Nomenclature, Synthesis and Applications of Spirocyclic Compounds

Nomenclature, Synthesis and Applications of Spirocyclic Compounds

By

Davood Habibi

and Masoumeh Beiranvand

Cambridge Scholars Publishing

Nomenclature, Synthesis and Applications of Spirocyclic Compounds

By Davood Habibi and Masoumeh Beiranvand

This book first published 2023

Cambridge Scholars Publishing

Lady Stephenson Library, Newcastle upon Tyne, NE6 2PA, UK

British Library Cataloguing in Publication Data
A catalogue record for this book is available from the British Library

ISBN (10): 1-5275-3941-5
ISBN (13): 978-1-5275-3941-9

CONTENTS

PREFACE

Spirocycle compounds are twisted structures with two or more rings that share a single atom and plays a significant role in life. Spiro molecules are exciting classes of chemicals because of their rigid conformational features and three-dimensional geometry, which are used in organic optoelectronic devices, pharmaceutical chemistry, and materials science, and as an intermediate in organic chemistry. Developing spirocycle compounds is challenging in modern organic synthesis because it includes creating a quaternary center, one of the most challenging works among synthetic transformations.

This book comprehensively overviews the tremendous and advanced works reported about spiro compounds during the last 20 years. The primary purpose of this comprehensive book is introduction spiro compounds (chapter 1), naming (chapter 2), presentation of the most common, new, and diverse synthetic methodologies (chapters 3-5), and applications of spiro compounds (chapter 6) to academic staff and industry experts to become familiar with.

In particular, this book presents a considerable tool for chemists working in this research area. Moreover, the increasing necessity and interest in efficient, versatile, and potential synthetic procedures to achieve spiro compounds quickly and economically and the corresponding libraries makes this book a vital reference instrument in organic synthesis. This wide-ranging collection can inspire academics and industrialists and help in their future progress. We hope to comply with the expectations of a significant part of the scientific community, organic chemistry faculties, and students.

In the end, we would like to thank the colleagues who helped us in making this great work with their valuable and exciting reports.

List of Abbreviations

AChE	Acetylcholine-esterase
AEMs	Anion-exchange membranes
AIBN	Azobisisobutyronitrile
AIDS	Acquired immunodeficiency syndrome
aza-MBH	Aza-Morita-Baylis–Hillman
BGTC	Bottom-gate and top-contact
BLHPC	Bent ladder-type hexaphenylene
Boc	*tert*-Butyloxy-carbonyl
BPI	Phenanthroimidazole
BuChE	Butyryl-cholinesterase
CAN	Ceric ammonium nitrate
CCR5	Chemokine receptor 5
CE	Current efficiency
CFA	Complete freund's adjuvant
CIP	Cahn-Ingold-Prelog
CM	Cross metathesis reaction
CNS	Central nervous system
CP	Cationic ring-opening polymerization
CPA	Chiral phosphoric acid
CTLs	Charge transport layers
CTMs	Charge transporting materials
DAB	1,4-Dideoxy-1,4-imino-*D*-arabinitol
DABCO	1,4-Diazabicyclo[2.2.2]octane
DAD	1,4-Diazabuta-1,3-diene
DAF	Diazafluorene
1,3-DC	1,3-Dipolar cycloaddition
1,3-DCRs	1,3-Dipolar cycloaddition reactions
DFT	Density functional theory
DIMCARB	*N*,*N*-Dimethyl ammonium *N'*,*N'*-dimethyl carbamate
DIPEA	*N*-Ethyldiisopropylamine
DLW	Direct laser writing
DMAD	Dimethyl acetylenedicarboxylate
DMAP	4-Di-methylaminopyridine

DMPA	Di-*p*-methoxy phenylamine
dr	Diastereoisomeric ratio
DSA-Ph	*p*-Bis(*p*-*N*,*N*-diphenyl-aminostyryl)benzene
DSCs	Dye-sensitized solar cells
D-Spiro-A	Donor-Spiro-Acceptor
DSX-LPP	Dispiroxanthene-ladderpenta phenylene
EBV	Epstein–Barr virus
ECD	Electrochromic devices
ee	Enantiomeric excess
EIS	Electrochemical impedance spectroscopy
EL	Electroluminescence
er	Enantiomeric ratio
ESE	Excess strain energy
ETL	Electron transporting layer
EQE	External quantum efficiency
FET	Field-effect transistor
FIB	The level of fibrinogen
FMO	Frontier molecular orbital
FOLEDs	Fluorescent OLEDs
FRAP	Ferric reducing anti-oxidant power
FRP	Free-radical polymerization
ΔG	Free binding energy
GER	Group equivalent reaction
GP	Glycogen phosphorylase
HAT	Hydrogen atom transfer
HAT	Human african trypanosomiasis
HDAc2	Histone deacetylase 2
HDF	Human dermal fibroblast
HGR	Human glutathione reductase
HOMO	Highest occupied molecular orbital
HTIB	[Hydroxy(tosyloxy)iodo]-benzene
HTL	Hole transport layer
HTM	Hole-transporting molecule
HWE	Horner–Wadsworth–Emmons
ICT	Intramolecular charge transfer
IE	Ionization energy
IEDDA	Inverse-electron-demand oxa-hetero-Diels–Alder
IEDDAR	Inverse-electron-demand Diels-Alder reaction
IL	Interlayers
IMDA	Intramolecular Diels-Alder
INH	Isoniazid

IRP	Intermolecular radical pair
J_{SC}	Short-circuit current density
K–DA-E	Knoevenagel/Diels–Alder/epimerization
LEDs	Light-emitting diodes
LPDE	Lithium perchlorate in diethyl ether
LPPP	Ladder-type poly(*p*-phenylene)s
LRRK2	Leucine-rich repeat kinase 2
LUMO	Lowest unoccupied molecular orbital
MBH	Morita-Baylis-Hillman
MC	Merocyanine
MC2	*Mycobacterium smegmatis*
MDR	Multidrug resistance phenotype
MDR-TB	Multidrug-resistant *Mycobacterium tuberculosis*
MED	Minimal effective dose
MES	Maximum electroshock seizure
MIC	Minimum inhibitory concentration
MIT	Molecular integration technology
mitoKATP	Mitochondrial ATP-dependent potassium channels
MP2	Møller–Plesset perturbation theory
MPP^+	1-Methyl-4-phenylpyridinium
MTB	*Mycobacterium tuberculosis* $H_{37}Rv$
NBS	*N*-Boromosuccimide
NCS	*N*-Chlorosuccinimide
NDA	Nitroso Diels-Alder
NF-κB	B Cells
NHC	*N*-Heterocyclic carbene
NIS	*N*-Iodosuccinimide
NMP	*N*-Methyl-2-pyrrolidone
NPS007994	*Streptomyces nodosus*
NPSGs	Nanopolyspirogrids
OFBA	*o*-Fluorobenzoic acid
OFETs	Organic field-effect transistors
OLEDs	Organic light-emitting diodes
OPVs	Organic photovoltaic devices
***o*-QDMs**	*Ortho*-quinodimethanes
***o*-QMs**	*Ortho*-quinomethanes
OSCs	Organic semiconductors
PA	Phenylacridine
PCs	Photoredox catalysts
PCEs	Power conversion efficiency
PCMs	Piezochromic materials

PCR	Peptide coupling reagent
PE	Power efficiency
PET	Photoinduced electron transfer
PFF	Pore filling fraction
PhOLEDs	Phosphorescent organic light-emitting diodes
PI	Photoinitiator
PIA	Introduced photoinduced absorption
PIFA	[Bis(trifluoroacetoxy)iodo]benzene
PL	Photoluminescence
PPV	Poly(*p*-phenylene vinylene)
***p*-QMs**	*Para*-quinone methides
PSCs	Perovskite solar cells
PTAA	Poly(triarylamine)
PTC	Phase transfer catalysis
PTSA	*p*-Toluenesulfonic acid
PTZ	Pentylenetetrazol
RCEM	Ring-closing enyne metathesis
RCM	Ring-closing metathesis
RFID	Radio frequency identification
RH	Relative humidity
RON	Reductive oxy-Nazarov
rr	Regioselective ratio
RRM	Ring-rearrangement metathesis
RSE	Ring strain energy
RT	Room temperature
RT-PCR	Reverse transcription and polymerase chain reaction
rubrene	5,6,11,12-Tetraphenylnaphthacene
RVD	Regulatory volume decrease
SAF	Spiro-acridine-fluorene
SAMs	Self-assembled monolayers
sarcKATP	Sarcolemmal ATP-dependent potassium channels
SBF	9,9′-Spirobifluorene
SBTF	Spiro[benzotetraphenefluorene]
SC	Solar cell
SCD1	Stearoyl-CoA desaturase-1
SE	Strain energy
SEI	Semiconductor-electrolyte interface
SET	Single-electron transfer
SFX	Spiro[fluorene-9,9′-xanthene]
SGR	Seyferth−Gilbert reagent

SI	Selectivity index
SIADH	Syndrome of inopportune anti-diuretic hormone
SMOLEDs	Small molecule organic light-cmitting diodcs
SOI	Secondary orbital interaction
SP	Spiropyran
SSDSCs	Solid-state dye-sensitized solar cells
SSYX	Shensong Yangxin capsule
STED	Stimulated-emission depletion
STP	Spirothiopyran
SZMC	Spirocyclic zwitterionic Meisenheimer
TADF	Thermally activated delayed fluorescence
TBAB	Tetra-*n*-butyl-ammonium bromide
TBAF	Tetra-*n*-butylammonium fluoride
TBAHS	Tetrabutylammonium hydrogen sulfate
TBHP	*tert*-Butyl hydroperoxide
TbTR	Trypanosoma brucei TR
TCO	Transparent conducting oxide
TDM	Transition density matrix
TFA	Trifluoroacetic acid
Th2	T Helper type 2
TMSCN	Trimethylsilyl cyanide
TMSOTf	Trimethylsilyl triflate
TNF-α	Tumor necrosis factor-alpha
TPD	*N*,*N*′-Diphenyl-*N*,*N*′-bis(3-methyl-phenyl)-1,1′-bi phenyl-4,4′-diamine
TPT	2,4,6-Triphenylpyrylium tetarfluoroborate
TR	Trypanothione reductase
TRPL	Time-resolved photoluminescence
TX	Thioxanthene
TXO_2	Dioxothioxanthene
U4CC	Ugi four-component condensation
UPS	Ultraviolet photoemission spectroscopy
V_{oc}	Open-circuit voltages
VECs	Vinylethylene carbonates
VOFET	Vertical organic field-effect transistor
V_2RA	V_2-receptor antagonists
YCK	Yuehchukene

1

SPIROCYCLIC COMPOUNDS

1.1. Introduction

Spirocyclic compounds, also known as spiro compounds, are twisted structures with two or more rings that share a single atom [1]. The connecting atom, also named the spiroatom, is most often a quaternary carbon (spiro carbon). Spirocyclic compounds can be broadly classified into two categories based on the nature of their rings: carbocyclic and heterocyclic. Carbocyclic spiro compounds contain only carbon atoms in their rings, while heterocyclic spiro compounds contain at least one heteroatom such as nitrogen, oxygen, or sulfur. These rings can be similar or different and the simplest spiro compounds are bicyclic. According to the number of spiro atoms, these compounds are classified as monospiro, dispiro, trispiro, etc.

In 1900, von Baeyer found the first spiro compound [2]. The spiro center in these compounds creates a naturally occurring three-dimensional structure which can reduce the conformational entropy penalty associated with target binding and produce diverse three-dimensional shapes [3]. Additionally, the perpendicular arrangement of spiro compounds results in the suppression of molecular interactions of π-systems, enhances solubility, and prevents the formation of excimers often observed in solid-state fluorescent dyes. Furthermore, the doubling of molecular weight, combined with the cross-shaped molecular structure and rigidity of spiro compounds, leads to entanglement in the amorphous solid state, inhibiting crystallization [4]. Spiro compounds can exhibit central or axial chirality, meaning spiroatoms can be chiral even without four different substituents (see Figure 1-1) [5].

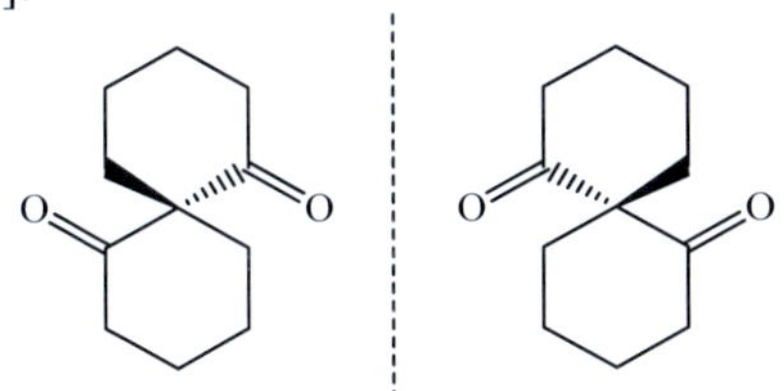

Fig. 1-1. Spirocycles with axial chirality.

Spiro molecules are an exciting class of chemicals due to their rigid conformational features and three-dimensional geometry which have numerous applications in organic optoelectronic devices, pharmaceutical chemistry, materials science, and as intermediates in organic synthesis [6-8]. Many natural products contain spiro motifs in their structure, such as Chitosenine **1**, Marcfortine B **2**, ACAT inhibitor **3**, Fredericamycin **4**, Elmenol G **5**, (-)-Horsfiline **6**, (+)-Elacomine **7**, Azaspiracid **8**, 6-Oxoleuconoxine **9**, among others (see Figure 1-2) [9-13].

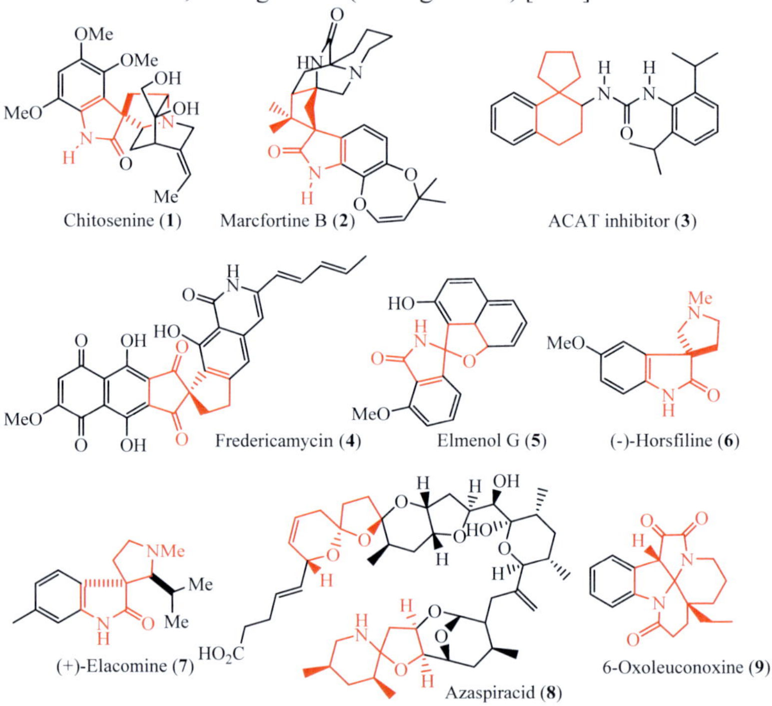

Fig. 1-2. Example of natural products containing spiro linkages.

Furthermore, spiro units form the core structures of essential commercial drugs like Spironolactone **10**, Drospirenone **11**, Ameinonide **12**, Buspirone **13**, Cevimeline **14**, Apalutamide **15**, Eplerenone **16**, Fenspiride **17**, and Fluspirilene **18**, as shown in Figure 1-3 [14].

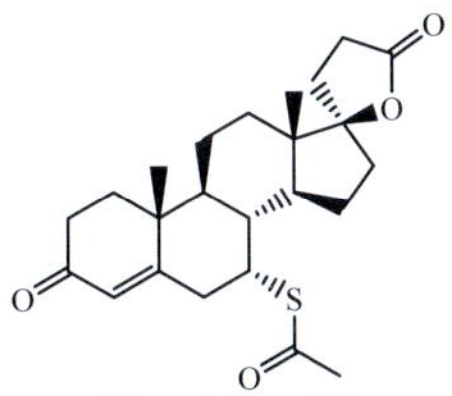

Spironolactone (**10**)
heart failure, edema in cirrhotic adults, treatment of primary hyperaldosteronism, add-on therapy in hypertension, and in nephroticsyndrome, off label uses: hirsutism, female pattern hair loss, and adult acne vulgaris

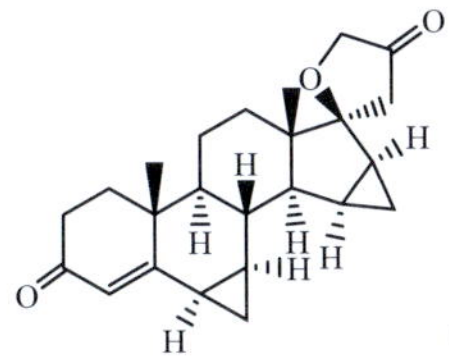

Drospirenone (**11**)
oral contraceptive (combination with ethinyl estradiol), menopause associated symptoms combination with estrogens)

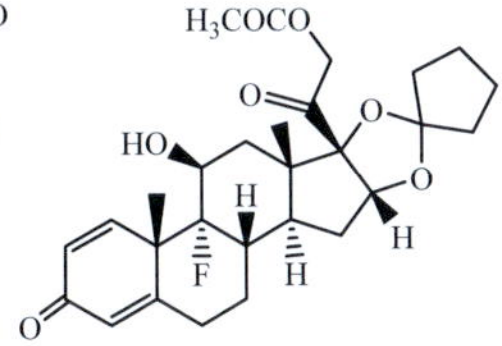

Amcinonide (**12**)
corticosteroid-responsive dermatoses

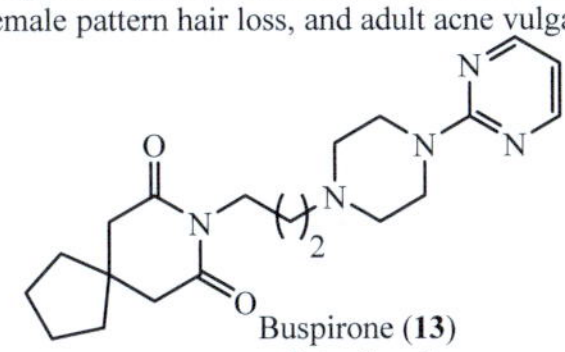

Buspirone (**13**)
anxiety disorders

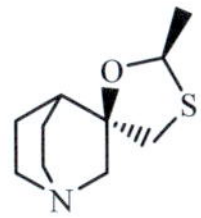

Cevimeline (**14**)
dry mouth in patients with Sjögren's Syndrome

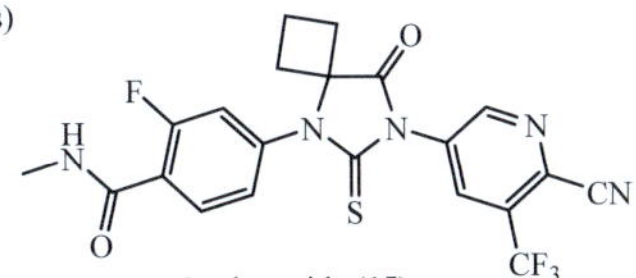

Apalutamide (**15**)
nonmetastatic, castration-resistant prostate cancer

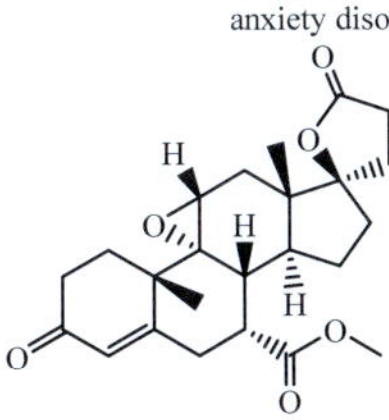

Eplerenone (**16**)
left ventricular systolic dysfunction

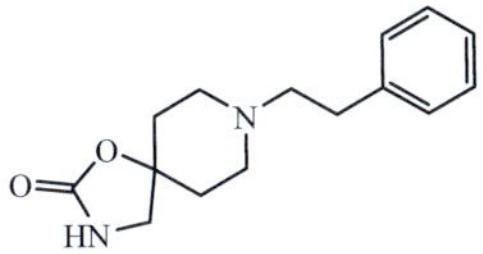

Fenspiride (**17**)
acute and chronic inflammatory diseases of ear, nose and throat organs and the respiratory tract

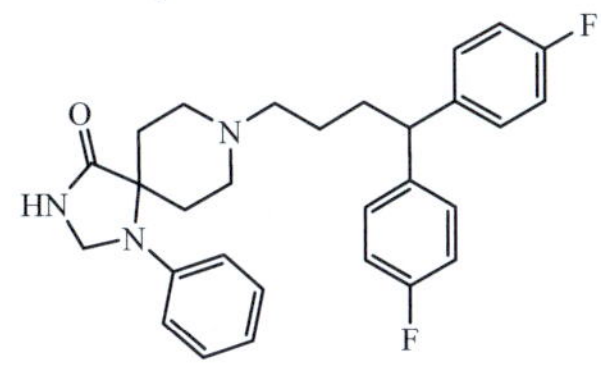

Fluspirilene (**18**)
schizophrenia

Fig. 1-3. Example of drugs containing spiro motifs.

1.2. Ring strain energy

One of the fundamental properties of spiro compounds is their ring strain energy (RSE), which refers to the increased energy of the molecule compared to its corresponding acyclic molecule. The concept of strain and strain energies (SEs) provides a basis for correlating molecular structures, stabilities, and reactivities. To quantitatively assess strain and SEs, one can take the difference between the enthalpy of formation, $\Delta H_f°(g)$, of the substance under consideration (either theoretically calculated or experimentally determined) and that of a hypothetical strain-free model. There are two approaches to this: (i) based on bond energies and (ii) based on group increments. Since spiro compounds contain multiple rings, it is important to determine whether the total RSE of a spiro compound is exclusively the sum of the RSE of the component rings. The excess strain energy (ESE) is the difference between the total RSE of a compound containing multiple rings and the sum of the RSE of the individual rings.

Hence, Rüchard et al. evaluated the enthalpies of formation for triangulanes and spiro-cyclopropanated cyclobutanes by measuring their heat of combustion in a micro calorimeter (Figure 1-4). They found that in triangulanes **19-22** (in which every ring is a three-membered hydrocarbon), there is an excess ring strain energy of 8.6 kcal/mol per spiro atom. Such an additional strain increment is virtually nonexistent for **25** (0.8 ± 0.4 kcal/mol), **26** (0.6 ± 0.2 kcal/mol), and **27** (0.3 ± 0.3 kcal/mol) but significant for **28** (2.4 ± 0.5 kcal/mol) [15, 16].

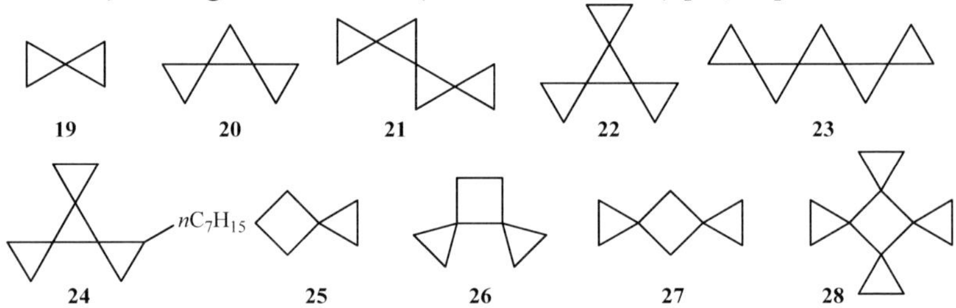

Fig. 1-4. Triangulanes and spirocyclopropanated cyclobutanes were studied by Rüchard et al.

Lammertsma et al. synthesized the first phospha[3]triangulane **36** as a $W(CO)_5$ complex which is a remarkably stable compound and possesses a Ph group (Scheme 1-1).

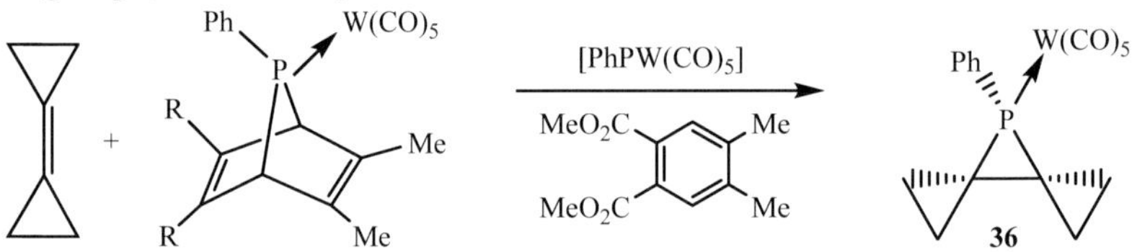

Scheme 1-1. Synthesis of phospha[3]triangulane **36**.

By comparing the X-ray structures of Ph-P substituted phospha-spiropentane **34** and phosphirane **35** complexes, researchers observed a denser structure. The heats of formation for the parent organophosphorus compounds and their hydrocarbon analogs were estimated using ring separation reactions and G2MP2 theory. The study found that the excess strain in phospha[*n*]triangulanes, which is the strain for spiro carbons over that of the three-membered rings, is approximately 5.3 kcal/mol per spiro carbon. This is around 40% less than the excess strain observed in linear [*n*]triangulane hydrocarbons (Figure 1-5) [17].

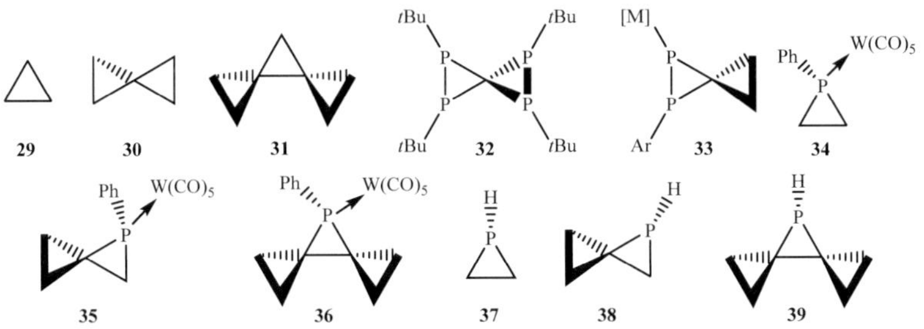

Fig. 1-5. Compounds **29-39** were studied by Rüchard et al.

The Lammertsma group investigated the ring strain in 2-aza-1-phospha-bicyclo[*n*.1.0]alkanes (n = 1-5) **40-44** using homodesmotic reactions at the G3(MP2) level. They compared the impact of cyclopropa(e)nation and heteroatom substitution with the corresponding bicyclic hydrocarbons and separate ring systems (Figure 1-6). The study revealed that the strain caused by fusion with cyclopropane is the sum of the individual rings. In contrast, the strain resulting from fusion with cyclopropene is much larger than the sum of rings due to the inverted nature of the bridgehead carbon. Substitution with nitrogen and phosphorus is favorable in all ring structures except in cyclohexane, and its effect is more pronounced in the more condensed structures. The evaluated SEs correlate very well with the experimental stability and reactivity of the bicyclic iron-amino phosphirane and phosphirene complexes [18].

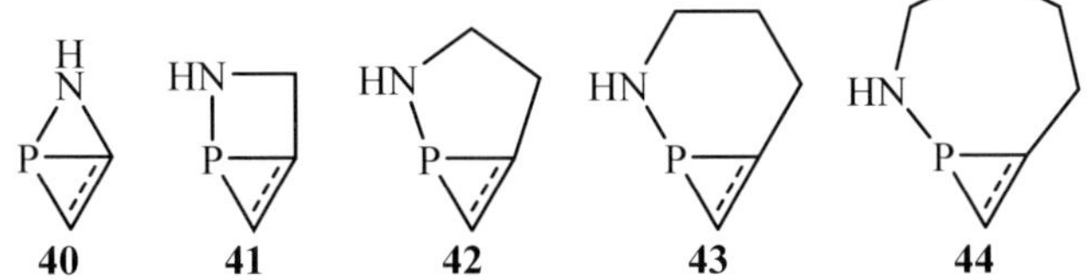

Fig. 1-6. Structure of 2-aza-1-phosphabicyclo[*n*.1.0]alka(e)nes **40-44**.

In 2014, Stedjan et al. calculated the RSEs for oxygen-containing spiro compounds **45-50** using the group equivalent reaction formalism (Figure 1-7). They found that the compounds containing two three-membered rings possess the highest RSEs and showed the most significant ESE of about 12 kcal/mol. The RSEs of cyclic lactones differ with ring size from those of cyclic ethers. Cyclic ethers' RSEs decrease by a small amount from the three- to four-membered rings, then decrease drastically as the ring increases to 5 atoms, and approach zero for the six-membered ring, representing the same unexpected behavior as seen in cycloalkanes. Cyclic lactones' RSEs reduce linearly to almost zero from the three- to the five-membered ring, then increase by 1-2 kcal/mol in the six-membered ring. Lactone-containing spiro compounds reveal regularly diminishing ESE as the size of the lactone ring increases, down to about 3 kcal/mol in the δ-lactone-containing spiro compound. Substitution of methyl group decreases RSE in these oxygen-containing spiro compounds, but fluorine substitution remarkably increases RSE, as has been reported in other compounds. But it has been shown that RSE alone is not fully related to the chemical reactivity of these spiro compounds [15].

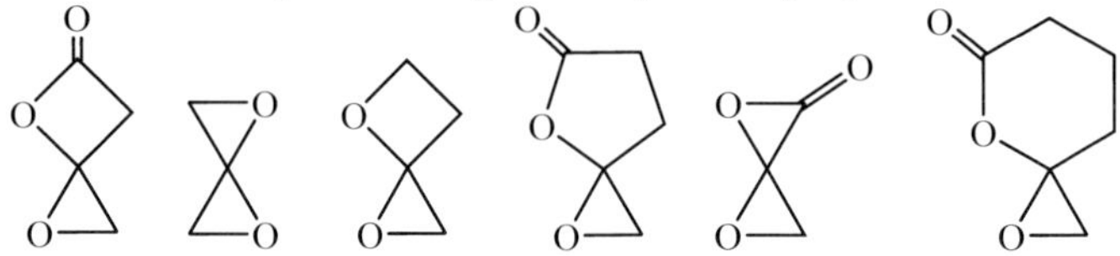

45 **46** **47** **48** **49** **50**

Fig. 1-7. Structure of oxygen-containing spiro compounds **45-50**.

1.3. Chirality

Chiral spiro compounds have attracted the attention of researchers and scientists owing to their potential applications in the pharmaceutical industry as either active pharmaceutical ingredients, catalysts in synthesizing active enantiomers, or surface modifiers on silica particles to resolve enantiomers [19].

The spatial orientation of molecules, i.e., their stereochemistry, has been one of the most important features of organic compounds since the tetrahedral structure of carbon atoms in organic molecules was introduced by van't Hoff and Lebel in the 19th century. In the pharmaceutical sciences, stereochemistry is of prime significance in the interaction of drugs and organisms since all receptors in human body are chiral and probably show different pharmacologic effects and pharmacokinetics between enantiomers. Hence, the US FDA required in 1992 that the properties of each enantiomer in a racemate should be explored individually before the drug is taken to the market as a pure enantiomer or as a racemate [20].

Absolute configuration is the spatial arrangement of the atoms of a chiral molecular entity (or group) and its stereochemical description, e.g., *R* or *S*. Neither the sign nor the magnitude for the rotation per se offers any information concerning the absolute configuration of a substance [21].

Historically, Cahn-Ingold-Prelog (CIP) sequence rules were first introduced to identify a molecule's spatial arrangement of atoms using simple, mostly atom- or bond-based stereo descriptors in 1956 [22].

Molecules with carbon and metal atoms can exhibit chirality in three ways: axial, central, and planar chirality [23]. Determining the absolute configuration of molecules with axial symmetry, which are dissymmetric but not asymmetric, presents intriguing challenges due to the absence of a formal asymmetric carbon that can serve as a basis for configurational correlations. Compounds with C_2 symmetry, such as allenes, spiranes, hindered biphenyls, hexahelicene, and *trans*-cyclooctene, are of particular interest. Both chemical and crystallographical methods have been used to address these cases, leading to ingenious solutions. Spiranes, due to their relatively rigid geometry, offer a unique opportunity to study interactions between functional groups held in fixed relative orientations. As a result, they are valuable substrates for chiroptical studies, including rotation of the sodium D line, ORD, and CD. However, assignments of absolute configuration to chiral C_2 spiranes were not made until 1968-1969, when configurations were assigned for spiranes **5l**, **52**, and **53** (Figure 1-8) [24].

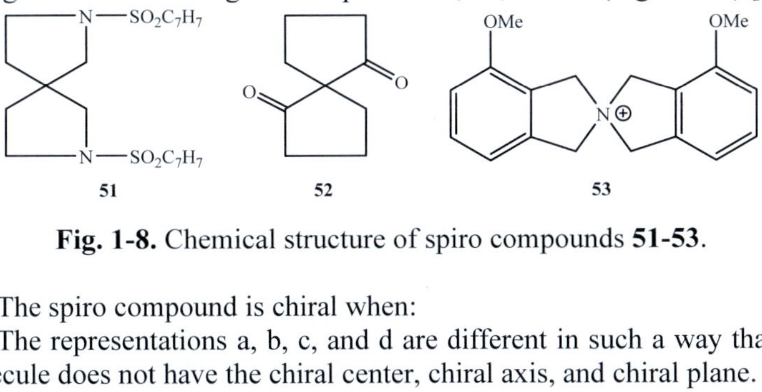

51 **52** **53**

Fig. 1-8. Chemical structure of spiro compounds **51-53**.

The spiro compound is chiral when:

The representations a, b, c, and d are different in such a way that the molecule does not have the chiral center, chiral axis, and chiral plane.

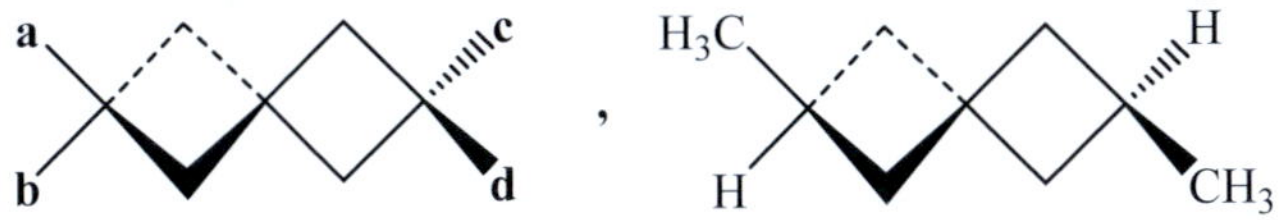

So, the eye can be put in the left or right of the molecule as below and the priority of representations a, b, c, and d is determined according to the CIP rule:

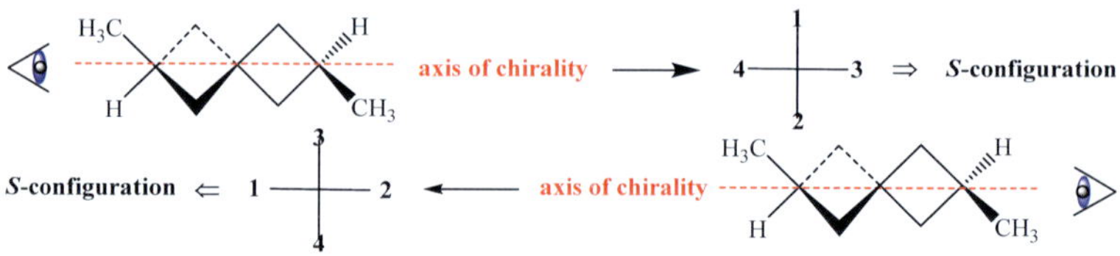

In the spiro-alkadiene compound, the eye can be put in left or right of the molecule as below and the priority of groups (double and single bonds) is determined according to the CIP rule:

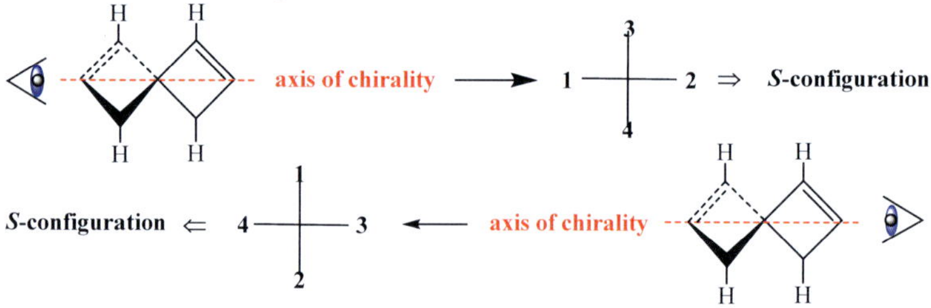

1.4. Spiro-aromaticity

Aromaticity is a fundamental concept in chemistry, but a precise definition has remained elusive [25]. Typically, aromatic compounds exhibit the following characteristics, although there are exceptions: (i) planarity with cyclic delocalized 4n+2 π electrons (Hückel's rule); (ii) equalization of bond lengths; (iii) deshielded chemical shifts of exocyclic protons in the ^{1}H NMR spectrum; (iv) the ability to undergo electrophilic substitution reactions; and (v) greater stability than their non-aromatic isomers [26]. According to van't Hoff's "tetrahedral carbon theory," organic compounds with a spiro carbon atom cannot be aromatic due to the sp^3 hybridization of the carbon atom. However, in 2002, Rzepa et al. suggested that the spiro atom itself could participate in conjugation to form a class of spiro-aromatic systems in which each ring could maintain aromaticity independently or join together to exhibit global aromaticity. When the spiro atom is carbon, spiro-aromaticity cannot be achieved because it has only four valence electrons. However, by using transition metals as the spiro atom, Huang et al. discovered two types of bis-spiro metallic-aromatics with square planar (Type I, Figure 1-9a) and tetrahedral (Type II, Figure 1-9b) geometries. The d electron configurations of the metal centers largely dictate their geometric and electronic structures. Huang et al. suggested that the metal center should have more empty d-orbitals to form more metal-aromatic tris-spiro structures. Rzepa et al.

evaluated some tris-spiro-aromatic candidates theoretically, using P, As, or V as the spiro atom, which demonstrated a certain degree of aromaticity, but none of them have been synthesized yet. Although Craig-Möbius type molecular orbitals were found in $Ta(DAD)^{3+}$ (DAD = 1,4-diazabuta-1,3-diene), there is not enough evidence to support its aromaticity. Additionally, the presence of Möbius-type orbitals in metallacycles is only a necessary condition, rather than sufficient, for Möbius aromaticity because Hückel aromatic systems with transition metals may also involve such orbitals, as revealed by Mauksch et al. The Huang group disclosed a hexalithio spiro vanada-cycle as a tris-spiro metalla-aromatic compound (Type III, Figure 1-9c). The V atom interacts with the π* orbital of the three biphenyl ligands via its two 3d-orbitals, resulting in three metallaaromatic rings that form a 40π Craig-Möbius aromatic system [27].

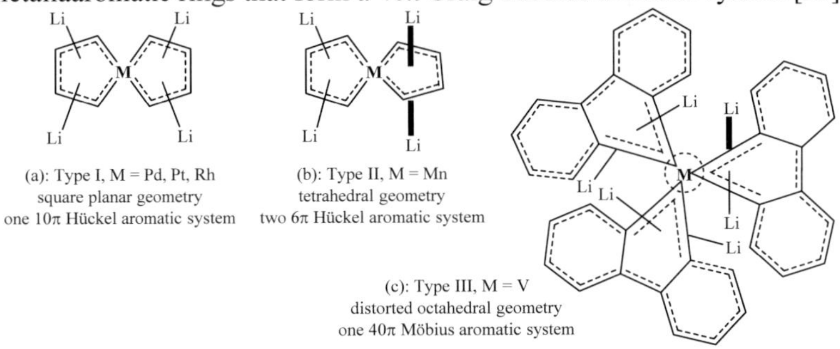

Fig. 1-9. Three types of spiro metallaaromatics.

References

[1] Tana Jingyun, Wanga Chunfei, Kuan Laoa Hio, Feng Gang, *et al.* 2019. "Efficient synthesis and facile functionalization of highly fluorescent spiro[pyrrol-pyran]". Dyes and Pigments. 107777.

[2] Saraswat Pankaj, Jeyabalan Govindasamy, Hassan Mohd. Zaheen, U. Rahman Mujeeb, K. Nyola Narendra. 2016. "Review of synthesis and various biological activities of spiro heterocyclic compounds comprising oxindole and pyrrolidine moities". Synthetic Communications. No. 20. 1643-1664.

[3] Sveiczer Attila, J. P. North Andrew, Mateu Natalia, L. Kidd Sarah, *et al.* 2019. "Spirocycles as rigidified sp^3-rich scaffolds for a fragment collection". Organic Letters. No. 12. 4600-4604.

[4] P. I. Saragi Tobat, Spehr Till, Siebert Achim, Fuhrmann-Lieker Thomas, Salbeck Josef. 2007. "Spiro compounds for organic optoelectronics". Chemical Reviews. No. 4. 1011-1065.

[5] Rios Ramon. 2012. "Enantioselective methodologies for the synthesis of spiro compounds". Chemical Society Reviews. No. 3. 1060-1074.

[6] Pakravan Narges, Shayani-Jam Hassan, Beiginejad Hadi, Tavafi Hadis, Paziresh Shadi. 2019. "A green method for the synthesis of novel spiro compounds: Enhancement of antibacterial properties of caffeic acid through electrooxidation in the presence of barbituric acid derivatives". Journal of Electroanalytical Chemistry. 113286.

[7] Sun Ming-Li, Zhang Feng, Qian Yan, Ou Chang-Jin, *et al.* 2018. "Catalyst-free photocyclization for the synthesis of spiro-fused aromatic organic semiconductor based on SFX". Tetrahedron. No. 16. 2063-2067.

[8] Pawlowski Robert, Skorka Patryk, Stodulski Maciej. 2020. "Radical-mediated non-dearomative strategies in construction of spiro compounds". Advanced Synthesis & Catalysis. No. 21. 4462-4486.

[9] Kotha Sambasivarao, Chandra Deb Ashoke, Lahiri Kakali, Manivannan Ethirajan. 2009. "Selected synthetic strategies to spirocyclics". Synthesis. No. 02. 165-193.

[10] V. Galliford Chris, A. Scheidt Karl. 2007. "Pyrrolidinyl-spiro-oxindole natural products as inspirations for the development of potential therapeutic agents". Angewandte Chemie International Edition. No. 46. 8748-8758.

[11] Marti Christiane, M. Carreira Erick. 2003. Construction of spiro-[pyrrolidine-3,3′-oxindoles] recent applications to the synthesis of

oxindole alkaloids". European Journal of Organic Chemistry. No. 12. 2209-2219.

[12] Sengoku Tetsuya, Shirai Anna, Takano Ayaka, Inuzuka Toshiyasu, *et al.* 2019. "Divergent synthesis of methylene lactone-and methylene lactam-based spiro compounds: Utility of amido-functionalized γ-hydroxylactam as a precursor for cytotoxic *N,O*-and *N,N*-spiro compounds". The Journal of Organic Chemistry. No. 19. 12532-12541.

[13] M. Trost Barry, Cramer Nicolai, Bernsmann Heiko. 2007. "Concise total synthesis of (±)-marcfortine B". Journal of the American Chemical Society. No. 11. 3086-3087.

[14] Zheng Yajun, M. Tice Colin, B. Singh Suresh. 2014. "The use of spirocyclic scaffolds in drug discovery". Bioorganic & Medicinal Chemistry Letters. No. 16. 3673-3682.

[15] K. Stedjan Matthew, D. Augspurger Joseph. 2015. "Ring strain energy in ether-and lactone-containing spiro compounds". Journal of Physical Organic Chemistry. No. 4. 298-303.

[16] Beckhaus Hans-Dieter, Rüchardt Christoph, I. Kozhushkov Sergei, N. Belov Vladimir, P. Verevkin Sergey, De Meijere Armin. 1995. "Strain energies in [*n*]triangulanes and spirocyclopropanated cyclobutanes: An experimental study". Journal of the American Chemical Society. No. 48. 11854-11860.

[17] Lammertsma Koop, Wang Bing, Hung Jui-Te, W. Ehlers Andreas, M. Gray Gary. 1999. "Synthesis, structure, strain energy, and excess strain of a phospha[3]triangulane". Journal of the American Chemical Society. No. 50. 11650-11655.

[18] L. G. Borst Mark, W. Ehlers Andreas, Lammertsma Koop. 2005. "G3(MP2) ring strain in bicyclic phosphorus heterocycles and their hydrocarbon analogues". The Journal of Organic Chemistry. No. 20. 8110-8116.

[19] Ya Liang, Jing-jihng Guo, Xiu-ming Liu, and Rong-bao Wei. 2008. "Chiral separation of spiro-compounds and determination configuration". Chemical Research in Chinese Universities. No. 4. 441-444.

[20] Ling-Yi Kong, Peng Wang. 2013. "Determination of the absolute configuration of natural products". Chinese Journal of Natural Medicines. No. 3. 193-198.

[21] P. Moss Gary. 1996. "Basic terminology of stereochemistry (IUPAC Recommendations 1996)". Pure and Applied Chemistry. No. 12. 2193-2222.

[22] M. Hanson Robert, Mayfield John, Vainio Mikko, Yerin Andrey, *et al.* 2018. "Algorithmic analysis of Cahn–Ingold–Prelog rules of stereochemistry: Proposals for revised rules and a guide for machine implementation". Journal of Chemical Information and Modeling. No. 9. 1755-1765.

[23] J. Coles Simon, B. Davies David, J. Eaton Robert, B. Hursthouse Michael, *et al.* 2007. "Stereogenic properties of spiranes combined with four equivalent conventional centres of chirality". Dalton Transactions. No. 20. 2040-2047.

[24] K. Hill Richard, A. Cullison David. 1973. "Dissymmetric spirans. II. Absolute configuration of 1,1'-spirobiindene and related compounds". Journal of the American Chemical Society, No. 4. 1229-1239.

[25] Yu Chao, Zhong Mingdong, Zhang Yongliang, Wei Junnian, *et al.* 2020. "Butadienyl diiron complexes: Nonplanar metalla-aromatics involving σ-type orbital overlap". Angewandte Chemie . No. 43. 19210-19215.

[26] Chen Dafa, Hua Yuhui, Xia Haiping. 2020. "Metallaaromatic chemistry: History and development". Chemical Reviews. No. 23. 12994-13086.

[27] Huang Zhe, Zhang Yongliang, Zhang Wen-Xiong, Wei Junnian, *et al.* 2021. "A tris-spiro metalla-aromatic system featuring Craig-Möbius aromaticity". Nature Communications. No. 1. 1-7.

2

NOMENCLATURE OF SPIRO COMPOUNDS

2.1. IUPAC Nomenclature of Spiro Compounds

The nomenclature and name "spirane" were first proposed by von Baeyer for bicyclic compounds with one common atom to both rings which intersect at a single point. Later, Radulescu extended this nomenclature to include spiro-fused ring systems and recognized that each ring must be named individually, along with specifying the spiro-fusion details. Patterson also utilized these two systems in his analysis of ring systems. The Chemical Society introduced a third method for naming spiro compounds which was later incorporated into the IUPAC rules along with the other two methods [1, 2].

2.2. Compounds with only monocyclic ring components

Monospiro hydrocarbons are composed of two saturated cycloalkane rings and are named using a specific nomenclature convention. The prefix "spiro" is added to a von Baeyer descriptor that indicates the number of carbon atoms connected to the spiro atom in each ring, arranged in ascending order and separated by a period and enclosed in square brackets. The name of the original hydrocarbon indicates the total number of skeletal atoms.

In non-substituted monospiro compounds, the carbon atoms are numbered continuously, beginning from a ring atom that is adjacent to the spiro atom, proceeding through the smaller ring (if present), then through the spiro atom, and finally, around the second ring. Therefore, the general format for writing IUPAC nomenclature for non-substituted spiro compounds is spiro[a.b]alkane.

For substituted spiro compounds, the nomenclature convention is slightly different. The name starts with the prefix indicating the position of the substituents followed by the spiro descriptor which includes the number of atoms in the smaller and larger rings, respectively, in the spiro ring system, enclosed in square brackets. The numbering of carbon atoms is done from the smaller ring to the larger ring, giving the substituents the lowest possible numbers. When there are two or more substitutions, the numbering is done in such a way that it gives the substitutions the lowest numbers [1, 2].

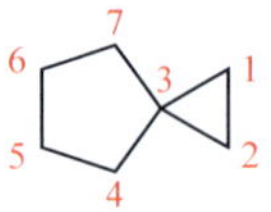

Spiro[2.4]heptane

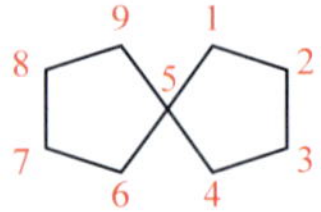

Spiro[4.4]nonane

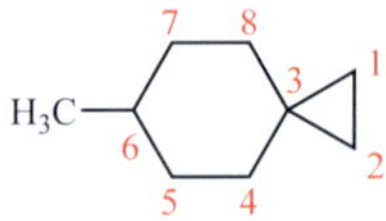

6-Methylspiro[2.5]octane

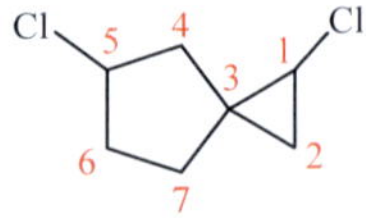

1.5-Dichlorospiro[2.4]heptane

The method mentioned for monocyclic spiro compounds is also used for radical monocyclic spiro rings, with the difference that they are named by replacing the "-ane" ending of the systematic name of the parent hydride with "-yl" [1].

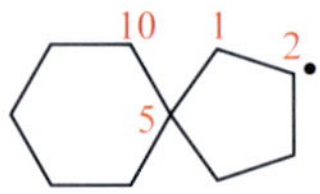

Spiro[4.5]decan-2-yl

When monospiro compounds possess a double bond, the same numbering pattern is retained, but in such a direction around the rings that the double bonds receive numbers as low as possible. The format of writing the IUPAC nomenclature for spiro compounds in the presence of a double bond is spiro[a.b]alkyene (y = position number of double bonds) [2].

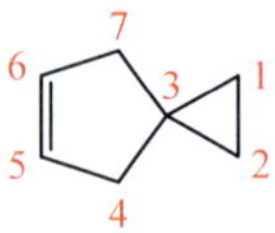

Spiro[2.4]hept-5-ene

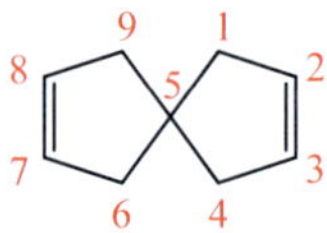

Spiro[4.4]nona-2,7-diene

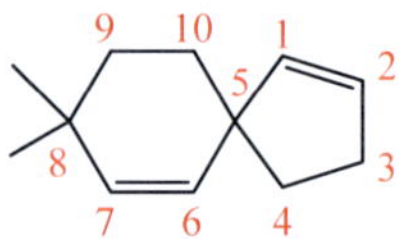

8,8-Dimethylspiro[4.5]deca-1.6-diene

The same numbering pattern is retained when monospiro compounds possess a functional group, but low locants are allocated for the principal functional group. Low locants are allocated for the double bond if there is a double bond [1].

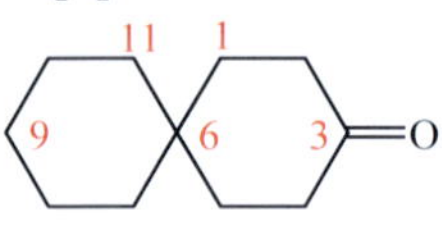

Spiro[5.5]undecan-3-one

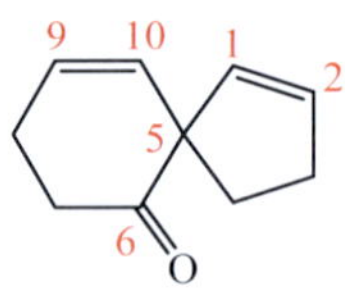

Spiro[4.5]deca-1,9-diene-6-one

When monospiro compounds contain a heteroatom, the naming convention for spiro heterocycles with one spiroatom is derived by prefixing "spiro" to the name of an alkane with the same number of carbon atoms. The number of atoms in each ring is indicated with Arabic numerals, separated by a period, enclosed in square brackets, arranged in ascending order, and placed between the spiro prefix and the name of the hydrocarbon. The "a" prefix indicates the position of the heteroatom, as shown in Table 1.1. The same numbering scheme is used for spiro heterocycles, regardless of whether the larger ring contains the heteroatom or not. It is essential to assign low numbers to the heteroatom.

Table 1.1. "a" prefixes

Valence	III		IV		III		II		I	
	B	bora	C	carba	N	Aza	O	oxa	F	fluora
	Al	alumina	Si	sila	P	phospha	S	thia	Cl	chlora
	Ga	galla	Ge	germa	As	arsa	Se	selena	Br	broma
	In	inda	Sn	stanna	Sb	stiba	Te	tellura	I	ioda
	Tl	thalla	Pb	plumba	Bi	bisma	Po	polona	At	astata

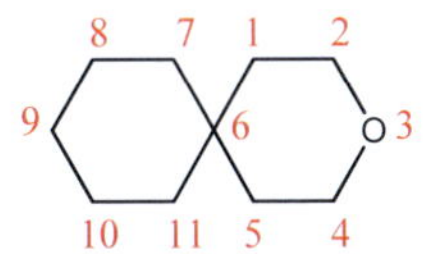

3-Oxaspiro[5.5]undecane

If there is a choice of name or numbering the heteroatoms, the following criteria are considered:

a) The lowest numbers are assigned to heteroatoms as a set, irrespective of the kind of heteroatom.

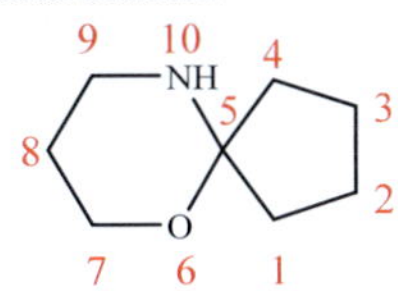

6-Oxa-10-azaspiro[4.5]decane

b) If there is still a choice, low locants are assigned in accordance with the following descending order of heteroatoms: F > Cl > Br > I > O > S > Se > Te > N > P > As > Sb > Bi > Si > Ge > Sn > Pb > B > Al > Ga > In > Tl.

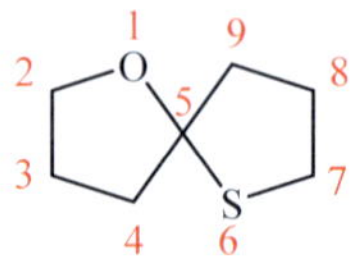

1-Oxa-6-thiaspiro[4.4]nonane

c) If unsaturation is present in a ring, the numbering remains the same, but the direction around the ring is such that the multiple bonds are given as low a number as possible.

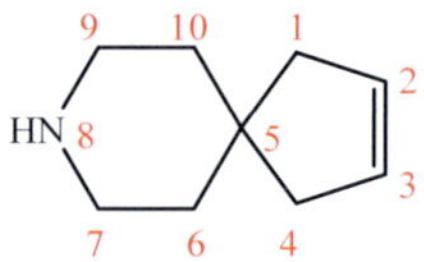

8-Azaspiro[4.5]dec-2-ene

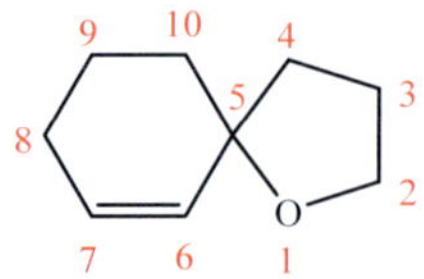

1-Oxaspiro[4.5]dec-6-ene

d) Suppose monospiro rings possess a non-standard spiro atom. In that case, they are named by placing the prefix spiro- before the von Baeyer descriptor, followed by the name corresponding to the hydrocarbon with the total number of skeletal atoms present in the ring system. The heteroatoms are indicated by "a" prefixes and non-standard valency by the lambda symbol "λ" (A non-standard bonding number of a natural skeletal atom of a parent hydride is indicated by the symbol "λ^n", cited in conjunction with a convenient locant). Note that the bonding number 'n' in the symbol "λ^n" is the sum of the total number of valence bonds of that skeletal atom to adjoining skeletal atoms, if any, in a parent hydride and the number of hydrogen atoms. The locant of the spiro atom is used as a superscript number to the corresponding number of the von Baeyer descriptor to indicate each time the spiro atom is revisited. If there is a choice, low locants are assigned to heteroatoms with higher bonding numbers. If necessary, to decrease the bonding number's numerical value, i.e., the lower number is assigned to λ^5 rather than λ^3. To indicate the locant of the spiro atom, a superscript number is used in conjunction with the corresponding number of the von Baeyer descriptor each time the spiro atom is revisited. When there is a choice, low locants are assigned to heteroatoms with higher bonding numbers. If necessary, to reduce the numerical value of the bonding number, the lower number is assigned to λ^5 instead of λ^3.

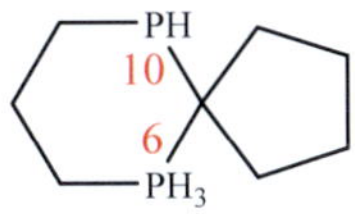

$6\lambda^5$,10-Diphosphaspiro[4.5]decane

e) Low locants are allocated for radical positions, or if the ring system is a substituent, its point of attachment.

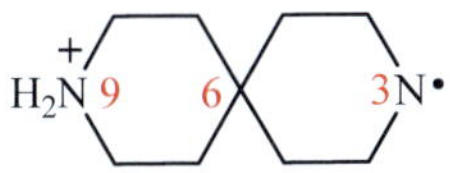

3,9-Diazaspiro[5.5]undecane-9-ium-3-yl

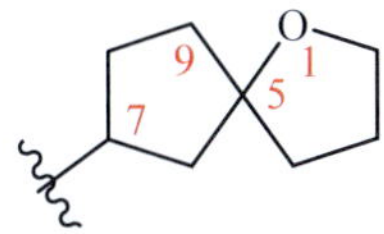

1-Oxaspiro[4.4]nonan-7-yl

f) Spiro compounds with an ionic spiro atom are to illustrate a non-standard valency for the spiro atom and produce the ion by the convenient subtractive or additive suffix. Alternative methods use the ionic replacement terms or describe the ion in the names of the components of the spiro system. Low locants are allocated for cationic sites.

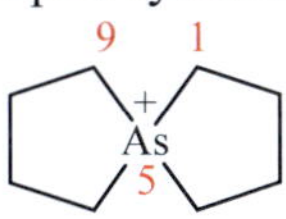

$5\lambda^5$-Arsaspiro[4.4]nonan-5-ylium (preferred name)
Alternative name 5-arsoniaspiro[4.4]nonane

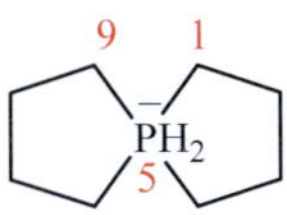

$5\lambda^5$-Phosphapiro[4.4]nonan-5-uide (preferred name)
Alternative name $5\lambda^5$-Phosphanuidaspiro[4.4]nonane

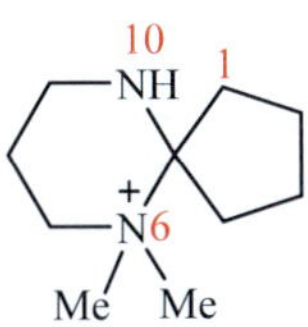

6,6-Dimethyl-6,10-diazaspiro[4.5]decan-6-ium (preferred name)
Alternative name 6,6-dimethy-10-aza-6-azoniaspiro[4.5]decane

g) Low locants are appropriated for substituents cited as prefixes in the order of citation [1, 2].

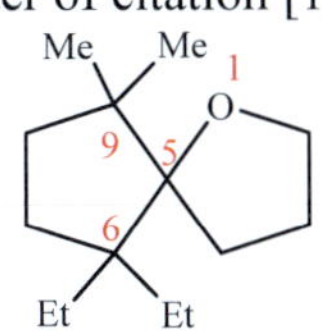

6,6-Diethyl-9,9-dimethyl-1-oxaspiro[4.4]nonane

2.3. Linear polyspiro alicyclic ring systems

Polyspiro hydrocarbons consisting of unbranched assemblies of three or more saturated cycloalkane rings are named using dispiro-, trispiro-, etc. to indicate the total number of spiro atoms present, followed by the name of the normal acyclic hydrocarbon with the same number of total carbon atoms. The von Baeyer descriptor is placed between the prefix and the hydrocarbon name, indicating the number of carbon atoms connected to the spiro atom or linking spiro atoms. The numbers are listed in order, starting at the smaller terminal ring (if one exists) and proceeding sequentially, always following the shorter path to the other terminal ring through each spiro atom and then back to the first spiro atom. The numbers are separated by a point and enclosed in square brackets. When a spiro atom is encountered for the second time, its locants are indicated as a superscript number to the corresponding number of linking atoms.

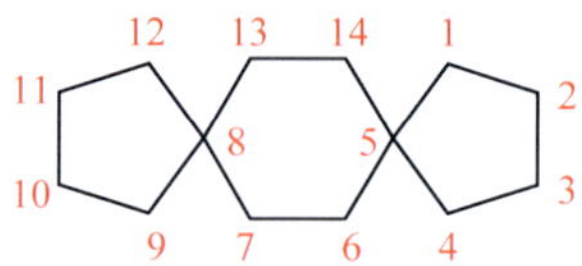

Dispiro[4.2.4^8.2^5]tetradecane

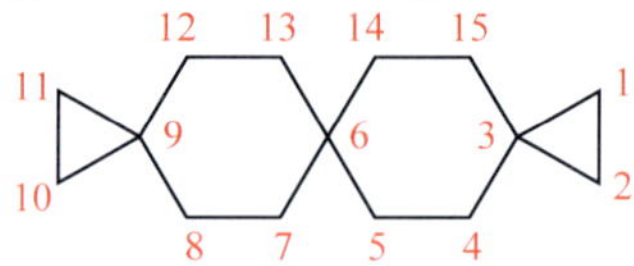

Trispiro[2.2.2.2^9.2^6.2^3]pentadecane

If there is a choice in the numbering, the numbers of the von Baeyer descriptor are considered in the order of citation and the name is selected with lower numbers at the first point of difference.

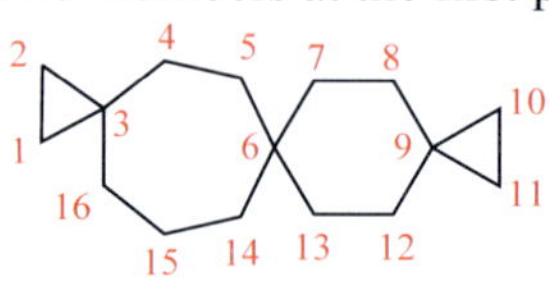

Trispiro[2.2.2.2^9.2^6.3^3]hexadecane

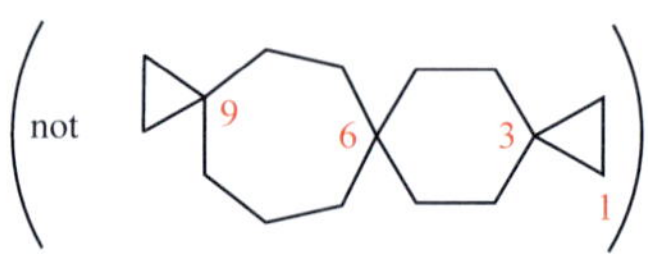

not Trispiro[2.2.2.2^9.3^6.2^3]hexadecane
(2.2.2.2^9.**2^6**.3^3 is lower than 2.2.2.2^9.**3^6**.2^3)

If there are chiral centers**,** low locants are assigned for chirality centers with the configuration *R* [1].

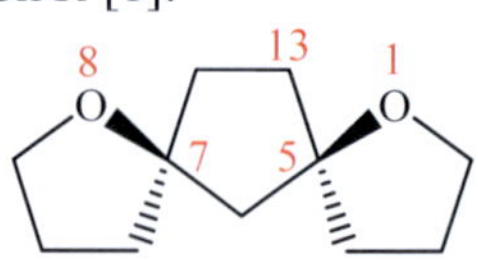

(5*R*,7*S*)-1,8-Dioxadispiro[4.1.4.2]tridecane
not (5*S*,7*R*)-1,8-dioxadispiro[4.1.4.2]tridecane (5*R* is lower than 5*S*)

2.4. Branched polyspiro alicyclic ring systems

The method mentioned for linear polyspiro alicyclic ring systems is also used for branched polyspiro alicyclic ring systems.

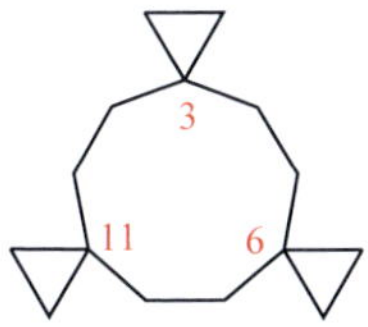

Trispiro[2.2.2^6.2.2^{11}.2^3]pentadecane

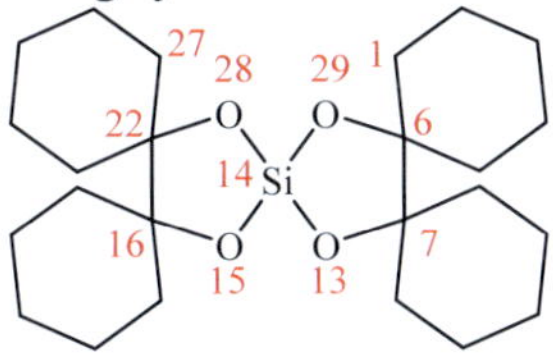

13,15,28,29-Tetraoxa-14-silapentaspiro [5.0.5^7.1.1.5^{16}.0.5^{22}.1^{14}.1^6]nonacosane

If there is a choice in numbering, the name that gives the lowest locants for spiro atoms is selected.

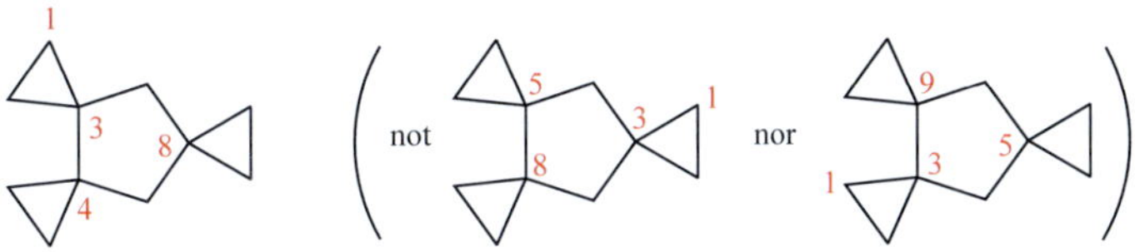

Trispiro[2.0.2^4.1.2^8.1^3]undecane, not trispiro[2.1.2^5.0.2^8.1^3]undecane, not trispiro[2.1.2^5.1.2^9.0^3]undecane, (3,**4**,8 is lower than 3,**5**,8 or 3,**5**,9)

If there is a choice in numbering, the numbers of the von Baeyer descriptor are considered in the order of citation. The name is selected with lower numbers at the first point of difference [1].

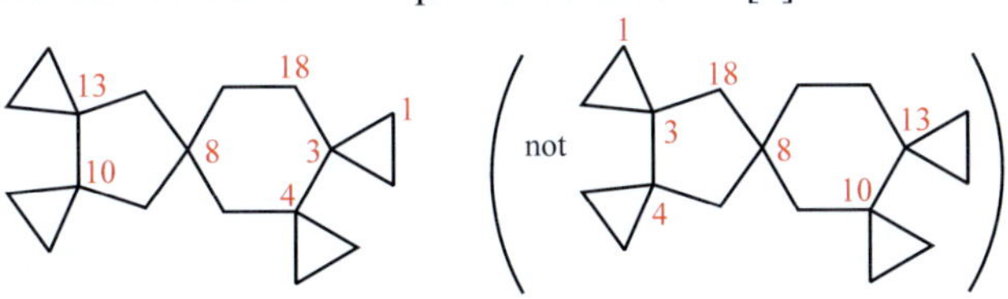

Pentaspiro[2.0.2^4.1.1.2^{10}.0.2^{13}.1^8.2^3]octadecane, not Pentaspiro[2.0.2^4.1.1.2^{10}.0.2^{13}.2^8.1^3] octadecane(2.0.2^4.1.1.2^{10}.0.2^{13}.1^8.2^3 is lower than 2.0.2^4.1.1.2^{10}.0.2^{13}.$\mathbf{2^8}$.1^3)

Ring systems consisting of three monocyclic rings with a single non-standard spiro atom are named by adding the prefix "spiro-" to the von Baeyer descriptor, followed by the name of the hydrocarbon containing the total number of skeletal atoms in the ring system. Heteroatoms are indicated by the prefix "a", and non-standard valences are denoted by the lambda symbol 'λ'. The locant of the spiro atom is represented by a superscript number to the corresponding von Baeyer descriptor number, indicating each time the spiro atom is revisited.

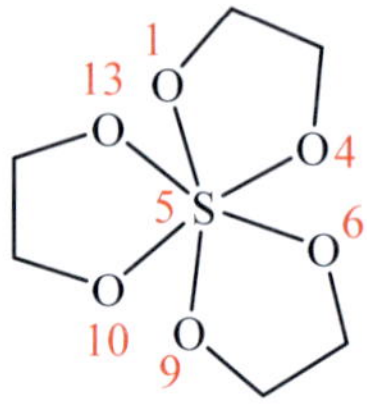

1,4,6,9,10,13-Hexaoxa-5λ^{6}-thiaspiro[4.4^{5}.4^{5}]tridecane

If there is a choice, a smaller ring is numbered before a larger ring.

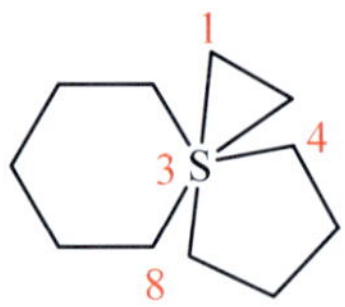

3λ^{6}-Thiaspiro[2.4^{3}.5^{3}]dodecane

Polyspiro ring systems that contain at least three monocyclic rings with one non-standard hetero spiro atom, as well as other spiro fusion, are named using a composition of methods for naming the polyspiro ring systems and for denoting heteroatoms with non-standard bonding numbers. If there is a choice for denoting, the following criteria are considered in order until a decision is made:

a) Low numbers are selected for spiro atoms.

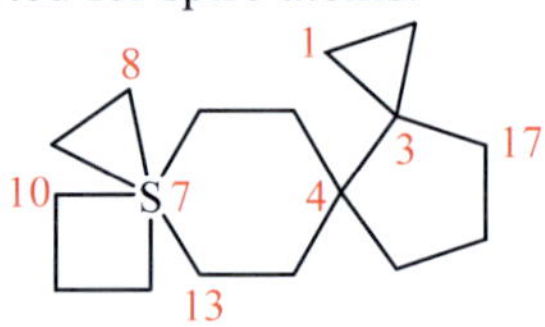

7λ^{6}-Thiatrispiro[2.0.2.2^{7}.3^{7}.2^{4}.3^{3}]heptadecane

b) Low numbers are selected for spiro atoms connecting three rings.

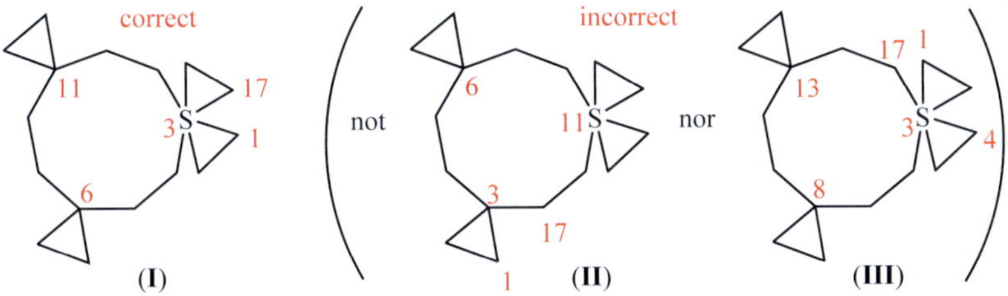

$3\lambda^6$-Thiatrispiro[$2.2.2^6.2.2^{11}.2^3.2^3$]heptadecane (**I**),
[not $11\lambda^6$-Thiatrispiro[$2.2.2^6.2.2^{11}.2^{11}.2^3$]heptadecane (**II**),
nor $3\lambda^6$-Thiatrispiro [$2.2^3.2.2^8.2.2^{13}.2^3$]heptadecane (**III**)]

c) Low numbers are selected for the von Baeyer descriptor in the order of citations [1, 2].

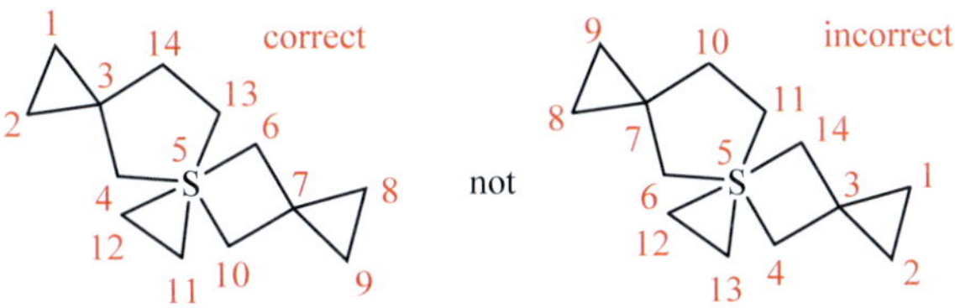

(**I**) (**II**)

$5\lambda^6$-Thiatrispiro[$2.1.1.2^7.1^5.2^5.2^3$]tetradecane (**I**),
[not $5\lambda^6$-Thiatrispiro[$2.1.1.2^7.2^5.2^5.1^3$]tetradecane (**II**)

2.5. Monospiro compounds composed of two equal polycyclic components

Monospiro compounds composed of two equal polycyclic components (excluding double bonds and heteroatoms named by von Baeyer nomenclature) are named by using the prefix "spirobi" followed by the name of the component ring system enclosed in square brackets. The locants needed to describe the component ring system are also placed in square brackets. The established numbering system of the polycyclic component system is retained, with one set of locants marked with a prime. The location of the spiro atom is indicated in the name by the appropriate locants (unprimed first), placed before the name.

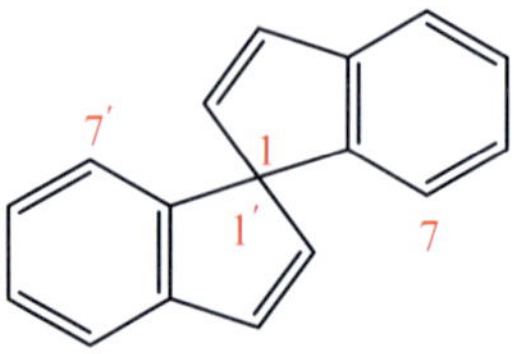
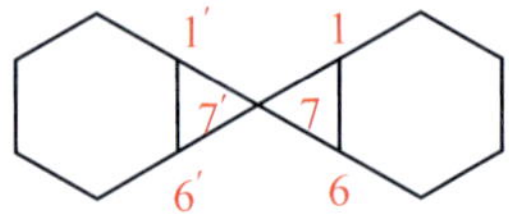

1,1′-Spirobi[indene] 7,7′-Spirobi[bicyclo[4.1.0]heptane]

a) If there is a choice to determine which component has primed locants, the lower number at the spiro atom is unprimed.

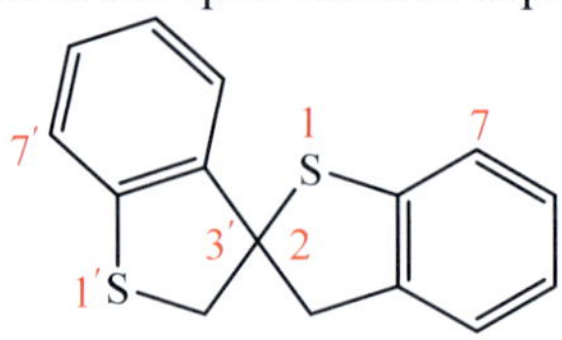

2′*H*,3*H*-2,3′-Spirobi[[1]benzothiophene]

b) The spiro system is named as the saturated hydrocarbon with replacement-term-prefixes in front of the completed spiro hydrocarbon name to indicate the heteroatoms. If there is a choice, low numbers are given to the spiro atom and then to the heteroatoms.

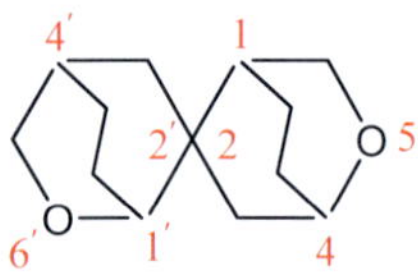
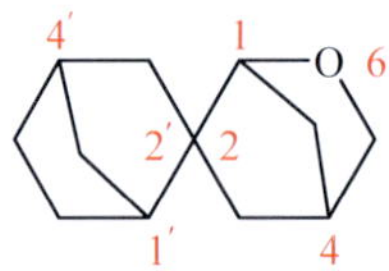

5,6′-Dioxa-2,2′-spirobi[bicyclo[2.2.2]octane] 6-Oxa-2,2′-spirobi[bicyclo[2.2.1]heptane]

c) Unsaturation in a component named by von Baeyer nomenclature is indicated by the endings -ene, -diene, etc.; only the final 'e' of the saturated hydrocarbon name is deleted if followed by a vowel. If there is a choice, low numbers are given to (i) spiro atoms, (ii) heteroatoms, (iii) principal functional groups, and (iv) double bonds.

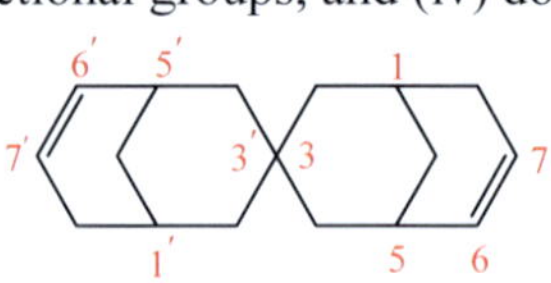
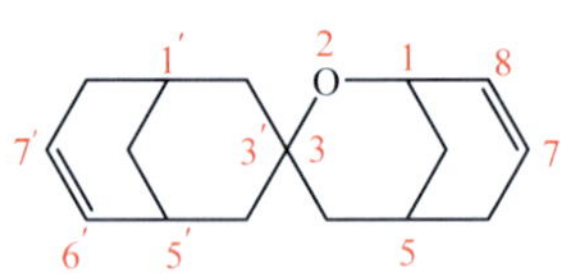

3,3′-Spirobi[bicyclo[3.3.1]nanane]-6,6′-diene 2-Oxa-3,3′-spirobi[bicyclo[3.3.1]nanane]-6′,7-diene

d) If a choice for numbering the components remains, the normal rules for numbering the components are used, and unprimed locants are selected in preference to identical locants with a prime [1].

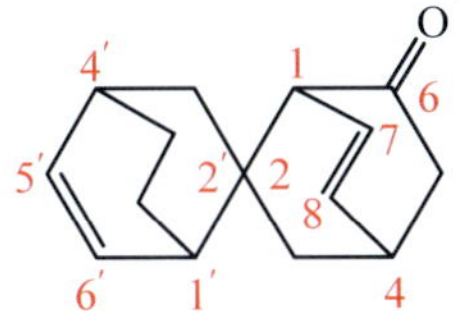

2,2′-Spirobi[bicyclo[2.2.2]octane]-5′,7-dien-6-one

2.6. Monospiro compounds with three equal polycyclic components

Monospiro compounds with three equal polycyclic components are named by placing the prefix "spiroter" before the name of the component ring system in square brackets. The multiplicative prefix "ter" indicates the replication of equal ring components. Locants required to define the component ring system are placed in square brackets. The middle component is numbered with primed locants, and the numbers of the third component are double-primed. The spiro atoms are indicated in front of the name by two pairs of locants separated by a colon. If necessary, indicated hydrogen is mentioned in front of these locants.

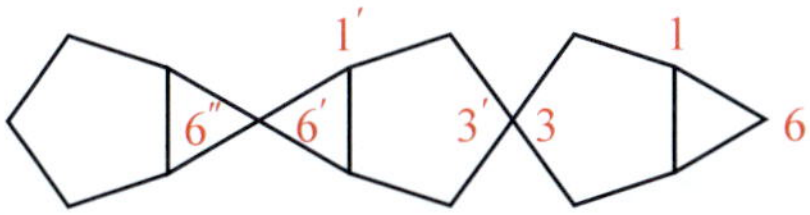

3,3′:6′,6″-Dispiroter[bicyclo[3.1.0]hexane]

a) If there is a choice of locants, the lowest setting for all spiro atoms is selected when contrasted as a set-in ascending order and, if still undecided, in citation order.

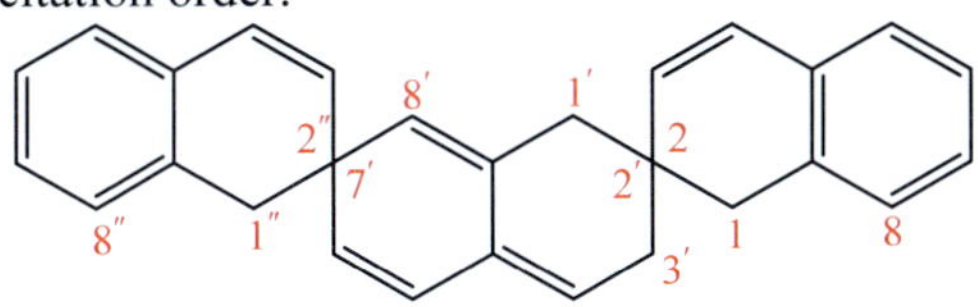

1*H*,1′,*H*,1″*H*,3′*H*-2,2′:7′,2″-Dispiroter[naphthalene]

b) Spiro ring systems composed of three identical components, each containing only one spiro atom, are named by placing the prefix "spiroter" before the name of the component ring system in square brackets. The three spiro locants are indicated before the name within the brackets [1].

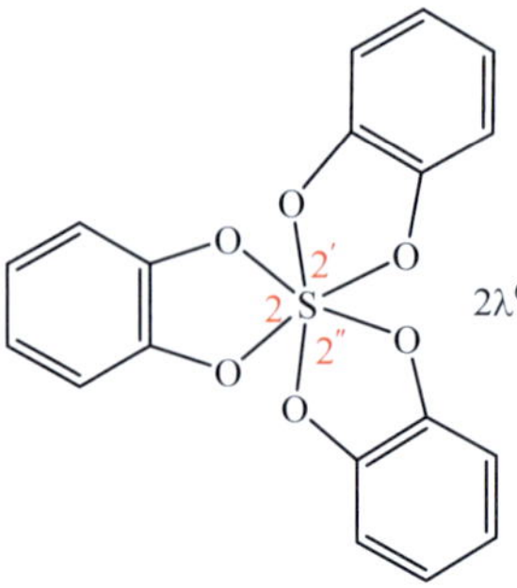

$2\lambda^6$,2′,2″-Spiroter[[1,3,2]benzodioxathiole]

2.7. Monospiro compounds with different components, at least one of which is polycyclic

Monospiro ring systems with different components, at least one of which is polycyclic, are named by placing the prefix "spiro" before the names of the components ordered in alphabetical sequence and enclosed in brackets. The spiro atom is assigned the lowest possible number, and its position is indicated by placing the appropriate numbers between the names of the two components. Locants of the second component are primed, and any locants needed to name this component ring system are not primed but placed in square brackets. Indicated hydrogen is cited, if necessary, in front of the complete name.

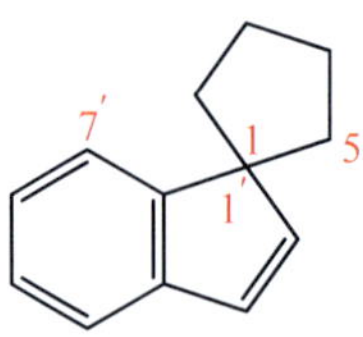

Spiro[cyclopentane-1,1′-indene]

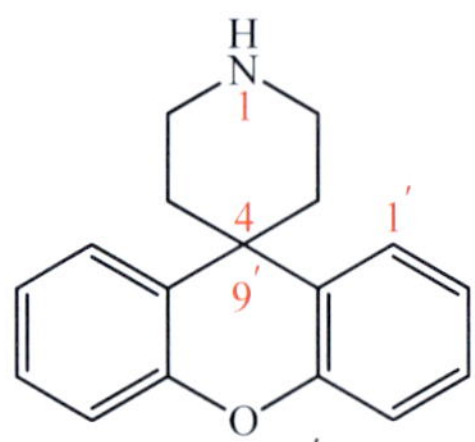

Spiro[piperidine-4,9′-xanthene]

a) If the alphabetic order of the component names is insufficient to determine the order, ring superiority principles based on lower italic fusion letters and numbers, heteroatom locants, and von Baeyer descriptor numbers are used. If the numbering of the bicyclic fused benzo ring components is consistent with the numbering of the spiro ring system, they are indicated before the component name, but if they are not consistent with the numbering of the spiro ring system, they are placed in square brackets.

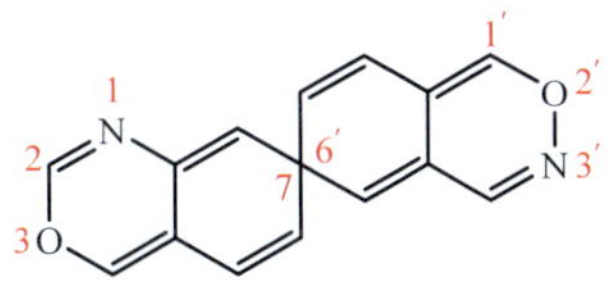

Spiro[3,1-benzoxazine-7,6′-[2,3]benzoxazine]

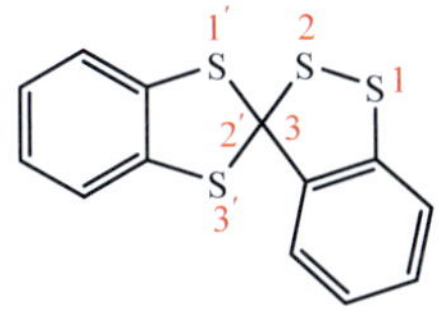

Spiro[1,2-benzodithiole-3,2′-3,2′-[1,3]benzodithiole]

b) Monospiro ring systems with different ring components, at least one polycyclic ring and at least one ring component, are named by placing the skeletal replacement "a" prefix before the "spiro" term.

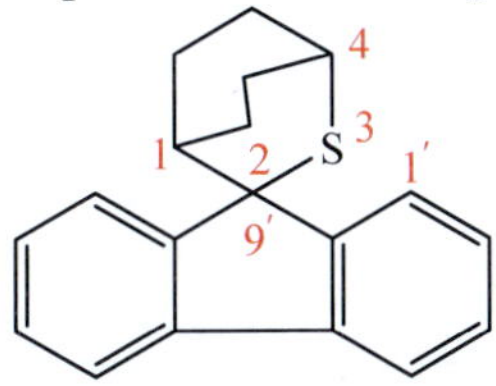

3-Thiaspiro[bicyclo[2.2.2]octane-2,9′-fluorene]

c) Spiro ring systems possess three components with only one spiro atom. At least two different components are named by placing the prefix "spiro" in front of the names of the components in alphabetic sequence with the convenient spiro locants. The same components use the bis-prefix. When three components are different, they are mentioned in the middle of the name , enclosed in parentheses to highlight this uncommon position [1].

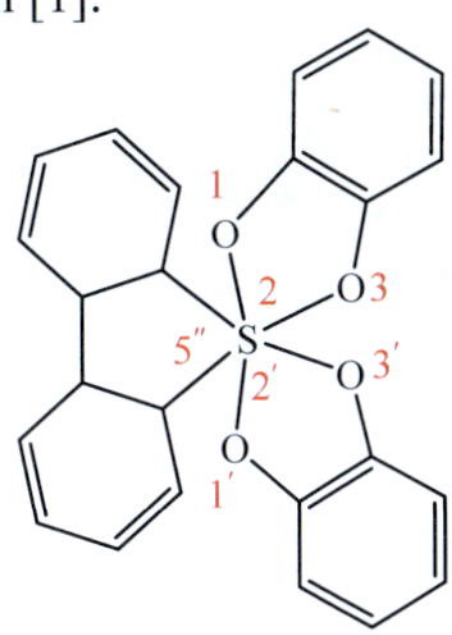

$2\lambda^6$-Spiro[1,3,2-benzodioxathiole-2′,2′–([1,2,3] benzoxadithiole)-2,5″-dibenzo[*b*,*d*]thiophene]

2.8. Unbranched polyspiro compounds with at least two different components and at least one of them is polycyclic

They are named by placing the name of each component in order of occurrence in square brackets, beginning with the terminal component earlier in alphabetical order. A prefix indicating the number of spiro atoms is placed in front of the component names (dispiro-, trispiro-, etc.). Locants required to describe the second and later components are not primed but placed in square brackets. The positions of the spiro atoms are illustrated by pairs of convenient locants separated by a comma between each pair of component ring system names. Indicated hydrogen, if required, is mentioned in front of the full name.

If there is a choice of names, the first mentioned is the one earlier in alphabetical sequence for the second component, and so on for the third component, and so forth. If both terminal ring systems are identical, the order of citation of ring system components is determined by comparing the pair of second ring components from the end of the structure, and so on. Multiplicative prefixes are not applied with dispiro systems having identical terminal components [1].

In cases where the alphabetical sequence is not sufficient to decide between two names, ring superiority principles based on lower italic fusion letters and numbers, heteroatom locants, and von Baeyer descriptor numbers are applied. If the locations in bicyclic fused benzo ring components correspond to the numbering of the spiro ring system, they are mentioned in front of the name of the ring. However, if they do not match the numbering of the spiro ring system, they are placed in brackets.

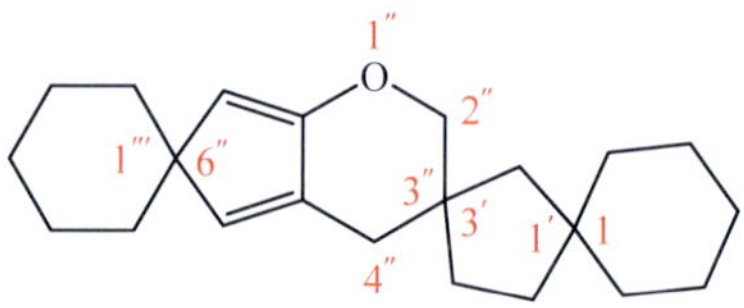

2″*H*,4″*H*-Trispiro[cyclohexane-1,1′-cyclopentane-3′,3″-cyclopenta[b]pyran-6″,1‴-cyclohexane]

If there is still a choice of numbering, the numbers of the von Baeyer descriptor are considered in the order of citation. The name is selected with lower numbers at the first point of difference.

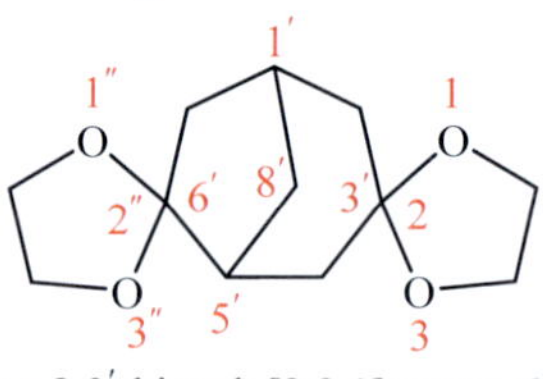

Dispiro[1,3-dioxolane-2,3′-bicyclo[3.2.1]octane-6′,2″-[1,3]dioxolane]

2.9. Branched polyspiro compounds with at least one polycyclic component

When three or more components are spiro-fused to an identical component, the resulting ring system is a branched spiro-fused ring system. Terminal components have only one spiro atom.

If a central component is spiro-fused to three or more identical terminal components, the central component is named first, and its locants are unprimed. The terminal components are cited with the appropriate multiplicative prefix (tris-, tetrakis-, etc.), and locants are primed, double-primed, etc. The lowest number is assigned to the spiro atom of the central ring. The spiro atoms are indicated by pairs of locants separated by colons. Indicated hydrogen is mentioned, if necessary, in front of the full name.

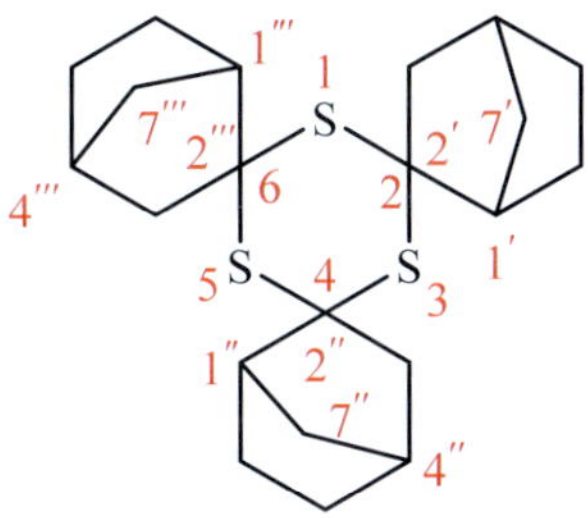

Trispiro[1,3,5-trithiane-2,2′:4,2″:6,2‴-tris(bicyclo[2.2.1]heptane]

Suppose two or more different terminal rings are spiro-fused to a central ring. In that case, if convenient, the alphabetically foremost is mentioned with a multiplicative prefix followed by the central ring and the remaining terminal rings in alphabetic sequence. The first said ring is spiro-fused to the lowest numbered spiro position of the central ring. Then, the lowest locants are given to the remaining terminal rings according to their alphabetic sequence.

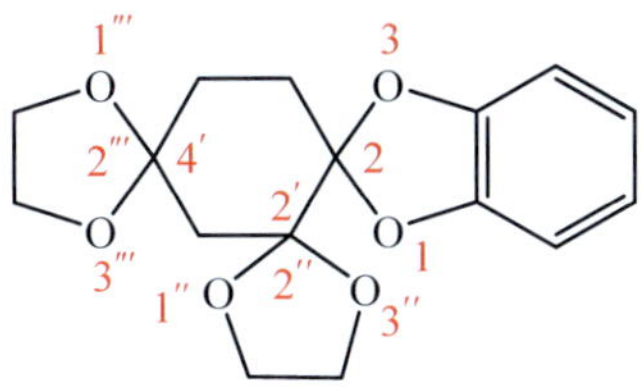

Trispiro[1,3]benzodioxole-2,1′-cyclohexane-2′,2″:4′,2‴-bis([1,3]dioxolane)]

a) When the application of the rules does not lead to a decision and there is a choice of locants, the lowest set of locants is assigned to the entire spiro system when compared as a set in increasing numerical order, and if still undecided, in the order of citation in the name. If a choice still needs to be made, heteroatoms and indicated hydrogen are taken into consideration.

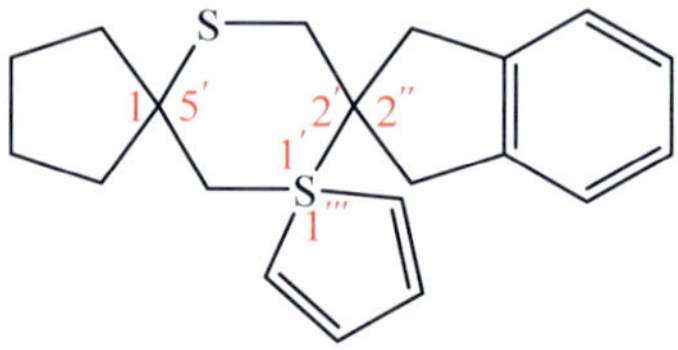

1′λ^4-Trispiro[cyclopentane-1,5′-[1,4]dithiane-2′,2″-indane-1′,1‴-thiophene]
not 1′λ^4-Trispiro[cyclopentane-1,3′-[1,4]dithiane-6′,2″-indane-1′,1‴-thiophene]
not 4′λ^4-Trispiro[cyclopentane-1,2′-[1,4]dithiane-5′,2″-indane-4′,1‴-thiophene]
(1,1′,1‴,2′,2″,5′ is lower than 1,1′,1‴,2″,3′,6′ or 1,1‴,2′,2″,4′,5′)

b) As cited before, if additional components are spiro-fused to a branched polyspiro compound, the following criteria are used in order:

i) Each monocyclic ring spiro-fused is named as a unit, including the maximum number of monocyclic rings. This unit is applied as a component for further spiro-fusion. Braces are used (instead of square brackets) after the initial polyspiro prefix to enclose components, at least one formerly spiro-fused, to assist in representing the composite nature of the name.

ii) If there is no such system or the system cannot be named by the standard spiro-fusion rules, the largest preferred spiro system is named and considered a unit for further spiro-fusion. The priming of this unit is continued in the residue of the name.

iii) After identification, the spiro-fused and the remaining components are treated in a standard manner [1, 2].

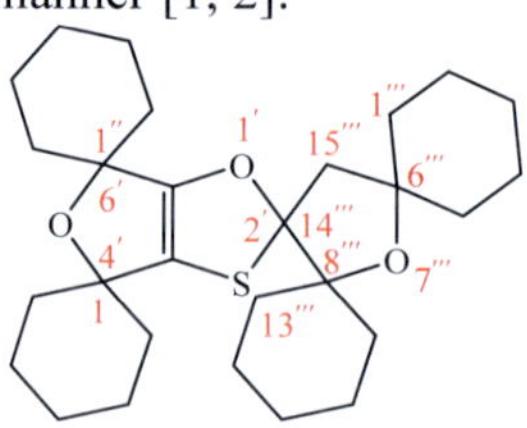

Trispiro{bis(cyclohexane)-1,4′:1″,6′-furo[3,4-*d*][1,3]oxathiole-2′,14‴-[7]oxadispiro-[5,1,5,2]pentadecane}

2.10. Chiral spiro compounds

Stereochemistry associated with the tetrahedral geometry is represented by the method of section E of the nomenclature of the organic chemistry book (Recommendations 1974) and R-7 of a guide to IUPAC nomenclature of organic compounds book (Recommendations 1993).

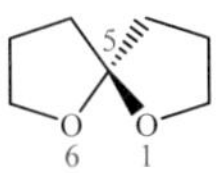

(*S*)-1,6-Dioxaspiro[4.4]nonane

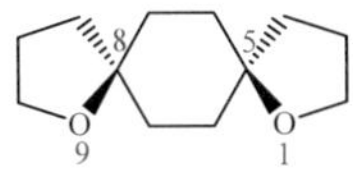

cis-1,9-Dioxadispiro[4.2.4.2]tetradecane

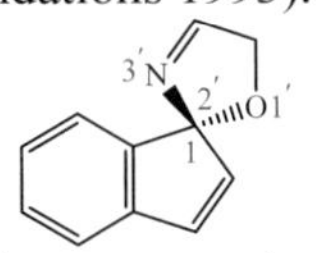

(*R*)-5′*H*-Spiro[indene-1,2′-[1,3]oxazole]

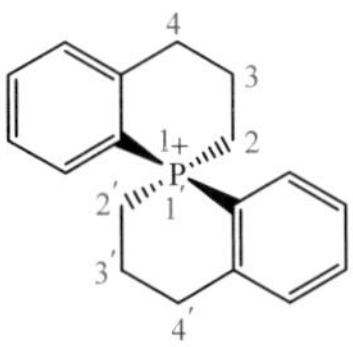

(*S*)-3,3′,4,4′-Tetrahydro-2*H*,2′*H*-1λ^5-1′-spirobi[phosphinolin]-1-ylium

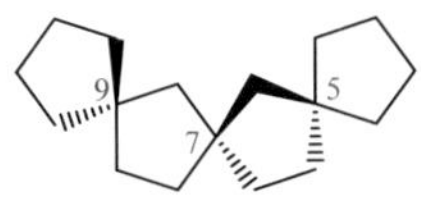

(*R*)-Trispiro[$4.1.1.4^9.2^7.2^5$]heptadecane

Stereochemistry associated with the non-tetrahedral geometry is indicated by the method of coordination chemistry [1].

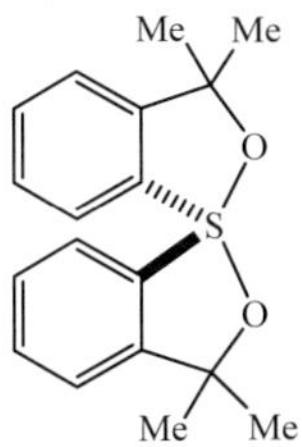

(*TBPY*-4-11′–A)-3,3,3′,3′-Tetramethyl-3*H*,3′*H*-1λ^4,1′-spirobi[2,1]benzoxathiole]

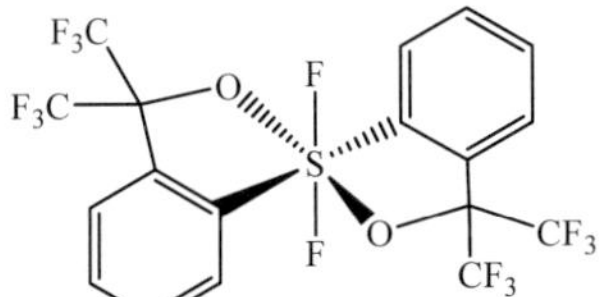

(*OC*-6-12)-1,1-Difluoro-3,3,3′,3′-tetrakis(trifluoromethyl)-3*H*,3′*H*-1λ^6,1′-spirobi[2,1]benzoxathiole]

2.11. Spiro macromolecule

A spiro macromolecule is a double-strand macromolecule in which adjacent constitutional units are joined to each other through three atoms, with two on one side and one on the other side of each constitutional unit. Alternatively, it is a double-strand macromolecule consisting of a continuous and uninterrupted sequence of rings, where adjacent rings have only one atom in common.

2.11.1. Bicyclic monospiro macromolecules

Both rings of a spiro macromolecule are named independently [3]. The spiro atom is considered a branch point in each ring, and a locant is added to the branch point term that refers to the less complex ring. The component names are placed in brackets in the order of choice, regardless of complexity, and the prefix "spiro" is placed in front of the names of the components.

The general format for the IUPAC nomenclature of bicyclic monospiro macromolecules is spiro[cyclo(ring name 1)-[Bn],[Bn′]-cyclo(ring name 2)], where [Bn] and [Bn′] are the locants for the spiro atom.

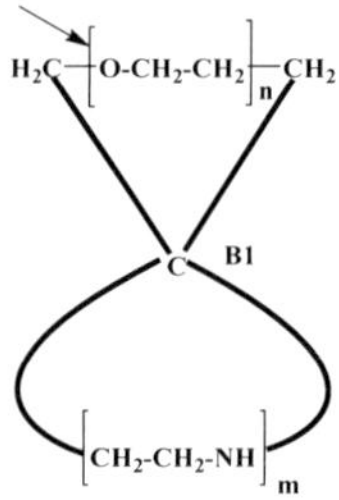

Spiro[cyclo[poly(oxyethylene)-[2:B1]propane-1,3-diyl]-[B1],[B1′]-cyclo [poly(iminoethylene)-[B1′] methylene]]

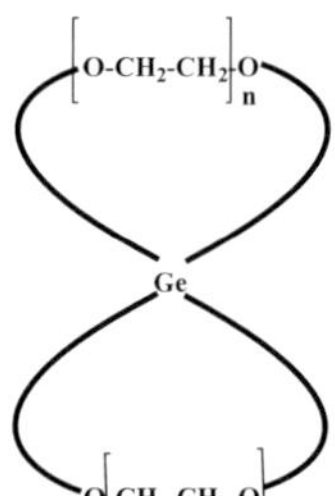

Spiro[cyclo[oxy[B1]germanediyl-poly(oxyethylene)]-[B1],[B1′]-cyclo [oxy[B1′]germanediyl-poly(oxy-ethylene)]] or [B1],[B1′]-Spirobi[cyclo [oxy [B1]germanediyl-poly(oxyethylene)]]

2.11.2. Spiro macromolecules with additional bridges

In a spiro macromolecule compound with a free spiro union, if there are additional bridges, the most complex spiro macromolecule that can be named according to the previous rules is selected and named accordingly. Branch point expressions are added as necessary, and the names of the bridges are placed in front of the name.

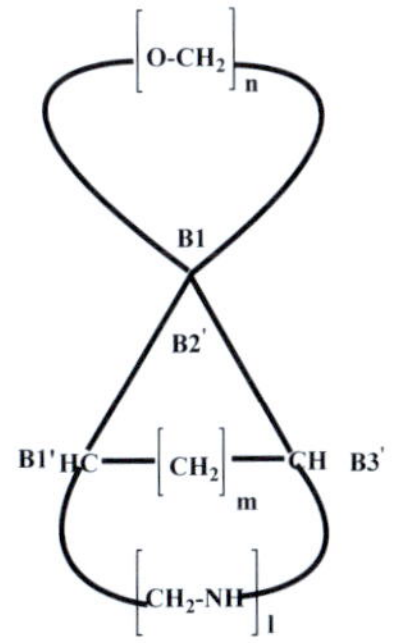

[B1′],[B3′]-[Poly(methylene)]-spiro[cyclo[poly(oxymethylene)-[B1] methylene]-[B1],[B2′]-cyclo[poly-(iminomethylene)-[1:B1′,2:B2′,3:B3′] propane-1,3-diyl]]

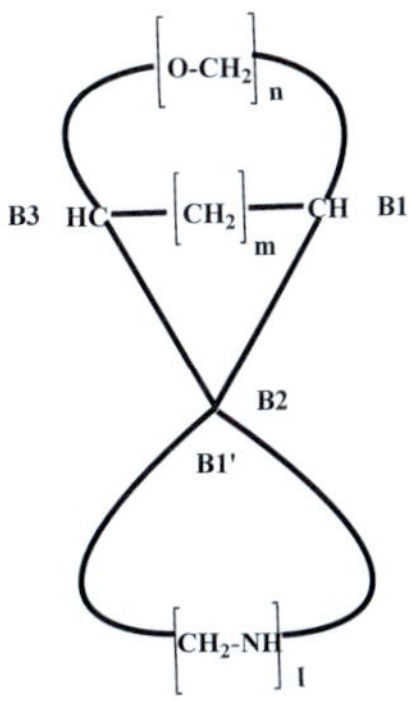

[B1],[B3]-[Poly(methylene)]-spiro[cyclo[poly(oxymethylene)-[1:B1,2:B2, 3:B3]propane-1,3-diyl]-[B2],[B1′]-cyclo[poly(iminomethylene)-[B1′] methylene]]

2.11.3. Polyspiro macromolecules

In a polymacrocyclic compound with more than one spiro atom, the numerical prefixes di-, tri-, etc., are added before the "spiro" prefix to indicate the number of spiro atoms shared by two macromolecular rings. The component rings of the macromolecule are named in the order of their occurrence, starting with the most superior terminal component. The locants for the spiro atoms represented by the respective branch point expression of each of the individual macrocycles are placed between the names of the respective components, with a prime added to the branch point expression referring to the second and two primes to those referring to the third component ring, etc. [3].

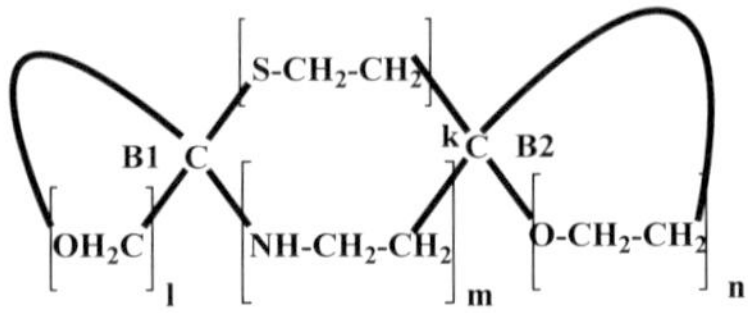

Dispiro[cyclo[poly(oxymethylene)-[B1]methylene]-[B1],[B2′]-cyclo[poly(sulfanediylethylene)-[B1′]methylene-poly(ethyleneimino)-[B2′]methylene]-[B1′],[B1″]-cyclo[poly(oxyethylene)-[B1″]methylene]]

2.12. Spiro polymer

A polymer, the molecules of which consist of an uninterrupted sequence of rings with adjacent rings having one atom in common, is named spiro polymer.

Examples:

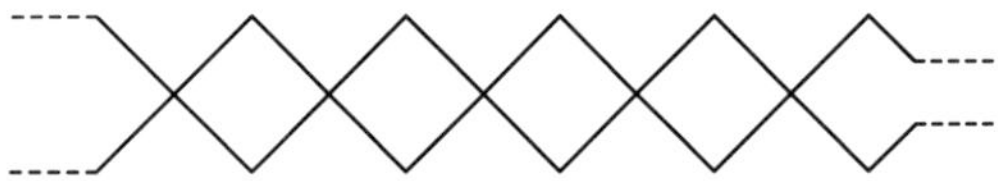

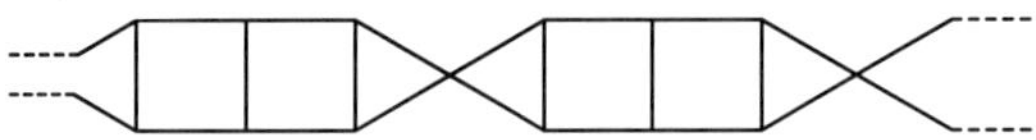

Structure-based nomenclature is based on the selection of a preferred constitutional repeating unit, which forms the basis for the naming of the polymer. The name of the polymer is obtained by adding the prefix "poly" to the name of the repeating unit, which is named according to the rules of organic nomenclature whenever possible. For double-strand polymers, such as spiro polymers and ladder polymers, the repeating unit is typically a tetravalent group that is attached to four atoms. However, since some of these attachments may involve double bonds, the unit may sometimes be hexavalent or octavalent. The steps to be followed for naming these polymers are similar to those for single-strand polymers and include:

- Identifying the preferred constitutional repeating unit,
- Orienting the repeating unit,
- Naming the repeating unit, and
- Naming the polymer.

2.12.1. Recognize the preferred constitutional repeating unit

Some ways to break the same structures into constitutional repeating units are illustrated below.

i) 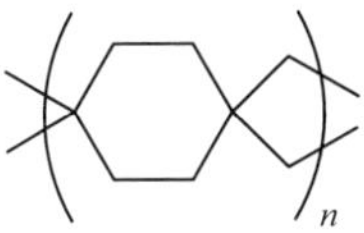 or 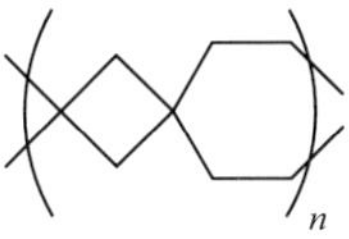

ii) 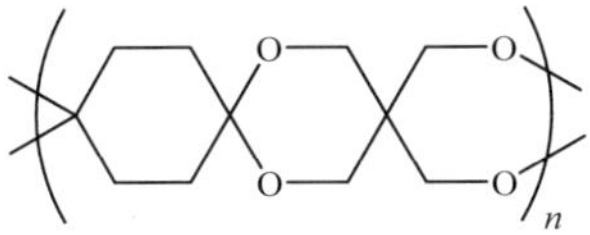 or 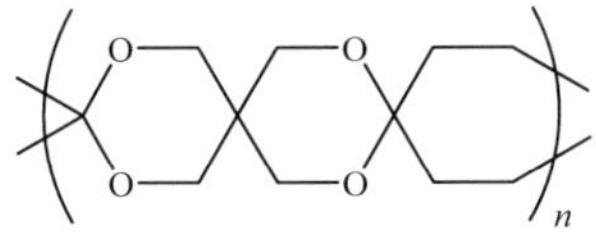

iii) 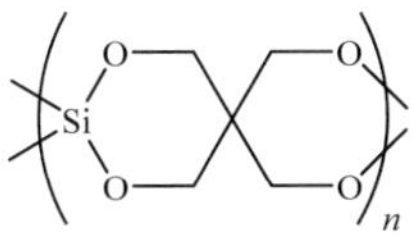 or 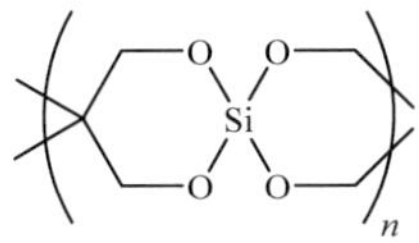

iv) 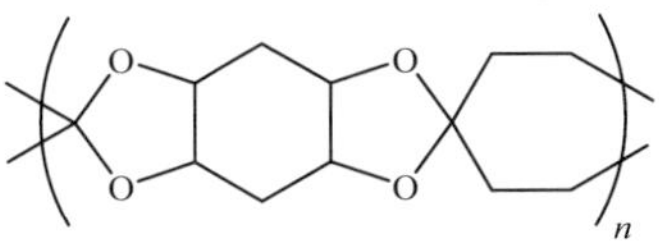 or 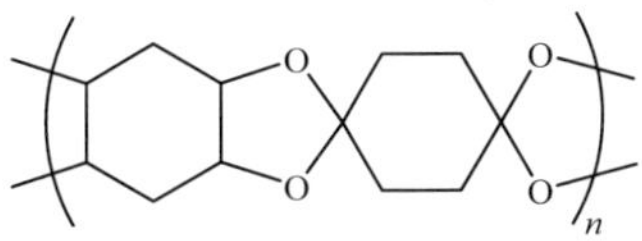

In order to create a unique name for a polymer, a single preferred constitutional repeating unit must be selected. This is achieved by following Rules:

Rule 1. The polymer chain should be terminated by breaking a non-aromatic ring. For aromatic rings (i.e., those with chemistry similar to benzene), the following criteria should be observed in decreasing order of preference:

- Minimize the number of free valences of the constitutional repeating unit.
- Maximize the number of the most preferred heteroatoms in the ring system.
- Maintain the most preferred ring system.
- Select the longest chain for the acyclic constitutional repeating unit.

The priority of ring systems is determined by using the following criteria consecutively until a decision is reached:

A heterocycle is superior to any carbocycle, regardless of the number of rings.

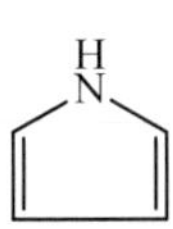 senior to 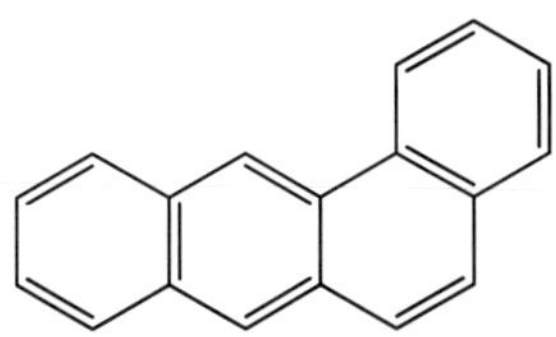

1. For heterocycles, the criteria are based on the nature and position of the heteroatom.

senior to
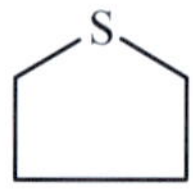

2. The larger number of rings.

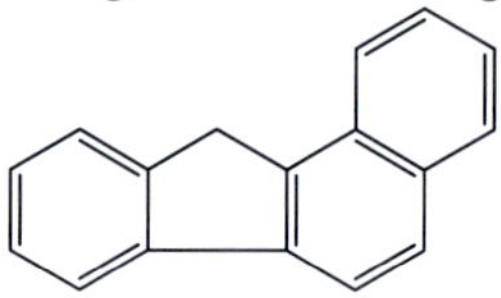
senior to
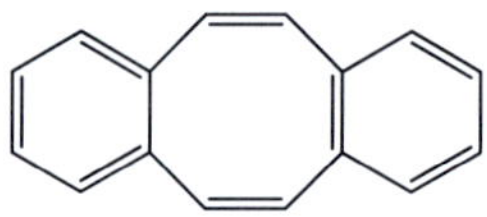

(four ring) (three reing)

3. Larger individual rings at the first point of disagreement.

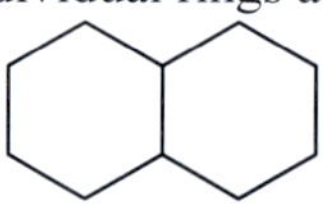
senior to
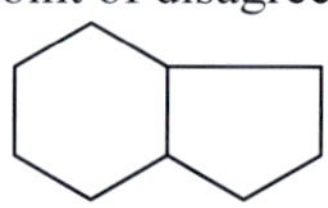

(6,6) (6,5)

4. The larger number of atoms in joint among rings.

senior to
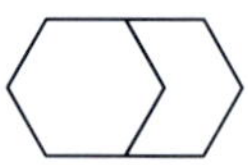
senior to
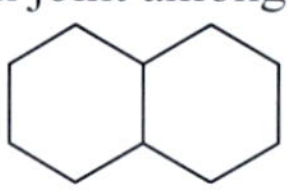
senior to

(bridged) (bridged) (*ortho*-fused) (spiro)

2.12.2. Orientate the constitutional repeating unit

Rule 2. The acyclic constitutional repeating unit that results from breaking the ring is oriented in such a way that the lowest free valence locant is positioned at the lower left.

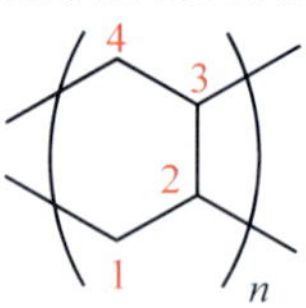

is preferred to
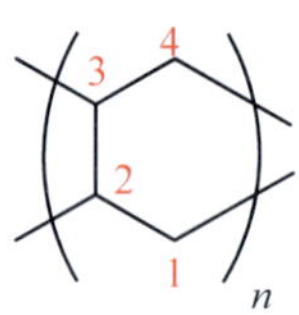

1,4:3,2– 2,3:4,1–

If there is a different choice, the acyclic unit is oriented to give the lower locant number on the cause of the first difference.

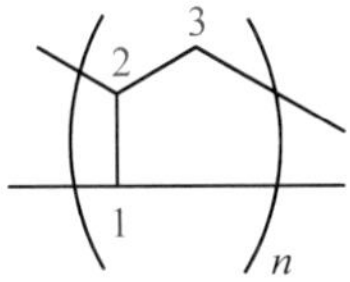

is preferred to

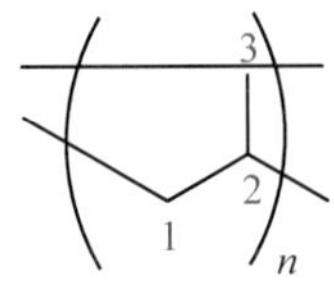

1,2:3,1– 1,3:3,2–

Rule 3: If the preferred constitutional repeating unit has a ring system with a fixed numbering of locants, it is oriented (e.g., by rotation) so that the lowest free valence locant is at the lower left, and the next locants in the ascending sequence follow a clockwise orientation.

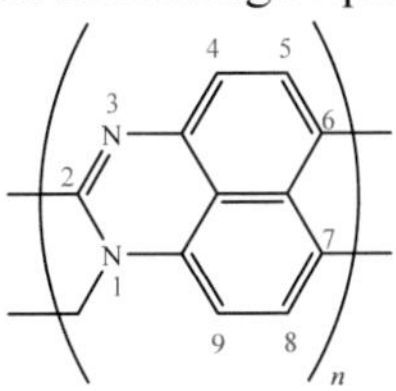

is preferred to

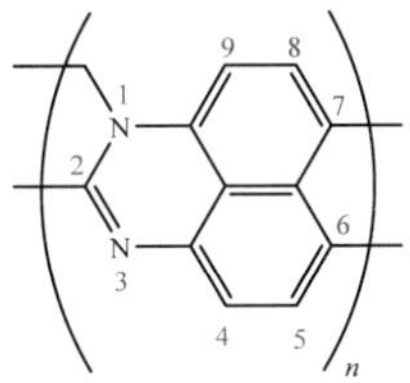

and

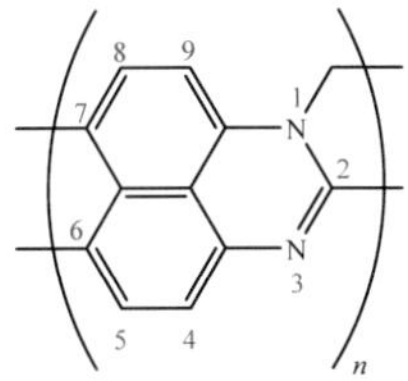

Whenever orienting the ring with the lowest free valence locant at the lower-left results in an anticlockwise direction of locants, the orientation is altered to the next lowest free valence locant at the lower left to maintain the rising sequence of locants in the clockwise direction.

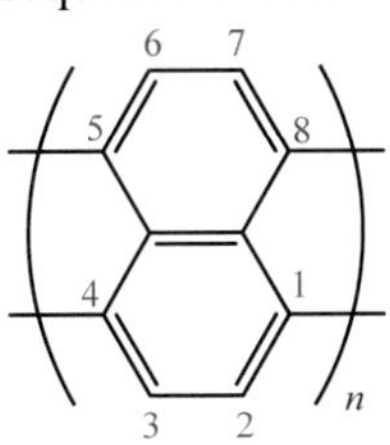

is preferred to

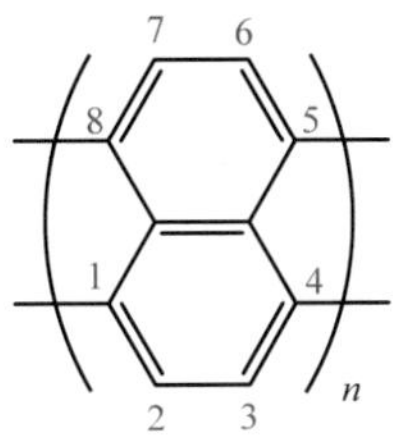

Rule 4. Suppose the preferred constitutional repeating unit contains a ring system and acyclic subunits. In that case, it is oriented in such a way that the ring system is on the left side, and the acyclic subunits are on the right of the ring system. The ring system is oriented according to the Rule 3.

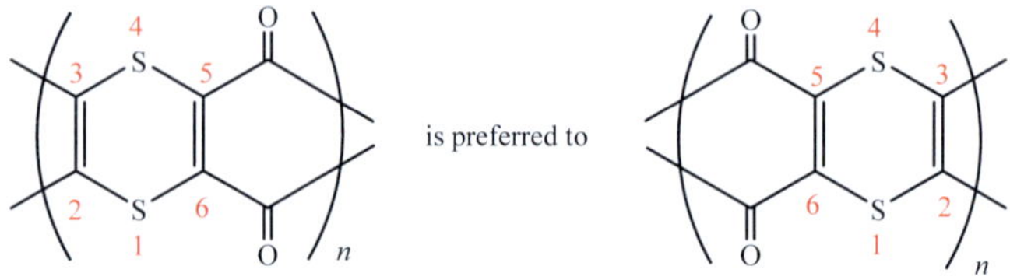

The naming of the preferred constitutional repeating unit follows a left-to-right order. The ring system is named according to the rules of naming carbocyclic and heterocyclic rings, followed by a suitable suffix, such as "2,3:5,6-tetrayl", "2,3:9-triyl-10-ylidene", or "2,3-diyl:6,7-diylidene". The locants for the free valences are always mentioned just before the corresponding ending and are presented in the following sequence: lower left, upper left, upper right, lower right. The left locants are separated from the right locants with a colon. Next, the names of acyclic subunits are listed in order of upper right and lower right, before the corresponding locants (which refer to the ring system positions where the acyclic subunits are joined).

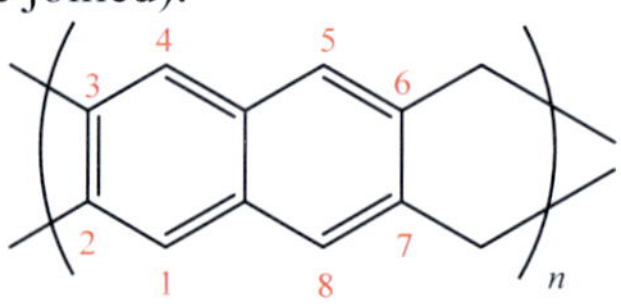

Poly(naphthalene-2,3:6,7-tetrayl-6,7-dimethylene)

2.12.3. Naming the polymer

Rule 5: The polymer is named with the prefix "poly" followed by the name of the preferred constitutional repeating unit in parentheses or brackets.

2.12.3.1. Polymers consisting of repeatedly spiro carbocycles

The preferred constitutional repeating unit resulting from a spiro polymer consisting of a repeating carbocycle is a tetravalent acyclic group.

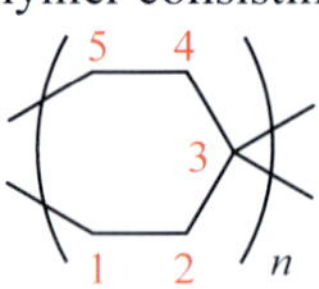

Poly(pentane-1,5:3,3-tetreyl)
(The longest straight chain is preferred)

not

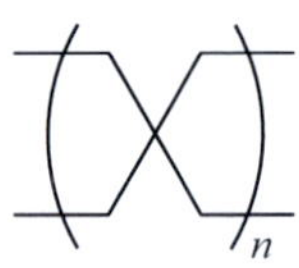

2.12.3.2. Polymers constituted of repeatedly spiro carbocyclic systems

The constitutional repeating unit resulting from a spiro polymer containing a repeating polycarbocyclic (multi-ring) system is a tetravalent group or a combination of a tetravalent (multivalent) cyclic group with one or two bivalent (multivalent) acyclic groups.

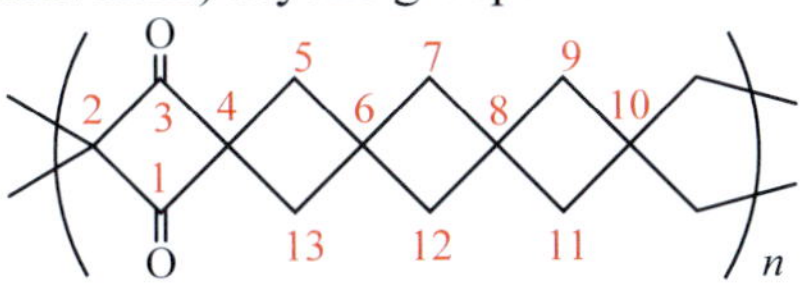

Poly [(l,3-dioxotrispiro[3.1.1.3.1.1]tridecane-2,2:10,l0-tetrayl)-10,10-dimethylene]

2.12.3.3. Polymers consisting of repeating spiro heterocycles

The preferred constitutional repeating unit of a spiro polymer containing a repeating heterocycle is an acyclic tetravalent carbon group in which some carbon atoms are replaced by heteroatoms. The orientation of the constitutional repeating unit is determined by assigning the lowest possible locants to the heteroatoms and ensuring that the heteroatom of highest seniority has a lower locant than the next most senior heteroatom. The constitutional repeating unit is then named using replacement nomenclature.

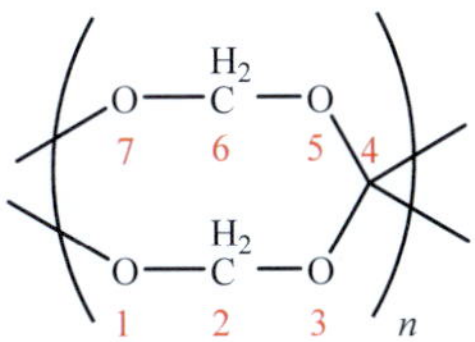

Poly(1,3,5,7-tetraoxaheptane-1,7:4,4-tetrayl)

2.12.3.4. Polymers constituted of repeatedly spiro heterocyclic systems

The constitutional repeating unit resulting from a spiro polymer, which includes a repeating heterocyclic ring system, is a combination of a tetravalent ring system with one or two bivalent acyclic groups. If there are multiple options for the constitutional repeating unit, it is oriented in such a way that the acyclic subunit with the terminal atom of higher seniority is positioned at the upper right [4].

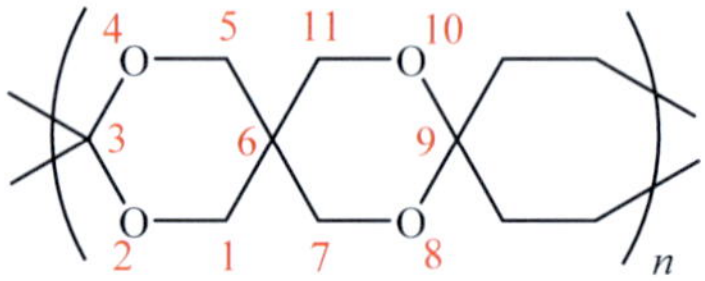

Poly(2,4,8,10-tetraoxaspiro[5.5]undecane-3,3:9,9-tetrayl-9,9-diethylene)
(all hetero atoms are in the ring system)

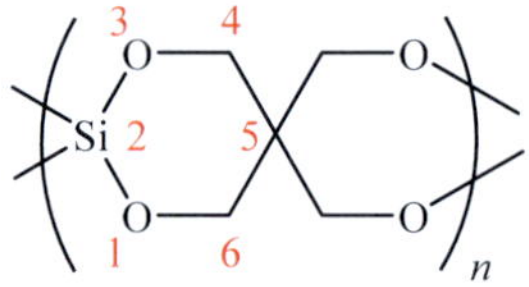

Poly [l,3-dioxa-2-silacyclohexane-2,2: 5,5-tetrayl-5,5-bis(methyleneoxy)]
(fixed ring numbering decides the orientation)

References

[1] P. Moss Gary. 1999. "Extension and revision of the nomenclature for spiro compounds". Pure and Applied Chemistry. No. 3. 531-558.

[2] Das Arijit. 2013. "New innovative methods for IUPAC nomenclature of bicyclo and spiro compounds in organic chemistry". Online Submission. No. 7. 596-597.

[3] Barón Matilde, B. Fox Richard, Fradet Anne, Hellwich Karl-Heinz, *et al.* 2008. "Structure-based nomenclature for cyclic organic macromolecules". Pure and Applied Chemistry. No. 2. 201-232.

[4] V. Metanomski Wladyslaw, E. Bareiss Rolf, Kahovec Jet, L. Loening Kurt, *et al.* 1993. "Nomenclature of regular double-strand (ladder and spiro) organic polymers". Pure and Applied Chemistry. No. 7. 1561-1580.

3

SYNTHESIS OF SPIROCYCLES I, GENERAL REACTIONS

3.1. Introduction

Spiranes are an exciting class of chemicals because of their interesting conformational features and their structural implications on biological systems. Developing spirocycle compounds is a challenging task in modern organic synthesis, as it involves creating a quaternary center, which is one of the most difficult synthetic transformations.

There are various synthetic methods for the synthesis of spiro compounds, including alkylation methods, Pericyclic-type reactions, cyclization reactions, rearrangement-based approaches, ring-contraction method, ring-enlargement method, ring closure of geminally di-substituted compounds, organometallic strategies, organocatalytic approaches, oxidative or reductive coupling reactions, metathesis approaches, Darzens reaction, and Pauson–Khand reaction.

3.2. Alkylation methods

Alkylation is a process in which an alkyl substitution migrates from one molecule to another via an alkyl carbocation, a carbanion, a free radical, or a carbene, which is produced from electrophilic alkylating agents like olefin and alkyl halide in the presence of a strong mineral acid or a Lewis acid such as metal oxide catalysts. This text describes reactions where the alkylation reaction is used as a key step in the preparation of spiro compounds.

In 1981, Eilerman et al. demonstrated intramolecular decarboxylative alkylation of β-keto-esters. They showed that regiospecific intramolecular alkylation of enolates, generated *in situ* by halide-induced non-hydrolytic decarboxylation of ω-halogeno-β-ketoesters **1**, could provide a convenient route to spirocyclic ketones **2** (Scheme 3-1, Eq. 1). The lactone carboxylate **3** also afforded the spirolactone **4** in a 69% yield (Scheme 3-1, Eq. 2). The reaction of ketoester **5** with 2-isopropylidene-l,4-dichloro butane (sodium hydride-DMF, 3 h) gave adduct **6** in a 65% yield.

Decarboxylative spirocyclization of **6** provided the ketone **7** in an 80% yield (9:1 mixture of epimers). Treatment of **7** with methyl-lithium furnished allylic alcohol **8** which was oxidized with pyridinium chlorochromate to give a mixture of (±)-β-vetivone **9a** and (±)-epi-β-vetivone **9b** (ratio 9:1) in a 72% yield. The alcohol **8** was also converted (*p*-toluenesulfonic acid at 25 °C) into a mixture of (±)-β-vetispirene **10a** and its epimer **10b** (ratio 9:1) in an 82% yield (Scheme 3-1, Eq. 3) [1].

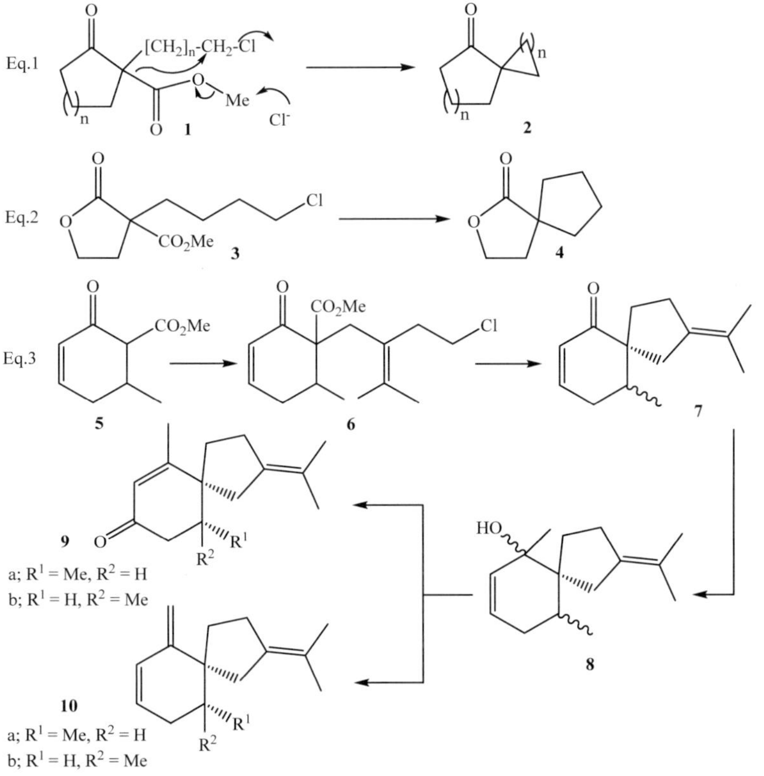

Scheme 3-1. Intramolecular decarboxylative alkylation of β-Keto-esters.

The synthesis of spirolactone **15** was reported by Posner et al. in 1988 via sequential intermolecular allylation/intramolecular alkylation reaction (Scheme 3-2). An allylic/homoallylic biselectrophile such as **12** was used to allylated lactone **11** to give compound 13. Then, the resulting compound **13** was produced *in situ* enolate ion intermediate **14**. Intramolecular C-alkylation of this enolate ion provided spirolactone **15**.

Compound **15** was converted into the natural product β-vetivone **16** by a series of synthetic transformations [2].

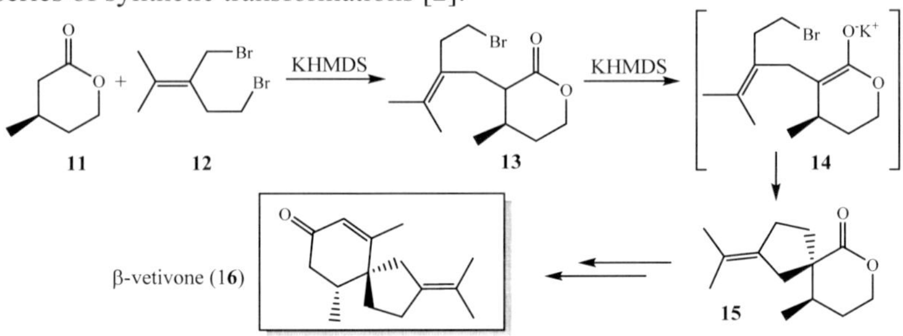

Scheme 3-2. Synthesis of β-vetivone 1**6**.

In 1989, Sakata et al. reported the synthesis of spiroketone **19** starting from several tribromo compounds **17** and 1-indanone **18** in the presence of sodium hydride which was reduced with Et_3SiH-CF_3CO_2H to give compound **20** in 69% yield. Treatment of **20** with BuLi and DMF gave **21** in 62% yield (Scheme 3-3) [3].

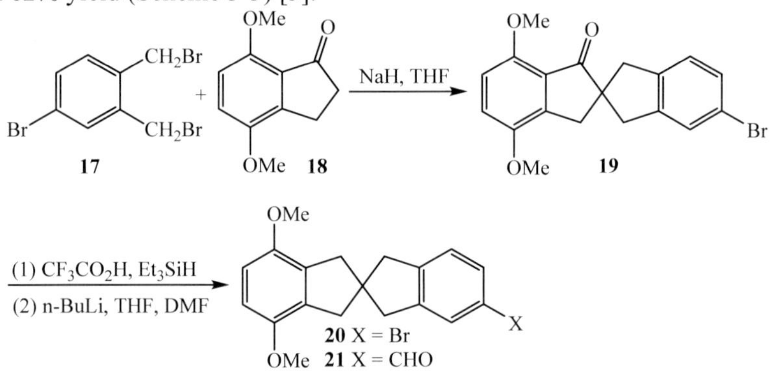

Scheme 3-3. Synthesis of spiroketone **19**.

Schulz et al. reported the synthesis of spiroionene **27** using a polyalkylation procedure (Scheme 3-4). First, tetrabromide **22** reacts with tosylamine **23** to give ditosylamide **24** which is then converted into diamine **25** through reductive cleavage with hydrobromic acid in *m*-cresol. The resulting diamine **25** then undergoes a reaction with 1,2-dibromo methylbenzene **26** to form spiro-conjugated ammonium salt **27** [4].

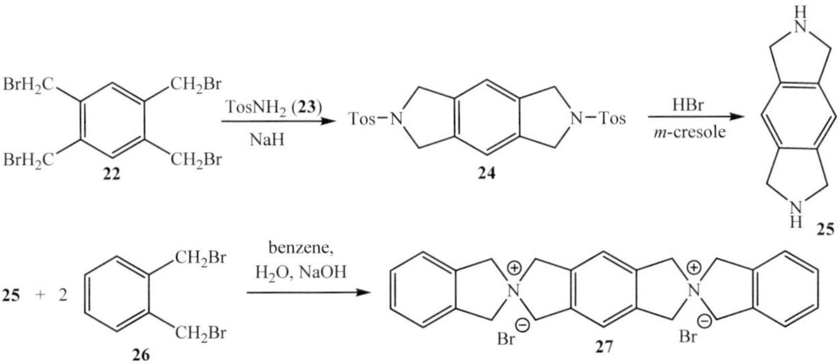

Scheme 3-4. The synthesis of polymeric ionenes with spirane structure.

Kotha et al. developed [4+2] and [2+2+2] cycloaddition reactions for the synthesis of diverse 2,2-spirobiindane-1,3-diones which are important precursors for the synthesis of unsymmetrical benzoannelated fenestranes. In this method, alkylation of indanedione **28** with 2,3-bis-(iodomethyl) buta-1,3-diene **29** and propargyl bromide **30** under phase transfer catalysis (PTC) conditions [K_2CO_3, dry CH_3CN, tetrabutylammonium hydrogen sulfate (TBAHS) at rt.] furnished key intermediates **31** and **32**, respectively. These intermediates then undergo [2+2+2] and [4+2] cycloaddition reactions with various dienophiles to generate various 2,2-spirobiindane-1,3-diones. For example, **31** reacts with dimethyl acetylene dicarboxylate **33** in anhydrous toluene under refluxing conditions for 24 hours to produce the aromatized product **34** directly (Scheme 3-5) [5].

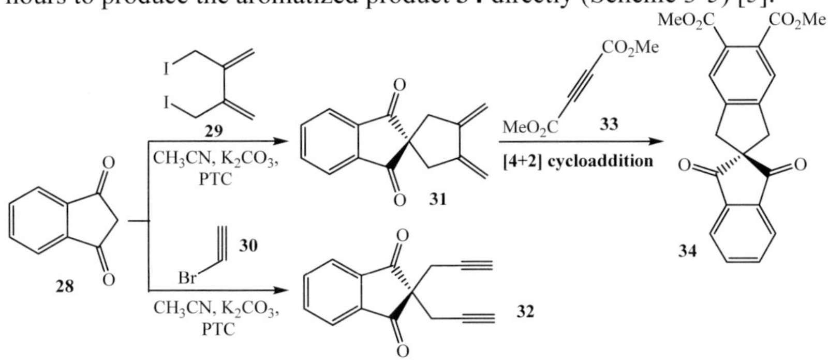

Scheme 3-5. Synthesis reported by Kota et al.

Edwards et al. showed that 6,6-spiroketal **39** might be prepared in an excellent yield by alkylation reaction of 3,4-dihydro-6-[(*para*-toluene-sulfonyl)methyl]-2*H*-pyran **35** and subsequent acid-catalyzed cyclization

(Scheme 3-6). 3-(*t*-Butyldimethylsilyloxy)-1-iodopropane **36** is the alkylating reagent required for this reaction, which is prepared by a reaction of propane-1,3-diol with tetrahydrofuran [6].

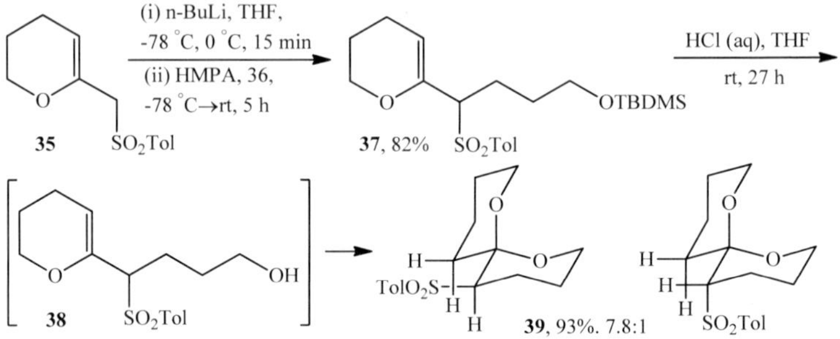

Scheme 3-6. Synthesis of 6,6-spiroketal **39**.

In 2006, Singh et al. reported a single-step reaction of *N,N′*-substituted and unsubstituted barbituric acids **40** with various alkyl dihalides **41** and **42** under phase transfer catalytic conditions using DMF–K_2CO_3 as a base and $TBAHSO_4$ as a catalyst to provide spirobarbituric acids **43** and **44** in moderate to high yields (Scheme 3-7). Although C_5-monoalkylated compounds in the enolic form were isolated, the second alkylation mainly occurred at C_5. The reaction mechanism for dialkylation at C_5 is described based on the tautomeric forms of the first alkylated intermediate [7].

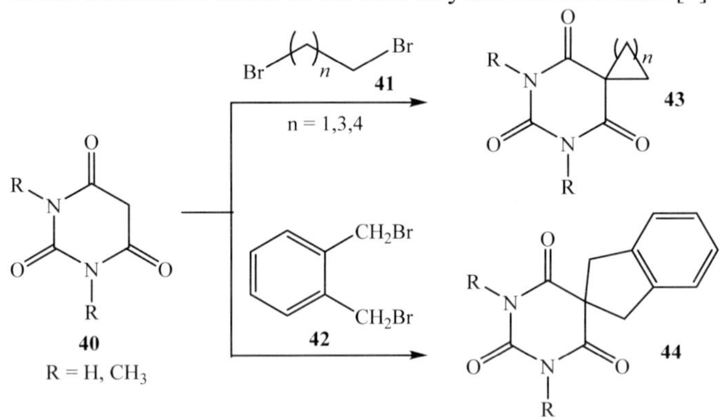

Scheme 3-7. Synthesis of spirobarbituric acids **43** and **44**.

To synthesize spirocyclic butenolides **48**, Marko et al. developed a two-step approach. Firstly, they used $ZnCl_2$ to mediate the condensation of 2-(trimethylsilyloxy)furan **45** and *ortho*-ester iodides **46**, yielding lactones **47** in yields of 95% and 83%, respectively. The lactones were then treated with a base to generate the corresponding dienolate which underwent intramolecular alkylation reaction to produce spirocyclic moieties **48** in good yield, whether the extra ring was five- or six-membered. In this short and efficient sequence, 2-(trimethylsilyloxy)furan acted as a dianion equivalent (Scheme 3-8) [8].

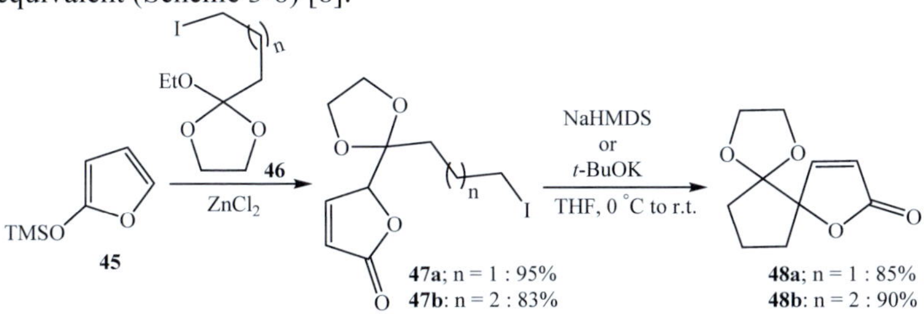

Scheme 3-8. Two-step synthesis of spirocyclic butenolides.

The Dussault group developed synthetic routes to prepare spiroketals **52a** and **52b** via intramolecular alkylations of *gem*-dihydroperoxides [9]. The synthesis of spirodioxolane **52a** commenced with the double reaction of heptanal with dilithiated isobutylene, resulting in the formation of diol **49a** as a mixture of syn- and anti- diastereomers that could not be resolved. Ozonolysis of the bismethanesulfonate in the presence of excess H_2O_2 did not lead to the desired *gem*-1,1-dihydroperoxide, but instead, the corresponding ketone was obtained, indicating the effect of the allylic sulfonates on carbonyl oxide formation selectivity.

The 3,3-bismesyloxy ketone was unstable towards β-elimination and rapidly reacted with H_2O_2/I_2 or H_2O_2/H_2SO_4 to yield a mixture of three diastereomeric spiro-bicyclic peroxyketals **52a**. The precursor for cyclization, **51b**, was produced from the ozonolysis of bismesyloxyalkene **50b** with excess H_2O_2 and used to synthesize spiro-bis-1,2-dioxane **52b**. The reaction of *gem*-dihydroperoxide **51b** with KO*t*-Bu in the presence of 18-crown-6 led to the formation of spiro-bis-1,2-dioxane **52b** in a short sequence as a mixture of three diastereomers (Scheme 3-9).

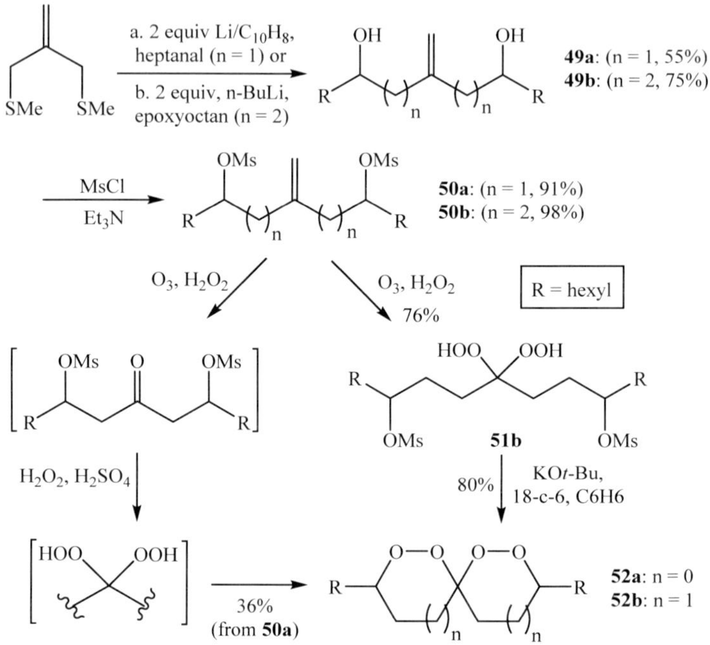

Scheme 3-9. Syntheses of spiro-**52a** and -**52b**.

Synthesis of spiro[3.3]heptane scaffold was developed by Radchenko et al. [10]. In this process, firstly, compound **54** was formed by the reaction of compound **53** with di-isopropyl-malonate. Subsequently, compound **54** was transformed into its (bis)tosyl derivative **55** which, in turn, was used to alkylate the di-isopropylmalonate to form diester **56**, possessing the spiro[3.3]heptane core. The synthesized spiro[3.3]heptane core can synthesize a library of isomeric glutamic acid analogs (Scheme 3-10).

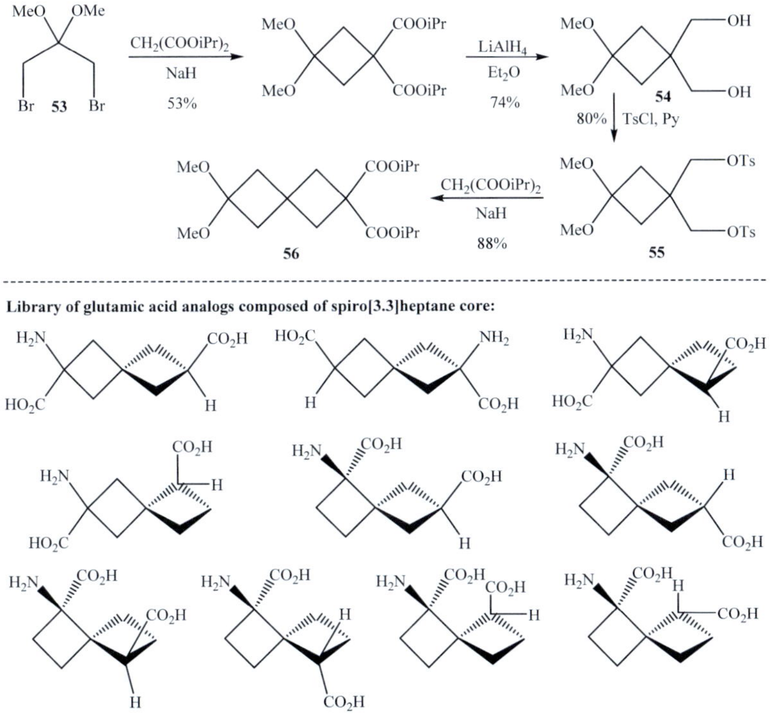

Scheme 3-10. Synthesis of spiro[3.3]heptane scaffold.

Miyaoka et al. reported on an enantioselective total synthesis of (–)-sinularianin B **61**, a unique sesquiterpenoid isolated from the Formosan soft coral Sinularia sp. [11]. One of the key transformations involved the formation of cyclopentane **58** containing a tertiary hydroxy group through a sulfone-mediated tandem intermolecular-intramolecular alkylation that involves a 5-*endo* cyclization (Scheme 3-11). This synthetic strategy also includes a ring-closing metathesis (RCM) to build the highly strained *trans*-5,6-fused ring **59** and an intramolecular Horner–Wadsworth–Emmons reaction to convert the 7-membered lactone **60** into γ-spirolactone **61** through an apparent acyl rearrangement. The total synthesis of sinularianin B was achieved in 16 steps starting from readily available allylic alcohol **57**, with a yield of 39.5%. The absolute configurations of the four stereocenters in (–)-sinularianin B were determined to be 1*R*, 4*S*, 8*S*, and 9*R*. The alkylation reaction is a remarkable example of a successful synthetic approach to this unique natural product.

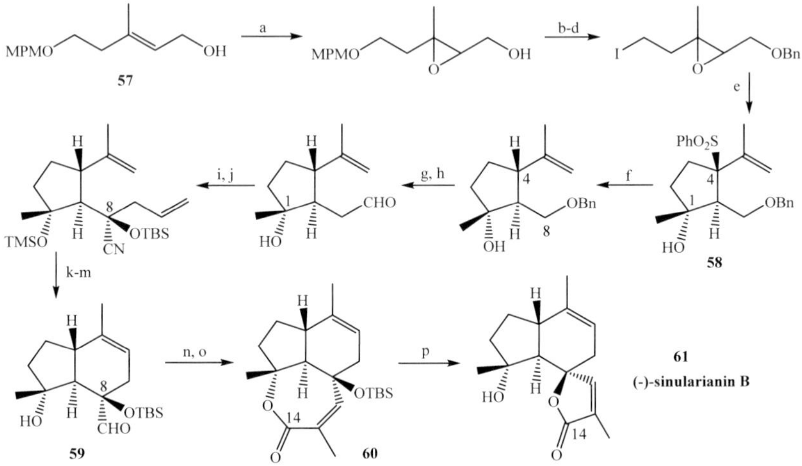

Scheme 3-11. Total synthesis of sinularianin B. (a) *tert*-butyl hydroperoxide (TBHP), diisopropyl D-tartrate, $Ti(O^iPr)4$, 4 Å MS, CH_2Cl_2, 20 °C, 97%, >95% ee; (b) BnBr, NaH, TBAI, THF, r.t.; (c) 2,3-dichloro-5,6-dicyano-1,4-benzoquinone (DDQ), CH_2Cl_2/sat. $NaHCO_3$ aq., r.t.; (d) I_2, Ph_3P, imidazole, CH_2Cl_2, r.t., 91%; (e) methallyl phenyl sulfone, *n*-BuLi, THF, -78 to -45 °C then *n*-BuLi, Me_3Al, -78 to -55 °C, 99%; (f) HCO_2H, Et_3N, *n*-Bu_3P, $Pd_2(dba)_3CHCl_3$, 1,4-dioxane, reflux, 91%; (g) Na, liq. NH_3, THF, -78 °C; (h) *o*-iodoxybenzoic acid, DMSO/ THF, r.t., 96%; (i) *tert*-butyldimethylsilyl cyanide, bis(triphenyl phosphoranylidene)ammonium chloride, THF, r.t. then trimethylsilyl cyanide (TMSCN), r.t.; (j) *n*-BuLi, *i*-Pr_2NH, THF, -78 to 0 °C then allyl bromide, hexamethylphosphoramide, -78 °C; (k) Grubbs 2nd generation catalyst, $ClCH_2CH_2Cl$, reflux; (l) di-isobutylaluminum hydride, toluene, -78 to 0 °C then sat. NH_4Cl aq., 0.5 M H_2SO_4 aq., Et_2O, r.t.; (m) tetra-*n*-butylammonium fluoride (TBAF), AcOH, THF, 40 °C, 67%; (n) 2-(diethoxy-phosphoryl)propanoic acid, carbodiimid, CH_2Cl_2, r.t.; (o) KO*t*Bu, THF, 60 °C, 78%; (p) TBAF, THF, r.t. then K_2CO_3, 4 Å MS, MeOH, 40 °C, 99%.

Likewise, Basavaiah et al. reported a facile, convenient, and two-step strategy for synthesizing spiro-oxindoles containing α-arylidene-γ-butyrolactone moiety. In this strategy, monoalkylation of 1-substituted-2-oxindoles **62** with Baylis-Hillman acetates **63** provides the tri-substituted alkenes **64** which on treatment with [bis(trifluoro acetoxy)iodo]benzene (PIFA) under appropriate conditions would lead to the formation of 2-oxindole-spiro-3-arylidene-γ-butyro lactone framework **65** (Scheme 3-12).

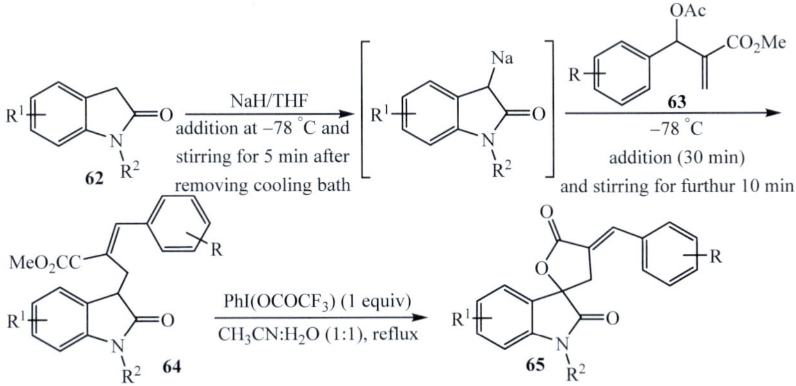

Scheme 3-12. Synthesis of spiro-oxindoles containing α-arylidene-γ-butyrolactone moiety.

The authors proposed a plausible mechanism for the transformation of compounds **64** into spirolactones **65**, which is depicted in Scheme 3-13. Initially, the carbonyl oxygen of **64** attacks the iodine of PIFA, leading to the formation of iminium ion intermediate **66** (with trifluoroacetate ion as a counter ion). The subsequent elimination of trifluoroacetic acid and PhI generates iminium ion intermediate **67**. Intramolecular nucleophilic attack of the carbonyl oxygen of the -CO_2Me group onto position 3 of intermediate **67**, followed by the release of methyl trifluoroacetate as shown in Scheme 3-13, provides the desired product **65** [12].

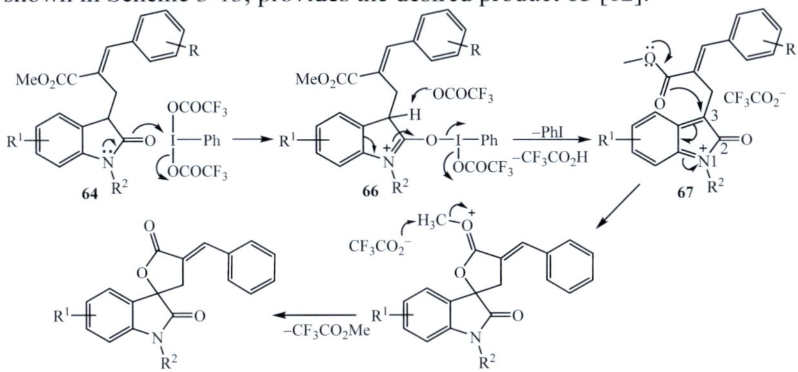

Scheme 3-13. Plausible mechanism.

3.2.1. Friedel–Crafts alkylation

The Friedel-Crafts alkylation reaction is a well-known classical method for the direct alkylation of an aromatic ring using a strong Lewis acid, such as aluminum chloride, ferric chloride, or other MX_n reagents, as a catalyst. In the following, some examples of Friedel-Crafts reactions are described to provide readers with more information on this topic.

Bo et al. synthesized a novel type of blue light-emitting spiro-bridged ladder-type poly(*p*-phenylene)s (LPPP) [13]. The synthesized spiro-bridged ladder-type polymer **68** represents the first LPPP exhibiting excellent thermal and optical stability (Scheme 3-14).

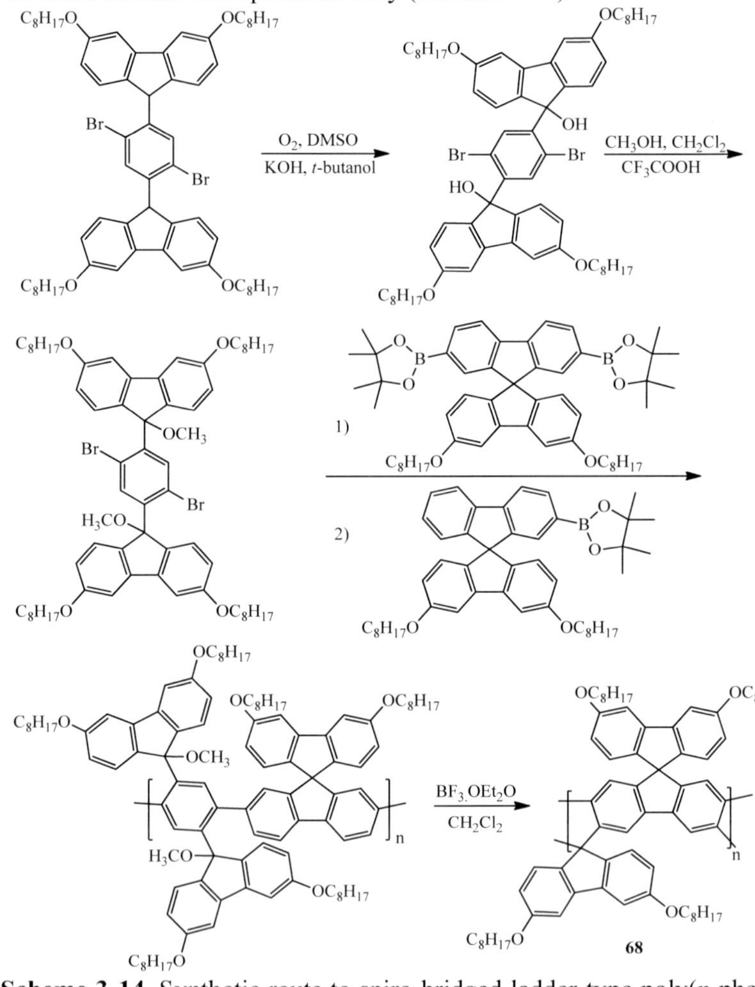

Scheme 3-14. Synthetic route to spiro-bridged ladder-type poly(*p*-phenylene)s.

LPPPs are traditionally synthesized through a two-step process: (1) a keto-functionalized precursor polymer is produced via Pd-catalyzed Suzuki polycondensation. (2) The ladder polymer is then produced in two steps: (i) the keto groups are reduced with lithium aluminum hydride or the carbonyl groups alkylated with alkyl lithium or an alkyl magnesium halide, followed by (ii) ring closure through a Friedel-Crafts alkylation. However, the authors noted that the defective keto reduction of the precursor polymer chain is the main source of ketonic defects in the final product. To avoid these defects, they combined the traditional two-step route into a single-step process, in which only a Friedel-Crafts-type reaction is required to produce an LPPP from a PPP precursor. Additionally, to prevent the oxidation of alkyl-substituted methylene bridges, they incorporated all aromatic spirobi-fluorene linkages into the polymer structure.

Bakthadoss et al. reported the first Friedel-Crafts reaction of Baylis-Hillman adducts derived from nitro-olefins in 2011. In this process, the Baylis-Hillman adducts derived from nitro-olefins **69** were reacted with isatin **70** and *N*-methyl glycine **71** in acetonitrile solvent without any catalyst for 10 hours at reflux to give the desired spiropyrrolidine oxindole compounds **72**, as shown in Scheme 3-15 [14].

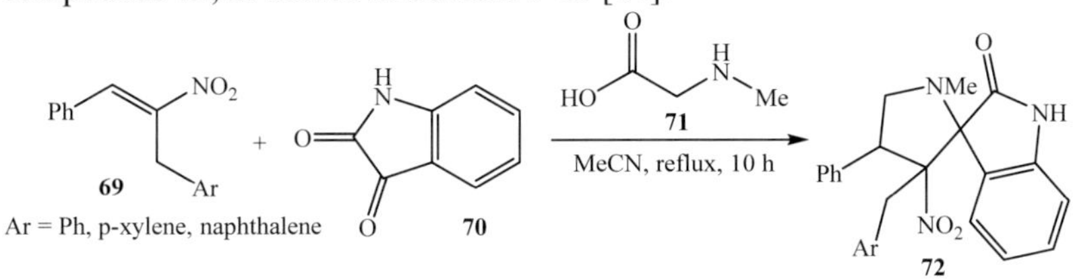

Scheme 3-15. Synthesis of 3-spiropyrrolidine compounds **72** from Baylis–Hillman adducts **70**.

In search of an efficient synthetic route for various indenoindene-fused α-methylene-γ-butyrolactones, the Kim group developed a tandem intra- and intermolecular Friedel–Crafts reaction from the spirolactone **75** which can be easily synthesized from ninhydrin **73** and Baylis–Hillman adduct **74** (Scheme 3-16). With compound **75** in hand, they investigated Friedel–Crafts reactions in benzene in the presence of H_2SO_4. The reactions proceeded via an initial intramolecular Friedel–Crafts reaction to form intermediate **76** via the corresponding oxonium intermediate. The subsequent intermolecular Friedel–Crafts reaction with benzene provided the final pentacyclic lactone **77** in good yield (95%) [15].

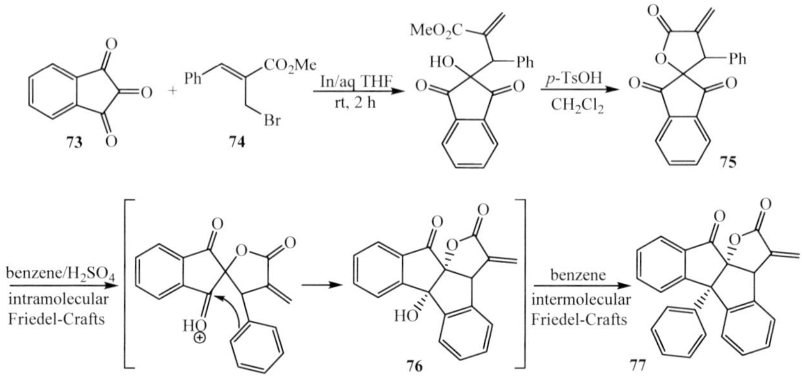

Scheme 3-16. Tandem intra- and intermolecular Friedel–Crafts reaction of the spirolactone **75**.

Hamada et al. reported a method for synthesizing spiro[4.5]cyclo hexadienones containing an allene motif via a base-promoted intramolecular *ipso*-Friedel-Crafts addition of phenols to propargyl bromides **78** (Scheme 3-17, Eq. 1). The desired allenylspiro[4.5]cyclohexadienones **79** were obtained in up to 99% yield when the spirocyclization was performed in a CH_2Cl_2-*tert*-BuOH mixed solvent system at room temperature, using potassium *tert*-butoxide as a base. The authors also demonstrated the feasibility of this type of allenyl-spirocycle **81** using a Pd-catalyzed intramolecular *ipso*-Friedel-Crafts alkylation when a propargyl carbonate with a naphthol unit **80** was utilized as a substrate (Scheme 3-17, Eq. 2) [16].

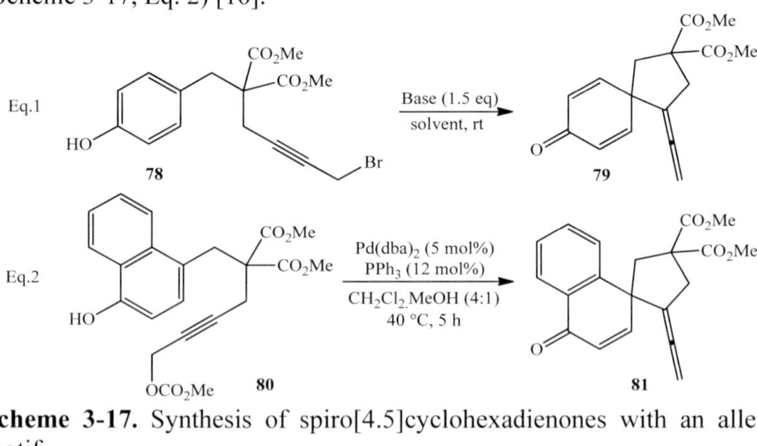

Scheme 3-17. Synthesis of spiro[4.5]cyclohexadienones with an allene motif.

Anion-exchange membranes (AEMs) based on polystyrene carrying benzyltrimethyl ammonium are crucial to produce alkaline fuel cells and water electrolyzers. To this end, Jannasch et al. developed a synthetic method to prepare highly alkali-stable mono- and spirocyclic piperidine-based cations on PS through a superacid-mediated Friedel-Crafts alkylation using 2-(piperidine-4-yl)propane-2-ol **82** [17]. The piperidine rings were then quaternized by either using iodomethane to generate *N,N*-dimethyl piperidinium cations or by cyclo-quaternizations using 1,5-dibromopentane and 1,4-dibromobutane, respectively, to provide *N*-spirocyclic quaternary ammonium cations (Scheme 3-18).

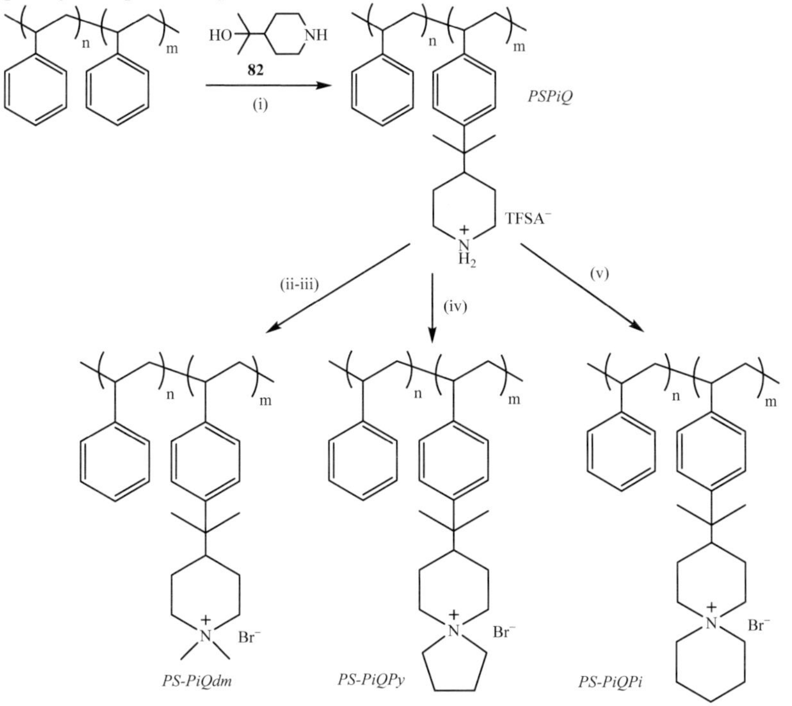

Scheme 3-18. Reagents and conditions: (i) triflic acid, CH_2Cl_2, 0 °C; (ii) iodomethane, K_2CO_3, *N*-methyl-2-pyrrolidone (NMP), room temperature; (iii) ion-exchange in 1 M NaBr, 50 °C; (iv) *N*-ethyldiisopropylamine (DIPEA), NMP, 80 °C; and (v) 1,5-dibromopentane, DIPEA, NMP, 80 °C.

This approach enables functionalization of up to 27% of the styrene units with piperidine rings and complete quaternization, with all the sensitive β-hydrogens of the cations present in ring structures, leading to high stability. The resulting AEMs based on these polymers exhibit excellent alkaline stability, with less than 5% ionic loss observed by ^1H NMR spectroscopy after 30 days in 2 M of aq NaOH at 90 °C.

3.3. Pericyclic-type reactions

Pericyclic reactions are concerted processes in which bonds are formed or broken in a cyclic transition state. These reactions are a powerful synthetic tool that offers a high degree of selectivity in both inter-and intramolecular versions. We will discuss some important examples of pericyclic reactions to synthesize spirocycle compounds.

3.3.1. 1,3-Dipolar cycloaddition reactions (1,3-DCRs)

Palmisano et al. developed a novel non-biomimetic synthesis of (-)-Horsfiline **86** in 1996 through a thermal intermolecular 1,3-dipolar cycloaddition (1,3-DC). The synthesis involved the access of (-)-horsfiline **86** in five steps from dipolarophile **83** which had a chiral auxiliary [*R*-(-)-menthyl]. Dipolarophile **83** was obtained by Wittig olefination of 5-methoxyisatin, followed by dipolar cycloaddition with *N*-methyl-azomethine ylide, which was produced *in situ* from formaldehyde and sarcosine **84**, to yield **85**. The ester was hydrolyzed and underwent formal decarboxylation to give (-)-horsfiline **86** (Scheme 3-19) with an overall yield of 37%. [18].

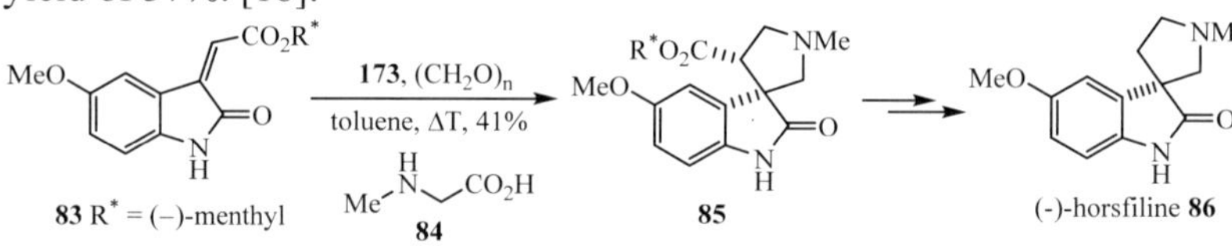

Scheme 3-19. Synthesis of (-)-Horsfiline **86** by 1,3-DC.

The William group also demonstrated the utility of 1,3-DC in the asymmetric, stereocontrolled total syntheses of (+) and (-)-spirotryprostatin B **92** [19]. They utilized a chiral azomethine ylide **89** which was generated *in situ* by adding 3-methoxy-3-methylbutanal **87** to 5,6-diphenyl-morpholin-2-one **88**. The resulting azomethine ylide **89** reacted with ethyl oxindolylidene acetate **90** to provide the desired cycloadduct **91**, possessing the correct relative and absolute stereochemistry of natural spirotryprostatin B. The key dipolar cycloaddition reaction sets four continuous stereogenic centers (Scheme 3-20).

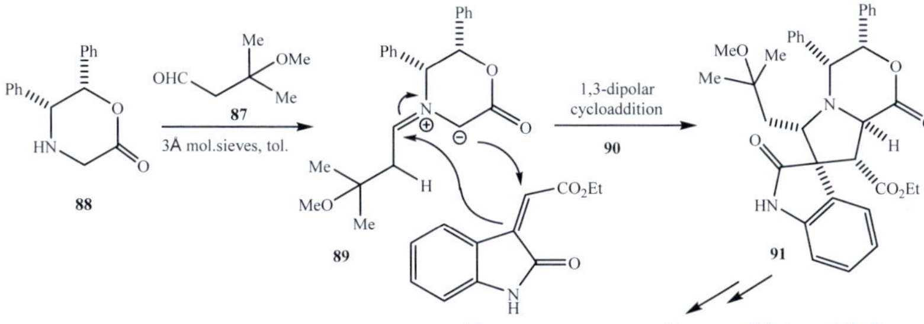

Scheme 3-20. The asymmetric, stereocontrolled total syntheses of (+) and (-)-spirotryprostatin B **92**.

Pardasani et al. conducted research on azomethine ylides generated through the decarboxylation of iminium species derived from cyclic secondary α-amino acids and isatin [20]. The simplicity of the procedure and the ability to control the regiochemistry of the cycloaddition step through the choice of dipolarophile are the main advantages of this reaction. Isatin **70** was reacted with (*R*)-(-)-thiazolidine-4-carboxylic acid **93** in a 1:1 molar ratio in refluxing acetonitrile for 22 hours, producing a single cycloadduct (2*S*,5*R*)-spiro{1-aza-3-oxa-7-thia-bicyclo[3,3,0]octane--2,3′-indol}-4,2′-dione **94** in 45% yield. The authors proposed a mechanism that involves the formation of an intermediate azomethine ylide **95** which undergoes 1,3-DC with either methyl acrylate or ethyl phenyl propiolate, resulting in spiro compounds **96** or **97** (Scheme 3-21).

In a similar study, Pardasani et al. proposed a synthetic and theoretical method to obtain novel spiro heterocycles through 1,3-DC between isatin derivative **70** and cyclic secondary α-amino acid, L-proline **98** (Scheme 3-22).

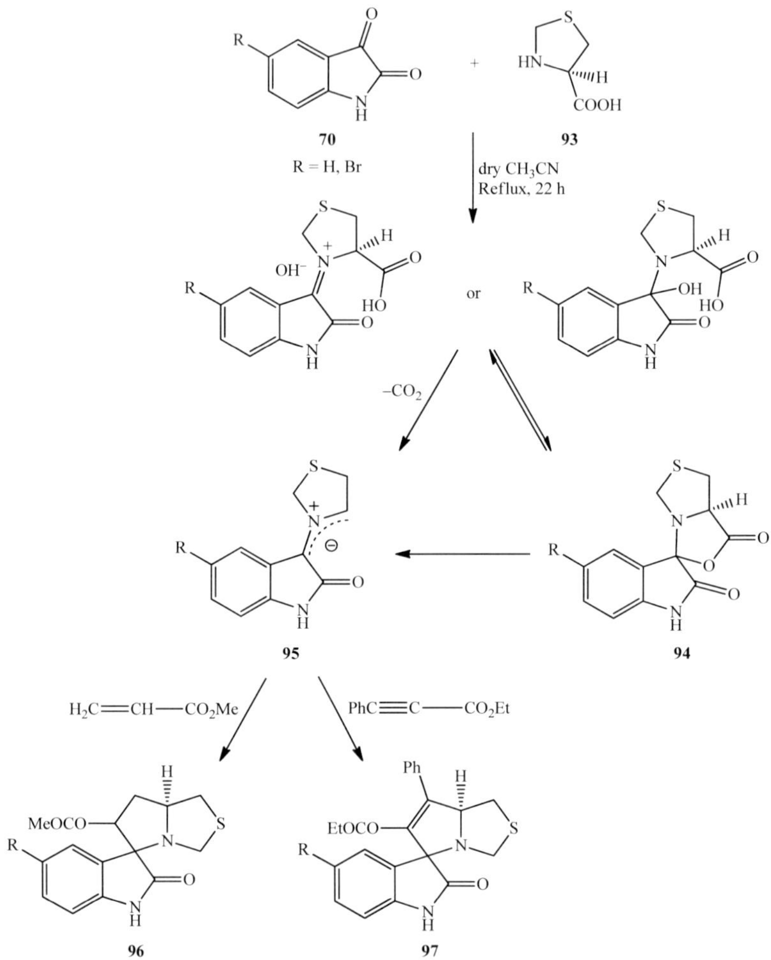

Scheme 3-21. Synthesis of substituted thiazolidines.

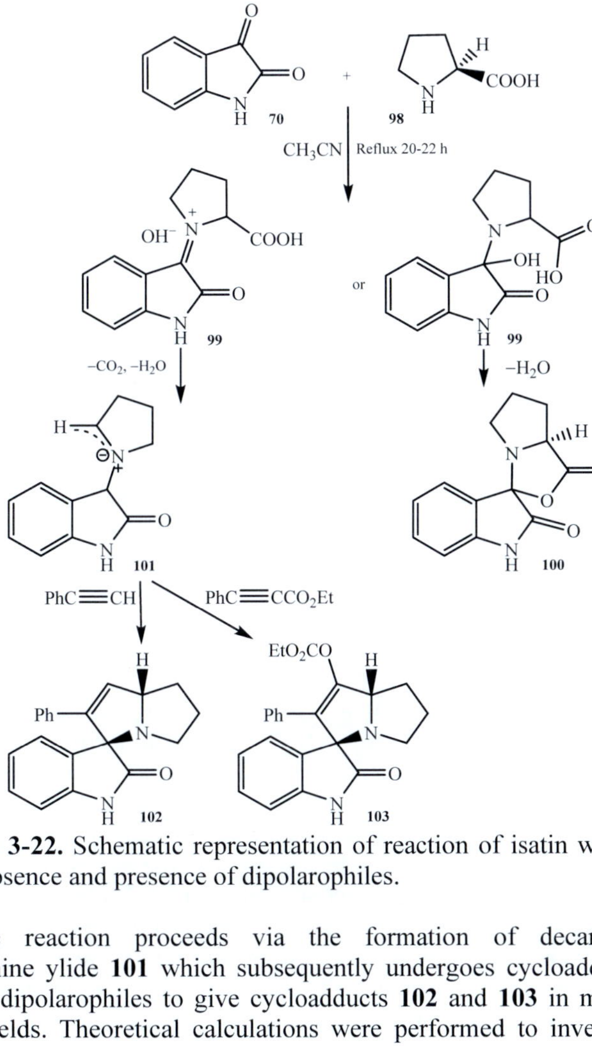

Scheme 3-22. Schematic representation of reaction of isatin with proline in the absence and presence of dipolarophiles.

The reaction proceeds via the formation of decarboxylative azomethine ylide **101** which subsequently undergoes cycloaddition with various dipolarophiles to give cycloadducts **102** and **103** in moderate to good yields. Theoretical calculations were performed to investigate the stereochemistry of the products, revealing that:

i. the azomethine ylide **101** has an almost planar structure, while the proline ring is planar and lies in the same plane as the isatin,
ii. the predominant frontier molecular orbital (FMO) approach is $HOMO_{dipole}$–$LUMO_{dipolarophile}$ due to its lower energy gap compared to $LUMO_{dipole}$–$HOMO_{dipolarophile}$, and

iii. the endo approach is preferred, with the phenyl group positioned towards the isatin ring [21].

In 2006, Raghunathan reported on the synthesis of a range of spiro pyrrolidines and polycyclic heterocycles through 1,3-DCRs using Baylis–Hillman adducts **104**. Various substituted Baylis–Hillman adducts **104** were synthesized based on literature procedures. The azomethine ylides generated by the reaction of sarcosine **84** with isatin **70**, ninhydrin **73**, or acenaphthenequinone **105** in boiling toluene for 48–78 h were used to react with the double bond of **104**, providing the corresponding cycloadducts as single regioisomers with overall yields of 40–55% (Scheme 3-23). This reaction allowed for the synthesis of a series of novel spiro heterocycles **106-108** in a regio- and stereocontrolled manner [22].

As illustrated in Schemes 3.24 and 3.25, Yavari et al. reported the efficient synthesis of functionalized spiro-2,5-dihydro-1,2-λ^5-oxa phospholes **111**. The reaction of Ph_3P with ethyl propiolate **109** in the presence of *N*-alkylisatins **110** allowed ethyl 2,2,2-triphenyl-2,5-dihydro-1,2-λ^5-oxaphosphole-4-carboxylate-spiro-1-alkyl-1,3-dihydro-2*H*-indol-2-ones **110** at 85–94% yield (Scheme 3-24). The assumed mechanistic pathway involves the initial formation of a 1,3-dipolar intermediate **112** between triphenylphosphine and ethyl propiolate **109** which reacts with the carbonyl group of *N*-alkylisatin **110** to generate **113**. Cyclization of this zwitterionic intermediate allows formation of the spiro compound **111** [23].

The reaction between Ph_3P and dialkyl acetylene dicarboxylates **114**, in the presence of *N*-alkylisatins **110**, proceeds smoothly in CH_2Cl_2 at ambient temperature to produce spiro compounds **115** and **116** in a 3:1 ratio. It is believed that the mechanism of the reaction involves the initial formation of a 1,3-dipolar intermediate **117** between Ph_3P and the acetylenic compound **114** which then reacts with the carbonyl group of *N*-alkylisatin **110** to produce intermediate **118**. Cyclization of this zwitterionic intermediate leads to the formation of spiro compound **115**. Another possible pathway involves intermediate **119** and its subsequent transformation to spirobutenolide **116** (Scheme 3-25). This method has the advantage of being accomplished under neutral conditions, and the starting material can be used without requiring any activation or modification.

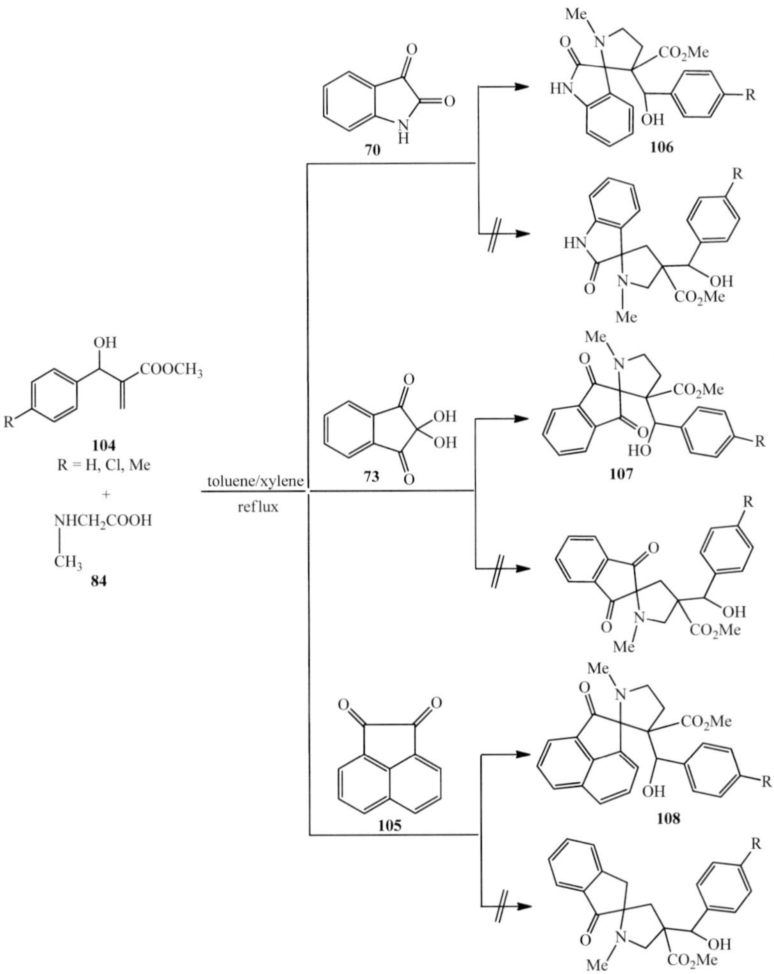

Scheme 3-23. 1,3-DCRs with Baylis–Hillman adducts.

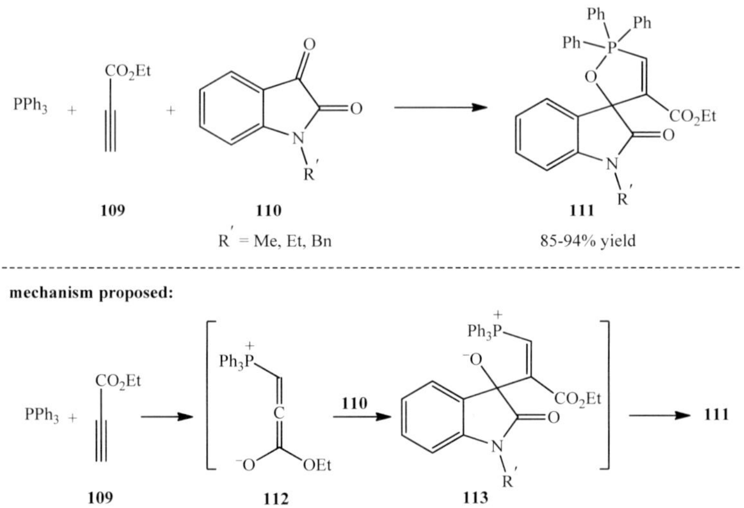

Scheme 3-24. The synthesis of spiro compound **111** and its postulated mechanism.

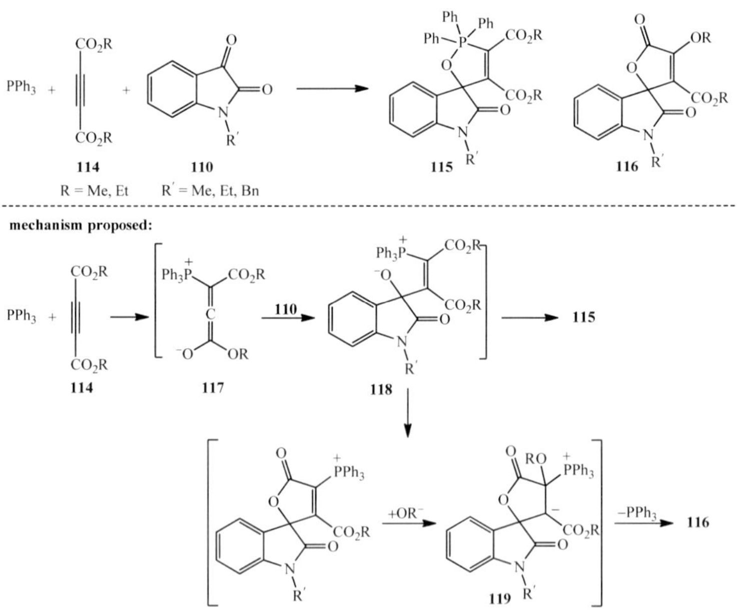

Scheme 3-25. The synthesis of spiro compounds **115/116** and its postulated mechanism.

The Raghunathan group developed a method using *N,N*-dimethyl ammonium *N′,N′*-dimethyl carbamate (DIMCARB) as a reusable reaction medium to synthesize several monoarylidene cyclopentanones [24]. These compounds were used as dipolarophiles in the 1,3-DCR with an azomethine ylide which was produced *in situ* by the decarboxylation method for the synthesis of spiropyrrolidines, using microwave-assisted methodology. By using a 1:1 ratio of ketone **120** and aldehyde **121/123**, monoarylidene cyclopentanones **122** and **125** were obtained in moderate to excellent yields, employing DIMCARB as a solvent (Scheme 3-26, Eq. 1). Subsequently, 2-arylidene-1-cyclopentanones **122** and **124** underwent 1,3-DCRs with a non-stabilized ylide, produced *in situ* by the decarboxylative condensation of isatin **70** and sarcosine **84**, to provide dispiropyrrolidinyl oxindoles **125** and **126** in a highly regioselective manner (Scheme 3-26, Eq. 2). The reactions were conducted in a solvent-free manner, using a microwave oven for 10 minutes.

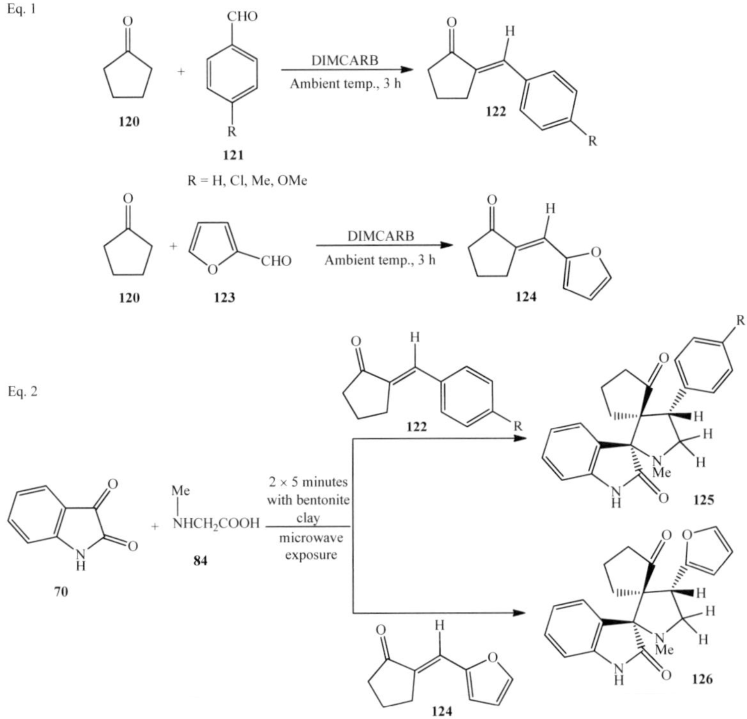

Scheme 3-26. The synthesis of spiropyrrolidines by DIMCARB as a reusable reaction medium.

In 2008, Perumal et al. reported on the high atom economy, chemo-, regio-, and stereoselective synthesis of novel spiro-pyrido-pyrrolizines and pyrrolidines which were discovered as potential antimycobacterial agents [25]. In this study, the 1,3-DC of an azomethine ylide, formed *in situ* from acenaphthenequinone **105** and α-amino acids [(i) proline, (ii) phenyl glycine, and (iii) sarcosine], with **127**, provided spiro[2.2″]acenaphthene-1″-one-spiro[3.3′]-5′-arylmethylidene-1′-methyltetra-hydro-4′(1*H*)-pyridinone-4-arylhexahydro-1*H*-pyrrolizines **128**, 4-aryl-5-phenyl-pyrrolo(spiro-[2.2″]acenaphthene-1″-one)-spiro[3.3′]-5′-aryl methylidene-1′-methyltetra-hydro-4′(1*H*)-pyridinones **129**, and 1-methyl-4-aryl pyrrolo-(spiro[2.2″] acenaphthene-1″-one)-spiro[3.3′]-5′-arylmethyl- idene-1′-methyltetrahydro -4′(1*H*)-pyridinones **130**, respectively, in a highly selective manner (Scheme 3-27). The reaction produced only water and carbon dioxide as byproducts.

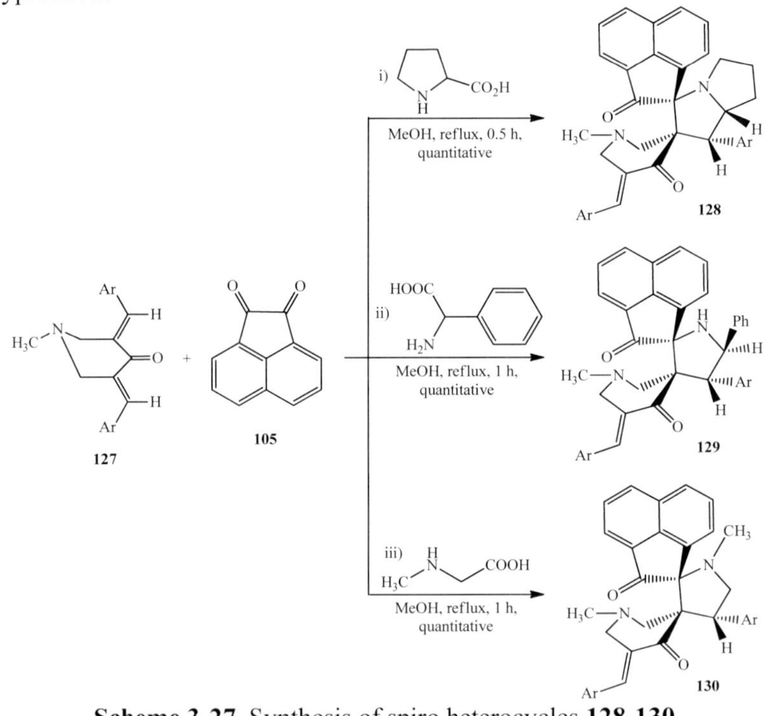

Scheme 3-27. Synthesis of spiro heterocycles **128-130**.

All these cycloadditions proceed:

i) Regioselectively, as the electron-rich carbon of the dipole adds to the β-carbon of the αβ–unsaturated moiety of **127**,

ii) Chemoselectively, as the cycloaddition takes place on only one C=C of **127** attributable to the steric hindrance applied by **128-130** for the second cycloaddition, and

iii) Stereoselectively, as only one diastereomer is generated individually in quantitative yields, although multiple stereocentres are present in each product **128-130**.

Based on the observed chemical shifts of H-2′eq (3.38 ppm) and H-2′ax (1.78 ppm) of **128**, the authors suggested that the two carbonyls linked to C-2 and C-3 of **128** are in *trans* relationship (Scheme 3-28). The reason for this stereoselectivity is that probably the transition state (**A**) would require less free energy of activation than the transition state (**B**), leading to **128′** as the latter would result in electrostatic expulsion between the *cis* carbonyls increasing the free energy of activation.

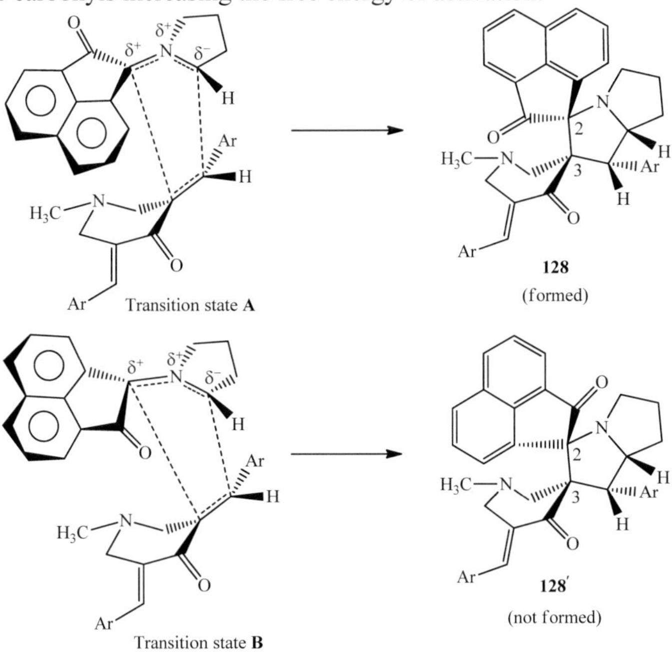

Scheme 3-28. Stereochemistry of cycloadducts differing in their relative configurations at C-2 and C-3.

The compounds were screened for their *in vitro* anti-mycobacterial activity against MTB and multidrug-resistant *Mycobacterium tuberculosis* (MDR-TB) by agar dilution to design MIC in duplicates. The MDR-TB clinical isolate was resistant to Isoniazid, Rifampicin, Ethambutol, and

Ciprofloxacin. All the compounds demonstrated excellent *in vitro* activity against MTB with an MIC of 12.5 mg/ml. Subsequently, some of the compounds were evaluated against MDR-TB, and all the compounds screened inhibited MDR-TB with MIC ranging from 0.4 to 3.13 mg/ml. They were more active than isoniazid (MIC:1.56 mg/ml) and Ciprofloxacin (MIC:12.5 mg/ml). The results of the structure-MTB activity relationship indicate that the weak electron-withdrawing halogens increase the antimycobacterial activity in the aryl rings. At the same time, electron-donating groups like methyl and methoxy decrease the activity.

In 2009, Perumal et al. synthesized a series of dispiropyrrolidine bisoxindoles **132**. They investigated a three-component reaction of isatin **70**, sarcosine **84,** and isatylidene malononitrile **131** in refluxing methanol for 100 min (Scheme 3-29) [26]. Initially, the reaction proceeds through the production of azomethine ylide (dipole **84**) through the decarboxylative condensation of isatin **70** with sarcosine **84**.

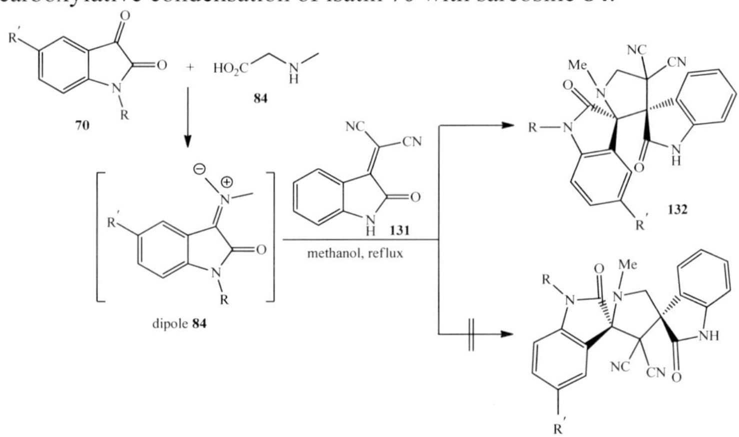

Scheme 3-29. Synthesis of dispiropyrrolidine bisoxindoles **132**.

The possible mode of the approach of azomethine ylide (dipole **84**) is shown in Figure 3-1. The regioselectivity in the product formation can be explained by considering the secondary interaction of the orbitals of the carbonyl group of dipolarophile **131** with those of the ylide **84** as shown in Figure 3-1. Accordingly, the observed regioisomer **132** via the path **A** is more favorable due to the secondary orbital interaction (SOI) which is impossible in the path **B**. Hence, only one regioisomer **132** was formed.

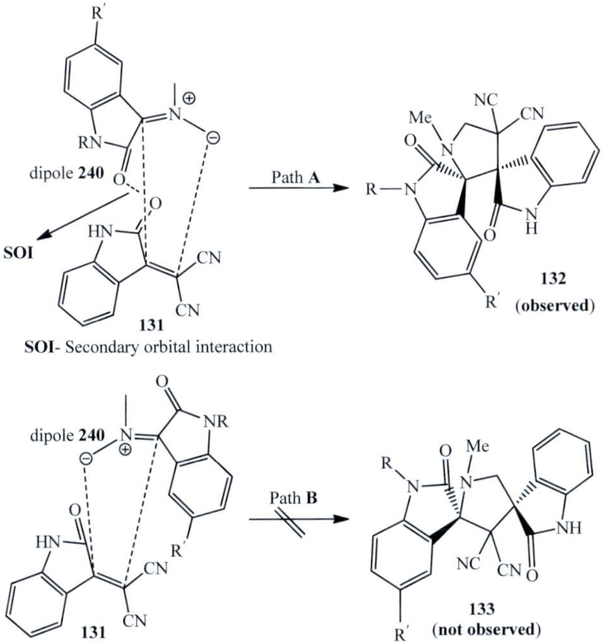

Fig. 3-1. Mode of the approach of azomethine ylide **84**.

The reaction was later extended using two other dipolarophiles such as 2-(1,3-dioxo-indan-2-ylidene)malononitrile **133** and 2-(1*H*-indole-3-carbonyl)-3-phenyl-acrylonitrile **135** under optimized conditions. The dipolarophiles **133** and **135** react with azomethine ylides (dipole **84**) produced *in situ* from isatin **70** and sarcosine **84** in methanol to afford spiroindane-1,3-diones **134** and spiropyrrolidine oxindoles **136**, respectively, as single products with good yields (Scheme 3-30).

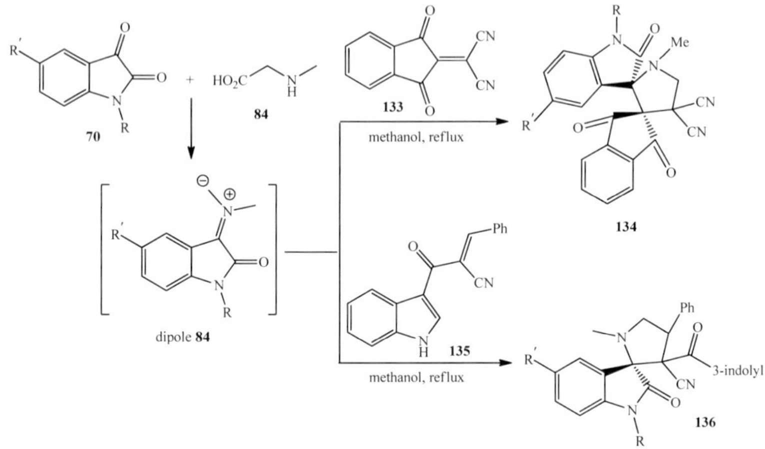

Scheme 3-30. Synthesis of spiroindane-1,3-diones **134** and spiro pyrrolidine oxindoles **136**.

Perumal et al. reported the synthesis of dispiro-oxindolylpyrrolothiazoles with high regio- and stereoselectivity in moderate to good yields (60–92%) in 2009. The reaction involved the use of 1,3-thiazolane-4-carboxylic acid, isatin, and 2-arylidene-1,3-indanediones **137**. The non-stabilized ylides produced *in situ* by the decarboxylative condensation of isatin **70** with amino acid, 1,3-thiazolane-4-carboxylic acid **138**, readily reacted with 2-arylidene-1,3-indanediones **137** to furnish dispirooxindolyl -pyrrolothiazoles **139-141** in a highly selective manner (Scheme 3-31).

The dispiro-oxindolylpyrrolothiazoles **139-141** synthesized were screened for their *in vitro* activity against MTB using the agar dilution method for the determination of minimum inhibitory concentration (MIC) in duplicate. The spiro heterocycles displayed MIC values in the range of 1.4-55.2 μM, with all the compounds being less potent than rifampicin and isoniazid. In the structure-MTB activity relationship, the presence of substituents in the isatin nucleus was found to be important for the activity of the dispirooxindolyl- pyrrolothiazoles, with the order of activity being NO_2>Cl>H. The presence of halogens (Br, F, and Cl) in the aryl ring increased the antimycobacterial activity of the dispiro compounds bearing the nitro group (series **141**) and chlorine (series **140**) in the isatin nucleus [27].

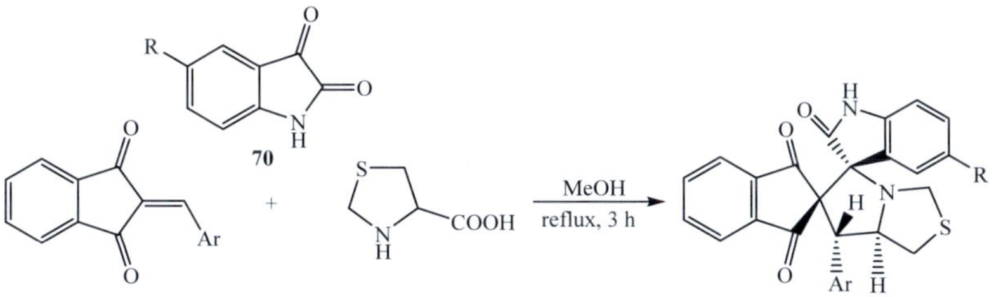

137 **138** **139** R = H, **140** R = Cl, **141** R = NO_2

Scheme 3-31. Synthesis of dispiro-oxindolylpyrrolothiazoles **139-141**.

Raghunathan et al. reported a highly regio- and stereoselective 1,3-DCR for the synthesis of tetraspiro-bisoxindolopyrrolidine **143-144** and tetraspiro-bispyrrolidines **145-146**. In this reaction, bis-dipolarophiles **142** reacts with 1,3-dipole generated from 1,2-diketones and secondary amino acids under solvent-free microwave conditions (Scheme 3-32) [28].

The Perumal group exploited benzo[*b*]thiophene 1,1-dioxides **147** as dipolarophiles under reflux conditions to give easy access to a series of spiro-oxindoles containing tri- and tetracyclic fused pyrrolobenzo[*b*] thiophene-1,1-dioxides **148-149** (Scheme 3-33) [29].

Bouhfid et al. reported the synthesis of novel spiro[1,4,2-dioxazole-5,3′-indolin]-2′-ones **154-157** via 1,3-DCRs of isatins **70**, **150-152** with aryl nitrile oxide **153**. The first step of the process (Scheme 3-34, Eq. 1) involved the alkylation of isatin **70** at room temperature under PTC conditions with benzyl chloride, allyl bromide, or propargyl bromide in the presence of dry K_2CO_3. Dimethylformamide was used as a solvent, and tetra-*n*-butyl-ammonium bromide (TBAB) was the catalyst. Next, the authors performed the 1,3-DC of isatin derivatives **150-153** with the nitrile oxide produced *in situ* from 4-methoxybenzaldoxime **153** and sodium hypochlorite (Scheme 3-34, Eq. 2). It was found that in addition to the C=O group participating in the formation of the 1,4,2-dioxazole ring, the C-C alkene (allyl) or alkyne (propargyl) bond underwent a 1,3-DC with the nitrile oxide to form the isoxazole or isoxazoline rings linked to the indole ring (**158** and **159**) via a methylene group [30].

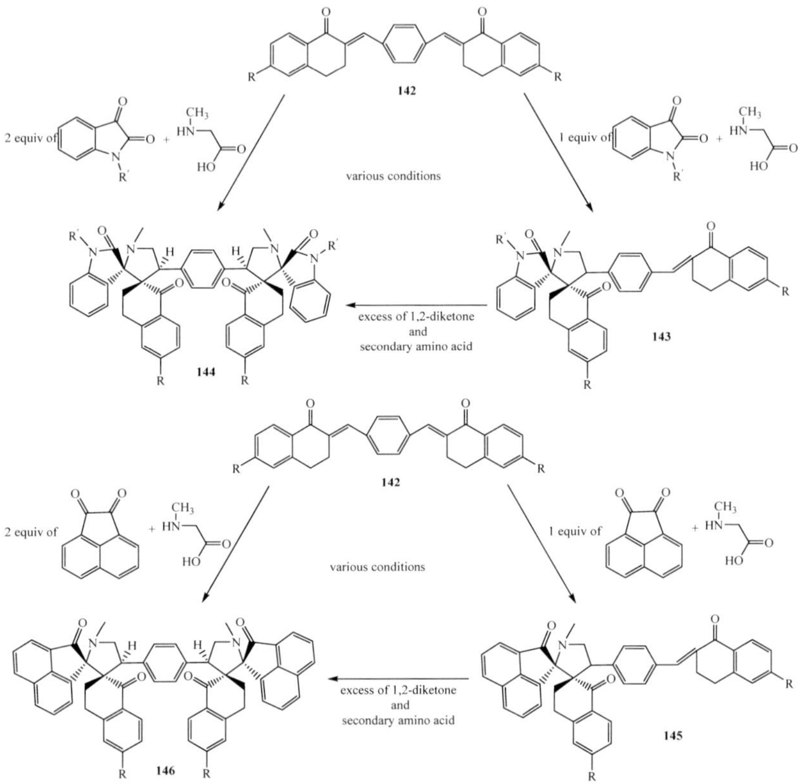

Scheme 3-32. Synthesis of tetraspiro-bisoxindolopyrrolidines **143-144** and tetraspiro-bispyrrolidines **145-146**.

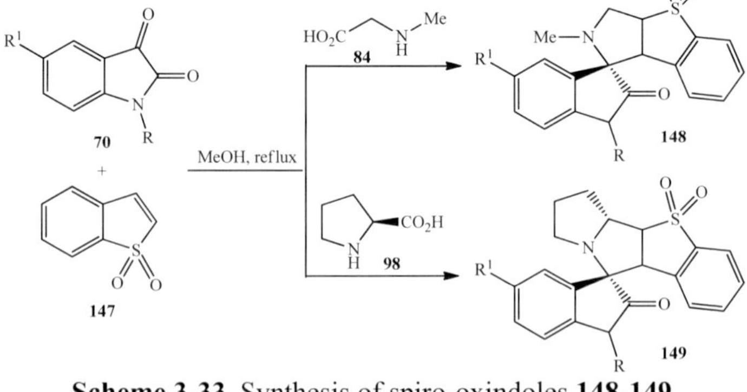

Scheme 3-33. Synthesis of spiro-oxindoles **148-149**.

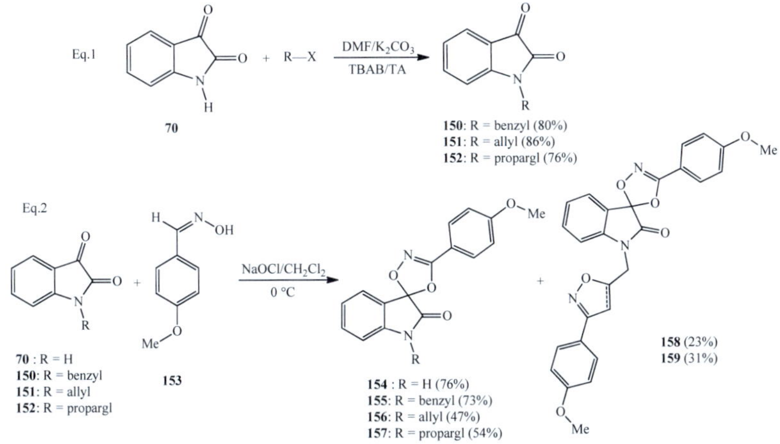

Scheme 3-34. *N*-Alkylation of isatin in PTC conditions (Eq. 1) and 1,3-DCR on *N*-alkylisatin (Eq. 2).

Coquerel et al. [31] showed that stereo-defined monocyclic, spiro-bicyclic, and bis-spiro-tricyclic pyrazolidin-3-ones **163** can be prepared by 1,3-DC of azomethine imines **161** obtained from hydrazones **160** with *R*-oxo-ketene **162** dipolarophiles produced *in situ*. The reaction brings about the formation of four covalent bonds and two contiguous chiral quaternary centers with excellent diastereoselectivity in a single catalyst/additive-free, highly atom-economic transformation (Scheme 3-35).

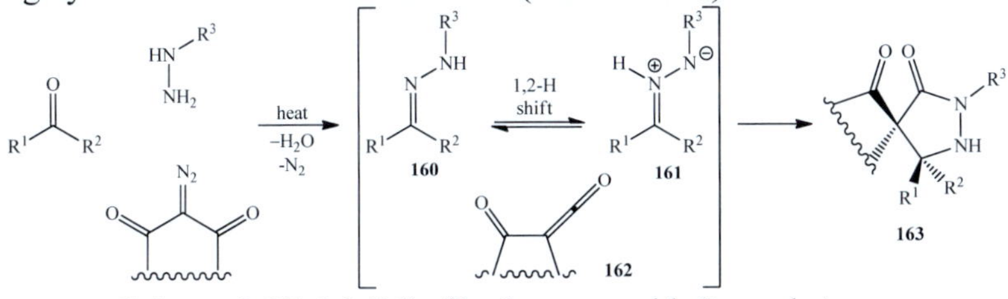

Scheme 3-35. 1,3-DC of hydrazones with *R*-oxo-ketenes.

Perumal et al. reported a three-component 1,3-DC to obtain spiro-oxindoles/acenaphthylen-1-ones containing benzo[1,4]thiazin-3-one ring **165-168** (Scheme 3-36) [32]. The reaction of isatin **70** or acenaphthene quinone **105** with 2-(4-methyl-benzylidene)-4*H*-benzo[1,4]thiazin-3-one **164** and sarcosine **84** or L-proline **98** in toluene under reflux condition furnished the expected spiro-oxindoles **165/166** and **167/168** in high yields.

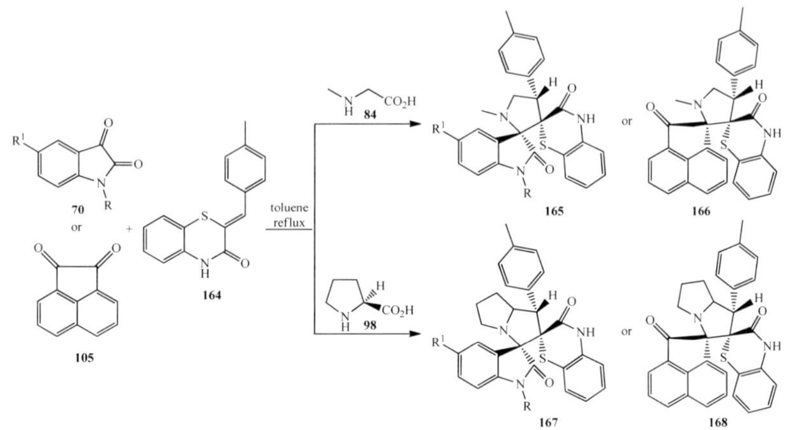

Scheme 3-36. Synthesis of spiro-oxindoles/acenaphthylen-1-ones containing benzo[1,4]thiazin-3-one ring **165-168**.

Later, Chen et al. reported the high regio- and stereoselectivity synthesis of spiro[pyrrolidine-2,3′-oxindole] derivatives via the Huisgen reaction isatin **70**, α-amino acids **98**, and (*E*)-β-phenyl nitro-olefins (Scheme 3-37).

Two regioisomers **170a** and **170b** were obtained in this reaction. They found that the regioselectivity of this reaction was affected by solvent and temperature, regardless of the ratio of the reactants. Low temperature and high solvent polarity result in high regioselectivity, and protic solvents lead to higher yield and regioselectivity [33].

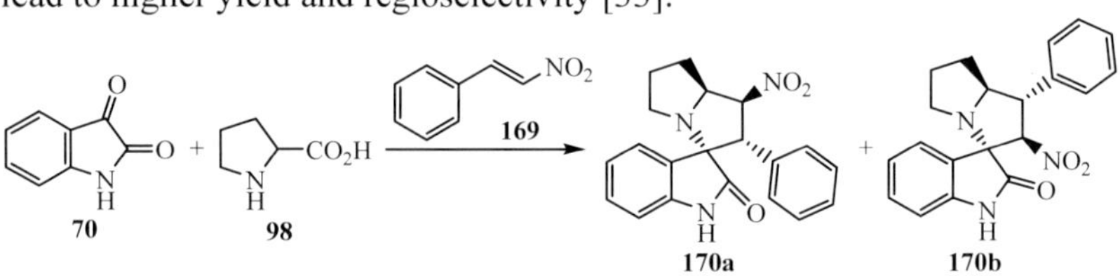

Scheme 3-37. Huisgen reaction between isatin, L-proline, and (*E*)-β-phenyl nitro-olefins.

Spiropyrrolidine oxindoles **173** can also be synthesized by multicomponent 1,3-DCR of isatin **70**, benzylamine **171**, and chalcones **172** [34]. Sarrafi et al. reported that [1,5]-H shift has a crucial role in forming the azomethine ylide and proceeds only through the sterically favorable iminium ion (Scheme 3-38).

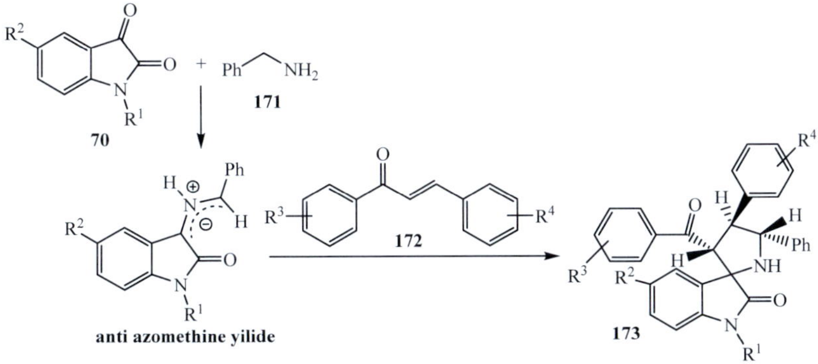

Scheme 3-38. Synthesis of spiropyrrolidine oxindoles **173**.

The synthesis of new enantiomerically pure spiro-oxindolo-pyrrolizidines **176** via a three-component asymmetric 1,3-DCR of azomethine ylides derived from isatin **70** was reported by Jadidi et al. The reaction proceeds by readily available isatin derivatives **70**, (*S*)-proline **174,** and chiral cinnamoyl oxazolidinone **175** at room temperature in aqueous ethanol as a green solvent and in the absence of any Lewis acid. The products **176** were achieved high to excellent yields and high optical purity. The chiral auxiliary was easily removed without racemization and/or by-product formation (Scheme 3-39) [35].

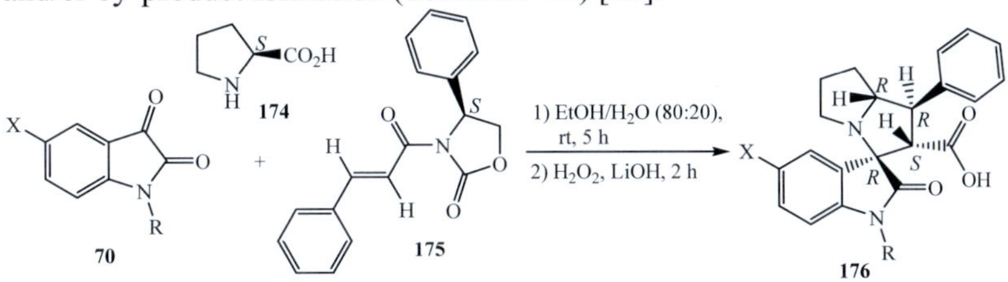

Scheme 3-39. Three-component asymmetric 1,3-DCR of azomethine ylides derived from isatin **70**.

Perumal et al. reported the chemo-, regio-, and stereoselective synthesis of novel hybrid spiro heterocycles through three- and four-component 1,3-DCRs in an ionic liquid [36]. In the three-component reaction, 1-methyl/benzyl-3,5-bis[(*E*)-arylmethylidene]-tetrahydro-4(1*H*)-pyridinones **177** were reacted with non-stabilized azomethine ylides produced *in situ* by the decarboxylative condensation of ninhydrin **73** and sarcosine **84** in the presence of [BMIm]Br ionic liquid to yield a library of pure 1-methyl-4-arylpyrrolo-(spiro[2.2′]indan-1′,3′-dione)-spiro[3.3″]-1″-

methyl/benzyl-5″-(arylmethylidene)piperidin-4″-ones **178-179** in excellent yields (Scheme 3-40, Eq. 1). The four-component 1,3-DCR of 1-methyl-3,5-bis[(*E*)-arylmethylidene]-tetrahydro-4(1*H*)-pyridinones **177**, ninhydrin **73**, sarcosine **84**, and *o*-phenylenediamine **180** in [BMIm]Br ionic liquid yielded 1-methyl-4-arylpyrrolo-(spiro[2.11′]-11*H*-indeno[1,2-*b*]quinoxaline)-spiro[3.3″]-1″-methyl/benzyl-5″-(arylmethylidene)piperi-din-4″-ones **181-182** in excellent yields (Scheme 3-40, Eq. 2).

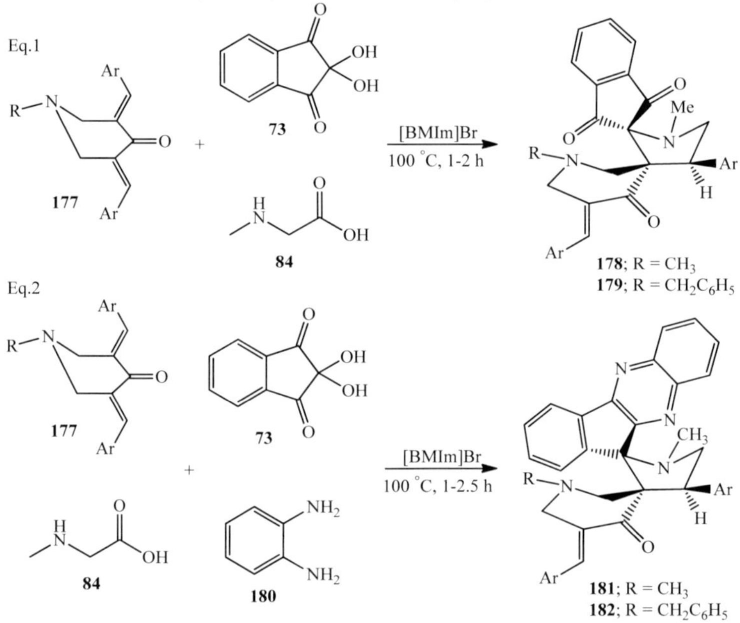

Scheme 3-40. Three and four-component 1,3-DCRs.

In another example, Lipson et al. investigated the synthesis and chemical properties of new derivatives of 3a′,6a′-dihydro-2′*H*-spiro-[indole-3,1′-pyrrolo[3,4-*c*]pyrrole]-2,4′,6′(1*H*,3′*H*,5′*H*)-triones **185** [37]. Isatin **70**, α-amino acids **183** (including sulfur-containing amino acids such as cysteine and ethionine), and maleimide **184** were heated in a medium of MeOH (2-PrOH)–H_2O (3:1) to afford the corresponding spiropyrrolidine oxindoles **185** in 41-90% yields (Scheme 3-41). For the synthesis of compounds involving cysteine, the heating was performed in a mixture of MeOH–H_2O (10:1) due to the rapid destruction of the amino acid and intense resinification. The interaction mechanism involves a 1,3-DC of an unstable azomethine ylide, generated *in situ* by the decarboxylation of the product from isatin **70** condensation with the amino acid **183**, to olefinic or acetylenic dipolarophiles.

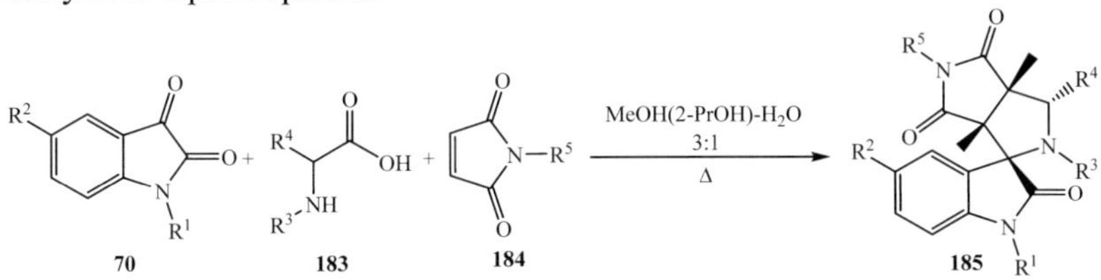

Scheme 3-41. Synthesis of 3a′,6a′-dihydro-2′*H*-spiro[indole-3,1′-pyrrolo-[3,4-*c*]pyrrole]-2,4′,6′ (1*H*,3′*H*,5′*H*)-triones **185**.

In 2013, Harrity et al. synthesized spiro-oxetane compounds **189-192** (Scheme 3-42). This process is initiated with the generation of oxetane-substituted sydnones **186a-c** (Eq. 1) which underwent a smooth base-mediated addition to propargyl bromide to form alkynes **187a** and **187b** in a good yield. For the generation of functionalized pyrazole products, these substrates were converted to the corresponding bromoalkynes **188a** and **188b** (Eq. 2). Subsequent intramolecular cycloaddition of alkynes delivers spiro-oxetane products **189-192** in an excellent yield and within a short reaction time (Eq. 3) [38].

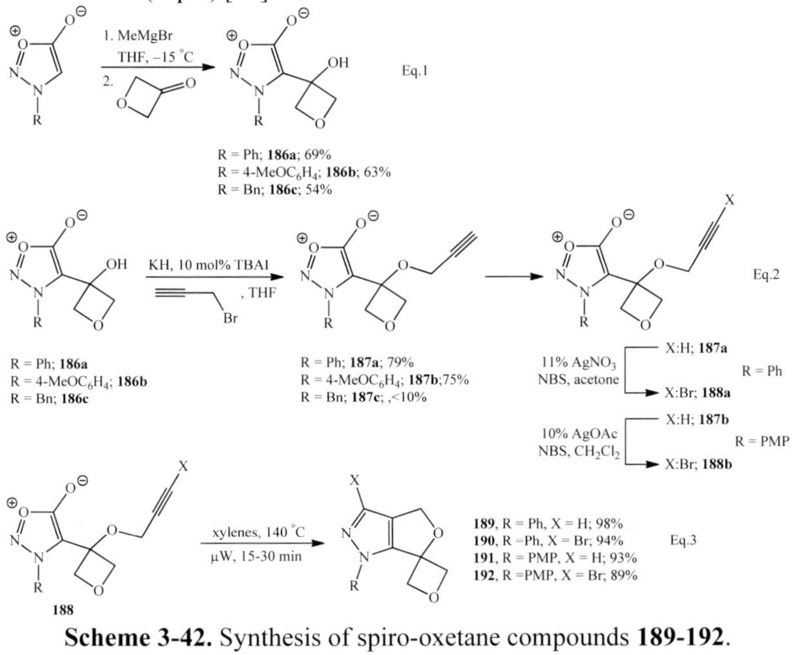

Scheme 3-42. Synthesis of spiro-oxetane compounds **189-192**.

Perumal et al. [39] reported the synthesis of bis- and tris-spiro-oxindoles containing both spiro[pyrrolidin-3,3′-oxindole] and spiro-[pyrrolidin-3,2′-oxindole] ring systems, and a chromeno-[3,4-*c*]pyrrole ring system. In this study, the azomethine ylide generated *in situ* with the dipolarophile **193** resulted in highly substituted bis-spiro-oxindoles **194** as the only product when equimolar quantities of isatin **70**, sarcosine **84**, and dipolarophile **193** were employed. Interestingly, when the same three-component reaction was performed with one equivalent of dipolarophile **193** and two equivalents of isatin **70** and sarcosine **84** under similar conditions, product **195** was obtained in a good yield. The chemo-selectivity and stereoselectivity observed with **195** are influenced by dipole-dipole repulsion and steric factors (Scheme 3-43).

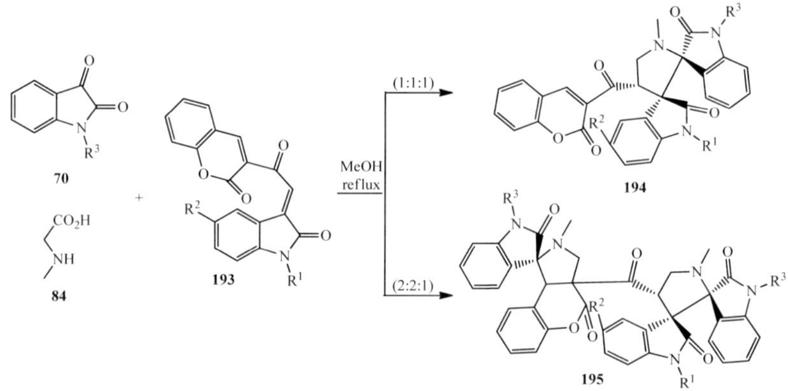

Scheme 3-43. Stoichiometry-controlled cycloaddition of azomethine ylide with dipolarophiles.

Mondal et al. demonstrated a successful transformation of curcumin **196** through a one-pot cycloaddition reaction. They used isatin and proline as the reagents in a 1:2:2 mol ratio in refluxing methanol (65 °C) [40]. The reaction proceeds through the formation of mono-cycloaddition regioisomers **197Aa** and **197Ab** which, after further reaction, generate products **198Aa−Ad** (Scheme 3-44).

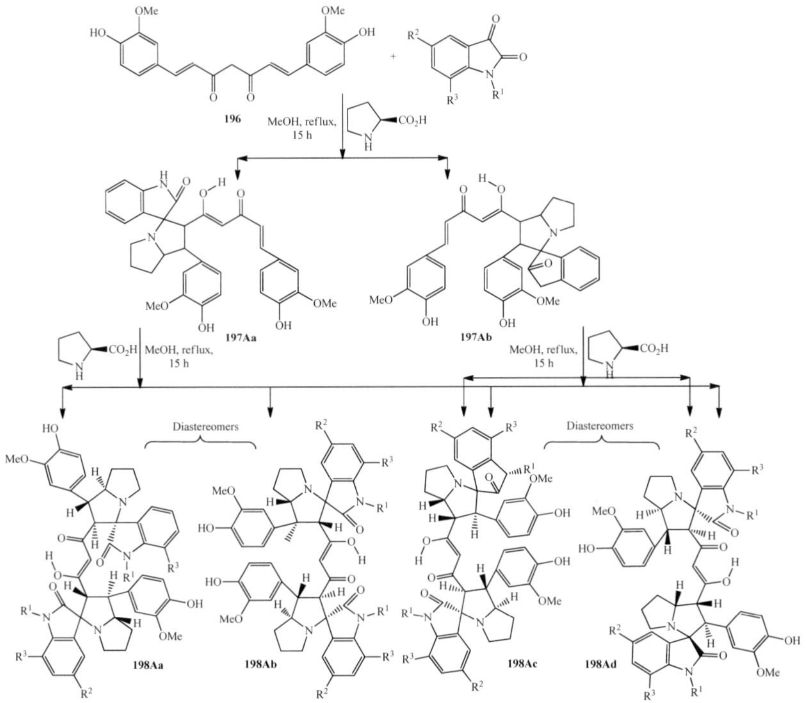

Scheme 3-44. Probable reaction pathway for dispiro-oxindolo curcuminoids.

They also performed the reaction with proline and acenaphthoquinone **105** as the 1,2-diketone compound. In this case, the bis-pyrrolizidine fused-dispiro-oxindole adducts **199a** and **199b** were obtained with around 76% yield (Scheme 3-45).

Rao et al. reported a tactical 1,3-DC approach for synthesizing carbohydrate derived polycyclic spiro heterocycles in 2015 [41]. A unique dipolarophile synthesized from D-glucose **200** reacted with azomethine ylide produced *in situ* from secondary α-amino acids (sarcosine, proline, or pipacolinic acid) and 1,2-diketone (isatin **70**/acenaphthoquinone **105**)/ indeno-quinoxalinone **201** to afford the corresponding cycloadducts **202-203** in a good yield (Scheme 3-46).

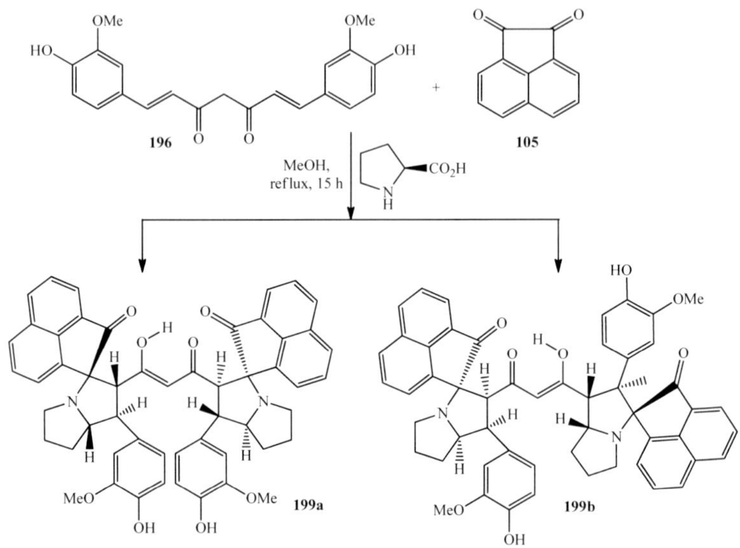

Scheme 3-45. Preparation of dispiro-acenaphthylen-2-one curcuminoids.

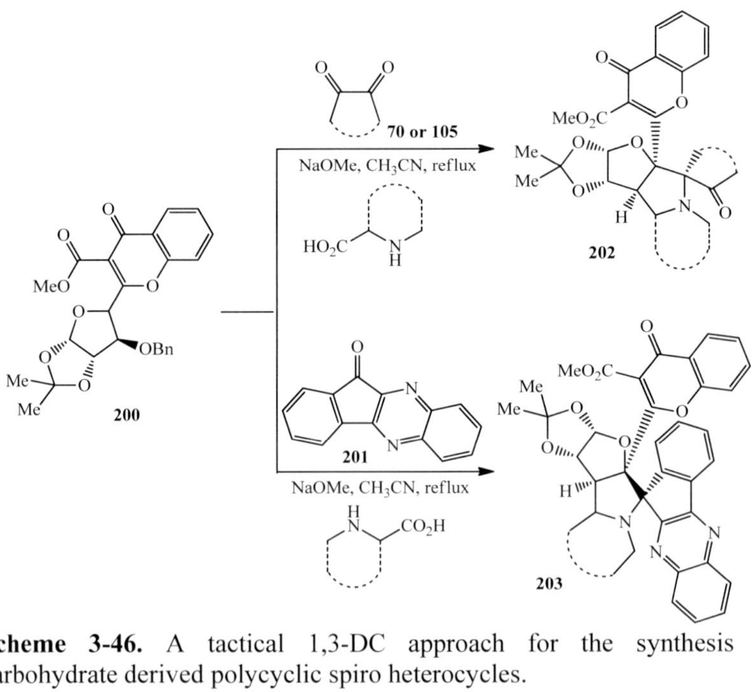

Scheme 3-46. A tactical 1,3-DC approach for the synthesis of carbohydrate derived polycyclic spiro heterocycles.

Dandia et al. demonstrated the significance of hydrophobicity in synthesizing spiro[acenaphthylene-1,5′-pyrrolo[1,2-*c*]thiazole] derivatives **206** through a 1,3-DCR of acenaphthenequinone **105**, 1,3-thiazole-4-carboxylic acid **204**, and Knoevenagel adduct **205** in an aqueous medium. The presence of NaCl played a crucial role in the reaction. In water alone, the reaction did not proceed due to the low magnitude of the hydrophobic effect. However, the addition of NaCl enhanced the hydrophobic effect of water, leading to the desired product **206** in high yield with excellent regio- and stereoselectivity. This process could generate four chiral centers containing one spiro center with one C-N and two C-C bonds in a single reaction, thereby enhancing the biocidal properties of the compound (Scheme 3-47) [42].

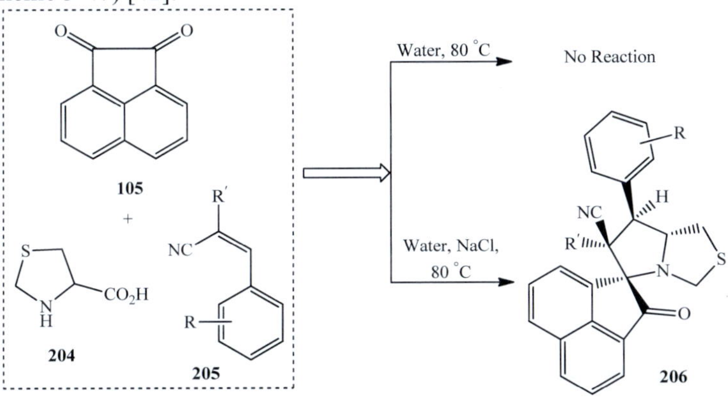

Scheme 3-47. Synthesis of spiro[acenaphthylene-1,5′-pyrrolo[1,2-*c*] thiazole] derivatives **206**.

In 2015, the Kumar group reported their efforts in search of novel potent cholinesterase inhibitors using a dipolar cycloaddition strategy [43]. For this purpose, a series of novel dimethoxyindanone embedded spiropyrrolidines were synthesized in an ionic liquid, [bmim]Br, and studied for their inhibitory activities towards cholinesterases. The target spiropyrrolidines **209a–j** were synthesized as shown in Scheme 3-48. The required starting precursors [2-arylmethylidene-5,6-dimethoxy-2,3-dihydro-1*H*-inden-1-one]s **207a–j** were synthesized by the reaction of 5,6-dimethoxy-1-indanone with convenient aromatic aldehydes in dilute methanolic sodium hydroxide. The three-component 1,3-DCR of **207a–j** with a relatively new azomethine ylide generated *in situ* from isatin **70** and phenylalanine **208** in [bmim]Br furnished the functionalized spiro pyrrolidines **209a–j** in good yields.

The proposed mechanism to explain the formation of final products **209** is depicted in Scheme 3-49. The hydrogen atom of [bmim]$^+$, which is electron-deficient, can form a hydrogen bond between the imidazole ring hydrogen of [bmim]$^+$ and the carbonyl group of isatin, facilitating the attack of the lone pair of phenylalanine on isatin and subsequent dehydration to generate azomethine ylide **210**. Similarly, the hydrogen atom of [bmim]$^+$ can form a hydrogen bond with the carbonyl group of 2-arylmethylidene-5,6-dimethoxy-2,3-dihydro-1*H*-inden-1-one which enables the easy addition of the azomethine ylide to the more electron-deficient β-carbon to produce spiropyrrolidine **209** in a regio- and stereo-selective manner.

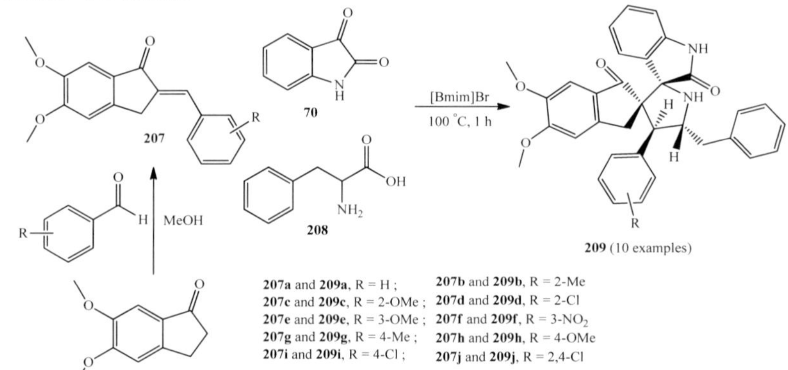

Scheme 3-48. Synthesis of spiropyrrolidines **209a–j**.

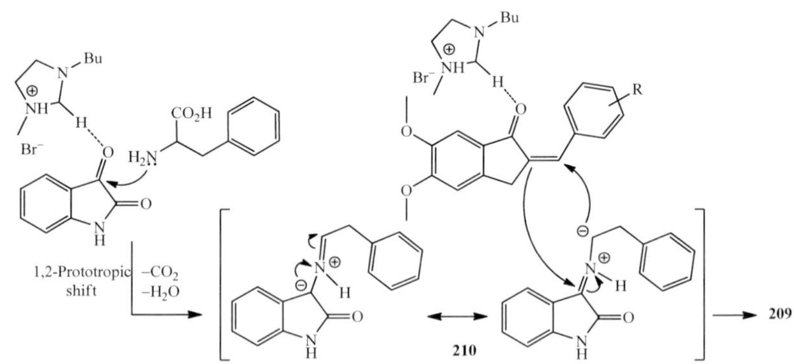

Scheme 3-49. A plausible mechanism for the formation of spiro pyrrolidine **209**.

The cholinesterase inhibitory activity of all synthesized compounds was evaluated *in vitro* against the acetylcholine-esterase (AChE) enzyme from electric eel and the BChE enzyme from equine serum. Compound **209f** (Ar=*m*-NO_2) exhibited the highest AChE inhibitory activity, with an IC_{50} value of 1.57 μM, followed by **209e** (Ar=*m*-OCH_3), **209h** (Ar=*p*-OCH_3), **209i** (Ar=*p*-Cl), **209j** (Ar=*o,p*-Cl), and **209g** (Ar=*p*-CH_3), all of which displayed considerable IC_{50} values below 10 μM. On the other hand, the other compounds **209a–d** exhibited mediocre AChE inhibitory activities, ranging from 10.15–15.17 μM. These results indicate that the electron-withdrawing nitro group had a greater impact on the AChE inhibitory potency than the electron-donating groups, such as methoxy (strong electron donor), methyl, and chloro (mild electron donor).

In another study, Javidan et al. developed a novel method for the synthesis of bis-spiro-oxindolopyrrolizidines and bis-spiro-oxindolo pyrrolidines via a highly regioselective pseudo-five-component 1,3-DCR of a bis-dipolarophile with a 1,3-dipole generated *in situ* from the reaction of isatin derivatives and secondary amino acids [44]. To achieve this, they prepared 3,3′-benzene-1,4-diylbis(1-phenylprop-2-en-1-one) **211** by reacting terephthalaldehyde and acetophenone. They then subjected one equivalent of the prepared bis-dipolarophile **211** to a pseudo-five-component reaction with two equivalents of isatin **70** and two equivalents of L-proline **98** in ethanol at room temperature to obtain bis-spirooxindolo pyrrolizidine **212** with a high yield (Scheme 3-50). Moreover, by replacing L-proline **98** with sarcosine **84**, a series of bis-spirooxindolopyrrolidines **213** were synthesized with yields ranging from 80% to 87%.

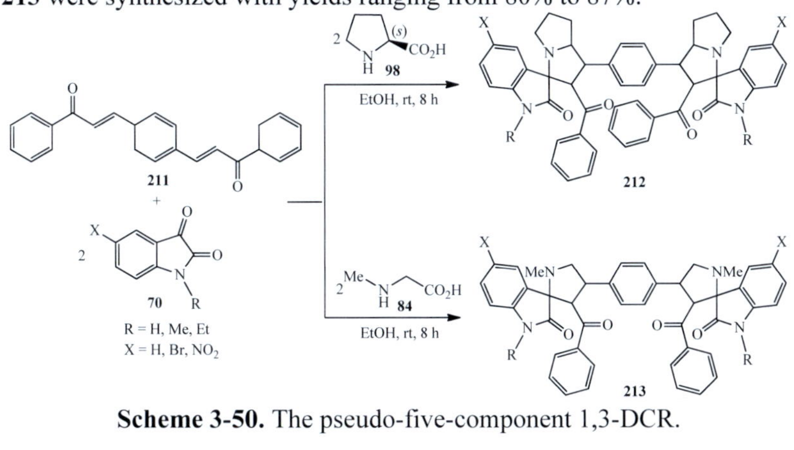

Scheme 3-50. The pseudo-five-component 1,3-DCR.

The Narayana group reported the synthesis of a novel series of spiro-oxindoles **215**. These compounds were discovered to exhibit inhibitory activity against methionine tRNA synthase (PDB ID:1PFV) and glucosamine-6-phosphate synthase (PDB ID:1JXA) enzymes through virtual screening. The synthesis of these compounds involved a three-component 1,3-DCR (domino-Knoevenagel-hetero-Diels-Alder reaction) approach. [45]. The reaction was performed by forming azomethine ylides produced *in situ* by the decarboxylative condensation of isatin **70** and amino acids **214** with the dipolarophile chalcones **172**. The synthetic route is demonstrated in Scheme 3-51.

All reactions take place chemoselectively since the addition of dipole occurs only to the available C=C bond rather than the C=O functional group of chalcones. Also, these reactions are regioselective due to the preferential attack of the nucleophilic carbon of the azomethine ylide to the end of the enone fragment of the dipolarophile chalcone. In addition, despite the presence of many stereocenters in the cycloadducts, all reactions proceeded through complete stereoselectivity leading to the generation of a single stereoisomer.

The synthesized compounds are found to be active against *Staphylococcus aureus*, *Escherichia coli*, *Aspergillus niger,* and *Aspergillus flavus* supporting the *in silico* screening. In addition, they are also appraised for antitubercular activity by the MABA method. The compounds **215a**, **215e,** and **215g** are found to be potent molecules with MIC of 0.8 μg/ml. The compounds **215b**, **215d,** and **215e** illustrate a remarkably radical scavenging capacity.

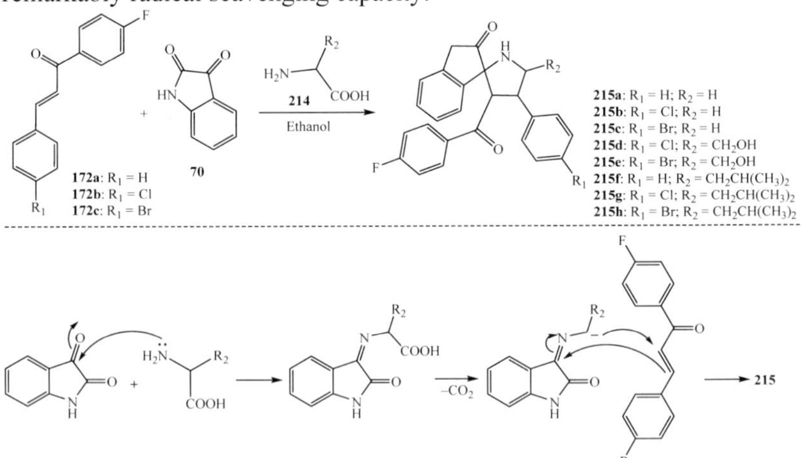

Scheme 3-51. Synthesis of spiro-oxindoles and their proposed mechanism.

Liu et al. reported the first 1,3-DCR of 2-alkylidene-1-indanones **216** with various azomethine ylides **217** promoted by imidazolium salts **218** for the synthesis of spiro heterocyclic compounds **219** containing highly substituted pyrrolidine and 1-indanone motifs with excellent yields (up to 99%) (Scheme 3-52) [46].

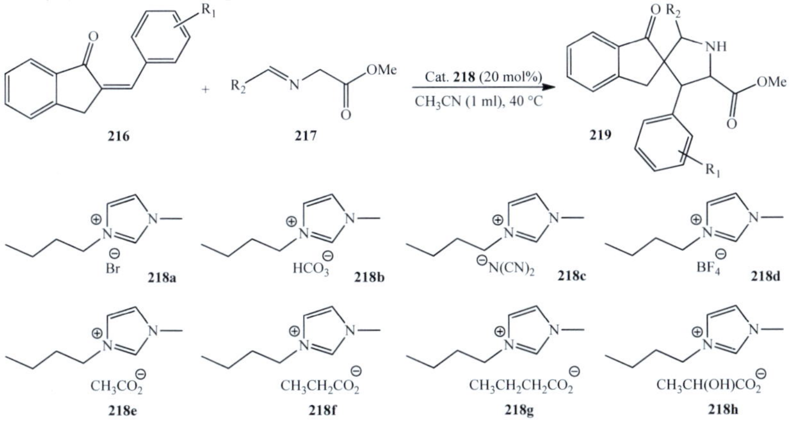

Scheme 3-52. The first 1,3-DCR of 2-alkylidene-1-indanones **216**.

Scheme 3-53 illustrates a plausible bifunctional catalytic mechanism for the mentioned reaction. According to this mechanism, the reaction involves the nucleophilic butyric anion as a base to capture a proton of azomethine ylides and an electrophilic imidazolium cation to accept the transferred electron to complete the reaction cycle. The butyric anion activates azomethine ylides, and the imidazolium cation might activate 2-alkylidene-1-indanone.

In 2016, Alizadeh et al. used an efficient method for the synthesis of spiro[indoline-3,2′-[1,3,4]-oxadiazol]-2-ones by treatment of various isatins **70** with hydrazonoyl chlorides **220** in the presence of Et_3N [47]. The mechanism of this reaction involves the formation of nitrile imine **A** from the reaction of hydrazonoyl chloride **220** and Et_3N. 1,3-DCR of intermediate **A** onto the C=O group of isatin **70** gives product **221**. Analogously, reaction of **A** with a C=O group of **105** and **223** generates the spiro compounds **222** and **224**, respectively (Scheme 3-54).

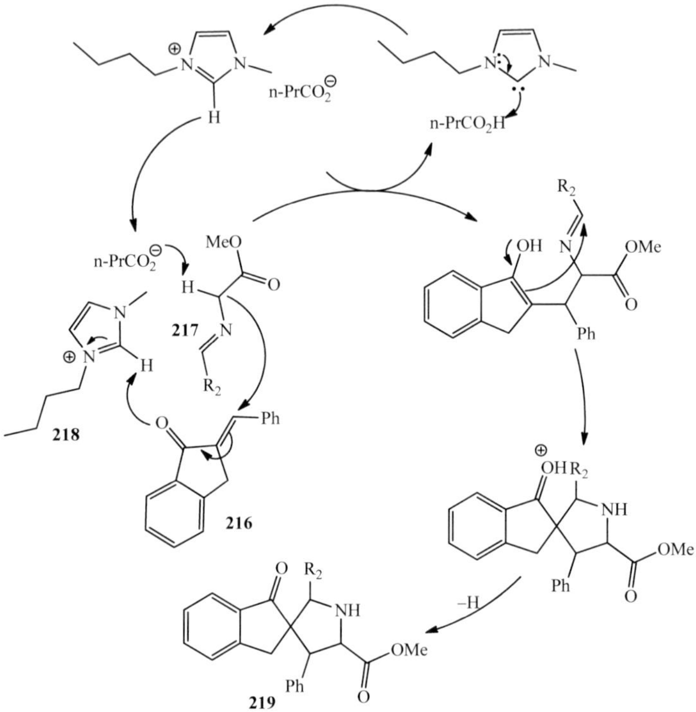

Scheme 3-53. Proposed mechanism.

Stepakov et al. reported a synthetic method to prepare 3-spiro[cyclo-propa[a]pyrrolizine]- and 3-spiro[3-azabicyclo[3.1.0]hexane] oxindoles (**227-228**) in moderate to high yields through a one-pot three-component reaction using substituted isatins **70**, α-amino acids **225**, and cyclopropenes **226** [48]. The key step is an intramolecular 1,3-cycloaddition reaction of an azomethine ylide generated *in situ* from isatins and peptide Gly-Gly or benzylamine onto a cyclopropene. It is noteworthy that the use of a peptide as an amine component for the generation of azomethine ylide is reported for the first time (Scheme 3-55).

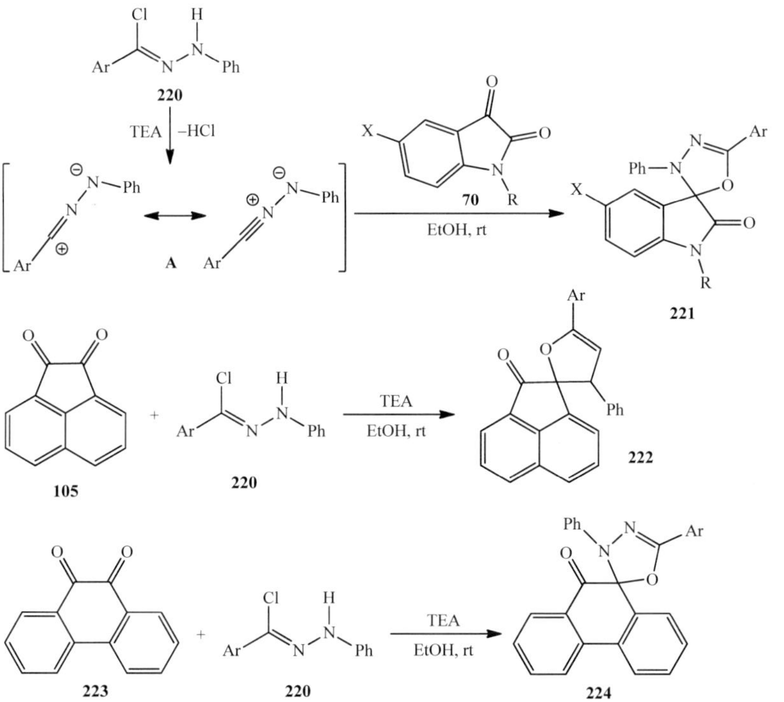

Scheme 3-54. Synthesis of spiro compounds **220, 222,** and **224**.

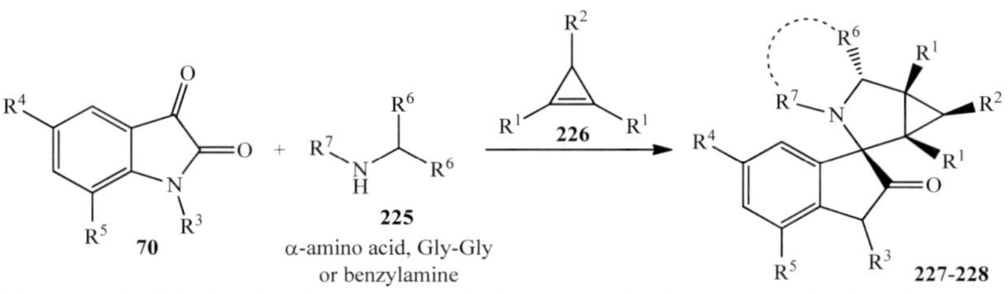

Scheme 3-55. Synthesis of a novel class of substituted 3-spiro[cyclopropa [*a*]pyrrolizine]- and 3-spiro[3-azabicyclo[3.1.0]hexane]-oxindoles **227-228**.

Li et al. conducted an elegant study in which they observed the Et_3N-catalyzed 1,3-DC of 2-arylidene-1,3-indandiones **229** with an *N*,*N*-cyclic azomethine imine **230**. This reaction resulted in the formation of spiroindane-1,3-dione-pyrazolidinones **231** in high yields with excellent diastereoselectivity under mild conditions (Scheme 3-56) [49]. The cycloaddition process, as depicted in Scheme 3-56, was proposed to occur via a concerted mechanism. Notably, the interaction between 2-arylidene-1,3-indandiones **229** and azomethine imine **230** was attributed to π-π stacking. The less hindered nitranion attacks the less substituted end of the double bond. In contrast, the more substituted end of the double bond reacts with the carbon atom of the imine ion to generate the desired cycloaddition product **231**. Introducing a substituent into the benzene ring of azomethine imine **230** weakens the interaction between 2-arylidene-1,3-indandiones **229** and azomethine imine **230** as a result of rising steric hindrance. This might be why azomethine imines with substituted aromatic rings is incompatible with the transformation.

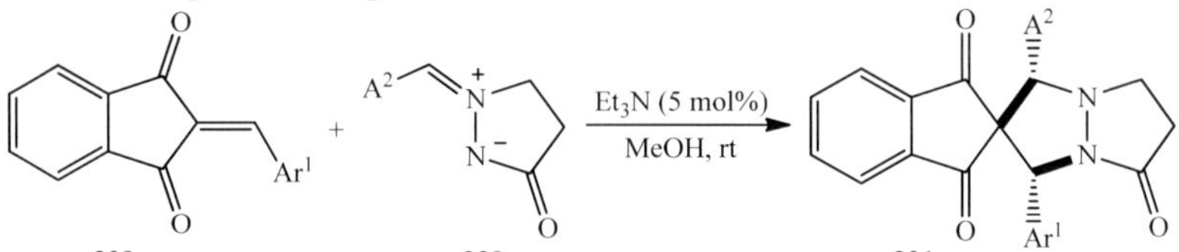

Scheme 3-56. Synthesis of spiroindane-1,3-dione skeletons **231**.

A novel one-pot 1,3-DC of 3-aminooxindoles **232**, aldehydes, and indolenines **233/234** was reported by Wang et al. [50]. Using substituted 2-alkenylindolenines **233**, the reaction is well proceeded in the presence of molecular sieves and the absence of catalyst, affording the desired indolenine-substituted spiro[pyrrolidin-2,3′-oxindole]s **235** containing four contiguous stereogenic centers in high yields (up to 99%) and excellent diastereoselectivities (up to >20:1 dr).

The reaction was also performed by 2-alkenylindolenines **234** which formed spiro compounds **235** and **236**. The cycloaddition proceeded smoothly with an excellent yield but with moderate diastereo-selectivity. The diastereomeric ratio could be switched to 1:6 by adding $NaHCO_3$ and catalyst DPP, favoring the formation of the epimer **236** as the primary product (Scheme 3-57).

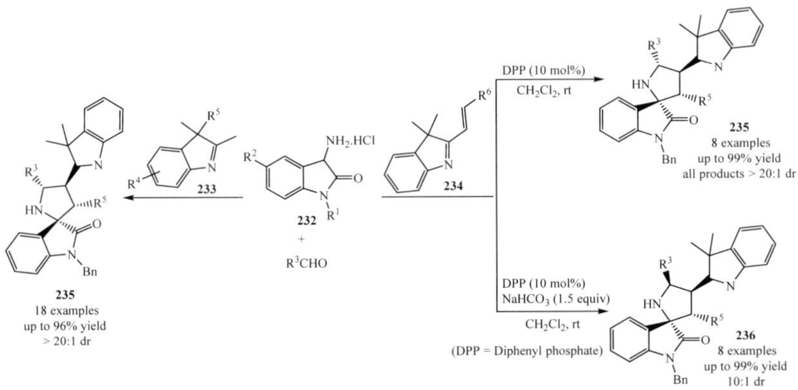

Scheme 3-57. 1,3-DC with divergent diastereoselectivities.

Another example to illustrate the synthesis of spiro-oxindoles was reported by Meshram et al. in 2017. Starting from isatin **70**, amino acid, but-2-ynedioates **237**, and phenacyl bromide **239**, the synthesis of spiro-oxindoles **238** and **240** was achieved in a one-pot 1,3-DCR using microwave irradiation under catalyst and base-free conditions in the aqueous medium (Scheme 3-58) [51].

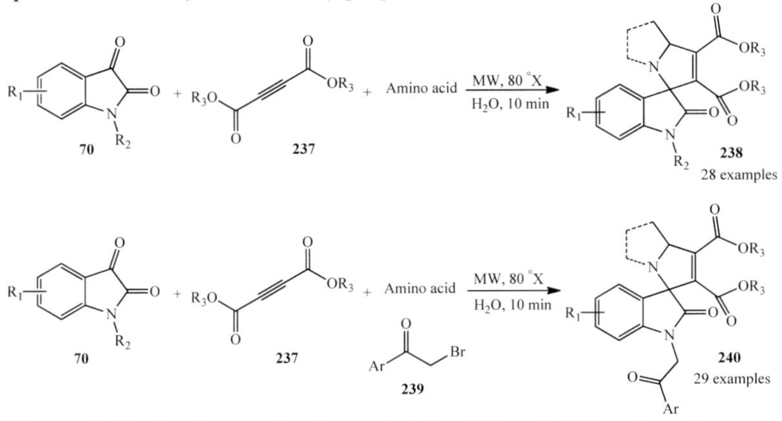

Scheme 3-58. Synthesis of spiro-oxindoles **238** and **240** under microwave irradiation in the aqueous medium.

In 2017, Barakat et al. reported the synthesis of diastereomerically pure spiro-oxindole, pyrrolidine, and thiazolidinone rings for the first time [52]. First, the Knoevenagel condensations of aldehydes with 2-thioxo-thiazolidin-4-one afforded the corresponding Knoevenagel adducts **241**.

Next, the three-component reaction between 5-arylidene-2-thioxothiazolidin-4-one **241**, isatin **70**, and L-proline **98** was performed at 60 °C in refluxed MeOH for 58 h to generate **242** with two spiro centers in a regioselective manner with 85% and 89% yields, respectively (Scheme 3-59).

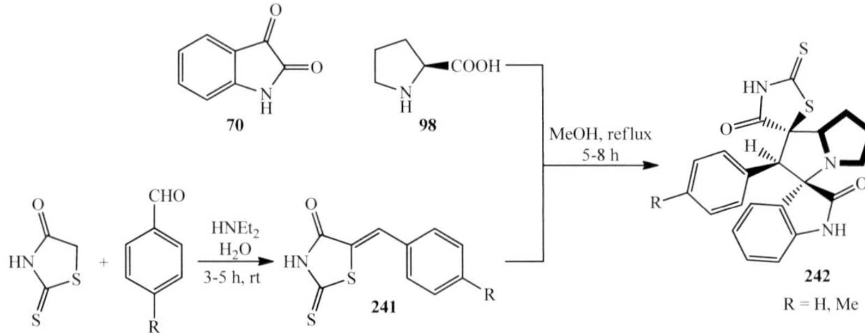

Scheme 3-59. Synthesis of spiro compound **242**.

The proposed mechanism for synthesizing spiro-oxindolepyrrolidine thioxothiazolidin-4-ones **242** is shown in Scheme 3-60. The regiochemistry of the cycloaddition reaction can be explained by SOI of the arylidine double bond of the dipolarophile **241** with ylides, as shown in Scheme 3-60. It is evident that the formation of regioisomer **242** via path **A** is more favorable; this could be attributed to the formation of *Z* isomer of **241** which allows the approach of azomethine from one side only. Hence, only one regioisomer **242** was achieved in the reaction. Moreover, the steric agent plays an additional role in the regioselective formation of products.

The compounds indicated antimicrobial and antifungal activity better than selected standards (ampicillin and gentamicin in case of anti-bacterial activity and Amphotericin A and Fluconazole in case of antifungal activity). A molecular docking study of the synthesized compounds indicated that the phenyl group plays a significant role in the designation of compound interaction inside the receptors.

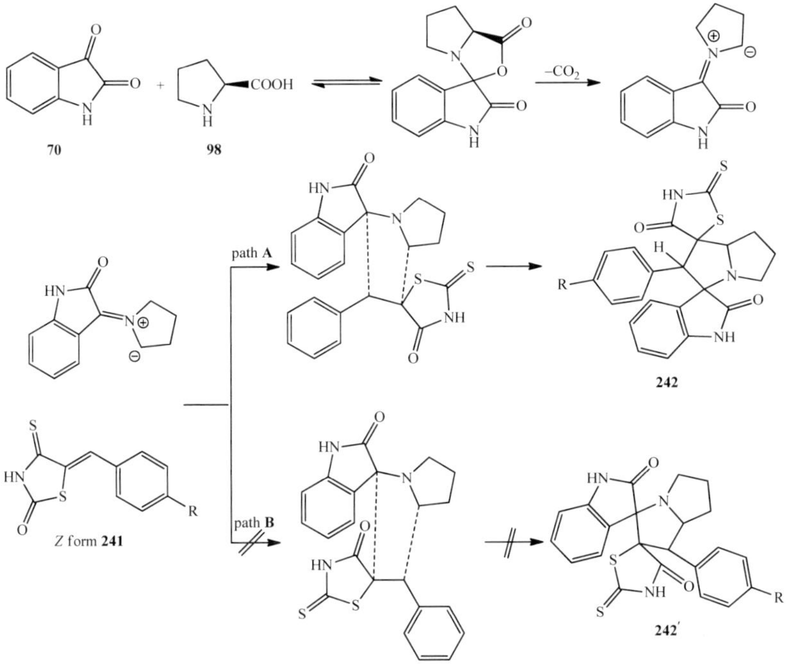

Scheme 3-60. Plausible approaches for the target compound.

Khodabakhshi et al. reported a simple procedure for the synthesis of dipyrazolo[1,5]diazocine-3,8-diones **244** via the one-pot reaction of 3-aminopyrazolone **243** with substituted benzaldehydes in the presence of *p*-toluenesulfonic acid (PTSA) as the catalyst. In this reaction, *para*-substituted benzaldehydes with electron-withdrawing groups gave higher yields than the *meta*/*ortho*-substituted benzaldehydes with electron-donating groups (Scheme 3-61) [53].

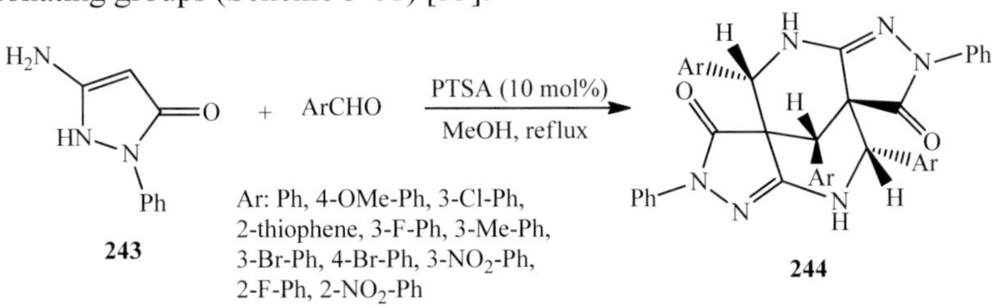

Scheme 3-61. Synthesis of dipyrazolo[1,5]diazocine-3,8-diones **244**.

Narayanarao et al. demonstrated the use of 1,3-DCRs for the regio- and stereoselective synthesis of novel spiro-indoline/indene analogs **246/247** via a one-pot three-component reaction [54]. The 1,3-DCR of azomethine ylides derived from ethynyl azaindole **245** under mild conditions is performed (Scheme 3-62).

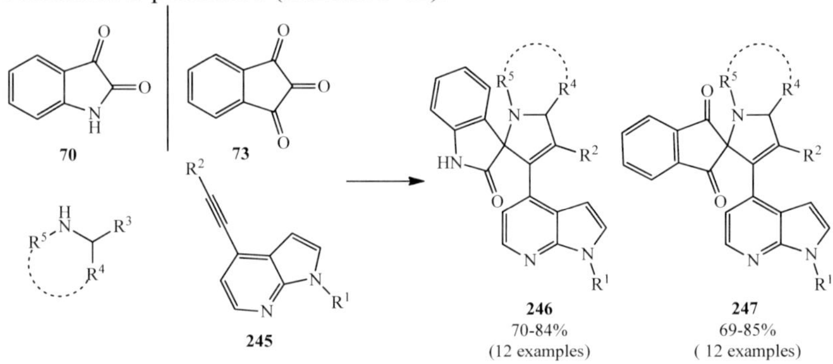

Scheme 3-62. Synthesis of novel spiro-indoline/indene analogs **246/247**.

Min Ko et al. applied an efficient synthetic strategy for the construction of spiro[oxindole-3,2′-pyrrolidine] derivatives containing a quaternary center from benzynes precursors **248**, isatin-derived ketimines **249**, and TBAT as a fluoride source. The proposed mechanism involves the conversion of imine **249** to azomethine ylide **249′** which reacts with the *in situ* generated benzyne **248′** to afford the product **250**. This process demonstrates that the hydrogen derived from **249** can be coordinated to the imine and carbonyl groups, followed by protonation of the nitrogen (Scheme 3-63) [55].

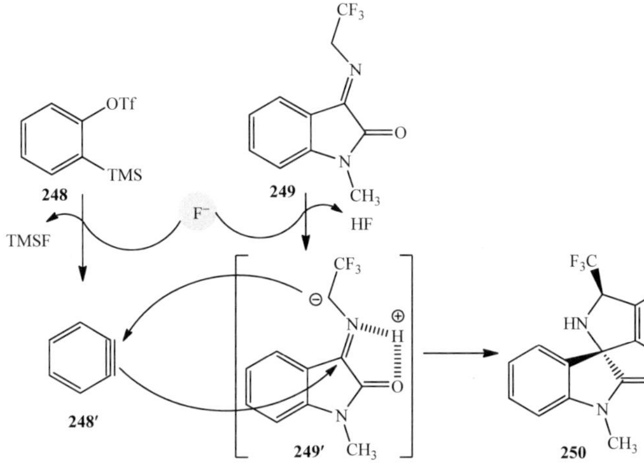

Scheme 3-63. Proposed reaction mechanism.

Stepakov et al. accomplished the successful development of a stereo-selective one-pot three-component 1,3-DC process, leading to the synthesis of various spiro[3-azabicyclo[3.1.0]hexane]s and spiro[cyclo propa[*a*]pyrrolizine]s with indeno[1,2-*b*]quinoxaline motifs. [56]. In this process, the reaction of azomethine ylides generated *in situ* from 11*H*-indeno[1,2-*b*]quinoxalin-11-ones **251** and amines **252**, with cyclopropenes **253,** furnished the desired complex spirocyclic compounds **254** in good to high yields (up to 96%) and excellent diastereoselectivity (as a single diastereomer) in most cases (Scheme 3-64).

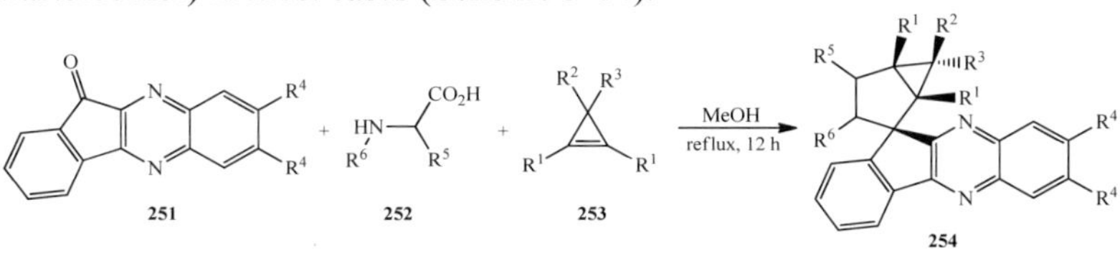

Scheme 3-64. Synthesis of complex spirocyclic compounds **254**.

Cheng et al. disclosed a 1,3-DCR of 3-diazoindolin-2-ones **255** with arynes **256** under mild conditions in excellent yields which easily gives access to a library of unstable spiro[indazole-3,3′-indolin]-2′-ones **257** and spiro[indazole-3,3′-indolin]-2′-imines **258** (Scheme 3-65). The isomerization reaction of spiro[indazole-3,3′-indolin]-2′-ones under thermal or acidic conditions affords indazolo[2,3-*c*]quinazolin-6(5*H*)-ones **259** whereas spiro[indazole-3,3′-indolin]-2′-imines could not undergo the same reaction [57].

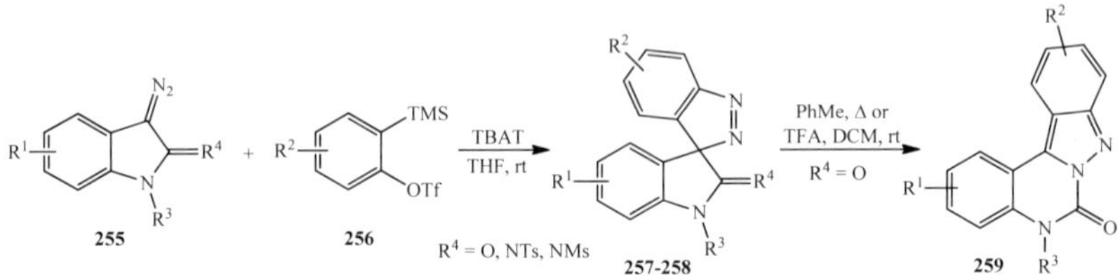

Scheme 3-65. 1,3-DCR of 3-diazoindolin-2-ones **255** with arynes **256**.

Four-component regio- and diastereoselective synthesis of pyrrolizidines incorporating spiro-oxindole/indanedione was carried out by 1,3-DCR of azomethine ylides, as reported by Alizadeh et al. in 2019 [58]. One-pot sequential four-component reactions give access to a library of diverse polycyclic pyrrolizidine fused spiro-1,3-indandione/oxindoles **262/262′** in excellent chemical yields (80-95%) and high diastereoselectivity employing 4-oxo-4*H*-chromene-3-carbaldehyde/2-chloro quinoline-3-carbaldehyde **260/260′**, 1-phenyl-2-(1,1,1-triphenyl-λ^5-phos-phanylidene) ethan-1-one **261**, ninhydrin or isatin **73/70**, and L-proline **98** in EtOH at room temperature (Scheme 3-66).

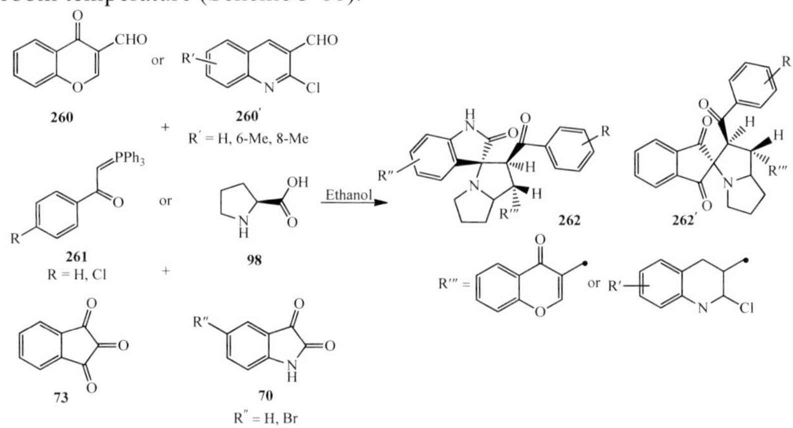

Scheme 3-66. Synthesis of pyrrolizidine fused spiro-1,3-indandion/ oxindoles.

The Periyasami group recently established the synthesis of a series of novel covalent cholesterol-spiro pyrrolidine/pyrrolizidine heterocyclic hybrids possessing biologically active oxindole, indanedione and acenaphthylene-1-one by the reaction of C3-β-cholestero-lacrylate with heterocyclic di- and tri-ketones [59]. Totally, 16 new C3-β-cholesterol tethered spiro pyrrolidine analogs were obtained as a single isomer in a good yield. All reactions, including diketone **105**/triketone **73** and sarcosine **84**/proline **98**, take place smoothly with novel C3-β-cholesterol acrylate **263** dipoloarphile in isopropyl alcohol medium under reflux for 1–2 h (Scheme 3-67).

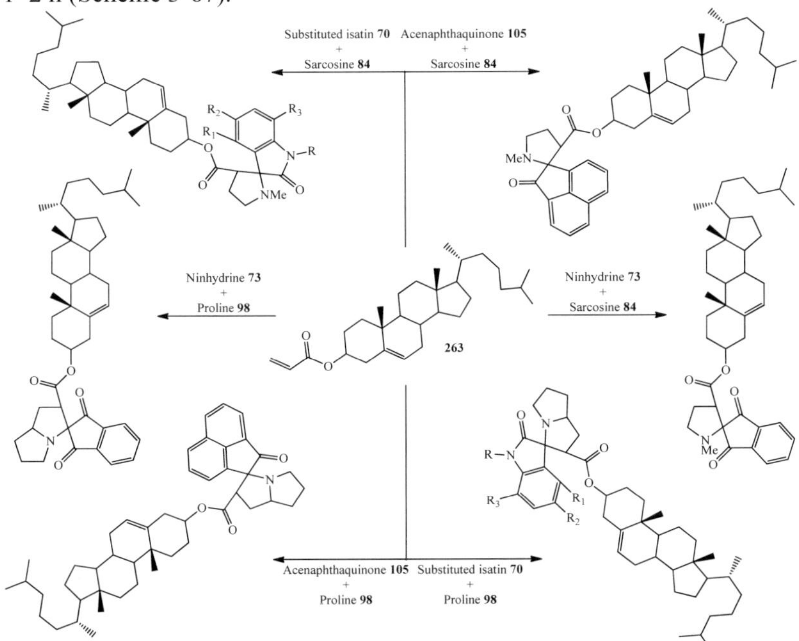

Scheme 3-67. Synthesis of novel C3-β-cholesterol tethered spiropyrrolidine/pyrrolizidine analogs from various substituted diketone/ triketone, cyclic, and/or acyclic amino acids with dipolarophile in the *i*-PrOH medium under reflux conditions for 1–2 h.

Additionally, among the 1,3-DCRs, a remarkable example was very recently reported by Kanizsai et al. toward the synthesis of aziridine-fused spiro[imidazolidine-4,3′-oxindole] framework **265** [60]. The reaction of 2*H*-azirines **264** with azomethine ylides generated from isatins **70** and α-amino acids led to the desired products **265** yields up to 81% with complete regio- and high diastereoselectivities, as indicated in Scheme 3-68.

In 2020, Sharma et al. demonstrated the reaction of *exo*-imines with 1,3-dipoles to give the 3D spirocycles **270**/**271** with a secondary amine (NH) in the spiro-ring [61]. Through a formal 1,3-DC strategy, the amine nitrogen of the cyanamide ion **266** was incorporated into the ring's core to provide *exo*-imine structures such as oxadiazol-imine **267**. These *exo*-imine dipolarophiles led to the creation of new rings by two new bonds via spiroannulation reactions with appropriate dipolar partners **268/269** (Scheme 3-69).

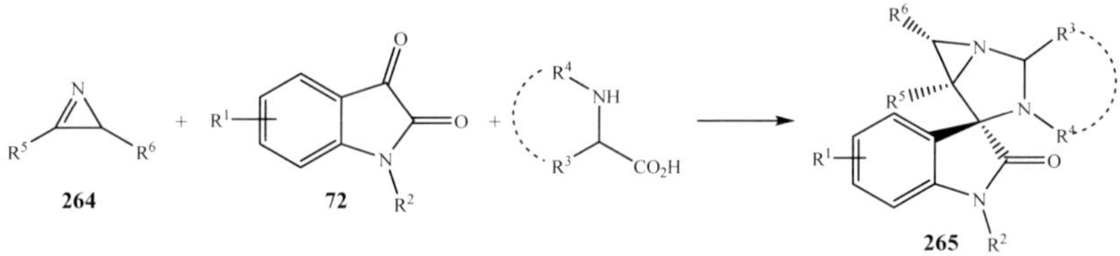

Scheme 3-68. Synthesis of aziridine-fused spiro[imidazolidine-4,3′-oxindole] framework **265**.

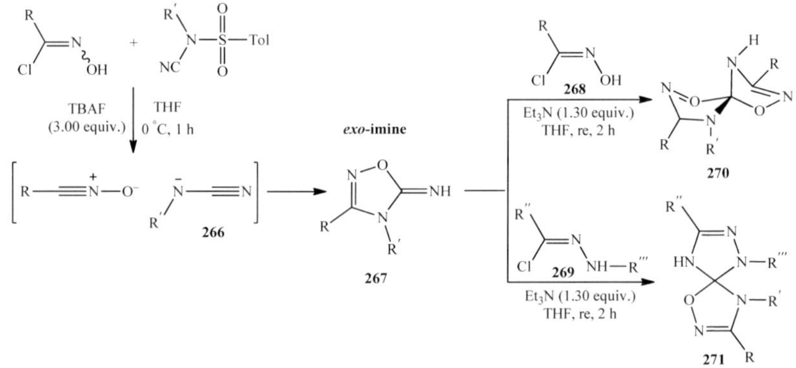

Scheme 3-69. Synthesis of biologically active heterospirocycles through iterative 1,3-DC pathways.

3.3.2. [2+1] Cycloaddition reactions

C_{60} spiro-compound **279** was synthesized by Xu et al. in 1997, starting from C_{60}, DL-valine **272,** and 4,4,5,5-tetramethylimidazolidine-2-thione **273** [62]. The nucleophilic amino group of α-amino acid **272** first attacks the carbon atom of thiocarbonyl group **273** to generate **274**. The acid-base reaction or acid-base equilibrium takes place between **274** and **277**. **276** or **277** first loses CO_2 and H_2S, then one of the C-N bonds breaks to form carbene **278**, and finally, **278** and C_{60} underdo [2+1] cycloaddition to provide **279** (Scheme 3-70).

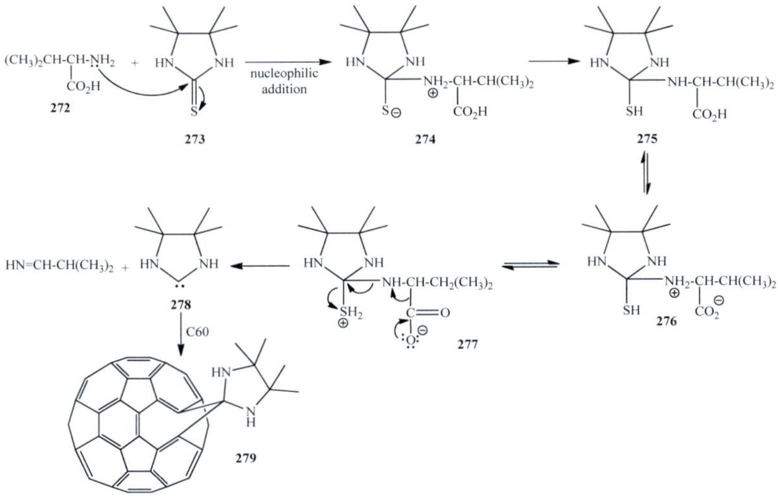

Scheme 3-70. [2+1] cycloaddition of C_{60}.

In 2015, Yuan et al. developed a new approach for the synthesis of spiro[2,5]octa-4,7-dien-6-one **282** via 1,6-conjugated addition mediated [2+1] annulation of *para*-quinone methides **280** (*p*-QMs) with sulfur ylides **281** (Scheme 3-71). This domino-type process was highly diastereoselective and represented good functional group tolerance and scalability without using metals and bases [63].

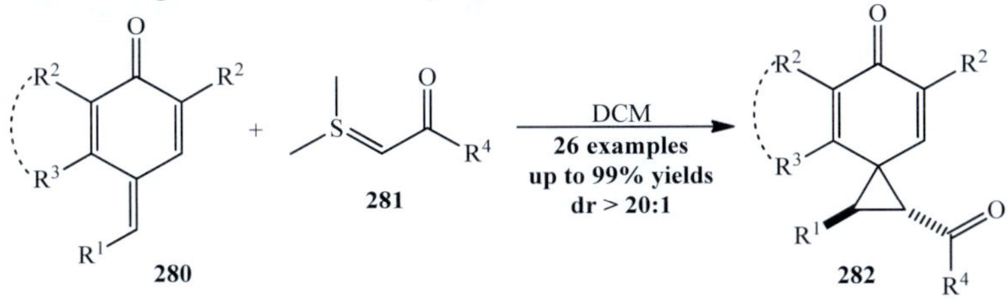

Scheme 3-71. 1,6-Conjugated addition mediated [2+1] annulation to synthrsis spiro[2,5]octa-4,7-dien-6-one **282.**

As illustrated in Scheme 3-72, Li et al. recently reported a base-promoted highly diastereoselective aziridination of cyclic electron-deficient alkenes **283** which leads to facile synthesis of novel spiro-[aziridine-pyrroline] (SAP) derivatives **284** [64]. They also explored the generality of the [2+1] cycloadditions. Due to the ketone group adjacent to the aziridine moiety, some products were found to be unstable. Thus, Horner–Wadsworth–Emmons (HWE) reaction was used to transform this ketone group to the corresponding α,β-unsaturated ester as the final bench-stable product **285** (Eq. 1). In addition, the authors indicated that the enantioenriched spiroaziridine **285** could be synthesized by employing the readily available and relatively inexpensive quinidine **286** as a chiral Brønsted base (Eq. 2).

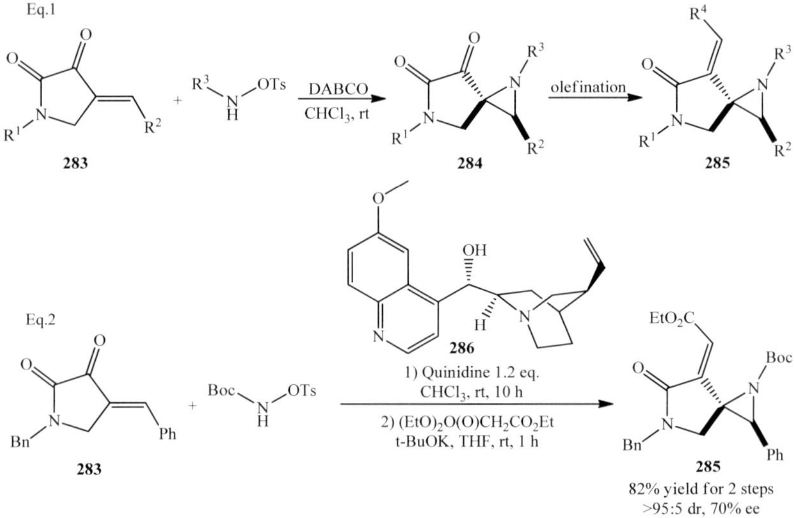

Scheme 3-72. Synthesis of spiro[aziridine-pyrroline] derivatives **284**.

In another interesting example, Roy et al. successfully employed sulfonium salts derived from the Baylis-Hillman bromides **287** and activated olefins (arylidene indane-1,3-diones **288**) to synthesize spiro-cyclopropanes **289** (Scheme 3-73) [65]. Spiro-cyclopropanation occurs through a Michael addition of the nucleophile (*S*-ylide) to the electron-deficient olefin, and subsequent ring-closing via elimination of dimethyl sulfide causes the generation of spiro-cyclopropanes **289** in a stereo-selective manner. Products were obtained mainly in a 1:1 diastereomeric ratio.

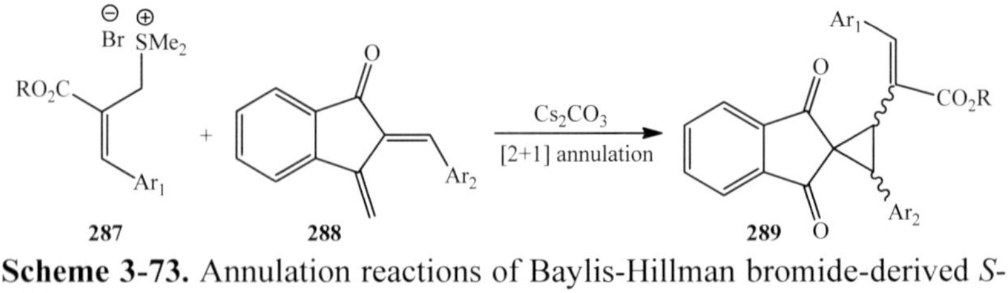

Scheme 3-73. Annulation reactions of Baylis-Hillman bromide-derived *S*-ylides.

Peng et al. showed an elegant example of a [2+1] annulation of (*Z*)-2-ylideneoxindoles **290** with sulfur ylides **291** in the synthesis of cyclopropane-fused spiro-pseudoindoxyl derivatives **292** (up to 88% yield and >20: 1 dr in all cases) (Scheme 3-74) [66].

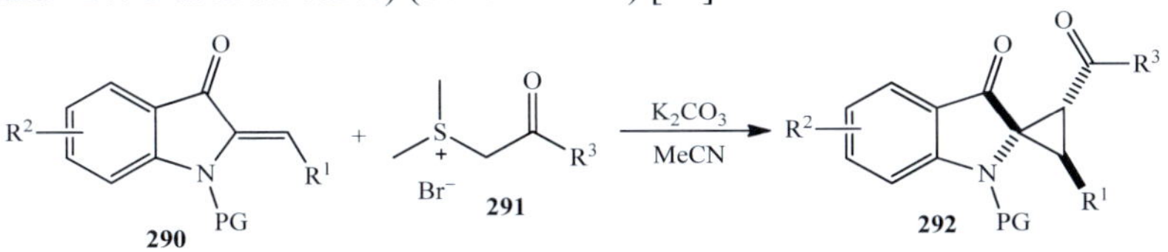

Scheme 3-74. The reaction of (*Z*)-2-ylideneoxindoles **290** with sulfonium salts **291**.

Recently, a diastereoselective 1,6-conjugate-addition-mediated [2+1] annulation reaction between *p*-quinone methides and pyrazolones was reported by Das et al. (Scheme 3-75). The corresponding bis-spiro-[cyclohexadienone−cyclopropane−pyrazolone] compounds **293** with contiguous quaternary spiro-centers were achieved in an excellent yield under mild reaction conditions [67].

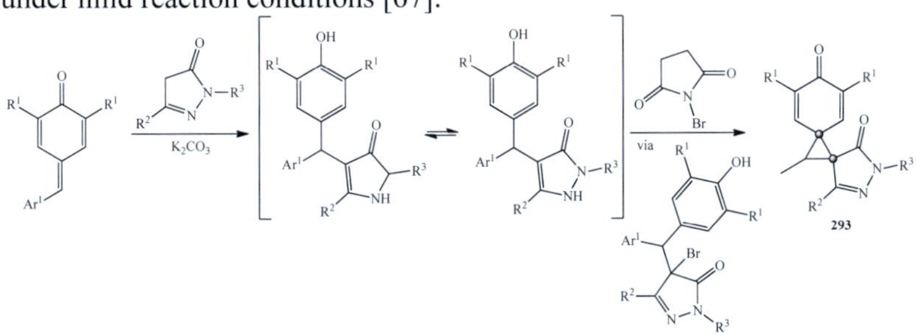

Scheme 3-75. Synthesis of bis-spiro compounds with contiguous quaternary spiro-centers **293**.

3.3.3. [2+2] Cycloaddition reactions

In 1984, Koft et al. applied the [2+2] photocycloaddition to the preparation of spiro compound **297**, as depicted in Scheme 3-76. Irradiation of **294** through uranium glass in methanol buffered with sodium acetate resulted in **295** within 48-72 h. Then, the reduction of the crude photolysate with sodium borohydride in methanol provided two products assumed to be alcohols **296a** and **296b**. Again without purification, ozonolysis followed by reductive workup with triphenyl-phosphine afforded a 2:1 mixture of aldehydes **297a** and **297b** in an overall yield of 25% based on **294** [68].

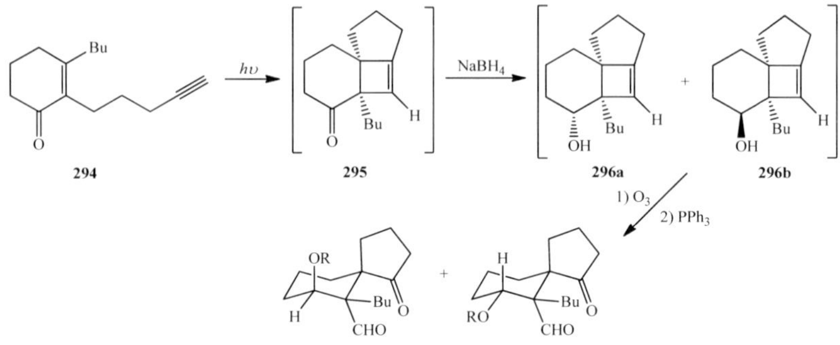

294 295 296a 296b

297a R = H, Ac 297b

Scheme 3-76. Koft et al. reported the [2+2] photocycloaddition.

Years later, Kraus et al. reported the [2+2] photocycloaddition strategy for the synthesis of diols containing the novel spiro[3.3]heptane units **301** and **302** by the irradiation of 2-ethoxy-l,4-naphthoquinone **298** with 3-butenol **299** and ally1 alcohol **300** (Scheme 3-77) [69].

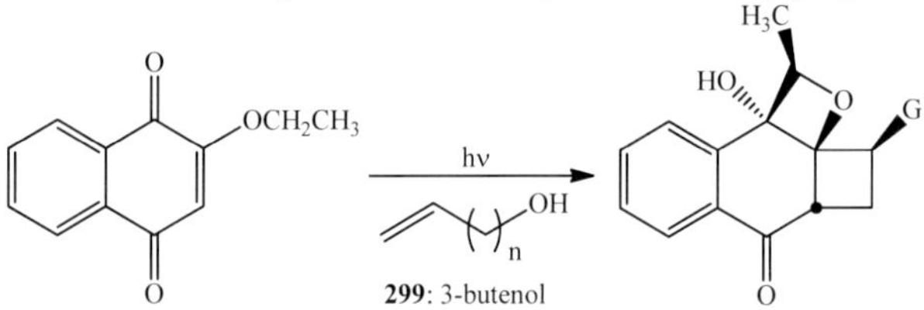

298

299: 3-butenol
300: allyl alcohol

301: G = CH_2CH_2OH
302: G = CH_2OH

Scheme 3-77. The synthesis of diols containing the novel spiro[3.3] heptane units **301** and **302**.

Baldwin et al. found that the reaction of methylenecyclopentane **303** with dichloroketene generated *in situ* from the dehalogenation of trichloroacetyl bromide **304** led to 1,1-dichloro- spiro[3.4]octan-2-one **305** with very high regioselectivity and fair yield 78% [70], The dichloro-ketene **305** was reduced with zinc in acetic acid to prepare spiro[3.4]octan-2-one **306** in 63% yield. A Wittig reaction converted this ketone in 80% yield to 2-methylenespiro[3.4]octane **307**. Finally, a Simmons–Smith reaction using activated zinc rather than the more traditional zinc–copper couple provided di-spiro[2.1.4.1]decane **308** (Scheme 3-78) [71].

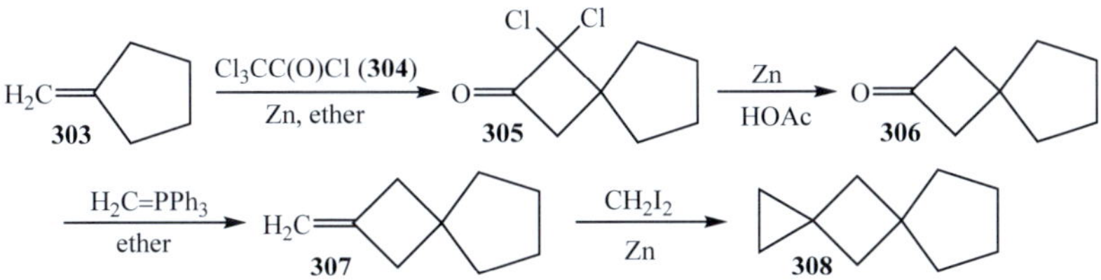

Scheme 3-78. The synthesis of di-spiro[2.1.4.1]decane **308**.

Keglevich et al. showed in 2000 that the unexpected [2+2] cycloaddition of the P=O group of cyclic 1-(2,4,6-triisopropyl phenyl)-phosphine oxides **309a-e** with the acetylene moiety of dimethyl acetylenedicarboxylate (DMAD) **33** gives easy access to spirocyclic oxaphosphetes **310a-d**, a completely new family of *P*-heterocycles (Scheme 3-79) [72].

Wendling et al. investigated the possibility of preparing unsaturated four-membered rings fused to polycyclic ring systems utilizing an intramolecular [2+2] cycloaddition reaction of electrophilic acetylenes linked to keto-derivatives. Their studies commenced with synthesizing the silyl enol ethers **312** starting from the corresponding acetylenic ω-ketoesters **311** and using conventional methods (TBDMSOTf, NEt_3, CH_2Cl_2, 20 °C) (Scheme 3-80).

To develop reaction conditions, they used TBAF as a deprotecting reagent. Surprisingly, the treatment of the silyl enol ethers **312** with TBAF led to a mixture of two easily separable products: the spiroketones **314** isolated in 15% yield and the allene **315** isolated in 45% yield. The formation of these two compounds could be explained through intermediates **A** and **B**. After deprotection of the silyl enol ether **312**, the resulting enolate **A** could evolve according to two different pathways: (a) an intramolecular Michael addition takes place, affording the spiroketone **314**, or (b) the enolate **A** is in equilibrium with the propargylic carbanion **B**; an intramolecular cyclization then takes place onto the carbonyl function, giving rise to the alcoolate **C**, which evolves to provide the allene **315**. The resulting allene **315** in the presence of *t*-BuOK produced corresponding oxetane **316** in a good yield (Scheme 3-81) [73].

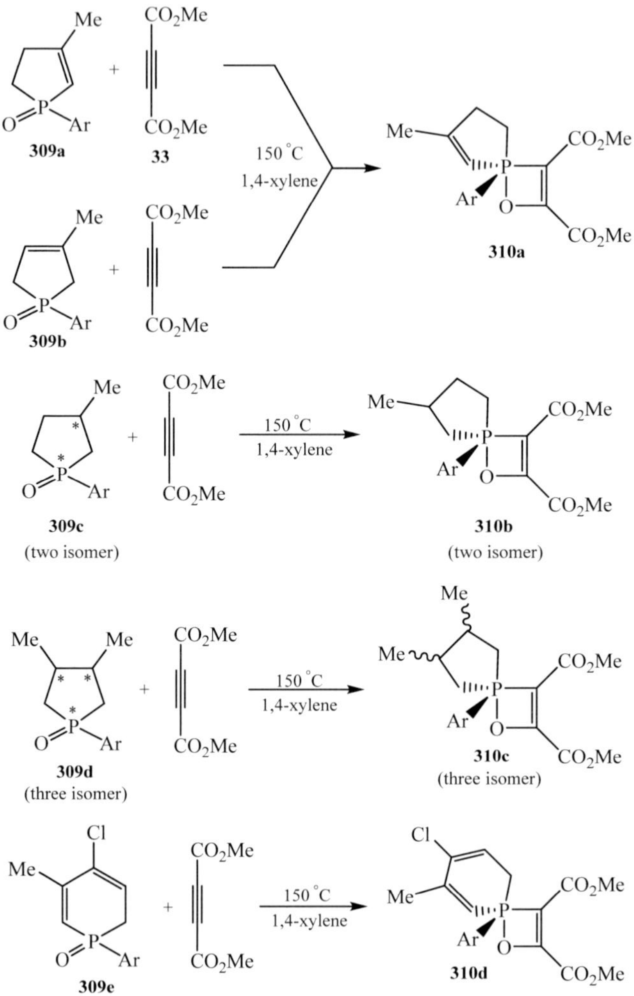

Scheme 3-79. Spirocyclic oxaphosphetes **310a-d**.

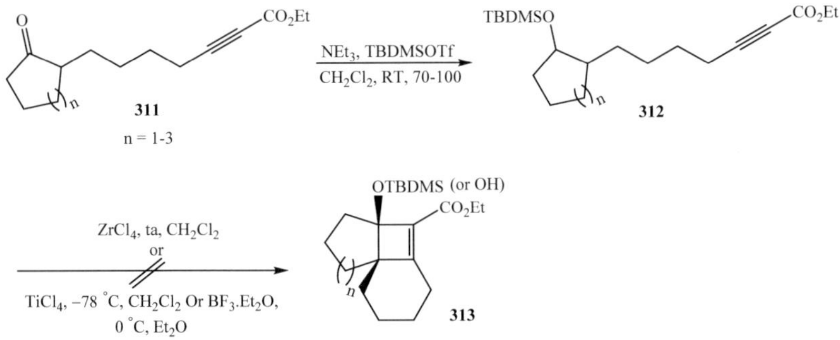

Scheme 3-80. Reactivity of acetylenic ω-ketoesters toward $ZrCl_4$.

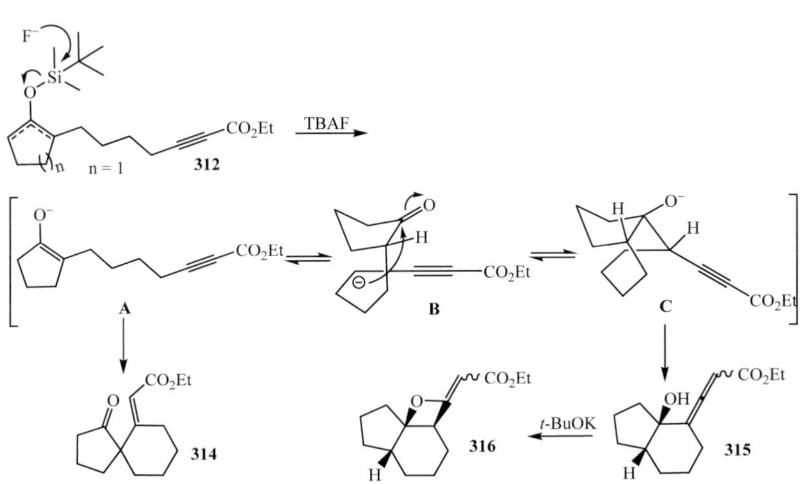

Scheme 3-81. Proposed mechanism for forming **314** and **315** and synthesis of oxetane **316**.

The rational design and synthesis of a novel type of β-turn mimetics **317** through a Staudinger reaction of unsymmetrical cyclic ketenes was reported by the Alonso group. This methodology led to the construction of a wide variety of this type of peptidomimetics, just by combining different imines and cyclic ketenes. To apply this strategy, they selected the *N*-benzyloxycarbonyl-L-proline acid chloride **318** as a precursor of the ketene involved in the Staudinger reaction with various imines (Scheme 3-82) [74].

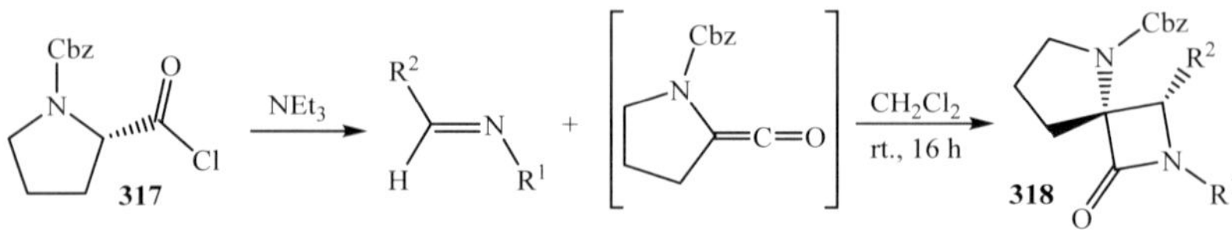

Scheme 3-82. Synthesis of spiro-β-lactams as β-turn mimetics.

In another study, (7-azaspiro[3.5]non-2-yl)carboxylate **319** was synthesized by Kaneko et al. via [2+2] cycloaddition reaction [75]. The cycloaddition reaction between olefin **319** and dichloroketene, generated *in situ* from trichloroacetyl chloride and zinc/copper and followed by reduction with zinc, provided the 7-azaspiro[3.5]nonane core of **320**. Treatment of this ketone **320** with tosylmethyl isocyanide (TosMIC) in the presence of potassium *tert*-butoxide provided the cyanide **321**. Deprotection of the piperidine amino group with trifluoroacetic acid (TFA) followed by reprotection afforded the benzylamine **322**. Hydrolysis of the cyano moiety afforded the carboxylic acid. Treatment with thienyl chloride in MeOH, then debenzylation using $Pd(OH)_2/C$, furnished the amine **323** (Scheme 3-83).

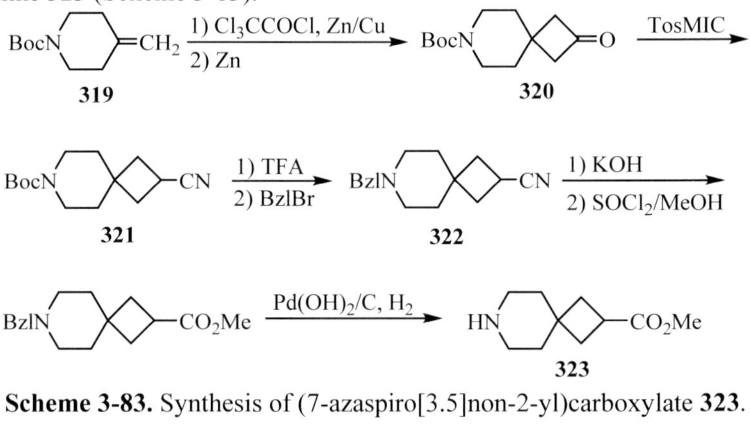

Scheme 3-83. Synthesis of (7-azaspiro[3.5]non-2-yl)carboxylate **323**.

As illustrated in Scheme 3-84, González et al. reported the synthesis of spiro-β-lactams **326** by a [2+2] cycloaddition between the unsymmetrical ketenes **325** derived from the dehydrohalogenation of *N*-benzyloxy carbonyl L-proline acid chlorides **324** with optically active imines [76].

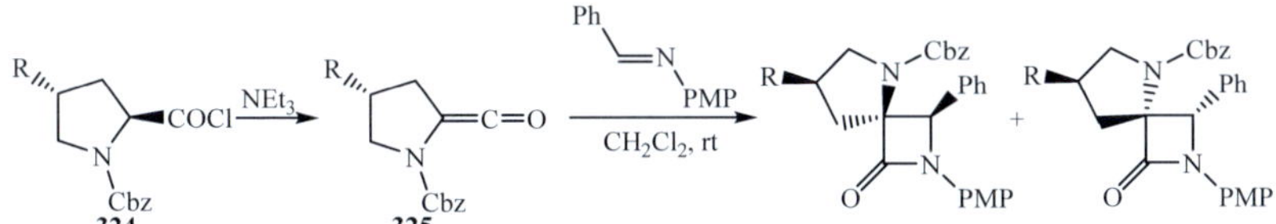

324 325 326

Scheme 3-84. Synthesis of spiro-β-lactams **326**.

The reaction of aryl and diaryl ketenes with isocyanides to furnish adducts comprising a 2:1 ketene: isocyanide stoichiometry has been known for more than 50 years. A structure **327** was suggested for these adducts based on analysis of the products formed from their sequential treatment with concentrated hydrochloric acid and sodium methoxide, in combination with IR data; this 4-iminocyclopenta-1,3-dione structure may be formed by consecutive *C*-alkylations of enolate-type intermediates (Scheme 3-85).

Robertson et al. [77] discovered a new ketene/isocyanide manifold pathway. They achieved the first X-ray crystallographic confirmation of the structures of compounds of general structure **328**. They showed that the products resulting from the reaction of diphenylketene with a variety of isocyanides depend heavily on the concentration of diphenylketene; a high concentration results in dioxolanes at much lower concentrations, the reactions follow an alternative route, and a cascade of formal pericyclic reactions produces polycyclic β-lactams.

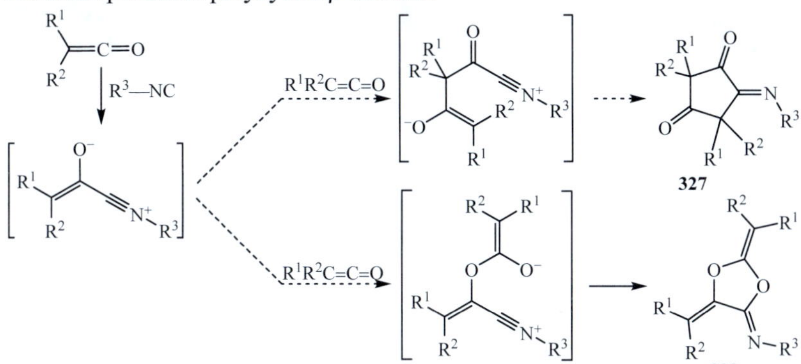

327

328

Scheme 3-85. The reaction of isocyanides with ketenes was first formulated (dashed arrows) and later modified (solid arrows).

In this work, the authors suggested that the structure of **329** is formed through electro-cyclization of the reactive zwitterion without a high concentration of diphenylketene, tautomerisation, and sequential [2+2] and ene reactions (Scheme 3-86).

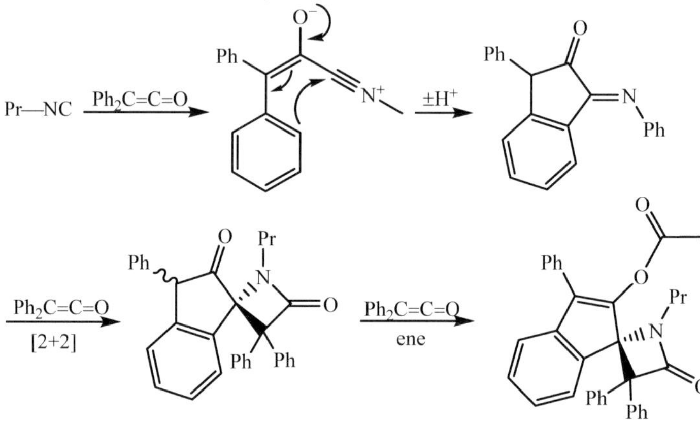

329

Scheme 3-86. Formation of the structure of **329** through a cascade reaction.

In an elegant work, the Delgado group [78] introduced the results of the photodimerization of *Z*-ligustilide **330** and analyzed experimental results using FMO methods. *Z*-Ligustilide **330** contains three different C–C double bonds and, under photochemical conditions, could provide several [π2s+π2s] photocyclodimers, considering the chemo-, regio- and stereochemical possibilities. Treatment of *Z*-ligustilide **330** under photochemical conditions using a mercury lamp furnished four main products: the natural product riligustilide **331** and three novel dimeric phthalides *endo*-*Z*,*Z*′-(3.8′,8.3′)-diligustilide **332**, *endo*-*Z*,*Z*′-(3a.7a′,7a.3a′) diligustilide **333** and *exo*-*Z*,*Z*′-(3a.7a′,7a.3a′)-diligustilide **334** (Figure 3-2). To further understand the formation of compounds **331**, **332**, **333,** and **334**, they performed molecular structure calculations at the semiempirical AM1 level. In this regard, the molecular structure of **330** was modeled and optimized for the ground and the first excited (triplet) states. They realized that for the triplet state, all the atoms in the chain are coplanar to the lactone, whereas this is not observed for the ground state. This difference between both molecular conformations causes a decrease in the steric hindrance for forming photocyclodimers.

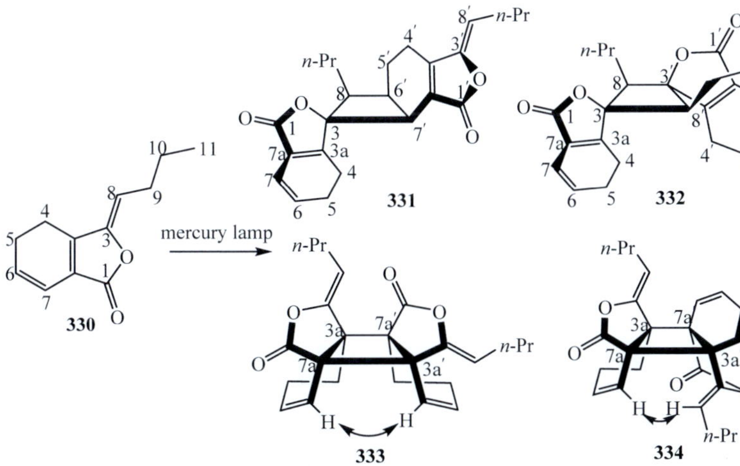

Fig. 3-2. *Z*-ligustilide products.

Mülliken's charge analysis expresses that for a charge-driven reactivity, the preferred double bond to form photocyclodimers is C(3)–C(8) since it displays opposite charge signs and the most considerable charge difference between the atoms.

The regiochemistry involved in the formation of the photocyclodimers **331**, **332**, **333**, and **334** may be explained through the analysis of coefficients and energies of the FMOs of the interacting monomers of *Z*-ligustilide **330**. The considered MOs are higher (HSOMO) and lower (LSOMO) solus-occupied MO for the ground state and the HOMO and LUMO for the triplet.

The tetracyclic core of the diterpene bielschowskysin was prepared by Sulikowski et al. as a single enantiomer via a stereoselective intramolecular [2+2] photocycloaddition of 5-alkylidene-2(5*H*)-furanone **336** [79]. In this reaction, irradiating a chloroform solution of **336** with a sun lamp for 2 h results in an approximately 3.6:1 mixture of geometric isomers **336** and **335**. Consecutive irradiation of a chloroform solution of **335** and **336** led to the consumption of both isomers and the generation of a complex mixture of products. Upon replacing chloroform with acetone, a similar equilibration-cyclization sequence was observed on irradiation with a sun lamp, but fewer by-products were observed. The intramolecular [2+2] photocycloaddition of butenolides **335** and **336** presumably follows the so-called rule of five, leading to intermediate 1,4-biradicals **335a** and **336a** (Scheme 3-87), respectively. Weedon and Maradyn trapped the triplet 1,4-biradicals by using hydrogen selenide which gave evidence for the intermediacy of 1,4-biradicals and the appropriateness of the intramolecular photocycloadditions. To extend this model to substrates **335** and **336**, the triplet 1,4-biradicals **335a** and **336a** could be interchanged by one of two mechanisms. First, the simple bond rotation could result in rapid interconversion of triplet biradicals **335a** and **336a**. A second pathway involves simple reversion of **335a** and **336a** to starting γ-alkylidene butenolides **335** and **336**, respectively. By either of these mechanisms, the favored generation of photoadduct **337** via the closure of **335a** may be explained by the creation of unfavorable dipole and/or electrostatic interactions in the alternative closure of 1,4-biradical **336a** to **338**.

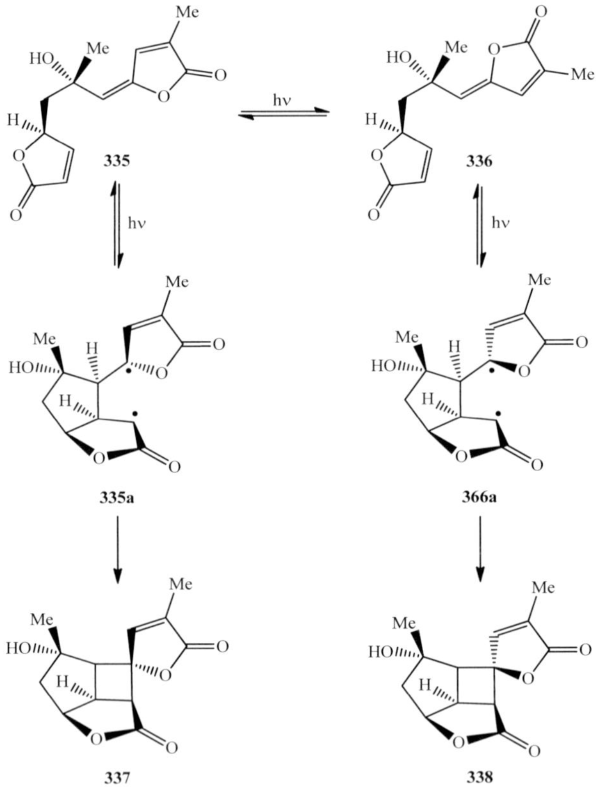

Scheme 3-87. Model for observed diastereoselectivity of [2+2] photocycloaddition.

Combining two spiroannelation methods allowed Fitjer et al. to facile access to spiroannelated five-membered rings. Their approach to the desired carbon frameworks relied on two ways for the construction of spiranes (Scheme 3-88):

i) a cycloaddition of dichloroketene to methylene-cyclopentane **339** with subsequent dehalogenation and ring enlargement to generate spiro [4.4]nonan-2-one **341** (**339**–**340**–**341**), and

ii) a cyclobutylidenation of **341** with subsequent epoxidation and oxaspiropentane to cyclopentanone rearrangement to generate dispiro-[4.1.4.2]tridecan-1-one **343** [80].

Experimentally, the addition of dichloroketene to methylene cyclopentane **339** proceeded regioselectively and generated, after dehalogenation, the β-quaternary cyclobutanone **340**. Upon treatment with diazomethane, this compound ring enlarged to produce the β-quaternary cyclopentanone **341**. The cyclobutylidenation of **341** to **342** proceeded smoothly, and subsequent epoxidation and boron trifluoride etherate catalyzed rearrangement of the resulting oxaspirohexane produced the desired α-quaternary cyclopentanone **343**. On the other hand, β-quaternary cyclobutanone **343** could be transformed into 2-oxaspiro[4.4]nona-3-one **344** by Bayer–Villiger oxidation.

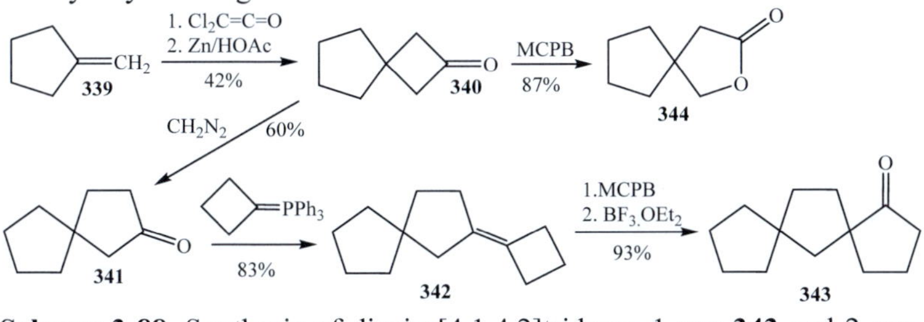

Scheme 3-88. Synthesis of dispiro[4.1.4.2]tridecan-1-one **343** and 2-oxa-spiro[4.4]nona-3-one **344**.

In 2009, *tert*-butyl 6-oxo-2-azaspiro[3.3]heptane-2-carboxylate **345** was prepared by Meyers et al. via a Wittig/[2+2] cycloaddition sequence (Scheme 3-89) [81]. To maintain efficient reaction conversion in the cycloaddition, it is essential to maintain the reaction temperature between room temperature and 30 °C. If the reaction temperature exceeds 30 °C, a significant loss of the Boc group is observed.

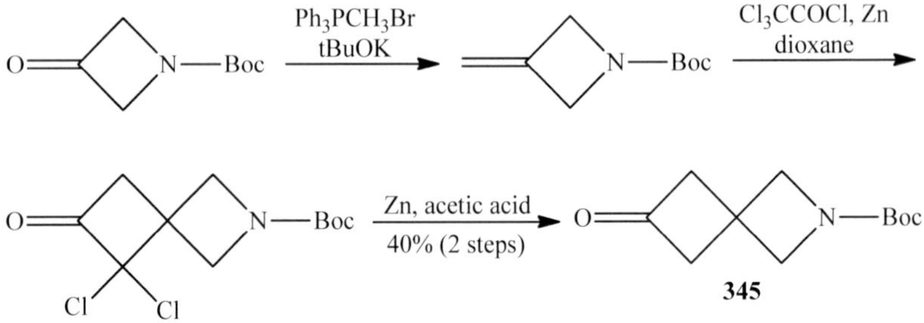

Scheme 3-89. Synthesis of **345** via [2+2] Cycloaddition.

In a report presented by Matsuo et al. in 2010, the spiro compounds 1-[(trimethylsilyl) methyl]spiro[3.4]octan-2-one **348** and 1-[(trimethylsilyl) methyl]spiro[3.5]nonan-2-one **349** were prepared by [2+2] cycloaddition of [2-(trimethylsilyl)ethylidene]cyclopentene **346** and [2-(trimethylsilyl) ethylidene]cyclohexane **347** with dichloroketene generated *in situ* from the dehalogenation of trichloroacetyl chloride with zinc which was activated with copper and reductive dechlorination of the dichlorocyclobutanone adducts (Scheme 3-90) [82].

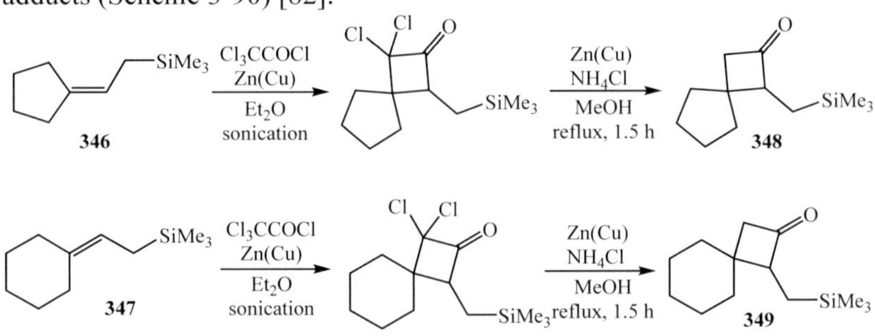

Scheme 3-90. Synthesis of spiro compounds **348** and **349**.

In other studies, the Jarrahpour group reported the synthesis of some new spirocyclic indeno[1,2-*b*]quinoxaline β-lactams **352** and **353** [83] in good yields following a [2+2] ketene-imine cycloaddition (Staudinger reaction) of *N*-phenyl-11*H*-indeno[1,2-*b*]quinoxalin-11-imines **350** and different phenoxyacetic acids **351** in the presence of triethylamine and *p*-toluenesulfonyl chloride (Scheme 3-91).

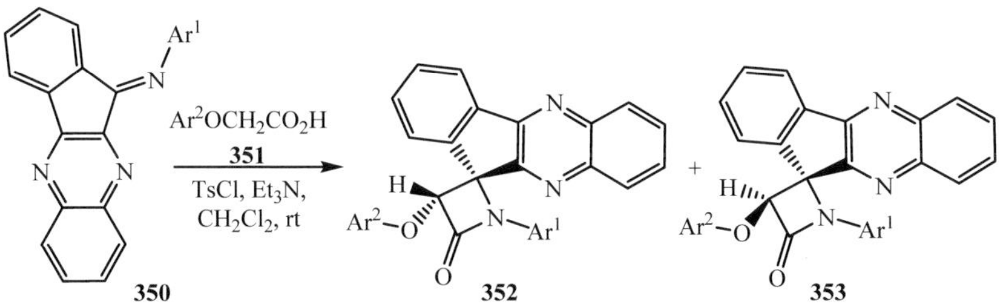

Scheme 3-91. Synthesis of indeno[1,2-*b*]quinoxaline spiro-β-lactams **352** and **353**.

Recently, Abe et al. investigated the regio- and stereoselectivity of photochemical [2+2] cycloaddition reaction of the pyrrole **354** with two types of enones **355** or **356** [84]. As shown in Scheme 3-92, in the reaction with cyclopentenone **355**, *exo*-selective formation of tricyclic cyclobutanes **357** and **358** was observed. In contrast, they used 1,4-naphthoquinone **356** as the enone led to the selective formation of oxetane **359**, although the reaction transformation was relatively low.

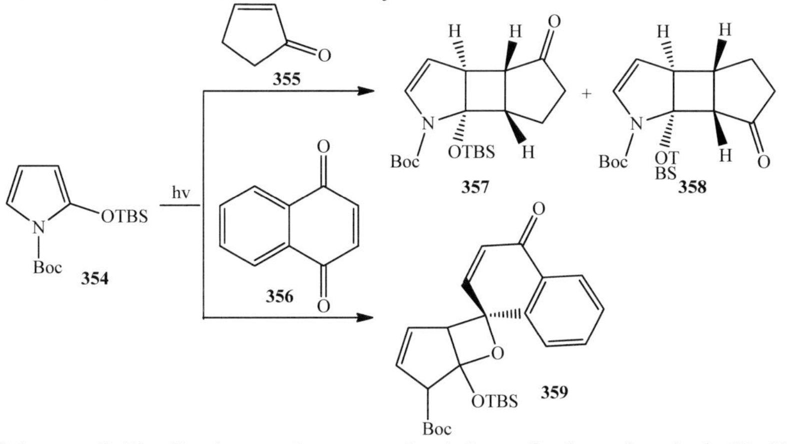

Scheme 3-92. Regio- and stereoselectivity of photochemical [2+2] cycloaddition reaction.

Photoinduced conversion of biomass-derived chemicals to high-value and advanced fuels has been a challenging topic in many fields of chemistry. Hence, the Zou group developed a green and efficient self-sensitized [2+2] cycloaddition process to convert biomass-derived isophorone **360** and β-pinene **361** to spirocyclic molecules **362** which cannot be obtained by thermal catalytic conversion (Scheme 3-93).

Specially, the cycloaddition of isophorone **360** and β-pinene **361** can take place smoothly with a quantum yield of 0.22 under 365 nm without additional photosensitizer because isophorone acts as an excellent photosensitizer. Triplet isophorone is involved in the reaction, and it will disorient with β-pinene **361** to produce a complex and trigger the photoaddition. The wavelength range of 300-375 nm absorption is the most effective in inducing photoreaction. Under optimal conditions, the final high-performance spiro-fuel **362** is obtained with an overall yield of 85%. Having a high density of 0.911 g/mL, a freezing point of -51 °C, and a volumetric NHOC of 38.67 MJ/L, the obtained spirocycloalkane **362** is very competitive as renewable high-density liquid fuels [85].

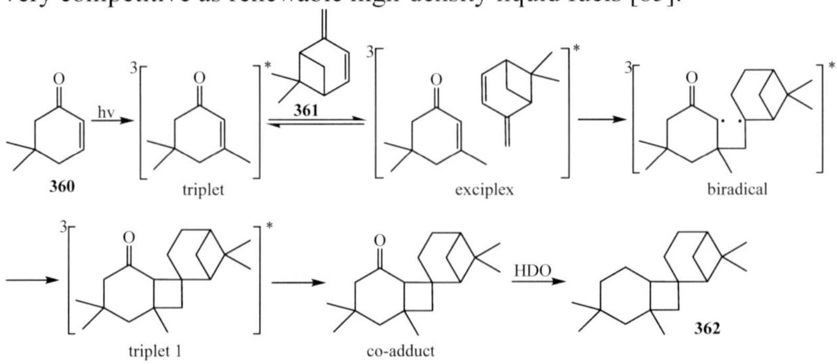

Scheme 3-93. Synthesis route of high-performance spiro-fuel **362** through self-photosensitized [2+2] cycloaddition and hydrodeoxygenation.

3.3.4. [3+2] Cycloaddition reactions

Over the last decades, excellent revisions have been reported concerning the reactivity and use of the [3+2] cycloaddition reaction. In this portion some important novel examples will be described concerning variations in [3+2] cycloaddition reactions that have been reported since 2003 to give a broad overview of the recent trends in this crucial protocol.

In 2003, Raghunathanet et al. [86] reported the formation of a new class of spiro[tetrahydro-naphthalen-one/butenolide]pyrrolidines **366/367** through formal [3+2] cycloaddition of azomethine ylide generated by thermal ring opening of *cis*-3-benzoyl-1-cyclohexyl-2-phenyl-aziridine **363** with (*E*)-2-arylidene-1,2,3,4-tetrahydronaphthalen-1-ones **364** and (*E*)-3-arylidene-5-phenyl-$\Delta^{4,5}$-butenolides **365** (Scheme 3-94).

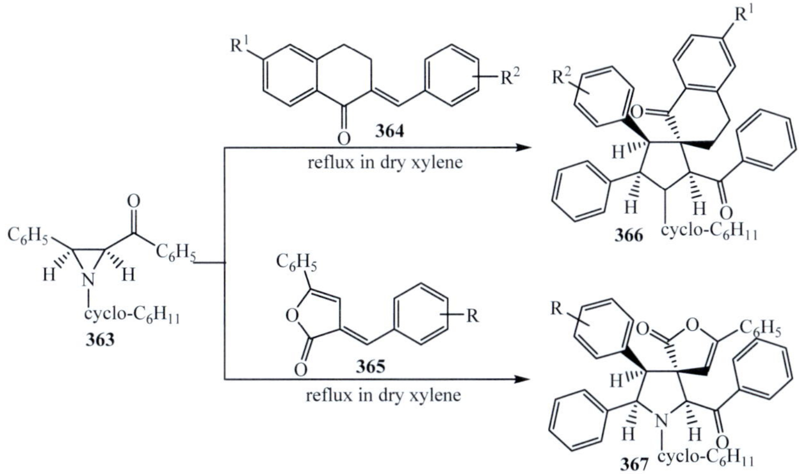

Scheme 3-94. Synthesis of spiro pyrrolidines through formal [3+2] cycloaddition **366/367**.

One year later, Raghunathan group reported the facile synthesis of novel dispiro heterocycles **369-370** through solvent-free microwave-assisted [3+2] cycloaddition of azomethine ylides generated by a decarboxylative route from sarcosine/ proline **84/98** and isatin **70** with the dipolarophile 9-arylidine-fluorene **368** (Scheme 3-95) [87].

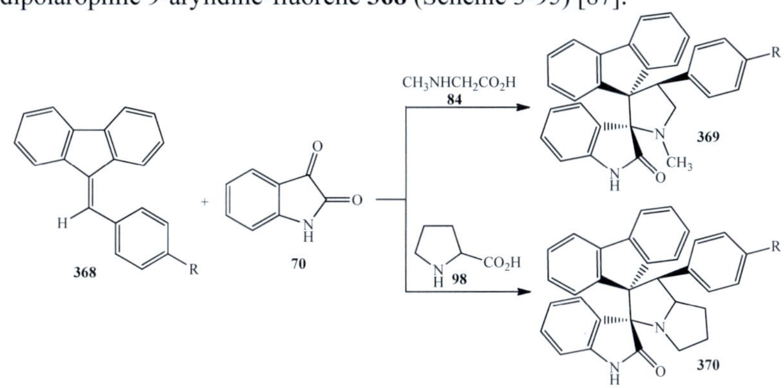

Scheme 3-95. Solvent-free microwave-assisted [3+2] cycloaddition of azomethine ylides.

In 2007, the Cheng et al. group explored [3+2] cycloaddition reaction of 2-arylthio-carbamoyl benzimidazolium, -imidazolinium, and –

thiazolium inner salts **371** (the ambident C-C-N and C-C-S 1,3-dipoles derived from carbenes) with ketenes that proceeded efficiently in a highly site-selective manner to generate the C-C-N cycloaddition products benzimidazoline-, imidazolidine-, or triazoline spiropyrrolidones **372** in 58-93% yields (Scheme 3-96) [88].

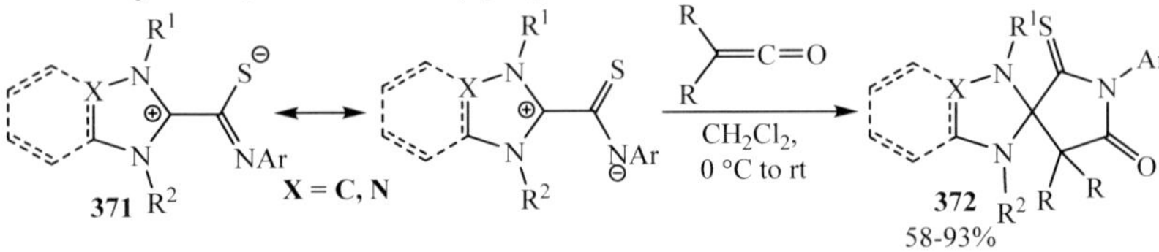

Scheme 3-96. Highly efficient and site-selective [3+2] cycloaddition.

Concurrently, Perumal et al. [89] reported an efficient one-pot synthesis of spirodihydrofuran oxindole **347** and spiro-2-hydroxy tetrahydrofuran oxindoles **375** through [3+2] oxidative cycloaddition of cyclic and acyclic 1,3-dicarbonyl compounds to 3-(phenyl-2-oxo ethylidene)-1-methyloxindoles **373** mediated by ceric ammonium nitrate (CAN) (Scheme 3-97).

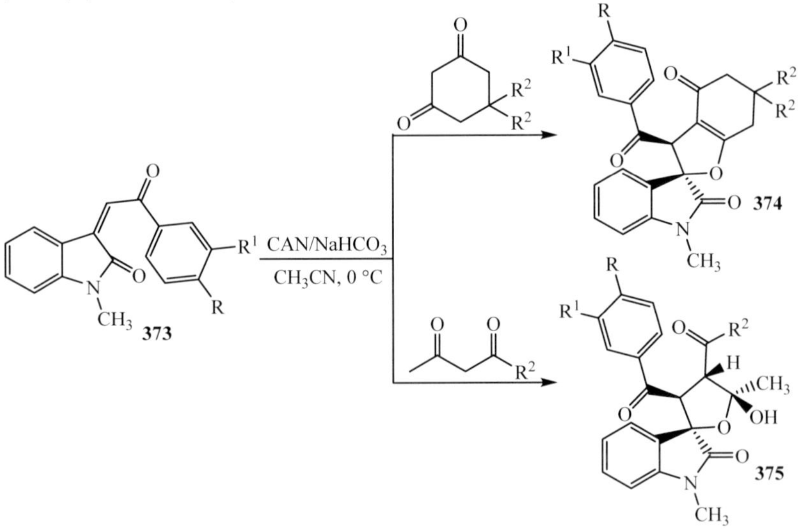

Scheme 3-97. Synthesis of spirodihydrofuran oxindole **374** and spiro-2-hydroxytetrahydrofuran oxindoles **375**.

Likewise, the Shanmugam group reported a novel regioselective synthesis of several functionalized 3-spiropyrrolizidine **377** and 3-

spiropyrrolidine oxindoles **378** from Baylis–Hillman adducts **376** of isatin and heteroaldehydes through [3+2] cycloaddition of azomethine ylides in excellent yields (Scheme 3-98) [90].

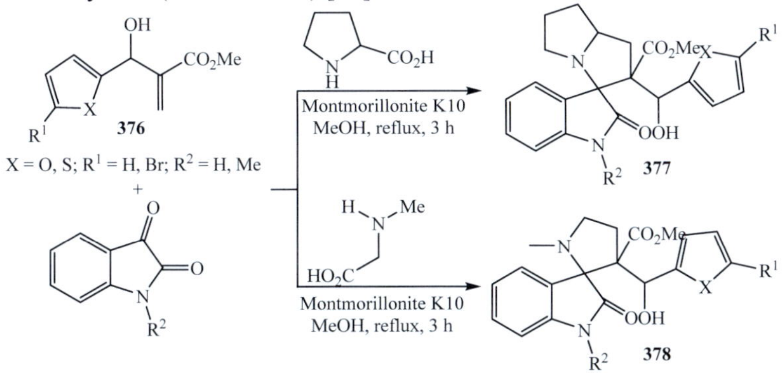

Scheme 3-98. Synthesis of 3-spiropyrrolizidine **377** and 3-spiropyrrolidine oxindoles **378** from Baylis–Hillman adducts **376**.

Spiro-oxindolopyrrolizidines **380** were obtained by Thangamani et al., using (*E*)-3-aryl-1-(thiophen-2-yl)prop-2-en-1-one dipolarophiles **379**, isatin **70**, and proline **98** as substrates under reflux condition (Scheme 3-99) [91]. The synthesized compounds were tested for their anti-bacterial and anti-fungal activities against various microbial organisms. Anti-bacterial and antifungal *in vitro* testing against a panel of microbial organisms disclosed that the compound containing fluorine and chlorine substituents was active against all the tested organisms.

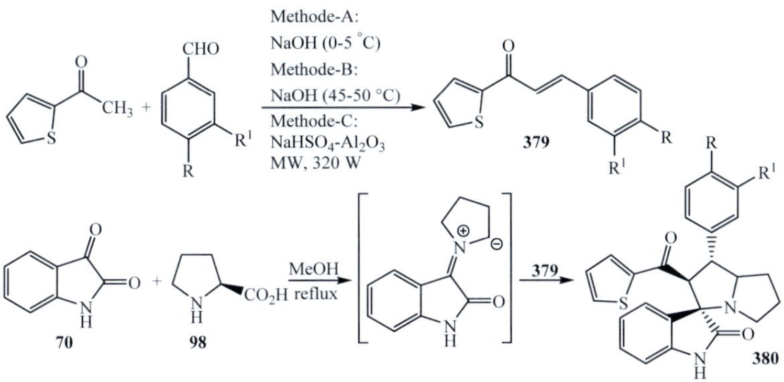

Scheme 3-99. Synthesis of spiro-oxindolopyrrolizidines **380**.

On the other hand, Raghunathan et al. also synthesized a series of hitherto unknown novel ferrocene-based monospirooxindolopyrrolizidine heterocycles **381**, following their developed procedure as shown in Scheme 3-100 [92].

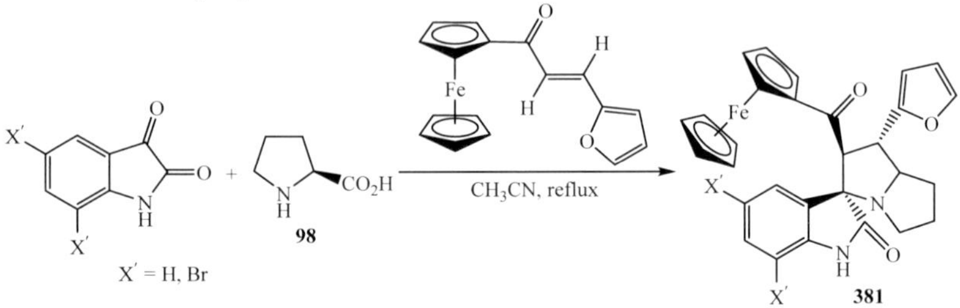

Scheme 3-100. Synthesis of ferrocene grafted spiro-oxindolo-pyrrolizidines **381**.

In 2011, Sarrafi et al. [93] investigated the regio- and stereochemical polar [3+2] cycloaddition of the azomethine ylides generated *in situ* from a reaction of isatin derivatives and proline **98**, with *trans*-β-nitrostyrene **382** and (*E*)-1-phenyl-2-nitropropene **383** for the synthesis of spiro-oxindolo-nitropyrrolizines **384-385** and **386-387** under reflux with ethanol (Scheme 3-101). Mechanism and the regiochemistry of two reactions have been studied in terms of global and local reactivity indexes, FMO analysis, and characterization of relevant transition states at the B3LYP/6-31G(d,p) level of theory. Investigation of theoretical results demonstrated that in the case of the reaction Eq. 1, the electronic factors control the reaction's regiochemistry, and the effects of FMO analysis agree with the experimentally favored product. While the reaction Eq. 2 (FMO analysis) failed to predict the significant development, the study of transition states energies successfully illustrated the experimental results.

One year later, Zhang et al. introduced an efficient procedure for the production of spiro[furan-3,3′-indolin]-2-ones **390a,b** and dispiro [cycloalkane-1,2′-furan-3′,3″-indolin]-2″-ones **391a,b** via photoinduced electron transfer (PET)-catalyzed [3+2] cycloaddition reactions of substituted spiro[indoline-3,2′-oxiran]-2-ones **388**/dispiro-oxiranes **389** with olefins [94]. The reactions proceed by ring opening of spiro[indoline-3,2′-oxiran]-2-ones **388** through $C_β$-O bond cleavage and subsequent cycloaddition with olefins by using 2,4,6-triphenylpyrylium tetra-fluoroborate (TPT) as a sensitizer (Scheme 3-102).

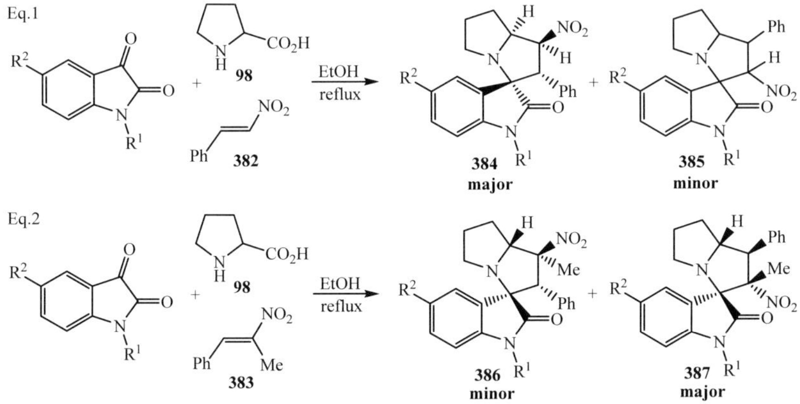

Scheme 3-101. The regio- and stereoselectivity of the polar [3+2] cycloaddition of azomethine ylides to nitrostyrenes **382** and **383**.

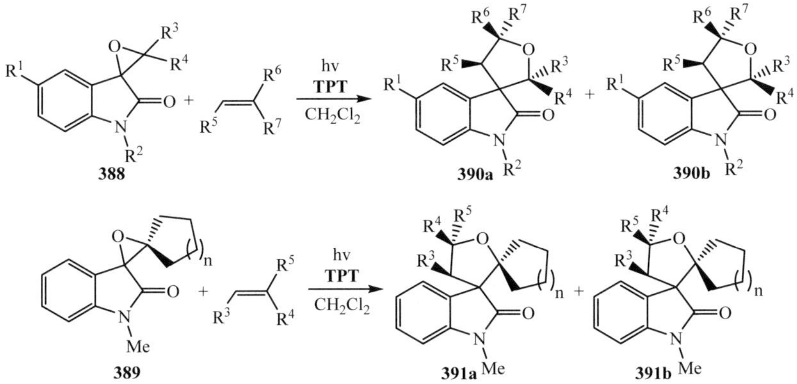

Scheme 3-102. PET-Catalyzed [3+2] cycloaddition reactions.

Raghunathan et al. reported for the first time a mild, efficient, and a facile four-component, one-pot synthesis of novel ferrocene grafted monospiroindenoquinoxaline pyrrolizidines **393** via [3+2] cycloaddition of azomethine ylide produced from 1,2-phenylenediamine **180**, ninhydrin **73**, and L-proline **98** with various unusual ferrocene derived dipolarophiles **392a-e** (Scheme 3-103) [95].

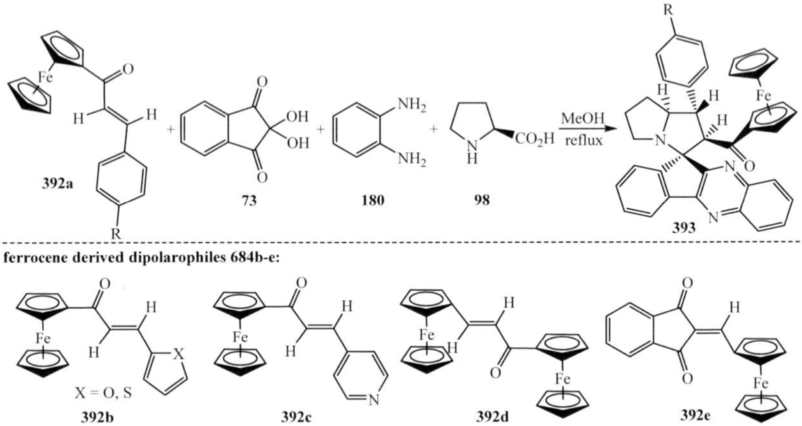

Scheme 3-103. A facile synthesis of novel ferrocene grafted spiroindeno-quinoxaline pyrrolizidines **393**.

In other examples, Ashraf Ali et al. employed indanones as dipolarophiles to develop a facile three-component [3+2] cycloaddition for the synthesis of dispiropyrrolidines **394** (Scheme 3-104). The synthesized compounds **394** were screened for their antimycobacterial activity against *Mycobacterium tuberculosis* $H_{37}Rv$ in HTS. Most of the compounds displayed a moderate to good activity with MIC of less than 20 μM [96].

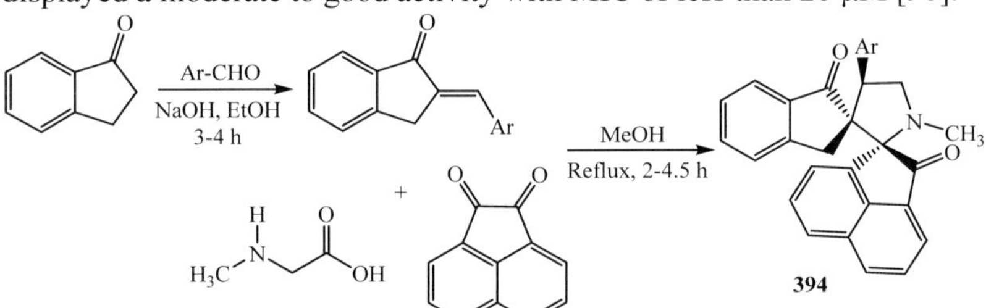

Scheme 3-104. Protocol for the synthesis of dispiropyrrolidines **394**.

Additionally, the Badru group used a cycloaddition reaction between *N*-cyclohexylidene *N*-phenyl nitrones **395** and cyclic dipolarophiles **396** to construct 2,5-diphenyl-3,3-spiropenta-methylene-5*H*-2,3,3a,6a-tetrahydro pyrrolo[3,4-*d*]isoxazole-4,6-diones **397** with one *endo*-diastereoisomer (Scheme 3-105) [97].

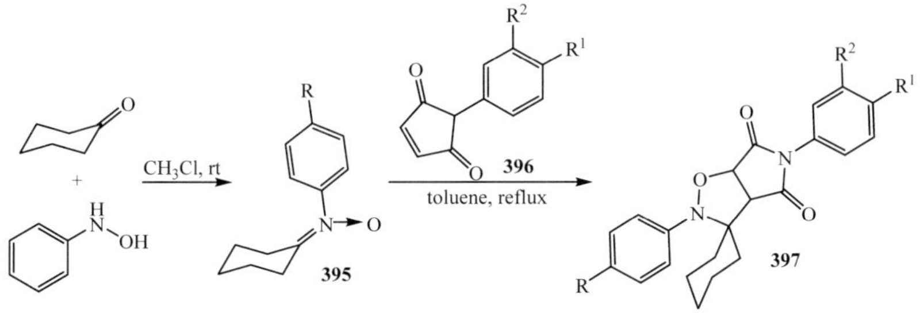

Scheme 3-105. Diastereoselective synthesis of novel spiroisoxazolidines **397**.

Shi et al. [98] developed the construction of a novel spiro-oxindole-based 2,5-dihydropyrrole scaffold **398** with potential bioactivity through efficient three-component reactions of isatins, amino-ester, and alkynes in the presence of 20 mol% trifluoroacetic acid (Scheme 3-106). The bioscreen of these new spirodihydropyrroles **398** allowed the finding of sixteen compounds with promising cytotoxicity to MCF-7 cells.

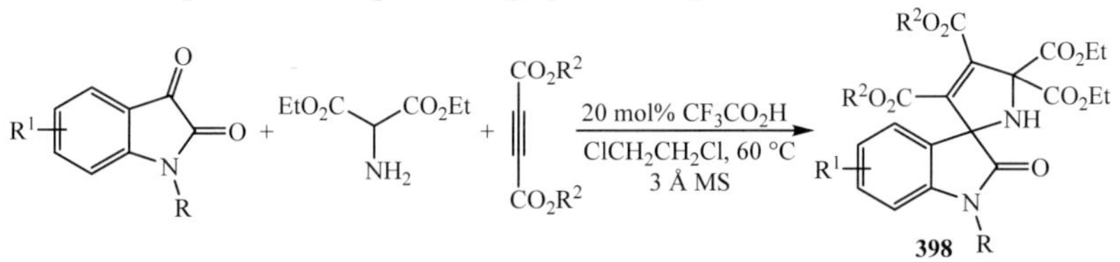

Scheme 3-106. Three-component reactions of isatins, amino-ester, and alkynes.

Ramana et al. elegantly presented a simple domino process for the construction of the tricyclic core existent in the spiro-pseudoindoxyl natural products **401** [99]. As shown in Scheme 3-107, this process involves two intramolecular events: the Au-catalyzed nitroalkyne **399** redox leading to isatogen **400** and its subsequent [3+2] cycloaddition with a suitably positioned olefin. Authors explored the option to modulate the size of the spiroannulated ring which is an essential variable in this class of natural products. Generally, this process converts a linear precursor into a tricyclic system with complete step, atom, and redox economy.

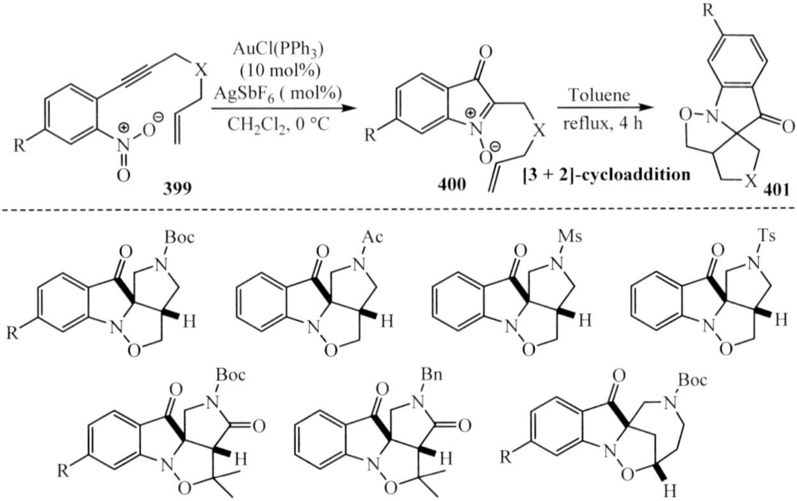

Scheme 3-107. Spiro-pseudoindoxyl skeleton construction **401** via nitroalkyne **399** redox and subsequent [3+2] cycloaddition.

Raghunathan et al. demonstrated an intermolecular [3+2] cycloaddition reaction to obtain functionalized spiro compounds [100]. In this reaction, the *O*-acryloylacridinediones **402** as dipolarophiles reacted with azomethine ylide derived from di-/tri-ketones and secondary-amino acids to provide acridinedione-derived monospiropyrrolidine/ pyrrolizidines **403**/**404** in good yield (Scheme 3-108).

In 2015, Rao et al. employed glyco 3-nitrochromanes **405** as dipolarophiles under reflux conditions to give easy access to biologically important glyco 3-nitrochromane hybrid pyrrolidinyl spiro heterocycles **406a,b** in good yields (Scheme 3-109) [101].

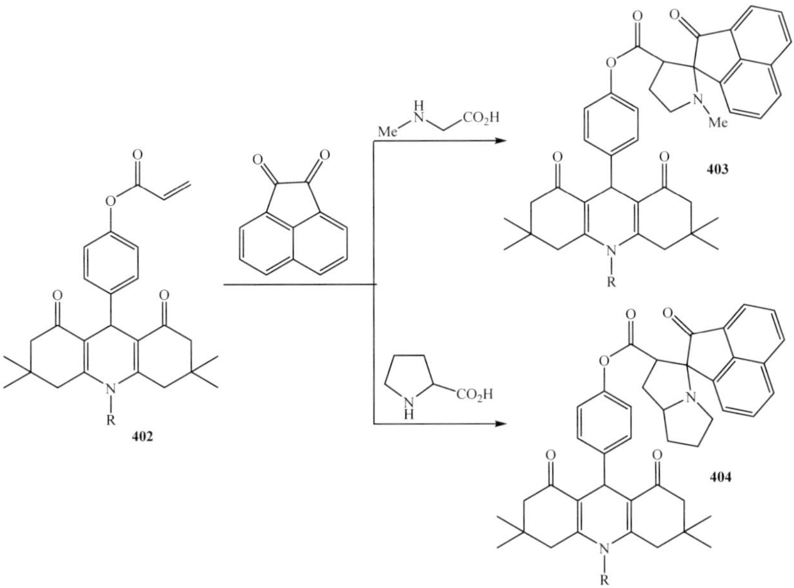

Scheme 3-108. Synthesis of monospirooxindolopyrrolidine/pyrrolizidines **403/404**.

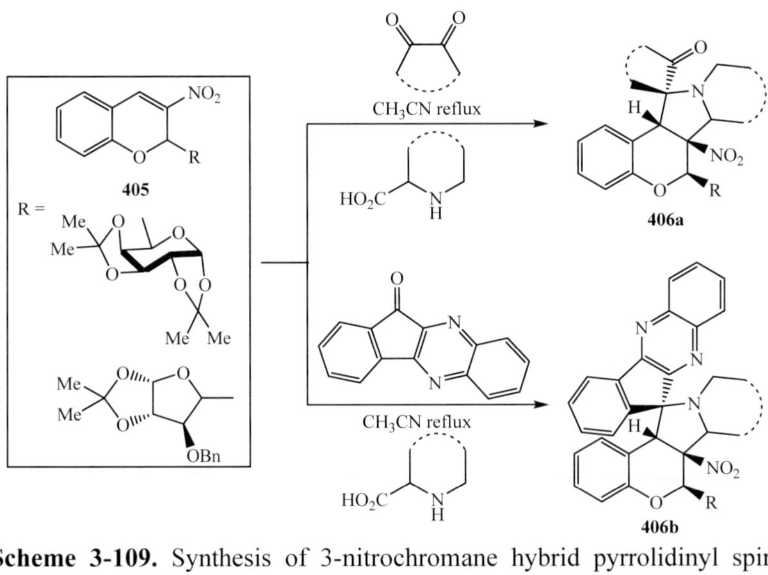

Scheme 3-109. Synthesis of 3-nitrochromane hybrid pyrrolidinyl spiro heterocycles **406a,b**.

The synthesis of novel structurally intriguing hexacyclic cage-like compounds **408** by the three-component tandem [3+2] cycloaddition–intramolecular annulation reactions of ninhydrin **73**, α-amino acids and (*E*)-3-arylidene-1-methylpiperidin-4-ones **407** was also achieved by the Kumar group (Scheme 3-110, Eq. 1) [102]. On the other hand, they obtained polycyclic dispiro compounds **409-410** through the four-component reactions of ninhydrin **73**, *o*-phenylenediamine **180**, α-amino acids, and (*E*)-3-arylidene-1-methylpiperidin-4-ones **407** (Scheme 3-110, Eq. 2).

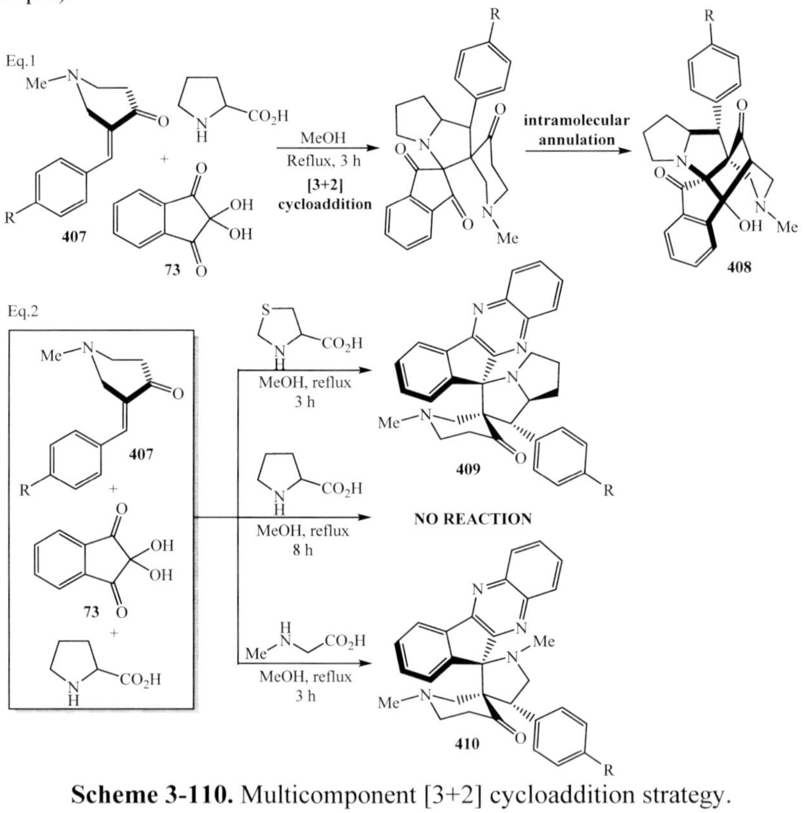

Scheme 3-110. Multicomponent [3+2] cycloaddition strategy.

In addition to the examples mentioned above, we can also cite the construction of spiro[benzo[*f*]imidazo[5,1,2-*cd*]-indolizine-4,2′-indene] derivatives **412** using triethylamine in CH_3CN through three-component tandem double [3+2] cycloaddition reactions at both C-1 and C-3 atoms of *N*-cyanomethylisoquinolinium ylide **411** (Scheme 3-111) [103].

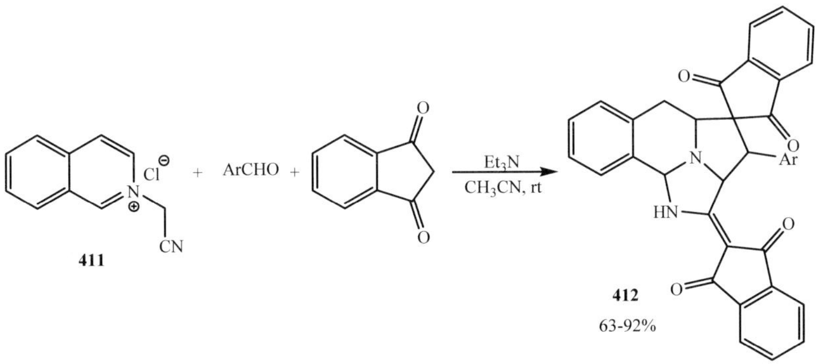

Scheme 3-111. Tandem double [3+2] cycloaddition reactions at C-1 and C-3 atoms of *N*-cyano-methylisoquinolinium ylide **411**.

The first formal [3+2] cycloaddition reaction of *in situ* generated azaoxyallyl cation **413** with cyclic ketones was elegantly reported in 2017 by Shao et al. [104], who performed an efficient procedure for the synthesis of spiro-4-oxazolidinones **414** in excellent yields (up to 99%) (Scheme 3-112).

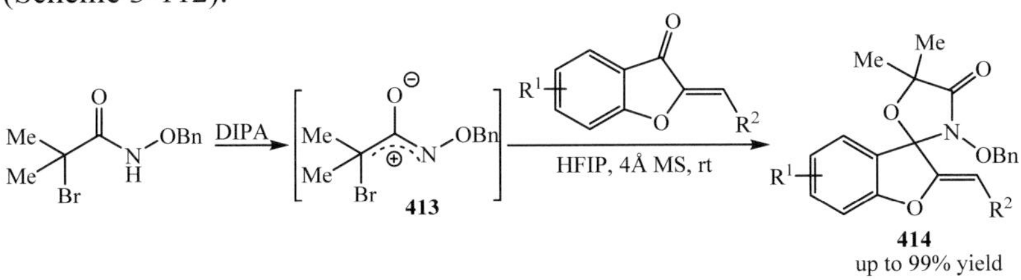

Scheme 3-112. [3+2] cycloaddition of azaoxyallyl cations **413**.

Likhar et al. showed in 2018 that a new series of spiro pyrrole/ pyrrolizine/thiazole compounds **417-419** was synthesized by using [3+2] dipolar cycloaddition through the three-component reaction of ninhydrin **73** and proline **98** with different dipolarophiles such as maleimide **184**, phenacyl bromide **415** and isothiocyanates **416** (Scheme 3-113) [105].

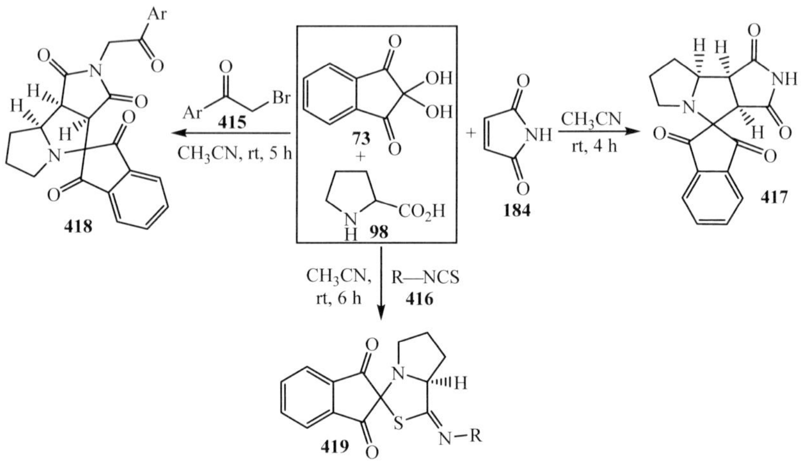

Scheme 3-113. Synthesis of new spiro pyrrole/pyrrolizine/thiazole compounds **417-419**.

Almost simultaneously, the other research group disclosed a catalyst-free mutually activated [3+2] cycloaddition reaction of isatin-derived α-(trifluoromethyl)imines **420** and vinylpyridines **421**, to give a series of 5′-trifluoromethyl-spiro[pyrrolidin-3,2′-oxindoles] **422** in excellent yields and diastereoselectivities (Scheme 3-114) [106].

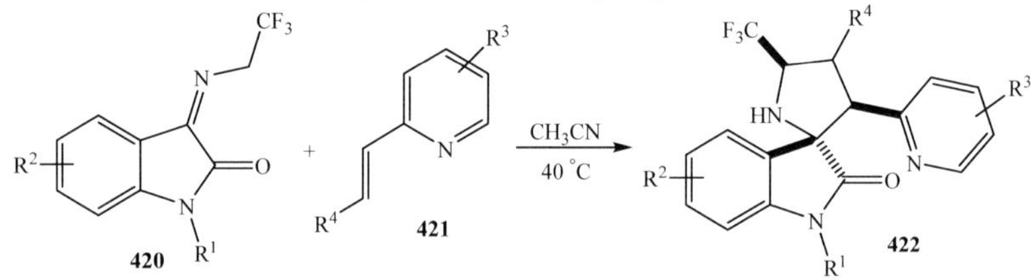

Scheme 3-114. A catalyst-free mutually activated [3+2] cycloaddition reaction.

Spiro-oxindole–pyrrolidine/pyrrolo[1,2-*c*]thiazole hybrids are essential structural motifs present in several natural alkaloids, and the synthesis of some was achieved starting with a one-pot three-component [3+2] cycloaddition reaction [107]. For example, forty novels spiro-linked quinolin-one-pyrrolidine/pyrrolo[1,2-*c*]thiazole oxindole/ acenaph-thalene hybrid hetero-cycles **424 -427** were synthesized using this method with ten (*E*)-6-(arylmethylidene)-2-aryl-7,8-dihydroquinolin-5(6*H*)-ones **423**, two 1,2-diketones and two α-amino acids (Scheme 3-115).

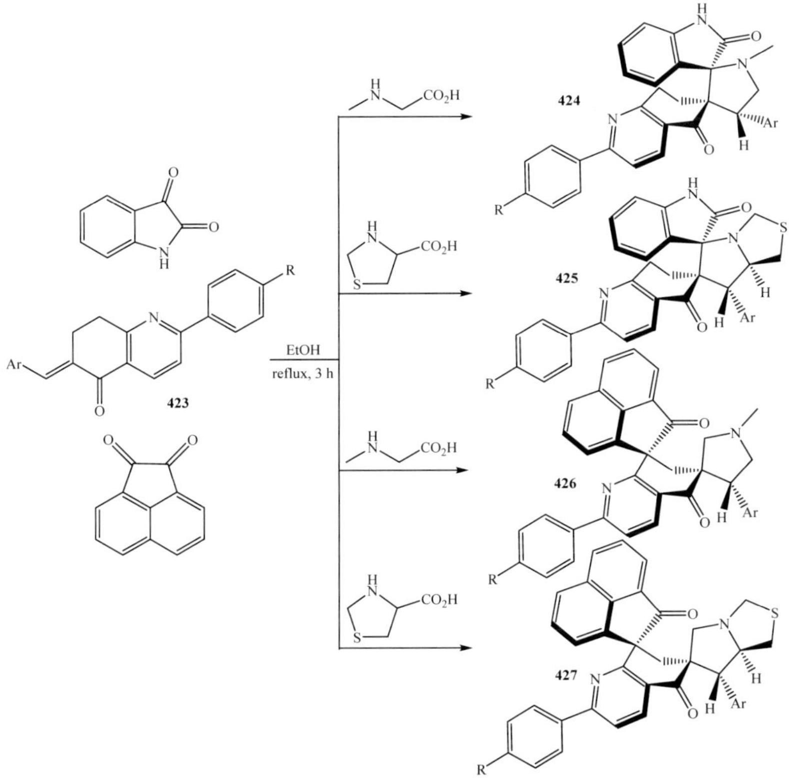

Scheme 3-115. Synthesis of spiro-linked quinolinone-pyrrolidine/pyrrolo-[1,2-*c*]thiazole oxindole/ acenaphthalene hybrid heterocycles **424-427**.

Likewise, Sun et al. explored the one-pot base-promoted cascade double [3+2] cycloaddition reaction of *N*-cyanomethylisoquinolinium salts **428** with various dipolarophiles such as (*E*)-3-arylideneindolin-2-ones **429**, 4-arylidenepyrazol-3-ones **430**, 2-arylidene-1,3-indanediones **431**, arylidenemalononitriles **432**, and (*E*)-*N*-hydroxy benzimidoyl chlorides **433** to create polycyclic spiro[indoline-3,8′-isoxazolo[5,4-*c*]pyrrolo[2,1-*a*]-isoquinolines **434-437** in good yields and with high diastereoselectivity (Scheme 3-116) [108].

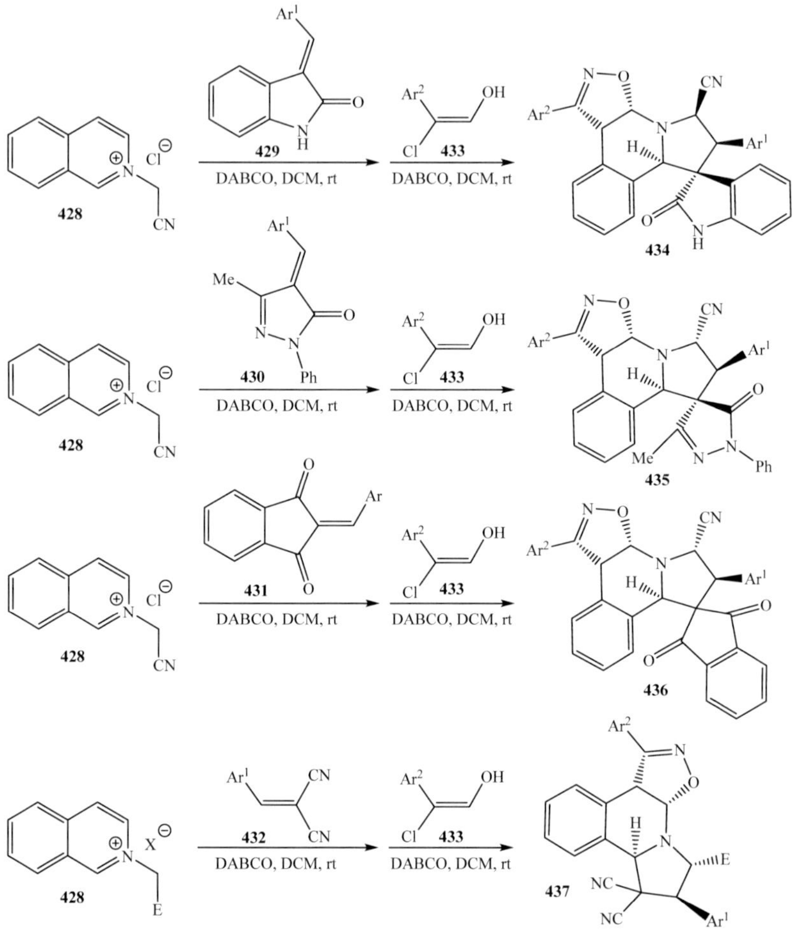

Scheme 3-116. One-pot base-promoted cascade double [3+2] cycloaddition reaction.

Later, [3+2] cycloaddition of *p*-QMs with nitrile imines under mild conditions leads to the formation of the corresponding spiro-pyrazoline-cyclohexadienone products **438** in good to excellent yields (up to 97% yield) with high regioselectivity (Scheme 3-117) [109].

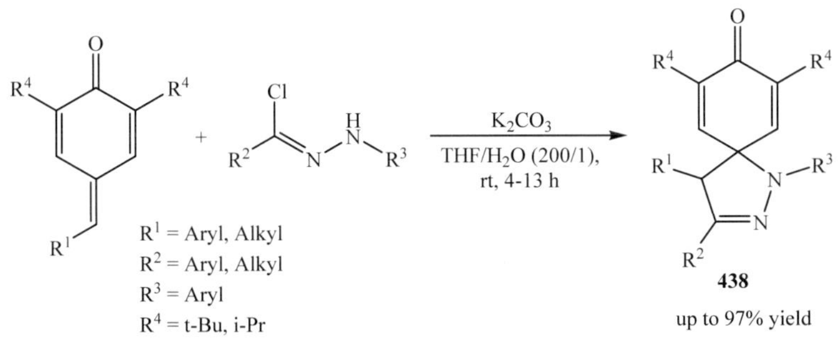

Scheme 3-117. Synthesis of spiro-pyrazoline-cyclohexadienones **438**.

Yan et al. introduced a new type of azomethine ylides **440** which was generated *in situ* by the reaction of ethyl glycinate hydrochloride **439** and dimethyl acetylenedicarboxylate **33** [110]. The reaction of produced azomethine ylides **440** with 3-phenacylideneoxindoline-2-ones **441** in ethanol afforded poly-substituted spiro[indoline-3,3′-pyrrolidine]s **442** in good yields (Scheme 3-118). The biological activities of the various spiro compounds obtained were investigated by *in vitro* evaluation against mouse breast cancer cells 4T1, mouse colon cancer cells CT26, human liver cancer cells HepG2, and human lung cancer cells A549 by MTT assay. The bioactivity detection of these spiro-oxindoles resulted in the discovery of eight compounds with specific cytotoxicity towards HepG2 cells and one compound with promising cytotoxicity towards CT26 cells.

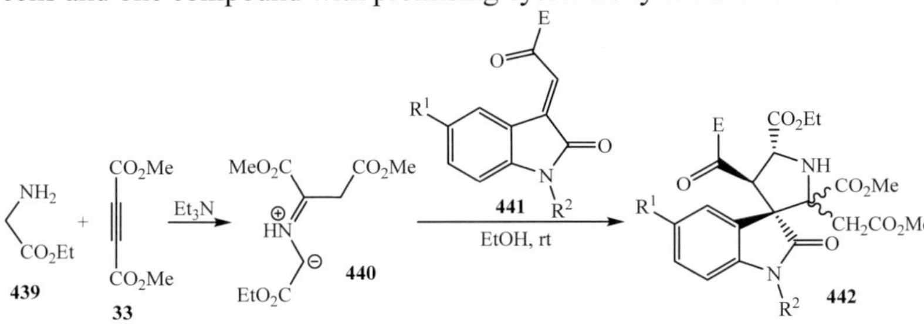

Scheme 3-118. Synthesis of poly-substituted spiro[indoline-3,3′-pyrrolidines] **442**.

Yan et al. successfully synthesized 7′-arylidenespiro[indoline-3,1′-pyrrolizine]s **446** by acetic acid-catalyzed three-component reaction of pyrrolidine, aromatic aldehydes, and 3-arylidene-oxindolin-2-ones **443** in refluxing toluene [111]. The similar three-component reaction with 2-arylidene-1,3-indanediones **444** also presented 7′-arylidenespiro[indene-

2,1′-pyrrolizine]s **447** in good yields. However, the reaction with 3-phenacylideneoxindoles **445** led to a mixture of 7′-arylidene-substituted spiro-oxindoles **448** and spiro[indoline-3,1′-pyrrolizine]s **449** in moderate yields. The reaction mechanism involved the generation of azomethine ylides, β-C−H functionalization of pyrrolidine, and successive [3+2] cycloaddition (Scheme 3-119).

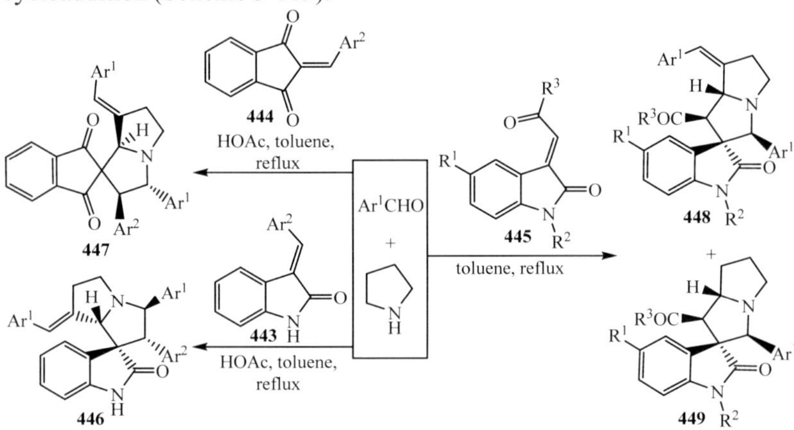

Scheme 3-119. Three-component reactions reported by Yan.

As an extension of this strategy, a regio- and stereocontrolled approach to the synthesis of 4-CF_3-substituted *exo*/*endo*-spiro[chromeno [3,4-*c*]pyrrolidine-3,3′-oxindol]s **451**/**451′** and *endo*-spiro[chromeno[3,4-*c*]pyrrolidine-1,3′-oxindole]s **452** through reversible [3+2] cycloaddition of azomethine ylides generated from isatins and sarcosine **84** to 3-nitro-2-(trifluoromethyl)-2*H*-chromenes **450** was also reported by Korotaev et al. (Scheme 3-120). The regio- and stereochemistry of the reaction products can be controlled by varying the temperature and solvent [112].

Recently, a three-component synthesis of novel spiro-oxindole tetrahydropyrrolizines **454** from secondary α-amino acids, isatins, and vinyl selenones **453** was reported by the Marini group (Scheme 3-121). Products were prepared in good yields and high diastereoselectivity by [3+2] cycloaddition of *in situ* generated azomethine ylides followed by the spontaneous elimination of benzeneseleninic acid [113].

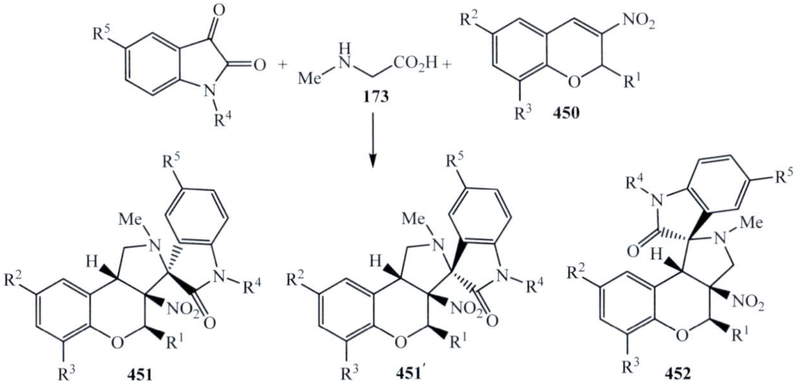

Scheme 3-120. Reversible [3+2] cycloaddition.

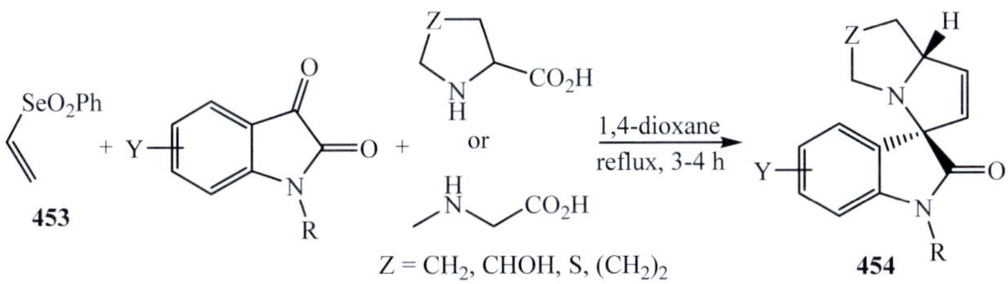

Scheme 3-121. Three-component synthesis of novel spiro-oxindole-tetra hydropyrrolizines **454**.

Hajra et al. nicely showed a highly efficient regio- and stereoselective spontaneous formal [3+2] cycloaddition of CO_2 in an aqueous medium for the one-pot synthesis of spiro[indoline-3,5′-oxazolidine]-2,2′-diones **455** in excellent enantiopuirity (ee up to 99%) under catalyst-free and ambient conditions (Scheme 3-122) [114].

Additionally, Sun et al. found [115] that the base promoted [3+2] cycloaddition reaction of MBH maleimides of isatins **456** with various cyclic 1,3-dicarbonyl compounds in the presence of mixed bases 1,4-diazabicyclo[2.2.2]octane (DABCO) and K_2CO_3 afforded functionalized dispiro[indoline-3,3′-furan-2′,3″-pyrrolidine]s **457** in satisfactory yields with high diastereoselectivity (Scheme 3-123, Eq. 1). The reaction proceeded with nucleophilic substitution of allylammonium ylide and oxa-Michael addition of the enloate. While the similar reaction of MBH adducts of isatins derived from acrylonitrile and methyl acrylate **458** led to functionalized spiro[indoline-3,4′-pyrane] derivatives **459** in good yields through formal [3+3] cycloaddition reaction (Scheme 3-123, Eq. 2).

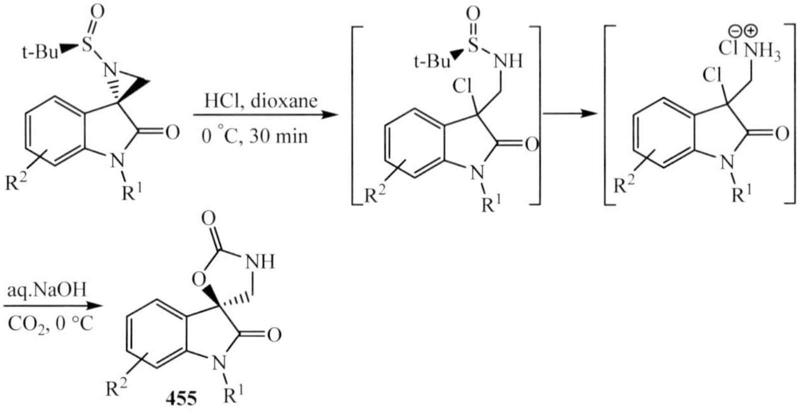

Scheme 3-122. Spontaneous formal [3+2] cycloaddition of CO_2.

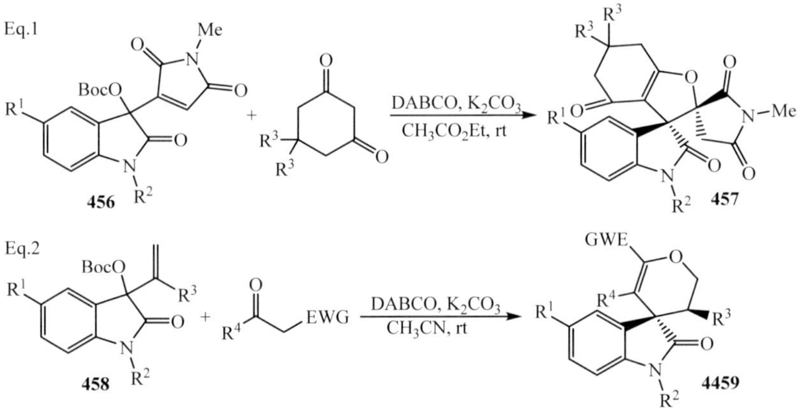

Scheme 3.123. Cycloaddition reactions of MBH adducts of isatins **456** and **458**.

Spironaphthofuranone-pyrazoline **461** was prepared via the regioselective [3+2] cycloaddition reaction of benzoaurones **460** with nitrile imines by Su et al. (Scheme 3-124). The desired products **461** were efficiently obtained in good to excellent yields (65-94% yield) with high regioselectivity [116].

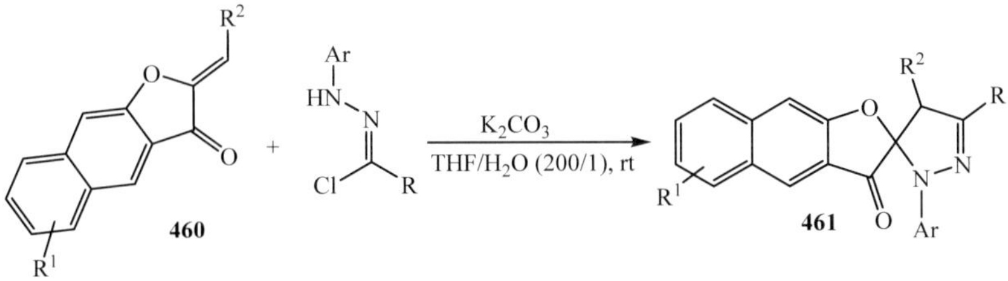

Scheme 3-124. Synthesis of the spiro-pyrazolines **461** with nitrile imines via [3+2] cycloaddition.

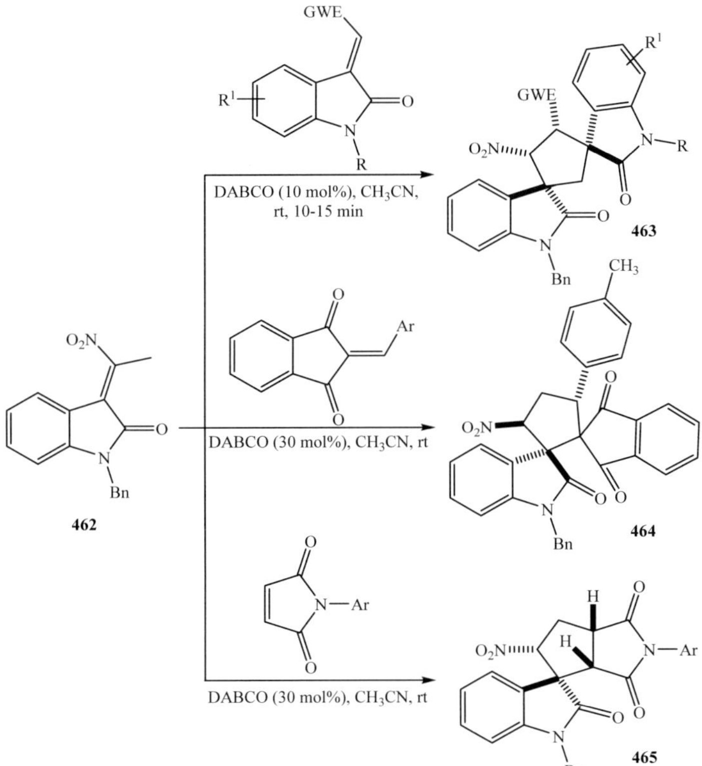

Scheme 3-125. Regio- and stereoselective synthesis of dispiro-bis oxindoles via [3+2] annulation involving nitroisatylidene **462** as a vinylogous Michael donor.

Very recently, Sankara et al. [117] developed a cascade [3+2] annulation involving a γ-selective vinylogous Michael addition of nitro-alkylidene oxindoles **462** to various electron-deficient alkenes followed by an intramolecular Michael addition which gives access to dispiro-bis-oxindoles **463-464** and spiro-oxindoles **465** (Scheme 3-125). This process leads to the formation of up to four contiguous chiral centers, including two quaternary spirocenters, in a high-yield regio- and diastereoselective transformation that also presents an appropriate entry into conformationally constrained γ-amino acid derivatives.

3.3.5. Diels–Alder cycloaddition reactions

The so-called Diels–Alder reaction (occasionally referred to as [4+2] cycloaddition) was first reported by Otto Diels and Kurt Alder in 1928 [118]. Diels–Alder reaction is a chemical reaction between a conjugated diene and a substituted alkene, commonly termed the dienophile, to form a substituted cyclohexene. Since the discovery of the Diels–Alder reaction, numerous attempts have been devoted to developing this applicable methodology in different features, aspects, and issues. This section will give a brief overview of the Diels–Alder reactions used to synthesize spiro compounds.

In 2002, Huang et al. presented the first examples of hydrogen-bond-promoted acceleration of hetero-Diels-Alder reactions and the use of such catalysis to carry out the hetero-Diels-Alder reactions of simple, unactivated ketones [119]. They explored the hetero-Diels-Alder reaction of cyclohexanone and diene **466** in various solvents (Scheme 3-126). The results showed that the hetero-Diels-Alder reactions are greatly accelerated in hydrogen-bonding solvents. This activation procedure demonstrates an attractive and operationally simple alternative to common Lewis acid catalysis.

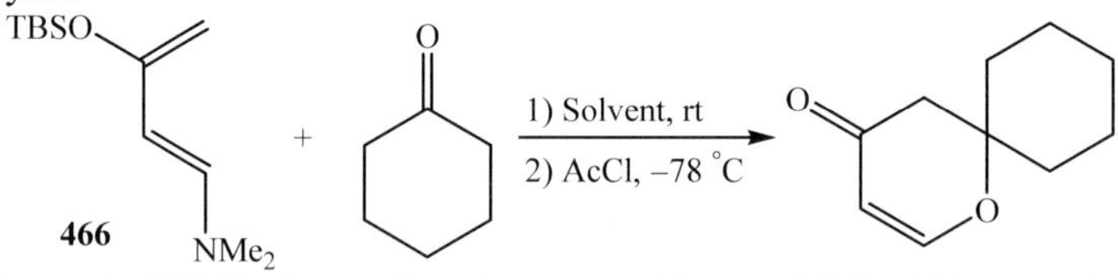

Scheme 3-126. Hydrogen-bond-promoted hetero-Diels-Alder reactions.

One year later, Kende et al. reported a novel aza-Diels−Alder reaction as the key step which was applied to the total synthesis of (±)-alantrypinone (a hexacyclic alkaloid) **469** and its 17-*epi* isomer **470** [120]. The reaction between azadiene **467** and 3-methyleneoxindole **468** proceeded smoothly in chloroform at room temperature to generate a chromate-graphically separable mixture of **469** in 55% yield and adduct **470** in 18% yield (Scheme 3-127).

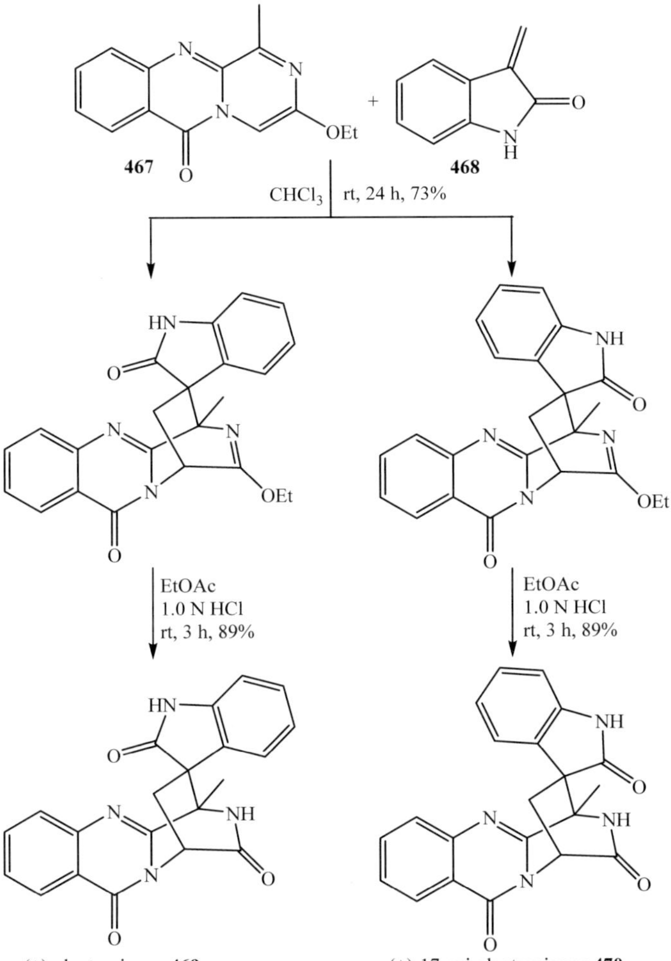

Scheme 3-127. Total synthesis of (±)-alantrypinone (a hexacyclic alkaloid) **469** and its 17-*epi* isomer **470**.

Miller et al. [121] showed that the Diels-Alder reaction of acylnitroso **471** with *N*-Cbz-protected spirocyclic diene **472** led to the valuable cycloadducts **473** for further derivatization and could be converted into the novel carbocyclic nucleoside spironoraristeromycin **474** (Scheme 3-128).

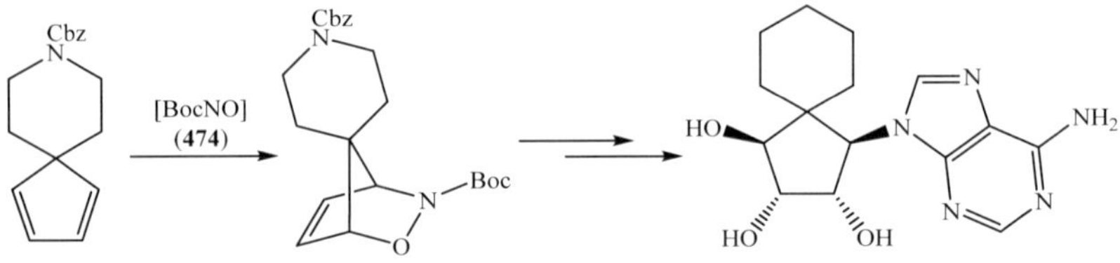

472 473 474

Scheme 3-128. Synthesis of a spironoraristeromycin using an acylnitroso Diels-Alder reaction.

A new highly selective and convergent approach to the synthesis of lambertellol core **477** or arthrinone **478** was reported by Commeiras et al. in 2010 through high chemo- and diastereoselective intermolecular Diels-Alder cycloaddition between *trans*-1,2-disiloxybenzo cyclobutenes **475** and 2-methylprotoanemonine **476** (Scheme 3-129). The chemoselectivity of the cycloaddition reaction can be modulated by ring substitution onto the starting lactone [122].

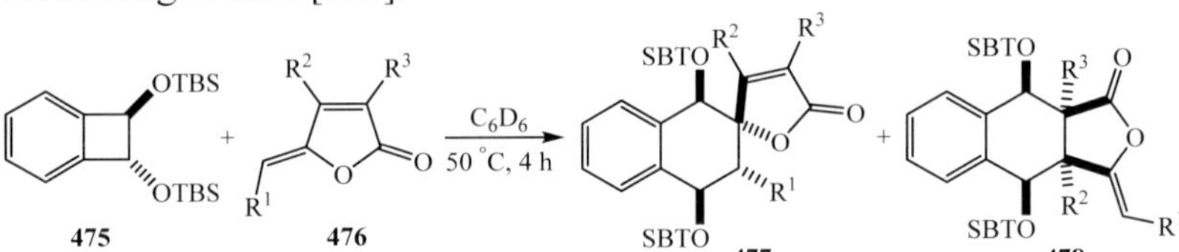

475 476 477 478

Scheme 3-129. Synthesis of lambertellol core **477** or arthrinone **478**.

Zipse et al. reported a concise total synthesis of variecolortide A and B (**481** and **482**) that is mainly devoid of protecting-group operations and is highly convergent [123]. The synthesis involves an unprecedented hetero-Diels-Alder reaction of a 1,4-anthraquinone **479** with a di-dehydro-diketopiperazine **480** to construct the central spirocyclic core of the natural products (Scheme 3-130).

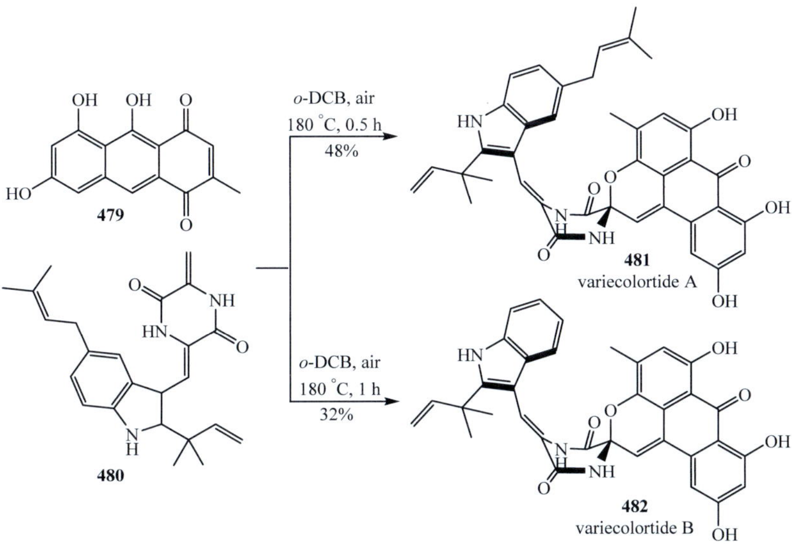

Scheme 3-130. The total synthesis of variecolortide A and B.

Liang et al. showed the dienone desymmetrization through a diastereoselective intramolecular Diels-Alder (IMDA) reaction which leads to densely functionalized intermediates **484** that are especially valuable for the assembly of the core unit of the structurally unique alkaloid, himandrine, **485** [124]. Desymmetrization of a "locally symmetrical" dienone was obtained with good levels of diastereo-selectivity by an oxidative cyclization of dienic sulfonamides **483** mediated by iodobenzene diacetate in TFA and followed by a tandem IMDA reaction (Scheme 3-131).

Contemporaneously, Moghaddam et al. reported the synthesis of highly functionalized spiro dihydrothiopyranes **487** from corresponding aldehydes and thio-oxindoles **486** in one step procedure through a tandem catalyst-free Aldol-Diels-Alder reaction accelerated in water without the isolation of intermediates (Scheme 3-132). The reaction was accomplished in water with no catalyst added [125]. The proposed tandem begins with the Aldol condensation between thio-oxindole and aldehyde to give α,β-unsaturated thio-oxindoles. The monomeric form of this kind of α,β-unsaturated thio-oxindole is inaccessible in water because it undergoes fast [4π+2π] dimerization reaction with one molecule as the diene partner (4π) and the other as the dienophile (2π). The C=C bond of one molecule acted as the dienophile, while the C=S was not involved in the reaction due to resonance with nitrogen lone pair electrons.

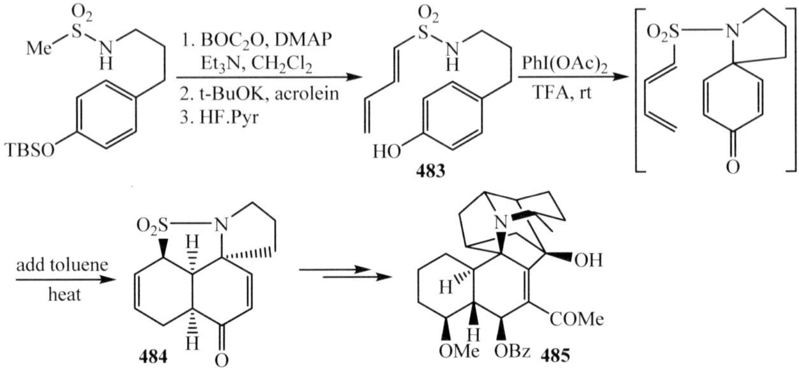

Scheme 3-131. Dienone desymmetrization through a diastereoselective IMDA reaction.

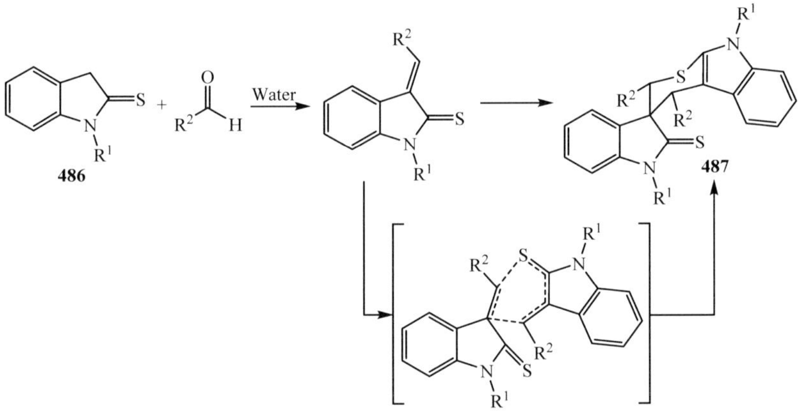

Scheme 3-132. A tandem Aldol-Diels-Alder reaction accelerated in water.

Another example was elegantly presented in 2010 by Presset et al. [126]. A series of α-spiro-δ-lactams **489** was synthesized efficiently and with high diastereoselectivity by microwave-assisted an imination (or aza-Wittig)/Wolff rearrangement/Diels-Alder cycloaddition domino sequence from simple starting materials (Scheme 3-133). These reactions feature an unprecedented reactivity of acylketenes **488** as dienophiles in 6π electrocyclic processes.

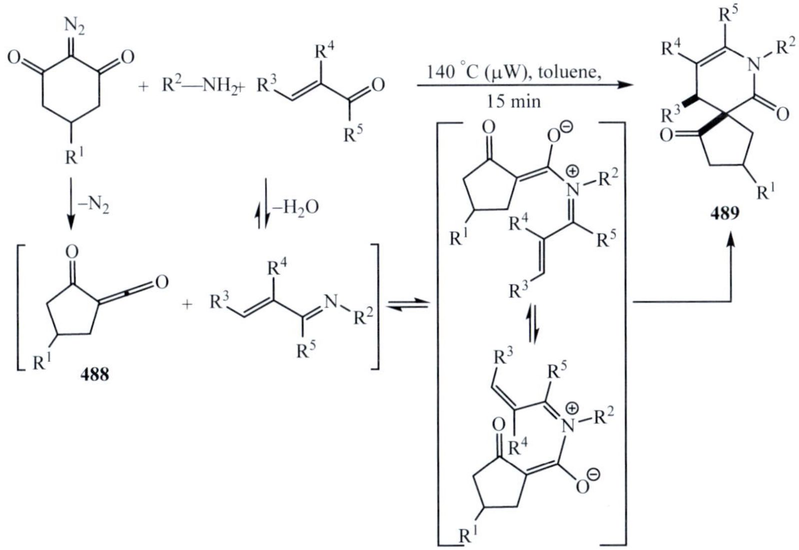

Scheme 3-133. Microwave-assisted an imination (or aza-Wittig)/Wolff rearrangement/Diels-Alder cycloaddition domino sequence.

In 2012, Miller et al. [127] prepared a series of novel oxazolidinone antibiotics **490** with a spiropiperazinyl substituent at the 4-position of the phenyl ring via nitroso Diels-Alder (NDA) chemistry and the *in vitro* antibacterial activities were explored against various gram-positive and gram-negative bacteria, and mycobacteria (Scheme 3-134). 1′-(2-Fluoro-4-((*R*)-5-(hydroxy-methyl)-2-oxooxazolidin-3-yl)phenyl)-2-oxa-3-azaspiro [bicyclo[2.2.1]heptane-7,4′-piperidine]-3-carboxylate **490a** and *tert*-butyl 1′-(4-((*S*)-5-(acetamidomethyl)-2-oxooxazolidin-3-yl)-2-fluorophenyl)-2-oxa-3-azaspiro[bicyclo[2.2.1]heptane-7,4′-piperidine]-3-carboxylate **490b** analogs were active against selected drug-resistant microbes, like methicillin-resistant Staphylococcus aureus and vancomycin-resistant enterococci and had no mammalian toxicity in a Hep-2 cellular assay (CC_{50} >100 μM).

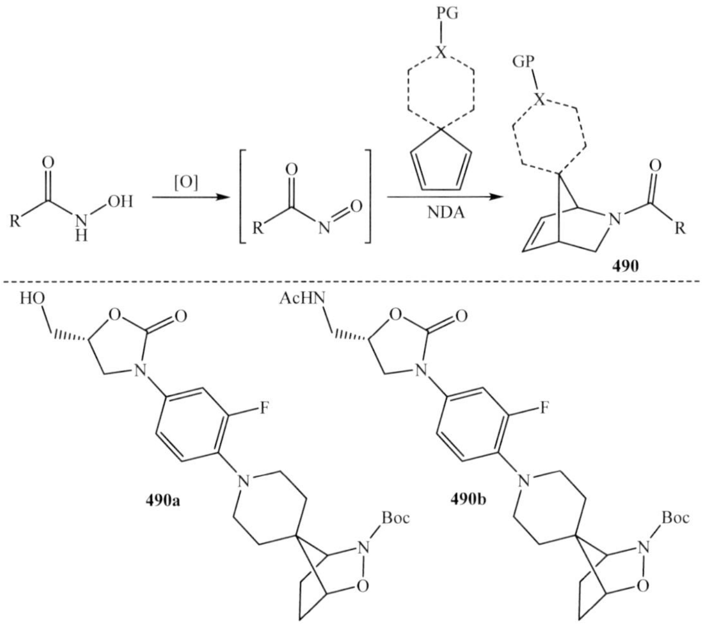

Scheme 3-134. Synthesis of novel oxazolidinone antibiotics **490**.

Likewise, an efficient Diels-Alder reaction of 3-hydroxy-2-pyrone **491** with a wide range of methyleneindolinones in the presence of 4-dimethylaminopyridine (DMAP) as a base catalyst under mild reaction conditions directly led to the construction of structurally diverse spirocyclic oxindoles **492** containing bicyclic lactone motif with moderate diastereoselectivities and good to excellent yields (Scheme 3-135) [128].

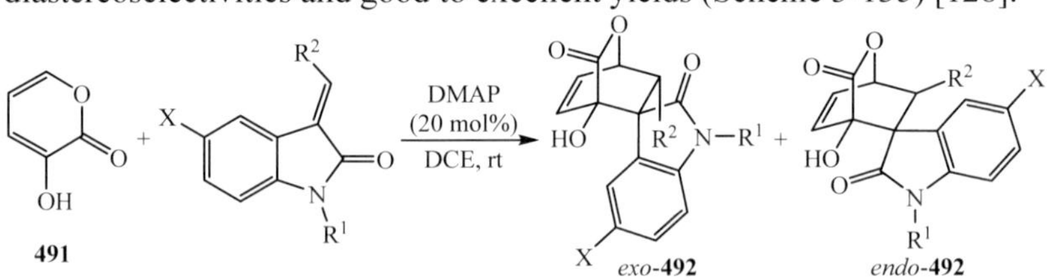

Scheme 3-135. DMAP-Catalyzed Diels-Alder reaction.

Callistemon viminalis (Myrtaceae) is a shrub native to Australia, which has also been cultivated in the south of China and applied for the treatment of cold and arthralgia in Chinese folk medicine. (±)-Calliviminones A (**495**) and B (**496**), two Diels-Alder adducts of polymethylated phloroglucinol and myrcene possessed spiro[5.5]undecene skeleton, were isolated from the fruits of *Callistemon viminalis*. Kong et al. reported the first example of carbon Diels-Alder adducts between phloroglucinol **493** and terpene **494** (Scheme 3-136). In this study, the biomimetic synthesis of compounds **495** and **496** confirmed the pivotal role of the Diels-Alder reaction in the plausible biosynthetic pathway. The bioactivity scan demonstrated that **495** and **496** displayed moderate inhibition of NO generation on lipopolysaccharide-induced RAW264.7 macrophages [129].

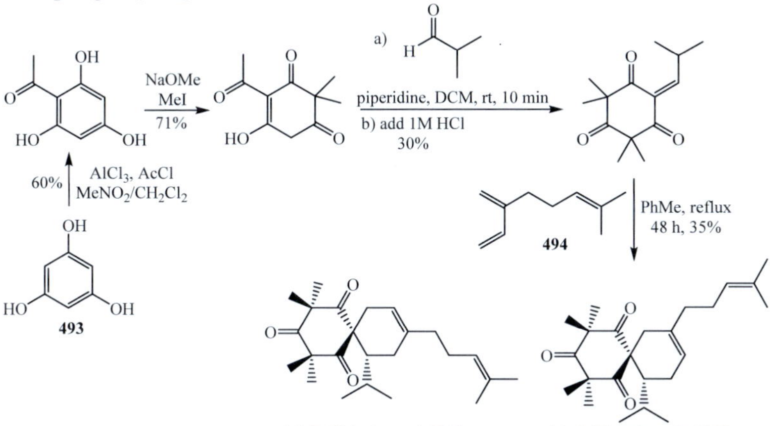

Scheme 3-136. Biomimetic synthesis of compounds **495** and **496**.

rel-(5′*R*,6′*R*,7′*R*)-5′-Carboxy-7′-aryl-1-aryl-3′,7′-dihydro-*2H*,*2′H*,*5H*-spiro[pyrrolidin-3,6′-thio-pyrano[2,3-*d*]thiazol]-2,2′,5-triones **498** were successfully synthesized by Lesyk et al. using *trans*-aconitic acid **497** in the hetero-Diels-Alder reaction (Scheme 3-137) [130].

Yan et al. explored *in situ* dienes formation from acid-mediated dehydration of various pinacoles in ionic liquid [Bmim]Br and a sequential Diels-Alder reaction with typical dienophiles such as 3-methylene-oxindolines **499** and 2-arylidene-1,3-indanediones **500** (Scheme 3-138). The reaction afforded spiro[cyclohexane-1,3′-indolin]-3-en-2′-ones **501** and spiro[cyclohexane-1,2′-inden]-3-ene-1′,3′-diones **502** in satisfactory yields, and high diastereoselectivity [131].

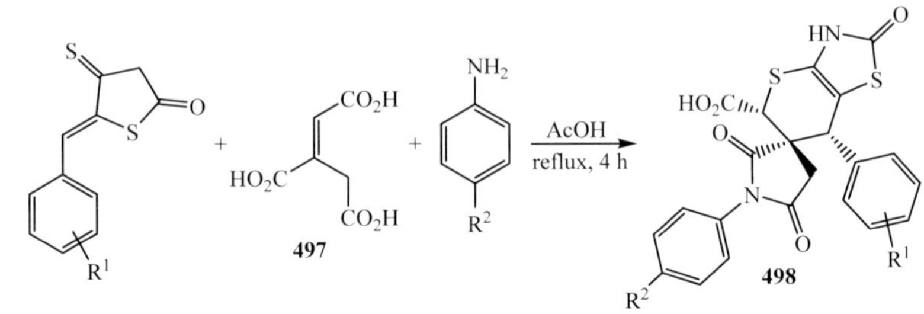

Scheme 3-137. Synthesis of *rel*-(5′*R*,6′*R*,7′*R*)-5′-Carboxy-7′-aryl-1-aryl-3′,7′-dihydro-*2H*,*2′H*,*5H*-spiro[pyrrolidin-3,6′-thio-pyrano[2,3-*d*]thiazol]-2,2′,5-triones **498**.

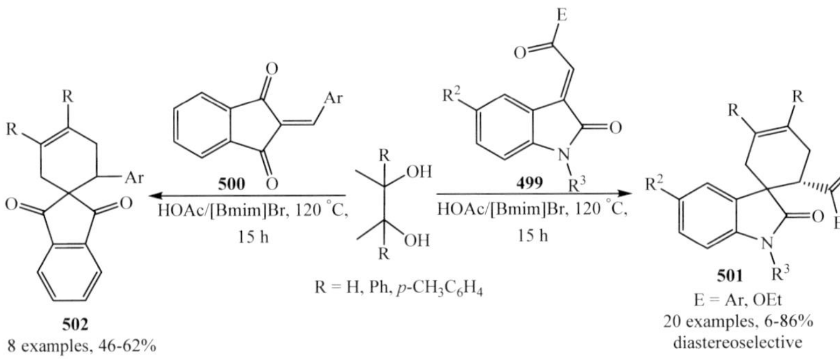

Scheme 3-138. HOAc-Mediated domino Diels−Alder reaction in ionic liquid [Bmim]Br.

In 2018, Nayak et al. presented a diastereoselective oxa-hetero-Diels–Alder reaction to prepare novel spiro indanone fused pyrano[3,2-*c*]chromenes **505** and studied their *in vitro* anticancer activities [132]. Using indane-1,3-dione **503** and 3-vinyl-2*H*-chromenes **504**, novel spiro heterocyclic frameworks **505** were obtained at the interphase between "drug-like" molecules and natural products (Scheme 3-139). *In vitro* cytotoxic studies of these compounds against breast cancer, colon cancer, oral cancer, and breast cancer cell lines were evaluated by MTT {3-(4,5-dimethylthiazol-2-yl)-2,5-diphenyl tetrazolium bromide} assay *in vitro*. The screening results disclosed that many of the compounds are showing moderate to high levels of anticancer activity against the tested cancer cell lines. Some exhibited potent inhibitory activities compared to the commercial anticancer drug 5-fluorouracil.

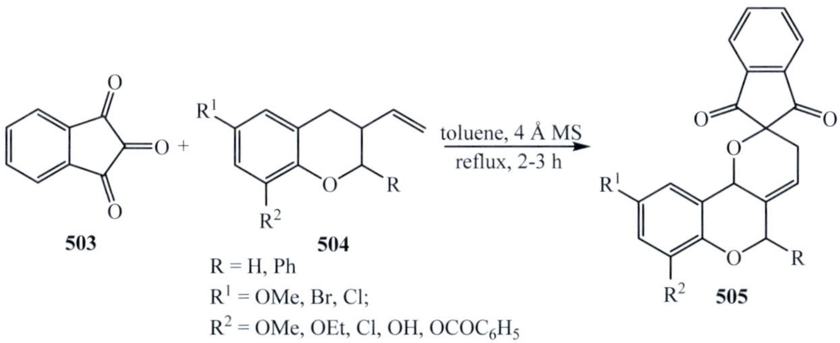

Scheme 3-139. Synthesis of spiro indanone fused pyrano[3,2-*c*]chromenes **505**.

Yan et al. reported the reaction of α,β-unsaturated *N*-arylaldimines **506** with two molecular 5-arylidene-1,3-dimethylbarbituric acids **507** in methylene dichloride at room temperature to afford unique spiro[pyrido [3′,2′:5,6]pyrano[2,3-*d*]pyrimidine-7,5′-pyrimidine]s **508** in moderate to good yields and with high diastereoselectivity (Scheme 3-140). The reaction proceeds through domino aza-/oxa-hetero-Diels–Alder reactions, in which the two substrates act as active aza-/oxa-1,4-dienes and reactive dienophiles [133].

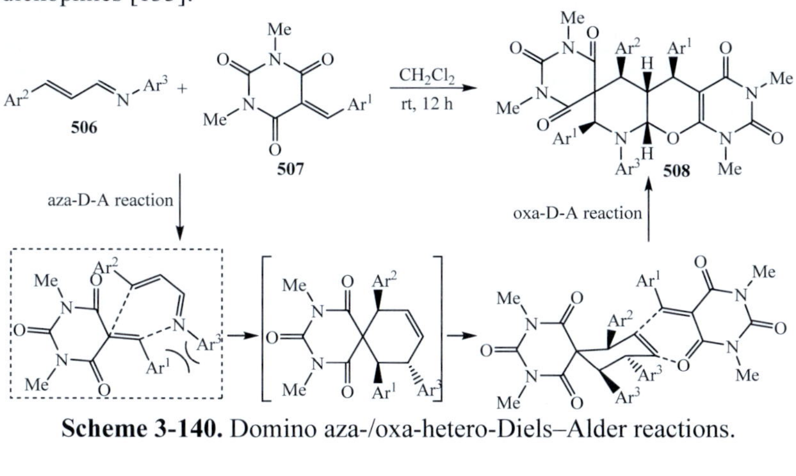

Scheme 3-140. Domino aza-/oxa-hetero-Diels–Alder reactions.

A straightforward procedure for synthesizing functionalized tetrahydro-spiropyran scaffolds **510** via a regioselective [4+2] cycloaddition reaction of MBH alcohol **509** and 2-arylidene-1,3-indanedione was reported by Anwar et al. (Scheme 3-141). MBH alcohol **509** as an ambident substrate exhibits a dual role of nucleophile &

electrophile with 2-arylidene-1,3-indanediones. A single isomer was formed exclusively in all the products through a cascade process that followed an oxa-Michael-Michael addition sequence using cesium carbonate as a base. Control experiments demonstrate that the reaction undergoes a stepwise pathway rather than a concerted mechanism [134].

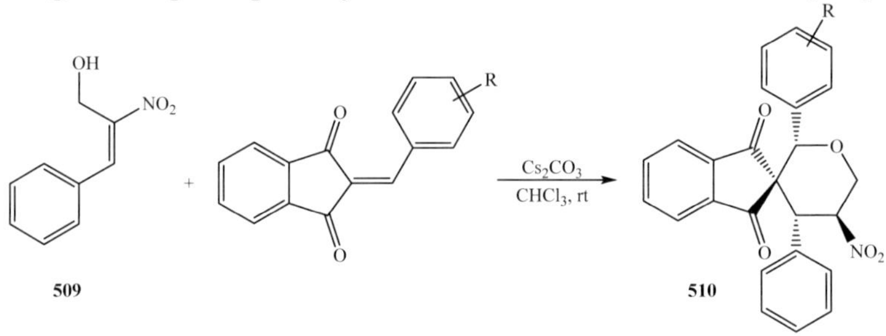

Scheme 3-141. Synthesis of functionalized tetrahydrospiropyran scaffolds **510** using MBH alcohol **509**.

Later, Yan et al. revealed an NH_4I-promoted three-component reaction for the synthesis of *cis*-2,4-diaryl-substituted dihydrospiro [carbazole-1,2′-indene]-1′,3′-diones **511** and dihydrospiro[carbazole-1,3′-indoline]s **512** from inexpensive starting materials (indole, acetophenones, and 2-aryli-dene-1,3-indanediones/(2-oxoindolin-3-ylidene) malono-nitriles) [135]. The reaction mechanism involved the acid-catalyzed Friedel-Crafts reaction between indole and acetophenones, dehydration, Diels-Alder reaction, and aromatization process (Scheme 3-142).

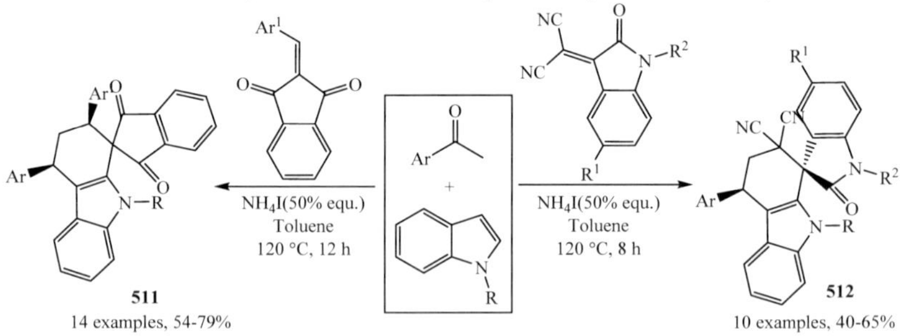

Scheme 3-142. NH_4I promoted a three-component reaction.

In 2019, the Zou group developed a one-pot Mannich-Diels-Alder reaction for the synthesis of high performance jet-fuel-range spirocycle-alkane **515** [136]. The reaction was catalyzed by HZSM-5 zeolite with a Si/Al molar ratio of 130, giving the target adduct **515** high selectivities

(81.6%) and yield (80.9%). Typically, in the first step of the Mannich reaction, the mixture of cyclopentanone, formaldehyde, dimethylamine hydrochloride, deionized water, and 0.8 g catalyst led to the formation of a β-amino-carbonyl compound **513**, then the subsequent Diels-Alder reaction between β-amino-carbonyl compound **513** and cyclopentadiene generates the adduct **514** (Scheme 3-143). Interestingly, after hydrodeoxygenation, the resultant spirocycloalkane **515** has much higher density and neat volumetric heat of ignition (0.952 g/ml, 40.18 MJ/L) than the wildly used JP-10 fuel (0.936 g/ml, 39.41 MJ/L).

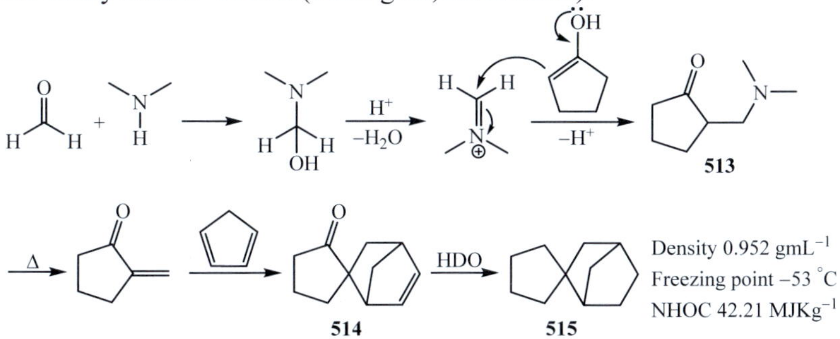

Scheme 3-143. Zeolite catalytic synthesis of high-performance jet-fuel-range spiro-fuel by one-pot Mannich–Diels–Alder reaction.

The synthesis of polycyclic triindanone-fused spiro[bicyclo[2.2.2] octane-2,3′-indoline] derivatives **517-517** by a domino reaction was recently studied by Yan et al. [137]. In this study, the domino reaction was performed with *in situ* generation of active cyclic dienes by triethylamine-promoted cyclotrimerization or cyclotetramerization of 1,3-indanediones and sequential Diels-Alder reaction of active cyclic dienes with dienophilic 3-methyleneoxindoles **516** (Scheme 3-144).

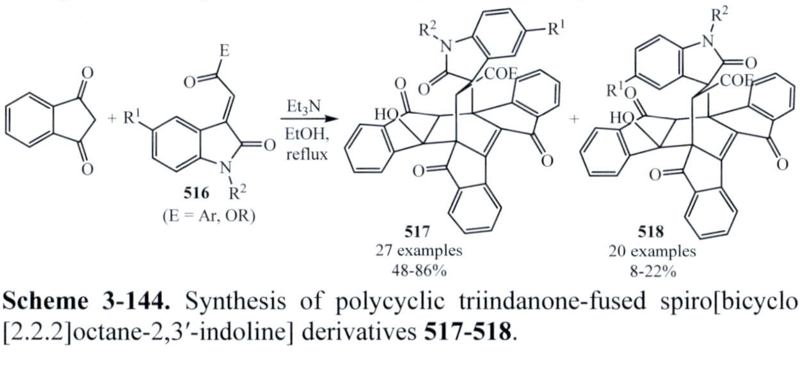

Scheme 3-144. Synthesis of polycyclic triindanone-fused spiro[bicyclo [2.2.2]octane-2,3′-indoline] derivatives **517-518**.

Maslivets et al. developed an original, facile, and highly efficient procedure for the *in situ* generation of cyclic enolate-iminium 1,4-dipoles **520** through thermal decomposition of dipyrazolodioxadiazocines **519** as shelf-stable "ready-to-use" precursors [138]. Various [4+2] cycloadducts **521** have been synthesized in high yields via unusual cycloconversion of dipyrazolo-dioxadiazocines **519** in the presence of ketones (Scheme 3-145).

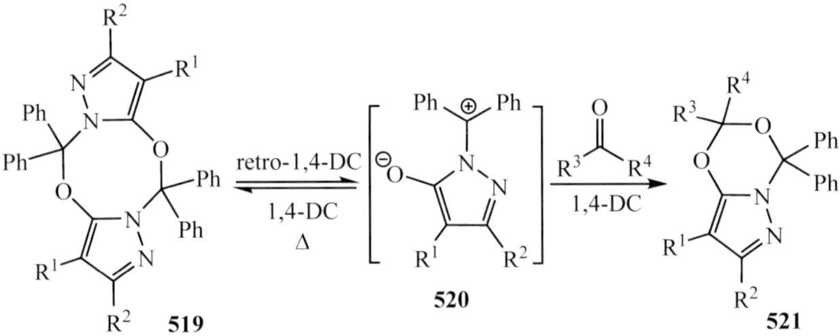

Scheme 3-145. Generation of cyclic enolate-iminium 1,4-dipoles **520** via thermolysis of dipyrazolo-dioxadiazocines **519**.

Sun et al. used *p*-TsOH catalyzed Diels-Alder reaction of 3-(indol-3-yl)maleimides **522** with 3-phenacylideneoxindoles **445** in toluene to provide *cis*/*trans* isomers of 3a′,4′,6′,1′c′-tetrahydrospiro[indoline-3,5′-pyrrolo[3,4-*c*]carbazole]s **523** which could be easily transformed to the corresponding 4′,6′-dihydrospiro[indoline-3,5′-pyrrolo[3,4-*c*]carbazole]s **524** in high yields and with high diastereoselectivity by further DDQ oxidation (Scheme 3-146) [139].

One year later, an unconventional base-mediated aza-Diels-Alder reaction of electron-deficient azoalkenes, which were *in situ* generated from base dehydrohalogenation of α-halogeno *N*-acyl-hydrazones **525**, with electron-deficient 3-phenacylideneoxindoles **526**/ 3-aryliminooxindol -2-ones **527**, was explored by the same group for the synthesis of functionalized spiro[indoline-3,3′-pyridazin]-2-ones **528** and spiro [indoline-3,3′-[1,2,4]triazin]-2-ones **529** in good yields (Scheme 3-147) [140].

An on-water [4+2] annulation reaction for the construction of spiro(indoline-2,3′-hydro-pyridazine) scaffolds **531** was reported by Zhan et al. in 2021 [141]. The reaction of 2-methyl-3*H*-indolium salt **530** with α-bromo *N*-acylhydrazone catalyzed by Cs_2CO_3 on water afforded the corresponding products **531** in high yields with high chemoselectivity under environmentally friendly conditions (Scheme 3-148). Water, as a solvent, can lead to the decrease of the energy barrier of the rate-determining step and a stable product with low energy.

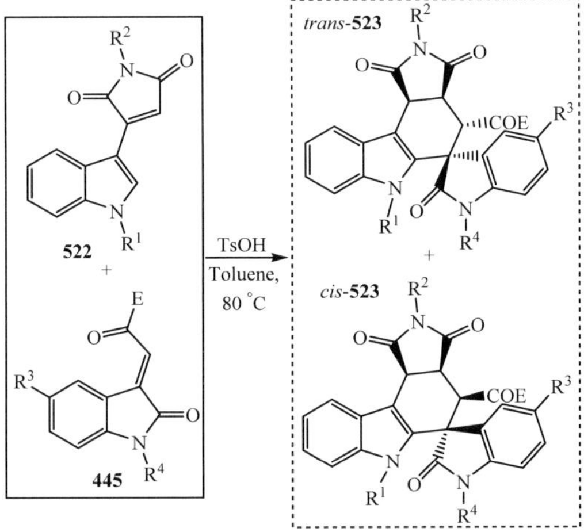

Scheme 3-146. *p*-TsOH catalyzed Diels-Alder reaction.

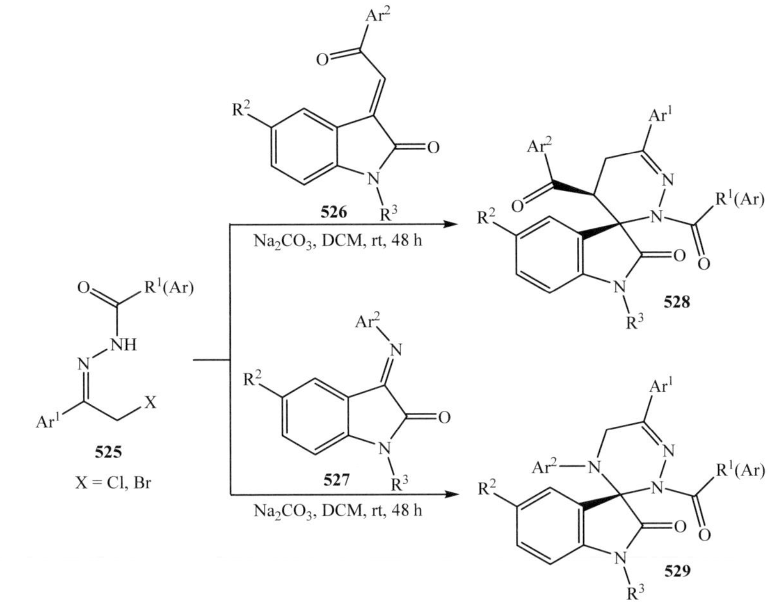

Scheme 3-147. Base-mediated aza-Diels-Alder reaction of electron-deficient azoalkenes.

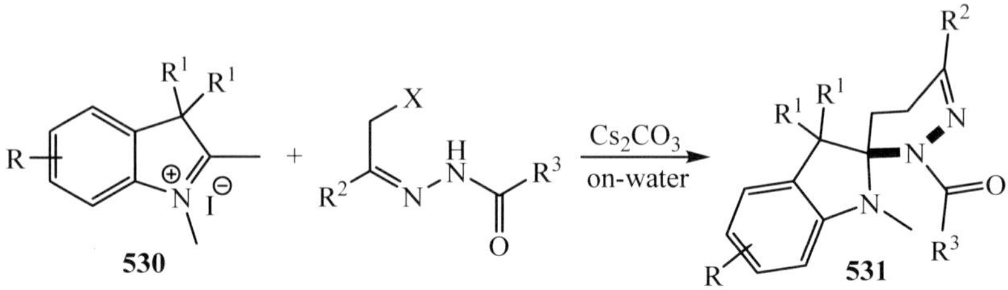

Scheme 3-148. Synthesis of spiro(indoline-2,3′-hydropyridazine)s **531** via an "on-water" [4+2] annulation reaction.

Krishna et al. reported acid-catalyzed intramolecular formal [4+2] cycloaddition reactions of isoindolinone-derived hydroxylactams **532** bearing enones or enal which furnish spiro isoindolinones **534** and related molecules **535** (Scheme 3-149). In these reactions, intramolecular [4+2] cycloadditions of *N*-acylenamides **533** generated *in situ* from the hydroxylactams with the enones or enals lead to the constitution of the fused ring systems and the spiro ring systems [142].

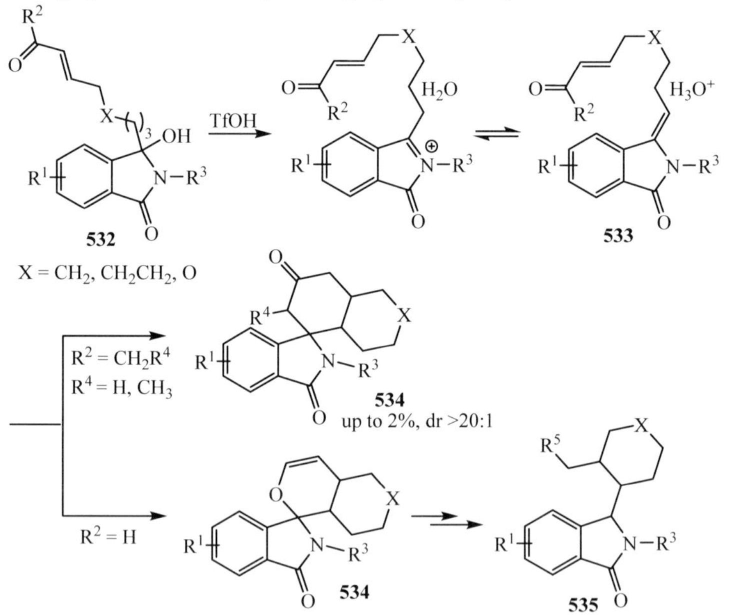

Scheme 3-149. Synthesis of spiro isoindolinones **534** and related molecules **535**.

DDQ oxidative dehydrogenative Diels-Alder reaction of 3-(indol-3-yl)-1,3-diphenylpropan-1-ones **536** with 3-methyleneoxindoles **516** in refluxing ethanol for the synthesis of functionalized dihydrospiro [carbazole-1,3′-indoline]s **537** was later reported by the Wang group, as shown in Scheme 3-150 [143].

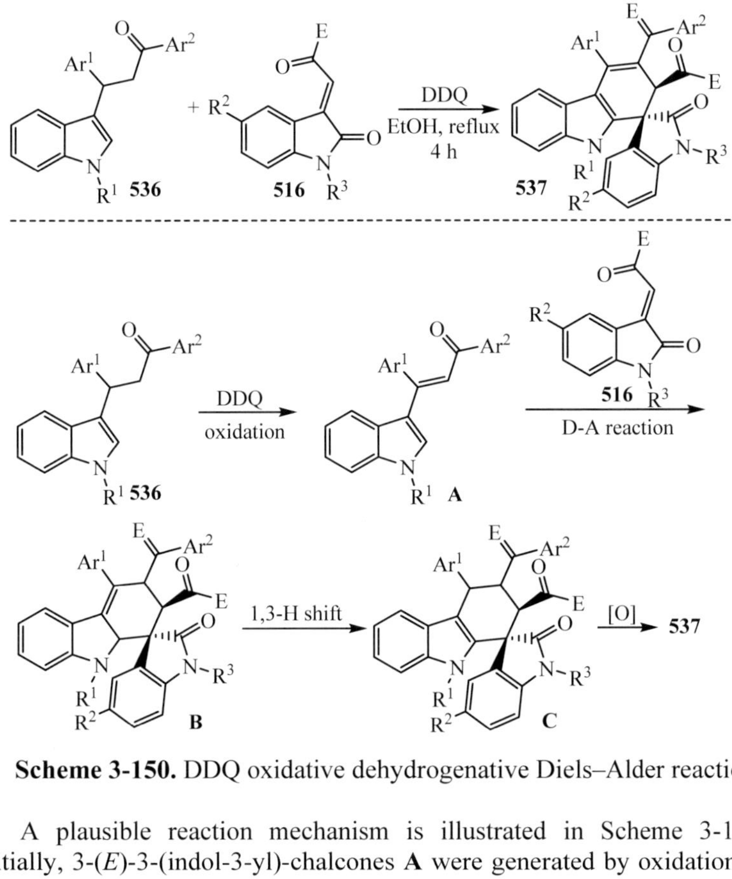

Scheme 3-150. DDQ oxidative dehydrogenative Diels–Alder reaction.

A plausible reaction mechanism is illustrated in Scheme 3-150. Initially, 3-(*E*)-3-(indol-3-yl)-chalcones **A** were generated by oxidation of the 3-(indol-3-yl)-1,3-diphenylpropan-1-ones **536** with DDQ. Later, a Diels-Alder reaction of 1,3-diene **A** with 3-methyleneoxindole **516** provided the intermediate **B** which was converted to intermediate **C** by the 1,3-proton shift process. Finally, the desired products were formed by the further oxidative dehydrogenation process in the air.

Recently, Swamy et al. showed that spiroannulation of δ-acetoxy allenoate **538** and alkyl benzoisothiazole dioxide (*N*-sulfonyl ketimine) triggered **539** by DABCO/MeCO$_2$H combination allows an essentially single diastereomer **540** through chemo- and regiospecific [4+2] carboannulation and a new hydroxyl group was introduced (Scheme 3-151).

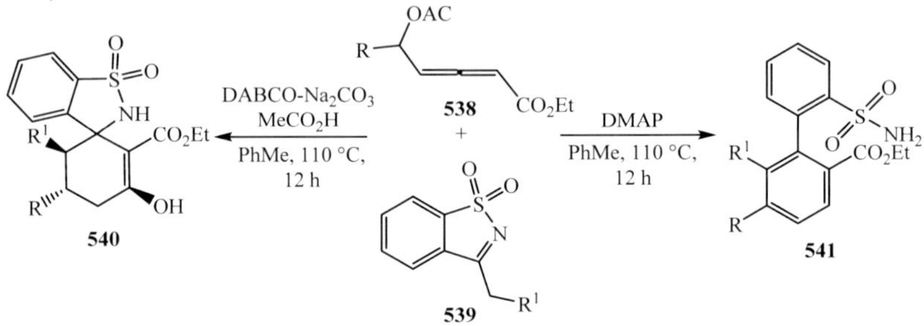

Scheme 3-151. Spiroannulation of δ-acetoxy allenoate **538**.

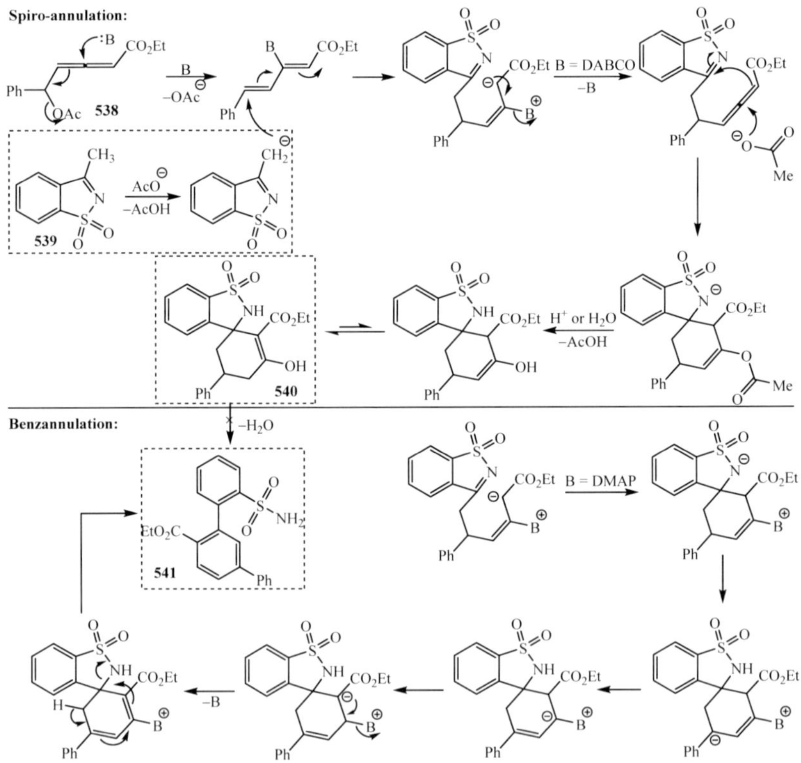

Scheme 3-152. Proposed pathways for [4+2] spiro- and benzannulations.

On the other hand, DMAP-catalyzed benzannulation provides unsymmetrical *m*-teraryls **541** through Mannich coupling, sequential proton transfers, and C−N bond cleavage. In this study, δ-acetoxy allenoate **538** is used as a 4C-synthon, and the carboannulation is completely base-dependent [144]. The proposed mechanism explaining the role of the base is detailed in Scheme 3-152.

3.3.6. Miscellaneous reactions

Other examples of pericyclic-type reactions besides those mentioned in previous sections are listed in the following.

Oxindoles and β-lactams are attractive structural motifs because of their unique biological importance. In 2021, the Li group designed a synthesis of 3,3-spiro[β-lactam]-oxindoles **544** through diastereoselective [3+1] cycloaddition reaction of oxindolyl azaoxyallyl cation precursors **542** with sulfur ylides **543** (Scheme 3-153). Using the Cs_2CO_3 base, the desired products were obtained with up to 94% yield with perfect diastereoselectivity [145].

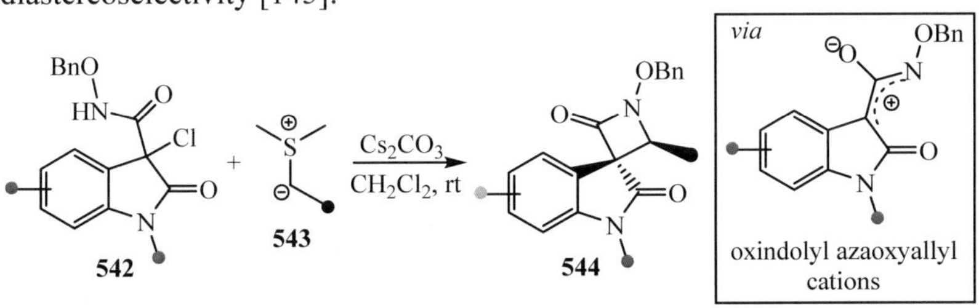

Scheme 3-153. Diastereoselective [3+1] cycloaddition reaction of oxindolyl azaoxyallyl cation precursors **542** with sulfur ylides **543**.

Additionally, spirocyclic piperidones **547** synthesized by Hayrapetyan et al. by [3+3] cycloaddition [146]. According to the proposed mechanism, in the first step, α-keto-lactams/α-ketolactones **545** were synthesized by Dieckmann-type condensation of lactams with convenient esters in refluxing toluene using NaH as a base. Additionally, the desired ketolactams/ketolactones **545** can also be furnished by deprotonating lactams with LDA at low temperatures and quenching the Li-enolates with convenient esters. Condensation between α-ketolactams/α-ketolactones **545** with primary amines proceeded readily in refluxing chloroform while using 4 Å molecular sieves as the dehydrating agent **546**. Then, the [3+3] aza-annulation reaction enamines **546** with itaconic anhydride **547**, resulting in spirocyclic piperidones **548** with excellent selectivity. Finally, the reduction of the enamide double bond with borohydride reagents, such as $NaBH_4$ and $NaBH_3CN$, in acetic acid was carried out with good yields (54–56%) and excellent selectivity (Scheme 3-154).

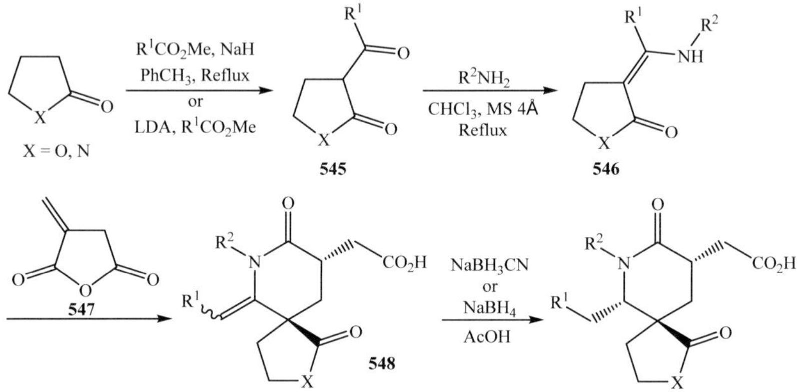

Scheme 3-154. Synthesis of spirocyclic piperidones **548** via [3+3] aza-annulation reaction.

A series of spiro[4,5]decanes bearing oxindoles containing three stereogenic centers **550** were also synthesized by Tanaka et al. via the formal [4+1] cycloaddition of enones **549** with cyclohexane-1,3-dione in the presence of CF_3SO_3H as a catalyst at 60 °C (Scheme 3-155) [147].

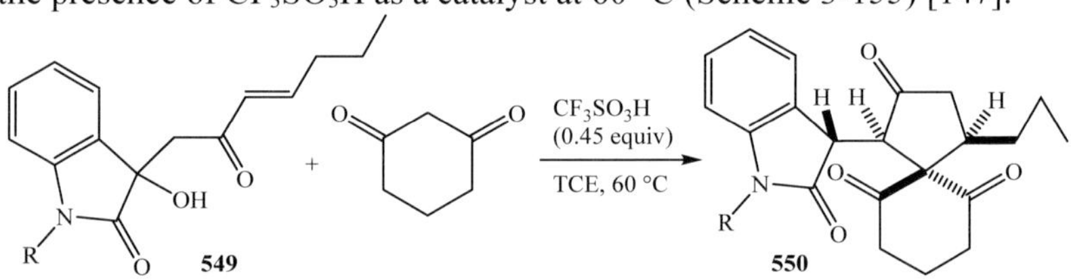

Scheme 3-155. CF_3SO_3H-Catalyzed [4+1] cycloaddition.

Likewise, Namboothiri et al. showed that Hauser-Kraus type annulation ([4+4] cycloaddition) of *p*-QMs **551** with sulfonylphthalide **552** involves an unprecedented 1,6-addition of sulfonylphthalide **552** to hydroxyaryl-*p*-quinone methide **551** followed by Dieckmann cyclization (Scheme 3-156) [148].

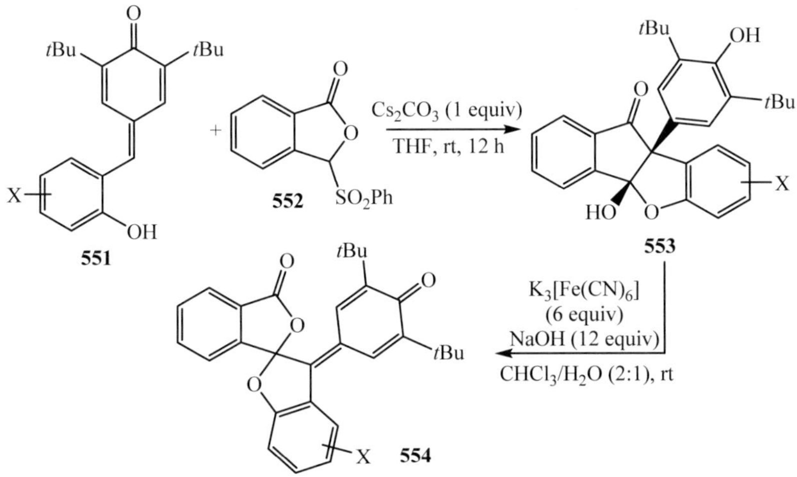

Scheme 3-156. Synthesis of spirolactones **554** via Hauser-Kraus type annulation.

In this reaction, spirolactones **554** are produced via a key indenofuran intermediate **553**. The authors' proposal for the reaction mechanism consists of two possible pathways (Scheme 3-157).

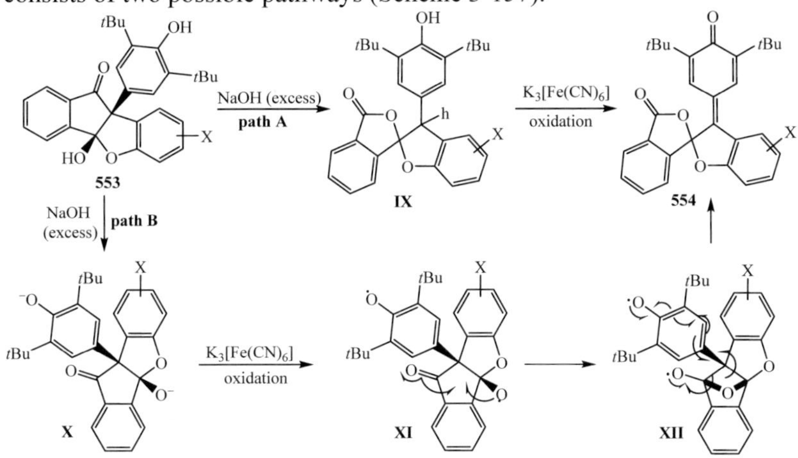

Scheme 3-157. Plausible mechanisms for the formation of spirolactone **554**.

The indenofuran intermediate **553** can undergo an intramolecular cyclization in excess NaOH to give an intermediate **IX** which upon oxidation conditions affords the quinone methide **554** (Scheme 3-157, pathway A). An equally probable alternative mechanism considers the generation of dianion **X** from **553** in the presence of an excess base and its oxidation to the diradical **XI** (Scheme 3-157, pathway B). Formation of oxetane diradical intermediate **XII** from **XI** and its ring fragmentation provides **554**.

The same group also applied a similar methodology to produce complex fused and spiro heterocycles **555-556** through a [4+4] cycloaddition reaction of sulfonylphthalides **552** with simple conjugated nitroalkenes [149]. Using nitroalkenes bearing an additional nucleophilic site, the desired products **555-556** were obtained in high yields with high selectivity through a cascade process involving Michael addition, Dieckmann cyclization, and a series of eliminations and rearrangements (Scheme 3-158).

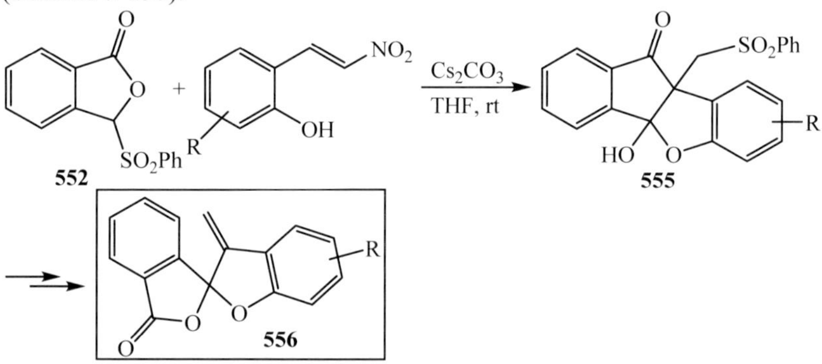

Scheme 3-158. Synthesis of complex fused and spiro heterocycles **555-556**.

In addition, they revealed an unusual reactivity of a Hauser donor such as 3-sulfonylphthalide **552** with *o*-hydroxynitrostyrene **557** as a Michael acceptor and developed an [4+4] annulation strategy toward the synthesis of functionalized benzofuran containing fused and spiro heterocycles **558-559** in high yields (Scheme 3-159) [150].

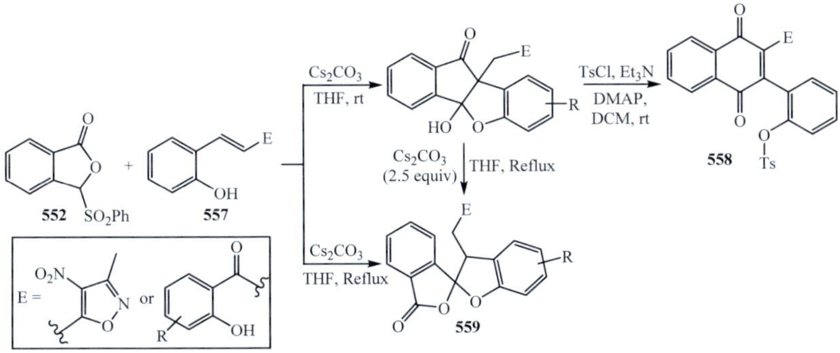

Scheme 3-159. Synthesis of functionalized benzofuran containing fused and spiroheterocycles **558-559**.

In 2015, Kutateladze et al. reported a method for synthesizing complex *N,O*-heterocycles with an additional spiro-connected nitrogen heterocyclic moiety [151]. Upon irradiation of cyclic imine containing *o*-amido group, reactive intermediate **560** generates amino azaxylylene **563**, which is capable of intramolecular cycloadditions to tethered unsaturated pendants to construction complex *N,O*-heterocycles **562-563** with additional spiro-connected nitrogen heterocyclic moiety (Scheme 3-160). The photocyclization and subsequent post-photochemical transformations are accompanied by the rise of Lovering's fsp^3 factor producing unprecedented three-dimensional molecular architectures.

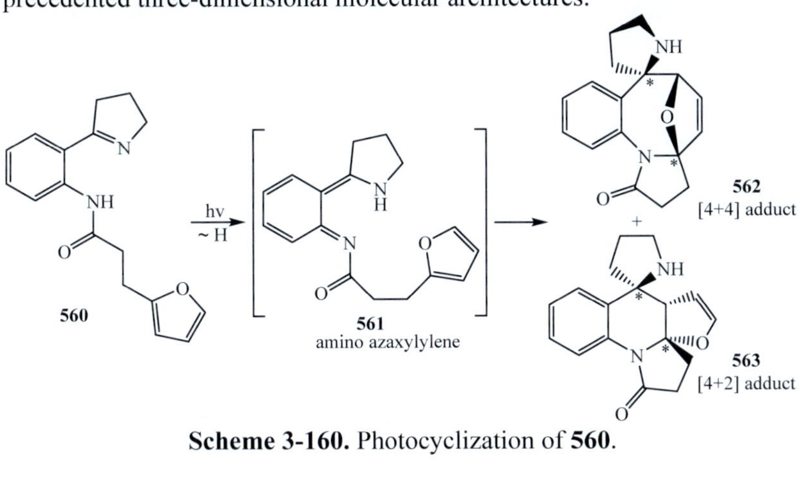

Scheme 3-160. Photocyclization of **560**.

The functionalized spirocyclic oxindole-butenolides **564** have been prepared in an intermolecular [2+2+1] multicomponent cycloaddition reaction from readily available isocyanides, activated alkynes and isatins [152]. This reaction proceeds via Michael addition-nucleophilic addition-intramolecular cyclization sequence, as shown in Scheme 3-161.

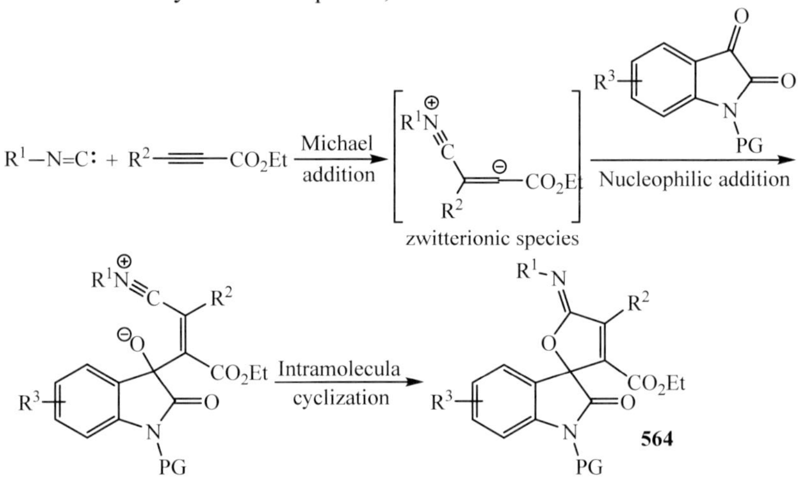

Scheme 3-161. Three-component [2+2+1] cycloaddition of isocyanide, ethyl propiolate, and substituted isatin.

3.4. Cyclization reactions

In general, any reaction that results in ring formation is called cyclization. Within this context, this section will outline the application of cyclization reactions to constructing spiro compounds. Common cyclization reactions for synthesizing spiro compounds are dearomatization reactions, *ipso*-cyclization, intramolecular cyclization, Nazarov reactions, Prins reactions, Pictet-Spengler reactions, radical cyclization, reductive cyclization, oxidative cyclization, ring-opening cyclization, and Conia-ene cyclization. Here, we will describe some important examples concerning some variations in cyclization reactions that have been reported since 2000 to give a broad overview of the recent trends in this cyclization protocol.

3.4.1. Dearomative cyclization

During the last decade, dearomative cyclization has emerged as a powerful strategy for constructing spirocycle compounds. It presents an alternative synthetic method without extra stoichiometric amounts of chemical oxidants or reductants. This section summarizes recent advances in this rapidly growing area as a review, in which the typical examples are listed.

In 2007, Xi et al. synthesized a wide variety of multiply substituted bicyclic/polycyclic spirocyclopentadienes **567a-d** and phenanthrenes **569** through a dearomatizing anionic cyclization and a novel skeletal rearrangement of the *ipso* attack intermediates followed by rearomatization [153].

4-Phenyl 1-lithio-1,3-butadienes **565** and 4-naphthyl 1-lithio-1,3-butadienes **566** were generated *in situ* from their corresponding iodo compounds, underwent highly selective intramolecular nucleophilic addition of the butadienyllithium to the aromatic rings, leading to full dearomatization of the phenyl rings and partial dearomatization of the naphthyl rings (Scheme 3-162).

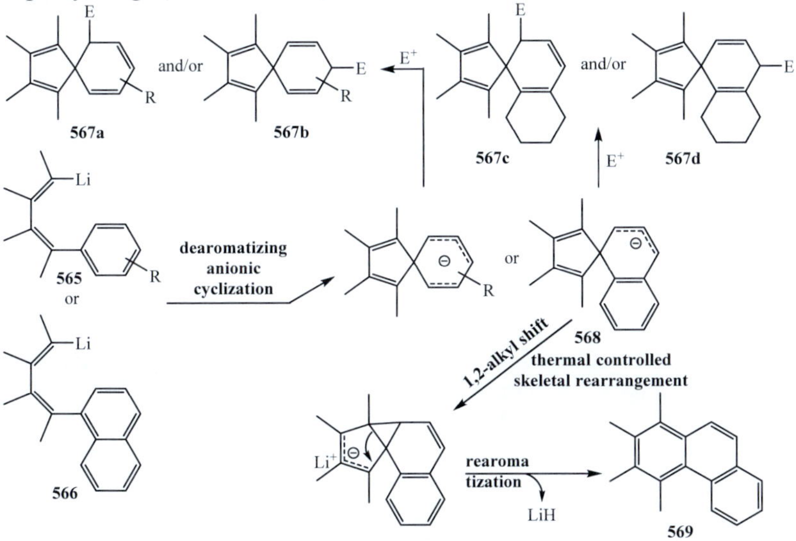

Scheme 3-162. Synthesis of multiply substituted bicyclic/polycyclic spiro cyclopentadienes **567a-d** and phenanthrenes **569**.

When the reactions were performed at lower temperatures, *ipso* intramolecular nucleophilic attack took place individually to provide the spirocyclopentadienes **567a-d** upon hydrolysis or further treatment with various electrophiles. However, at temperatures higher than room temperature, such as 75 °C, 4-naphthyl 1-lithio-1,3-butadienes **566** undergo a novel skeletal rearrangement through the 1,2-alkyl shift to provide the formally vicinal attack lithiated dearomatized intermediates **568** which are thermally unstable and produce the stable phenanthrenes **569** through the rearomatization process.

Later, an efficient synthesis of spiroindolenines **572′**, an important class of heterocycles with pharmaceutical activities, was disclosed through Brønsted acid catalysis [154]. The process goes through a [1,5]-hydride shift/cyclization sequence involving *p*-TsOH·H_2O-catalyzed reaction of 2-substituted indoles **570** with 2-(pyrrolidin-1-yl)benzaldehydes **571** (Scheme 3-163). Spiro-indolenines **572** were formed in good to excellent yields and moderate diastereoselectivity (3.5/1 dr). When the inseparable mixture of spiroindolenine isomers **572** was washed with isopropyl ether after flash chromatography, the major isomers **572′** could be achieved up to >20/1 dr.

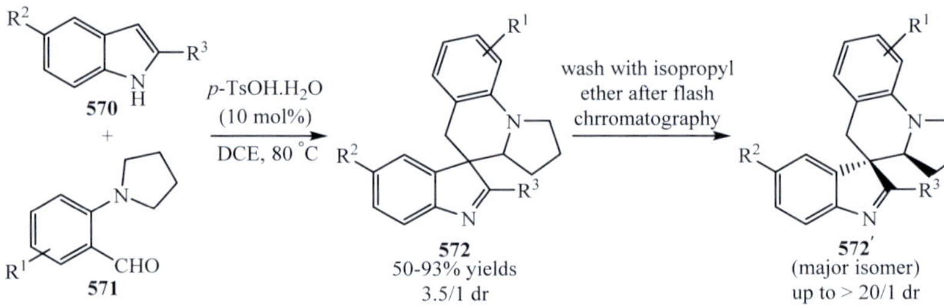

Scheme 3-163. *p*-TsOH·H_2O-Catalyzed [1,5]-hydride shift/cyclization sequence.

The Li group presented a chemoselective N-H or C-2 arylation of indole-2-carboxamides **573** for the preparation of indolo[1,2-*a*]quinoxalin-6-ones **574** and 2,3′-spirobi[indolin]-2′-ones **575** [155]. When ZnI_2/Ag_2CO_3 was used, the reaction afforded indole-2-carboxamides **574** in up to 86% yields via intramolecular N-H/C-H coupling whereas the dearomative cyclization reactions were carried out in the presence of TfOH, resulting in C-2 arylation of indole via C-H functionalization or a Fridel-Crafts alkylation reaction to generate 2,3′-spirobi[indolin]-2′-ones **575** as the only products with good functional group tolerance (Scheme 3-164).

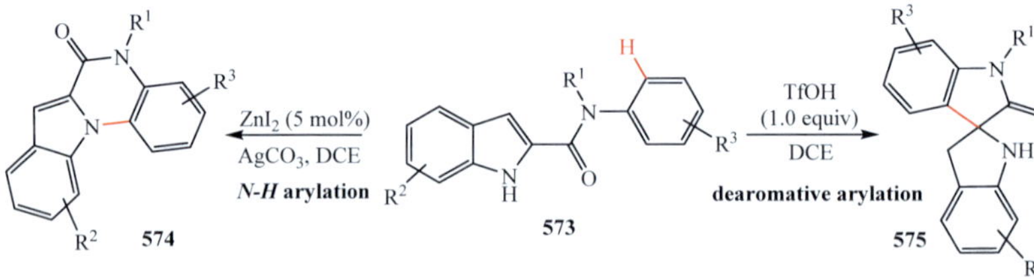

Scheme 3-164. Chemoselective N−H or C-2 arylation of indole-2-carboxamides **573**.

A radical-induced dearomative cascade cyclization to sulfonated spiro[cyclopentenone-1,3′-indole] derivatives **578** was developed by Cui et al. which involves three-component sulfonylative spirocyclization of indolyl ynones **576** with aryldiazonium salts **577** and a sulfur dioxide surrogate of DABCO·$(SO_2)_2$. The suggested mechanism is shown in Scheme 3-165 [156].

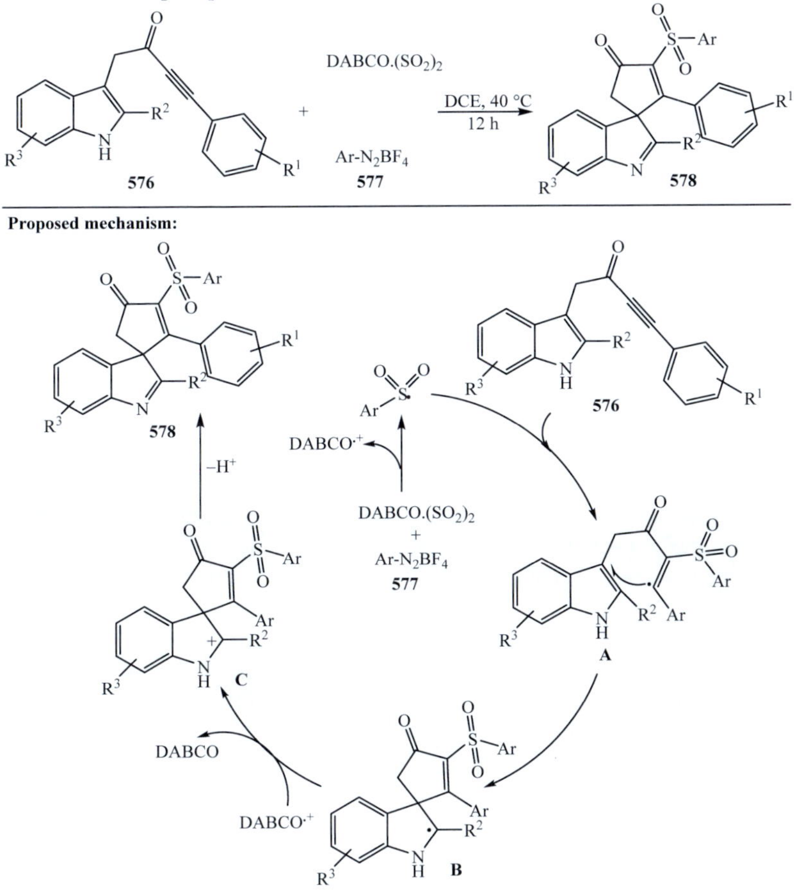

Scheme 3-165. Radical-induced dearomative cascade cyclization and proposed mechanism.

Initially, an arysulfonyl radical and a radical cation $DABCO^{\bullet+}$ were generated *in situ* from aryldiazonium tetrafluoroborate **577** and DABCO·$(SO_2)_2$. Afterward, the addition of the arylsulfonyl radical to indolyl ynone **576** gave a vinyl radical **A** followed by the cyclization of the indole at C3 to form a spirocyclic radical intermediate **B**, in which the radical cation DABCO$^{\bullet+}$ could be oxidized to give the intermediate **C**. Finally, the intermediate **C** undergoes dehydrogenation to form the desired product **578**.

Likewise, Zhang et al. presented spiroamination and spirolactonization of anisoles **579** via anodic oxidation dearomatization iodine reagents (Scheme 3-166).

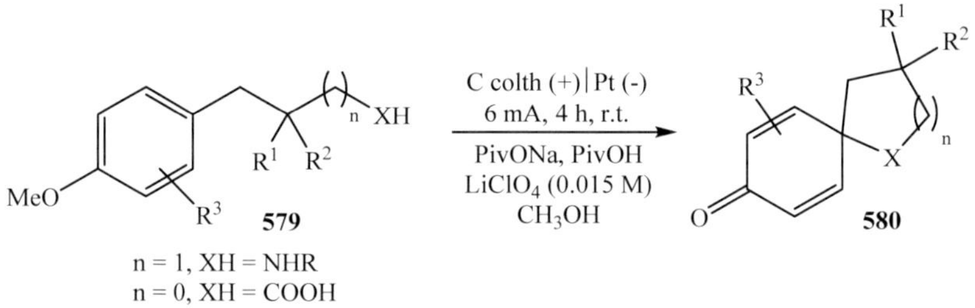

Scheme 3-166. Electrochemical oxidation dearomatization of anisoles **579**.

A wide variety of spiropyrrolidines/spirolactones **580** with diverse functional groups was obtained with yields up to 97%. Based on the experimental results, as shown in Scheme 3-167, the aromatic ring **579** was first oxidized at the anode to give the intermediate **I** with radical cation focused on the aromatic core. Then, with the base's assistance, the nucleophilic amido group attacked the aromatic core to generate the radical intermediate **II** which could be oxidized at the anode to produce cation intermediate **III**. Subsequently, intermediate **III** was captured by MeOH and then hydrolyzed to generate the target product **580**. Meantime, the concomitant reduction of protons realized H_2 at the cathode. However, radical intermediate **II** formed by adding N radical intermediate **V** to the aromatic core, could not be excluded [157].

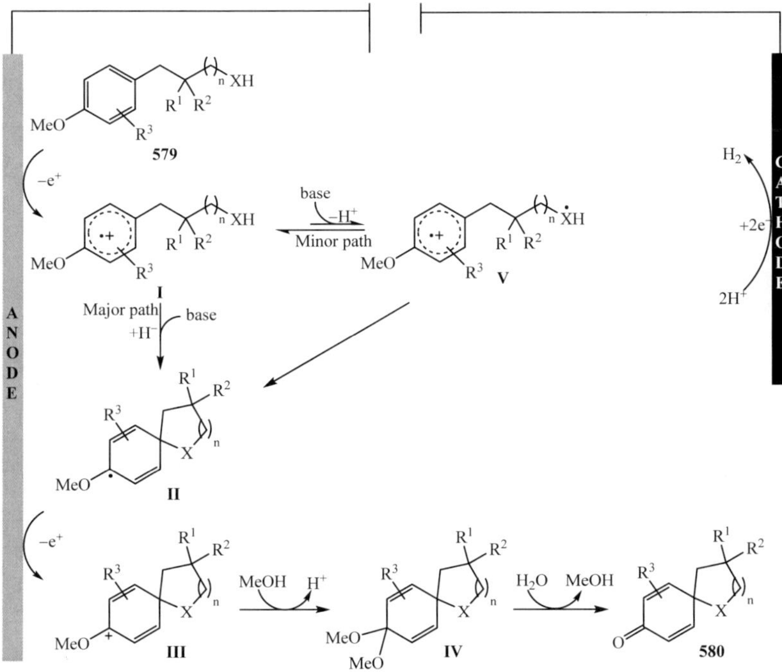

Scheme 3-167. Possible mechanism.

3.4.2. Intramolecular *ipso*-cyclization

Recently, the intramolecular *ipso*-cyclization approach has emerged as an attractive tool for synthesizing spirocycles as it provides an efficient way to form new carbon-carbon/heteroatom bonds in a single step [158]. Here, we present an overview of works reported in this field over the past 20 years.

In 2003, the Kikugawa group prepared spirodienones **583** bearing the 1-azaspiro[4.5]decane ring system from *N*-methoxy-(4-halogenophenyl) amides **581** by the intramolecular *ipso* attack of a nitrenium ion generated with [hydroxy(tosyloxy)iodo]-benzene (HTIB) **582** in trifluoroethanol (TFEA) [159]. The presence of a halogen atom may stabilize the cation, making it an excellent leaving group (Scheme 3-168).

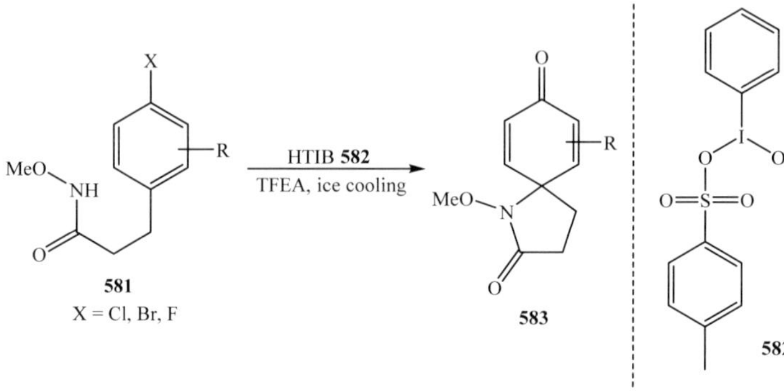

581
X = Cl, Br, F

583

582

Scheme 3-168. Intramolecular *ipso*-cyclization of *N*-methoxy-(4-halogeno phenyl)amides **581**.

In addition, 3-halospirotrienones **585** bearing a spiro[4.5]decane ring system were also synthesized by an iodonium ion-induced intramolecular *ipso*-cyclization of methoxy-substituted 4-aryl-1-alkynes **584** in the presence of ICl under highly mild reaction conditions (Scheme 3-169) [160]

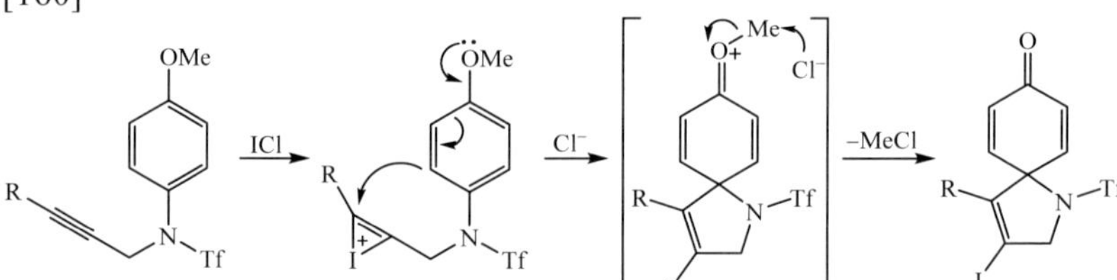

584 585

Scheme 3-169. ICl-induced intramolecular *ipso*-cyclization of methoxy-substituted 4-aryl-1-alkynes **584**.

Li et al. reported the selective synthesis of 8-methylene azaspiro [4,5]trienes **587** by intramolecular electrophilic *ipso*-cyclization reaction of *N*-(4-methylphenyl) propiolamides **586** with ICl or I_2 (Scheme 3-170). The products could be isolated in moderate to excellent yields [161].

The same group presented the synthesis of various 8-(polyfluoro alkoxy)azaspiro[4.5]trienes **590** by the reaction of *N*-arylpropynamides **588** with polyfluoroalkyl alcohols **589** in the presence of *N*-iodo-succinimide (NIS) (Scheme 3-171) [162].

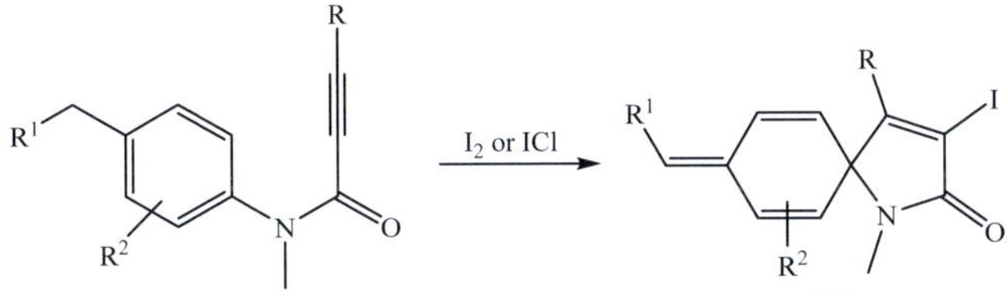

586 587

Scheme 3-170. Intramolecular electrophilic *ipso*-cyclization reaction *N*-(4-methylphenyl)propiolamides **586**.

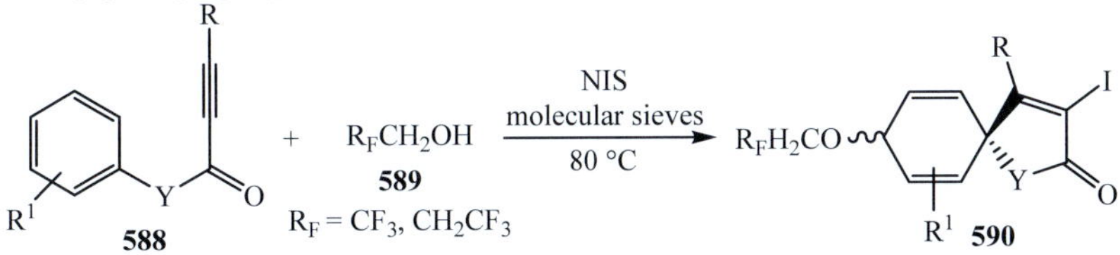

588
Y = N, O
589
$R_F = CF_3, CH_2CF_3$
590

Scheme 3-171. Synthesis of various 8-(polyfluoroalkoxy)azaspiro[4.5] trienes **590**.

They also reported an efficient intramolecular electrophilic *ipso*-iodocyclization for the synthesis of spiro[4,5]trienyl acetates **592** from *para*-unactivated arylalkynes **591**, including *N*-arylpropiolamides and phenyl 3-phenylpropiolate, as well as NIS and HOAc (Scheme 3-172) [163].

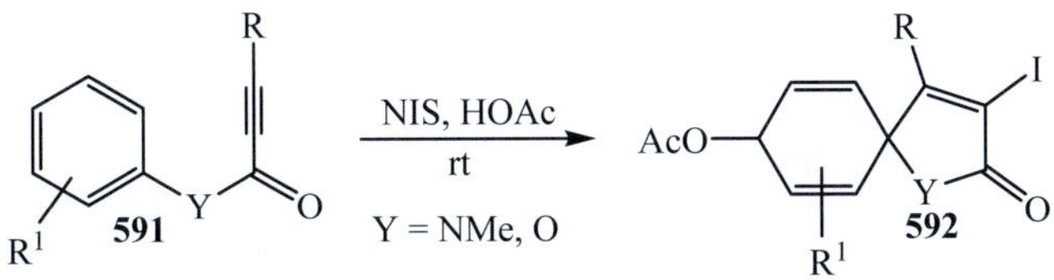

591 592

Scheme 3-172. Intramolecular electrophilic *ipso*-iodocyclization of *para*-unactivated arylalkynes **591**.

Moreover, the halo-substituted spiro[4,5]trienones **594** were synthesized by the intramolecular *ipso*-halocyclization of 4-(*p*-unsubstituted-aryl)-1-alkynes **593** in the presence of *N*-boromosuccimide (NBS) which can be converted into azaquaternary tricyclic products **595** via Pd-catalyzed Heck reaction (Scheme 3-173) [164].

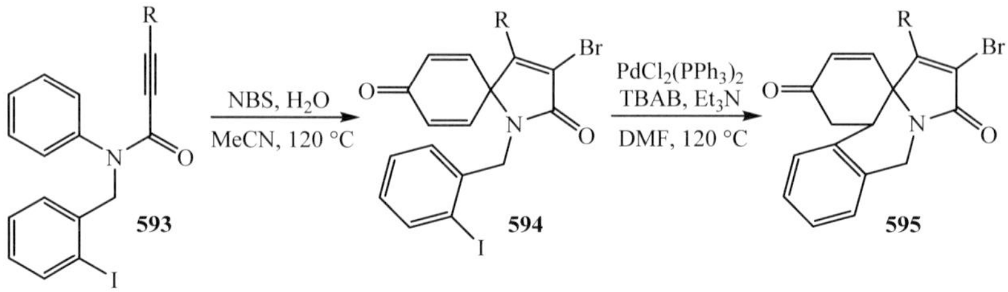

Scheme 3-173. Intramolecular *ipso*-halocyclization of 4-(*p*-unsubstituted-aryl)-1-alkynes **593**.

In 2011, an interesting example starting from *N*-aryl-3-phenyl-propiolamidesin **596** was addressed by Likhar et al. for the selective synthesis of spiro[4.5]trienes **597** using iodine/cerium(IV) ammonium nitrate and acetic acid (Scheme 3-174) [165].

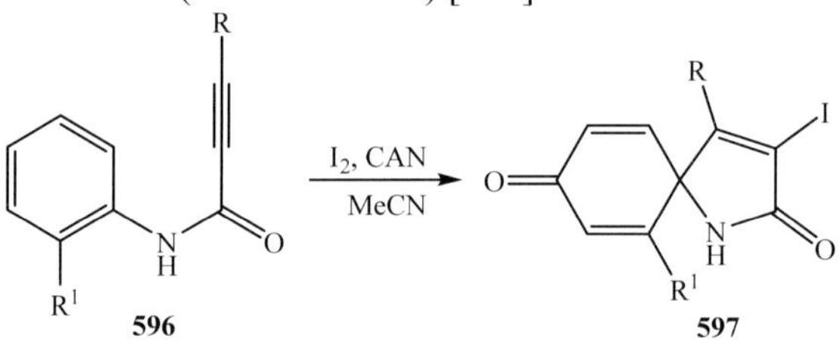

Scheme 3-174. Iodocyclization of *N*-aryl-3-phenylpropiolamides **596** by I_2/CAN.

Li et al. showed that a variety of *N*-arylacrylamides **598** in the presence of iodine mono-chloride, water, and sodium bicarbonate can lead to the intramolecular electrophilic *ipso*-cyclization reaction that provides the corresponding 3-iodo-1-azaspiro[4.5]deca-6,9-diene-2,8-diones **599** in moderate to good yield (Scheme 3-175) [166].

Scheme 3-175. Synthesis of 3-iodo-1-azaspiro[4.5]deca-6,9-diene-2,8-diones **599**.

Chen et al. extended the ICl-induced intramolecular electrophilic cyclization to the preparation of spiro-conjugated molecules **601** by 1-[4′-methoxy(1,1′-biphenyl)-2-yl]alkynones **600** (Scheme 3-176) [167]. They observed an interesting substituent effect during their experimentation and found that with minor modification to the pattern of reacting substrate, both *ipso* and *ortho*-cyclization can be, regiospecifically, "switched on" or "off".

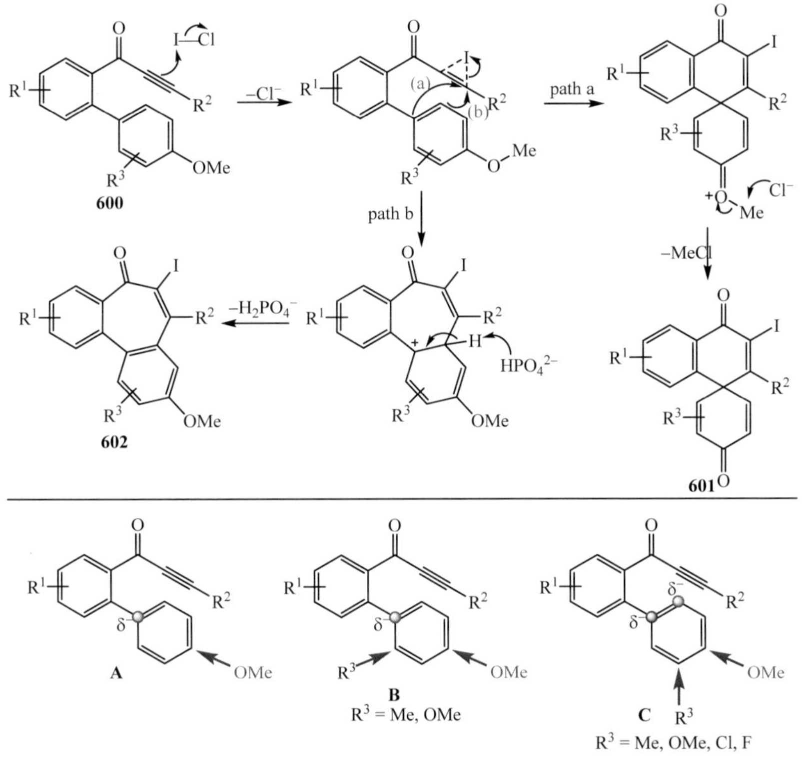

Scheme 3-176. ICl-induced iodocyclization of 1-[4′-methoxy(1,1′-biphenyl)-2-yl]alkynones **600**.

The *ipso*-electrophilic cyclization can be amplified by the electronic effect of the *para*-methoxy group on the distal phenyl group through resonance (**A**), as shown in Scheme 3-176. The presence of a second electron-donating substituent at the ortho position of the distal phenyl ring reinforces this effect, leading to further *ipso*-cyclization (**B**). However, when a second substituent, such as methoxy, methyl, or a halogen, exists at the meta position of the distal phenyl group, the electron density is directed to the ortho positions to the 1,1′-diphenyl bond (**C**). As a result, the ipso and *ortho*-positions become activated, leading to a mixture of spiro compound **601** and dibenzo[7]annulen-5-one **602**. When three substituents are present on the distal phenyl group, the reaction outcome depends on the identity and positions of the substituents.

In another study, Kesharwani et al. reported the synthesis of 1*H*-pyrrolo-[2,1-*j*]-quinolone frameworks **604**, which are present in the family of tricyclic marine alkaloids such as Lepadiformine and Cylindricine, by employing *N*-(alkynoyl)-6-methoxytetrahydroquinolines **603** (Scheme 3-177) [168].

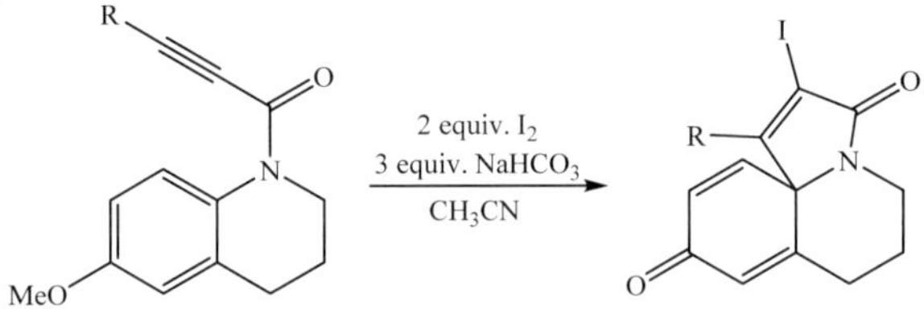

Scheme 3-177. Synthesis of 1*H*-pyrrolo-[2,1-*j*]quinolone frameworks **604**.

Efficient construction of azaspiro[4.5]trienones **607** libraries was accomplished in a highly effective one-pot reaction tandem that consists of an Ugi four-component condensation (U4CC) of substituted *p*-anisidines **605**, aldehydes, 3-alkyl/aryl-propiolic acids **606** and isocyanides, combined with an iodine mediated *ipso*-iodocyclization (Scheme 3-178) [169].

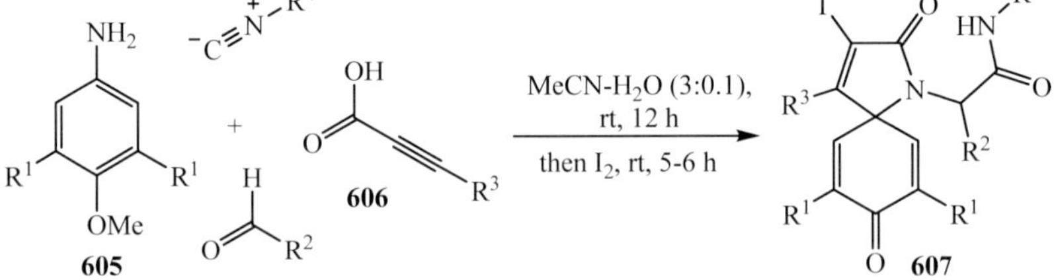

Scheme 3-178. Tandem U4CC/electrophilic *ipso*-iodocyclization in one-pot.

In 2019, Baidya et al. showed an elegant example of a selenium radical-triggered switchable *ortho*/*ipso* oxidative cyclization cascade of *N*-aryl alkynamides **608** using diaryl diselenides **609** as the selenium precursors and inexpensive $K_2S_2O_8$ as an oxidant (Scheme 3-179). The reaction furnished a variety of 3-selenyl quinolin-2-ones **610** and 3-seleno-spiro[4,5]trienones **611/611′** in good to excellent yields [170].

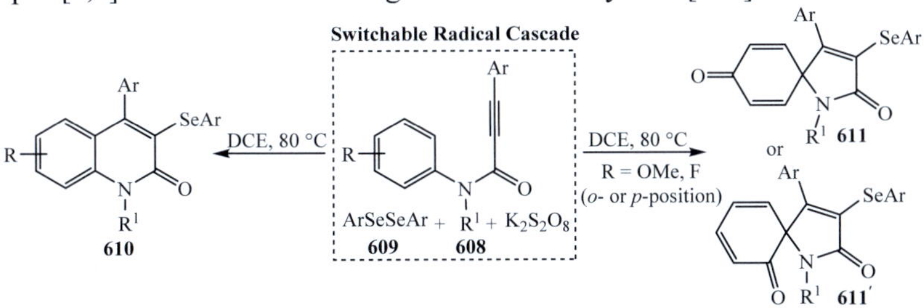

Scheme 3-179. Metal-free switchable *ortho*/*ipso*-cyclization of *N*-aryl alkynamides **608**.

A new electrochemical three-component annulation-iodosulfonylation of 1,5-enyne-containing *p*-QMs **612** was developed by Hao et al. using available arylsulfonyl hydrazides **613** and potassium iodide under environmentally benign conditions (Scheme 3-180). Potassium iodide acts as the triple role of an electrolyte, a redox catalyst, and an iodination reagent [171]. Iodosulfonated spirocyclo hexadienone containing (*E*)-indenes **614** were formed in moderate to high yields.

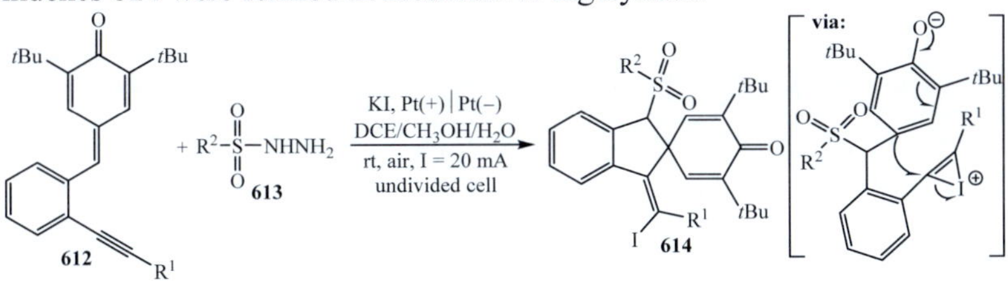

Scheme 3-180. Electrochemical annulation-iodosulfonylation of 1,5-enyne containing *p*-QMs **612**.

Reddy et al. disclosed a CAN-mediated domino thiocyanation/*ipso*-cyclization/dearomatization of *N*-(*p*-methoxyphenyl)-propiolamides **615** to give access to diversely substituted thiocyanato-spirocyclohexadienones **616** (Scheme 3-181). Further, conversions of the thiocyanated spirocyclo hexadienones to their corresponding derivatives **617** with unique structural motifs were established for the first time [172].

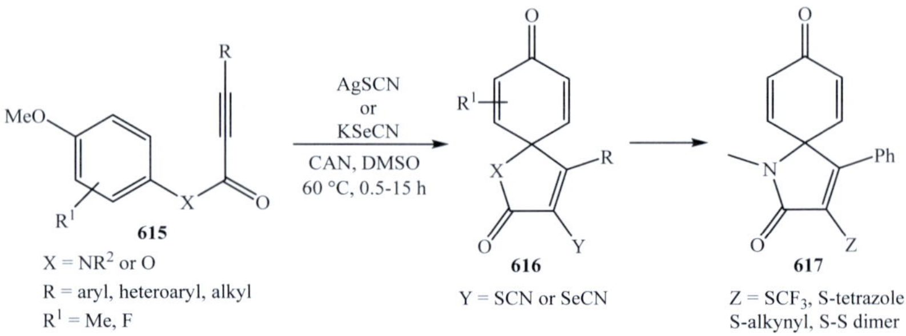

Scheme 3-181. CAN-Mediated domino thiocyanation/*ipso*-cyclization/ dearomatization of *N*-(*p*-methoxyphenyl)-propiolamides **615**.

Yang et al. developed an efficient method for trifluoromethylthiolation and dearomatization of activated alkynes **618** using stable and readily available $AgSCF_3$ (Scheme 3-182).

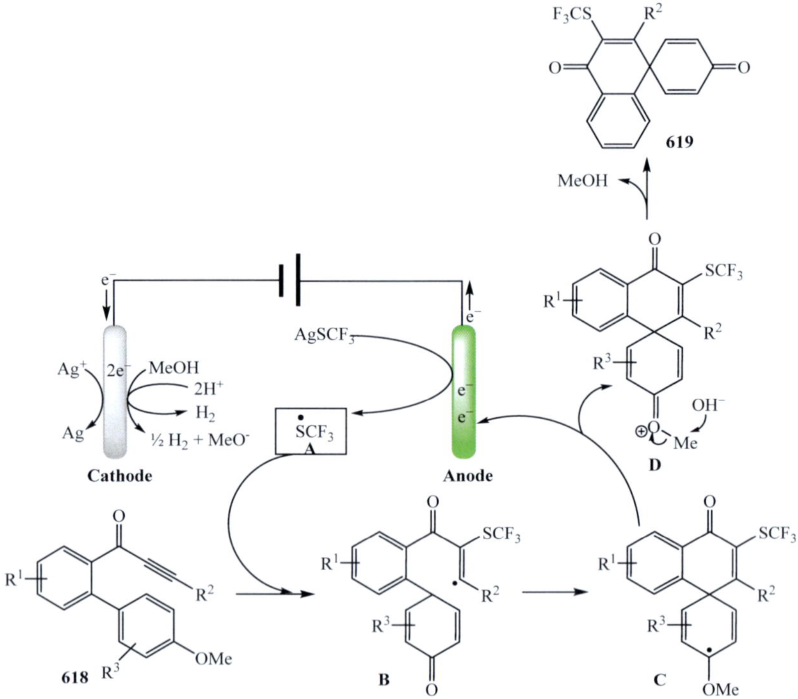

Scheme 3-182. Electrochemical trifluoromethylthiolation and spirocyclization of alkynes with $AgSCF_3$.

According to the experimental results, the SCF_3 radical **A** is generated *in situ* through the anodic oxidation of $AgSCF_3$. Then, the radical intermediate is added to the C≡C bond of alkyne **618** to form vinyl radical intermediate **B** which is further converted into intermediate **C** through an intramolecular radical *ipso*-cyclization. The anodic oxidation of **C** generates intermediate **D** which undergoes elimination of MeOH to yield SCF_3-containing spiro[5,5]-trienones **619**. During the reaction, the electro-reduction of a proton from MeOH or H_2O in the system releases hydrogen gas at the cathode. However, reducing Ag(I) to Ag cannot be prevented [173].

3.4.3. Miscellaneous reactions

Other examples of cyclization besides those mentioned in previous sections are listed in the following.

In 2003, spirodienones bearing 1-azaspiro[4.5]decane skeleton **623/623′** was synthesized by an intramolecular electrophilic substitution reaction of *N*-phthalimido-*N*-acylnitrenium ions **622** generated from *N*-acylaminophthalimides **620**, with hypervalent iodine compounds (PIFA **621** and HTIB **582**) in TFEA (Scheme 3-183) [174].

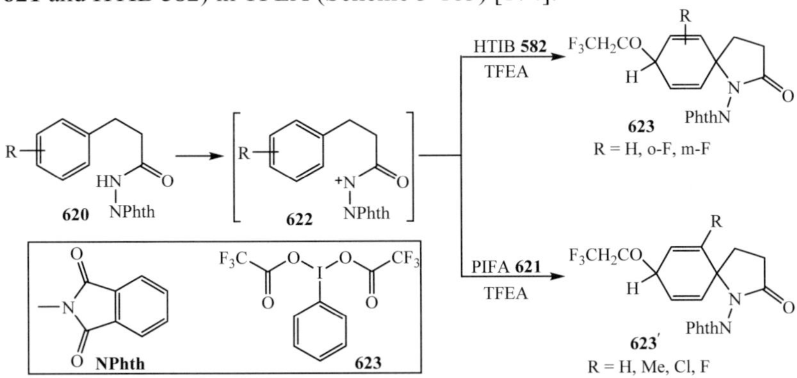

Scheme 3-183. Intramolecular cyclization with nitrenium ions.

Zhang et al. developed an oxidative intramolecular cyclization for the synthesis of spirocycle-propane quinolinediones **625** via the PIFA-mediated cyclization of 2,2-di-substituted 2-benzoylacetamides **624**. This reaction is carried out by any metal catalyst at room temperature (Scheme 3-184).

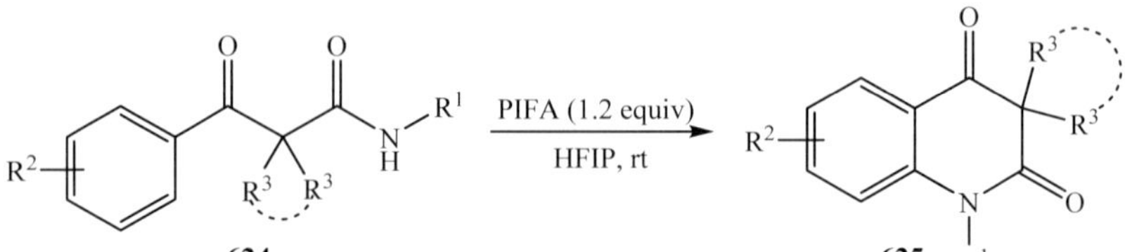

624 **625**

Scheme 3-184. PIFA-Mediated oxidative intramolecular cyclization of **624**.

The authors suggested a plausible mechanism for the present one-pot oxidative cyclization, which is disclosed in Scheme 3-185 [175].

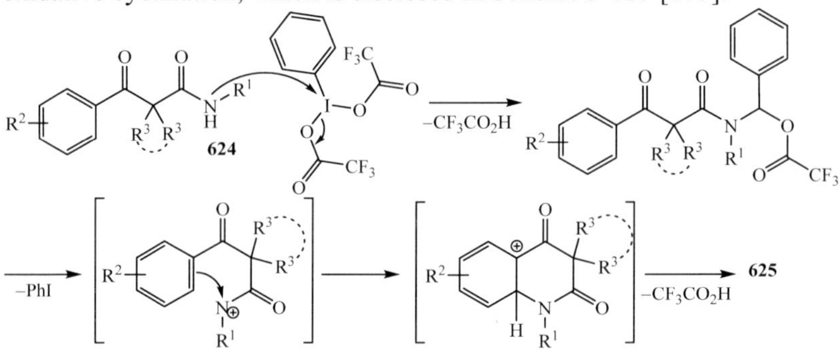

Scheme 3-185. Proposed mechanism.

Teimouri et al. applied an efficient, eco-friendly catalyst-free three-component domino enamination/aldol-like reaction/Stork enamine annulation/intramolecular cyclization to synthesize new spirobicyclic oxazolidinedione **627** containing cyclopentenone moieties of the type shown in Scheme 3-186. Starting from readily available primary or secondary amines, β-enaminocarbonyl compounds and alloxan **626** under catalyst-free conditions in water furnished the desired products **627** in moderate to good yields [176].

In a study by Li et al. in 2011, the reductive oxy-Nazarov (RON) cyclization was introduced as the key step in the total synthesis of (±)-Cephalotaxine **632** [177]. The RON cyclization is an effective, regio-, and stereospecific cyclopentannulation of vinylalkylidene dioxolanone **628** to hydroxycyclopentenone **631** under mild hydride reduction conditions. The reaction proceeds via the thermally conrotatory ring closure of a tethered 1,2-oxidopentadienyl cation species **629** or resonance hybrid form **630**. By using this RON cyclization as a key step, a successful total synthesis of (±)-cephalotaxine **632** was achieved, as shown in Scheme 3-187.

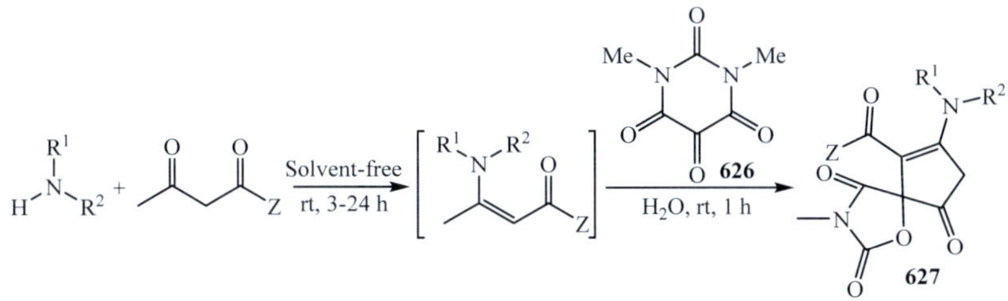

Scheme 3-186. Three-component domino enamination/aldol-like reaction/ Stork enamine annulation/ intramolecular cyclization.

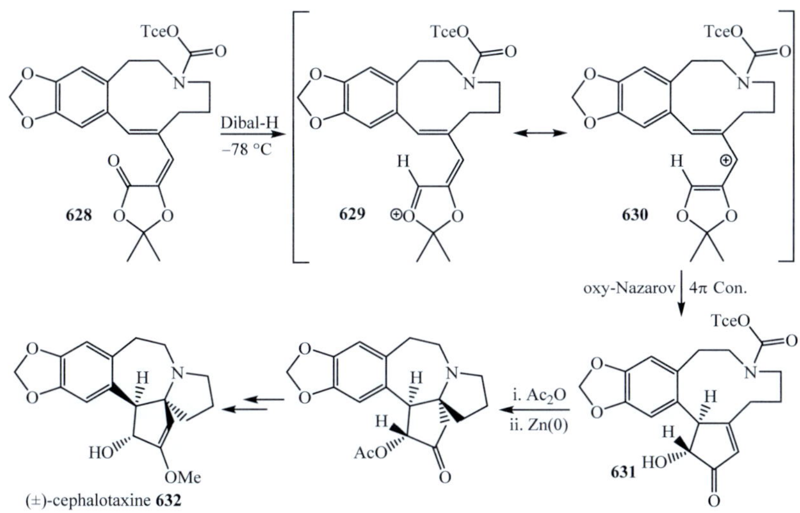

Scheme 3-187. Total synthesis of (±)-cephalotaxine **632**.

The TfOH-promoted classical Nazarov-type cyclization of benzofulvenes **633** led to 5*H*,10′*H*-spiro[benzo[*k*]phenanthridine-5,6′-dibenzopentalene] derivatives **634**/**634**′ through novel C-C and C-N bond formation (Scheme 3-188). This approach affords a new molecular framework with reasonable to excellent yields and tolerates various electron-withdrawing/donating substituents. Moreover, it is free of bases, oxidants, and metals, which is favorable for synthetic organic chemistry [178].

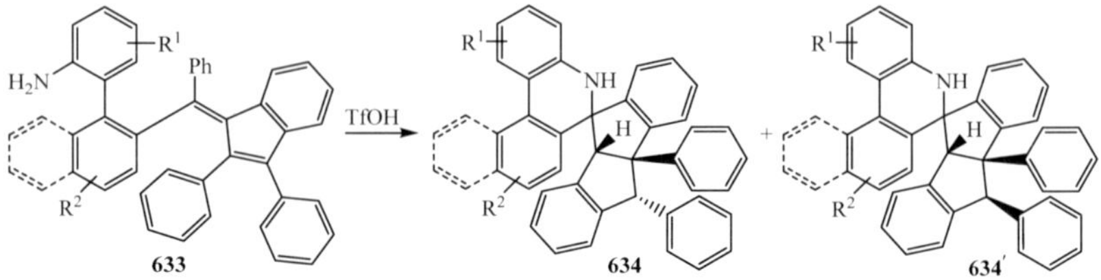

Scheme 3-188. TfOH-Promoted classical Nazarov-type cyclization of benzofulvenes **633**.

In addition, spirolactams **638** and poly-substituted pyrroles **639** were prepared by CAN–mediated oxidative cyclization of furfurylamines **635** with ynones **636** (Scheme 3-189). The protocol involved *in situ* generation of *N*-furan-2-ylmethyl-β-enaminones **637** and their subsequent oxidation by CAN (6 equivalents for spiro-lactam formation, 3 equivalents for pyrrole formation) [179].

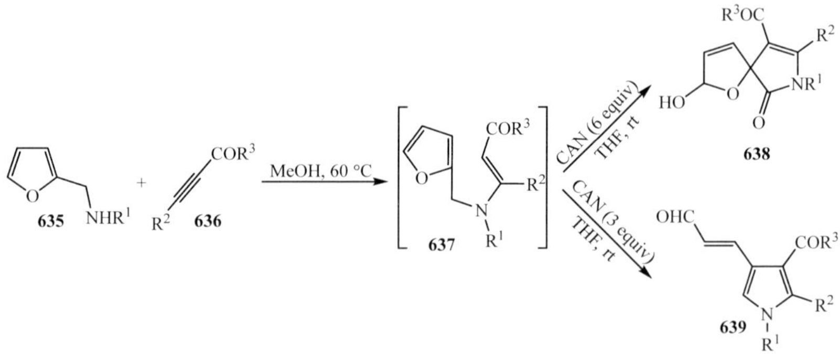

Scheme 3-189. CAN-Mediated oxidative cyclization of *N*-furan-2-ylmethyl-beta-enaminones **637.**

The Pictet-Spengler reaction is a direct, efficient, and versatile synthetic method in which a β-aryl ethylamine reacts with an aldehyde or ketone, followed by ring closure, resulting in the formation of spirocyclic compounds [180].

In 2017, Gámez-Montaño et al. reported the first strategy based on I-MCR's, in which two one-pot processes were combined to obtain spiro[pyrrolidine-3,3′-oxindoles] linked to another heterocycle [181]. They synthesized six novel tris-heterocycles 1′-tetrazolylmethyl-spiro[pyrrolidine-3,3′-oxindoles] **640** in moderate to good yields in two steps: i) a one-pot Ugi-azide/ Pictet-Spengler process, followed by ii) a one-pot oxidative spiro-rearrangement (Scheme 3-190).

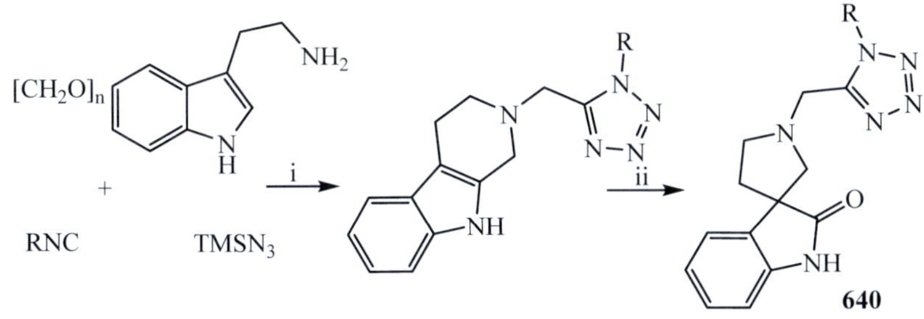

640

i) MeOH/PhMe (1:1), 90 °C, ii) NBS, THF/AcOH/H_2O (3:2:2), -10 °C

Scheme 3-190. One-pot Ugi-azide/Pictet-Spengler process.

Yoshimi et al. used a new radical cyclization in the one-step synthesis of spiro dihydro-isoquinolinones **642** via PET-promoted decarboxylation of *N*-(2-phenyl)benzoyl five-membered alicyclic amino acids **641** (Scheme 3-191). Reductive *ipso*-radical cyclization onto a benzene ring by an alkyl radical was achieved for the first time under mild conditions.

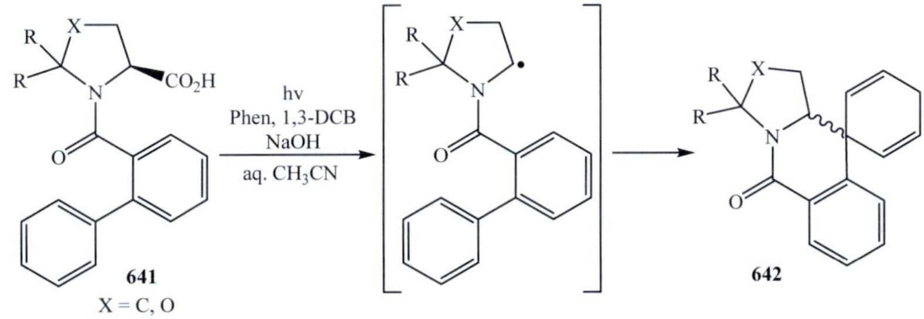

641
642
X = C, O
R = H, CH_3

Scheme 3-191. PET-Promoted decarboxylation of *N*-(2-phenyl)benzoyl five-membered alicyclic amino acids **641**.

Scheme 3-192 shows a plausible mechanism for the reductive *ipso*-radical cyclization of **641** by PET-promoted decarboxylation. Radical **643** was generated by photoinduced decarboxylation and attacked at the *ipso*-position to generate radical **644**. Subsequent back electron transfer from the radical anion of 1,3-dicyanobenzene (1,3-DCB) (as an electron-acceptor) gives anion intermediate **645** which undergoes protonation to produce spirocyclic product **642** [182].

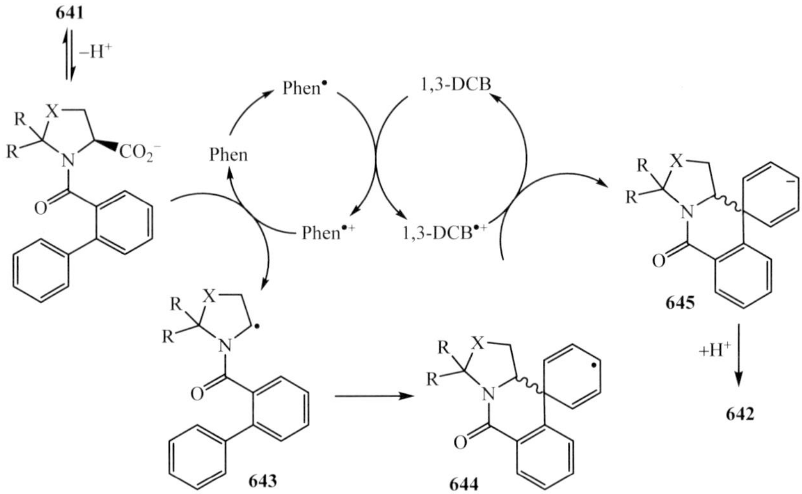

Scheme 3-192. A plausible mechanism for the photoreaction of **641**.

Moreover, the Reddy group reported $Na_2S_2O_8$-promoted radical cyclization of *N*-(4-methoxy-phenyl)-*N*-methyl-3-phenylpropiolamides **646** with 1-phenylcyclopropan-1-ols **647** for the synthesis of azaspiro[4.5]-deca-3,6,9-triene-2,8-diones **648** in good yields with high selectivity (Scheme 3-193). The radical reactions are started by β-carbonyl radicals, generated *in situ* from cyclopropanols using $Na_2S_2O_8$ [183].

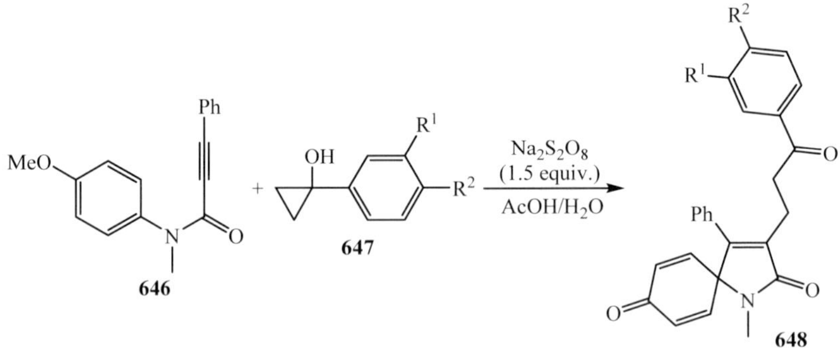

Scheme 3-193. $Na_2S_2O_8$-Promoted radical cyclization.

Pohmakotr et al. also applied $PhSCF_2TMS$ as a useful *gem*-difluoromethylene building block for the construction of *gem*-difluoromethylenated spiro-γ-butyrolactones **651** through cyclization of γ-alkenyl-γ-*gem*-difluoro(phenylsulfanyl)methyl-γ-butyrolactones **649** and

650 to Bu_3SnH in the presence of a catalytic amount of azobisisobutyronitrile (AIBN) in toluene, as illustrated in Scheme 3-194 [184].

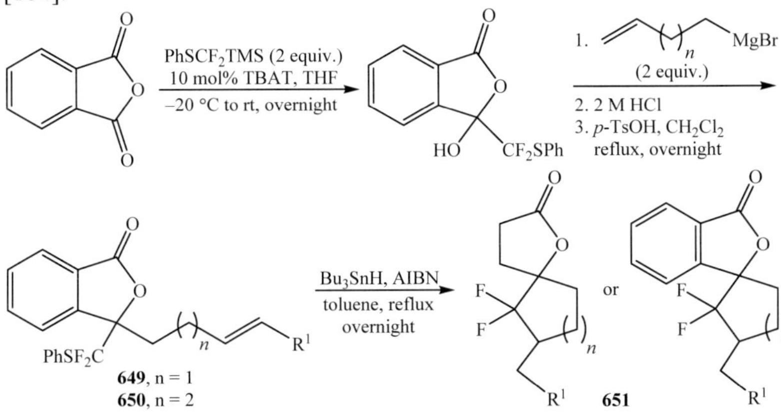

Scheme 3-194. Synthesis of *gem*-difluoromethylenated spiro-γ-butyro lactones **651**.

References

[1] G. Eilerman Robert, J. Willis Brian. 1981. "A new spiro-annelation procedure: Intramolecular decarboxylative alkylation of β-keto-esters". Journal of the Chemical Society, Chemical Communications. No. 1. 30-32.

[2] H. Posner Gary, G. Hamill Terence. 1988. "An asymmetric total synthesis of fragrant spiro[4.5]decane sesquiterpene (-)-beta-vetivone via an enantiomerically pure vinylic sulfoxide". The Journal of Organic Chemistry. No. 26. 6031-6035.

[3] Sakata Yoshiteru, Nakashima Shuichi, Goto Yasushi, Tatemitsu Hitoshi, Misumi Soichi. 1989. "Synthesis of completely fixed porphyrin-quinone compounds and the mutual orientation effect on electron transfer". Journal of the American Chemical Society. No. 24. 8979-8981.

[4] MiJllen Klaus, Lex Johann, C. Schulz Rolf, Walter Frank. 1990. "An ionene with spirane structure (spiroionene)". Polymer Bulletin. No. 3. 263-269.

[5] Kotha Sambasivarao, Manivannan Ethirajan. 2001. "Synthesis of spiro-indanes by cycloaddition strategy". Journal of the Chemical Society, Perkin Transactions 1. No. 20. 2543-2547.

[6] C. Craig Donald, L. Edwards Gavin, J. Sinclair David. 2001. "A new synthesis of spiroketals via sulfonyl-substituted dihydro pyrans". Tetrahedron. No. 3. 563-570.

[7] Singh Palwinder, Paul Kamaldeep. 2006. "A practical approach for spiro- and 5-monoalkylated barbituric acids". Journal of Heterocyclic Chemistry. No. 3. 607-612.

[8] Maulide Nuno, E. Markó István. 2006. "2-(Trimethyl silyloxy) furan as a dianion equivalent: A two-step synthesis of functionalized spirocyclic butenolides". Organic Letters. No. 17. 3705-3707.

[9] Ghorai Prasanta, H. Dussault Patrick, Hu Chunhua. 2008. "Synthesis of spiro-bisperoxyketals". Organic Letters. No. 12. 2401 -2404.

[10] S. Radchenko Dmytro, O. Grygorenko Oleksandr, V. Komarov Igor. 2008. "Synthesis of conformationally restricted glutamic acid analogs based on the spiro[3.3]heptane scaffold". Tetrahedron: Asymmetry. No. 24. 2924-2930.

[11] Ota Koichiro, Hiroaki Miyaoka. 2013. "First total synthesis of (-)-sinularianin B". Chemical Communications. No. 74. 8148-8150.

[12] Basavaiah Deevi, Lingam Harathi, Hari Babu Thelagathoti. 2018. "Baylis-Hillman acetates in organic synthesis: A simple two-step strategy for oxindole-spiro-α-arylidene-γ-butyrolactone framework". Tetrahedron. No. 19. 2306-2313.

[13] Wu Yonggang, Zhang Jianyuan, Fei Zhuping, Bo Zhishan. 2008. "Spiro-bridged ladder-type poly(*p*-phenylene)s: Towards structurally perfect light-emitting materials". Journal of the American Chemical Society. No. 23. 7192-7193.

[14] Bakthadoss Manickam, Sivakumar Nagappan. 2011. "First Friedel-Crafts reaction of the Baylis-Hillman adducts derived from nitroolefins: Application towards synthesis of pyrrolidines and spiropyrrolidines". Synlett. No. 09. 1296-1302.

[15] Bo Ram Park, Se Hee Kim, Yu Mi Kim, Jae Nyoung Kim. 2011. "Synthesis of indenoindene-fused α-methylene-γ-butyrolactones via a tandem intra-and intermolecular Friedel–Crafts reaction". Tetrahedron Letters. No. 14. 1700-1704.

[16] Nemoto Tetsuhiro, Wu Riliga, Zhao Zengduo, Yokosaka Takuya, Hamada Yasumasa. 2013. "Synthesis of spiro[4.5]cyclohexadienones with an allene motif via a base-promoted intramolecular *ipso*-Friedel-Crafts addition of phenols to propargyl bromides". Tetrahedron. No. 16. 3403-3409.

[17] S. Olsson Joel, Huong Pham Thanh, Jannasch Patric. 2020. "Functionalizing polystyrene with N-alicyclic piperidine-based cations via Friedel-Crafts alkylation for highly alkali-stable anion -exchange membranes". Macromolecules. No. 12. 4722-4732.

[18] Palmisano Giovanni, Annunziata Rita, Papeo Gianluca, Sisti Massimo. 1996. "Oxindole alkaloids. A novel non-biomimetic entry to (−)-Horsfiline". Tetrahedron: Asymmetry. No. 1. 1-4.

[19] R. Sebahar Paul, Osada Hiroyuki, Usui Takeo, M. Williams Robert. 2002. "Asymmetric, stereocontrolled total synthesis of (+) and (-)-spirotryprostatin B via a diastereoselective azomethine ylide [1,3]-dipolar cycloaddition reaction". Tetrahedron. No. 32. 6311-6322.

[20] Pardasani Pushpa, T. Pardasani Ram, Sherry Densely, Chaturvedi Vivek. 2002. "Stereoselective synthesis of substituted thiazolidine derivatives". Synthetic Communications. No. 3. 435-44.

[21] T. Pardasani Ram, Pardasani Pushpa, Chaturvedi Vaibhav, Yadav Shashit Kumar, *et al.* 2003. "Theoretical and synthetic approach to novel spiroheterocycles derived from isatin derivatives and L-proline via 1,3-dipolar cycloaddition". Heteroatom Chemistry: An International Journal of Main Group Elements. No. 1. 36-41.

[22] Jayashankaran Jayadevan, R. S. Manian Rathna Durga, Sivaguru Manickasamy, Raghunathan Raghavachary. 2006. "Synthesis of novel spiropyrrolidines through [3+2] cycloaddition reactions with Baylis-Hillman adducts as dipolarophiles". Tetrahedron Letters. No. 31. 5535-5538.

[23] Yavari Issa, Hossaini Zinatossadat, Sabbaghan Maryam, Ghazanfarpour-Darjani Majid. 2007. "Efficient synthesis of functionalized spiro-2,5-dihydro-1,2-λ^5-oxaphospholes". Tetrahedron. No. 38. 9423-9428.

[24] Sridhar Gowri, Gunasundari Thanikachalam, Raghunathan Raghavachary. 2007. "A greener approach for the synthesis of 1-*N*-methyl-(spiro[2.3′]oxindolespiro[3.2″]/spiro[2.3′]indan-1,3-di onespiro[2.2″])cyclopentanone-4-aryl pyrrolidines". Tetrahedron Letters. No. 2. 319-322, 2007.

[25] Kumar Raju Ranjith, Perumal Subbu, Senthilkumar Palaniappan, Yogeeswari Perumal, Sriram Dharmarajan. 2008. "A highly atom economic, chemo-, regio- and stereoselective synthesis, and discovery of spiro-pyrido-pyrrolizines and pyrrolidines as antimycobacterial agents". Tetrahedron. No. 13. 2962-2971.

[26] Vidhya Lakshmi Neelakandan, Thirumurugan Prakasam, T. Perumal Paramasivan. 2010. "An expedient approach for the synthesis of dispiropyrrolidine bisoxindoles, spiropyrrolidine oxindoles and spiroindane-1,3-diones through 1,3-dipolar cycloaddition reactions". Tetrahedron Letters. No. 7. 1064-1068.

[27] Uma Maheswari Shanmugavel, Balamurugan Kamaraj, Perumal Subbu, Yogeeswari Perumal, Sriram Dharmarajan. 2010. "A facile 1,3-dipolar cycloaddition of azomethine ylides to 2-arylidene-1,3-indanediones: Synthesis of dispiro-oxindolyl pyrrolothiazoles and their antimycobacterial evaluation". Bioorganic & Medicinal Chemistry Letters. No. 24. 7278-7282.

[28] Rajesh Raju, Raghavachary Raghunathan. 2010. "Regio- and stereoselective synthesis of novel tetraspiro-bispyrrolidine and bisoxindolopyrrolidines through 1,3-dipolar cycloaddition reaction". Tetrahedron Letter. No. 44. 5845-5848.

[29] Vidhya Lakshmi Neelakandan, Thirumurugan Prakasam, Jayakumar Chinnappan, T. Perumal Paramasivan. 2010. "An easy access to novel spiro-fused pyrrolobenzo[*b*]thiophene 1,1-dioxide derivatives via 1,3-dipolar cycloaddition using benzo[*b*]thiophene 1,1-dioxide". Synlett. No. 06. 955-961.

[30] Bouhfid Rachid, Joly Nicolas, Essassi El Mokhtar, Lequart Vincent, *et al.* 2011. "Synthesis of new spiro[1,4,2-dioxazole-

5,3′-indolin]-2′-one by 1,3-dipolar cycloaddition". Synthetic Communications. No. 14. 2096-2102.

[31] Presset Marc, Mohanan Kishor, Hamann Marie, Coquerel Yoann, Rodriguez Jean. 2011. "1,3-Dipolar cycloaddition of hydrazones with α-oxo-ketenes: A three-component stereoselective entry to pyrazolidinones and an original class of spiro-oxindoles". Organic Letters. No. 15. 4124-4127.

[32] Vidhya Lakshmi Neelakandan, Tamilisai Rajendren, T. Perumal Paramasivan. 2011. "A facile synthetic approach to novel spiro-oxindoles/acenaphthylen-1-ones containing benzo[1,4]thiazin-3-one ring via 1,3-dipolar cycloaddition". Tetrahedron Letters. No. 41. 5301-5307.

[33] Chen Gang, Miao Yan-qing, Zhou Rui, Zhang Li, *et al.* 2013. "Investigation of regioselectivity in the synthesis of spiro-[pyrrolidine-2,3′-oxindoles] by use of the Huisgen reaction". Research on Chemical Intermediates. No. 6. 2445-2450.

[34] Sarrafi Yaghoub, Hamzehloueian Mahshid, Alimohammadi Kamal, Yeganegi Saeid. 2012. "Experimental and theoretical approaches to [1,5]-prototropic generation of an azomethine ylide and a 1,3-dipolar cycloaddition for novel spiropyrrolidine oxindoles synthesis". Journal of Molecular Structure. 168-176.

[35] Taghizadeh Mohammad Javad, Arvinnezhad Hamid, Samadi Saadi, Jadidi Khosrow, *et al.* 2012. "Synthesis of new enantiomerically pure spirooxindolopyrrolizidines via a three-component asymmetric 1,3-dipolar cycloaddition reaction of azomethine ylides derived from isatin". Tetrahedron Letters. No. 38. 5148-5150.

[36] Michael Rajesh Stephen, Devi Bala Balasubramainan, Perumal Subbu. 2012. "Multi-component, 1,3-dipolar cycloaddition reactions for the chemo-, regio-and stereoselective synthesis of novel hybrid spiroheterocycles in ionic liquid". Tetrahedron Letters. No. 40. 5367-5371.

[37] L. Pavlovskaya Tatyana, Red′kin Ruslan Gr, G. Yaremenko Fedor, V. Shishkina Svetlana, *et al.* 2013. "Synthesis and chemical properties of new derivatives of 3a′,6a′-dihydro-2′*H*-spiro-[indole-3,1′-pyrrolo[3,4-*c*]pyrrole]-2,4′,6′(1*H*,3′*H*,5′*H*)-trione". Chemistry of Heterocyclic Compounds. No. 6. 882-896.

[38] Nassoy Anne-Chloé, Raubo Piotr, P. A. Harrity Joseph. 2013. "Synthesis and cycloaddition chemistry of oxetanyl-substituted sydnones". Tetrahedron Letters. No. 24. 3094-3096.

[39] Lanka Srinu, Thennarasu Sathiah, T. Perumal Paramasivan. 2014. “Stoichiometry-controlled cycloaddition of azomethine ylide with dipolarophiles: Chemoselective and regioselective synthesis of bis-and tris-spiro-oxindole derivatives”. Tetrahedron Letters. No. 16. 2585-2588.

[40] P. Bharitkar Yogesh, Das Mohua, Kumari Neha, Kumari M. Padma, *et al.* 2015. “Synthesis of bis-pyrrolizidine-fused dispiro-oxindole analogues of curcumin via one-pot azomethine ylide cycloaddition: Experimental and computational approach toward regio-and diastereoselection”. Organic Letters. No. 18. 4440-4443.

[41] Naga Siva Rao Jonnalagadda, Raghunathan Raghavachary. 2015. “A tactical 1,3-dipolar cycloaddition approach for the synthesis of carbohydrate derived polycyclic spiro heterocycles”. Tetrahedron Letters. No. 12. 1539-1544.

[42] Dandia Anshu, Parewa Vijay, Kumari Sukhbeer, Bansal Sarika, Sharma Amit. 2016. “Imposed hydrophobic interactions by NaCl: Accountable attribute for the synthesis of spiro[acenaphthylene-1,5′-pyrrolo[1,2-*c*]thiazole] derivatives via 1,3-dipolar cyclo-addition reaction in aqueous medium”. Green Chemistry. No. 8. 2488-2499.

[43] I. Almansour Abdulrahman, Suresh Kumar Raju, Arumugam Natarajan, Basiri Alireza, *et al.* 2015. “A facile ionic liquid promoted synthesis, cholinesterase inhibitory activity and molecular modeling study of novel highly functionalized spiropyrrolidines”. Molecules. No. 2. 2296-2309.

[44] Taghizadeh Mohammad Javad, Javidan Abdollah, Keshipour Sajjad. 2015. “Synthesis of new bis-spirooxindolo pyrrolizidine (pyrrolidine) derivatives via a pseudo five-component 1,3-dipolar cycloaddition reaction”. Chemistry of Heterocyclic Compounds. No. 5. 467-471.

[45] Sapnakumari Majal, Narayana Badiadka, Shashidhara Ks-Kn, Balladka Sarojini. 2017. “Multicomponent synthesis, biological evaluation and molecular docking of new spiro-oxindole derivatives”. Journal of Taibah University for Science. No. 6. 1008-1018.

[46] Hu Wei-Qiang, Cui Yan-Su, Wu Zhi-Jun, Zhang Chuan-Bao, et al. 2015. “Construction of spirocycles containing highly substituted pyrroli-dine and 1-indanone motifs with spiro quaternary stereogenic centers via 1,3-dipolar cycloaddition of 2-

alkylidene-1-indanone and azomethine ylides promoted by simple imidazolium salts". RSC Advances. No. 87. 70910-70914.
[47] Alizadeh Abdolali, Moafi Leila. 2016. "A convenient 1,3-dipolar cycloaddition reaction for the synthesis of spiro-oxindoles and some other spirocompounds containing the 1,3,4-oxadiazole moiety". Helvetica Chimica Acta. No. 6. 457-461.
[48] S. Filatov Alexander, A. Knyazev Nickolay, P. Molchanov Alexander, L. Panikorovsky Taras, *et al.* 2017. "Synthesis of functionalized 3-spiro[cyclopropa[*a*]pyrrolizine] and 3-spiro[3-azabicyclo[3.1.0]hexane]oxindoles from cyclopropenes and azomethine ylides via [3+2]-cycloaddition". The Journal of Organic Chemistry. No. 2. 959-975.
[49] Duan Jindian, Cheng Jing, Cheng Yuyu, Li Pengfei. 2016. "Synthesis of dinitrogen-fused spirocyclic heterocycles via organocatalytic 1,3-dipolar cycloaddition of 2-arylidene-1,3-indandiones and an azomethine imine". Asian Journal of Organic Chemistry. No. 4. 477-480.
[50] Zhu Guodong, Liu Siyuan, Wu Shiqi, Peng Lianghong, *et al.* 2017. "Assembly of indolenines, 3-amino-oxindoles, and aldehydes into indolenine-substituted spiro[pyrrolidin-2,3′-oxindoles] via 1,3-dipolar cycloaddition with divergent diastereoselectivities". The Journal of Organic Chemistry. No. 8. 4317-4327.
[51] Mali Prakash Rambhau, Shirsat Prashishkumar Kisan, Khomane Navnath Bapurao, Nayak Lakshama, *et al.* 2017. "1,3-Dipolar cycloaddition reactions for the synthesis of novel oxindole derivatives and their cytotoxic properties". ACS Combinatorial Science. No. 10. 633-639.
[52] Barakat Assem, M. Soliman Saied, Mohammed Al-Majid Abdullah, Shahidul Islam Mohammad, *et al.* 2018. "New spiro-oxindole constructed with pyrrolidine/thioxothiazolidin-4-one derivatives: Regioselective synthesis, x-ray crystal structures, Hirshfeld surface analysis, DFT, docking and antimicrobial studies". Journal of Molecular Structure. 101-114.
[53] Khodabakhshi Mohammad Reza, Matloubi Moghaddam Firouz, Kiamehr Mostafa. 2018. "PTSA Catalyzed synthesis of functionalized dipyrazolo[1,5]diazocine-3,8-diones via the reaction of 3-aminopyrazolone with benzaldehyde derivatives". Tetrahedron Letters. No. 51. 4503-4508.
[54] Narayanarao Manjunatha, Koodlur Lokesh, Gopal Subramanya, Y. Reddy Suman, Kamila Susmita. 2018. "one-pot three-

component synthesis of fused spiro-indoline/indene derivatives derived from ethynyl azaindole by 1,3-dipolar cycloaddition reaction". Synthetic Communications. No. 18. 2441-2451.

[55] Ryu Heesun, Seo Jeongseob, Min Ko Haye. 2018. "Synthesis of spiro[oxindole-3,2′-pyrrolidine] derivatives from benzynes and azomethine ylides through 1,3-dipolar cycloaddition reactions". The Journal of Organic Chemistry. No. 22. 14102-14109.

[56] S. Filatov Alexander, A. Knyazev Nickolay, N. Ryazantsev Mikhail, Natalia Soldatova, V. Suslonov Vitalii, *et al.* 2018. "A highly diastereoselective one-pot three-component 1,3-dipolar cycloaddition of cyclopropenes with azomethine ylides generated from 11*H*-indeno[1,2-*b*]-quinoxalin-11-ones". Organic Chemistry Frontiers. No. 4. 595-605.

[57] Cheng Bin, Li Yuntong, Zu Bing, Wang Taimin, *et al.* 2019. "Syntheses of spiro[indazole-3,3′-indolin]-2′-ones and spiro [indazole-3,3′-indolin]-2′-imines via 1,3-dipolar cycloadditions of arynes and studies on their isomerization reactions". Tetrahedron. No. 52. 130775.

[58] Alizadeh Abdolali, Roosta Atefeh, Halvagar Mohammadreza. 2019. "Four-component regio-and diastereoselective synthesis of pyrrolizidines incorporating spiro-oxindole/indanedione via 1,3-dipolar cycloaddition reaction of azomethine ylides". ChemistrySelect. No. 1. 71-74.

[59] Periyasami Govindasami, Kamalraj Subban, Padmanaban Ramanathan, Yeswanth Kumar Santhakumar, *et al.* 2019. "Glucosamine-6-phosphate synthase inhibiting C3-β-cholesterol tethered spiro heterocyclic conjugates: Synthesis and their insight of DFT and docking study". Bioorganic Chemistry. 102920.

[60] Angyal Anikó, Demjén Andras, Harmat Veronika, Wölfling Janos, *et al.* 2019. "1,3-Dipolar cycloaddition of isatin-derived azomethine ylides with 2*H*-azirines: Stereoselective synthesis of 1,3-diazaspiro [bicyclo[3.1.0]hexane]oxindoles". The Journal of Organic Chemistry. No. 7. 4273-4281.

[61] Sharma Pallavi, Prabhath M. R. Ranga, Wong Derek, Ampem-Lassen Maame Adjoa, *et al.* 2020. "Synthesis of biologically active heterospirocycles through iterative 1,3-dipolar cyclo-addition pathways". The Journal of Organic Chemistry. No. 1. 1223-1230.

[62] Xu Ju-Hua, Li Yu-Liang, Zheng Da-Gui, Yang Jing-Kui, *et al.* 1997. "An unexpected reaction: [2+1] Cycloaddition of [60]

fullerene with 4,4,5,5-tetramethylimidazolidine-2-thione and DL-valine". Tetrahedron Letters. No. 37. 6613-6616.
[63] Yuan Zhenbo, Fang Xinxin, Li Xuanyi, Wu Jie, *et al.* 2015. "1,6-Conjugated addition-mediated [2+1] annulation: Approach to spiro[2.5]octa-4,7-dien-6-one". The Journal of Organic Chemistry. No. 21. 11123-11130.
[64] Yang Kai-Chuan, Li Jun-Long, Shen Xu-Dong, Li Qiang, *et al.* 2017. "Facile synthesis of novel spiroheterocycles via diastereo-selective aziridination of cyclic enones". RSC Advances. No. 34. 21175-21179.
[65] Roy Suparna, Piradhi Venkatesh. 2017. "Formal [2+1] annulation (spiro-cyclopropanation) reaction between sulfur ylides derived from Baylis-Hillman bromides and arylidene indane-1,3-dione". ChemistrySelect. No. 21. 6159-6162.
[66] Tang Xue, Zhu Hong-Ping, Zhou Jin, Chen Yang, *et al.* 2018. "Highly diastereoselective synthesis of cyclopropane-fused spiro-pseudoindoxyl derivatives through [2+1] annulation of 2-ylidene-oxindoles and sulfonium bromides". Organic & Biomolecular Chemistry. No. 43. 8169-8174.
[67] B. Kale Someshwar, K. Jori Popat, Thatikonda Thanusha, G. Gonnade Rajesh, Das Utpal. 2019. "1,6-Conjugate-addition-induced [2+1] annulation of *para*-quinone methides and pyrazolones: Synthesis of bis-spiro compounds with contiguous quaternary spiro-centers". Organic Letters. No. 19. 7736-7740.
[68] R. Koft Emil, B. Smith Amos. 1984. Intramolecular [2+2] photochemical cycloadditions. 3. Perhydrohistrionicotoxin synthetic studies: Synthesis of spiro[4.5]decanones via intramolecular [2+2] photocycloaddition. The Journal of Organic Chemistry. No. 5. 832-836.
[69] A. Kraus George, Shi Jianmin, Reynolds Don. 1990. "A reinvestigation of the photochemistry of 2-alkoxy-1,4-naphtho quinones". Synthetic Communications. No. 12. 1837-1841.
[70] T. Brady William, G. Liddell Harold, L. Vaughn Walter. 1966. "Halogenated ketenes. I. dichloroketene[1,2]". The Journal of Organic Chemistry. No. 2. 626-628.
[71] E. Baldwin John, E. Keene Maryann, Shukla Rajesh. 2000. "Syntheses of dispiro[2.1.4.1]decane and dispiro[2.0.3.3]decane". Tetrahedron Letters. No. 49. 9441-9443.
[72] Keglevich György, Forintos Henrietta, Miklós Keserű György, Hegedűs László, Tőke László. 2000. "Synthesis of the spiro derivatives of 1,2-oxaphosphetes by [2+2] cycloaddition of cyclic

1-(2,4,6-triisopropylphenyl)phosphine oxides with dimethyl acetylenedicarboxylate". Tetrahedron. No. 27. 4823-4828.

[73] Wendling Florence, Michel MieScheme 2001. "Unexpected reactivity of acetylenic ω-ketoesters toward TBAF and *t*-BuOK; New cascade reactions affording allene and oxetane derivatives". Organic Letters. No. 17. 2689-2691.

[74] Alonso Eduardo, López-Ortiz Fernando, del Pozo Carlos, Peralta Emma, *et al.* 2001. "Spiro-β-lactams as β-turn mimetics. Design, synthesis, and NMR conformational analysis". The Journal of Organic Chemistry. No. 19. 6333-6338.

[75] Kaneko Toshihiko, S. J. Clark Richard, Ohi Norihito, Ozaki Fumihiro, *et al.* 2004. "Piperidine carboxylic acid derivatives of 10*H*-pyrazino[2,3-*b*][1,4]benzothiazine as orally-active adhesion molecule inhibitors". Chemical and Pharmaceutical Bulletin. No. 6. 675-687.

[76] Macías Alberto, Alonso Eduardo, Del Pozo Carlos, Venturini Alessandro, González Javier. 2004. "Diastereoselective [2+2] cyclo-addition reactions of unsymmetrical cyclic ketenes with imines: Synthesis of modified prolines and theoretical study of the reaction mechanism". The Journal of Organic Chemistry. No. 21. 7004-7012.

[77] Robertson Jeremy, J. Bell Stephen, Krivokapic Alexander. 2005. "On the reaction of diphenylketene with isocyanides". Organic & Biomolecular Chemistry. No. 23. 4246-4251.

[78] Quiroz-García Beatriz, Figueroa Ricardo, Cogordan J. Antonio, Delgado Guillermo. 2005. "Photo-cyclodimers from *Z*-ligustilide. Experimental results and FMO analysis". Tetrahedron Letters. No. 17. 3003-3006.

[79] Doroh Brandon, A. Sulikowski Gary. 2006. "Progress toward the total synthesis of bielschowskysin: A stereoselective [2+2] photocycloaddition". Organic Letters. No. 5. 903-906.

[80] Meyer-Wilmes Imelda, Gerke Ralf, Fitjer Lutz. 2009. "Helical primary structures of 1, 3-spiroannelated five-membered rings: (±)-Trispiro[4.1.1.4.2.2]heptadecane and (±)-tetraspiro[4.1.1.1.4.2.2.2]heneicosane". Tetrahedron. No. 8. 1689-1696.

[81] J. Meyers Marvin, Muizebelt Inouk, v. Wiltenburg Jim, L. Brown David, Thorarensen Atli. 2009. "Synthesis of *tert*-butyl 6-oxo-2-azaspiro[3.3]heptane-2-carboxylate". Organic Letters. No. 16. 3523-3525.

[82] Matsuo Jun-ichi, Kawano Mizuki, Okuno Ryosuke, Ishibashi Hiroyuki. 2010. "Titanium(IV) chloride-mediated carbon−carbon

bond-forming reaction between 3,3-dialkylcyclobutanones and aldehydes". Organic Letters. No. 17. 3960-3962.

[83] Ameri Rad Javad, Jarrahpour Aliasghar, Cüneyt Ersanl Cem, Atioğlu Zeliha, *et al.* 2017. "Synthesis of some novel indeno[1,2-*b*]quinoxalin spiro-β-lactam conjugates". Tetrahedron. No. 8. 1135-1142.

[84] Xue Jianfei, Takagi Ryukichi, Abe Manabu. 2018. "Photochemical [2+2] cycloaddition reaction of enone derivatives with 2-siloxy-1*H*-pyrrole derivatives". Arkivoc. 192-204.

[85] Xie Junjian, Pan Lun, Nie Genkuo, Xie Jiawei, *et al.* 2019. "Photoinduced cycloaddition of biomass derivatives to obtain high-performance spiro-fuel". Green Chemistry. No. 21.5886-5895.

[86] Amal Raj Alex, R. Raghunathan Raghavachary. 2003. "Synthesis of spiro pyrrolidines via formal [3+2] cycloaddition of unusual enones and *cis*-3-benzoyl-1-cyclohexyl-2-phenylaziridine". Tetrahedron. No. 16. 2907-2911.

[87] Jayashankaran Jayadevan, Durga R. S. Manian Rathna, Raghunathan Raghavachary. 2004. "A facile synthesis of novel dispiroheterocycles through solvent-free microwave-assisted [3+2] cycloaddition of azomethine ylides". Tetrahedron Letters. No. 39. 7303-7305.

[88] Li Jia-Qi, Liao Rong-Zhen, Ding Wan-Jian, Cheng Ying. 2007. "Highly efficient and site-selective [3+2] cycloaddition of carbene-derived ambident dipoles with ketenes for a straightforward synthesis of spiro-pyrrolidones". The Journal of Organic Chemistry. No. 16. 6266-6269.

[89] Savitha Govardhan, K. Niveditha Krishna, Muralidharan Dhanya, T. Perumal Paramasivan. 2007. "An efficient one-pot synthesis of spiro dihydrofuran oxindole and spiro-2-hydroxytetrahydro furanoxindole derivatives via (3+2) oxidative cycloaddition mediated by CAN". Tetrahedron Letters. No. 16. 2943-2947.

[90] Shanmugam Ponnusamy, Viswambharan Baby, Madhavan Suchithra. 2007. "Synthesis of novel functionalized 3-spiro pyrrolizidine and 3-spiropyrrolidineoxindoles from Baylis–Hillman adducts of isatin and heteroaldehydes with azomethine ylides via [3+2] cycloaddition". Organic Letters. No. 21. 4095-4098.

[91] Thangamani Arumugam. 2010. "Regiospecific synthesis and biological evaluation of spirooxindolo-pyrrolizidines via [3+2]

cycloaddition of azomethine ylide". European Journal of Medicinal Chemistry. No. 12. 6120-6126.
[92] Suresh Babu Adukamparai Rajukrishnan, Raghunathan Raghavachary, Baskaran Sundari. 2009. "An expedient synthesis of ferrocene grafted spirooxindolopyrrolizidines via [3+2]-cycloaddition of azomethine ylides". Tetrahedron. No. 11. 2239-2243.
[93] Alimohammadi Kamal, Sarrafi Yaghoub, Tajbakhsh Mahmood, Yeganegi Saeed, Hamzehloueian Mahshid. 2011. "An experimental and theoretical investigation of the regio-and stereoselectivity of the polar [3+2] cycloaddition of azomethine ylides to nitrostyrenes". Tetrahedron. No. 8. 1589-1597.
[94] Wang Lihong, Li Zhanshan, Lu Lianhong, Zhang Wei. 2012. "Synthesis of spiro[furan-3,3′-indolin]-2′-ones by PET-catalyzed [3+2] reactions of spiro[indoline-3,2′-oxiran]-2-ones with electron-rich olefins". Tetrahedron. No. 5. 1483-1491.
[95] Suresh Babu Adukamparai Rajukrishnan, Gavaskar Dasari Sunil, Raghunathan Raghavachary. 2012. "A facile synthesis of novel ferrocene grafted spiro-indenoquinoxaline pyrrolizidines via one-pot multicomponent [3+2] cycloaddition of azomethine ylides". Tetrahedron Letters. No. 49. 6676-6681.
[96] Chee Wei Ang, Ashraf Ali Mohamed, Keng Yoon Yeong, Ismail Rusli, *et al.* 2013. "A facile three-component [3+2]-cycloaddition for the regioselective synthesis of highly functionalised dispiro pyrrolidines acting as antimycobacterial agents". Bioorganic & Medicinal Chemistry Letters. No. 5. 1383-1386.
[97] Shah Sakshi, Badru Rahul, Singh Baldev. 2013. "Diastereo-selective synthesis of novel spiro-isoxazolidines via [3+2] cycloaddition". Synthetic Communications. No. 7. 1073-1082.
[98] Tan Wei, Zhu Xiao-Tong, Zhang Shu, Xing Gui-Juan, *et al.* 2013. "Diversity-oriented synthesis of spiro-oxindole-based 2,5-dihydropyrroles via three-component cycloadditions and evaluation on their cytotoxicity". RSC Advances. No. 27. 10875-10886.
[99] V. Suneel Kumar Chepuri, V. Ramana Chepuri. 2014. "Tuning the regioselectivity of gold-catalyzed internal nitroalkyne redox: A cycloisomerization and [3+2]-cycloaddition cascade for the construction of spiro-pseudoindoxyl skeleton". Organic Letters. No. 18. 4766-4769.
[100] Lanka Srinu, Thennarasu Sathiah, T. Perumal Paramasivan. 2014. "Synthesis of acridinedione derived mono spiro-pyrrolidine/

pyrrolizidines-A facile approach via intermolecular [3+2] cycloaddition reaction". Tetrahedron Letters. No. 30. 4047-4053.
[101] Naga Siva Rao Jonnalagadda, Raghunathan Raghavachary. 2015. "A facile synthesis of glyco 3-nitrochromane hybrid pyrrolidinyl spiro heterocycles via [3+2] cycloaddition of azomethine ylides". Tetrahedron Letters. No. 17. 2276-2279.
[102] Malathi Karuppiah, Kanchithalaivan Selvaraj, Ranjith Kumar Raju, I. Almansour Abdulrahman. *et al.* 2015. "Multicomponent [3+2] cycloaddition strategy: Stereoselective synthesis of novel polycyclic cage-like systems and dispiro compounds". Tetrahedron Letters. No. 44. 6132-6135.
[103] Shi Rong-Guo, Sun Jing, Yan Chao-Guo. 2017. "Tandem double [3+2] cycloaddition reactions at both C-1 and C-3 atoms of *N*-cyanomethylisoquinolinium ylide". ACS Omega. No. 11. 7820-7830.
[104] Shao Pan-Lin, Li Zi-Rui, Wang Zhi-Peng, Zhou Ming-Hui, *et al.* 2017. "[3+2] Cycloaddition of azaoxyallyl cations with cyclic ketones: Access to spiro-4-oxazolidinones". The Journal of Organic Chemistry. No. 19. 10680-10686.
[105] R. Mali Prakash, B. Khomane Navnath, Sridhar Banavar, M. Meshram Harshadas, R. Likhar Pravin. 2018. "Synthesis of new spiro pyrrole/pyrrolizine/thiazole derivatives via (3+2) cycloaddition reactions". New Journal of Chemistry. No. 16. 13819-13827.
[106] Gui Hou-Ze, Gao Yu-Ning, Wei Yin, Shi Min. 2018. "Highly efficient and diastereoselective construction of trifluoro methyl-containing spiro[pyrrolidin-3,2′-oxindole] by a catalyst-free mutually activated [3+2] cycloaddition reaction". Chemistry–A European Journal. No. 40. 10038-10043.
[107] Vasudevan Sumesh Remani, Shylaja Adaikalam, Ranjith Kumar Raju, I. Almansour Abdulrahman, Suresh Kumar Raju. 2018. "Synthesis of spiro-linked quinolinone-pyrrolidine/pyrrolo[1,2-*c*]thiazole-oxindole/acenaphthalene hybrids via multi-component [3+2] cycloaddition". Tetrahedron Letters. No. 46. 4086-4089.
[108] Liu Dan, Sun Jing, Zhang Yu, Yan Chao-Guo. 2019. "Diastereoselective synthesis of spirocyclic isoxazolo[5,4-*c*]pyrrolo [2,1-*a*]isoquinolines via cascade double [3+2] cycloadditions". Organic & Biomolecular Chemistry. No. 34. 8008-8013.
[109] Su Yingpeng, Zhao Yanan, Chang Bingbing, Zhao Xiaolong, *et al.* 2019. "[3+2] Cycloaddition of *para*-quinone methides with

nitrile imines: Approach to spiro-pyrazoline-cyclohexadienones". The Journal of Organic Chemistry. No. 11. 6719-6728.

[110] Huang Ying, Huang Yi-Xin, Sun Jing, Yan Chao-Guo. 2019. "A [3+2] cycloaddition reaction for the synthesis of spiro[indoline-3,3′-pyrrolidines] and evaluation of cytotoxicity towards cancer cells". New Journal of Chemistry. No. 23. 8903-8910.

[111] Huang Ying, Fang Hui-Lin, Huang Yi-Xin, Sun Jing, Yan Chao-Guo. 2019. "Synthesis of 7′-arylidenespiro[indoline-3,1′-pyrrolizines] and 7′-Arylidenespiro[indene-2,1′-pyrrolizines] via [3+2] cycloaddition and β-C–H functionalized pyrrolidine". The Journal of Organic Chemistry. No. 19. 12437-12451.

[112] B. Kutyashev Igor, V. Ulitko Maria, Yu Barkov Alexey, S. Zimnitskiy Nikolay, *et al.* 2019. "A regio-and stereocontrolled approach to the synthesis of 4-CF_3-substituted spiro[chromeno [3,4-*c*]pyrrolidine-oxindoles] via reversible [3+2] cycloaddition of azomethine ylides generated from isatins and sarcosine to 3-nitro-2-(trifluoromethyl)-2*H*-chromenes". New Journal of Chemistry. No. 47. 18495-18504.

[113] Palomba Martina, De Monte Emanuela, Mambrini Andrea, Bagnoli Luana, *et al.* 2021. "A three-component [3+2]-cycloaddition/elimination cascade for the synthesis of spiro-oxindole-pyrrolizines". Organic & Biomolecular Chemistry. No. 3. 667-676.

[114] Hajra Saumen, Biswas Anurag. 2020. "Catalyst-free stereo controlled formal [3+2] cycloaddition of CO_2 for the synthesis of enantiopure spiro[indoline-3,5′-oxazolidine]-2,2′-diones under aqueous and ambient conditions". Organic Letters. No. 13. 4990-4994.

[115] Pan Liu-Na, Sun Jing, Shi Rong-Guo, Yan Chao-Guo. 2020. "Diastereoselective synthesis of dispiro[indoline-3,3′-furan-2′,3″-pyrrolidine] via [3+2] cycloaddition reaction of MBH maleimides of isatins and 1,3-dicarbonyl compounds". Organic Chemistry Frontiers. No. 20. 3202-3208.

[116] Su Yingpeng, Ma Chenglong, Zhao Yanan, Yang Caixia, *et al.* 2020. "Regioselective synthesis of spiro naphthofuranone-pyrazoline via a [3+2] cycloaddition of benzoaurones with nitrile imines". Tetrahedron. No. 31-32. 131355.

[117] Siva Sankara Chenikkayala, N. N. Namboothiri Irishi. 2021. "Regio-and stereoselective synthesis of dispiro-bisoxindoles via [3+2] annulation involving nitroisatylidene as a vinylogous Michael donor". Organic Letters. No. 12. 4618-4623.

[118] Diels von Otto, Alder Kurt. 1928. "Synthesen in der hydroaromatischen Reihe". Justus Liebigs Annalen der Chemie. No. 1. 98-122.
[119] Huang Yong, H. Rawal Viresh, 2002. "Hydrogen-bond-promoted hetero-Diels−Alder reactions of unactivated ketones". Journal of the American Chemical Society. No. 33. 9662-9663.
[120] S. Kende Andrew, Fan Junfa, Chen Zecheng. 2003. "A concise total synthesis of (±)-Alantrypinone by a novel hetero-Diels−Alder reaction". Organic Letters. No. 18. 3205-3208.
[121] Lin Weimin, G. Virga Kristopher, Kim Kyung-Hee, Zajicek Jaroslav, *et al.* 2009. "Diastereoselective synthesis of a spiro-noraristeromycin using an acylnitroso Diels−Alder reaction". The Journal of Organic Chemistry. No. 16. 5941-5946.
[122] Blanc Romain, Heran Virginie, Rahmani Raphaël, Commeiras Laurent, Parrain Jean-Luc. 2010. "Diels–Alder cycloaddition of *o*-quinonedimethides and alkylidene-5*H*-furan-2-ones: New and rapid access to lambertellol cores and arthrinone derivatives". Organic & Biomolecular Chemistry. No. 24. 5490-5494.
[123] A. Kuttruff Christian, Zipse Hendrik, Trauner Dirk. 2011. "Concise total syntheses of variecolortides A and B through an unusual hetero-Diels–Alder reaction". Angewandte Chemie International Edition, . No. 6. 1402-1405.
[124] Liang Huan, A. Ciufolini Marco. 2010. "Tandem phenolic oxidative amidation−intramolecular Diels−Alder reaction: An approach to the himandrine core". Organic Letters. No. 8. 1760-1763.
[125] Matloubi Moghaddam Firouz, Hojabri Leila, Taheri Salman, Pirani Parisa. 2010. "A tandem Aldol-Diels-Alder reaction accelerated in water: An approach to the catalyst-free one-pot synthesis of spirothio-oxindoles". Journal of the Iranian Chemical Society. No. 3. 781-790.
[126] Presset Marc, Coquerel Yoann, Rodriguez Jean. 2010. "Periselectivity switch of acylketenes in cycloadditions with 1-azadienes: Microwave-assisted diastereoselective domino three-component synthesis of α-spiro-δ-lactams". Organic Letters. No. 18. 4212-4215.
[127] Ji Cheng. Lin Weimin, C. Moraski Garrett, A. Thanassi Jane, *et al.* 2012. "Syntheses and biological studies of novel spiropiperazinyl oxazolidinone antibacterial agents using a spirocyclic diene derived acylnitroso Diels−Alder reaction". Bioorganic & Medicinal Chemistry. No. 11. 3422-3428.

[128] Zhou Ming-Qiang, Zhao Jian-Qiang, You Yong, Xu Xiao-Ying, *et al.* 2015. "DMAP-Catalyzed Diels–Alder reaction of 3-hydroxy-2-pyrone and methyleneindolinones for the synthesis of spirocyclic oxindoles". Tetrahedron. No. 23. 3903-3908.

[129] Wu Lin, Luo Jun, Zhang Yalong, Zhu Mengdi, *et al.* 2015. "Isolation and biomimetic synthesis of (±)-calliviminones A and B, two novel Diels–Alder adducts, from Callistemon viminalis". Tetrahedron Letters. No. 1. 229-232.

[130] Zelisko Nataliya, Karpenko Olexandr, Muzychenko Volodymyr, Gzella Andrzej, *et al.* 2017. "*trans*-Aconitic acid-based hetero-Diels-Alder reaction in the synthesis of thiopyrano[2,3-*d*][1,3]thiazole derivatives". Tetrahedron Letters. No. 18. 1751-1754.

[131] Yang Ren-Yin, Sun Jing, Yan Chao-Guo. 2018. "HOAc-Mediated domino Diels–Alder reaction for synthesis of spiro[cyclohexane-1,3′-indolines] in ionic liquid [Bmim]Br". ACS Omega. No. 5. 5406-5416.

[132] Panda Pravati, Nayak Sabita, Ku Sahoo Susanta, Mohapatra Seetaram, *et al.* 2018. "Diastereoselective synthesis of novel spiro indanone fused pyrano[3,2-*c*]chromene derivatives following hetero-Diels–Alder reaction and in vitro anticancer studies". RSC Advances. No. 30. 16802-16814.

[133] Yang Wen-Juan, Sun Qiu, Sun Jing, Yan Chao-Guo. 2018. "Domino aza/oxa-hetero-Diels–Alder reaction for construction of novel spiro[pyrido[3′,2′:5,6]pyrano[2,3-*d*]pyrimidine-7,5′-pyrimidine]". Organic Chemistry Frontiers. No. 18. 2754-2758.

[134] Babu Gudise Veera, Chandrasekhar Settipalli Poorna, Koti Reddy Eeda, Anwar Shaik. 2019. "Oxa-Michael–Michael reaction of MBH alcohol and 2-arylidene-1,3-indanedione: Regioselective formal [4+2] cycloaddition towards tetrahydrospiropyran scaffolds". European Journal of Organic Chemistry. No. 12. 2234-2242.

[135] Sun Jing, Yang Ren-Yin, Zhan Shao-Cong, Yan Chao-Guo. 2019. "Construction of tetrahydrospiro[carbazole-1,2′-indenes] and dihydrospiro[carbazole-1,3′-indolines] via NH_4I promoted three-component reaction". ChemistrySelect. No. 34. 10100-10103.

[136] Pan Lun, Xie Junjian, Nie Genkuo, Li Zheng, *et al.* 2020. "Zeolite catalytic synthesis of high-performance jet-fuel-range spiro-fuel by one-pot Mannich–Diels–Alder reaction". AIChE Journal. No. 1. e16789.

[137] Liu Dan, Cao Jun, Sun Jing, Liu Ru-Zhang, Yan Chao-Guo. 2020. "Convergent synthesis of triindanone-fused spiro[bicyclo [2.2.2]octane-2,3′-indolines] via domino reaction of 1,3-indanedione and 3-methyleneoxindoles". Organic Letters. No. 22. 8931-8936.

[138] E. Zhulanov Vladimir, A. Vigovskaya Valeria, V. Dmitriev Maksim, S. Silaichev Pavel, *et al.* 2020. "Dipyrazolodioxa diazocines as shelf-stable "ready-to-use" precursors for an in situ generation of enolate-iminium 1,4-dipoles: A straightforward atom-economical approach to pyrazolo[5,1-*d*][1,3,5]dioxazines". Organic & Biomolecular Chemistry. No. 17. 3382-3391.

[139] Yan Chen, Sun Jing, Yan Chao-Guo. 2021. "Convenient construction of spiro[indoline-3,5′-pyrrolo[3,4-*c*]carbazole] and spiro[indene-2,5′-pyrrolo[3,4-*c*]carbazole] via acid-catalyzed Diels-Alder reaction". Chinese Chemical Letters. No. 3. 1253-1256.

[140] Shi Wenjing, Sun Jing, Yan Chao-Guo. 2021. "Aza-Diels-Alder reaction of both electron-deficient azoalkenes with electron-deficient 3-phencaylideneoxindoles and 3-aryliminooxindol-2-ones". Green Synthesis and Catalysis. No. 4. 362-366.

[141] Zuo Wei-Fang, Zhou Jin, Wu Yu-Ling, Fang Hua-Ying, *et al.* 2021. "Synthesis of spiro(indoline-2,3′-hydropyridazine) via an "on-water" [4+2] annulation reaction". Organic Chemistry Frontiers. No. 5. 922-927.

[142] Krishna Yarkali, Tanaka Fujie. 2021. "Intramolecular formal [4+2] cycloadditions: Synthesis of spiro isoindolinone derivatives and related molecules". Organic Letters. No. 5. 1874-1879.

[143] Zhan Shao-Cong, Fang Ren-Jie, Yang Ren-Yin, Zhao Ru-Fang, *et al.* 2021. "DDQ Dehydrogenative Diels–Alder reaction for the synthesis of functionalized spiro[carbazole-1,3′-indolines] and spiro[carbazole-1,5′-pyrimidines]". New Journal of Chemistry. No. 34. 15423-15428.

[144] Sanjeeva Kumar Avvari, Chauhan Sachin, Swamy K. Kumara. 2021. "Contrasting carboannulation involving δ-acetoxy allenoate as a four-carbon synthon using DABCO and DMAP: Access to spiro-carbocyclic and *m*-teraryl scaffolds". Organic Letters. No. 3. 1123-1129.

[145] Leng Hai-Jun, Li Qing-Zhu, Xiang Peng, Qi Ting, *et al.* 2021. "Diastereoselective [3+1] cyclization reaction of oxindolyl azaoxyallyl cations with sulfur ylides: Assembly of 3,3′-spiro[β-lactam]oxindoles". Organic Letters. No. 4. 1451-1456.

[146] Hayrapetyan Davit, Stepanova Valeriya. 2021. “Facile synthesis of spirocyclic lactams via [3+2] and [3+3] aza-annulation reactions”. European Journal of Organic Chemistry. No. 14. 2121-2125.
[147] Huang Ji-Rong, Sohail Muhammad, Taniguchi Tohru, Monde Kenji, Tanaka Fujie. 2017. “Formal (4+1) cycloaddition and enantioselective Michael–Henry cascade reactions to synthesize spiro[4,5]decanes and spiro-oxindole polycycles”. Angewandte Chemie. No. 21. 5947-5951.
[148] Basu Pallabita, Satam Nishikant, N. N. Namboothiri Irishi. 2020. “Synthesis of indenofurans, benzofurans and spirolactones via Hauser–Kraus annulation involving 1,6-addition of phthalide to quinone methides”. Organic & Biomolecular Chemistry. No. 29. 5677-5687.
[149] Kumar Tarun, Satam Nishikant, N. N. Namboothiri Irishi. 2016. “Hauser–Kraus annulation of phthalides with nitroalkenes for the synthesis of fused and spiro heterocycles”. European Journal of Organic Chemistry. No. 20. 3316-3321.
[150] Suresh Alati, V. Baiju Thekke, Kumar Tarun, N. N. Namboothiri Irishi. 2019. “Synthesis of spiro- and fused heterocycles via (4+4) annulation of sulfonylphthalide with *o*-hydroxystyrenyls”. The Journal of Organic Chemistry. No. 6. 3158-3168.
[151] A. Mukhina Olga, M. Kuznetsov Dmitry, M. Cowger Teresa, G. Kutateladze Andrei. 2015. “Amino azaxylylenes photogenerated from *o*-amido imines: Photoassisted access to complex spiro-poly-heterocycles”. Angewandte Chemie International Edition. No. 39. 11516-11520.
[152] Li Jian, Liu Yuejin, Li Chunju, Jie Haohua, Jia Xueshun. 2012. “Atom-economical synthesis of the functionalized spirocyclic oxindole-butenolide via three-component [2+2+1] cycloaddition strategy”. Green Chemistry. No. 5. 1314-1321.
[153] Liu Lantao, Wang Zhihui, Zhao Fei, Xi Zhenfeng. 2007. “Dearomatizing anionic cyclization and novel skeletal rearrangement: High yield formation of multiply substituted bicyclic or polycyclic spirocyclopentadienes and phenanthrene derivatives from 4-aryl-1-lithio-1,3-butadienes”. The Journal of Organic Chemistry. No. 9. 3484-3491.
[154] Wang Peng-Fei, Jiang Chun-Huan, Wen Xiaoan, Xu Qing-Long, Sun Hongbin. 2015. “C–H Bond functionalization via [1,5]-hydride shift/cyclization sequence: Approach to spiro indolenines”. The Journal of Organic Chemistry. No. 2. 1155-1162.

[155] Kong Lingkai, Sun Yue, Zheng Zhong, Tang Rong, *et al.* 2018., "Chemoselective N–H or C-2 arylation of indole-2-carbox amides: Controllable synthesis of indolo[1,2-*a*]quinoxalin-6-ones and 2,3′-spirobi[indolin]-2′-ones". Organic Letters. No. 17. 5251-5255.

[156] Chen Zhichao, Zhang Hong, Zhou Shu-Feng, Cui Xiuling. 2021. "Metal-free sulfonylative spirocyclization of indolyl-ynones via insertion of sulfur dioxide: Access to sulfonated spiro[cyclo pentenone-1,3′-indoles]". Organic Letters. No. 20. 7992-7995.

[157] Zhang Cui, Bu Faxiang, Zeng Chulin, Wang Dan, *et al.* 2022. "Electrochemical oxidation dearomatization of anisol derivatives toward spiropyrrolidines and spirolactones". Chinese Chemical Society. No. 4. 1199-1207.

[158] Raji Reddy Chada, Kumar Prajapti Santosh, Warudikar Kamal-kishor, Ranjan Ravi, Bhargava Rao Billa. 2017. "Ipsocyclization: An emerging tool for multifunctional spirocyclo hexadienones". Organic & Biomolecular Chemistry. No. 15. 3130-3151.

[159] Miyazawa Etsuko, Sakamoto Takeshi, Kikugawa Yasuo. 2003. "Synthesis of spirodienones by intramolecular *ipso*-cyclization of *N*-methoxy-(4-halogenophenyl)amides using [hydroxy(tosyloxy) iodo]benzene in trifluoroethanol". The Journal of Organic Chemistry. No. 13. 5429-5432.

[160] Zhang Xiaoxia, C. Larock Richard. 2005. "Synthesis of spiro[4.5]trienones by intramolecular *ipso*-halocyclization of 4-(*p*-methoxyaryl)-1-alkynes". Journal of the American Chemical Society. No. 35. 12230-12231.

[161] Yu Quan-Fu, Zhang Yue-Hua, Yin Qin, Tang Bo-Xiao, *et al.* 2008. "Electrophilic ipso-iodocyclization of *N*-(4-methylphenyl) propiolamides: Selective synthesis of 8-methylene azaspiro [4,5]-trienes". The Journal of Organic Chemistry. No. 9. 3658-3661.

[162] Wang Zhi-Qiang, Tang Bo-Xiao, Zhang Hong-Ping, Wang Feng, Li Jin-Heng. 2009. "Electrophilic *ipso*-halocyclization of *N*-aryl-propynamides with polyfluoroalkyl alcohols: Selective synthesis of 8-(polyfluoroalkoxy)azaspiro[4.5]trienes". Synthesis. No. 06. 891-902.

[163] Tang Bo-Xiao, Tang Dong-Jun, Tang Shi, Yu Quan-Fu, *et al.* 2008. "Selective synthesis of spiro[4,5]trienylacetates via an intramolecular electrophilic *ipso*-iodocyclization process". Organic Letters. No. 6. 1063-1066.

[164] Tang Bo-Xiao, Zhang Yue-Hua, Song Ren-Jie, Tang Dong-Jun, *et al.* 2012. "Intramolecular *ipso*-halocyclization of 4-(*p*-un-

substituted-aryl)-1-alkynes leading to spiro[4,5]trienones: Scope, application, and mechanistic investigations". The Journal of Organic Chemistry. No. 6. 2837-2849.

[165] R. Likhar Pravin, S. Racharlawar Shailesh, V. Karkhelikar Manjusha, SubhasMadavu Salian, Sridhar Banavar. 2011. "Iodocyclization of *N*-aryl-3-phenylpropiolamides by I_2/CAN: A convenient route for the selective synthesis of quinolin-2-ones". Synthesis. No. 15. 2407-2414.

[166] Wu Cui-Yan, Deng Guo-Bo, Hu Ming, Wei Wen-Ting, Li Jin-Heng. 2014. "*ipso*-Iodocyclization of *para*-substituted 4-aryl-1-alkenes leading to 3-iodo-1-azaspiro[4.5]deca-6,9-diene-2,8-diones". Synthesis. No. 19. 2585-2590.

[167] Chen Yu, Liu Xiaochen, Lee Minwook, Huang Chenlong, *et al.* 2013. "ICl-Induced intramolecular electrophilic cyclization of 1-[4′-methoxy(1,1′-biphenyl)2-yl]alkynones—a facile approach to spiroconjugated molecules". Chemistry–A European Journal. No. 30. 9795-9799.

[168] Low-Beinart Lila, Sun Xiaohan, Sidman Eli, Kesharwani Tanay. 2013. "Synthesis of pyrrolo[2,1-*j*]quinolone framework via intramolecular electrophilic *ipso*-cyclization". Tetrahedron Letters. No. 11. 1344-1347.

[169] Yugandhar Doddapaneni, Srivastava Ajay Kumar. 2015. "Efficient construction of azaspiro[4.5]trienone libraries via tandem Ugi 4CC/electrophilic *ipso*-iodocyclization in one-pot". ACS Combinatorial Science. No. 8. 474-481.

[170] Sahoo Harekrishna, Sankar Grandhi Gowri, Ramakrishna Isai, Baidya Mahiuddin. 2019. "Metal-free switchable *ortho*/*ipso*-cyclization of *N*-aryl alkynamides: Divergent synthesis of 3-selenyl quinolin-2-ones and azaspiro[4,5]trienones". Organic & Biomolecular Chemistry. No. 48. 10163-10166.

[171] Zuo Hang-Dong, Hao Wen-Juan, Zhu Chi-Fan, Guo Cheng, *et al.* 2020. "Electrochemical annulation–iodosulfonylation of 1,5-enyne-containing para-quinone methides (*p*-QMs) to access (*E*)-spiroindenes". Organic Letters. No. 11. 4471-4477.

[172] Raji Reddy Chada, Ajaykumar Uprety, H. Kolgave Dattahari. 2020. "Expeditious access to spiro-fused 2,5-cyclohexadienones via thio(seleno)cyanative *ipso*-cyclization". The Journal of Organic Chemistry. No. 23. 15521-15531.

[173] Yang Wen-Chao, Zhang Ming-Ming, Sun Yu, Chen Cai-Yun, Wang Lei. 2021. "Electrochemical trifluoromethylthiolation and spirocyclization of alkynes with $AgSCF_3$: Access to SCF_3-

containing spiro[5,5]trienones". Organic Letters. No. 17. 6691-6696.

[174] Kikugawa Yasuo, Nagashima Akira, Sakamoto Takeshi, Miyazawa Etsuko, Shiiya Megumi. 2003. "Intramolecular cyclization with nitrenium ions generated by treatment of *N*-acylaminophthalimides with hypervalent iodine compounds: Formation of lactams and spiro-fused lactams". The Journal of Organic Chemistry. No. 17. 6739-6744.

[175] Zhang Zhiguo, Gao Xiaolong, Li Zhonglian, Zhang Guisheng, *et al.* 2017. "PIFA-Mediated oxidative cyclization of 1-aroyl-*N*-arylcyclopropane-1-carboxamides and their application in the synthesis of pyrrolo[3,2-*c*]quinolinones". Organic Chemistry Frontiers. No. 3. 404-408.

[176] Teimouri Mohammad Bagher, Heydari Mahdi, Mohammadi Kazem. 2020. "Substrate-controlled selectivity switch in a three-component reaction: Sequential synthesis of spiro-oxazolidine dione-cyclopentenones and hydroxy enamino barbiturates in water". RSC Advances. No. 23. 13601-13610.

[177] Z. Li Wei-Dong, Duo Wei-Guo, Zhuang Cheng-Han. 2011. "Concise total synthesis of (±)-Cephalotaxine via a trans-annulation strategy: Development of a facile reductive oxy-Nazarov cyclization". Organic Letters. No. 13. 3538-3541.

[178] Sun Shang-You, Raju Selvam, Vedarethinam Guganchandar, Chen Pei-Lin, Chuang Shih-Ching. 2021. "TfOH-Promoted classical Nazarov-type cyclization of benzofulvenes: Synthesis of polycyclic 5H,10′*H*-spiro[benzo[*k*]phenanthridine-5,6′-dibenzo pentalenes]". Organic Letters. No. 16. 6212-6216.

[179] Peng Hui, Li Jiuyi, Wang Furong, Liu Bo, Yin Biaolin. 2016. "Synthesis of spiro-lactams and polysubstituted pyrroles via ceric ammonium nitrate-mediated oxidative cyclization of *N*-furan-2-ylmethyl-β-enaminones". The Journal of Organic Chemistry. No. 12. 4939-4946.

[180] Calcaterra Andrea, Mangiardi Laura, Delle Monache Giuliano, Quaglio Deborah, *et al.* 2020. "The Pictet-Spengler reaction updates its habits". Molecules. No. 2. 414.

[181] V. Alvarez-Rodríguez Nancy, Islas-Jácome Alejandro, Rentería-Gómez Angel, E. Cárdenas-Galindo Luis, *et al.* 2018. "Synthesis of 1′-tetrazolylmethyl-spiro[pyrrolidine-3,3′-oxindoles] via two coupled one-pot processes Ugi-azide/Pictet–Spengler and oxidative spiro-rearrangement". New Journal of Chemistry. No. 3. 1600-1603.

[182] Yamada Tomoaki, Ozaki Yui, Yamawaki Mugen, Sugiura Yoshihiko, *et al.* 2017. "Reductive *ipso*-radical cyclization onto aromatic rings of five-membered alicyclic amino acids bearing *N*-(2-phenyl)benzoyl groups by photoinduced electron transfer promoted decarboxylation". Tetrahedron Letters. No. 9. 835-838.

[183] Reddy C. Ravikumar, Yarlagadda Suresh, Ramesh Boora, Reddy M. Rajashekhar, *et al.* 2017. "$Na_2S_2O_8$-Promoted radical cyclization for the synthesis of azaspiro[4.5]deca-3,6,9-triene -2,8-dione and pyrrolo[2,1-*j*]quinolone derivatives". European Journal of Organic Chemistry. No. 16. 2332-2337.

[184] Chatupheeraphat Adisak, Soorukram Darunee, Kuhakarn Chutima, Tuchinda Patoomratana, *et al.* 2013. "Synthesis of *gem*-difluoromethylenated spiro-γ-butyrolactones by employing $PhSCF_2Si(CH_3)_3$ as a *gem*-difluoromethylenating agent". European Journal of Organic Chemistry. No. 30. 6844-6858.

4

SYNTHESIS OF SPIROCYCLES II: METAL-CATALYZED SYNTHESIS OF SPIRO COMPOUNDS

4.1. Introduction

The enantioselective construction of spirocyclic centers presents an intriguing synthetic challenge, which continues to fuel the exploration of innovative catalysts and intricate molecules. Transition-metal-mediated processes have been employed to synthesize a diverse range of spirocyclic compounds. In this chapter, we will explore the various metals utilized in these transformations, including palladium, nickel, group 11 metals, rhodium, iron, and ruthenium, each discussed in dedicated sections.

4.2. Alkylation reactions

4.2.1. Pd- Catalyzed alkylation reactions

The Pérez-Gómez group developed a novel synthetic pathway for the synthesis of spiro-oxoindoles **3** using α-diazocarbonyl compounds in a Pd-catalyzed remote C–H alkylation reaction [1]. This protocol relies on the activation of remote C–H bonds by σ-alkyl Pd(II) species generated *in situ* via intramolecular carbopalladation of alkenes, followed by the insertion of a carbenoid coupling partner. The group conducted a series of experiments using acrylamide **90** and α-diazocarbonyl compound **2** as substrates with $Pd(OAc)_2/PPh_3$ as a catalyst and Cs_2CO_3 (1 equiv) at 100 °C (Scheme 4-1). The group found that dry MeCN as a solvent provided a complete transformation of the starting material **1**, producing a 96% NMR yield of **3a** as a 2.2:1 diastereomeric mixture. Other ligands such as dppe, dppf, or BINAP did not significantly improve the diastereoselectivity, and reducing the $Pd(OAc)_2$ loading from 10 to 5 mol% did not affect the product yield. The group further investigated the scope of this reaction by varying the substitution patterns of both coupling partners. They observed that increasing the steric bulk of the ester function from $-CO_2Me$ to -

CO_2Et or -CO_2Bn of the diazo compound **2** did not significantly affect the yield of the spirocycles **3**, but some products' diastereoisomeric ratio (dr) decreased. The reaction of acrylamide **1** with various diazocarbonyls **2** led to moderate to good yields for substrates containing aryl groups with electron-donating and -withdrawing substituents.

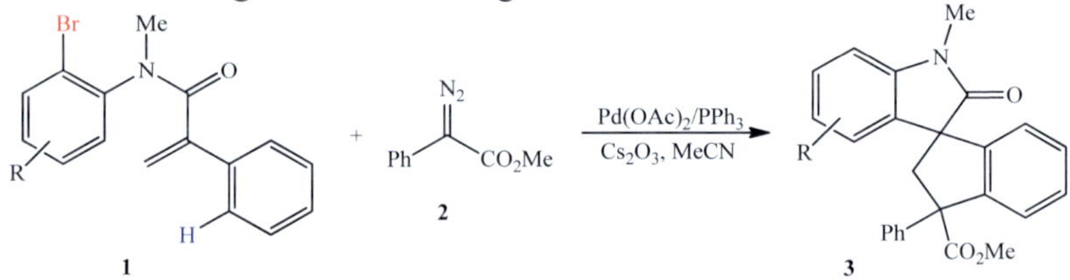

Scheme 4-1. Pd-Catalyzed remote C–H alkylation.

Scheme 4-2 depicts a proposed mechanistic pathway which presumably leads to the formation of product **3**. The mechanical path leading to product **3** is thought to proceed through the intramolecular 5-*exo*-trig carbopalladation of the alkenyl substrate **1**, resulting in the formation of the σ-alkyl Pd(II) intermediate of type A (Scheme 4-2). Next, a C–H activation step would generate the five-membered C(sp^2), C(sp^3)-palladacycle of type B (path a, Scheme 4-2). This intermediate would undergo carbene transfer from the diazo compound **2**, followed by the N_2 release. Finally, a C–C coupling step with reductive elimination of Pd(0) would lead to the formation of the spiro-oxoindole **3**.

Lautens et al. reported the use of an *in situ* generated palladacycle to perform remote-selective C–H alkylation reactions of arenes **4** without using a covalently bonded directing group or the addition of an external mediator [2]. Based on the proposed catalytic cycle and selecting an appropriate ligand, benzofurans and indole **6** via a C-H alkylation/aryl migration sequence, or spiro-fused benzocyclobutenes **5** via reductive elimination of the key spiropalladacycle could be constructed selectively, as indicated in Scheme 4-3.

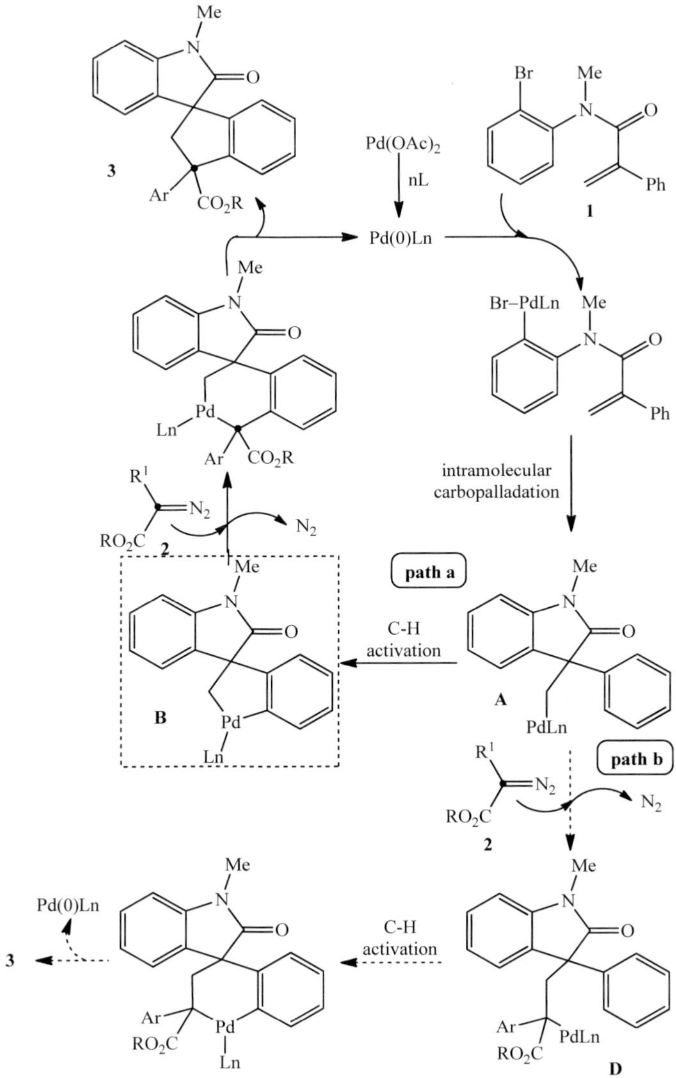

Scheme 4-2. Proposed reaction pathway for the remote alkylation.

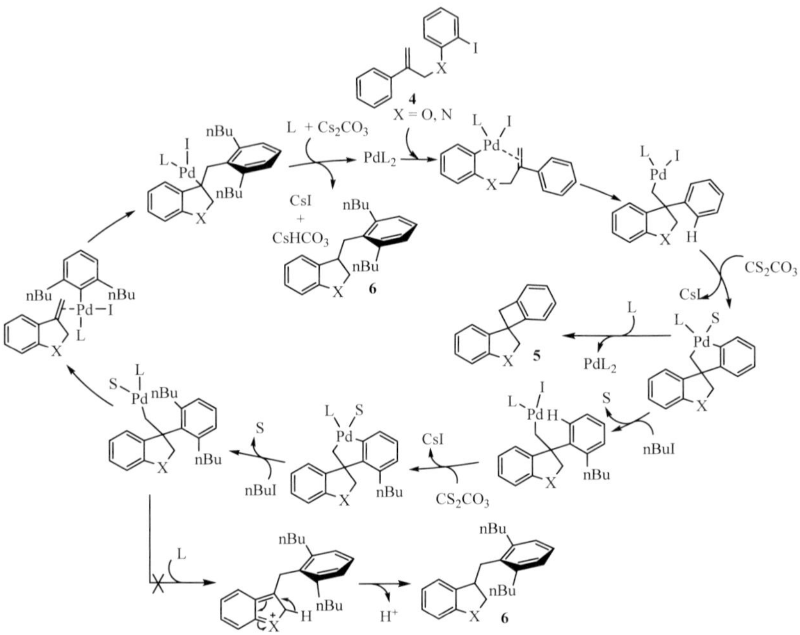

Scheme 4-3. Proposed catalytic cycle (L = P^tBu_3, S = $CsHCO_3$).

In 2010, Hamada et al. reported the first successful Pd-catalyzed intramolecular *ipso*-Friedel-Crafts allylic alkylation of *para*-substituted phenols **7** which provided a novel pathway for the synthesis of spiro[4.5]cyclohexadienones **8** frameworks (Scheme 4-4) [3]. The reaction exhibited the following characteristics:

i. The use of monodentate phosphorus ligands with electron-withdrawing properties improved the reactivity of the reaction,
ii. Substrates containing a methyl carbonate as a leaving group were suitable for the reaction, and
iii. $Pd(dba)_2$ was identified as a best catalyst for the reaction.

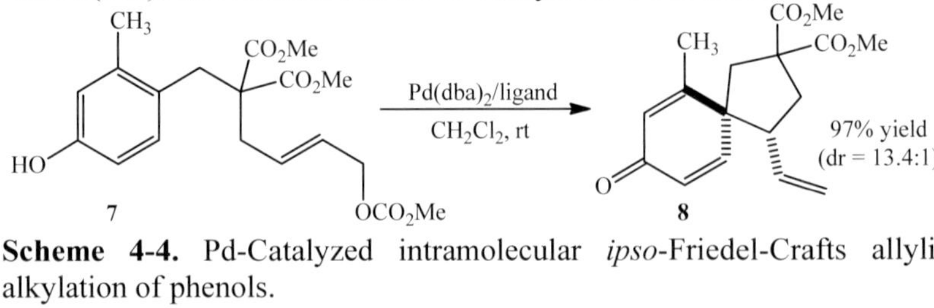

Scheme 4-4. Pd-Catalyzed intramolecular *ipso*-Friedel-Crafts allylic alkylation of phenols.

Figure 4-1 shows an asymmetric version of this reaction which was later reported by the same group using chiral phosphorus ligands. With the use of 5 mol% of the Pd catalyst and 12 mol% of (-)-9-NapBN (-)-**9e**, the spirocyclic adduct was obtained with up to 93% enantiomeric excess (ee), although with a low chemical yield. However, when using 6 mol% of (*R*,*R*)-**9k**, the spirocyclic adducts were obtained in good yields with up to 89% ee (diastereoselectivity = 9.2:1) [4].

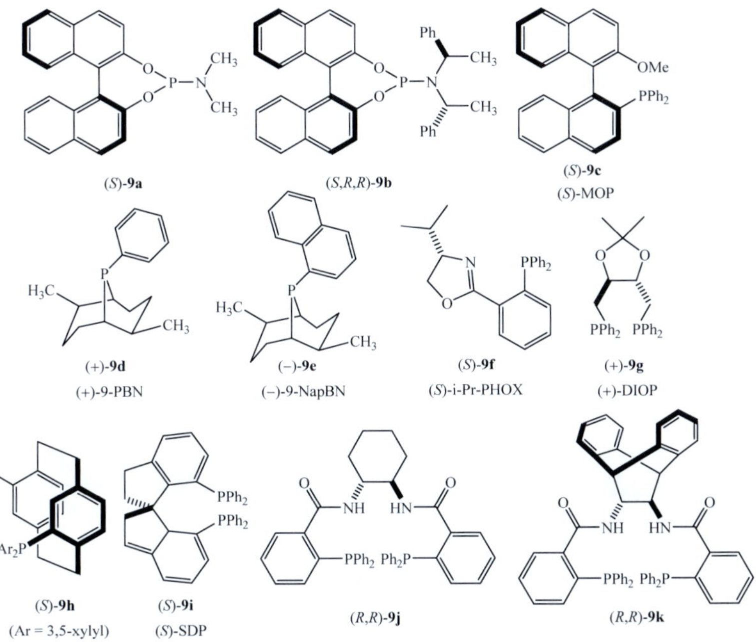

Fig. 4-1. Chiral phosphorus ligands.

The transition state of (*R*,*R*)-**9k** can be explained using a general working model of Trost's catalyst. Considering the relative configuration of the reaction adduct, two possible transition states can be assumed, in which the malonate ester unit is in the open space created by a raised flap. The absolute configuration of **8** proposes that the major enantiomer is generated via transition state **TS-A**. Compared with transition state **TS-A**, the transition state **TS-B** possesses violent steric expulsion between the methyl group on the aromatic ring and the lowered flap. As a result, the (*S*,*S*)-adduct is preferentially formed through **TS-A** (Scheme 4-5).

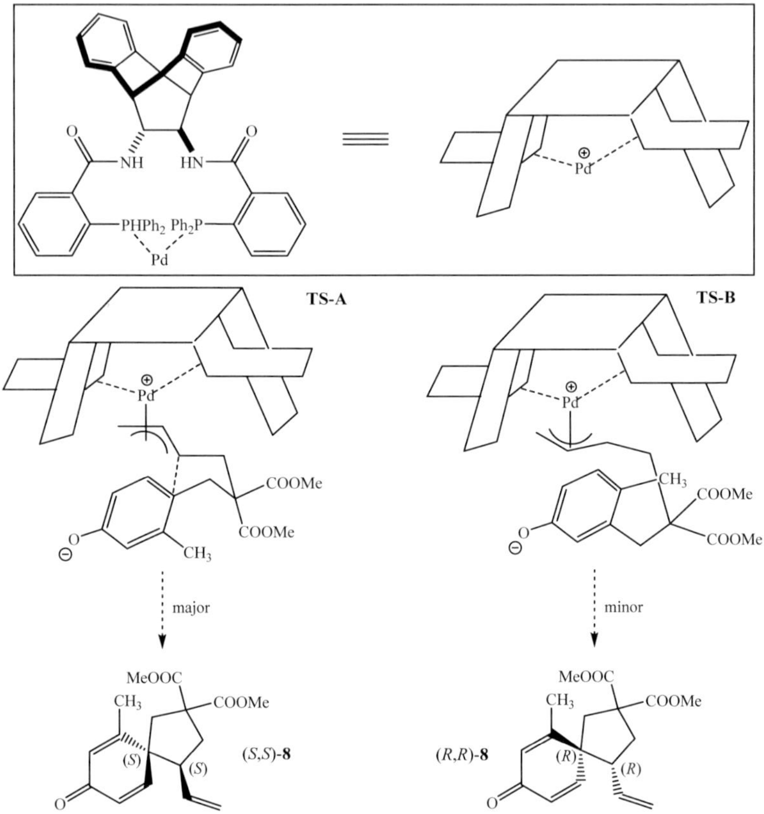

Scheme 4-5. Working model.

One year later, they reported Pd-catalyzed intramolecular *ipso*-Friedel-Crafts allylic alkylation of phenols via arylative activation of allenes. According to this, functionalized spiro[4.5]cyclohexadienones **8** was synthesized through a Pd-catalyzed sequential process: Heck insertion to an allene–intramolecular *ipso*-Friedel-Crafts allylic alkylation cascade (Scheme 4-6).

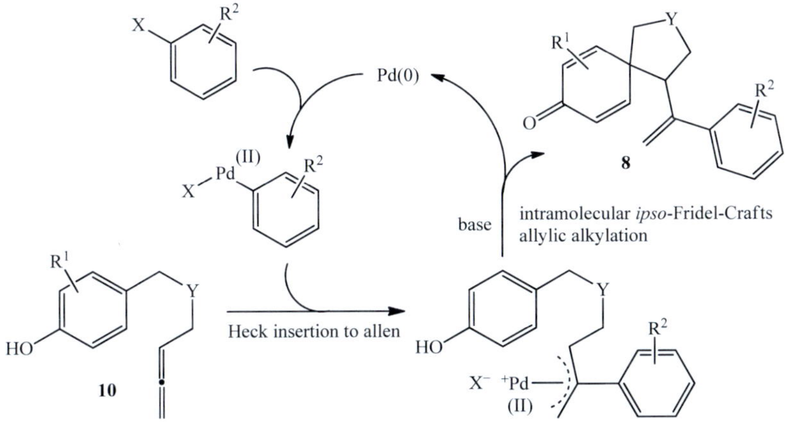

Scheme 4-6. Plan to synthesize spiro[4.5]cyclohexadienones **8** through the Heck insertion/intramolecular *ipso*-Friedel–Crafts allylic alkylation cascade.

The reaction involves the use of substrate **10** and various aryl iodides in the presence of 5 mol% of the catalyst (Scheme 4-7). The reaction tolerates *ortho*-, *meta*-, and *para*-substituted aryl iodides, resulting in good yields of spirocyclic adducts **8**. Additionally, the reaction proceeds smoothly under the same conditions when using 2-iodonaphthalene as a substrate [5].

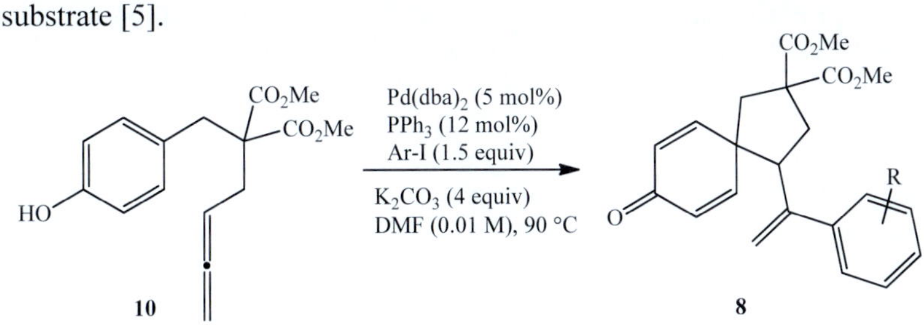

Scheme 4-7. Pd-Catalyzed intramolecular *ipso*-Friedel–Crafts allylic alkylation of phenols via arylative activation of allenes.

Nemoto et al. investigated transition-metal-catalyzed intramolecular Friedel–Crafts-type reactions using phenols as substrates [6]. Their findings are as follows:

i) When phenols **11**, which have one additional CH_2 unit compared to compound **7**, were used with the optimized reaction conditions, conventional *O*-alkylation was the dominant reaction pathway, leading to dimeric compound **12** in 16% yield (Scheme 4-8, Eq. 1). Spiro[5.5] cyclohexadienone **8** was only obtained in 5% yield, indicating that the tether length between the aromatic-ring unit and the π-allylpalladium unit played a critical role in this reaction.

ii) Activation of propargyl carbonate with a Pd catalyst led to the dearomatization of phenol **13**, producing spirocyclic products **14** and **15**. Pd(II) complexes could serve as electrophilic partners in the intramolecular nucleophilic dearomatization of phenols: η^1-allenyl-palladium(II) complex **A** and η^3-propargylpalladium(II) complex **B** provided functionalized spirocyclohexadienones via the corresponding palladacycle intermediate and π-allylpalladium intermediate, respectively (Scheme 4-8, Eq. 2).

iii) The Au-catalyzed intramolecular *ipso*-Friedel–Crafts alkenylation of phenol **16** yielded the desired product **17**. A plausible mechanism of this reaction is presented in Scheme 4-8, Eq. 3. The reaction is initiated by the interaction of the triple bond in **16** with the Au(I) complex, followed by *ipso*-Friedel–Crafts alkenylation of the phenol with the Au-activated alkyne to form vinyl-Au(I) intermediate **A**. Protodeauration occurs in the presence of a Brønsted acid buffered with 2,6-di-*tert*-butylpyridine, yielding spirocyclic adduct **17**. Dihydronaphthalene **18** is produced by the dienone–phenol rearrangement of **17**, followed by isomerization of the exocyclic olefin in excess MsOH.

Wang et al. reported on the development of an efficient cascade asymmetric reaction that involved Friedel-Crafts alkylation/*N*-hemiketalization/Friedel-Crafts alkylation of oxindolyl β,γ-unsaturated α-ketoesters **19** with 3-alkylindoles **20**. This reaction allows for the construction of multifunctionalized, optically active spiro[oxindolepyrrolo(1,2-*a*)indole] **21** derivatives in high yields with excellent enantio- and diastereoselectivities. The reaction is catalyzed by a chiral diphosphine-Pd(II) catalyst $L_6/Pd(PPh_3)_2Cl_2$ (5 mol %) and $AgSbF_6$ (10 mol %) as a co-catalyst, using a substrate ratio of **20**/**19** to 3:1, and a reaction temperature ranging from 25 to 90 °C in $PhCF_3$ (Scheme 4-9) [7].

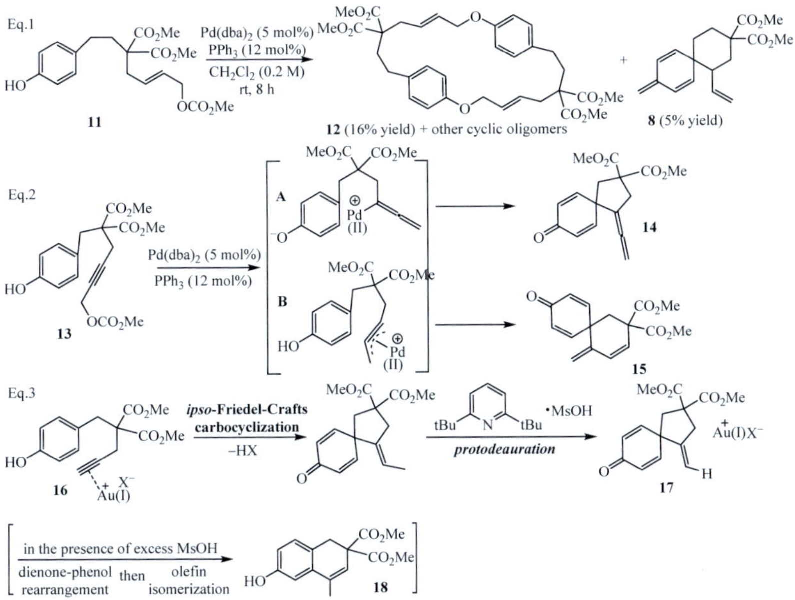

Scheme 4-8. Transition-metal-Catalyzed intramolecular Friedel-Crafts-type processes.

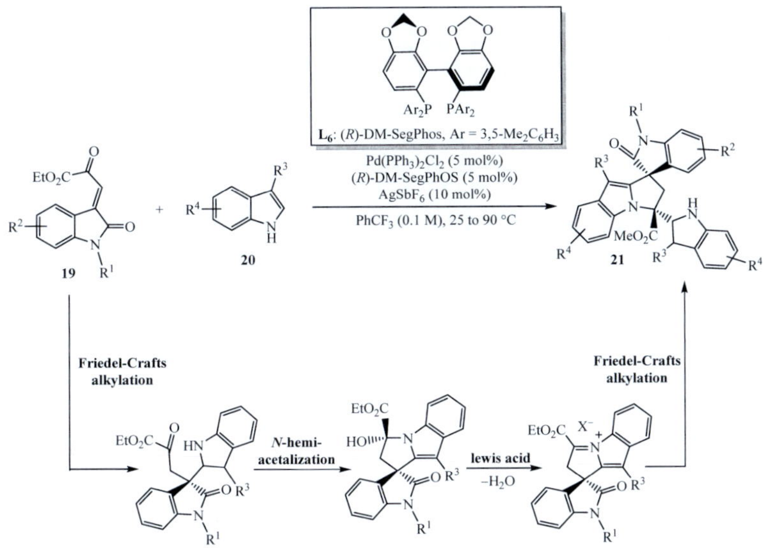

Scheme 4-9. Proposed cascade process.

4.2.2. Other metal-catalyzed alkylation reactions

You et al. reported the enantioselective synthesis of spiroindolenines through the Ir-catalyzed intramolecular C-3 allylic alkylation of indoles **22** [8]. When the reaction was carried out with a 2-methyl-1,2,3,4-tetra hydroquinoline-derived phosphoramidite ligand (*R*,*R*$_a$)-**L6**, highly enantio-enriched spiroindolenines **23** were produced in 92-98% yield with about 99:1 dr and 97% ee (Scheme 4-10).

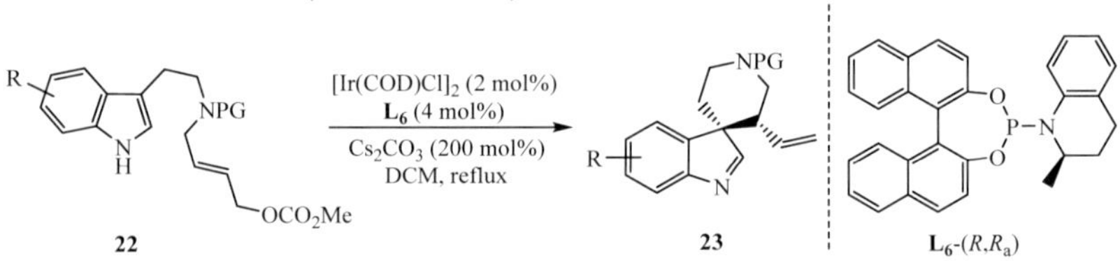

Scheme 4-10. Ir-Catalyzed intramolecular C-3 allylic alkylation of indoles **22**.

The same group reported the first enantioselective synthesis of spiro-2*H*-pyrroles **25** using asymmetric dearomatization of pyrroles via Ir-catalyzed intramolecular asymmetric allylic alkylation reactions. Reactions of allylic carbonate tethered pyrroles **24** in the presence of $[Ir(cod)Cl]_2$ as a catalyst and a BINOL-derived phosphoramidite ligand lead to the efficient production of spiro-2*H*-pyrroles **25** in good yields with up to >99/1 dr and 96% ee (Scheme 4-11) [9].

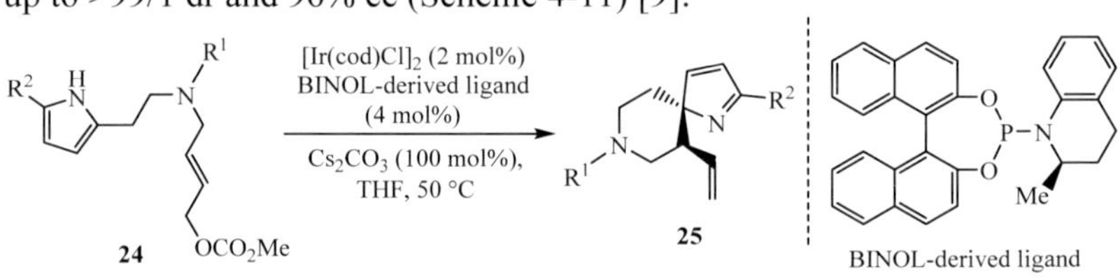

Scheme 4-11. The enantioselective synthesis of spiro-2*H*-pyrroles.

Recently, Cresswell et al. reported a practical and scalable solution to a valid, catalytic entry to α,α,α-tri-substituted (α-tertiary) primary amines **28** by C–H functionalization. For this purpose, they began their examinations using cyclohexylamine **26** ($E_{p/2}$ = +1.53 V vs. SCE, in MeCN) and butyl acrylate **27** as model substrates (Scheme 4-12-**A**). By irradiating **26** and **27** in MeCN with 34 W Kessil blue light-emitting diodes (LEDs) in the presence of various photoredox catalysts (PCs) and hydrogen atom transfer (HAT) catalysts (Scheme 4-12-**B**, **C**), they found that α-C–H alkylation could cleanly result in 85% NMR yield using

4CzIPN as a PC and tetrabutyl-ammonium azide ($Bu_4N^+N_3^-$) as a HAT catalyst ($E_{p/2}$ of N_3^- = +0.87 V vs. SCE in MeCN).

After optimizing the reaction conditions, the researchers proceeded to explore the scope of the reaction with different amine partners **29** (Scheme 4-13). In all cases, except for **29f** and **29p**, a 1:1 stoichiometry of amine to alkylating agent, either methyl acrylate **30** or 2-methoxyethyl acrylate **31**, was used. The resulting γ-amino esters **32** were then converted into the desired γ-lactams **33** through a subsequent lactamization step [10].

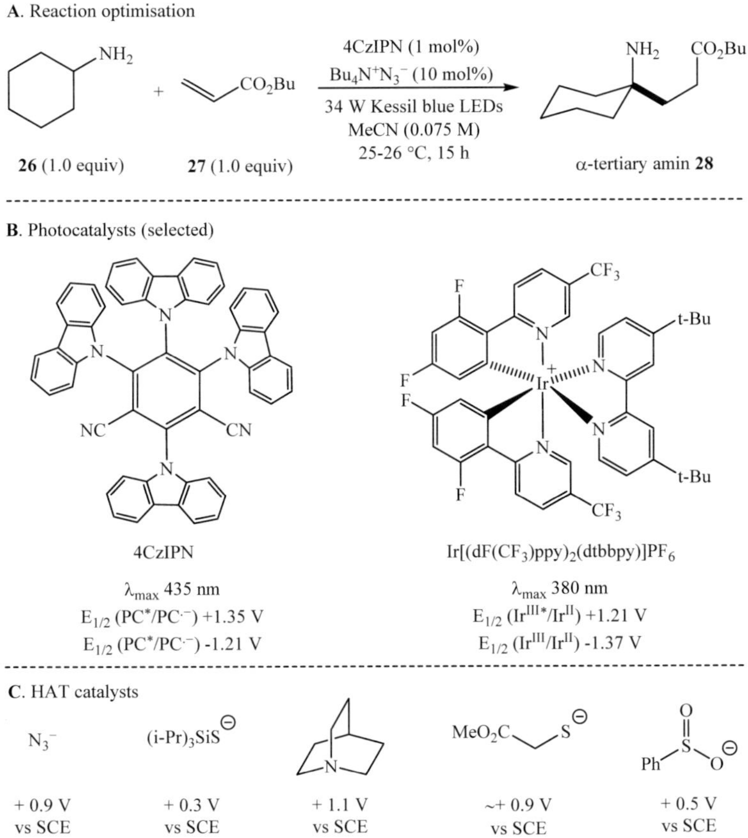

Scheme 4-12. Reaction model.

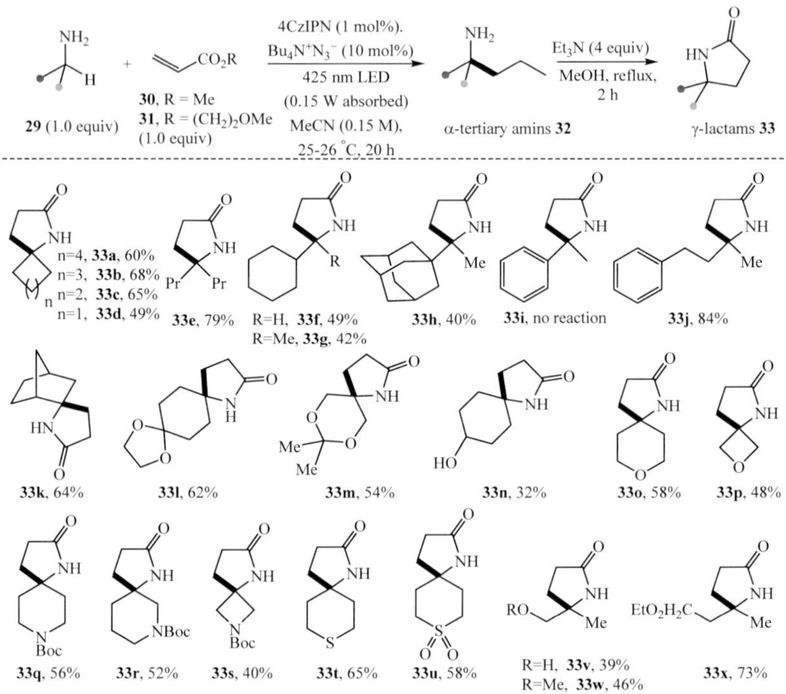

Scheme 4-13. α-C–H alkylation primary amines.

Prabhu et al. in 2012 reported a novel method for chemoselective di-functionalization of activated alkynes for access to 3-alkylated spiro[4,5]-trienones, thiaspiro[4,5]trienones and azaspiro[4,5]trienones using PC under visible light conditions [11]. The used tricarbonyl compounds in PC were applied as the alkylating source. In the process, aryl alkynoate **34** reacts with tricarbonyl compound **35** as a tertiary alkyl radical source under visible light irradiation to give its corresponding spiro products **36** in good to moderate yields (Scheme 4-14).

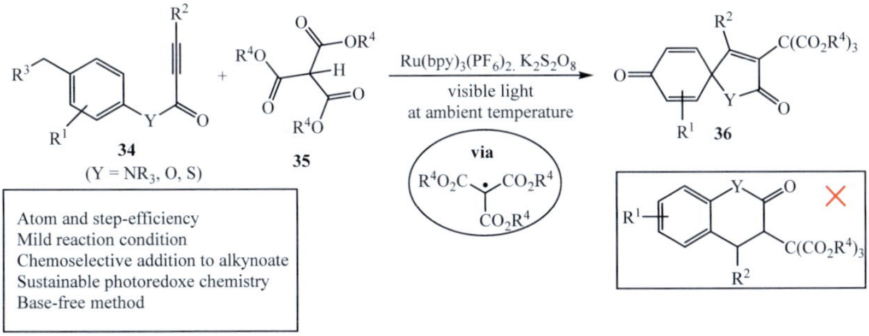

Scheme 4-14. Visible light-mediated-*ipso*-annulation of activated alkynes.

Notably, the reaction demonstrated tolerance to a broad range of functional groups. Likewise, the method allowed authors to access their corresponding mono-carbonylated products by transforming tri-carbonylated spirolactams, which are difficult to access using traditional methods (Scheme 4-15).

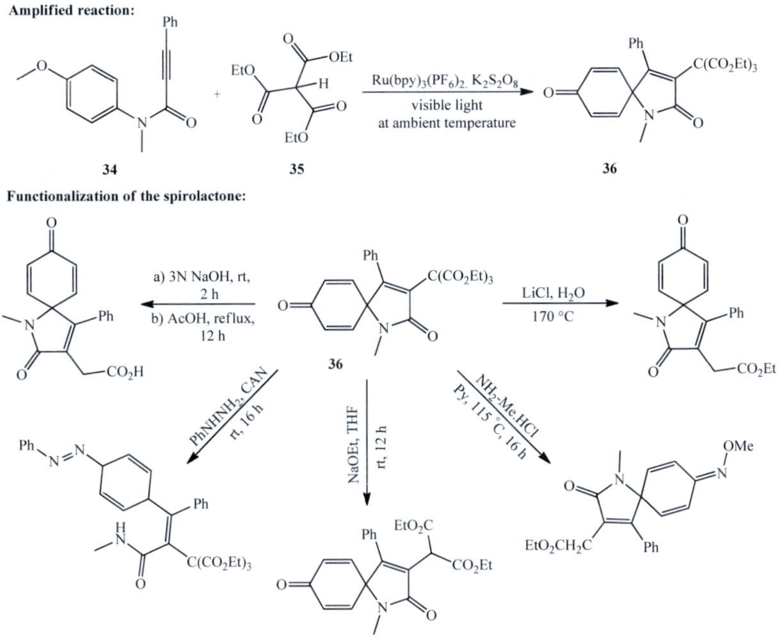

Scheme 4-15. Synthetic utility and product modification.

Zhang et al. proposed a method for the diversity-oriented synthesis of spirocyclic pyrrolidines through late-stage cascade reactions of *o*-succinimide-substituted benzaldehydes. The reaction of benzaldehydes **37** with *N*-ethyl maleimide **38** in the presence of [Ru(*p*-cymene)Cl_2]$_2$, $AgSbF_6$, and 4-chlorobenzoic acid produced *o*-succinimide-substituted benzaldehydes **39** in good yields. A catalytic amount of aniline **T6** acted as a transient directing group and was effective for the Ru-catalyzed *ortho*-C(sp^2)-H alkylation of benzaldehyde with maleimide (Scheme 4-16).

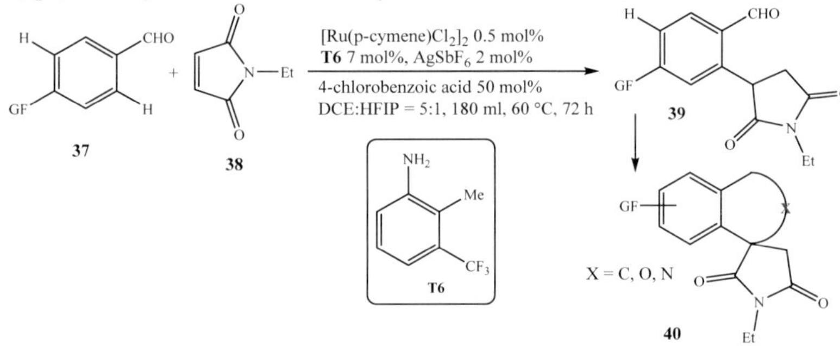

Scheme 4-16. Assembly of diverse spirocyclic pyrrolidines.

The proposed mechanism for the reaction involves the condensation of benzaldehyde with **T6** to form imine **A**. The Ru dimer catalyst, upon treatment with Ag salt, forms the active species **B**. The *ortho*-C(sp^2)-H activation of benzaldehyde leads to the formation of the key metallacycle **C**. The coordination of *N*-ethyl maleimide **38** and the following alkene migratory insertion provide a seven-membered intermediate **E**. Due to the lack of *syn*-planar β-hydrogen atom concerning the metal center, the β-hydride elimination process is impossible. Therefore, proto-demetallation by carboxylic acid leads to the regeneration of the active catalyst **B** along with product **F** (Scheme 4-17).

Through the consecutive cascade reactions, the *o*-succinimide-substituted benzaldehydes **39** could form diverse useful spiro heterocycles **40** in good yields [12].

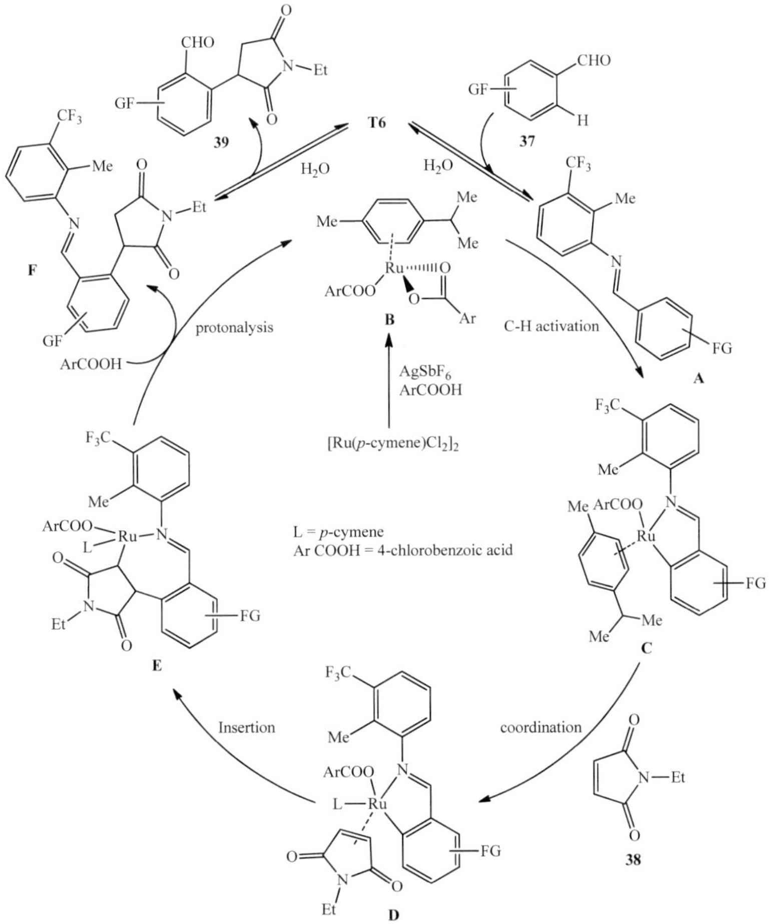

Scheme 4-17. Proposed mechanism.

9,9′-Spirobifluorenes **42** have attracted considerable attention as promising materials for use in organic electronic devices due to their high morphological stability and higher solubility compared to the corresponding non-spiro-linked parent compounds. The Jana group has reported an efficient synthesis of 9,9′-spirobifluorene via Fe-catalyzed intramolecular Friedel–Crafts alkylation [13]. Typically, these compounds are obtained by cyclization of the corresponding tertiary alcohols in a boiling mixture of acetic acid and hydrochloric acid. Under similar reaction conditions, tertiary carbinol **41** underwent smooth electrophilic alkylation in the presence of $FeCl_3$ (5 mol%) at room temperature and furnished the desired 9,9′-spirobifluorene **42** in 95% yield within 5 min, as outlined in Scheme 4-18.

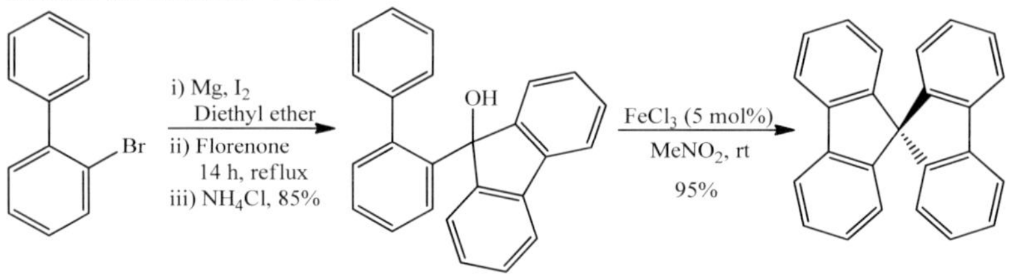

Scheme 4-18. Fe-Catalyzed synthesis of 9,9′-spirobifluorene **42**.

Yang et al. developed a Fe-catalyzed decarbonylative alkylative spirocyclization of *N*-arylcinnamamides to introduce simple linear/branched alkyl groups into biologically important aza-spirocyclohexadienones **45** [14]. In the presence of $Fe(acac)_2$, *N*-arylcinnamamides **43** reacted with aliphatic aldehydes **44** to give alkylated 1-aza-spirocyclohexadienones **45** in good yields (Scheme 4-19). The reaction involved oxidative decarbonylation of aldehydes into primary, secondary, and tertiary alkyl radicals, which underwent subsequent radical addition, spirocyclization, and oxidation to construct dual $C(sp^3)$-$C(sp^3)$ and C=O bonds. The authors proposed a possible mechanism for the reaction which involved the reaction of *N*-(4-hydroxyphenyl)-*N*-methyl cinnamamide **43** with *iso*-butyraldehyde **44** as an example.

Gade et al. developed a Cu-catalyzed enantioselective alkylation of β-ketoesters using alcohols for the in-situ generation of alkylating reagents. The chiral allylation products **47** obtained from 2-substituted allylic alcohols **46** could be transformed into bi-spirolactones **48** in a one-pot procedure by adding **8** equivalents of $BF_3 \cdot Et_2O$, as shown in Scheme 4-20. The Cu complex plays a key role in cyclization [15].

The proposed mechanism involves the coordination of two carbonyl oxygen atoms of the β-ketoester to the Cu complex to form intermediate **A**, followed by deprotonation with DIPEA to give the key ester enolate intermediate **B**. Then, it reacts with the *in situ* generated benzyl or allyl iodide from the alcohol to afford intermediate **C**. Finally, de-coordination from intermediate **C** releases the product and regenerates the catalyst (Scheme 4-20). In intermediate **B**, the *re*-face attack is much more accessible than the *si*-face because the phenyl group blocks the *si*-face in the oxazolinyl unit.

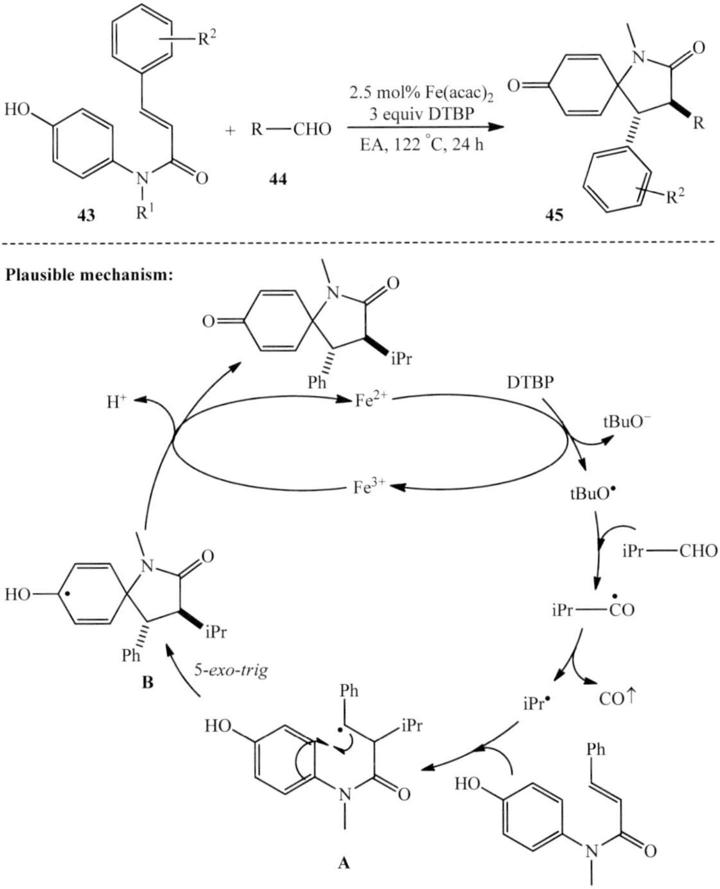

Scheme 4-19. Fe-Catalyzed decarbonylative alkylative spirocyclization.

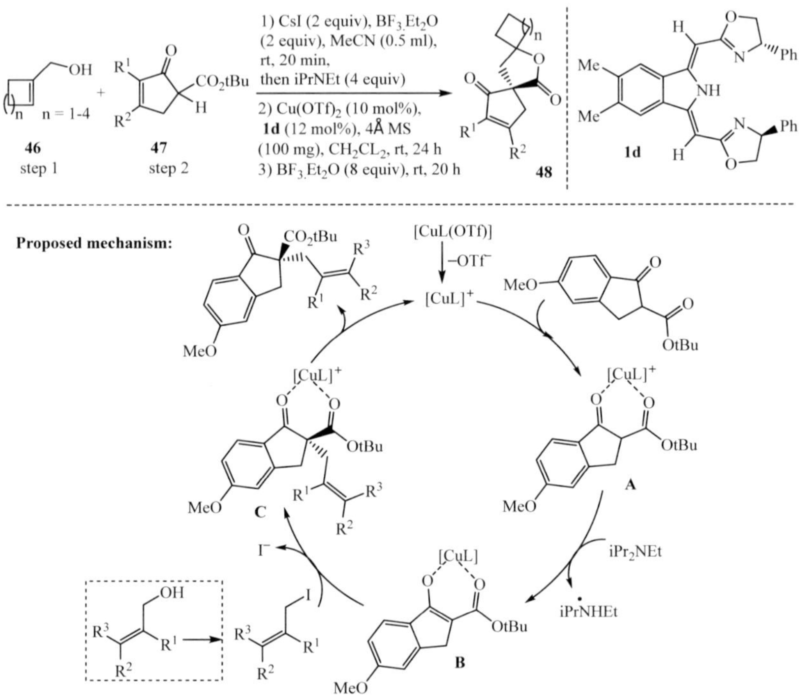

Scheme 4-20. Enantioselective synthesis of bi-spirolactones **48** and proposed mechanism.

Bjørnstad et al. established a synthetic route for preparing rigidly bridged bidentate cyclopentadienyl-alkoxide complexes of titanium [16]. The starting material for spiroannulation is tetrahydrofluoren-1-one **50** which is organized through the alkylation/cyclization sequence of indene **49**. For the spiroannulation, the fluorene C1 ketone is first converted into a spiro-carbon through Martin's efficient tandem acylation-alkylation, and subsequent Rh(I)-catalyzed hydroacylation. The generated spiroketone is converted into a stereochemical mixture of alcohols **51** in a mixture of acetone and dichloromethane which is, in turn, dehydrated by 4-toluenesulfonic acid catalysis and heating in tetrahydrofuran isomerized to the desired spiroketone **52** in 79% yield (Scheme 4-21).

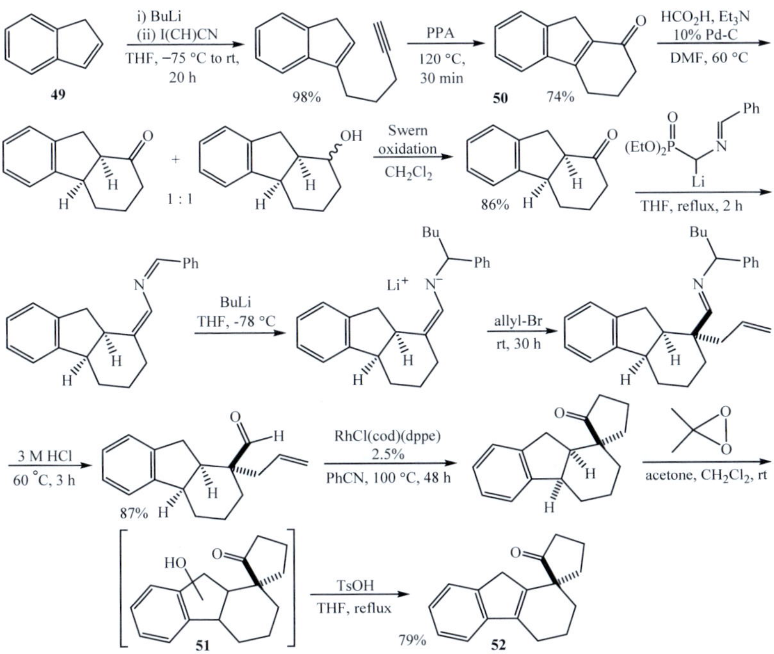

Scheme 4-21. Preparation of spiroketone **52**.

Upon treatment with L-Selectride, spiroketone **52** undergoes conversion into proligand **53**. Additionally, spiroketone **52** can yield the methylenated product **54** through Ti(IV)-promoted methylenation using bis(iodozincio)methane. This product reacts with 9-borabicyclo[3.3.1] nonane and hydrogen peroxide under alkaline conditions, resulting in the formation of proligand **55**. (Scheme 4-22, Eq. 1). Finally, the proligands **53** and **55** were dilithiated by treatment with two equivalents of butyllithium in diethyl ether at 0 °C. The respective dilithiated species **56** and **57** were precipitated from the diethyl ether solution and reacted with $TiCl_3$ and $PbCl_2$ to give semi-metallocenes **58** and **59** (Scheme 4-22, Eq. 2).

In 2020, Reddy et al. established a strategy to functionalize spiro[4.5]trienones **63**/**64** through a domino Ag-catalyzed decarboxylative *ipso*-carboacylation/alkylation reaction with *N*-aryl-propiolamides **60** using α-keto acids/alkyl carboxylic acids **61**/**62** (Scheme 4-23, Eq. 1). The reaction proceeds by $AgNO_3$ (10 mol%) as a catalyst and $K_2S_2O_8$ as an oxidant in CH_3CN:H_2O. The strategy requires the generation of radicals via the decarboxylation of acids to start. Additionally, the reaction extended for the first time to generate acyl/alkyl tricyclic pyrrolo[2,1-*j*]-quinolone **66** (a challenging framework found in tricyclic marine alkaloids) from *N*-(alkynoyl)-6-methoxy tetrahydroquinolines **65** (Scheme 4-23, Eq. 2) [17].

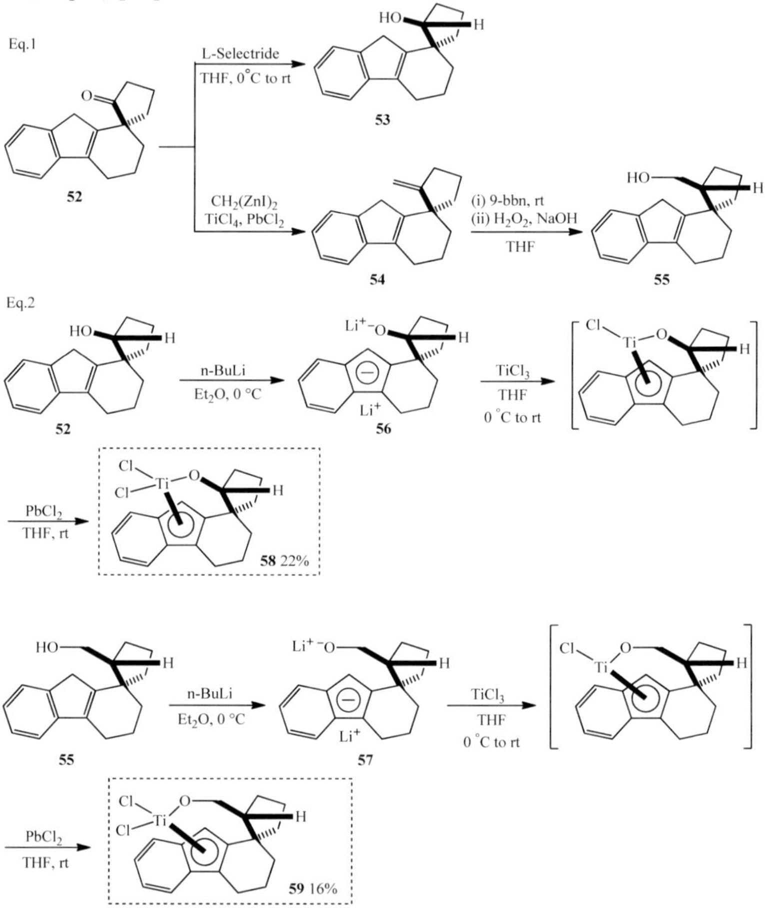

Scheme 4-22. Preparation of the proligands **53** and **55** and semi-metallocenes **58** and **59**.

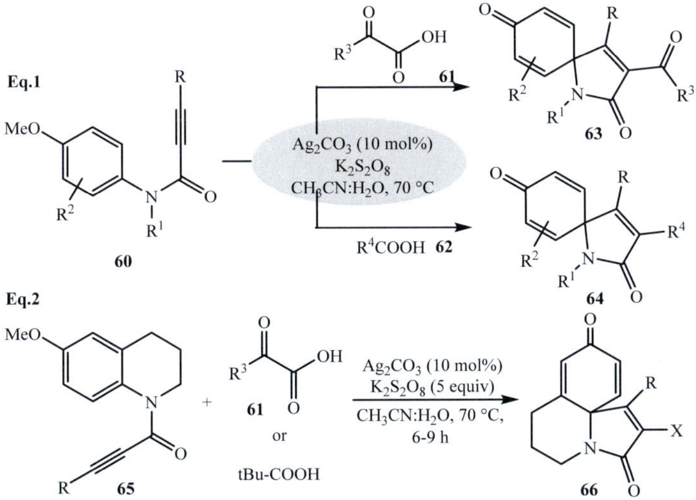

Scheme 4-23. Ag-Catalyzed oxidative *ipso*-cyclization.

The proposed reaction pathway involves the oxidation of Ag(I) with stoichiometric $K_2S_2O_8$ to give Ag(I) which engages the α-keto acid **61** to generate the radical acyl **A** involving decarboxylation. Next, adding the radical acyl **A** onto alkyne **60** gives the radical vinyl **B** which undergoes intramolecular *ipso*-cyclization and dearomatization to generate the intermediate **C**. Then, oxidation of **C** to oxocarbonium **D** followed by nucleophilic attack of hydroxide (produced from H_2O) affords the intermediate **E**. Finally, hydrolysis of hemiacetal **E** provides the product **63** (Scheme 4-24).

Dhotare et al. used $SbCl_3$-catalyzed solvent-free Friedel–Crafts reaction of phenols **67** with mandelic acids **68** to construct the 3-aryl benzofuran-2(3*H*)-ones **69** in good to high yields. The synthetic utility of this method was exemplified by the synthesis of a new class of potential bioactive spirocyclic benzofuran-2-ones **70** through alkylation of 3-arylbenzofuran-2(3*H*)-ones **69** at 3-position with *tert*-butyl bromoacetate as a key step (Scheme 4-25) [18].

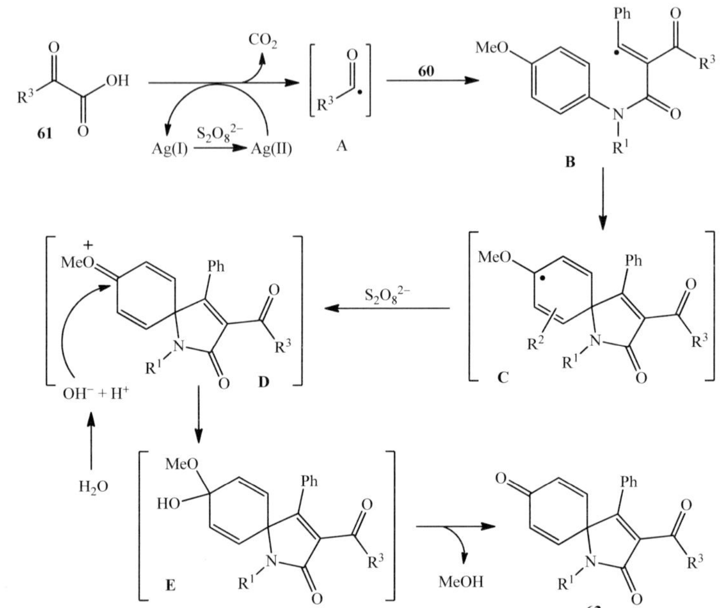

Scheme 4-24. Plausible mechanism.

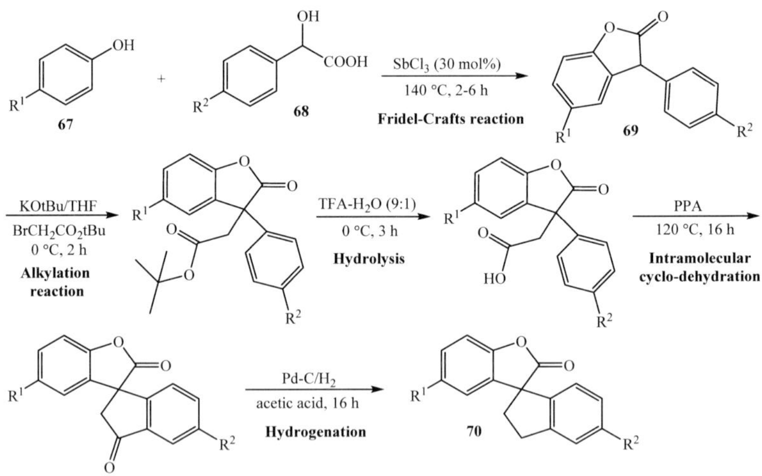

Scheme 4-25. $SbCl_3$-Catalyzed solvent-free Friedel–Crafts reaction.

4.3. Pericyclic-type reactions
4.3.1. 1,3-DCRs
4.3.1.1. Ag-Catalyzed 1,3-DCRs

In this context, Raghunathan et al. reported the efficient regioselective synthesis of highly substituted spiropyrrolidines **73** (Scheme 4-26). 1,3-DCR of (*E*)-3-arylidene-4-chromanones **71** with *N*-metalated azomethine ylides **72** derived from imines of the glycine methyl ester with various aromatic aldehydes, in the presence of silver acetate and triethylamine afforded the spiropyrrolidines **73** in a good yield [19].

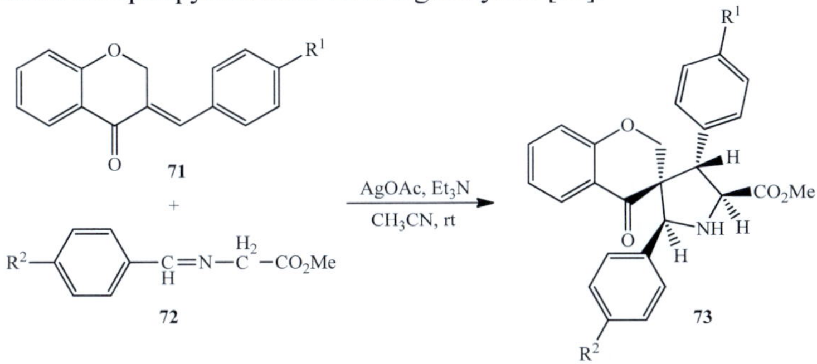

Scheme 4-26. 1,3-DCR of *N*-metalated azomethine ylides **72**.

Later in 2006, Grigg et al. reported a Ag-catalyzed 1,3-DCR of imines derived from 2-amino-γ-lactone/thiolactone **74/75**, utilizing NEt_3 or DBU as a base [20]. The imines **76/77** reacted regio- and stereospecifically with methyl acrylate to provide spiro lactones/thiolactones analogs **78/79** in a good yield via the *syn* dipoles and an *endo*-transition states (Scheme 4-27).

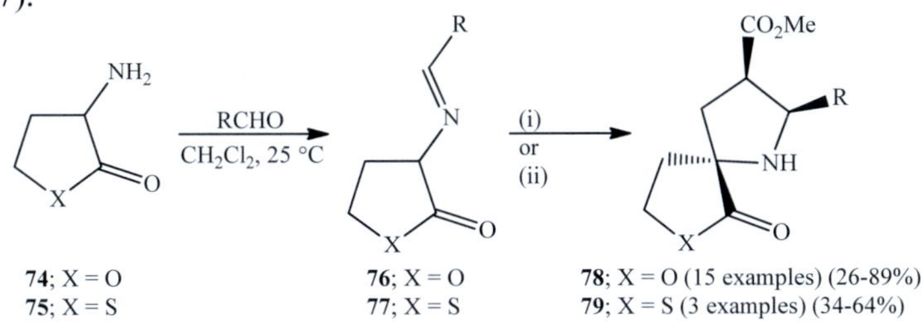

Scheme 4-27. Reagents and conditions: (i) NEt_3, AgOAc in MeCN, 25 °C; (ii) NEt_3 or DBU, AgO in toluene, 25 °C.

The synthesis of spiro nitropyrrolidines employing metallo-azomethine ylide formation from aldimines of cyclic α-amino ester **80** using a combination of AgOAc in MeCN or Ag_2O in toluene with NEt_3

was reported by Grigg et al. [21]. The authors investigated imines of a range of aldehydes (aryl, heteroaryl, aliphatic) in order to explore the diversification of the metallo-azomethine ylide cycloaddition.

In some cases, imines of long-chain aliphatic aldehydes were used to enhance the lipophilicity of the cycloadducts. The aryl and aliphatic imines **81** and **82** were used in cycloadditions with a range of nitrostyrenes. Aryl imines **81** reacted with various nitrostyrenes in acetonitrile in the presence of NEt_3 and AgOAc to provide mixtures of **84** (major) and **85** (minor) cycloadducts in 59–83% yield. In aryl imines, the isomer ratio varied from 4.5:1 to 2:1 depending on the aryl group present in the imines. Cycloadducts **85** generated by the base catalyzed epimerisation of **84**.

Aliphatic aldimines **82** undergo Ag_2O-catalyzed cycloaddition with *trans*-nitrostyrene in toluene in the presence of NEt_3 to provide the corresponding *endo*-cycloadducts **84** in 42–85% yield. In all cases, the cycloaddition was regio- and stereoselective and included only the *E*,*E*-dipole **83** (Scheme 4-28).

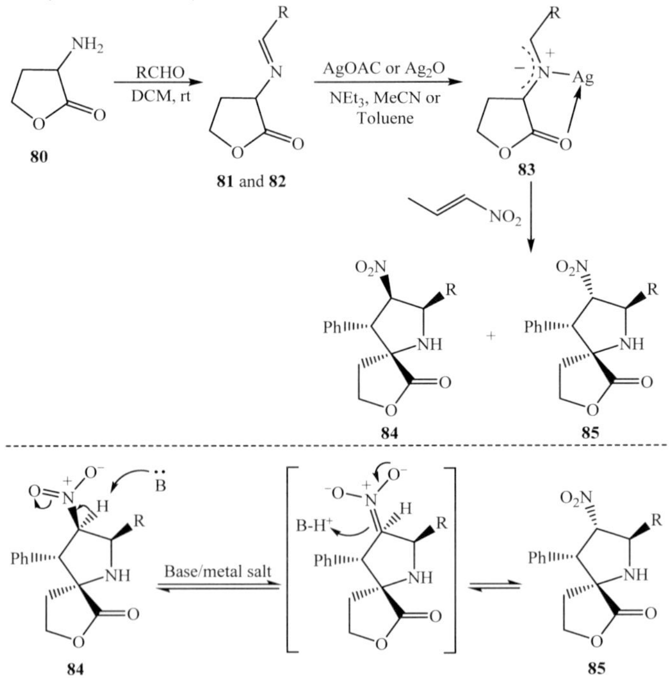

Scheme 4-28. Synthesis of spiro nitropyrrolidines via Ag(I)-catalyzed azomethine ylide cycloadditions.

Mernyák et al. investigated the stereoselective synthesis of spiro pyrazolines **88** of steroidal α,β-unsaturated ketones through 1,3-DC [22]. This process uses steroidal α,β-unsaturated ketones **86** as dipolarophiles and hydrazonoyl chlorides **87** as nitrilimines precursors (Scheme 4-29). The reaction was carried out under both homogeneous and heterogeneous conditions. Silver acetate as a base proved to be more efficient than its NEt_3 counterpart.

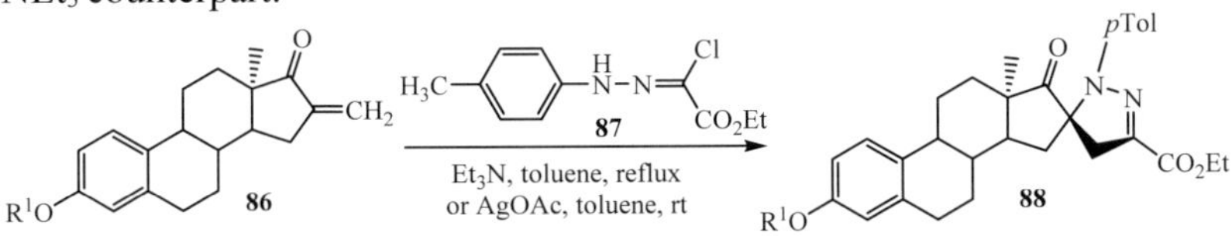

Scheme 4-29. Stereoselective synthesis of spiro pyrazolines **88**.

In 2016, Sansano et al. developed an efficient method for the enantioselective synthesis of poly-substituted spiro nitroprolinates **92** which was successfully achieved by 1,3-DCs employing α-imino γ-lactones **89** as azomethine ylide precursors and nitroalkenes **90** as dipolarophiles (Scheme 4-30) [23]. In this procedure, (*R*,*R*)-MeDuPhos **91**· AgF acts as an efficient bifunctional catalyst because the fluoride behaves as a base. The Ag(I) atom offers a tetrahedral geometry where the coordination sphere is saturated by the carbonylic oxygen atom, the nitrogen atom of the azomethine ylide, and the two phosphorus atoms of the ligand **91**.

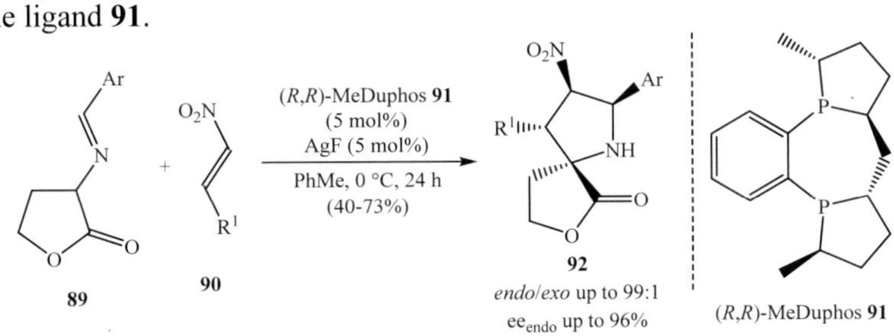

Scheme 4-30. Enantioselective synthesis of poly-substituted spiro-nitroprolinates **92**.

This coordination pattern results in blockage of the (1*si*, 3*re*) prochiral face, thus prompting the preferential formation of cycloadducts *endo*-**92** and *exo*-**92** over its enantiomeric counterparts *ent-endo*-**92** and *ent-exo*-**92** (Figure 4-2).

In the same year, highly regio- and diastereoselective syntheses of a new set of functionalized pyrrolidines, spiropyrrolidine/pyrrolizidine oxindoles, spiroacenaphthylenolyl-pyrrolidine/pyrrolizidine and spiro-1,3-indandionolyl-pyrrolidine/pyrrolizidine through azomethine ylides cycloaddition reaction were reported by the Babu group [24]. The Ag-catalyzed cycloaddition reactions of azomethine ylides derived from *N*-benzylideneiminoglycinates **93** with a variety of arylidene/heteroarylidene malononitriles **94** furnished a new set of C-3, C-5-aryl/heteroaryl substituted C-4, C-4-dicyanopyrrolidine-2-carboxylate scaffolds with very high regio- and diastereoselectivity. The three-component 1,3-DCRs of azomethine ylides **95** derived from 1,2-dicarbonyls and α-amino acids with different indole/pyrrole-based 2π components **96** (dipolarophiles) undergo very high regio- and stereocontrol (Scheme 4-31).

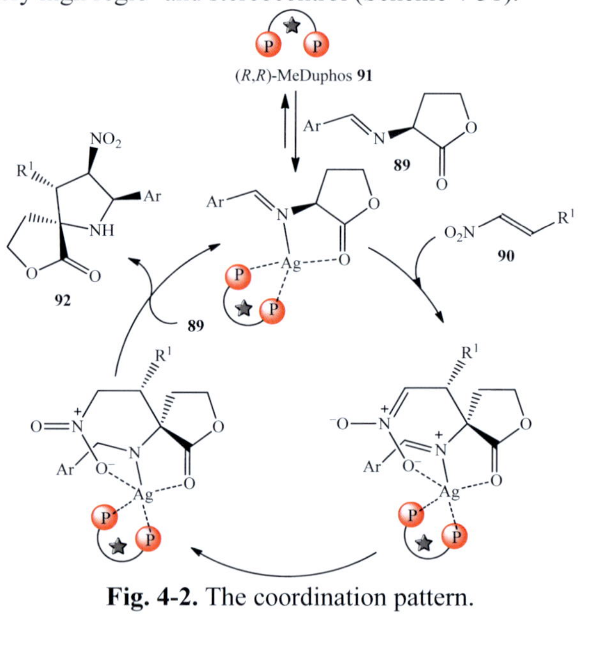

Fig. 4-2. The coordination pattern.

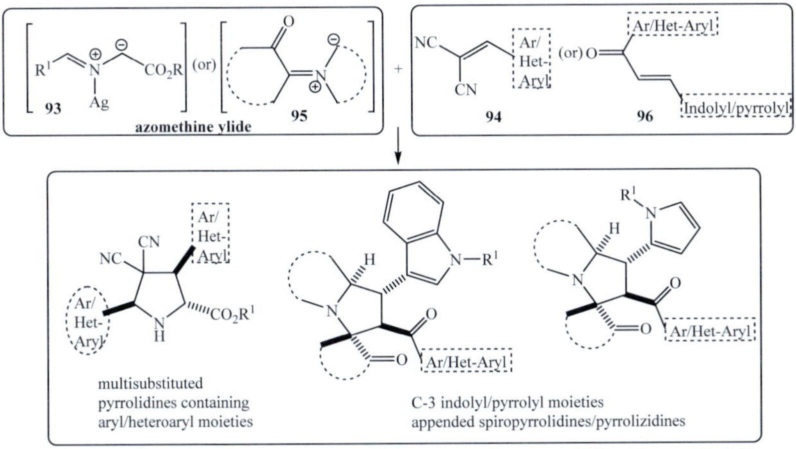

Scheme 4-31. Synthesis of pyrrolidines, spiropyrrolidine/pyrrolizidines-appended with aryl/heteroaryl moieties.

4.3.1.2. Cu-Catalyzed 1,3-DCRs

In this field, a Cu(I)/(*S*)-TF–BiphamPhos-catalyzed asymmetric 1,3-DC employing various imino ester **97** and ethyl cyclopropylidene acetate **98** for preparation of various optically active 5-azaspiro[2,4]heptanes **99** was disclosed by Wang et al. for the first time in 2011 (Scheme 4-32) [25].

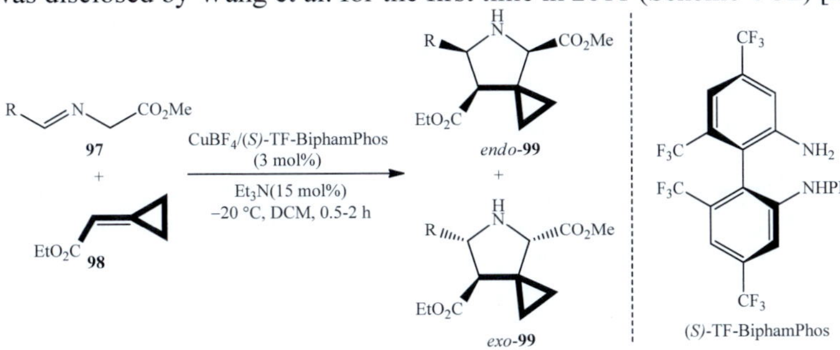

Scheme 4-32. Asymmetric 1,3-DC of various imino ester **97** with 2-cyclopropylideneacetate **98**.

Shortly after, the same group presented asymmetric synthesis of spiro-(butyrolactone-pyrrolidines) **102** bearing a unique spiro quaternary stereogenic center by a Cu(I)/(*S*)-TF–BiphamPhos catalyzed 1,3-DC. In this process, homoserine lactone derived cyclic imino esters **100** reacted as a dipole with dimethyl maleate **101** (Scheme 4-33) [26].

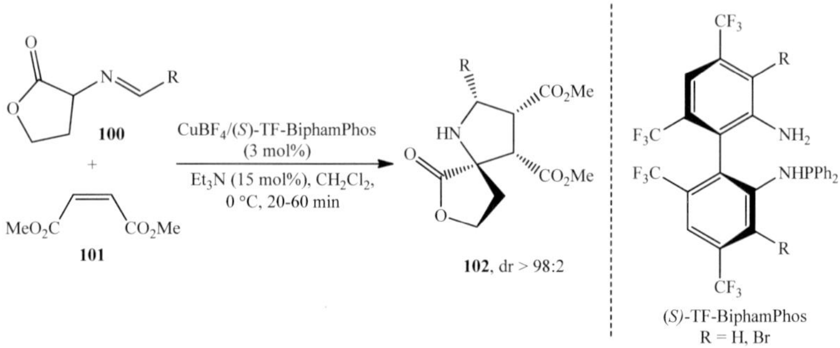

Scheme 4-33. Cu(I)-Catalyzed asymmetric 1,3-DC of various homoserine lactone-derived cyclic imino esters **100** with dimethyl maleate **101.**

One year later, the same group presented the *exo*-selective construction of spiro[butyrolactone-pyrrolidine] **105** bearing one to two quaternary stereogenic centers via Cu(I)/(*R*)-DTBM-BIPHEP-catalyzed 1,3-DC of various imino esters derived from glycinate **103** with α-methylene-γ-butyrolactone **104** (Scheme 4-34) [27].

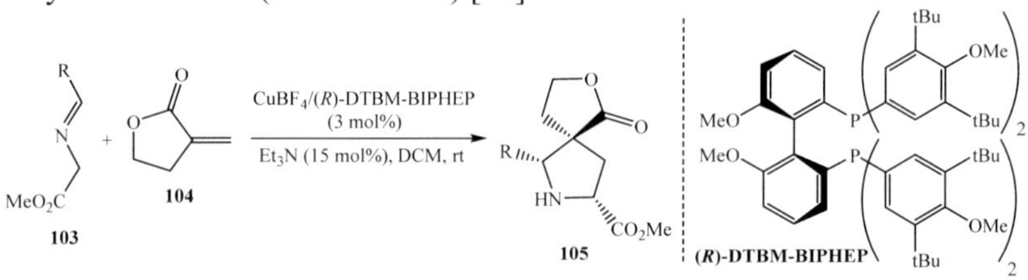

Scheme 4-34. Cu(I)/(*R*)-DTBM-BIPHEP-Catalyzed 1,3-DC.

Waldmann et al. reported the synthesis of pentacyclic spirotryprostatin scaffold bearing one all-carbon quaternary stereogenic center [28]. Final products **109** were synthesized in high yields following a one-pot reaction sequence consisting of a highly enantioselective 1,3-DC of *N*-benzylidene glycine methyl esters **106** with 3-methyleneoxindoles **107** using ferrocene-based ligand **108** and $Cu(CH_3CN)_4PF_6$, acylation of the hindered amine with (L)-FmocProCl and spontaneous ring closure upon *N*-deprotection (Scheme 4-35).

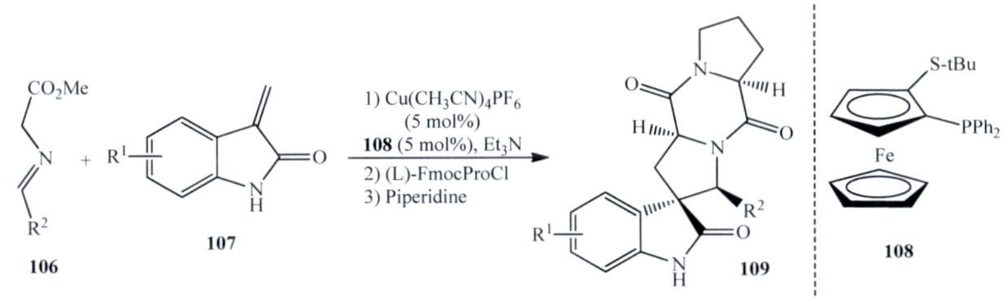

Scheme 4-35. The one-pot procedure obtains spirotryprostatin A analogs.

A three-component reaction between isatin derivatives **70**, L-proline **98,** and terminal alkynes **110** containing an amide or ester functional group reported by Pal et al. gives access to spiro-oxindoles **111**, catalyzed by CuI in good yields (75–92%) (Scheme 4-36). The spiro-oxindole framework **111** demonstrates a new template for identifying novel inhibitors of PDE4 [29].

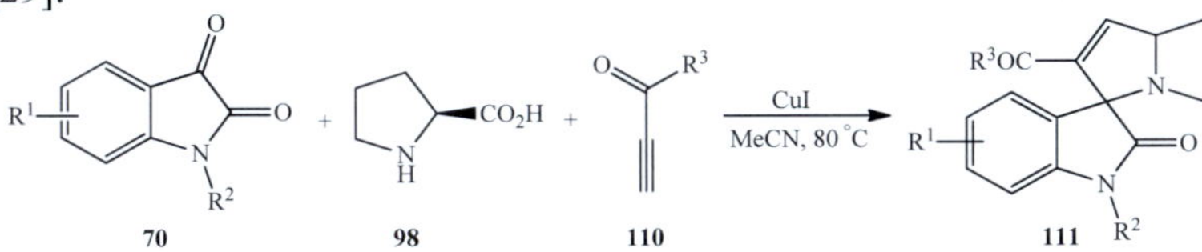

Scheme 4-36. Cu-Mediated 1,3-DC.

The Deng group developed the first examples of the highly stereoselective construction of structurally novel spirocyclic pyrrolidine-thia(oxa)zolidinediones bearing spiro heteroquaternary stereocenters [30]. A highly efficient asymmetric 1,3-DC of azomethine ylides with 5-alkylidene thia(oxa)zolidine-2,4-diones **112** catalyzed by a chiral *N*,*O*-ligand/$Cu(CH_3CN)_4BF_4$ system afforded spirocyclic pyrrolidinethia(oxa)-zolidinediones **113** (Scheme 4-37) in good to excellent yields, with excellent levels of diastereo- and enantioselectivity (dr up to 99:1; ee up to 98%).

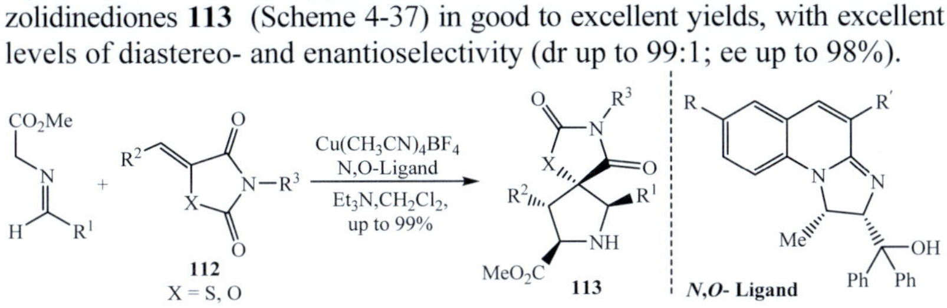

Scheme 4-37. The asymmetric construction of spirocyclic pyrrolidinethia-(oxa)zolidinediones **113**.

The spiro[pyrazolidin-3,3′-oxindole] derivatives **116** was synthesized by Sun et al. [31] via reacting of methyleneindolinones **114** with *N,N′*-cyclic azomethine imines **115** in the presence of 5 mol% of $Cu(OAc)_2$ at room temperature (Scheme 4-38). The desired product **116** could be obtained in moderate to high yields (up to 82%) with good stereoselectivities (up to 15:1). The proposed transition state to explain the role of the catalyst is depicted in Figure 4-3.

The electron-deficient methyleneindolinone is activated by $Cu(OAc)_2$ and stabilized forward. On the other hand, azomethine imine is also activated simultaneously by $Cu(OAc)_2$ and the 1,3-dipole may attack from the behind to give the desired product in good yields.

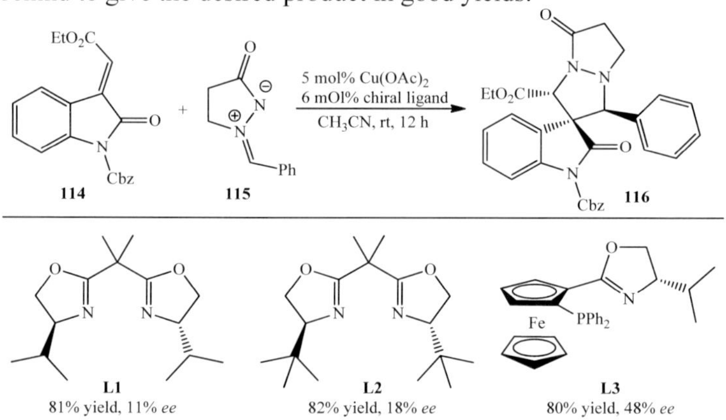

Scheme 4-38. Cu-Catalyzed 1,3-DC.

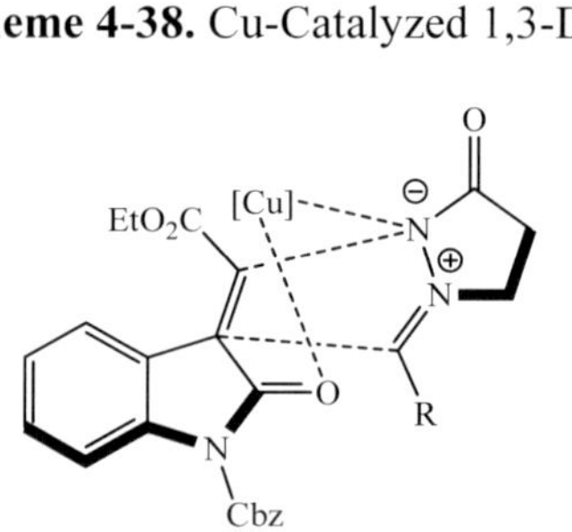

Fig. 4-3. Proposed transition state.

Zhou et al. reported a diastereoselective 1,3-DCR using homoserine lactone **117** derived cyclic imino esters as dipoles and α,α,β-tri-substituted olefins **118** as dipolarophlies catalyzed by Cu(II)/PPFA (Scheme 4-39). The *exo*-spiropyrrolidines **119** were obtained in good yields and high diastereoselectivities (up to >98:2 dr) [32].

Scheme 4-39. Cu(II)/PPFA catalyzed 1,3-DC.

Later on, the Wang group introduced a unique ligand-controlled Cu(I)-catalyzed umpolung-type 1,3-DC of azomethine ylide generated *in situ* from imino ester **120** to realize the efficient kinetic resolution of racemic alkylidene norcamphors **121** with the participant construction of previously inaccessible spiro *N*-heterocycles with high levels of regio- and stereoselectivity [33]. When the phenyl group on the phosphorus atom of ligand **L1** was replaced by bulky electron-donating 3,5-bis(methyl)phenyl (**L2**), or electron-withdrawing 3,5-bis(trifluoromethyl)phenyl group (**L3**), the conventional cycloadduct *endo*-**122** was formed mainly in high yields with good to excellent regioselective ration (rr) and good enantiomeric excess. Chiral ligand **L4** containing cyclohexyl groups also demonstrated normal regioselectivity control with a deleterious effect on the enantioselectivity. Ligand **L5** incorporating two bromine atoms at the 3,3′-positions of the biphenyl scaffold completely reversed the regioselectivity, providing *endo*-**123** in 90% yield with exclusive diastereoselectivity and 92% ee (Scheme 4-40). The achievement of this methodology relies on the strategy of the kinetic resolution and the serendipitous discovery of a unique ligand-enabled regiospecific cycloaddition which not only furnishes evidence for the existence of the minor zwitterionic resonance form in metalated azomethine ylide, but also diversifies the existing chemistry of azomethine ylide-involved 1,3-DCs with rare polarity inversion.

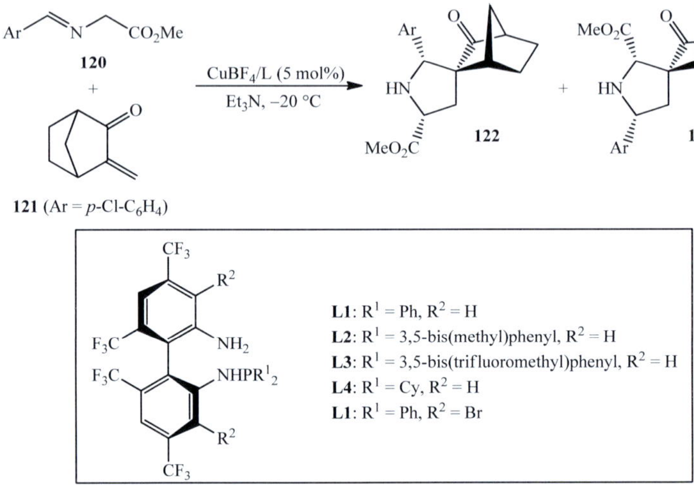

Scheme 4-40. Ligand-controlled umpolung-type 1,3-DC.

In 2019, Muthusamy et al. reported an atom-economical diastereoselective synthesis of a thermodynamically controlled product, spiro-indolofurobenzopyrans **126**, from diazoamides **124** and *O*-propargyl salicylaldehydes **125** in the presence of Cu(I) thiophene-2-carboxylate (Cu(I)TC) as a catalyst in a tandem manner (Scheme 4-41). This process leads to forming two carbon-carbon bonds, a carbon-oxygen bond, and two stereogenic centers in a single synthetic step [34].

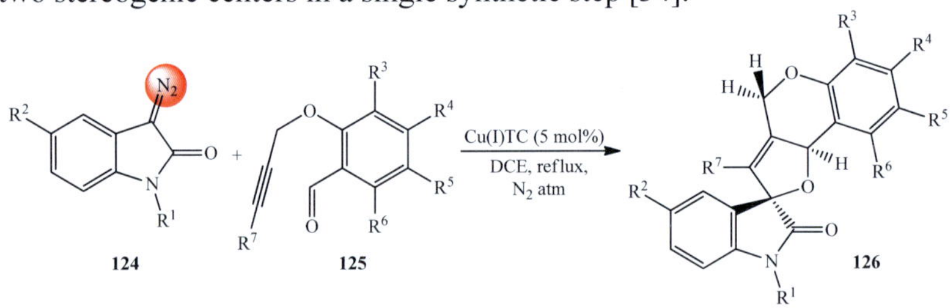

Scheme 4-41. Synthesis of a thermodynamically controlled product, spiro-indolofurobenzopyrans **126**.

Recently, Shi et al. nicely showed $CuCO_3$-catalyzed 1,3-DC of isatin *N,N′*-cyclic azomethine imine **127** with electron-deficient alkynes **128** to create spiro (indoline-3,10-pyrazolo[1,2-a]pyrazole) derivatives **129** in a good yield (Scheme 4-42) [35].

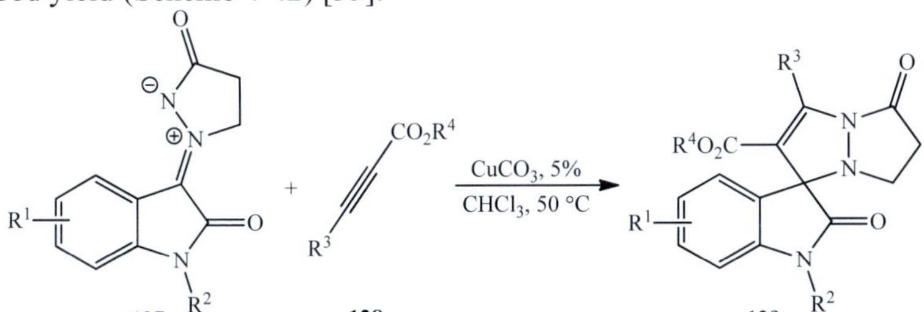

Scheme 4-42. $CuCO_3$-Catalyzed 1,3-DC of isatin *N,N′*-cyclic azomethine imine **127**.

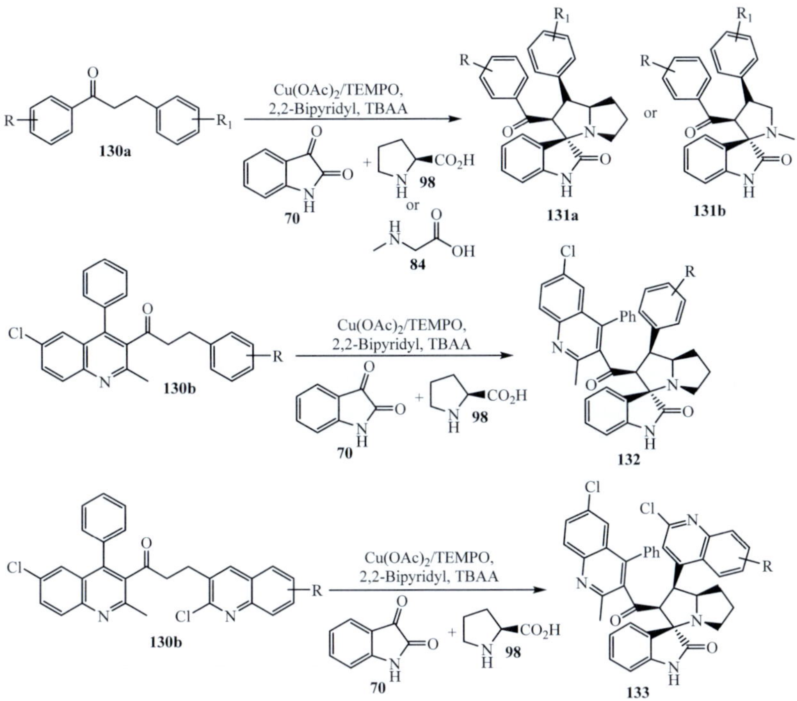

Scheme 4-43. Cu/TEMPO catalyzed dehydrogenative 1,3-DC in synthesizing spiro-oxindoles as potential antidiabetic agents.

Likewise, in 2020, Nawaz Khan et al. reported [36] the preparation of analogs spiro-[indoline-3,3′-pyrrolizin/pyrrolidin]-2-ones **131**, **132** and **133** through the Cu–TEMPO catalyzed dehydrogenation of alkylated ketones **130a-c**, followed by 1,3-DC of azomethine ylides via decarboxylative condensation of isatin **72** and L-proline/sarcosine **98/84**, in high regioselectivities and yields (Scheme 4-43). The synthesized derivatives were investigated further with molecular docking, *in vitro* anti-oxidant, and anti-diabetic activities and their QSAR studies indicate the good biodegradability and bioavailability, including drug transport and ADMET properties.

4.3.1.3. Other metal-catalyzed 1,3-DCRs

The preparation of oxapolycycles **136** through one-pot chemo-selective cross-metathesis/carbonyl ylide formation-intramolecular cycloaddition of α-diazo-β-keto esters was introduced by Hodgson et al. [37]. Cross-metathesis of unsaturated 2-diazo-3,6-diketoester **134** with various other olefins using Grubbs' 2nd generation catalyst **135**, followed by $Rh_2(OAc)_4$-catalyzed tandem carbonyl ylide formation-intramolecular cycloaddition led to the cycloadducts **136** in good to excellent yields (Scheme 4-44).

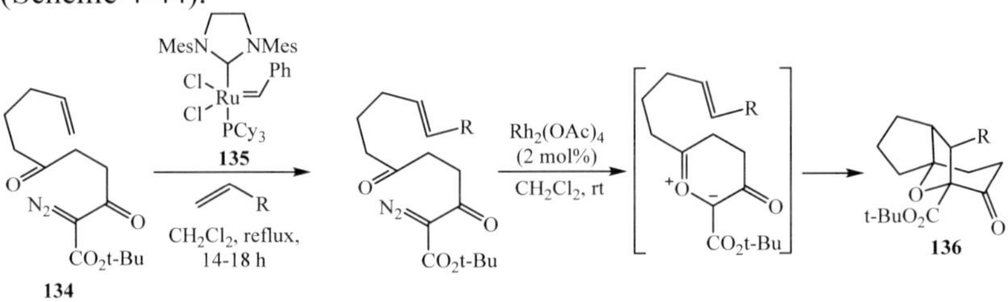

Scheme 4-44. Cross-metathesis/carbonyl-ylide formation/intramolecular cycloaddition using diazo ester **134**.

Commeiras et al. demonstrated the synthetic utility of γ-alkylidene-butenolides **137-138** as highly competent dipolarophile partners in both intra- and intermolecular Ir(II)-catalyzed 1,3-DCRs (Scheme 4-45). The strength of this approach lies in the formation of spiro[6.4]lactone moieties **139-140** with the simultaneous construction of quaternary spiro stereo-centers [38].

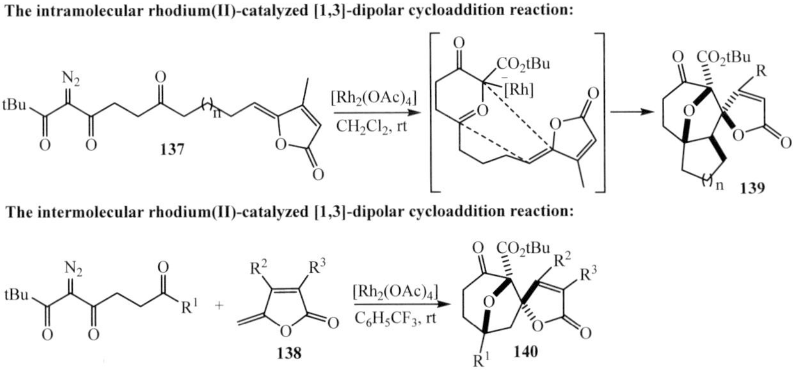

Scheme 4-45. Intra- and intermolecular Ir(II)-catalyzed 1,3-DCRs.

In 2014, a straightforward procedure for the synthesis of a series of spiro-isoxazoline-oxindoles **144** via chiral *N,N′*-dioxide–nickel(II) complex **143** catalyzed enantioselective 1,3-DCR of 3-arylidene-oxindoles **141** with nitrile oxides **142** was reported by Feng et al. (Scheme 4-46) [39]. Spiro-isoxazoline-oxindoles **144** were provided in moderate yields (up to 65%) with good regioselectivities (up to 99:1), excellent enantioselectivities (up to 99% ee), and exclusive diastereoselectivity as well.

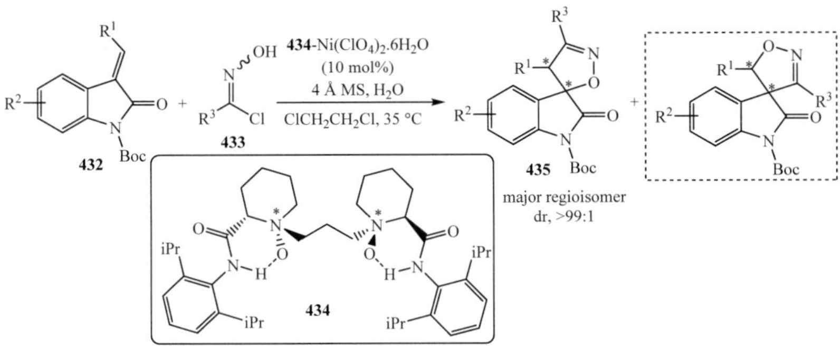

Scheme 4-46. 1,3-DCR between 3-arylideneoxindoles **141** and nitrile oxides **142** and chiral *N,N′*-dioxide–nickel(II) complex **143**.

Cobalt(L-phenylalanine)$_2$ for the first time was used in 2017 by Bdiri et al. [40] in a catalytic asymmetric 1,3-DCR of azomethine ylides **145** with steric exocyclic olefins **146** for the preparation of a series of novel spiropyrrolidines **147** with good to high yields (up to 90%), excellent diastereoselectivities, and good enantioselectivities (up to 87% ee) (Scheme 4-47).

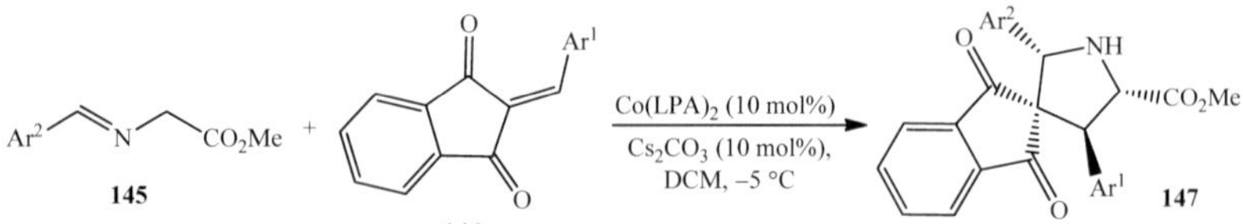

Scheme 4-47. Cobalt(L-phenylalanine)$_2$ catalyzed asymmetric 1,3-DCR.

In 2019, Bazgir et al. reported a Lewis acid-catalyzed regio- and diastereoselective synthesis of spiroisoxazolines via a one-pot sequential Knoevenagel condensation/1,3-DCR [41]. This involved using $ZnCl_2$ as a catalyst to selectively synthesize spiro-isoxazolines **152-154** from CH acids (oxindole **148**, 1,3-indandione **149**, and 3*H*-pyrazol-3-one **150**), aldehydes, and bromonitrile oxide precursor **151**, under mild reaction conditions and in good yields. The authors also prepared spiroindene-isoxazole-1,3-diones containing benzoimidazole or benzothiazole moieties **157** as new ligands by a sequential 1,3-DC/nucleophilic substitution of arylideneindandione **155**, dibromo-formaldoxime **151**, and the amino or mercapto heterocycles **156** for metal-catalyzed coupling reactions and C–H activation, as depicted in Scheme 4-48.

Catalytic, enantioselective 1,3-DCRs of nitrile imines **158** generated from hydrazonoyl bromides with a variety of methyleneindolinones **159** developed by the Stanley group in 2012. The nitrile imine cycloadditions occur in the presence of a catalyst generated *in situ* from $Mg(NTf_2)_2$ and an aminoindanol-derived bisoxazoline ligand **160** (Scheme 4-49). The spiro[pyrazolin[3,3′-oxindole] derivatives **161** are produced in good yields (up to 98%) and high enantioselectivity (up to 99% ee) [42].

Feng et al. reported a new 1,3-DC of nitrile imines **163** with 3-alkenyl-oxindoles **164**, catalyzed by a new chiral $Mg(ClO_4)_2$ complex of an *N,N′*-dioxide ligand **165** [43]. The authors proposed a mechanism to explain the obtained products **166** in this approach (Scheme 4-50). The nitrile imine was generated from the dehydrochlorination of hydrazonyl chloride **162** assisted by the organic base reacted with the alkylidene oxindole **164** to furnish spiro[pyrazolin-3,3′-oxindole] derivatives **166**. The 1,3-DCR is controlled by the $HOMO_{dipolar}$-$LOMO_{dipolarophile}$ interaction.

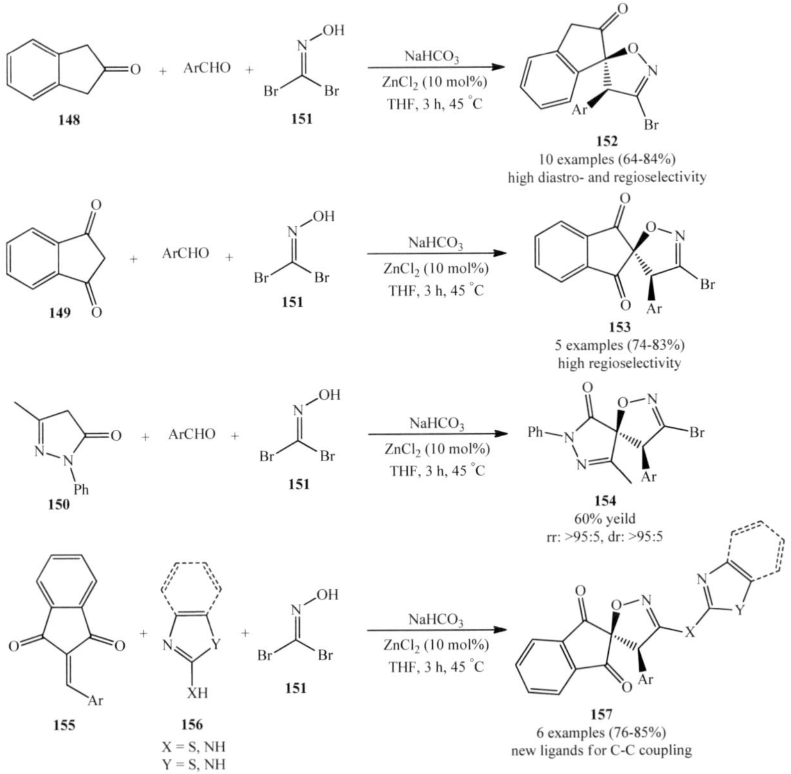

Scheme 4-48. The Lewis acid catalyzed one-pot sequential Knoevenagel condensation/1,3-DCR.

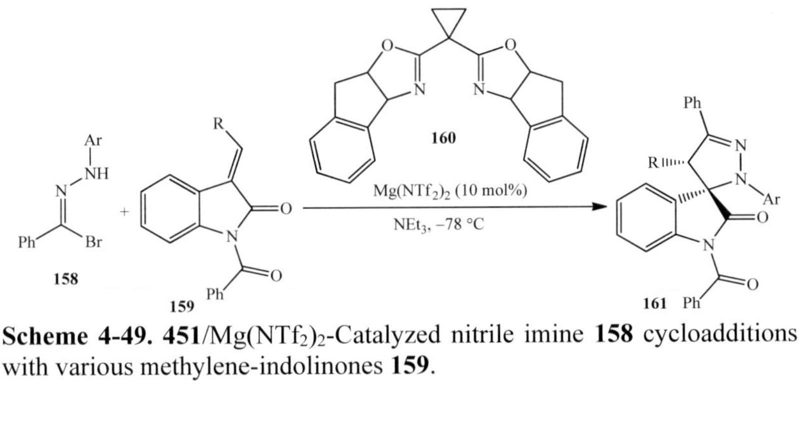

Scheme 4-49. **451**/Mg(NTf$_2$)$_2$-Catalyzed nitrile imine **158** cycloadditions with various methylene-indolinones **159**.

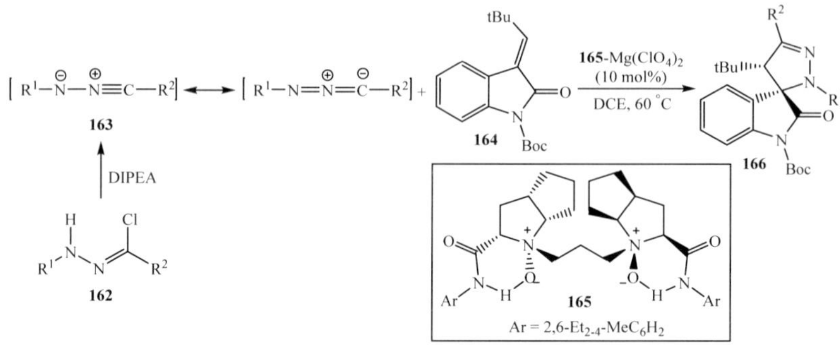

Scheme 4-50. Proposed mechanism for the 1,3-DCR.

4.3.2. [3+2] Cycloaddition reactions

4.3.2.1. Pd-Catalyzed [3+2] cycloaddition reactions

The Liu group reported a highly enantioselective [3+2] cycloaddition of vinyl cyclopropane derived from 1,3-indanedione **166** with nitro-alkenes, which catalyzed by Pd(0) and a chiral bis(*tert*-amine) ligand (Scheme 4-51). Spiro[cyclopentane-1,2′-indene]-1′,3′-dione adducts **167** were formed in high yields with good diastereoselectivities and excellent enantioselectivities [44]. The proposed mechanism is summarized in Scheme 4-51.

One year later, He et al. disclosed a Pd-catalyzed [3+2] cycloaddition of vinyl cyclopropane **169** and α,β-unsaturated imines **170** formed *in situ* from aryl sulfonyl indoles **168** [45]. The process affords spirocyclopentane -1,3′-indolenines **171** in up to 74% yield and up to 97% ee. It involves a first conjugate addition of the carbon anion of the zwitterionic π-allylpalladium complex from vinyl cyclopropane to the *in situ* generated unsaturated imine followed by a Pd-catalyzed intramolecular C3-allylation of indole (Scheme 4-52).

In 2016, Lu et al. demonstrated that the vinyl aziridine **173** could also be used as a 1,3-dipole precursor in the [3+2] cycloaddition reaction of methyleneindolinones **172** to give the 3,3′-pyrrolinyl spiro-oxindoles **174** with high yields and excellent stereoselectivities (Scheme 4-53) [46].

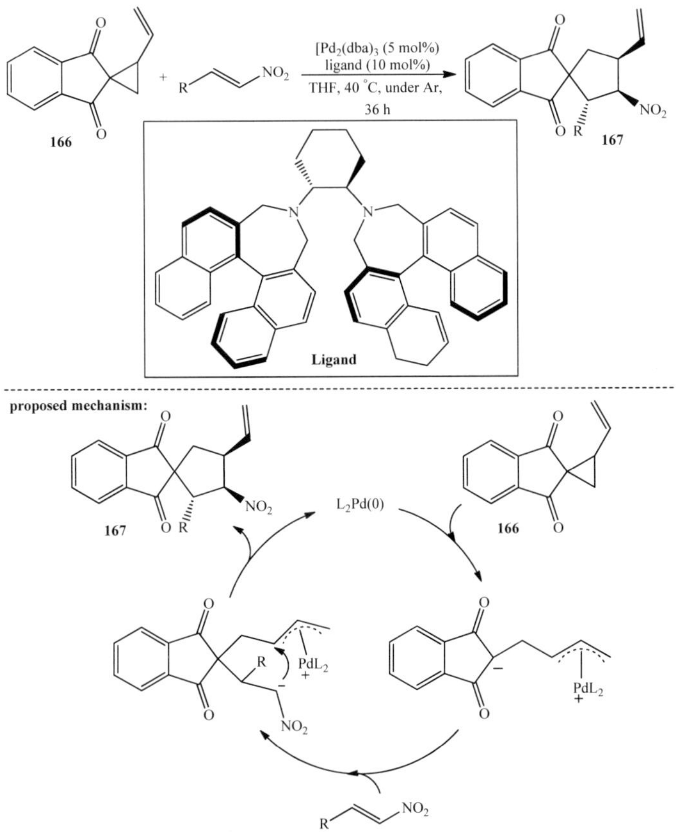

Scheme 4-51. Pd-catalyzes synthesis of spiro[cyclopentane-1,2′-indene]-1′,3′-dione adducts **167** and proposed mechanism.

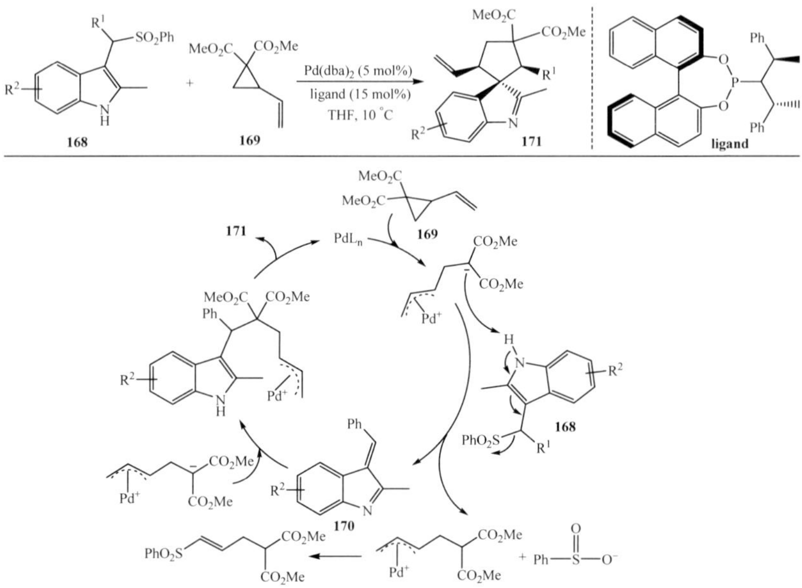

Scheme 4-52. Construction of spirocyclopentane-1,3′-indolenines **171**.

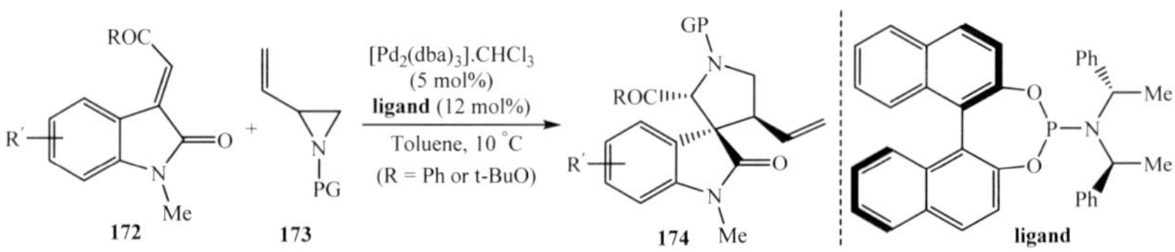

Scheme 4-53. Synthesis of 3,3′-pyrrolinyl spiro-oxindoles **174**.

Ding et al. designed an elegant method for the utility of allenic amides **175** as dipolarophiles in the reaction with vinyl epoxides to yield **176** with Pd-catalysis (Scheme 4-54). The chiral *N*-heterocyclic carbene (NHC) displayed its unique role as a ligand in this asymmetric [3+2] cycloaddition. Chiral tetrahydro-furans **177** bearing three functional groups (mono-substituted alkene, tetra-substituted enolether, and amide) and vicinal tertiary and quaternary stereocenters were formed in high yields with high diastereo- and enantioselectivities [47].

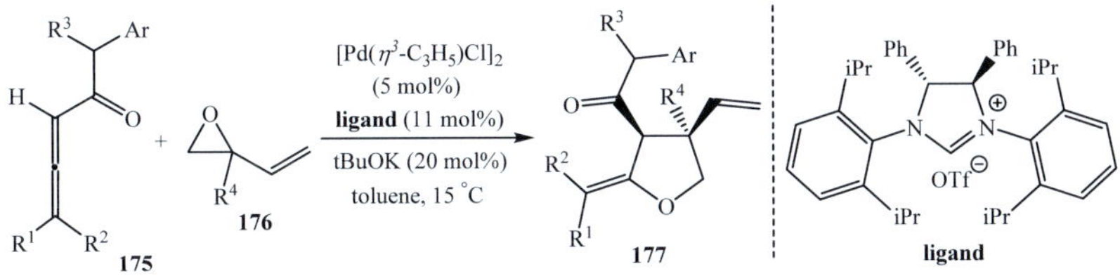

Scheme 4-54. Pd/NHC-Catalyzed asymmetric [3+2] cycloaddition reaction of vinyl epoxides **176**.

Additionally, a series of tetrahydrofuran-fused spirocyclic 1,3-indandiones **180** [48] was synthesized by Guo et al. via an asymmetric [3+2] cycloaddition of 2-arylidene-1,3-indandiones **178** with vinylethylene carbonates **179** in the presence of $Pd_2dba_3.CHCl_3$ and axially chiral phosphoramidite ligand (Scheme 4-55).

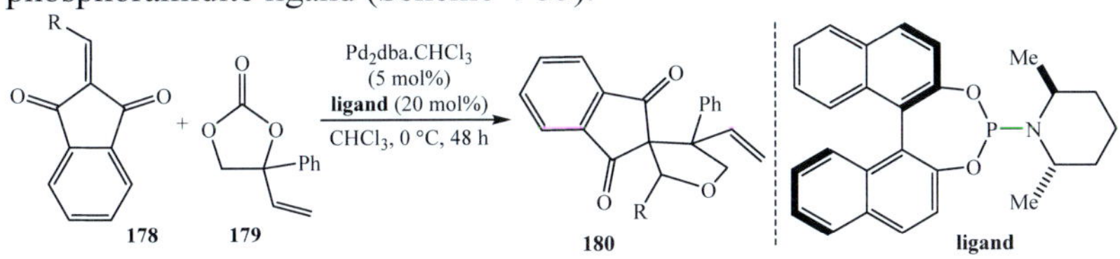

Scheme 4-55. Synthesis of tetrahydrofuran-fused spirocyclic 1,3-indandiones **180**.

Wang et al. presented another example demonstrating the synthesis of tetrahydrofurans in 2020 [49]. Starting from methyleneindolinones and vinylethylene carbonates upon Pd catalysis, the synthesis of functional group-enriched 3,3′-tetrahydrofuryl spiro-oxindoles **181** was achieved in a [3+2] cycloaddition reaction in toluene (Scheme 4-56).

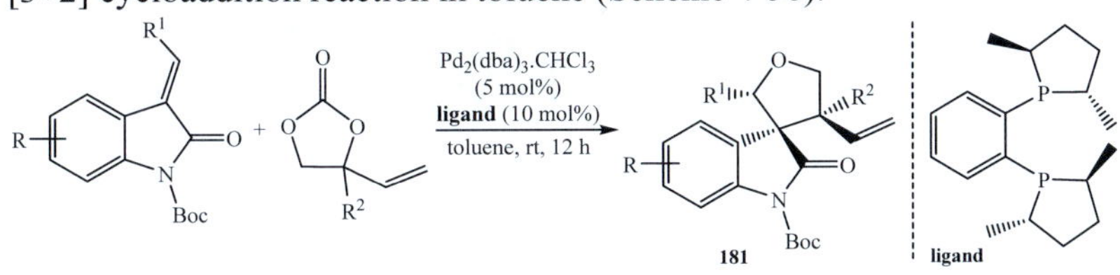

Scheme 4-56. Asymmetric synthesis of 3,3′-tetrahydrofuryl spiro-oxindoles **181**.

In the same year, a mild and efficient Pd-catalyzed [3+2] cycloaddition of vinyl-aziridine **182** and indane-1,3-diones **183** was reported by Vetica et al. [50]. The synthesized spiro-pyrrolidines **185** afforded excellent yields and, with the introduction of the leucine-derived phosphine ligand **184**, moderate to good enantio- and diastereoselectivities (Scheme 4-57).

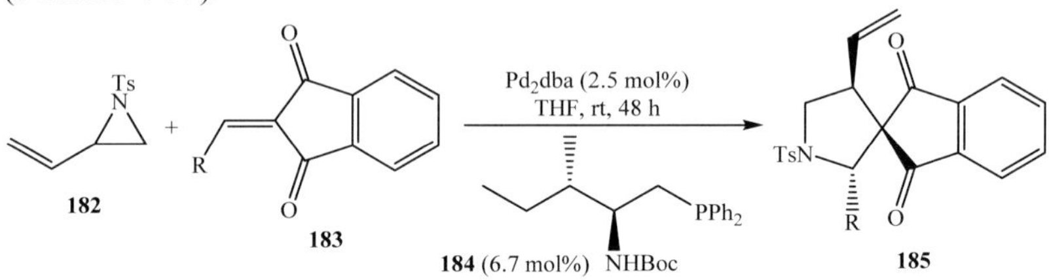

Scheme 4-57. Pd-Catalyzed [3+2] cycloaddition of vinylaziridine **182** and indane-1,3-diones **183**.

4.3.2.2. Rh-Catalyzed [3+2] cycloaddition reactions

In this context, Dong et al. reported a Rh(III)-catalyzed [3+2] annulation of cyclic *N*-sulfonyl ketimines **186** with internal alkynes in the presence of catalytic amounts of $AgSbF_6$ as an additive to form synthetically challenging spirocyclic sultams **187** in a high yield under mild conditions (Scheme 4-58) [51].

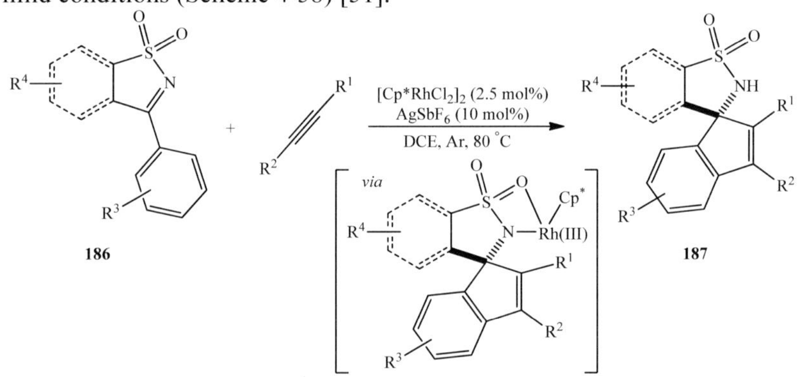

Scheme 4-58. Rh-Catalyzed spirocyclic sultam synthesis **187**.

Pham et al. showed that spirocyclic sultams **187** can also be prepared by chiral Cp^x–Rh(III)-catalyzed enantioselective annulation of *N*-sulfonyl ketimines and alkynes using protected l-phenylalanine **188** as an addictive (Scheme 4-59). Using Rh(III) complexes equipped with a convenient atropchiral cyclopentadienyl ligand and a carboxylic acid additive allows enantioselective and high-yielding access to such spirocyclic sultams [52].

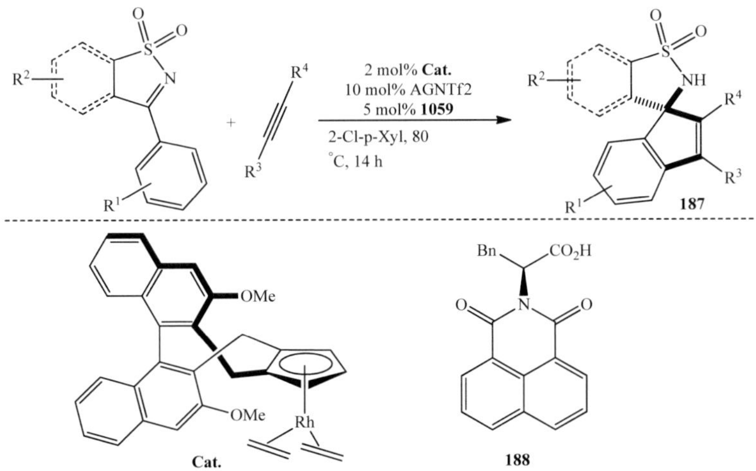

Scheme 4-59. Enantioselective spirocyclic sultam synthesis.

In 2014, Xia et al. [53] disclosed the first Rh(III)-catalyzed [3+2] annulation of 5-aryl-2,3-dihydro-1*H*-pyrroles **189** with internal alkynes using a $Cu(OAc)_2$ oxidant for building spiro[indene-1,2′-pyrrolidine]s **190** in good yields with excellent regioselectivities. This process involves the functionalization of an aryl C(sp^2)-H bond and the addition/protonolysis of an alkene C=C bond (Scheme 4-60).

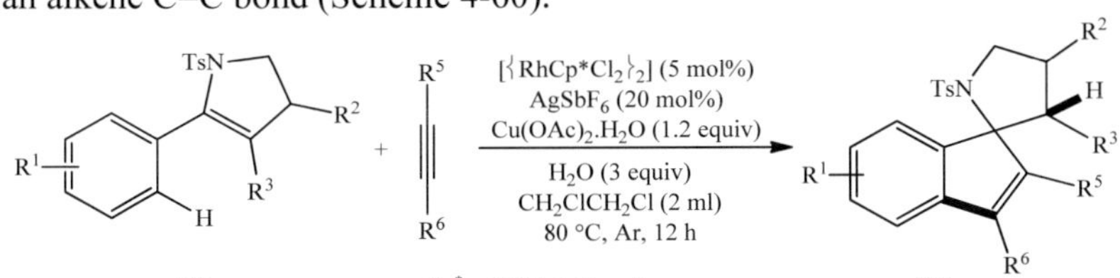

189 Cp* = C5Me5, Ts = 4-toluenesulfonyl 190

Scheme 4-60. Rh(III)-Catalyzed [3+2] annulation of 5-aryl-2,3-dihydro-1*H*-pyrroles **189**.

Another example of a Rh-catalyzed cycloaddition reaction was reported in 2016 by Reddy et al. [54]. In this reaction, the intramolecular cycloaddition of oxonium ylides **193** generated from 3-diazooxindole **191** and a bifunctional substrate i.e., 2-methyl-2-(4-methylpent-3-en-1-yl)-2*H*-chromene-3-carbaldehyde **192** in the presence 5 mol% of $Rh_2(OAc)_4$ lead to produce a highly substituted 3,3,5a-trimethyl-3a,5,5a,11b-tetrahydro-3*H*,4*H*-spiro[furo[2,3-*a*]-xanthene-2,3'-indolin]-2'-ones **194** in good yields with high diastereoselectivity (Scheme 4-61).

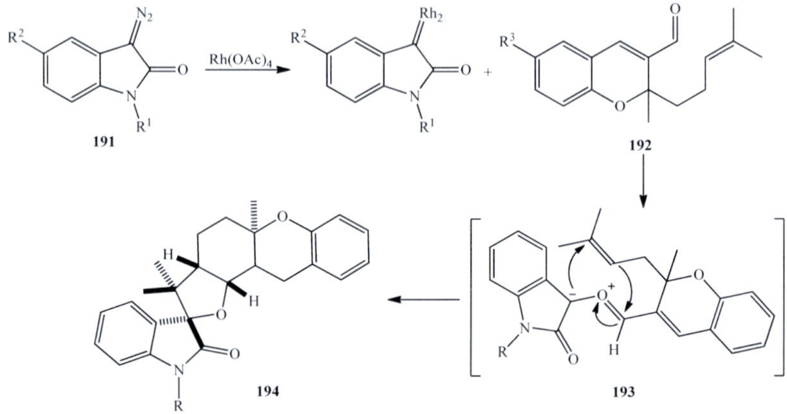

Scheme 4-61. Intramolecular [3+2] cycloaddition of oxonium ylides **193**.

In the same year, the Kim group noted the Rh(III)-catalyzed [3+2] annulations reaction of a range of *N*-acyl ketimines **196** produced *in situ* from 3-hydroxyisoindolinones **195** with various activated olefins to provide biologically important spiroindanes/spiroindenes **197** (Scheme 4-62). The synthesized compounds were evaluated for *in vitro* anti-cancer activity against different human cancer cell lines. Notably, quinone-containing spiroindenes demonstrated potent anti-cancer activity about two to three times stronger than the anti-cancer agent doxorubicin [55].

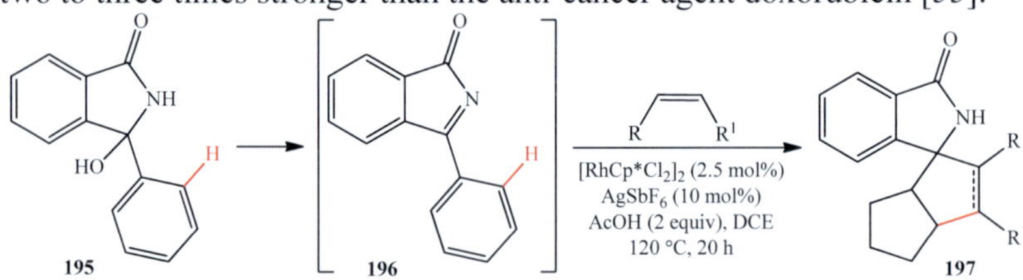

Scheme 4-62. Rh-Catalyzed [3+2] annulation of cyclic *N*-acyl ketimines **196** with activated olefins.

4.3.2.3. Other transition metal-catalyzed [3+2] cycloaddition reactions

Yan et al. conducted a study in 2017 investigating the synthesis of functionalized spiro[indene-2,1′-pyrrolo[2,1-*a*]isoquinoline]s **200** using a Ru-catalyzed oxidative [3+2] cycloaddition reaction under visible-light irradiation [56]. In this study, alkyl 2-(3,4-dihydroisoquinolin-2-yl) acetates **198** with 2-arylidene-1,3-indandiones **199** in the presence of [Ru(bpy$_3$)Cl$_2$] (10 mol%) as a sensitizer was irradiated with visible light to afford the corresponding functionalized spiro[indene-2,1′-pyrrolo[2,1-*a*]-

isoquinoline]s **200** in satisfactory yields and high diastereoselectivities (Scheme 4-63).

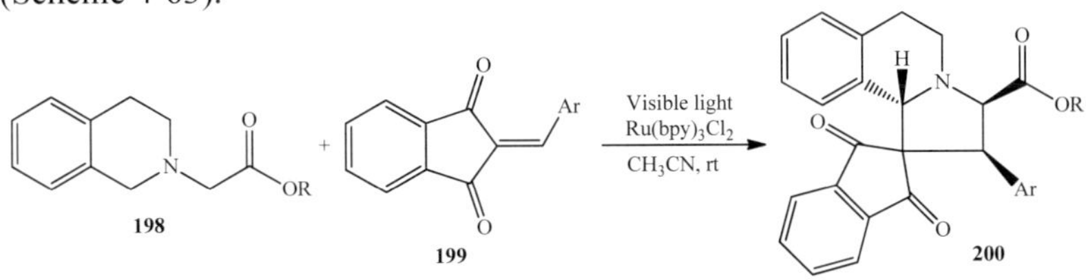

Scheme 4-63. Visible-light-catalyzed oxidative [3+2] cycloaddition reaction.

In another study, the catalytic asymmetric *exo′*-selective [3+2] cycloaddition of methylene-indolinones **172** with imino esters for the construction of new diastereomers of spiro[pyrrolidin-3,3′-oxindole] **201** was reported by Awata et al. (Scheme 4-64). By employing the imidazoline–aminophenol ligand [Ni(OAc)$_2$–(**L**)] complex, the reaction proceeded in the stepwise Michael–Mannich reaction to provide the *exo′* adducts as stable isomers [57].

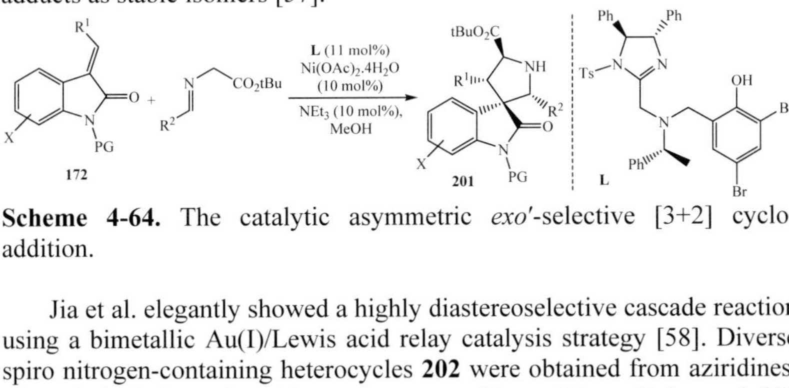

Scheme 4-64. The catalytic asymmetric *exo′*-selective [3+2] cycloaddition.

Jia et al. elegantly showed a highly diastereoselective cascade reaction using a bimetallic Au(I)/Lewis acid relay catalysis strategy [58]. Diverse spiro nitrogen-containing heterocycles **202** were obtained from aziridines, alkynyl alcohols, and amides under very mild conditions (Scheme 4-65), indicating this strategy's advantage in building up molecular complexity from simple starting materials.

Bis(imidazolidine)pyridine (PyBidine)-Cu(OTf)$_2$ complex was used by Arai et al. in an asymmetric [3+2] cycloaddition of methylene-indolinones with various imino esters to furnish spiro[pyrrolidin-3,3′-oxindole]s **203** in up to 98% ee (Scheme 4-66). PyBidine-Cu(OTf)$_2$ complex displays that imidazolidine-NH can harmonize with the Cu^{2+} cation for constructing a new dual activation system [59].

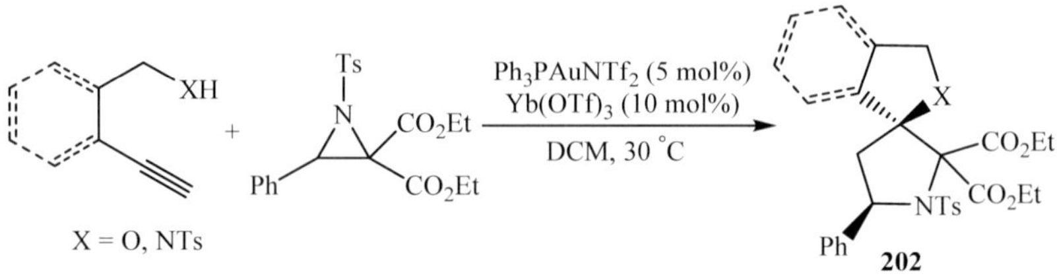

Scheme 4-65. Synthesis of diverse nitrogen containing spiro heterocycles **202**.

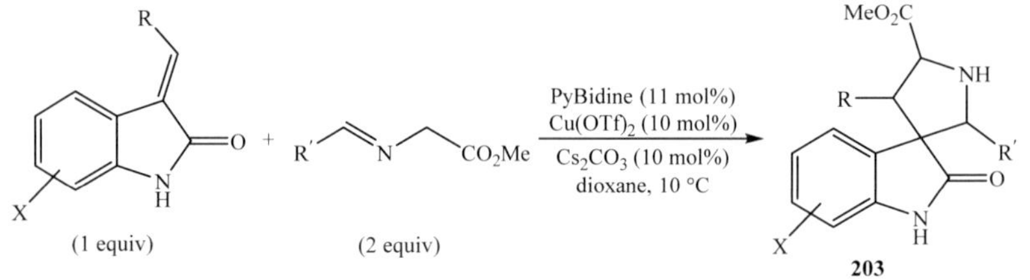

Scheme 4-66. PyBidine–Cu(OTf)$_2$-Catalyzed asymmetric [3+2] cycloaddition.

Additionally, the synthesis of diversely substituted spirooxindolo-2-iminothiazolidine scaffolds **205** is accomplished by employing $FeCl_3$ as a catalyst [60]. This reaction proceeds through the Lewis acid-mediated activation of spiro-oxindole aziridines **204** to produce a 1,3-dipole which undergoes intramolecular cycloaddition with different aryl/alkyl isothiocyantes (Scheme 4-67).

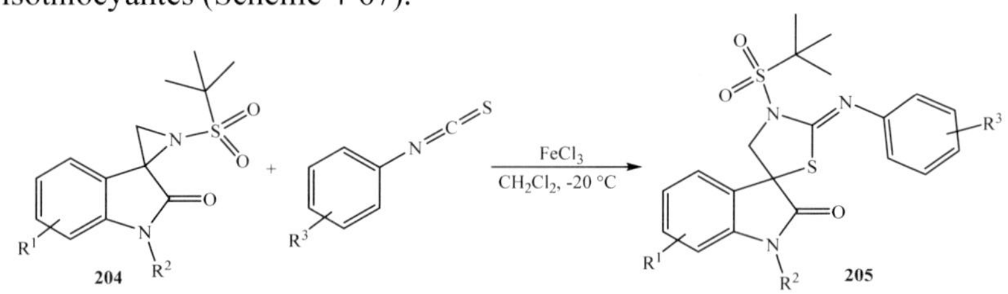

Scheme 4-67. $FeCl_3$-Catalyzed [3+ 2] cycloaddition reaction.

Raghunathan et al. [61] introduced the use of a rare mineral Ball-clay as a solid support and $ZrOCl_2.8H_2O$ as a catalyst for [3+2] cycloaddition reactions azomethine ylides derived from acenaphthenequinone and sarcosine/L-proline with (*E*)-2-oxoindolino-3-ylidene acetophenones **206** which led to the formation of dispiro-oxindolo-pyrrolidines and pyrrolizidines **207-210** (Scheme 4-68).

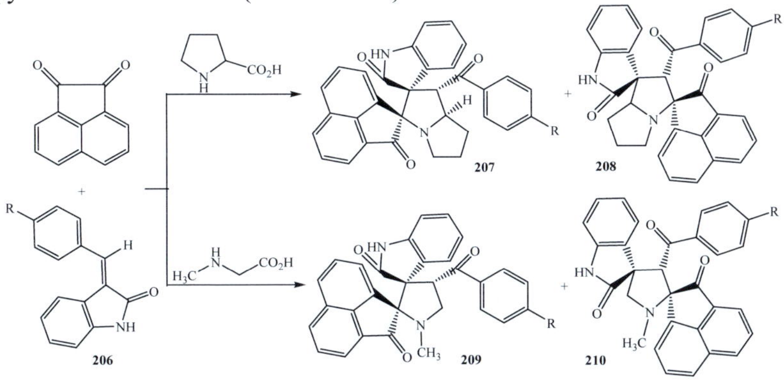

Scheme 4-68. Synthesis of dispiro-oxindolopyrrolidines and –pyrrolizidines **207-210**.

4.3.3. Diels–Alder cycloaddition reactions

4.3.3.1. Pd-Catalyzed Diels–Alder cycloaddition reactions

In 2002, Meijere et al. reported a new highly efficient three-component domino Heck-Diels-Alder reaction with bicyclopropylidene to give access to spiro[2.5]oct-4-enes [62]. Bicyclo-propylidene **211** was found to surpass even methyl acrylate in its rate of undergoing carbopalladation with aryl- or alkenylpalladium species, which led to substituted allylidene-cyclopropanes **212**, mainly in high yields. This diene **212** reacts with dienophiles in a Diels-Alder fashion to give spiro[2.5]octenes **213** in good yields (Scheme 4-69).

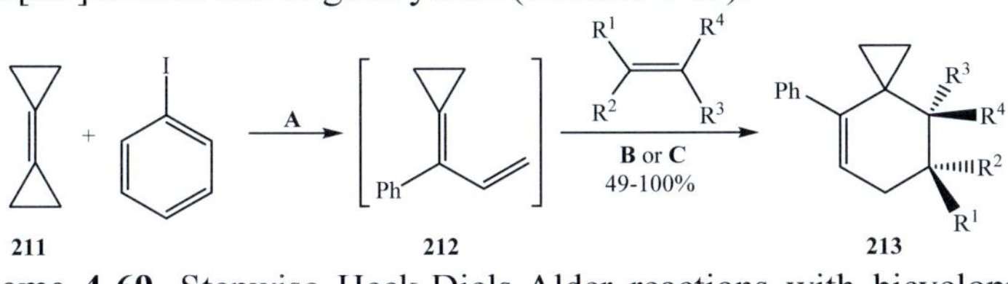

Scheme 4-69. Stepwise Heck-Diels-Alder reactions with bicyclopropylidene **211**. **A**: $CDCl_3$, 12 h; **B**: $Pd(OAc)_2$, PPh_3, C_6D_6; **C**: $Pd(OAc)_2$, PPh_3, Et_3N, DMF.

Later, Shintani et al. presented a [4+2] cycloaddition reaction of γ-methylidene-δ-valero-lactones **214** with methyl acrylate, an electron deficient olefin, in the presence of 5 mol% of $PdCp(\eta^3\text{-}C_3H_5)/P(O\textit{i}\text{-}Pr)_3$

catalyst at 40 °C, that affords spiro[2.4]heptanes **215** in high yield (87-97%) (Scheme 4-70) [63].

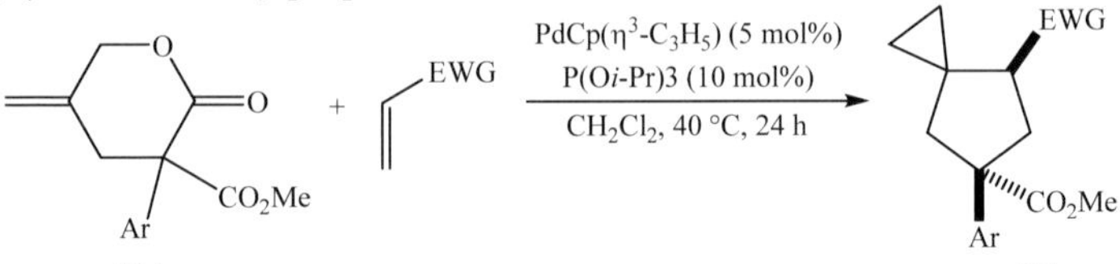

Scheme 4-70. Pd-Catalyzed synthesis of spiro[2.4]heptanes **215**.

In 2008, Müller developed a new domino reaction, involving insertion, coupling, isomerization, and Diels-Alder steps for the synthesis of spirocyclic benzofuranones **219** and dihydroindolones **220**. This reaction utilizes propargyl allyl ethers **216** and either alkynoyl *ortho*-iodo phenolester **217** or alkynoyl *ortho*-iodo anilides **218** as substrates, and is catalyzed by palladium (Scheme 4-71). Density functional theory (DFT) and Møller–Plesset perturbation theory (MP2) calculations show that the Diels-Alder step is thermodynamically and kinetically favored over a Claisen rearrangement in the product-forming step. The solid-state structure of these poly(hetero)cyclic frameworks reveals significant substituent effects on the torsional angles between the central butadiene chromophore and the terminal (hetero)aryl substituents [64].

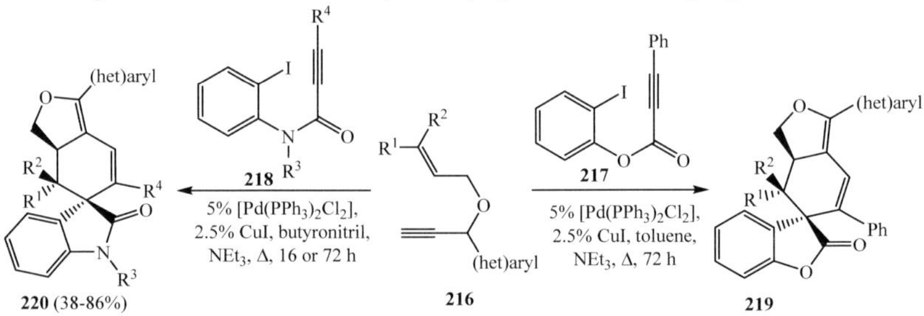

Scheme 4-71. Insertion–coupling–isomerization–Diels–Alder domino reaction to spirocyclic benzo-furanones **219** and dihydroindolones **220**.

Muthukrishnan et al. reported the successful synthesis of spiro[5.5] undeca-1,4-dien-3-ones **222** through the 1,6-conjugate addition initiated formal [4+2] annulation reaction between sulfonyl allenols **221** and *p*-QMs [65]. The reaction was carried out in DMF in the presence of $Pd(PPh_3)_4$ and K_2CO_3 as a base, providing the expected products in good to excellent yields (Scheme 4-72).

A plausible mechanism for this reaction is depicted in Scheme 4-72. Firstly, the oxidative addition of sulfonyl allenol **221** to Pd(0) generates the Pd-alkoxide **A** which subsequently undergoes 4-*exo*-dig metal-catalyzed oxycyclization to form the π-allyl oxetane intermediate **B**. The ring opening of intermediate **B** produces the reactive diene Pd-complex **C**. Then, the isomerization of intermediate **C** generates intermediate **D** which on deprotonation under basic conditions undergoes 1,6-conjugate addition with *p*-QM to form intermediate **E**. The subsequent ring closure and dearomatization of intermediate **E** afford the desired spirocyclic-framework **222**.

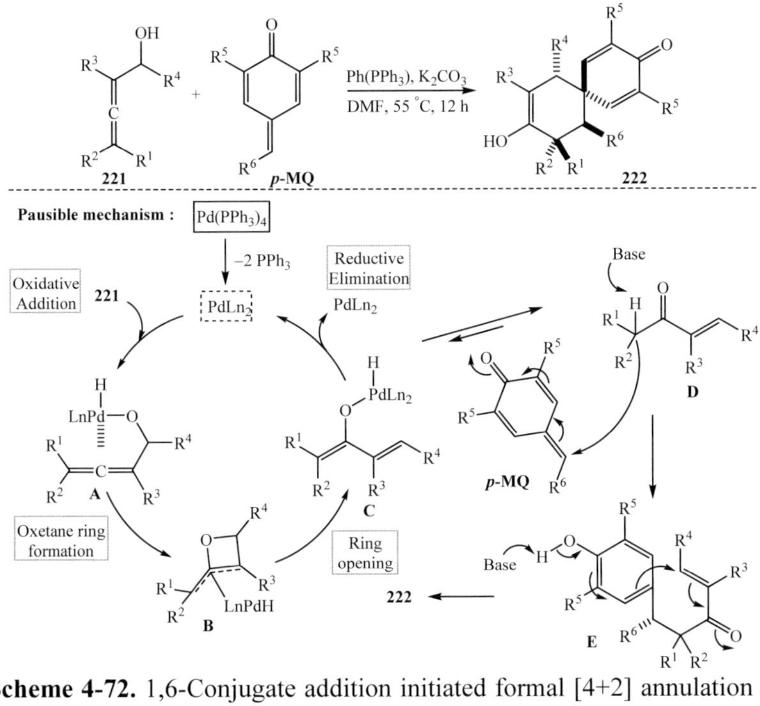

Scheme 4-72. 1,6-Conjugate addition initiated formal [4+2] annulation of *p*-QM with sulfonyl allenols **221**.

Trost et al. developed a Pd-catalyzed enantioselective cycloaddition method for carbonylogous 1,4-dipoles **225**. This was achieved through the *in-situ* deprotonation of allylic carbonates **223** [66]. By using C_2-unsymmetric phosphoramidite as a supporting ligand **224**, this dipole was used for the asymmetric synthesis of chiral cyclohexanones **226** through a catalytic [4+2] cycloaddition. Cyclic acceptors, such as barbiturate

(**226A**), Meldrum's acid (**226B**), and 1*H*-indene-1,3(2*H*)-dione (**226C**), also successfully delivered the desired spiro product in 85% yield with an excellent 93% ee (Scheme 4-73).

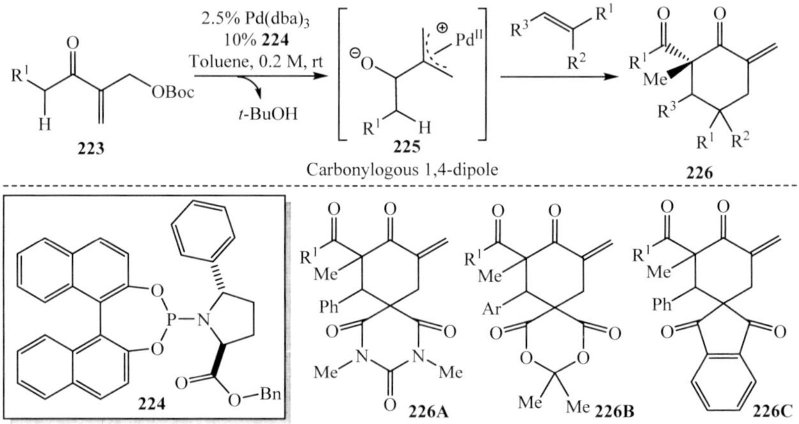

Scheme 4-73. Pd-Catalyzed enantioselective cycloaddition of carbonylogous 1,4-dipoles **225**.

In the same year, Guo et al. [67] reported the preparation of pharmacologically attractive chiral tetrahydropyran-fused spirocyclic scaffolds **232-234** by a Pd-catalyzed asymmetric [4+2] cycloaddition of 2-methylidenetrimethylene carbonate **227** with alkenes derived from pyrazolones, indandione or barbiturate **228-230** (Scheme 4-74). Under the catalysis of a complex formed from $Pd_2(dba)_3.CHCl_3$ (2.5 mol%) and axially chiral phosphoramidite ligand **231**, the desired compounds were obtained in good to excellent yields and with high enantioselectivities (up to 99% ee).

Compared to the well-established 1,3-DCs, Pd-catalyzed 1,4-DCs have limitations in their application to broad substrates due to the intrinsic thermodynamic properties of the intramolecular transformations and undesired isomerizations. To address this issue, Trost et al. developed a method to control and modulate the intermolecular reactivity of aliphatic 1,4-dipoles **235** by employing a ligated Pd-catalyst which enables two distinct cycloaddition pathways with a wide range of acceptors [68]. This study demonstrated that a diverse array of chiral 6-membered rings **236** and spiro[2.4]heptanes **237** could be formed with high yield and selectivity (Scheme 4-75).

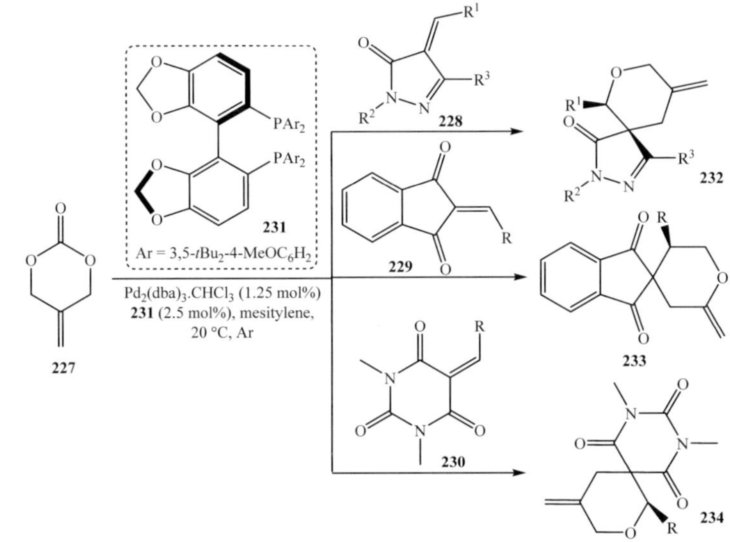

Scheme 4-74. Access to chiral tetrahydropyran-fused spirocyclic scaffolds **232-234**.

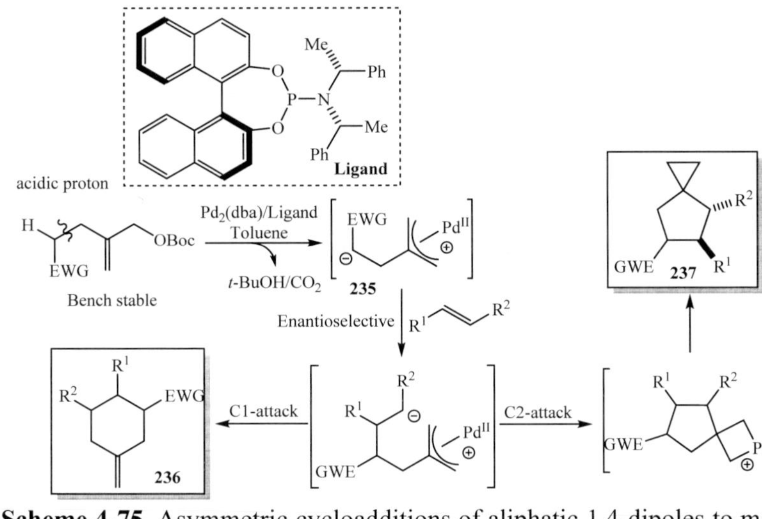

Scheme 4-75. Asymmetric cycloadditions of aliphatic 1,4-dipoles to make all-carbon rings **236-237**.

4.3.3.2. Other metal-catalyzed Diels–Alder cycloaddition reactions

In 2008, Tanaka et al. [69] reported the preparation of enantioenriched spirocyclic benzopyranones and isatin derivatives **239-240** by cationic Rh(I)/(*R*,*R*)-walphos-catalyzed highly enantioselective [4+2] annulation of 2-alkynyl benzaldehydes **238** with cyclic electron-deficient carbonyl compounds (Scheme 4-76).

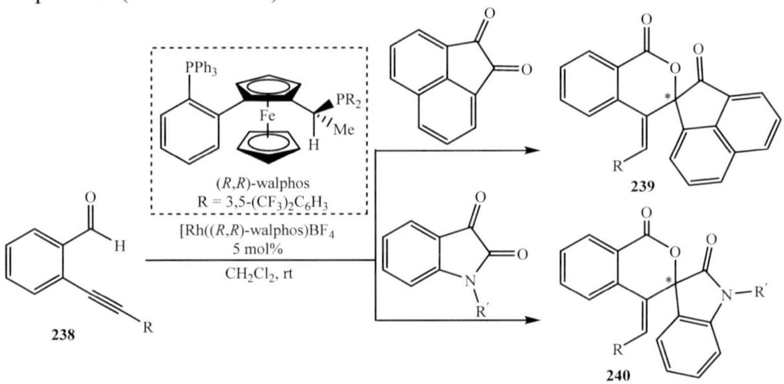

Scheme 4-76. Cationic Rh(I)/(*R*,*R*)-walphos-catalyzed highly enantio-selective [4+2] annulation.

An efficient $AuCl_3$-catalysed cascade reaction based on an initial cycloisomerization reaction of enynol or enynamines **241** to generate 1,3-butadienes **242/242′** that subsequently undergo an *in-situ* Diels–Alder reaction with suitable dienophiles was developed by the Barluenga group (Scheme 4-77). This process gives interesting fused or spirocyclic compounds **243-244** in good to excellent yields [70].

Shankaraiah et al. synthesized tetrahydrospiro[carbazole-4,3′-indoline] derivatives **246** through an AgOTf catalyzed alkyne-annulation/Diels–Alder cascade using *N*-tosyl-2-(but-3-en-1-yn-1-yl) aniline **245** in the presence of methyleneindolinones under mild reaction conditions as depicted in Scheme 4-78 [71].

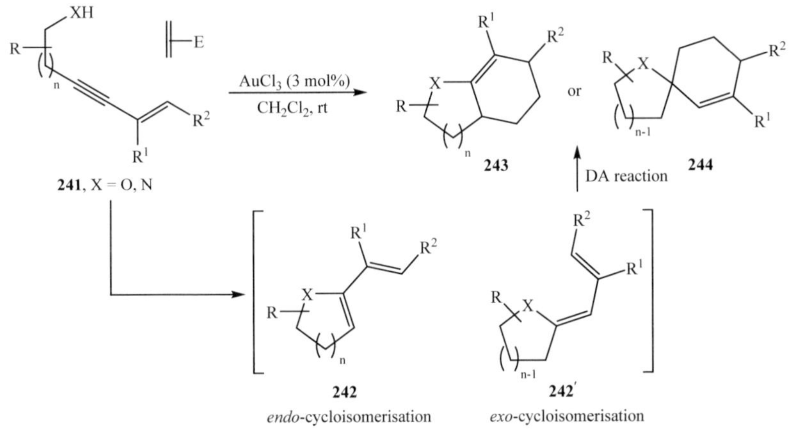

Scheme 4-77. The cascade cycloisomerization/Diels–Alder cycloaddition reaction catalyzed by $AuCl_3$.

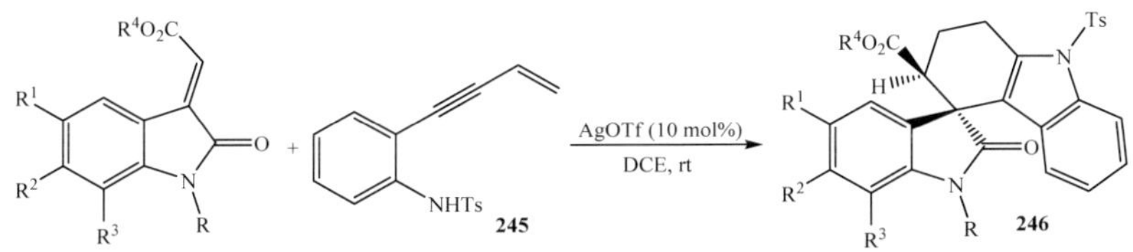

Scheme 4-78. Ag(I)-Catalyzed domino alkyne-annulation/Diels–Alder reaction.

Another elegant example of a Ag-catalyzed cascade reaction was performed by Tu et al. in 2017 [72]. They reported the new Ag-catalyzed dimerization and trimerization of β-alkynyl ketones **247** to yield spiro isochromenes **248** in water. Without water, spiro isochromenes **249** with methylene moiety were obtained through catalytic 6-*endo*-dig oxo-cyclization/[4+2] cycloaddition cascades by using pyridine as an additive, while, dispiro trimerization products **250** were prepared using $BiNPO_4H$ as a Brønsted acid catalyst through double C(sp^3)-H functionalization. The proposed mechanism is shown in Scheme 4-79.

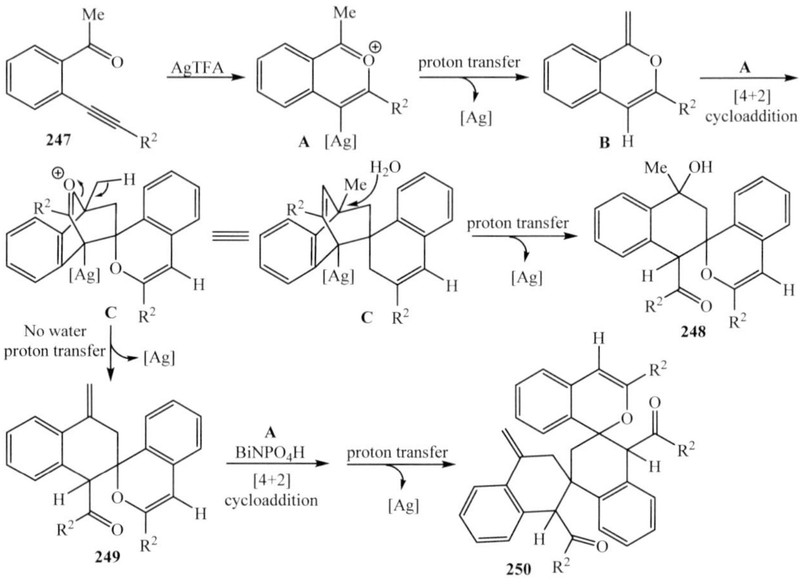

Scheme 4-79. Plausible reaction pathway.

Pahari et al. also found that the combination of the three-component domino Knoevenagel-hetero-Diels–Alder reaction allowed the construction of spirocyclic chromenones **251** [73]. Hence, 1,3-dicarbonyl compounds with ketones and alkynes or alkenes in the presence of a magnetically separable silica (Fe_3O_4@SiO_2) catalyst produced corresponding spiro compounds in good yields (Scheme 4-80).

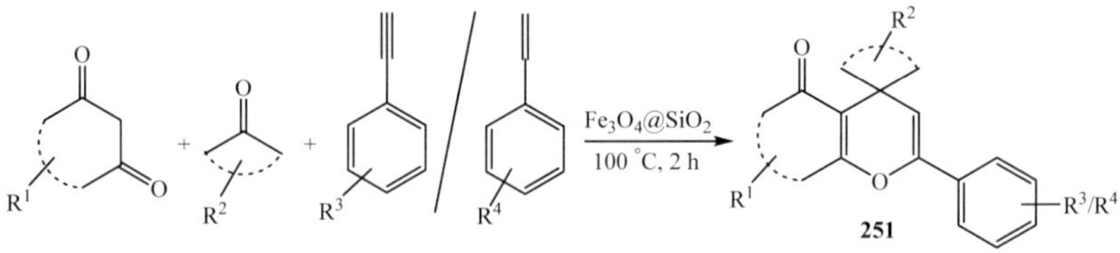

Scheme 4-80. Synthesis of spirochromenones **251** from 1,3-diketones, ketones, and alkynes or alkenes.

The Romo group developed a highly diastereo- and enantioselective Diels−Alder reaction catalyzed by an Evans' copper−bis(oxazoline) **254** complex to construct a highly functionalized spirolactam **255**, a key intermediate in the total synthesis of the marine toxin, gymnodimine [74]. The Evans' copper-bis(oxazoline)hexafluoroantimonate complex promoted the Diels-Alder reaction of α-methylene lactam **252** and diene

(*E*)-**253** leads to spirolactam **255** as a single diastereomer in 85% yield, as shown in Scheme 4-81.

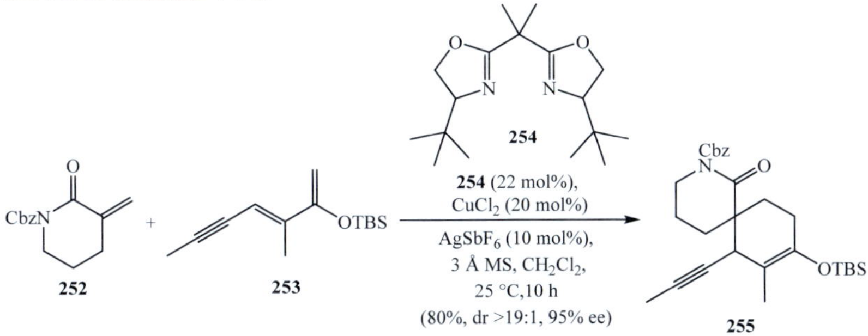

Scheme 4-81. Diastereo- and enantioselective Diels-Alder reaction leading to spirolactam **255**.

Recently, Wang et al. presented the inverse-electron-demand oxa-hetero-Diels–Alder (IEDDA) reactions of isatin-derived β,γ-unsaturated α-ketoesters **256** in the presence of 2.5–10 mol% of chiral bisoxazoline (BOX) **257**/Cu(II) complexes as Lewis acid catalysts for the synthesis of the corresponding chiral spiro oxindole-pyrans **258** in excellent yields with the excellent diastereo- and enantioselectivities (Scheme 4-82). These spiro oxindolepyran backbones are found in natural products and bioactive synthetic compounds [75].

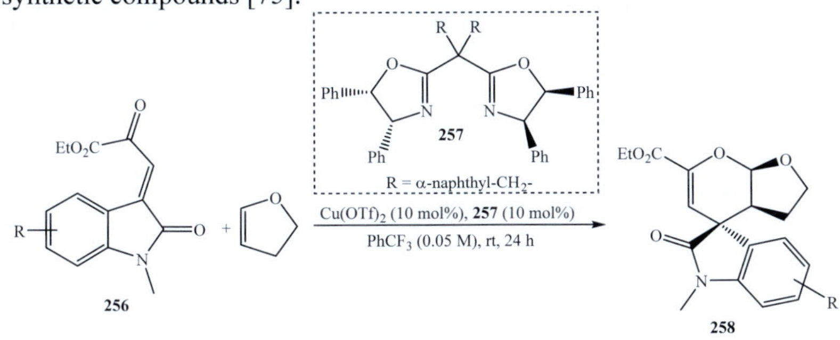

Scheme 4-82. Box-copper catalyzed asymmetric IEDDA.

4.3.3.3. Miscellaneous reactions

The Danheiser group [76] reported in 2000 that the treatment of thiol esters **259** with 1.05 equiv of lithium hexamethyldisilazide in THF at -78 °C gives the corresponding lithium enolates which are acylated by exposure to 1.2 equiv of trifluoroethy trifluoroacetate at -78 °C for 10 min. Treatment of the resulting *R*-trifluoroacetyl thiol esters with 1.5 equiv of

MsN_3 in the presence of 1.0 equiv of water and 1.5 equiv of Et_3N in acetonitrile (25 °C, 2.5 h), then, provides the desired diazo compounds **260** in good to excellent yield which subsequently adds to methylene cyclohexane **261** in the presence of Rh(II)acetate (0.02 equiv) to generate cycloadduct **262** (Scheme 4-83).

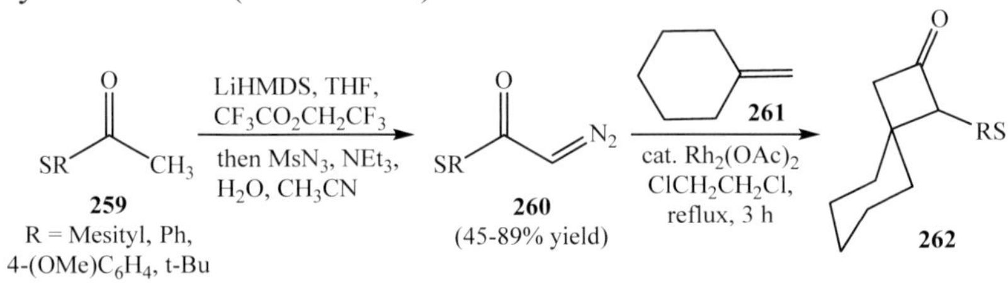

Scheme 4-83. The Danheiser group reported synthesis.

Rh(II) catalyzes the [4+2]/[2+2] cycloaddition reaction using *N*-protected isatin-3-arylimine and acyl ketene derived from α-diazocarbonyl compounds for the preparation of a novel class of spiro(oxindolyl)-oxazinones **267** and spiro(oxindolyl)-β-lactams **268**, as reported by Reddy et al. in 2012 [77]. Scheme 4-84 shows the mechanism of these reactions, starting with the formation of ketene intermediate **265** from 2-diazo-1,3-diketone **263** and its monoester counterpart **264** generated from the rhodium carbenoid through a Wolff rearrangement (Eq. 1). Acyl ketene **265** may then undergo either [4+2] or [2+2] cycloaddition depending on the diazodicarbonyl compound used. The reaction between 2-diazo-1,3-diketone **263** and isatin-3-arylimine **266** furnishes spiro(oxindolyl)-oxazinone **267** through [4+2] cycloaddition, while 2-diazoalkyl-aceto-acetate **264** (alkyl = methyl and ethyl) gives spiro(oxindolyl)-β-lactam **268** through [2+2] cycloaddition (Eq. 2).

The synthesis of spiro[2,2]dimethyl-benzofuran, bicyclo[4.2.0]octane]-7′-one **271**, with a fused and spirocyclic oxygen-containing rigid skeleton structure, was made possible by Ma et al. in 2017 [78] through a Rh(II)-catalyzed Wolff rearrangement/[2+2] cycloaddition reaction of 3-diazochroman-4-one **269** with an alkene in dichloromethane. Scheme 4-85 shows a plausible mechanism for this transformation. The presence of Rh(II) facilitates the initial loss of nitrogen gas from **269**, leading to the formation of benzofuran ketene intermediate **270** through the Wolff rearrangement. This intermediate is then captured by an alkene through the [2+2] cycloaddition to generate **271**. The cycloaddition occurs with regio-, stereo-, and chemoselectivity due to the orbital-controlled transformation in a concerted asynchronous mechanism. The configuration is dependent on the ketene substituents. Finally, **271** can easily be oxidized to spiro-

[2,2-dimethyl-benzofuran, hexahydroisobenzofuran-2′-one] **272** through Baeyer-Villiger oxidation.

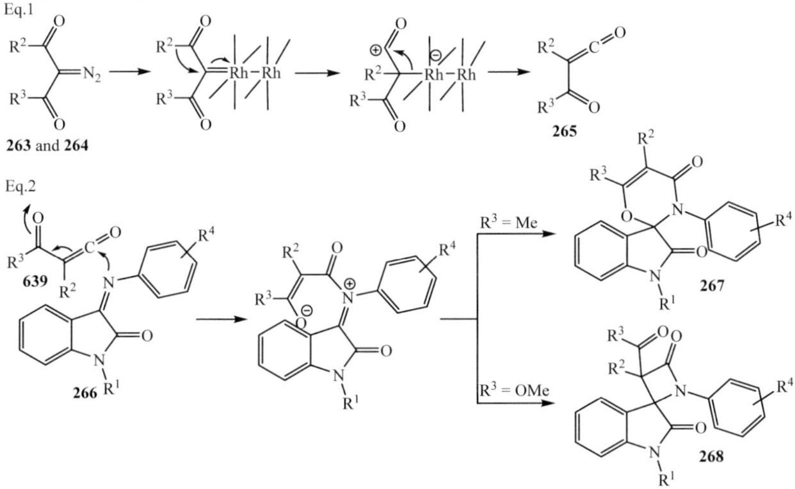

Scheme 4-84. A plausible mechanism of ketene generation through rhodium carbenoid (Eq.1) and a plausible mechanism for [4+2] and [2+2] cycloaddition (Eq. 2).

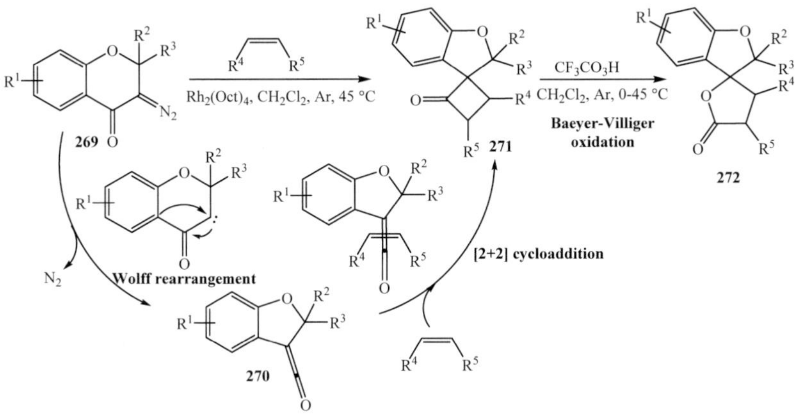

Scheme 4-85. The synthesis of spiro[2,2-dimethyl-benzofuran, hexahydro-isobenzofuran-2′-one] **272**.

In 2014, access to spiro cyclopentadiene-chroman-2,4-dione complexes **274** was reported by Wang et al. via a Pd-catalyzed [2+2+1] oxidative annulation of 4-hydroxycoumarins **273** with unactivated internal alkynes in the presence of oxidant $Cu(OAc)_2$ (Scheme 4-86) [79].

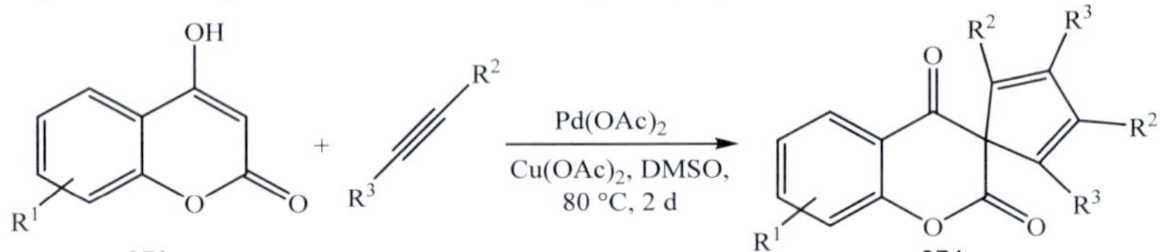

Scheme 4-86. Pd-Catalyzed [2+2+1] oxidative annulation.

The Luan group developed a novel Pd(0)-catalyzed norbornene (NBE)-mediated dearomatizing [2+2+1] spiroannulation of 1-bromo-2-naphthols **275** with aryl iodides and alkynes which leads to the rapid assembly of spirocarbocyclic scaffolds **276-277** (Scheme 4-87). With NBE as a transient mediator, this three-component cascade reaction was accomplished through successive Catellani-type C-H activation, unsymmetrical biaryl coupling, alkyne migratory insertion, and arene dearomatization [80].

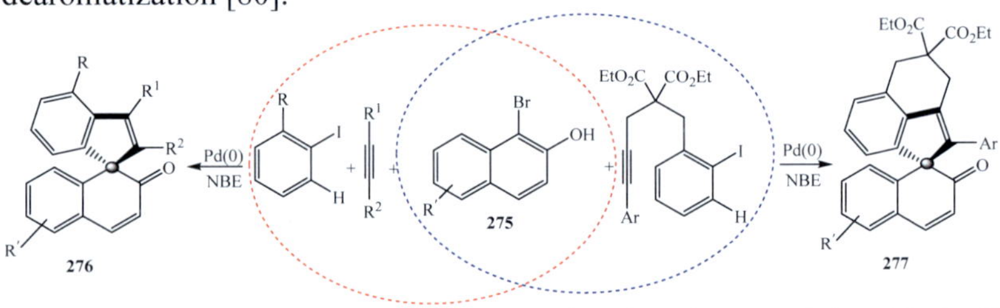

Scheme 4-87. Pd(0)-Catalyzed intermolecular dearomatizing [2+2+1] annulation.

Another domino reaction for the synthesis of fused/spiro/bridged ring-containing polycyclic compounds **280** was recently developed by Chen et al. which consists of the Pd-catalyzed [2+2+1] annulation of 3-iodo-chromones **278**, cyclic α-haloketones, and tetracyclododecene **279** (TCD) (Scheme 4-88) [81].

As illustrated in Scheme 4-89, Tanaka et al. reported the synthesis of enantioenriched C_2-symmetric spirobipyridine ligands **282** based on the cationic Rh(I)/(*R*)-Segphos or (*R*)-H_8-BINAP complex-catalyzed enantio-selective intramolecular double [2+2+2] cycloaddition of bis-diynenitriles **281** [82].

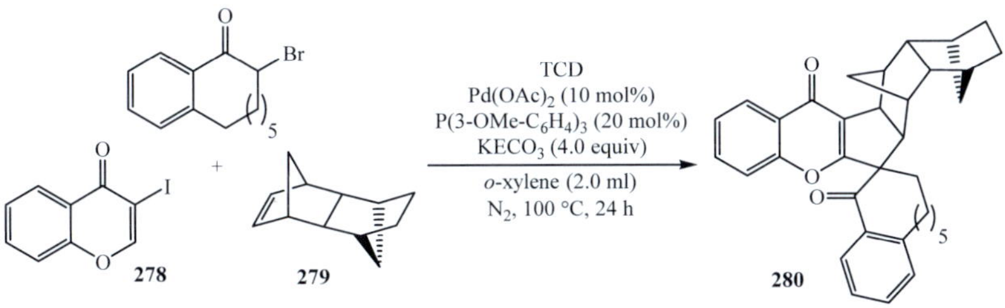

Scheme 4-88. Synthesis of fused/spiro/bridged ring-containing polycyclic compounds **280**.

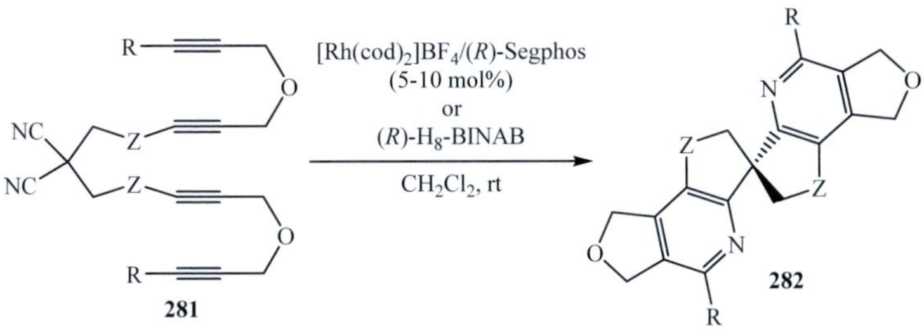

Scheme 4-89. Cationic Rh(I)/modified-BINAP catalyzed double [2+2+2] cycloaddition.

In 2010, Ramana et al. presented the synthesis of (spiro)-annulated tricyclic homochiral scaffolds **286-288** employing a ([RhCl(PPh₃)₃])-catalyzed [2+2+2] cyclotrimerization of diynes **283-285** with various symmetric di-substituted and mono-substituted alkynes (Scheme 4-90) [83].

A domino synthesis of dispiro-fused *N*-vinyl β-lactams **293** from diazo-Meldrum's acid **289** and 2*H*-azirines **290** or 5-alkoxyisoxazoles **291** through the "2-azabuta-1,3-diene **292** formation/ Staudinger ketene-imine cycloaddition" sequence was disclosed by Novikov et al. in 2019 (Scheme 4-91) [84].

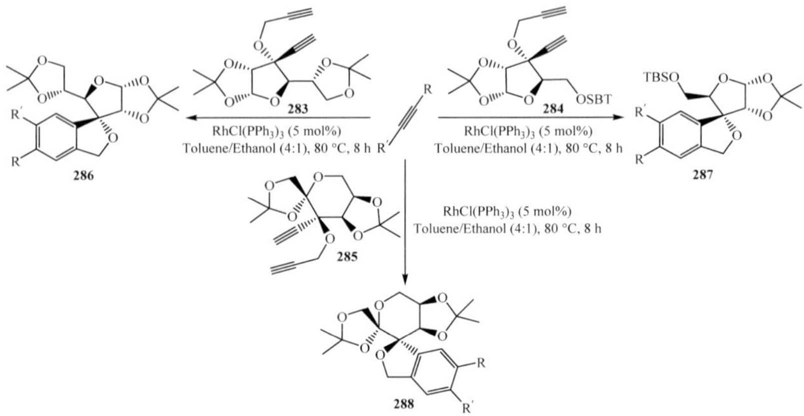

Scheme 4-90. Synthesis of (spiro)-annulated tricyclic homochiral scaffolds **286-288**.

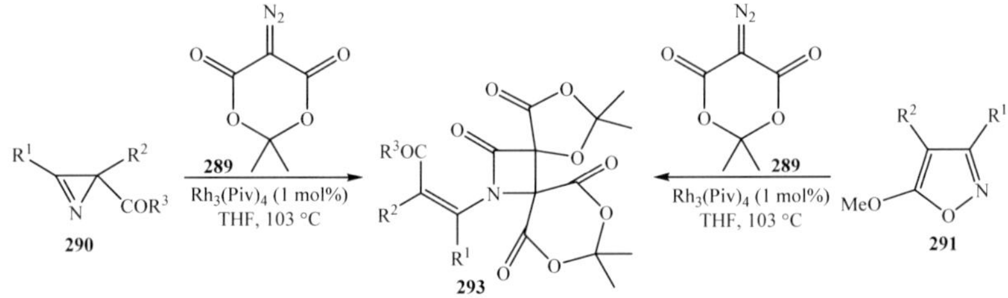

Scheme 4-91. Spiro β-lactams by Rh(II)-catalyzed reaction of diazocarbonyl compounds **289** with azirines/isoxazoles **290-291**.

The proposed mechanism to explain the formation of 2-azabuta-1,3-diene **292** intermediates is described in Scheme 4-92. The $Rh_2(Piv)_4$-catalyzed formation of the 2-azabuta-1,3-diene **292** intermediates in the first step of the reaction sequence proceeds via the addition of the Rh carbenoid to the azirines/isoxazoles **290/291** and affords the first example of the production of iminium-type ylides from diazo-Meldrum's acid **289**.

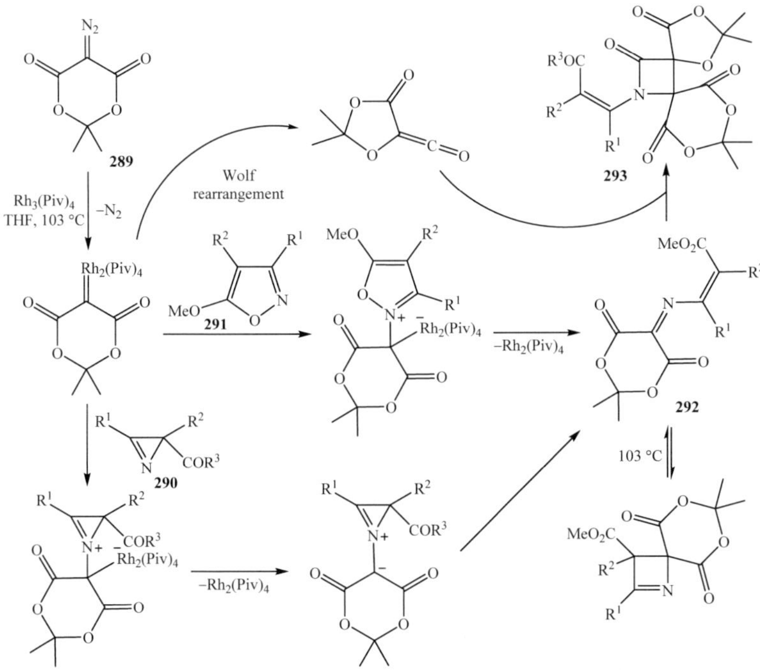

Scheme 4-92. A plausible mechanism for the formation of 2-azabuta-1,3-diene **292** intermediates.

Xu et al. reported a photocatalytic decarboxylative [2+2+1] cascade annulation of 1,6-enynes **294** with *N*-hydroxyphthalimide esters **295** to prepare indene-containing spiropolycyclic frameworks **296** through a sequence of radical addition, 5-*exo*-dig cyclization, 1,n-H shift and 5-*endo*-trig cyclization [85]. The reaction uses *fac*-Ir(ppy)$_3$ as a PC and TFA as an acidic additive under irradiation with blue LEDs in DMF (Scheme 4-93).

The proposed mechanism involves the irradiation of the *fac*-Ir(ppy)$_3$ PC to the excited state which initiates the reaction. The protonated cationic intermediate **294'** undergoes fragmentation and decarboxylation via a single-electron transfer (SET) with the excited species [*fac*-Ir(ppy)$_3$]*, leading to carbon-centered radical **A**. Radical **A** then adds to the C–C double bond of **293**, and after 5-*exo*-dig cyclization, an intramolecular 1,5-H atom shift occurs due to the reactivity and instability of vinylic radical **B**. Subsequently, 5-*endo*-trig cyclization occurs, providing intermediate **D**. A single-electron oxidation event of intermediate **D** generates cationic **E** and regenerates the photocatalyst. Alternatively, the reaction may proceed by a chain propagation process, where intermediate **D** can also oxidize **294′** to A. Finally, intermediate **E** is converted into the spiropolycyclic target **296** through further deprotonation.

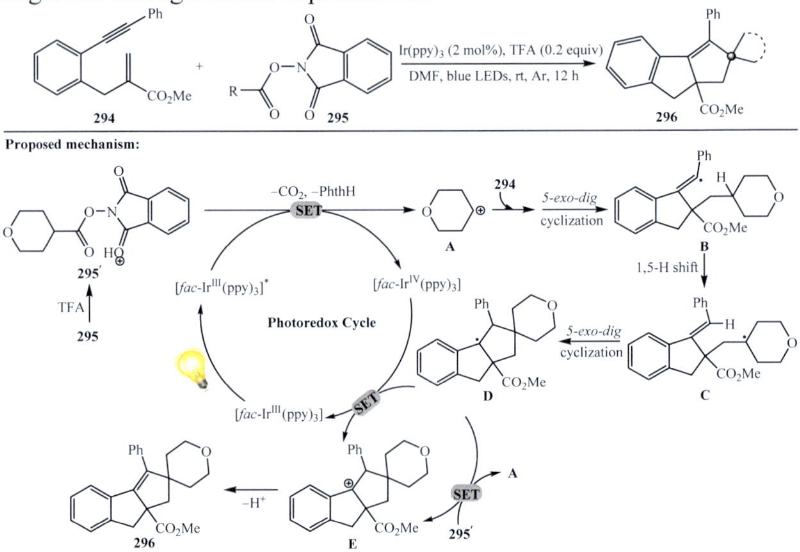

Scheme 4-93. Photocatalytic decarboxylative [2+2+1] cascade annulation and proposed mechanism.

Additionally, Göbel et al. [86] showed that the reaction of chiral *N,N′*-bis(aryl)-tetrahydro-pyrrolo[2,1-*c*][1,4]oxazine-3,4-diylidenediamines **297** with carbon monoxide and ethylene in the presence of $Ru_3(CO)_{12}$ will form spiro-lactams **298** by a [2+2+1] cycloaddition reaction, whereas the imine double bond next to the oxazine nitrogen atom remains unreacted. Also, the same spiro-lactams can be synthesized if iron carbonyl complexes of the *N,N′*-bis(aryl)tetrahydropyrrolo[2,1-*c*][1,4]oxazine-3,4-diylidenediamines **297** are introduced as starting compounds (Scheme 4-94).

Recently, the synthesis of spiro[imidazolidine-4,3′-indolin]-2′-imines **301** through Cu(I)-catalyzed [2+2+1] cycloaddition of 3-diazoindolin-2-imines **299** and triazines **300** was reported by Lu et al. (Scheme 4-95). The cascade process involves the formation of a copper–carbene intermediate and a [2+2+1] cycloaddition [87].

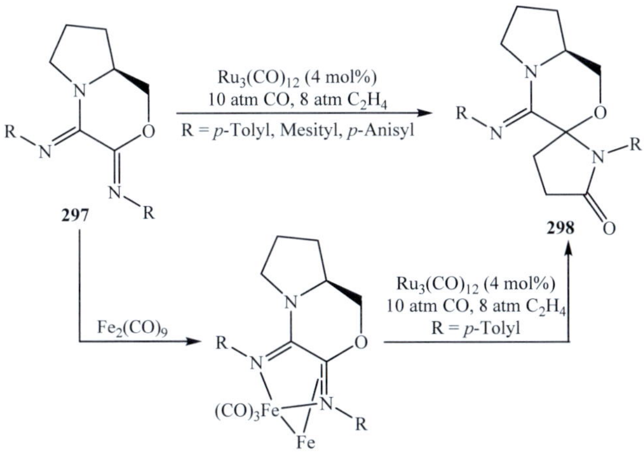

Scheme 4-94. One-pot Ru-catalyzed synthesis of spiro[pyrrolidin-2-one] derivatives **298**.

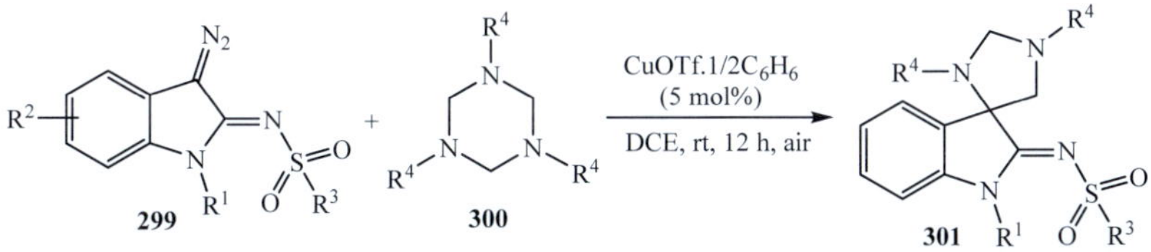

Scheme 4-95. Cu(I)-Catalyzed [2+2+1] cycloaddition of 3-diazoindolin-2-imines **299** and triazines **300**.

In 2020, Ouyang et al. exploited oximes **302** as five-atom units for the iminyl radical-triggered cascade 1,5-HAT/ [5+2] or [5+1] cycloaddition reactions with various electron-deficient alkenes for the synthesis of azepine **303** and spirotetrahydro-pyridines **304** (Scheme 4-96). This procure applies $FeCl_2$ as a catalyst and exhibits broad substrate scope and good functional group compatibility [88].

Scheme 4-97 depicts a plausible mechanism for the cascade 1,5-HAT /[5+2] annulation. Initially, SETs from Fe(II) to the oxime **302** leads to N−O bond cleavage, generating Fe(III) and an iminyl radical **A**. Subsequently, the generated radical **A** undergoes 1,5-HAT to form an alkyl carbon radical **B** which adds to the alkene to develop a new carbon radical **C**. The radical carbon **C** then adds to the intramolecular imine group to give intermediate **D** which can be oxidized by Fe(III) to generate an iminium ion **E** with simultaneous regeneration of the Fe(II) catalyst. After deprotonation, the azeping **302** is formed. Fe(III) salts can also induce the reaction, possibly because of the reduction from trivalent to divalent salts under the reaction conditions.

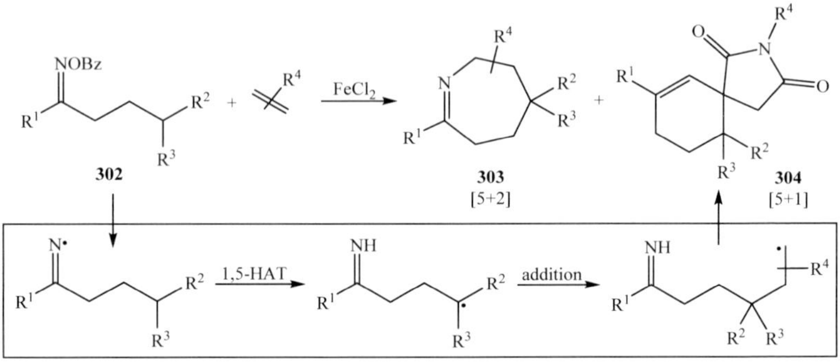

Scheme 4-96. Iminyl radical-triggered cascade 1,5-HAT/[5+2] or [5+1] annulation.

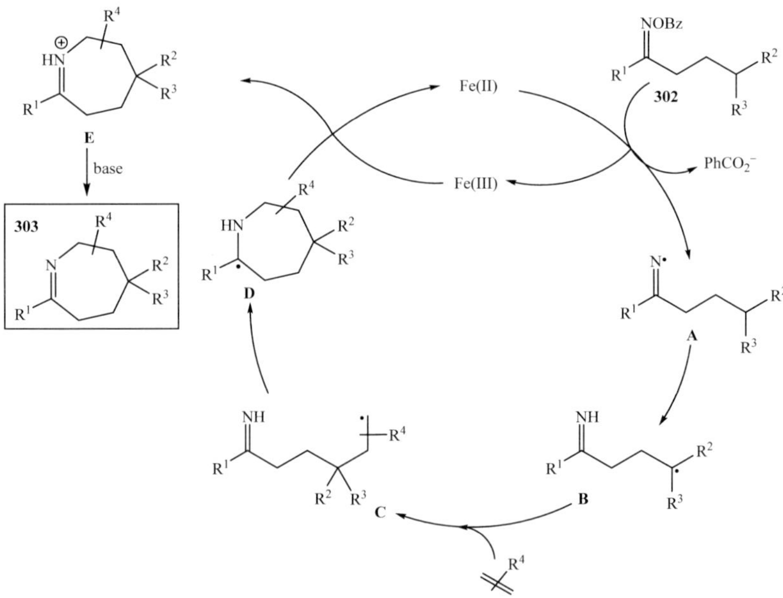

Scheme 4-97. A plausible mechanism for cascade 1,5-HAT/[5+2] annulation.

4.4. Cyclization reactions

In Chapter 3, different cyclization reactions have been discussed; and we will open a new vision of the metal contributions in this chapter.

4.4.1. Transition metal-catalyzed Conia-ene cyclization

The Conia-ene reaction, a type of ene reaction first reported by Conia et al. in 1975 [89], involves an intramolecular cyclization reaction between an enolizable carbonyl, such as an ester or ketone, and an alkyne or alkene, resulting in a cyclic product with a new carbon-carbon bond. Over the past two decades, the rapid advancement of transition-metal catalysis has had a profound impact on the asymmetric Conia-ene reaction. In this context, several transition metal-catalyzed Conia-ene reactions have been developed for the synthesis of spiranes.

One such example is the Ag(I)-catalyzed intramolecular addition of silyl enol ethers to alkynes which allows for the diastereoselective synthesis of a range of spiro compounds **305/305′** (Scheme 4-98), as demonstrated by the Miesch group in 2012.

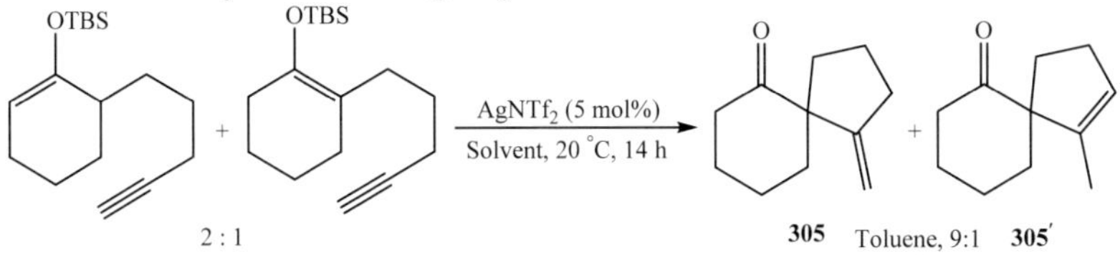

Scheme 4-98. An Ag-catalyzed spirocyclization of alkynyl silyl enol ethers.

A plausible mechanism for the Ag-catalyzed cycloisomerization reaction is outlined in Scheme 4-99 [90].

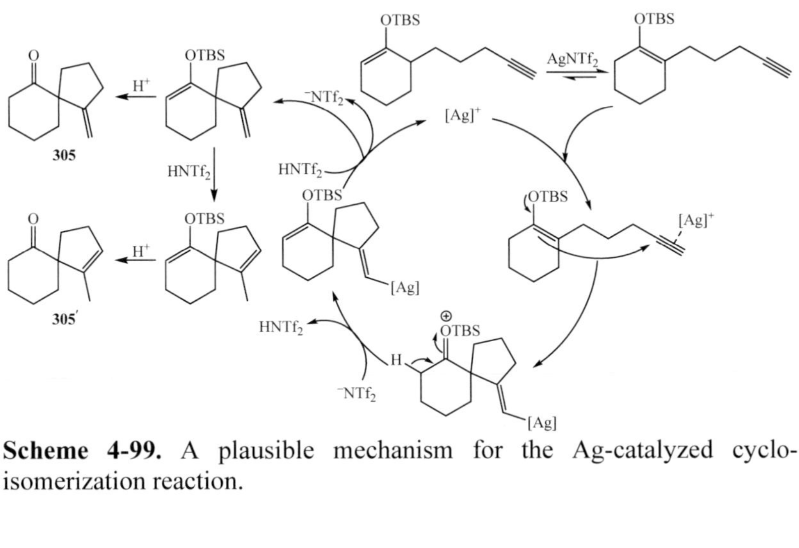

Scheme 4-99. A plausible mechanism for the Ag-catalyzed cyclo-isomerization reaction.

Liu et al. reported the chemo- and enantioselective synthesis of spiroketals **312** and spiroaminals **310** by using a combination of chiral Lewis acid *N,N′*-dioxide Ni(II) catalysis with achiral π-acid Au(I) catalysis [91]. The reaction between keto esters **306** and alkynyl alcohols **307** and amides **308** afforded the corresponding spiroketals **309** and spiroaminals **310** in up to 99% yield, 19:1 *endo*:*exo*, and more than 99% ee under mild reaction conditions (Scheme 4-100). Control experiments illustrate that the *N,N′*-dioxide ligand was essential for forming the spiro products. The tetradentate *N,N′*-dioxides, and the bidentate keto ester coordinate to the Ni^{2+} center to establish a complex with octahedral geometry as the intermediate TS. The ligand's neighboring amide group shields the *re*-face of the keto ester, and the enamide attacks from the *si*-face of the keto ester to generate the *endo*-**310**.

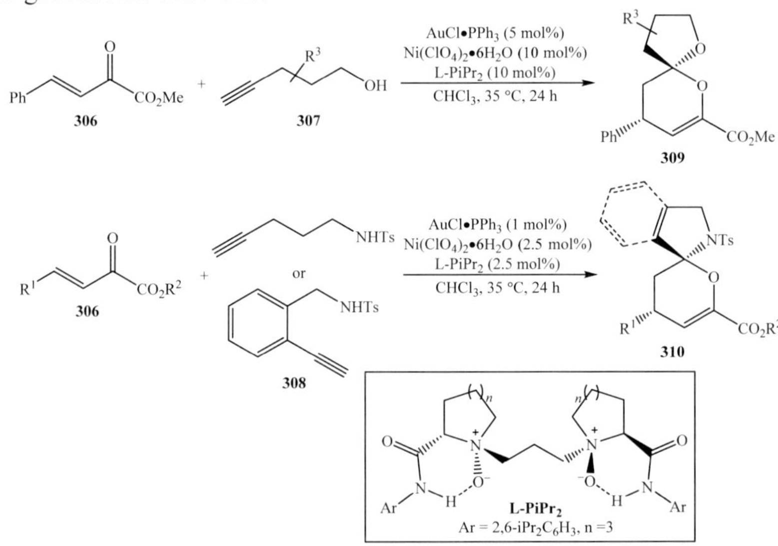

Scheme 4-100. Bimetallic Au(I)/chiral *N,N′*-dioxide Ni(II) asymmetric relay catalysis.

Another elegant example of the Conia-ene reaction was reported by Reddy et al. in 2013 [92]. In this case, an MBH-acetate of an acetylenic aldehyde bearing an alkyl group on the alkyne functionality **311** was used as a substrate for the construction of spirocyclic cyclopentenes **314**. In this process, the mentioned MBH-acetate **311** reacts with active methylene compounds **312** in the presence of K_2CO_3, forming mono-alkylated compounds **313**, which undergo 5-*exo*-dig carbocyclization to give the desired products **314**. CuI promotes the K_2CO_3-mediated 5-*exo*-dig carbocyclization through **I** and **II** involving anti-addition on the alkyne (Scheme 4-101).

This method is a new variation of the Conia-ene type reaction which affords cyclopentenes having a *Z*-geometry on the exocyclic olefin. In contrast, in the Conia-ene reaction, the products are cyclopentanes with *E*-geometry as it proceeds via a concerted transition state leading to *syn*-addition to the alkyne.

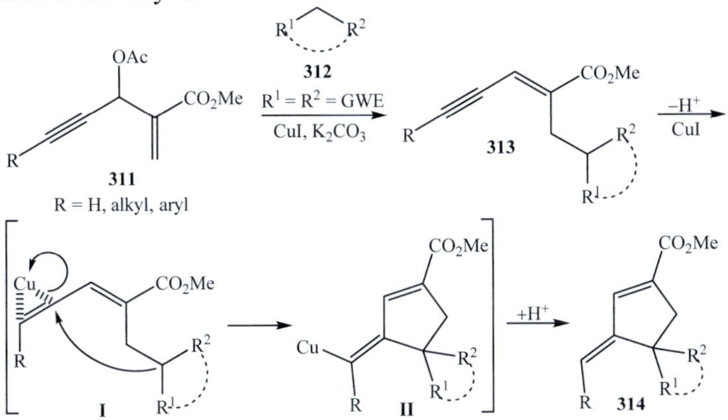

Scheme 4-101. Synthesis construction of spiro cyclic-cyclopentenes **314**.

The Miesch group nicely presented a spontaneous spirocyclization of keto-sulfonamides **315** via ynamides through a one-pot process for the preparation of azaspiro compounds **316**, which are building blocks for indole alkaloids. Push–pull ynamides were formed through Michael addition/elimination without Cu (Scheme 4-102) [93].

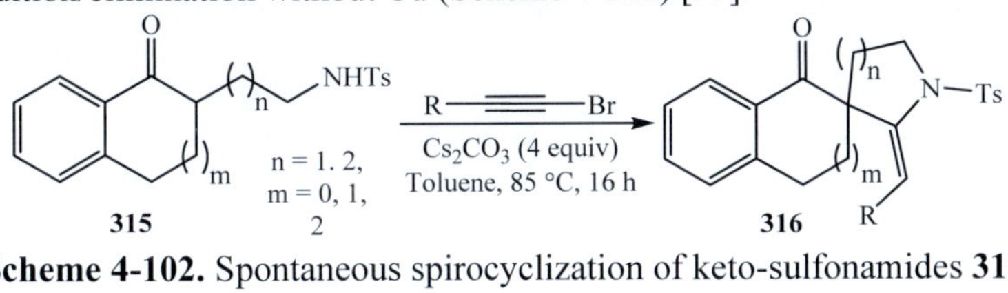

Scheme 4-102. Spontaneous spirocyclization of keto-sulfonamides **315**.

An efficient Pd-catalyzed enantioselective cyclization of 1,6-enamidynes **317-318** was developed to synthesize spirocyclic ring systems **319-320** [94]. Exploration of enamide ring size disclosed that five- and seven-membered rings were tolerated, with the seven-membered ring system providing a desired product with exceptional enantioselectivity (Scheme 4-103).

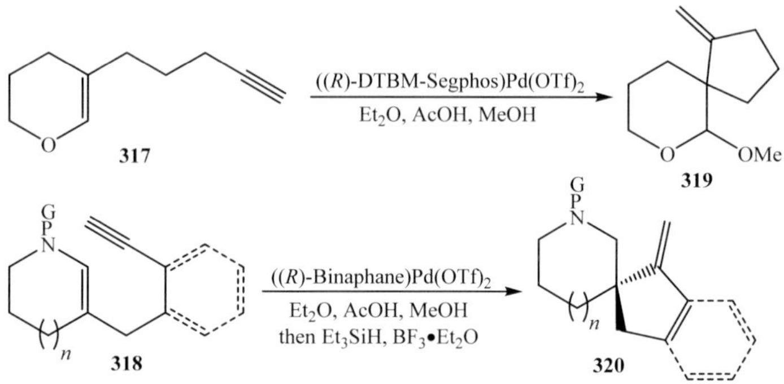

Scheme 4-103. Pd-catalyzed enantioselective cyclization of 1,6-enamidynes **317-318**.

In 2015, Córdova revealed for the first time that porous Pd(II)-polyimines are excellent catalysts for cooperatively catalyzed and enantioselective cascade reactions [95]. The synthesized Lewis acidic catalyst Pd^{2+}/**PP-1** by complexation (Scheme 4-104) was shown as a perfect heterogeneous catalyst for the dynamic kinetic asymmetric *trans*-formations between enals and propargylic nucleophiles together with a chiral amine co-catalyst **321**.

The Pd^{2+}/**PP-1**/chiral amine system delivered an excellent catalysis platform for synthesizing highly functionalized spirocyclic oxindoles **322** in a high yield with up to 97:3 dr and 99.5:0.5 enantiomeric ratio (er) (Scheme 4-105).

Additionally, Herrera et al. developed a new synthetic strategy that combined organocatalysis and metal catalysis to construct synergistic catalysis using prolines and Pd catalysts [96]. They obtained a variety of spiro pyrazolones **325** through the reaction of alkyne tethered pyrazolones **323** with enals under synergistic catalysis **324**. Proline bulky substituents **324** ($-CPh_2OSiMe_3$) consecutively act as steric hindrances and directing groups in a Michael/Conia-ene cascade reaction under synergistic catalysis **324** (Scheme 4-106).

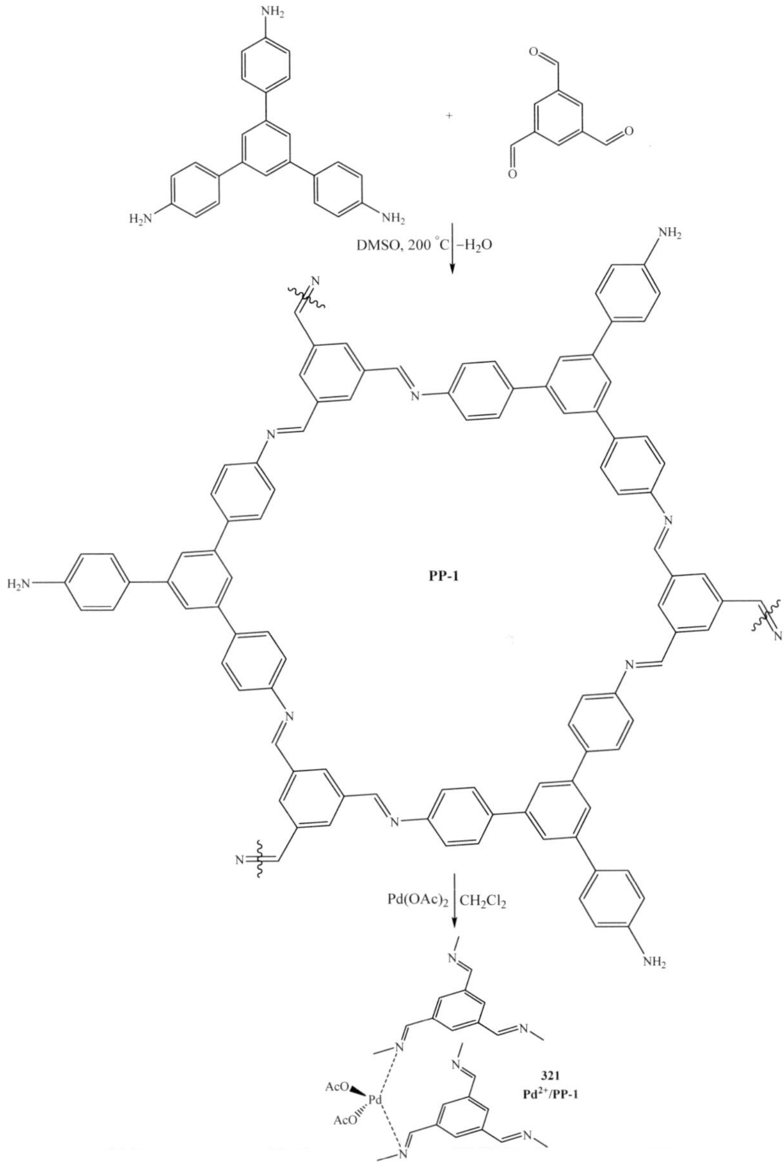

Scheme 4-104. Synthesis of Pd^{2+}/PP-1 **321**.

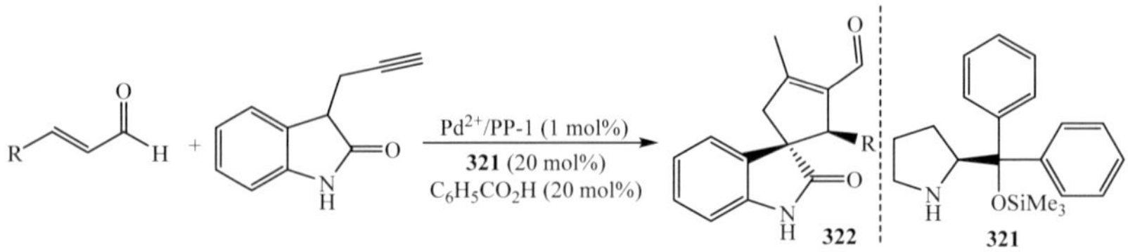

Scheme 4-105. Synthesis of highly functionalized spirocyclic oxindoles **322**.

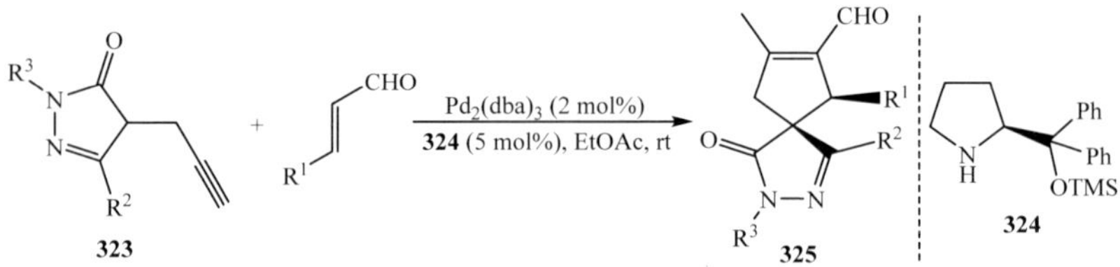

Scheme 4-106. Michael/Conia-ene cascade reaction under synergistic catalysis **324**.

On the other hand, $PtCl_2$ and $PPh_3AuCl/AgOTf$ catalysts have been successfully applied to synthesizing benzopyrones **327** and tricyclic spiro ketones **328** using hydrative-carbocyclization of oxodiyne substrates **326** in wet dioxane (Scheme 4-107). Carbocyclization with Pt and Au catalysts is distinct due to their altered regioselectivity in the oxo-assisted hydration of the neighboring alkyne carbons [97].

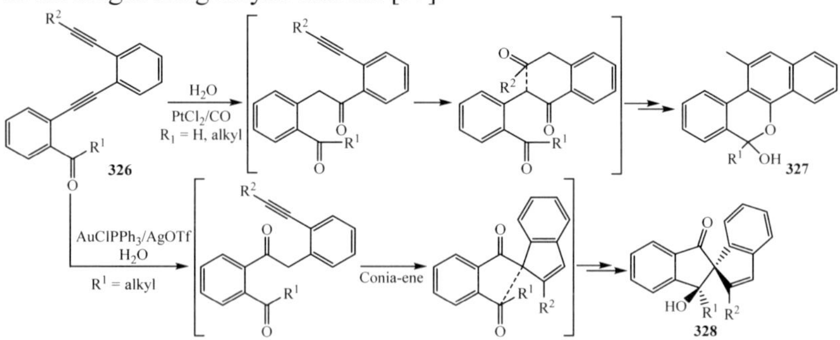

Scheme 4-107. Pt- and Au-catalyzed hydrative carbocyclization of oxo diynes **326**.

A convergent approach for the synthesis of diverse *N*-heterocycles was reported by the Sharma group [98], that involves trapping of diazo-derived rhodium carbenoids with Au-activated aminoalkynes and incorporates both the donor/acceptor (D/A) as well as acceptor/acceptor

(A/A) diazo carbonyls (Scheme 4-108). In this procedure, aminoalkynes **329** react with Meldrum's acids diazo **330** in the presence of $Rh_2(esp)_2$ to furnish the desired spiro compounds **331** in 65% yield.

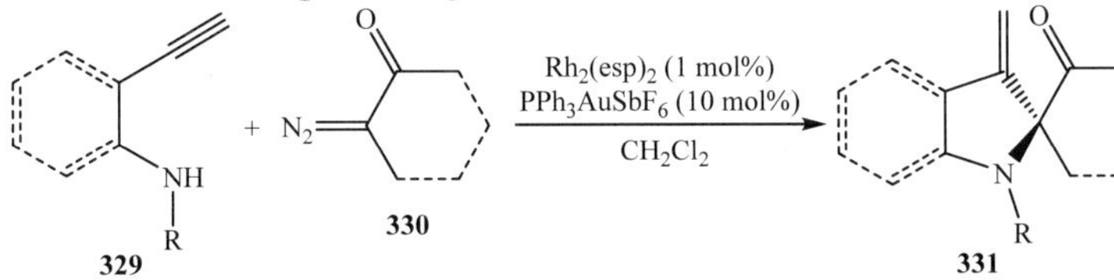

Scheme 4-108. Synthesis of spiro compounds **331** using aminoalkynes **329**.

Mechanistic investigations represented that the Rh(II)/Au(I) catalyzed reaction of amino-alkynes with D/A diazos is concerted, while the reaction with A/A diazo is stepwise and goes with carbene N-H insertion and a subsequent Conia-ene cyclization (Scheme 4-109).

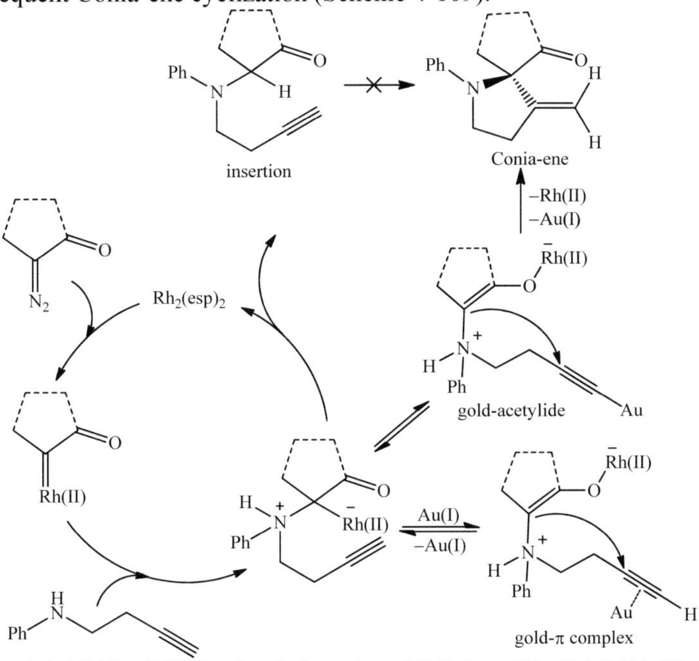

Scheme 4-109. Plausible reaction mechanism for the stepwise and synergistic Rh(II/Au(I)-catalyzed diazo N-H insertion/Conia-ene cascade.

Sharma et al. developed in 2019 an Rh(II)/Au(I) catalyzed carbene cascade approach [99]. The stereoselective synthesis of diverse spirocarbocycles **332** was carried out using a stepwise sp^2 C-H functionalization/Conia-ene cascade cyclization (Scheme 4-110).

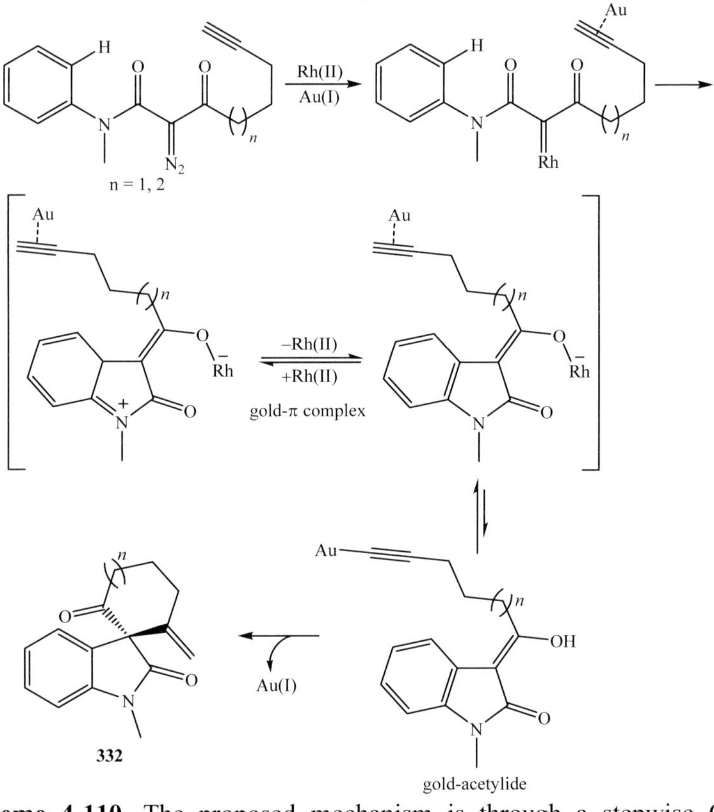

Scheme 4-110. The proposed mechanism is through a stepwise C–H functionalization and Conia-ene cascade cyclization.

4.4.2. Transition metal-catalyzed dearomative cyclization

As mentioned in Section 3.4.1, dearomative cyclization has emerged as a powerful strategy for constructing spirocycle compounds. As a result of extensive research over the past few decades, numerous elegant and efficient protocols for transition-metal-catalyzed dearomative cyclization to afford spiro compounds have been reported. Recent developments in this area are discussed in detail in this section.

In this context, functionalizing trifluoromethylation of activated alkynes **333** with the cheap reagent sodium trifluoromethanesulfinate ($NaSO_2CF_3$ or Langlois' reagent) and TBHP catalyzed by CuI (10%) has been demonstrated to furnish 3-(trifluoro-methyl)-spiro[4.5]trienones **334** in moderate yields [100]. This strategy involves an attachment of the CF_3 group with the electron-deficient carbon–carbon triple bond to form the spiro[4.5]decane skeleton through a tandem cyclization/dearomatization process, as shown in Scheme 4-111. The addition of MnO_2 as an oxidant improved the yield of final spiro products **334**.

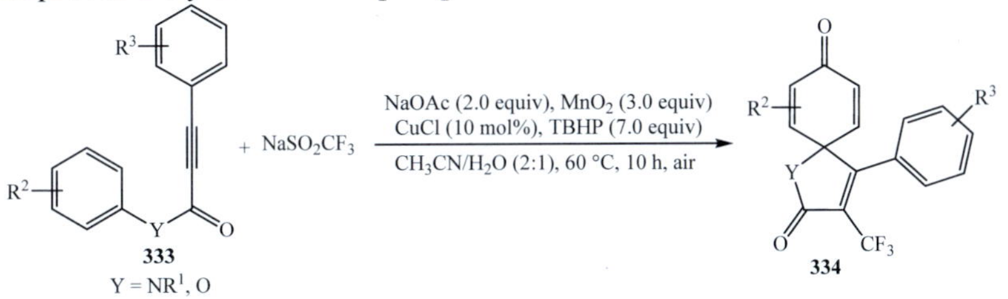

Scheme 4-111. Cu-Catalyzed difunctionalization of activated alkynes **333**.

The first Pd-catalyzed dearomative cyclization through modified Catellani-type C-H functionalization was elegantly reported in 2015 by Liang et al. who performed an efficient procedure for synthesizing spirocyclic indolenines **335** [101]. In this strategy, Pd-catalyzed norbornene-mediated tandem *ortho*-alkylation dearomative cyclization reaction of simple aryl halides with substituted indoles, leading to a new class of spirocyclic indolenines **335** bearing an all-carbon quaternary (Scheme 4-112).

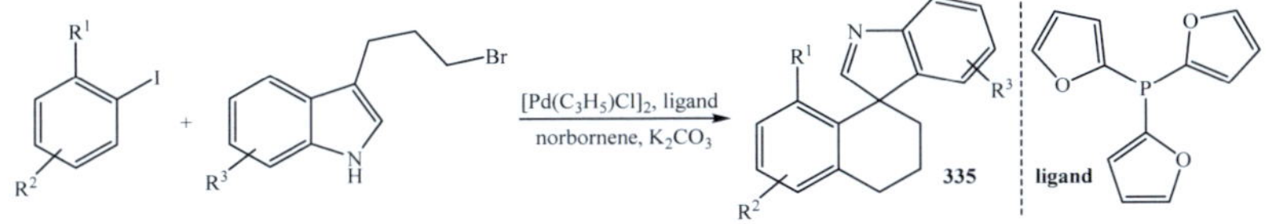

Scheme 4-112. Pd-Catalyzed dearomative cyclization.

A reaction pathway for Pd-catalyzed dearomative cyclization by norbornene-mediated sequence is represented in Scheme 4-113.

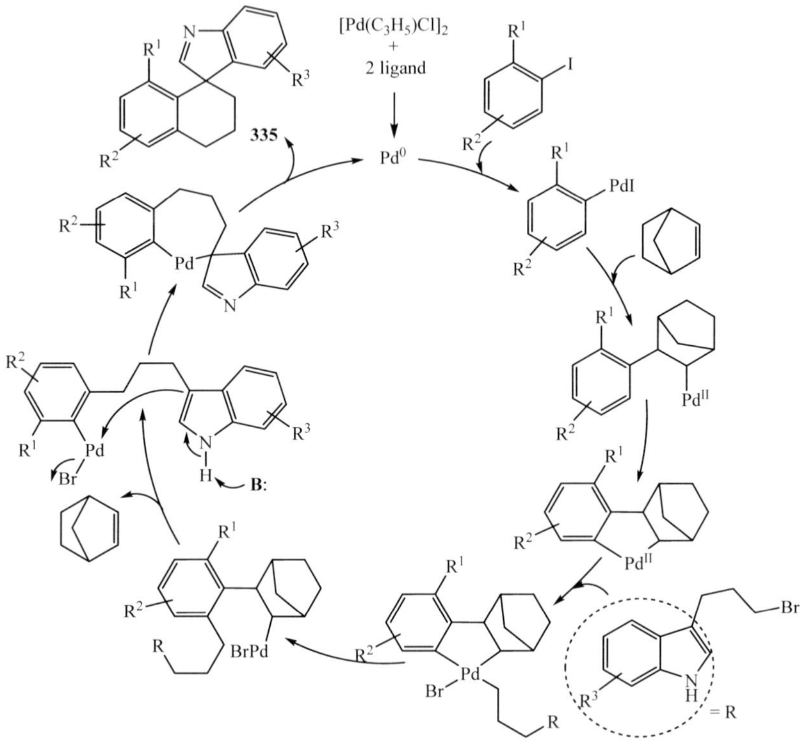

Scheme 4-113. Proposed reaction mechanism.

Zhang et al. developed an efficient visible light-induced intramolecular dearomative cyclization of α-bromo-*N*-benzyl-alkyl amides **336** to construct 2-azaspiro[4.5]decanes **337** in the presence of an Ir catalyst (Scheme 4-114) [102]. The proposed mechanism starts with the excitation of the Ir catalyst under visible light which undergoes a single electron transfer process with α-bromo-*N*-benzyl-alkylamide **336** to afford the electrophilic radical **I** and Ir(IV) metal complex. The key C-C bond is formed through a rapid regioselective addition of radical **I** to the *para*-position of the R^1O substituted electron-rich aromatic ring, thus forming radical **II**. Given that radical **II** has a lower barrier of oxidization due to the existence of the electron-rich R^1O group, a second single electron transfer event with Ir(IV) metal complex would afford the cation intermediate **III** and finally provide the photoredox catalytic cycle. The desired product **337** is produced from intermediate **III** in the presence of the base. The reaction proceeds smoothly under mild reaction conditions and affords the products in good to excellent yields (up to 98%).

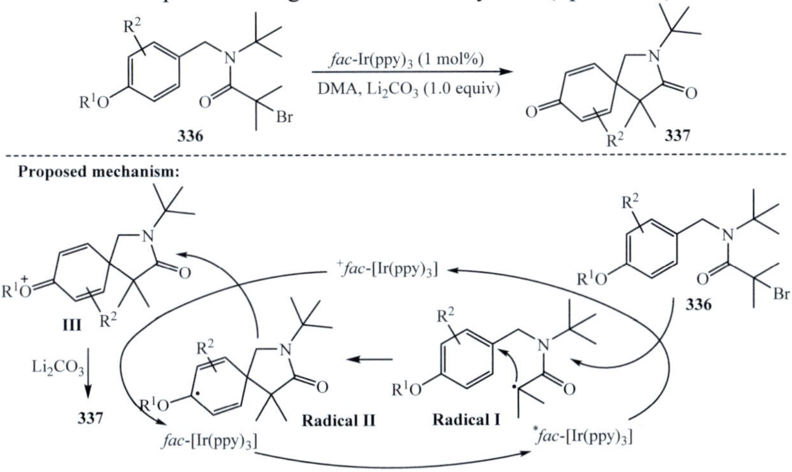

Scheme 4-114. Efficient construction of 2-azaspiro[4.5]decanes **337.**

Additionally, this group presented a visible-light-induced photo-catalytic intermolecular dearomative cyclization, starting from 2-bromo-1,3-dicarbonyl compounds **338** and alkynes using *fac*-Ir(ppy)₃ as a photocatalyst in the presence of an external base (Scheme 4-115). To complete the dearomative cyclization, an external base is required to neutralize the hydrobromic acid produced thereof [103].

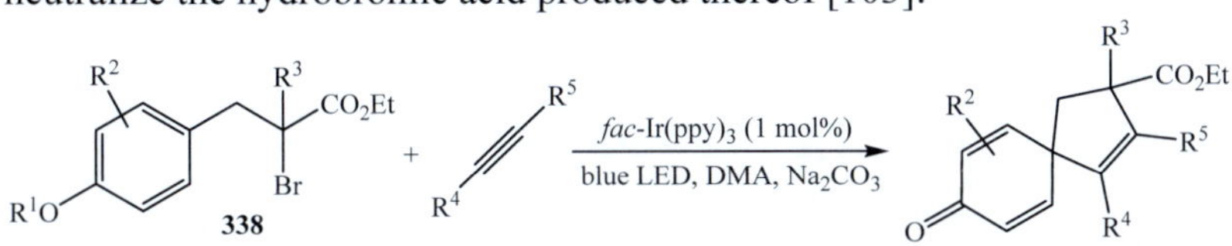

Scheme 4-115. Visible-light-induced intermolecular dearomative cyclization of **338**.

Another visible-light-driven regioselective dearomative cyclization strategy led to the formation of spiro[4.5]deca-1,6,9-trien-8-ones **340**. In the presence of H_2O, a variety of 2-benzyl-2-bromomalonates **339** smoothly undergo 5-*exo*-dig radical dearomative cyclization with alkynes to furnish the corresponding spiro[4,5]decanes **340** in moderate to good yield under oxidant-free conditions (Scheme 4-116) [104].

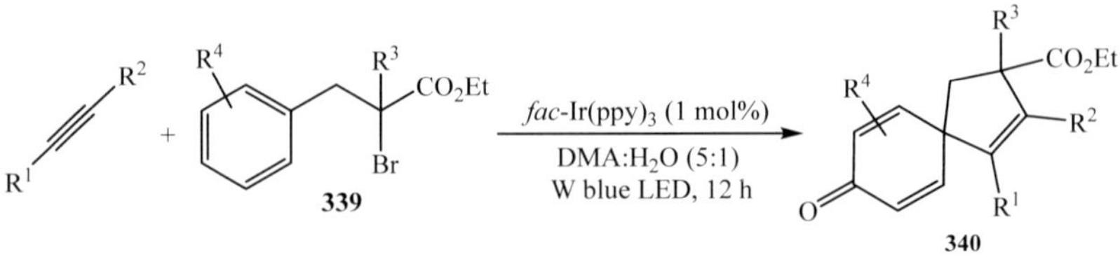

Scheme 4-116. Synthesis of spiro[4,5]decanes **340**.

Using water as an external oxygen source, they also reported a *fac*-Ir(ppy)$_3$-catalyzed intermolecular dearomative cyclization of 2-bromo-2-((5-bromofuran-2-yl)methyl)malonate **341** and alkynes providing substituted spirolactones **342** in yields of 19-91% through a 5-*exo*-dig radical cyclization under visible light (Scheme 4-117) [105].

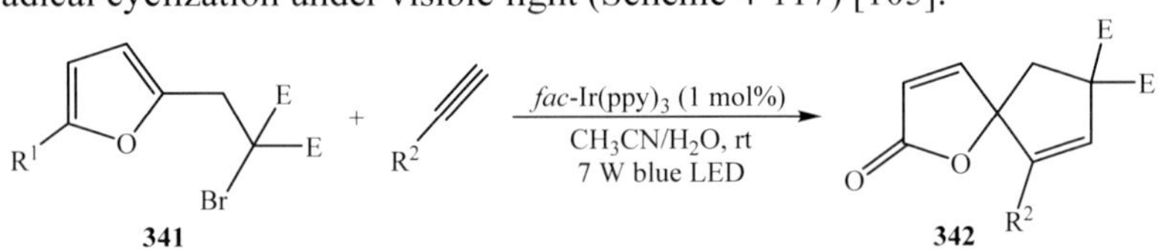

Scheme 4-117. Synthesis of substituted spirolactones **342**.

Later, they found that spiro[4.5]deca-1,7,9-trien-6-ones **344** could be obtained from alkynes and 2-bromo-2-(2-methoxybenzyl)-malonate **343** in the presence of 1 mol% *fac*-Ir(ppy)$_3$, as illustrated in Scheme 4-118 [106].

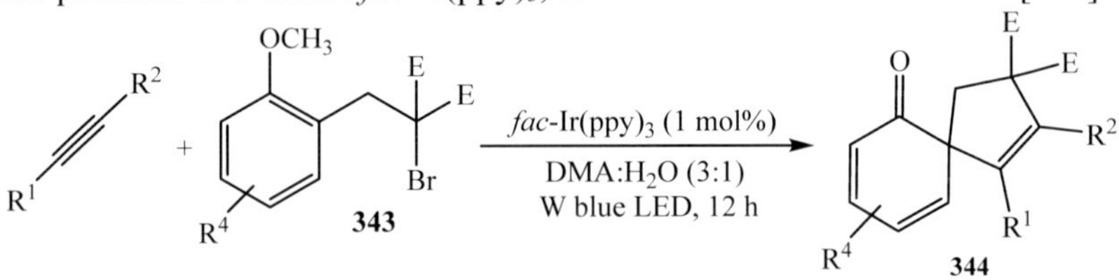

Scheme 4-118. Photocatalytic radical *ortho*-dearomative cyclization.

The French group presented a diastereoselective reaction, which admirably combines an Au-catalyzed dearomatization of *N*-propargyl- and *N*-homoallenyl-2-bromo-β-tryptamines **345-346** and an acidic hydrolysis (Scheme 4-119). Spiro-oxindoles **347/347′** presenting a spiro[indoline-3,4′-piperidin]-2-one scaffold was assembled in excellent diastereoselectivities [107].

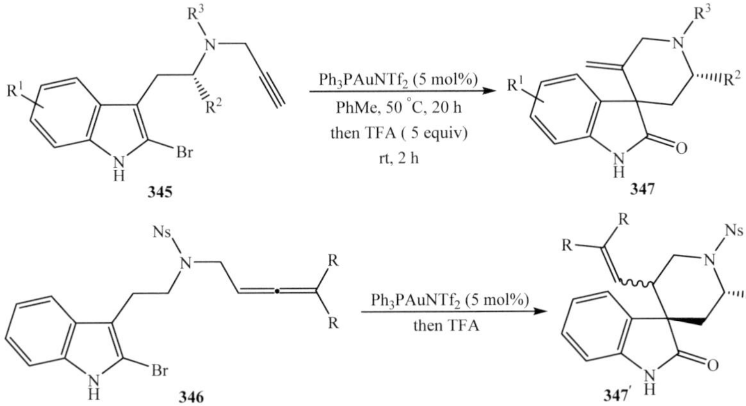

Scheme 4-119. Au-Catalyzed dearomatization of *N*-propargyl- and *N*-homoallenyl-2-bromo-β-tryptamines **345-346.**

Zhu et al. in 2018 reported a novel cascade radical addition/ dearomative spirocyclization between biaryl ynones **348** and fluoroalkyl bromides or diethylphosphite as radical sources, to give access to difluoromethylated or monofluoromethylated and phosphorated spiro[5.5]-trienones **349-350**, catalyzed by copper or silver in moderate to high yields (Scheme 4-120) [108].

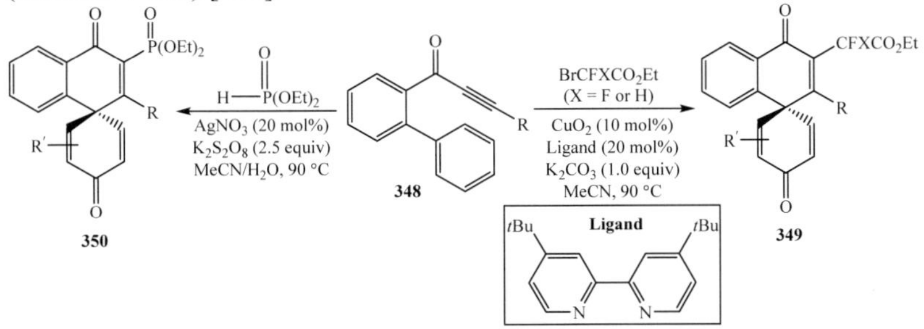

Scheme 4-120. Dearomative spirocyclization of biaryl ynones **348.**

Sarkar et al. successfully developed the Ru(VIII)-catalyzed synthesis of spiro-oxacycles **353** and spiro-azacycles **354** using oxidative de-aromatization as the key step from easily available phenols **351** and naphthols **352** (Scheme 4-121). The catalysis is accomplished in the open air at room temperature with a yield of ≤95% [109].

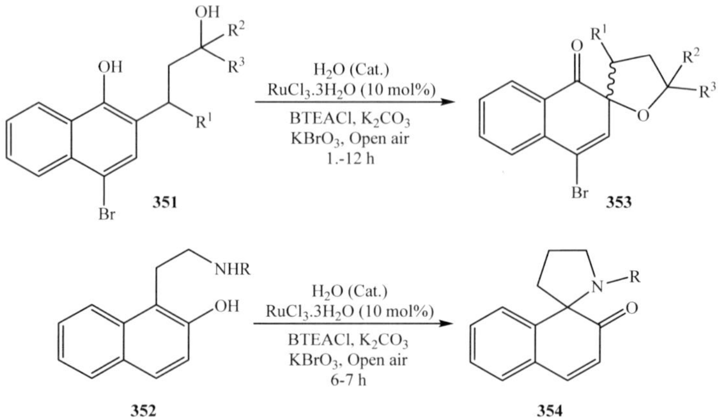

Scheme 4-121. Ru(VIII)-Catalyzed synthesis of spiro-oxacycles **353** and spiro-azacycles **354**.

A plausible mechanistic hypothesis for this reaction is that the catalytic cycle initiates with the oxidation of $RuCl_3 \cdot 3H_2O$ to Ru(VIII) in the presence of $KBrO_3$. The active species Ru(VIII) generates a hydrogen ruthenate complex with the alcoholic naphthoxide **350** and forms a spirofuran **352** and Ru(VI) oxide that is further oxidized to Ru(VIII) by $KBrO_3$. The chiral ammonium tribromides of the additive maintain a coordinated sphere through the catalytic cycle that delivers moderate diastereoselectivity (Scheme 4-122).

In 2019, You et al. reported an efficient synthesis of spiroindolenines **356** via a dearomatization reaction of indoles **355** using JohnPhosAuCl/AgOMs as a catalyst in excellent yields (≤99%) (Scheme 4-123). Additionally, the use of Hantzsch ester (HEH) as the hydrogen transfer reagent in the reaction resulted in the formation of spiroindolines **357** in a cascade fashion, starting from readily available indoles, with yields ranging from modest to good (≤79%) [110].

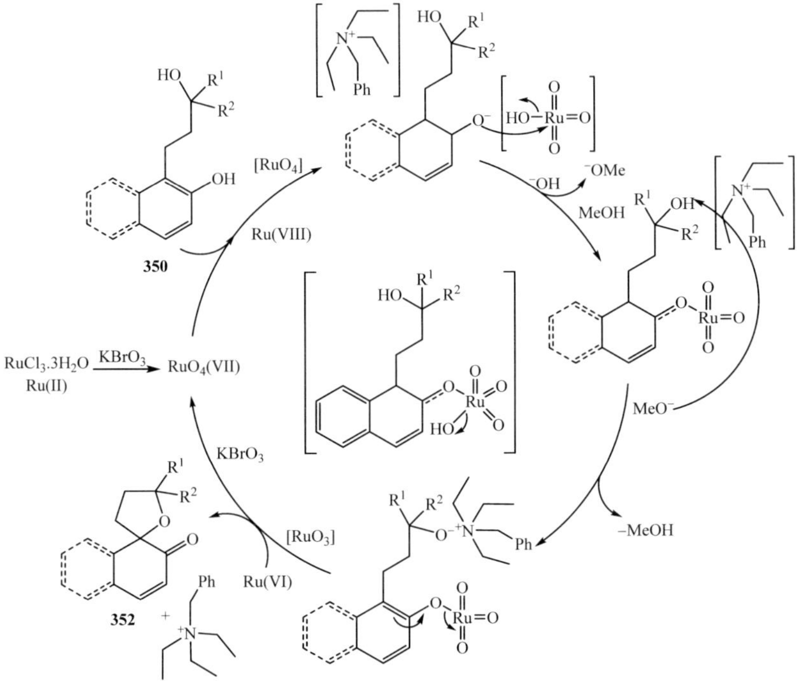

Scheme 4-122. Catalytic cycle.

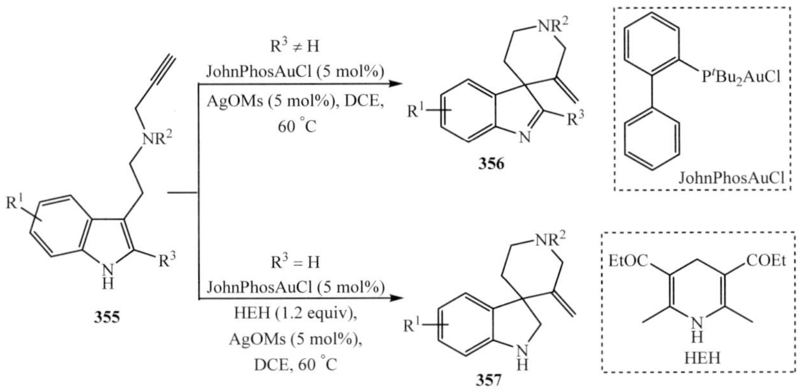

Scheme 4-123. Au-Catalyzed intramolecular dearomatization reactions of indoles.

A catalytic cycle was suggested, as shown in Scheme 4-124.

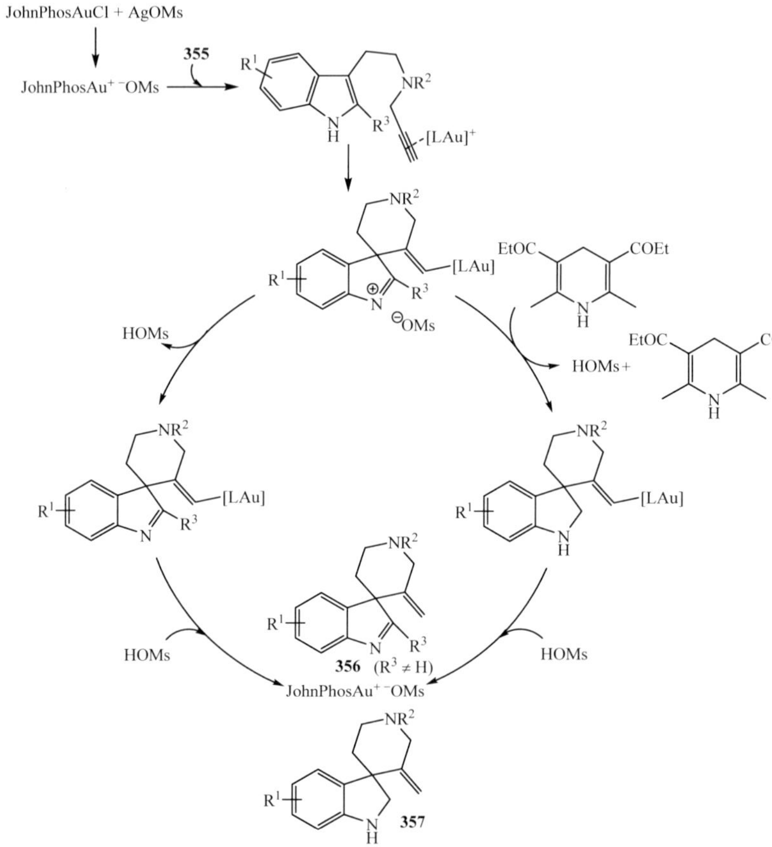

Scheme 4-124. Proposed reaction mechanism.

Various sulfonylated spiro[indole-3,3′-pyrrolidine] derivatives **359** were synthesized via a dearomative tandem spirocyclization reaction of *N*-[(1*H*-indol-3-yl)methyl]methacrylamides **358** with sulfinate sodium in the presence of $AgNO_3$ and K_2CO_3 by Li et al. (Scheme 4-125). The resulting sulfonylated spiroindolines **359** are of growing interest to the agrochemical and pharmaceutical industry [111].

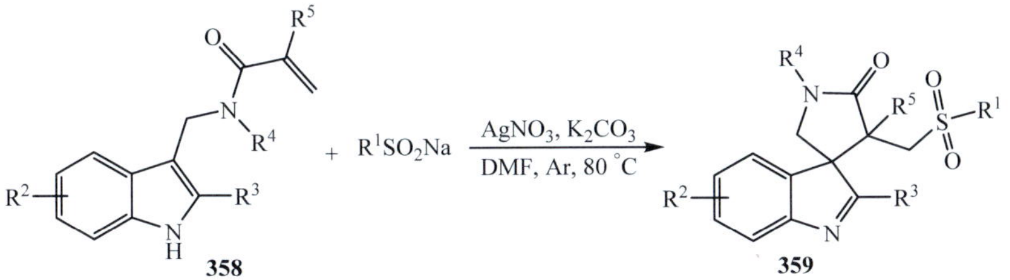

Scheme 4-125. Synthesis of sulfonylated spiro[indole-3,3′-pyrrolidines] **359**.

Recently, Xia et al. reported a novel double C−H functionalization acylative dearomatization of biaryl ynones **360** with aldehydes using environmentally friendly iron salts ($FeCl_2 \cdot 4H_2O$) as catalysts and $EtOH/H_2O$ as solvents with 4.0 equiv of TBHP as oxidant at 90 °C, where the regiospecific remote *para*-C-H functionalization on the Ar^3 ring affords the desired spiro[5.5]trienones **361** in high yields [112]. A plausible mechanism for this transformation was proposed, as shown in Scheme 4-126.

The use of readily available compounds (toluene and analogs, alkanes, ethers, ketones, etc.) as alkylating reagents was successfully applied by the same group in the synthesis of alkylated spiro[5.5]trienones **362** [113]. They showed that the cleavage of an unactivated aryl nitro group triggered by alkyl radicals resulted in dearomative cyclization, providing diversified alkylated spiro[5.5]trienones **362** in good yields (Scheme 4-127). The reaction was performed with 5 mol% CuBr and 4 equiv of TBHP at 120 °C.

In an elegant work, the Yang group introduced a novel Pd-catalyzed cascade dearomative spirocyclization and C−H annulation of aryl halides **363/364** with diaryl acetylenes **365**, giving straightforward access to diverse functionalized spiroembedded polycyclic aromatic compounds **366/367** (Scheme 4-128). The cascade process is performed by sequential oxidative addition of Pd(0) to the C−X bond, alkyne migratory insertion, Heck-type de-aromatization spiroannulation, and C−H bond annulation [114].

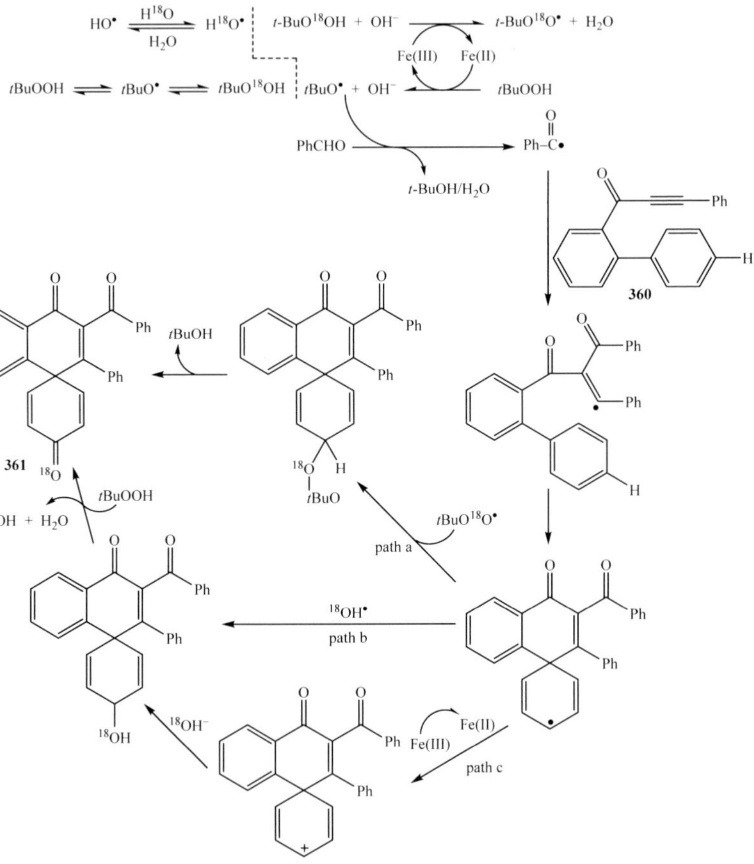

Scheme 4-126. A plausible mechanism for Fe-catalyzed de-aromatization of biaryl ynones **360** with aldehydes.

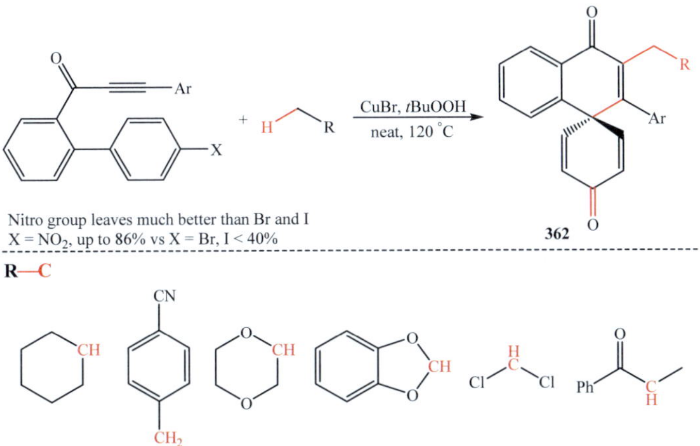

Scheme 4-127. Access to diversified alkylated spiro[5.5]trienones **362**.

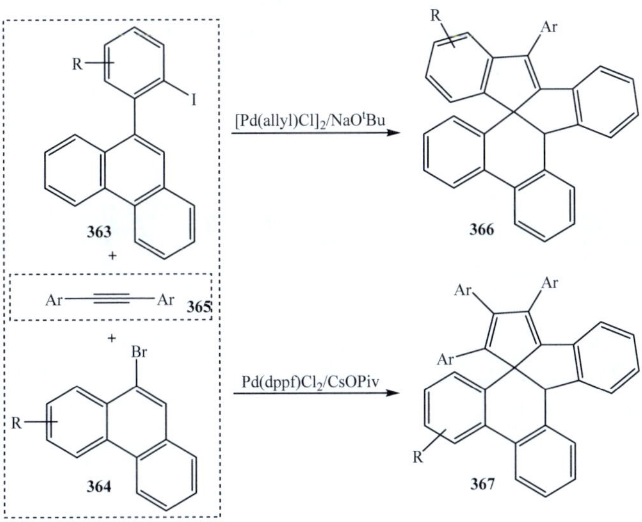

Scheme 4-128. Pd-Catalyzed intermolecular dearomative annulation.

4.4.3. Transition metal-catalyzed *ipso*-cyclization

Ipso-annulation reactions have attracted much attention due to their extensive applications in organic synthesis. We herein summarize recent progress in the transition metal-catalyzed *ipso*-cyclization for constructing diversely substituted spirocycles.

Spirobacillene A (**369**) is an indole alkaloid isolated from Lysinibacillus fusiformis, and the first total synthesis of spirobacillene A was reported by Unsworth et al. in 2013 [115]. A Lewis acid-mediated spirocyclization of an anisole onto a tethered ynone **368** was used as a key step (Scheme 4-129).

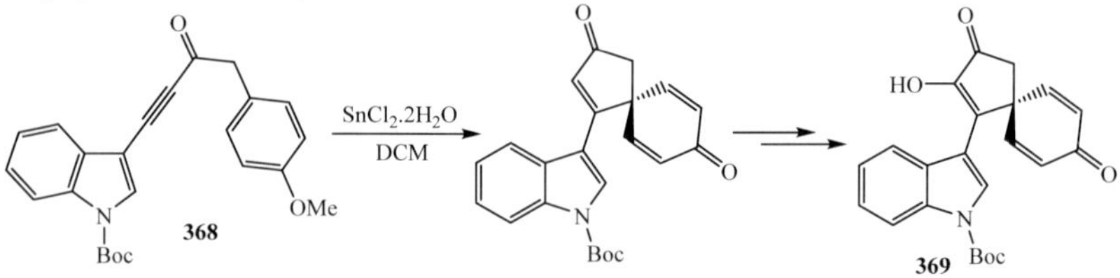

Scheme 4-129. Total synthesis of spirobacillene A **369**.

N-arylpropiolamide is a versatile building block that is often employed as a candidate to explore radical *ipso*-cyclizations. In 2015, Song et al. developed a new oxidative *ipso*-cyclization of *N*-(*p*-methoxy-aryl)propiol-amides **370** with disulfides and H_2O using inexpensive $CuCl_2$ catalyst and O_2 oxidant, leading to 3-(arylthio)-1-azaspiro[4.5]deca-3,6,9-triene-2,8-diones **371** through sequential oxidative C-S bond formation, *ipso*-cyclization, and dearomatization (Scheme 4-130) [116].

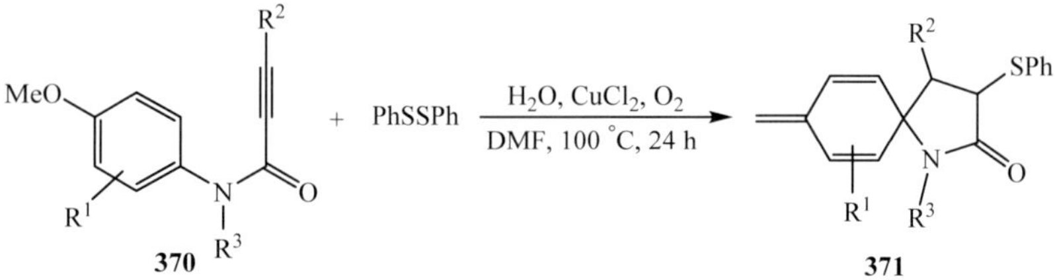

Scheme 4-130. Oxidative *ipso*-cyclization of *N*-(*p*-methoxyaryl)propiol-amides **370** with disulfides.

Shortly after, the same group extended the scope of this oxidative *ipso*-annulation with the Cu-catalyzed annulation of activated alkynes **372** with unactivated alkanes **373** which allows the preparation of 3-alkyl spiro[4,5]trienones **374** via a sequence of C-H oxidative coupling, *ipso*-carbocyclization and dearomatization (Scheme 4-131) [117].

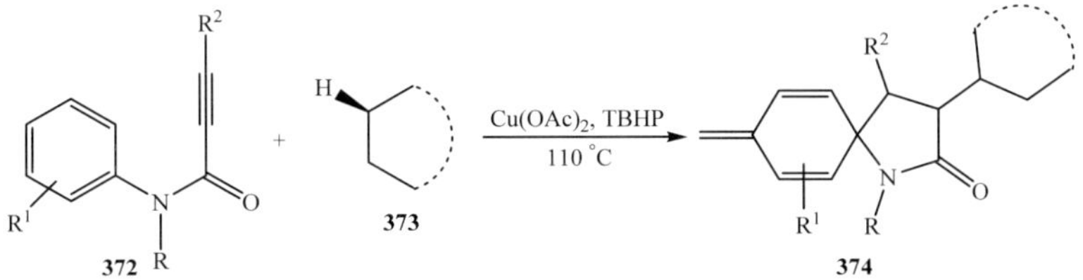

Scheme 4-131. Cu-Catalyzed synthesis of 3-alkyl spiro[4,5]trienones **374**.

A plausible mechanism for this system was proposed, as indicated in Scheme 4-132.

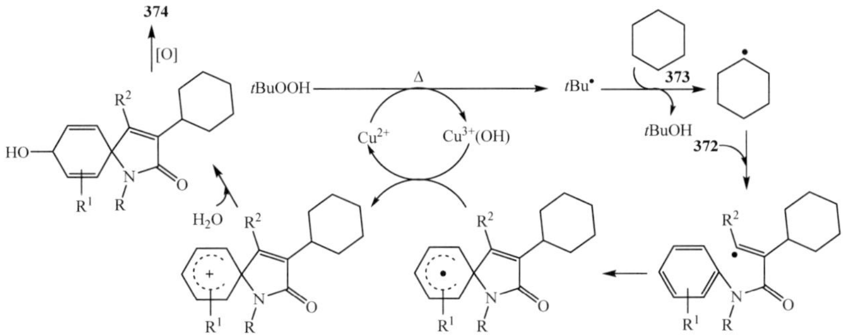

Scheme 4-132. Possible mechanism.

Qiu et al. also reported the synthesis of 3-bromo-1-azaspiro[4.5]deca-3,6,9-triene-2,8-dione **376** through $ZnBr_2$-mediated oxidative spiro-bromocyclization of *N*-arylpropiolamide **375** (Scheme 4-133). The reaction was efficiently performed at room temperature, and excellent tolerance of functional groups was represented. In this annulation, the water is proposed to play a dual role, both as a reactant to form a keto group in the final product and as a solvent to improve the solubility of both $ZnBr_2$ and Oxone [118].

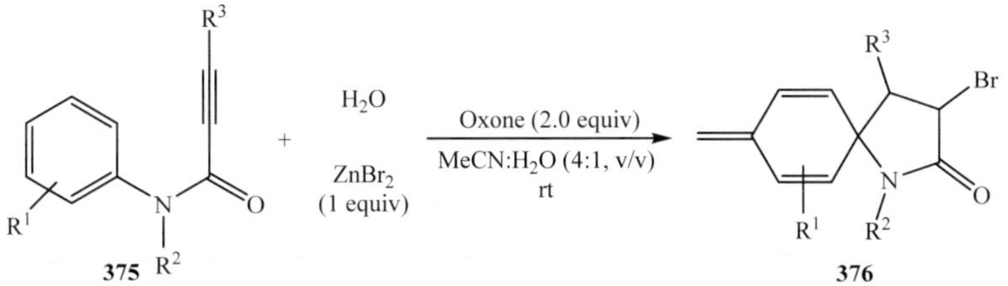

Scheme 4-133. $ZnBr_2$-Mediated oxidative spiro-bromocyclization of propiolamides **375**.

Later on, the Van der Eycken group showed that Ugi adducts **377** derived from alkynoic acids and 2-aminophenols could readily be applied to Au-catalyzed *ipso*-cyclization reactions producing spirocyclic pyrrol-2-one-dienones **378/378′** [119]. A mechanistic model for the Au(I)-catalyzed *ipso*-cyclization of Ugi adducts is given in Scheme 4-134.

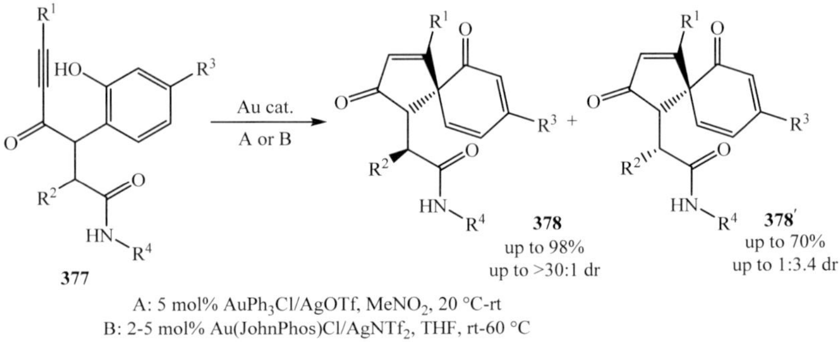

Scheme 4-134. Au-Catalyzed post-Ugi *ipso*-cyclization.

Diastereoselectivity for this annulation can be controlled by choosing the appropriate catalytic system, as indicated in Scheme 4-135.

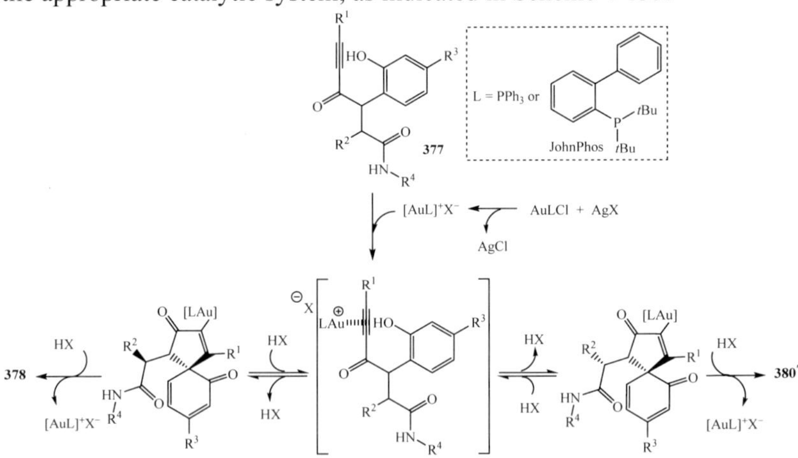

Scheme 4-135. A mechanistic model for a Au(I)-catalyzed *ipso*-cyclization of Ugi adducts **377**.

In 2019, Qiu et al. also reported a tunable protocol for the synthesis of 10-hydroxy-1-azaspiro[4.5]deca-3,6,8-trien-2-ones **379** and benzo[*b*]pyrolo[2,1-*c*][1,4]oxazin-3-ones **380** by the use of tetra-*n*-butyl ammonium bromide (TBAB)/oxone and $ZnBr_2$/oxone (Scheme 4-136) [120]. Based on the proposed mechanism, authors believed that species **B**, generated from α-addition, *ipso*-cyclization, and single electron oxidation, is a key intermediate. After intramolecular trapping by the carbonyl group, the intermediate **A** is transformed into spiro-tricyclic cation intermediate **B**.

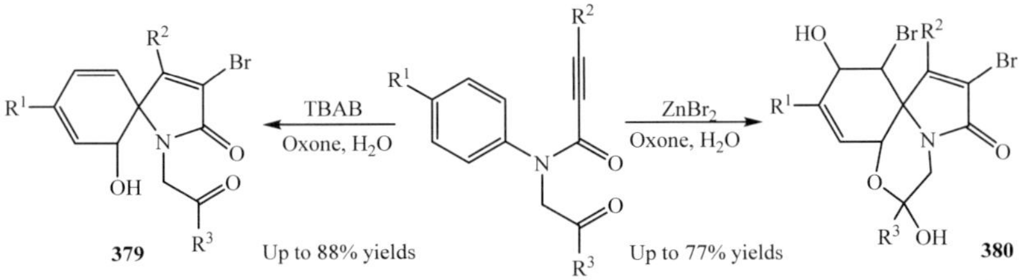

Scheme 4-136. *Ortho*-hydroxylative *ipso*-cyclization of *N*-arylpropiol amide.

Subsequently, nucleophilic addition of water to intermediate **B** affords the desired spiro-tricycles **381** which can be generated the products **379** or **380** depending on the bromide salt (Scheme 4-137).

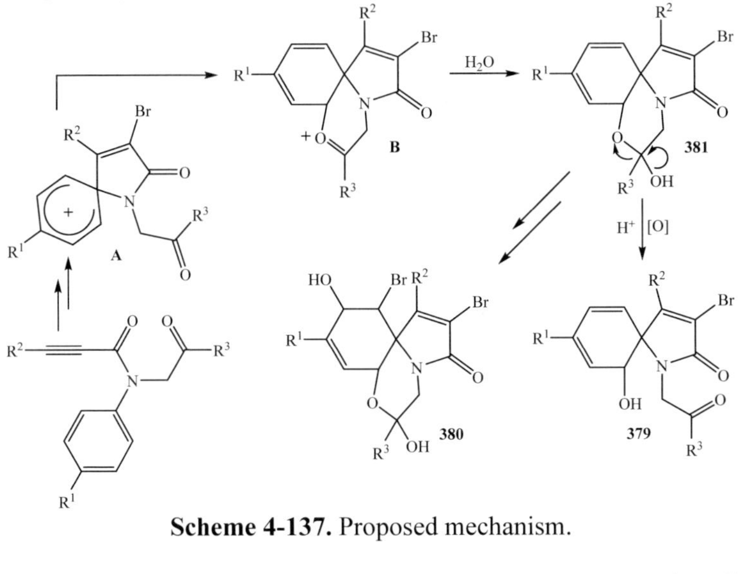

Scheme 4-137. Proposed mechanism.

Recently, a quite elegant example of nitrogen-centered radical strategy via visible-light-mediated acylation/*ipso*-cyclization of propiol-amides **60** with acyl oxime esters **382** was presented by Liu et al. to provide 3-acyl-substituted spiro[4,5]trienones **383** [121]. Scheme 4-138 shows a plausible mechanism for visible-light-mediated nitrogen-centered radical strategy. The alkyl- and aryl-substituted acyl radicals generated from the cleavage of carbon-carbon σ-bonds in acyl oxime esters **382** attacked the carbon-carbon triple bonds in propiolamides and subsequently underwent *ipso*-cyclization.

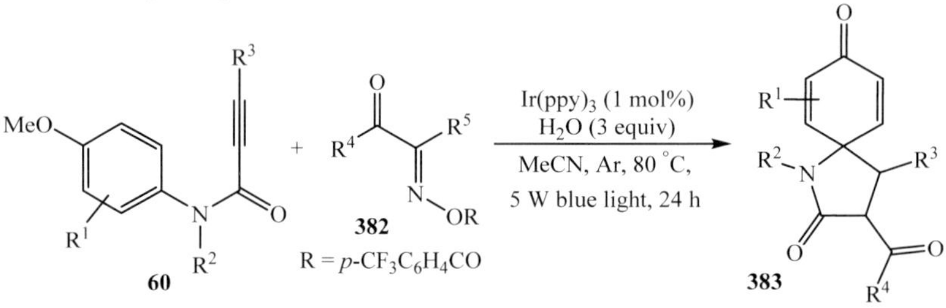

Scheme 4-138. Visible-light-mediated nitrogen-centered radical strategy.

4.4.4. Other transition metal-catalyzed cyclization reactions

Metal-catalyzed cyclization is one of the most important strategies for ring construction. This section has been divided according to the different transition metals used and the described specific cyclization reactions.

In 2018, Rong et al. demonstrated an efficient *in situ* reduction and intramolecular cyclization reaction to obtain nitrogen-containing spiro compounds **387-389** [122]. 5-Nitro-1*H*-indazole **384**, 6-nitro-1*H*-indazole **385**, and 5-nitroindole **386** were used to synthesize spiro compounds in Fe-H_2O-AcOH medium (Scheme 4-139).

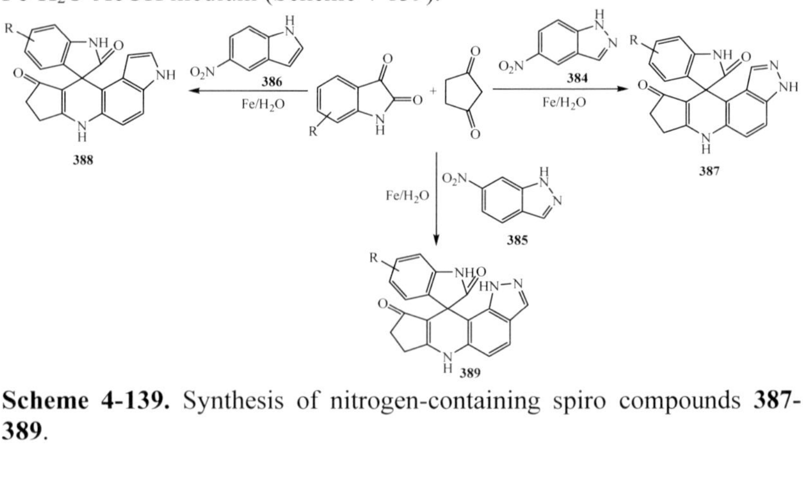

Scheme 4-139. Synthesis of nitrogen-containing spiro compounds **387-389**.

The switchable synthesis of fused or spiro polyheterocyclic compounds **391**-**392** through the dehydrogenative annulation reactions of 2-arylindazoles **390** with maleimides presented by Zhang et al. (Scheme 4-140). Mechanistically, the reaction involves an Rh(III)-catalyzed C-H metalation of 2-arylindazole, followed by maleimide insertion and intramolecular cyclization [123].

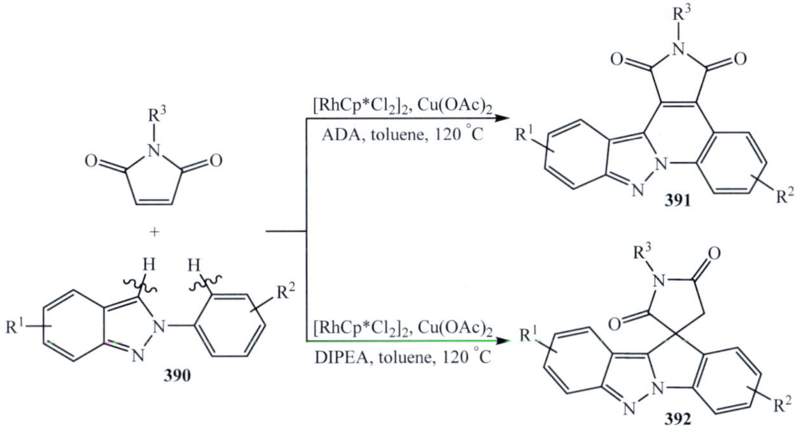

Scheme 4-140. Synthesis of fused or spiro polyheterocyclic compounds **391**-**392**.

In 2019, the You group suggested an Au-catalyzed dearomatization reaction of β-naphthols **393** through intramolecular hydroalkylation of alkynes to spironaphthalenones **394** that were achieved in good to excellent yields (76–99%) under open-flask conditions (Scheme 4-141) [124].

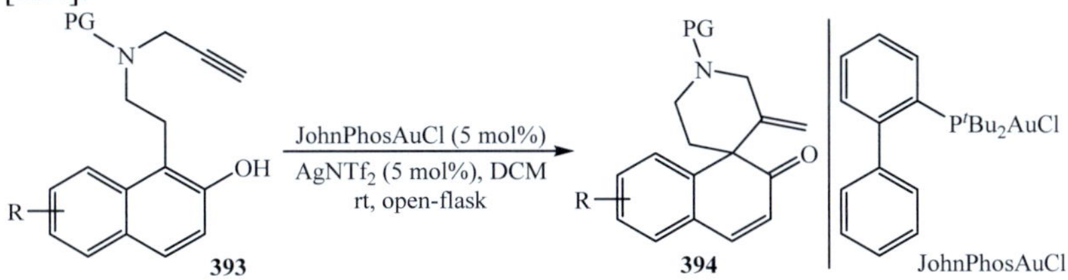

Scheme 4-141. Au-Catalyzed dearomatization reaction of β-naphthols **393**.

Likewise, Wang et al. showed that the Nazarov cyclization/hydroamination cascade gives access to spiro[indene-1,4′-quinoline] frameworks **396** in good yields, employing various *N*-sulfonyl

3-(2-sulfonamidoaryl)-1,1-diphenylpenta-1,4-dien-3-ols **395** as substrates and $FeBr_3$ as a catalyst (Scheme 4-142) [125].

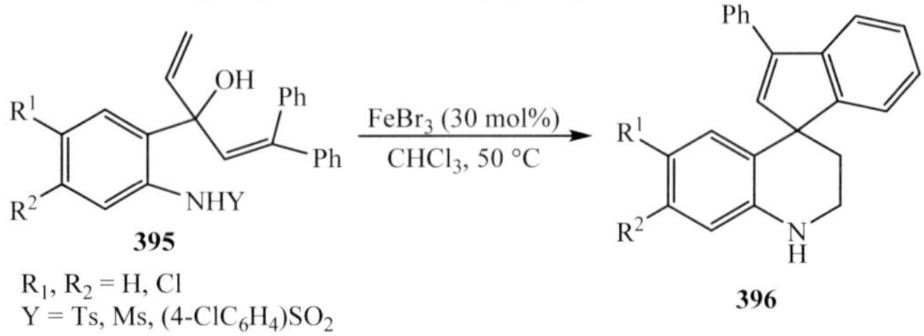

Scheme 4-142. Nazarov cyclization of 1,4-pentadien-3-ols **395**.

Additionally, Xie et al. reported a $FeCl_3$-promoted synthesis of spiro[4.5]decanes **398** from 2-(5,5-dimethoxypentyl)-1-substituted cyclopentanols **397** (Scheme 4-143). The reaction involves the $FeCl_3$-promoted dehydration of alcohol followed by $FeCl_3$-mediated Prins cyclization [126].

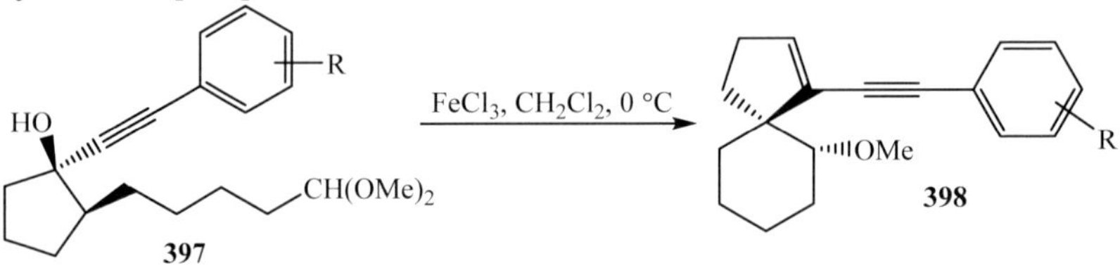

Scheme 4-143. $FeCl_3$-Promoted one-step synthesis of spiro[4.5]decanes **398**.

1,8-Dioxa-3-azaspiro[4.5]dec-2-enes **399** were successfully prepared by a tandem Prins strategy using 5 mol% $Cu(OTf)_2$ in dichloromethane at 0 °C under mild reaction conditions in short reaction times (Scheme 4-144) [127].

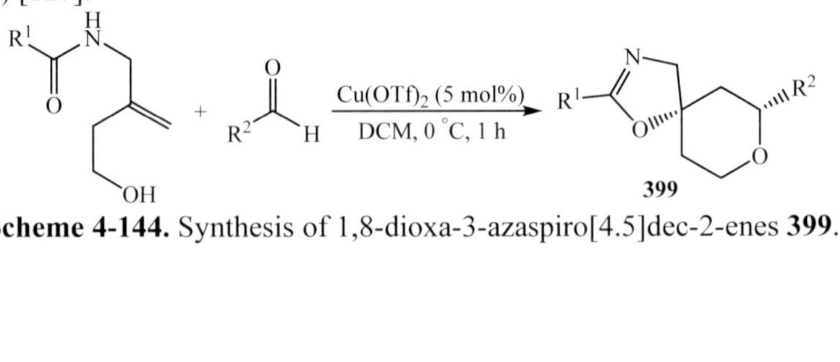

Scheme 4-144. Synthesis of 1,8-dioxa-3-azaspiro[4.5]dec-2-enes **399**.

4.5. Metathesis reactions

Metathesis, also known as olefin scrambling, was discovered in the mid-1950s and has become one of the most important tools in target-oriented synthesis, leading to significant advances in organic synthesis, catalysis, polymer science, and material science in the last 15 years. The term "metathesis" comes from the Greek word "metatithenai," meaning transposition, as it involves the redistribution of multiple carbon-carbon bonds by a scission-recombination process catalyzed by a transition metal complex. Olefin metathesis (Scheme 4-145) is one of the most widely studied metathesis reactions due to the abundance of olefins and their availability in both terminal and di-substituted forms. However, recent developments have expanded the scope of metathesis to include enyne and alkyne metathesis (Schemes 4-146 and 4-147), which have also shown great promise in organic synthesis [128-136].

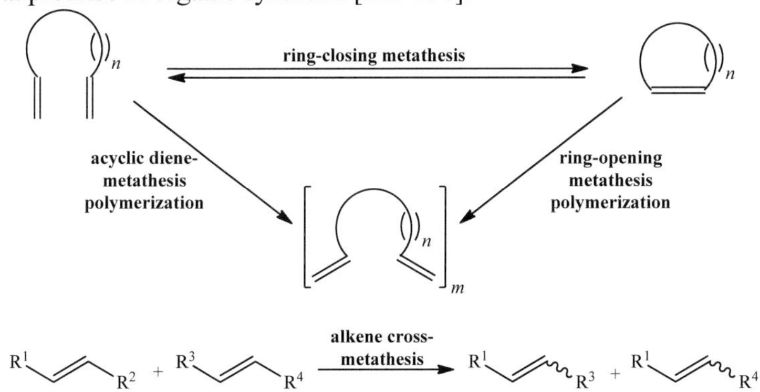

Scheme 4-145. Olefin metathesis reactions.

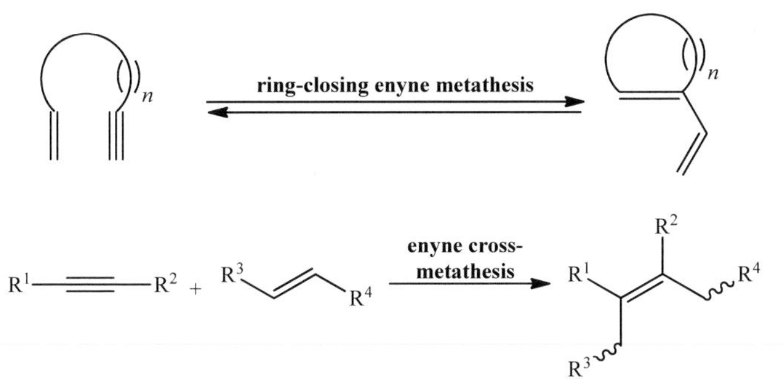

Scheme 4.146. Enyne metathesis.

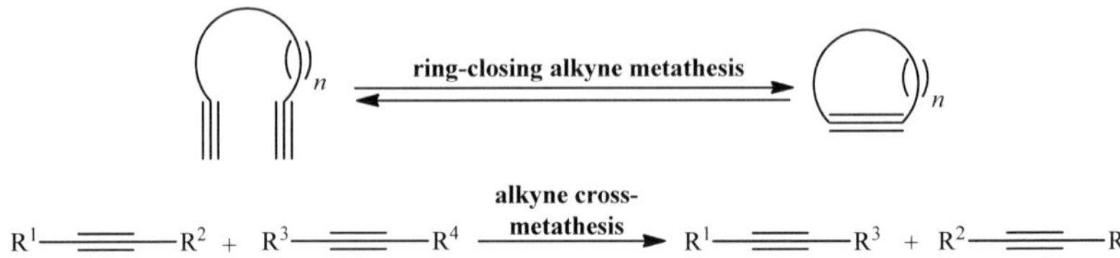

Scheme 4.147. Alkyne metathesis.

Various catalysts used in metathesis include Schrock catalyst **400**, the first-generation Grubbs catalyst (G-I, **401**), the second-generation Grubbs catalyst (G-II, **402**), the first-generation Hoveyda-Grubbs catalyst (GH-I, **403**), and the second-generation Hoveyda–Grubbs catalyst (GH-II, **404**) (Figure 4-4) [137, 138].

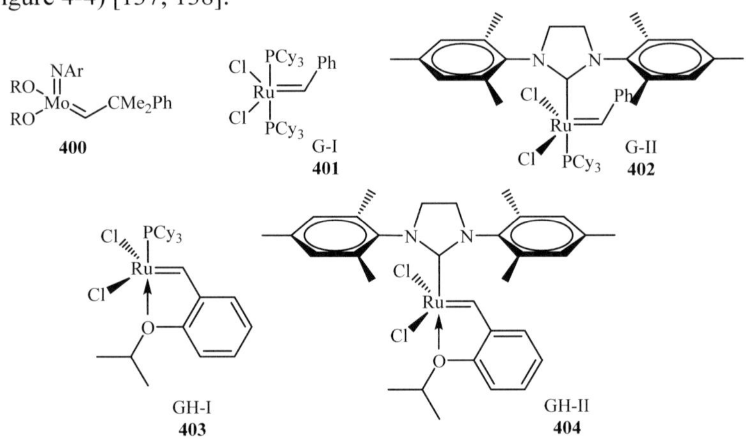

Fig. 4-4. Metathesis catalysts.

Several interesting reviews covering this topic have been reported until 2020. In this regard, some metathesis reactions will be discussed for the synthesis of spiro compounds.

In this context, spirocyclic butenolides were efficiently prepared via a RCM strategy by Langer et al. in 2002 [139]. Using the Grubbs catalyst **401**, the desired spirocyclic butenolides **405** could be obtained in good yields (57-80%) when the reaction was carried out at 35 °C in CH_2Cl_2 in the presence of catalytic amounts of Ti(*i*PrO)$_4$. The Lewis acid was used to avoid interruption of the catalytic cycle by chelation of the substrate carbonyl to the metal (Scheme 4-148).

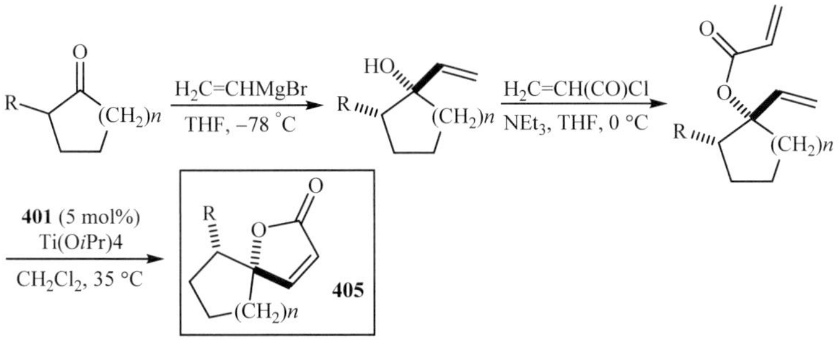

405

Scheme 4-148. Synthesis of spirocyclic butenolides **405** by RCM.

To demonstrate the applicability of the methodology, as mentioned earlier, Kotha et al. predicted that various spirocyclic systems **407** could be synthesized by applying RCM as a key step in Scheme 4-149 [140]. The required dialkylated starting materials **406** were prepared from various 1,3-diketones or substrates involving an active methylene group.

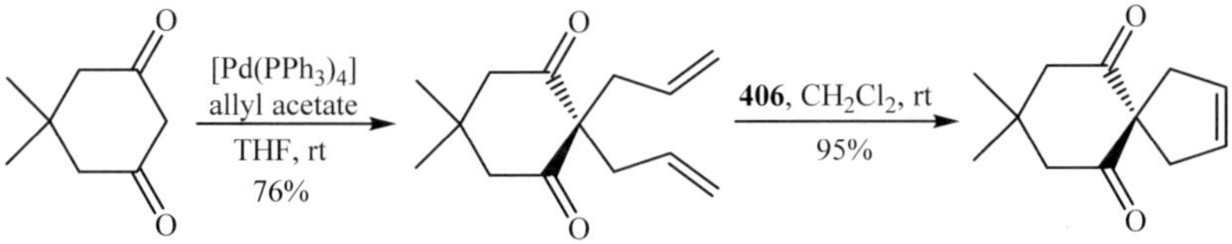

406 407

Scheme 4-149. Synthesis of spiro-cyclic systems **407** via RCM.

Remarkably, the Plumet group reported the stereoselective synthesis of spiro-β-lactams tethered to tetrahydrofuran rings **411-413** via Ru-catalyzed metathesis sequences with oxa norbornene precursors obtained by the Staudinger [2+2] cycloaddition of related imines **408-410** (Scheme 4-150) [141].

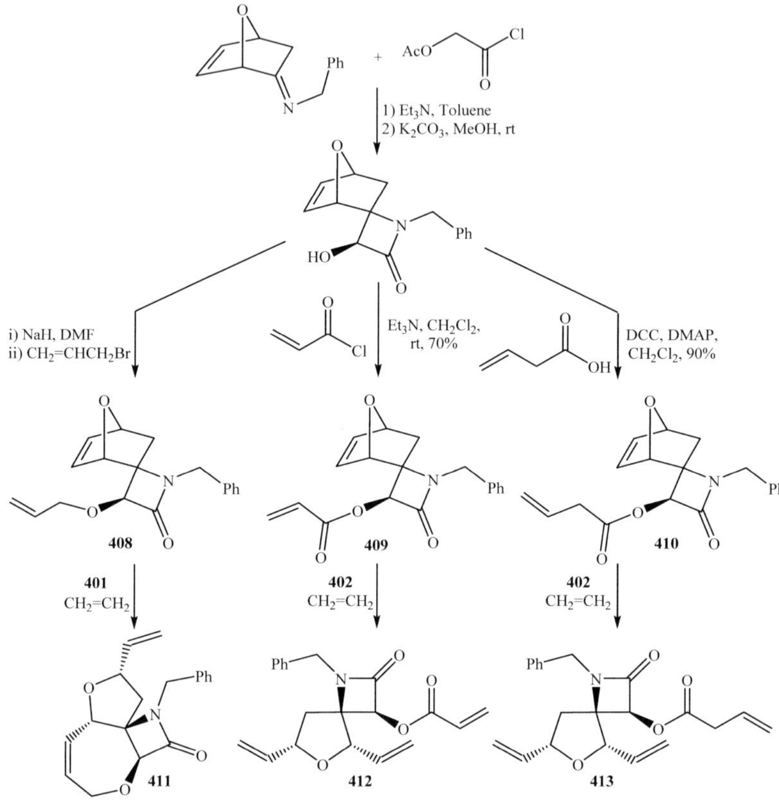

Scheme 4-150. Stereoselective synthesis of spiro-β-lactams **411-413**.

The combination of the diallylboration with RCM has also been applied to synthesize spiro-β-amino alcohols **415**, which are valuable reagents in natural product synthesis [142]. As shown in Scheme 4-151, the procedure is based on the transformation of Boc-protected azaspiro-olefins **414** to amino alcohols **415** by the halocyclocarmabation–dehydrohalogenation–isomerization–hydrolysis sequence.

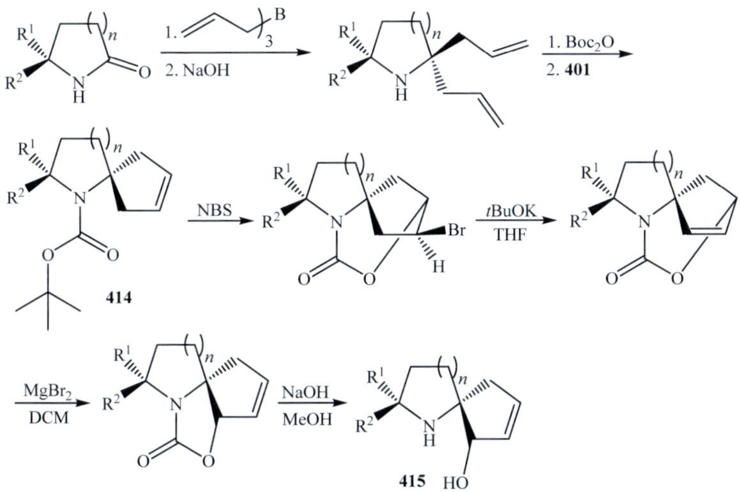

Scheme 4-151. The combination of diallylboration and RCM in the synthesis of spiro-β-amino alcohols **415**.

On the other hand, Mol et al. explored the combination of diallylboration of lactams and the subsequent RCM of the resulting α,α-diallyl *N*-heterocyclic compounds **416** to synthesize a series of nitrogen-containing spiro[4.*n*]alkenes **417** (Scheme 4-152). The authors found that free amine protection allowed the neutralization of the catalyst-deteriorating basicity of the nitrogen atom and facilitated RCM [143].

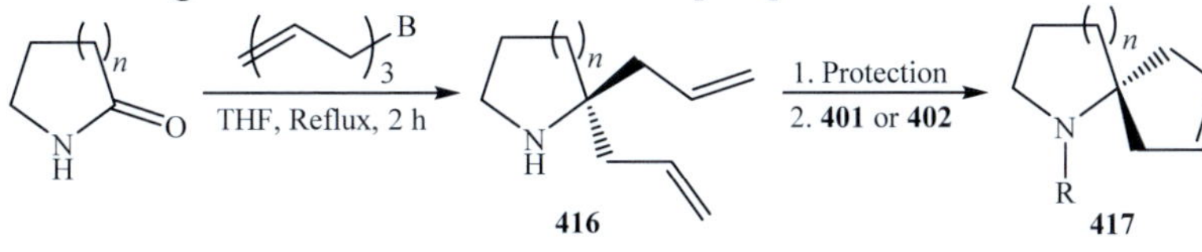

Scheme 4-152. Synthesis of nitrogen-containing spiro compounds **417** from lactams by allylboration and subsequent RCM.

Another exciting example reported by the Majumdar group is the combination of the Claisen rearrangement and RCM/ring-closing enyne metathesis (RCEM) process [144]. In this work, the analogous substituted spironaphthyridinones **418-419** were formed by conventional RCM or RCEM process in the presence of Grubbs' first-generation catalyst **401** at room temperature under a nitrogen atmosphere (Scheme 4-153).

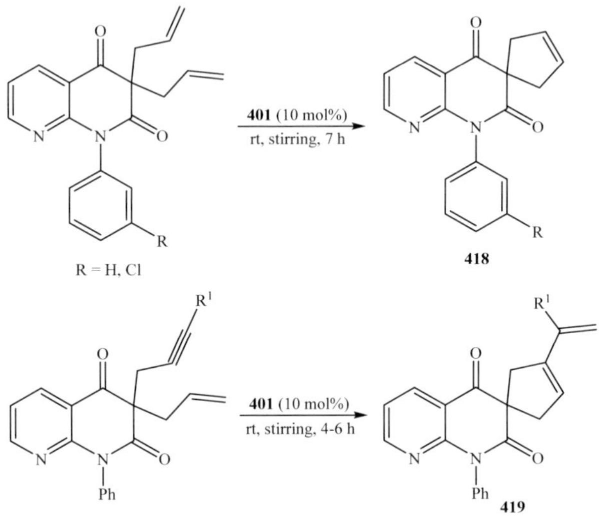

Scheme 4-153. Grubbs' catalyst mediated RCM or RCEM reaction.

Felpin et al. introduced a rapid and efficient strategy utilizing sequential Pd(0)/C-catalyzed allylation and RCM reactions for the construction of spiropyridones and spiroquinolones **420** (Scheme 4-154) [145].

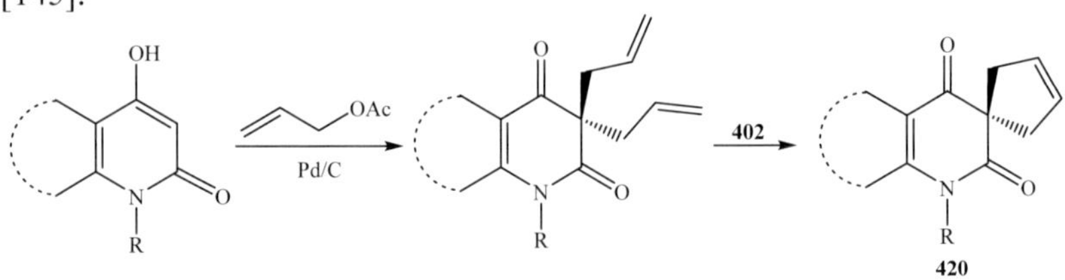

Scheme 4-154. Synthesis of spiropyridones and spiroquinolones **420**.

The syntheses were achieved using the developed protocol to the following a regioselective allylation at C3, taking advantage of the unusual reactivity of the Pd(0)/C catalyst. As illustrated in Scheme 4-155, the authors also explored the synthetic utility of this methodology in nucleoside chemistry.

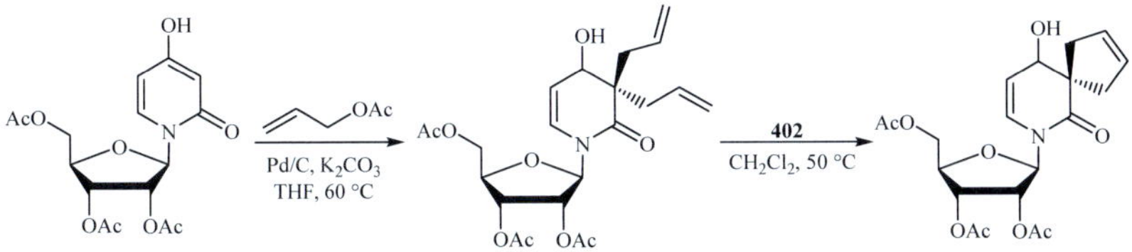

Scheme 4-155. Preparation of a spiro nucleoside.

Later, Dash et al. showed an elegant example of an efficient and general method based on selective alkylation (Scheme 4-156) and olefin metathesis (Scheme 4-157) by exploiting the four possible points of diversity in the hydantoin ring to prepare a diverse series of 5,5-spirocyclic and 1,5-, 4,5-, and 3,4-fused bicyclic imidazolidinones, that represent alkaloid-like scaffolds for a lead generation [146].

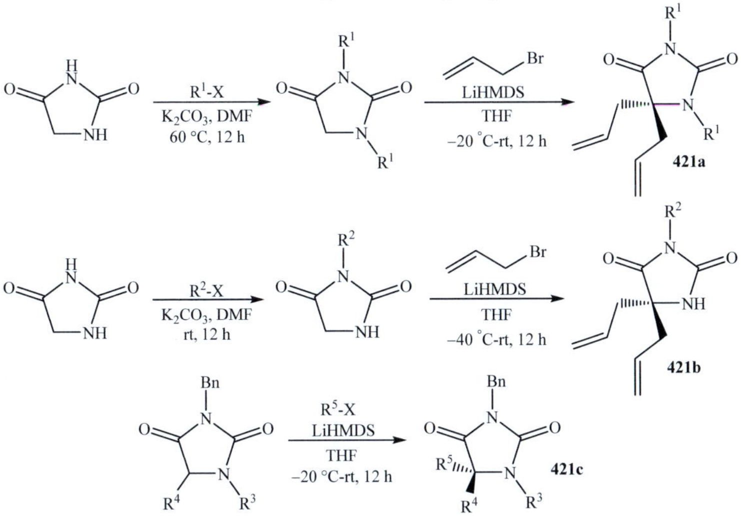

Scheme 4-156. Selective alkylation of hydantoin.

When the reaction was performed with hydantoins containing trienes and tetraenes **421a-c**, functionalized spirohydantoins **422** were obtained via selective RCM and cross metathesis (CM) reaction (Scheme 4-157, Eq. 1). The hydantoin containing triene can undergo a tandem metathesis sequence involving ring-closing-ring-opening and cross metathesis (RC-RO-RC-CM) that lead to construct either the functionalized spiro compound **423** or bicyclic hydantoin dimer **424** (Scheme 4-157, Eq. 2).

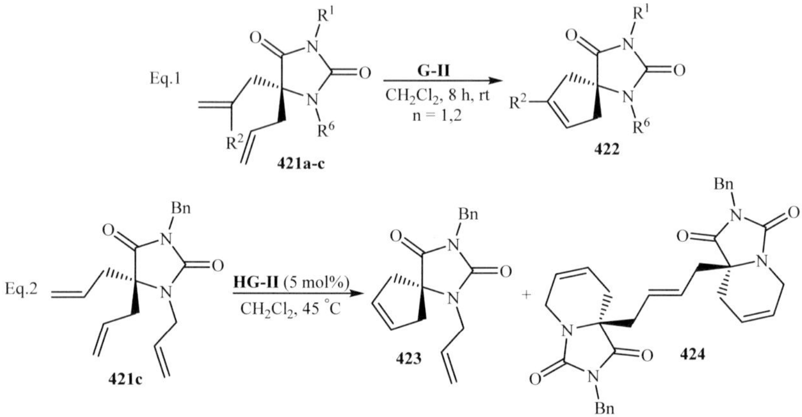

Scheme 4-157. Synthesis of functionalized spirohydantoins **422-423**.

In 2013, Siddiqui et al. reported an efficient method for the synthesis of functionalized spirodihydrofuran oxindoles **427** via intramolecular alkyne carbonyl metathesis [147]. The products **427** were easily obtained through a one-pot reaction of *N*-methyl isatin **425**, alkynes, and phenacyl bromides **426** in the presence of indium bromide. Scheme 4-158 illustrates a plausible mechanism for this transformation. The authors reported various examples with yields ranging from 76 to 92%.

Interestingly, Kotha et al. demonstrated a new approach to twisted C_2-symmetric cage bis-spirocyclic pyran **431** via an allyl Grignard reaction and an RCM sequence [148]. In this process, the diallyldiol **430** was formed by adding the etheral solution of the hexacyclic dione **429** to a freshly prepared allyl Grignard reagent **428** at 0 °C. Then, the diallyldiol **430** could be converted into the desired product **431** in the presence of Grubbs' first-generation catalyst **401** in refluxing toluene (Scheme 4-159).

The same group reported a simple and helpful protocol using a diversity-oriented strategy for the synthesis of spirocycles **434-435** starting from inexpensive and commercially available monocyclic ketones **432** and bicyclic diones **433** which undergo ring-closing metathesis with the aid of Grubbs' first generation catalyst (Scheme 4-160) [149].

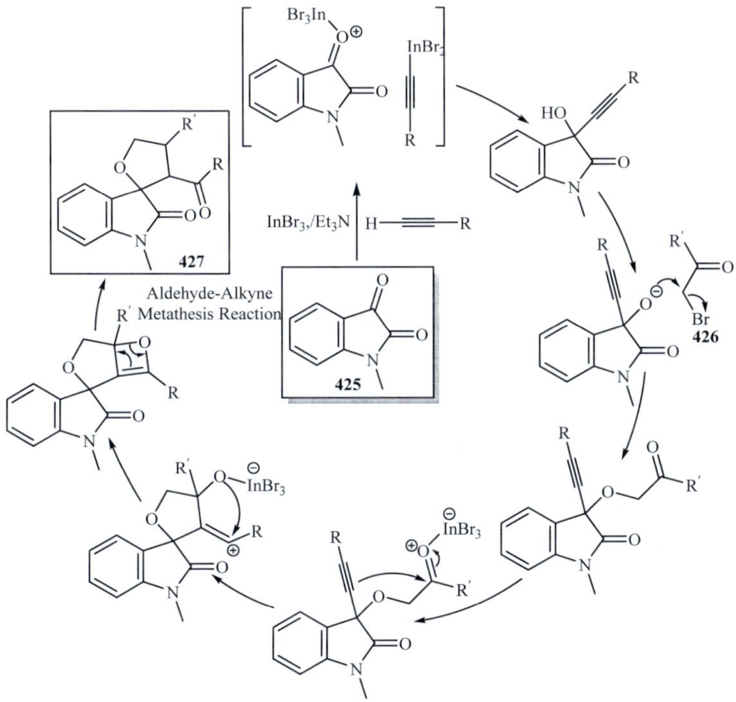

Scheme 4-158. Plausible mechanism for the formation of spiro dehydrofuran oxindoles **427**.

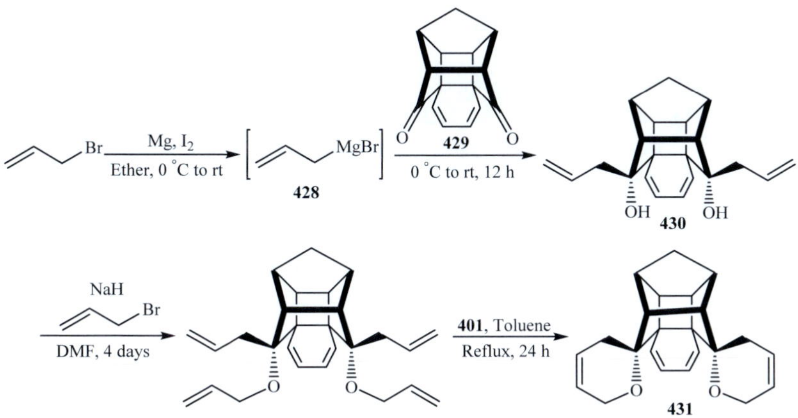

Scheme 4-159. Bis-spirocyclic pyran **431** via an allyl Grignard reaction and an RCM sequence.

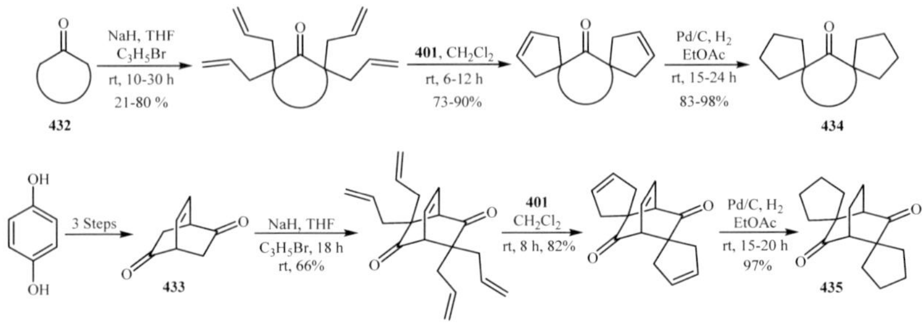

Scheme 4-160. Diversity-oriented approach to spirocycles **434-435** via RCM.

Building upon their previous work on the use of ring-rearrangement metathesis (RRM) for the synthesis of oxa-spirocycles **340-341**, the group developed a new approach to prepare oxygenated norbornenes as suitable precursors for the RRM sequence [150]. The synthesis involved the oxidation of diol **436** with the Swern reagent to give the oxidized product which was then treated with allyl magnesium bromide to afford Grignard addition product **437**. Further reaction with allyl bromide in the presence of NaH provided di *O*-allylated compound **438** in 63% yield which upon treatment with Grubbs first generation catalyst **401** under ethylene atmosphere generated the rearranged spirocyclic compound **440** in 75% yield (Scheme 4-161). Compound **437** was also treated with propargyl bromide in the presence of NaH to provide the di *O*-propargylated compound **439** (61%) which was then subjected to enyne RRM in the presence of G-I catalyst under an ethylene atmosphere to generate 1,3-diene **441** in 65% yield.

Martinková et al. developed a strategy for the synthesis of spiro-β-lactone-γ-lactam synthons **442** and ent-**442**, which are similar in structure to natural products such as oxazolomycins [151]. The synthesis employs two chirons, the protected 3-isothiocyanato-3-C-vinyl-α-d-xylofuranose and its antipode **443**. The key transformations involved in the synthesis include RCM, highly diastereoselective reduction of a double bond, and formation of a β-lactone skeleton to afford a lactone-pyrrolidine cyclic framework. The retrosynthetic strategy for **442** and ent-**442** is outlined in Figure 4-5.

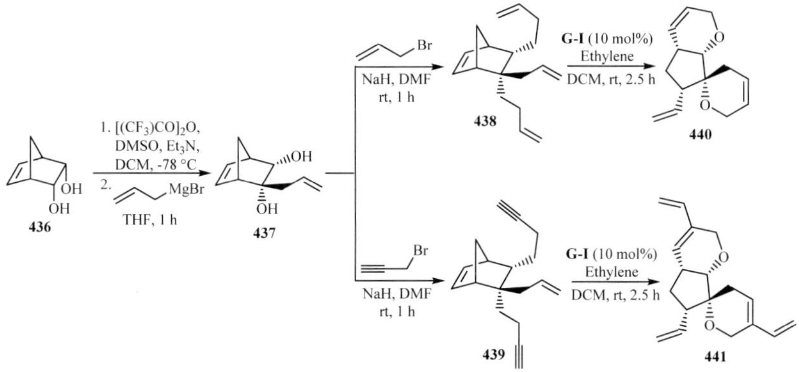

Scheme 4-161. Synthesis of oxa-spirocyclic compounds **440-441** by RRM route.

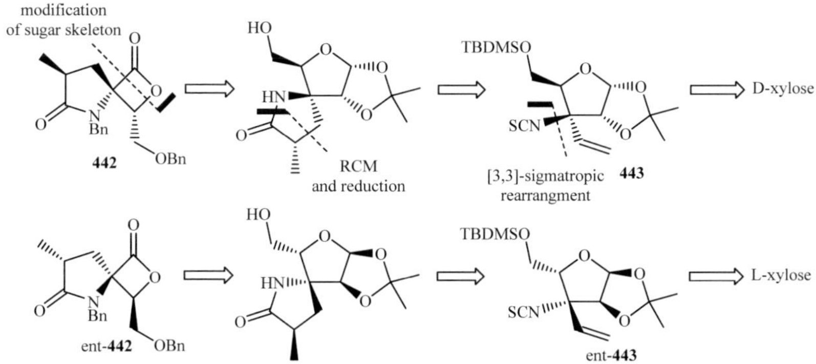

Fig. 4-5. Retrosynthetic strategy toward **442** and ent-**442**.

References

[1] Pérez-Gómez Marta, Hernández-Ponte Sergio, Bautista Delia, García-López José-Antonio. 2017. "Synthesis of spiro-oxoindoles through Pd-catalyzed remote C-H alkylation using α-diazo carbonyl compounds". Chemical Communications. No. 19. 2842-2845.

[2] Ye Juntao, Shi Zhihao, Sperger Theresa, Yasukawa Yoshifumi, *et al.* 2017. "Remote C-H alkylation and C-C bond cleavage enabled by an *in situ* generated palladacycle". Nature Chemistry. No. 4. 361-368.

[3] Nemoto Tetsuhiro, Ishige Yuta, Yoshida Mariko, Kohno Yuta, *et al.* 2010. "Novel method for synthesizing spiro[4.5]cyclohexa dienones through a Pd-catalyzed intramolecular *ipso*-Friedel–Crafts allylic alkylation of phenols". Organic Letters. No. 21. 5020-5023.

[4] Yoshida Mariko, Nemoto Tetsuhiro, Zhao Zengduo, Ishige Yuta, Hamada Yasumasa. 2012. "Enantioselective construction of all-carbon quaternary spirocenters through a Pd-catalyzed asymmetric intramolecular *ipso*-Friedel-Crafts allylic alkylation of phenols". Tetrahedron: Asymmetry. No. 11-12. 859-866.

[5] Nemoto Tetsuhiro, Nozaki Tomoyuki, Yoshida Mariko, Hamada Yasumasa. 2013. "Palladium-catalyzed intramolecular *ipso*-Friedel–Crafts allylic alkylation of phenols via arylative activation of allenes". Advanced Synthesis & Catalysis. No. 13. 2693-2700.

[6] Nemoto Tetsuhiro, Hamada Yasumasa. 2016. "Synthesis of spirocyclic and Fused cyclic compounds by transition-metal-catalyzed intramolecular Friedel-Crafts-type reactions of phenol derivatives". Synlett. No. 16. 2301-2313.

[7] Li Nai-Kai, Zhang Jun-Qi, Sun Bing-Bing, Li Hai-Yan, Wang Xing-Wang. 2017. "Chiral diphosphine–palladium-catalyzed sequential asymmetric double-Friedel-Crafts alkylation and *N*-hemiketalization for spiropolycyclic indole derivatives". Organic Letters. No. 8. 1954-1957.

[8] Wu Qing-Feng, He Hu, Liu Wen-Bo, You Shu-Li. 2010. "Enantioselective construction of spiroindolenines by Ir-catalyzed allylic alkylation reactions". Journal of the American Chemical Society. No. 33. 11418-11419.

[9] Zhuo Chun-Xiang, Liu Wen-Bo, Wu Qing-Feng, You Shu-Li. 2012. "Asymmetric dearomatization of pyrroles via Ir-catalyzed

allylic substitution reaction: Enantioselective synthesis of spiro-2*H*-pyrroles". Chemical Science. No. 1. 205-208.
[10] S. H. Ryder Alison, B. Cunningham William, Ballantyne George, Mules Tom, *et al.* 2020. "Photocatalytic α-tertiary amine synthesis via C-H alkylation of unmasked primary amines". Angewandte Chemie. No. 35. 15096-15101.
[11] Manna Sabyasachi, Kumar Someswara Ashwathappa Puneeth, Ramaiah Prabhu Kandikere. 2020. "Visible light-mediated *ipso*-annulation of activated alkynes: Access to 3-alkylated spiro[4,5] trienones, thiaspiro[4,5]trienones and azaspiro[4,5]trienones". Chemical Communications. No. 86. 13165-13168.
[12] Li Feng, Zhou Yirong, Yang Heng, Liu Dandan, *et al.* 2018. "Assembly of diverse spirocyclic pyrrolidines via transient directing group enabled *ortho*-C(sp^2)–H alkylation of benzaldehydes". Organic Letters. No. 1. 146-149.
[13] Sarkar Soumen, Maiti Sukhendu, Bera Krishnendu, Jalal Swapnadeep, Jana Umasish. 2012. "Highly efficient synthesis of polysubstituted fluorene via iron-catalyzed intramolecular Friedel –Crafts alkylation of biaryl alcohols". Tetrahedron Letters. No. 41. 5544-5547.
[14] Peng Xiang, Liu Ren-Xiang, Xiao Xiang-Yan, Yang Luo. 2020. "Fe-Catalyzed decarbonylative alkylative spirocyclization of *N*-arylcinnamamides: Access to alkylated 1-azaspirocyclohexa dienones". Molecules. No. 3. 432.
[15] Deng Qing-Hai, Wadepohl Hubert, H. Gade Lutz. 2012. "Highly enantioselective copper-catalyzed alkylation of β-ketoesters and subsequent cyclization to spirolactones/bispirolactones". Journal of the American Chemical Society. No. 6. 2946-2949.
[16] Bjørnstad Vidar, Undheim Kjell. 2008. "Stereoselective syntheses of spirane bridged semi-titanocenes". Synthesis. No. 06. 962-970.
[17] Raji Reddy Chada, H. Kolgave Dattahari, Subbarao Muppidi, Aila Mounika, Kumar Prajapti Santosh. 2020. "Ag-Catalyzed oxidative *ipso*-cyclization via decarboxylative acylation/ alkylation: Access to 3-acyl/alkyl-spiro[4.5]trienones". Organic Letters. No. 14. 5342-5346.
[18] B. Dhotare Bhaskar, K. Choudhary Manoj, K. Nayak Sandip. 2016. "$SbCl_3$-Catalyzed solvent-free Friedel-Crafts reaction of phenols with mandelic acids to 3-aryl benzofuran-2(3*H*)-ones: Synthesis of spirocyclic 2,3-dihydrobenzofuran-2-ones". Synthetic Communications. No. 21. 1772-1780.

[19] Subramaniyan Gowtham, Raghunathan Raghavachary. 2001. "Synthesis of highly substituted spiropyrrolidines via 1,3-dipolar cycloaddition reaction of *N*-metalated azomethine ylides". Tetrahedron. No. 14. 2909-2913.
[20] Grigg Ronald, Abul Basher Sarker Mohammed. 2006. "XY-ZH Compounds as potential 1,3-dipoles. Part 63: Silver catalysed azomethine ylide cycloaddition-the synthesis of spiro homoserine lactone analogues". Tetrahedron. No. 44. 10332-10343.
[21] Grigg Ronald, Kilner Colin, Abul Basher Sarker Mohammed, Orgaz de la Cierva Cecilia, Dondas H. Ali. 2008. "X = Y-ZH Compounds as potential 1,3-dipoles. Part 64: Synthesis of highly substituted conformationally restricted and spiro nitropyrrolidines via Ag(I) catalysed azomethine ylide cycloadditions". Tetrahedron. No. 37. 8974-8991.
[22] Mernyák Erzsébet, Kozma Eszter, Hetényi Anasztázia, Márk László, *et al.* 2009. "Stereoselective synthesis of spiro and condensed pyrazolines of steroidal α,β-unsaturated ketones and nitril-imines by 1,3-dipolar cycloaddition". Steroids. No. 6. 520-525.
[23] Cayuelas Alberto, Ortiz Ricardo, Najera Carmen, M. Sansano Jose, *et al.* 2016. "Enantioselective synthesis of polysubstituted spiro-nitroprolinates mediated by a (*R*,*R*)-Me-DuPhos·AgF-catalyzed 1,3-dipolar cycloaddition". Organic Letters. No. 12. 2926-2929.
[24] Rajkumar Vadla, Arulananda Babu Srinivasarao, Padmavathi Rayavarapu. 2016. "Regio-and diastereoselective construction of a new set of functionalized pyrrolidine, spiropyrrolidine and spiropyrrolizidine scaffolds appended with aryl-and heteroaryl moieties via the azomethine ylide cycloadditions". Tetrahedron. No. 36. 5578-5594.
[25] Liu Tang-Lin, He Zhao-Lin, Tao Hai-Yan, Cai Yue-Peng, Wang Chun-Jiang. 2011. "Stereoselective construction of a 5-aza-spiro [2,4]heptane motif via catalytic asymmetric 1,3-dipolar cycloaddition of azomethine ylides and ethyl cyclopropylidene acetate". Chemical Communications. No. 9. 2616-2618.
[26] Liu Tang-Lin, He Zhao-Lin, Tao Hai-Yan, Wang Chun-Jiang. 2012. "Stereoselective construction of spiro(butyrolactone pyrrolidines) by highly efficient copper(I)/TF-biphamPhos-catalyzed asymmetric 1,3-dipolar cycloaddition". Chemistry-A European Journal. No. 26. 8042-8046.

[27] Li Qing-Hua, Liu Tang-Lin, Wei Liang, Zhou Xiang, *et al.* 2013. "*exo*-Selective construction of spiro[butyrolactone-pyrrolidine] via 1,3-dipolar cycloaddition of azomethine ylides with α-methylene-γ-butyrolactone catalyzed by Cu(I)/DTBM-BIPHEP" . Chemical Communications. No. 83. 9642-9644.

[28] P. Antonchick Andrey, Schuster Hannah, Bruss Hanna, Schürmann Markus, *et al.* 2011. "Enantioselective synthesis of the spirotryprostatin A scaffold". Tetrahedron. No. 52. 10195-10202.

[29] Nath Singh Shambhu, Regati Sridhar, Paul Abir Kumar, Layek Mohosin, *et al.* 2013. "Cu-mediated 1,3-dipolar cycloaddition of azomethine ylides with dipolarophiles: A faster access to spiro-oxindoles of potential pharmacological interest". Tetrahedron Letters. No. 40. 5448-5452.

[30] Yang Wu-Lin, Tang Fei-Fei, He Fu-Sheng, Li Chun-Yan, *et al.* 2015. "Asymmetric construction of spirocyclic pyrrolidine-thia (oxa)zolidinediones via *N,O*-ligand/Cu(I) catalyzed 1,3-dipolar cycloaddition of azomethine ylides with 5-alkylidenethia(oxa) zolidine-2,4-diones". Organic Letters. No. 19. 4822-4825.

[31] Zhang Di, Zhang Da-Ming, Xu Guang-Yang, Sun Jiang-Tao. 2015. "Copper-catalyzed 1,3-dipolar cycloaddition of methylene indolinones and *N,N′*-cyclic azomethine imines". Chinese Chemical Letters. No. 3. 301-303.

[32] Tang Li-Wei, Li Chuan, Zhao Bao-Jing, Lan Li, *et al.* 2017. "Diastereoselective construction of spirocyclic pyrrolidines bearing two quaternary centers via Cu^{II}-P,*N*-Ligand catalyzed 1,3-dipolar cycloaddition". Tetrahedron. No. 7. 923-930.

[33] Shen Chong, Yang Yuhong, Wei Liang, Dong Wu-Wei, *et al.* 2019. "Kinetic resolution of alkylidene norcamphors via a ligand-controlled umpolung-type 1,3-dipolar cycloaddition". Iscience. 146-159.

[34] Muthusamy Sengodagounder, Prabu Ammasi, Suresh Eringath-odi. 2019. "Copper-catalyzed synthesis of spiro indolofuro-benzo-pyrans: Tandem reactions of diazoamides and *O*-propargyl Sali-cylaldehydes". Organic & Biomolecular Chemistry. No. 35. 8088 -8093.

[35] Shi Yang, Wang Gangqiang, Chen Zhiyu, Wu Minghu, *et al.* 2020. "Synthesis of spiro(indoline-3,1′-pyrazolo[1,2-*a*]pyrazoles) by 1,3-dipolar cycloadditions of isatin *N,N′*-cyclic azomethine imines with alkynes". Journal of Heterocyclic Chemistry. No. 4. 2044-2047.

[36] Teja Chitrala, N. Babu Spoorthy, Noor Ayesha, Daniel J. Arul, *et al.* 2020. "Cu/TEMPO Catalyzed dehydrogenative 1,3-dipolar cycloaddition in the synthesis of spiro-oxindoles as potential anti-diabetic agents". RSC Advances. No. 21. 12262-12271.

[37] M. Hodgson David, Angrish Deepshikha. 2006. "Oxapolycycles from one-pot cross-metathesis/carbonyl ylide formation intra-molecular cycloaddition of α-diazo-β-keto esters". Advanced Synthesis & Catalysis. No. 16-17. 2509-2514.

[38] Rodier Fabien, Rajzmann Michel, Parrain Jean-Luc, Chouraqui Galle, Commeiras Laurent. 2013. "Diastereoselective access to polyoxygenated polycyclic spirolactones through a Rhodium-catalyzed [3+2] cycloaddition reaction: Experimental and theoretical studies". Chemistry–A European Journal. No. 7. 2467-2477.

[39] Lian Xiangjin, Guo Songsong, Wang Gang, Lin Lili, *et al.* 2014. "Asymmetric synthesis of spiro[isoxazolin-3,3′-oxindoles] via the catalytic 1,3-dipolar cycloaddition reaction of nitrile oxides". The Journal of Organic Chemistry. No. 16. 7703-7710.

[40] Bdiri Bilel, Zhou Zhi-Ming. 2017. "Novel asymmetric synthesis of spiroindene-1,3dione-pyrrolidines via Co(II)/amino acids complex catalysed asymmetric 1,3-dipolar cycloaddition of azomethine ylides and 2-arylidenindane-1,3-diones". Tetrahedron Letters. No. 49. 4600-4608.

[41] Yazdani Hossein, Bazgir Ayoob. 2019. "Lewis acid catalyzed regio-and diastereoselective synthesis of spiroisoxazolines via one-pot sequential Knoevenagel condensation/1,3-dipolar cyclo-addition reaction". Synthesis. No. 07. 1669-1679.

[42] L. Gerten Anthony, C. Slade Michael, M. Pugh Kelsie, M. Stanley Levi. 2013. "Catalytic, enantioselective 1,3-dipolar cycloadditions of nitrile imines with methyleneindolinones". Organic & Biomolecular Chemistry. No. 45. 7834-7837.

[43] Wang Gang, Liu Xiaohua, Huang Tianyu, Kuang Yulong, *et al.* 2013. "Asymmetric catalytic 1,3-dipolar cycloaddition reaction of nitrile imines for the synthesis of chiral spiropyrazoline oxindoles". Organic Letters. No. 1. 76-79.

[44] Wei Feng, Ren Chuan-Li, Wang Dong, Liu Li. 2015. "Highly enantioselective [3+2] cycloaddition of vinylcyclopropane with nitroalkenes catalyzed by palladium(0) with a chiral bis(*tert*-amine) ligand". Chemistry–A European Journal. No. 6. 2335-2338.

[45] Liu Ze-Shui, Li Wen-Ke, Kang Tai-Ran, He Long, Liu Quan-Zhong. 2015. “Palladium-catalyzed asymmetric cycloadditions of vinylcyclopropanes and in situ formed unsaturated imines: Construction of structurally and optically enriched spiro indolenines”. Organic Letters. No. 1. 150-153.

[46] Li Tian-Ren, Cheng Bei-Yi, Fan Si-Qi, Wang Ya-Ni, *et al.* 2016. “Highly stereoselective [3+2] cycloadditions of chiral palladium-containing N^1-1,3-dipoles: A divergent approach to enantio-enriched spiro-oxindoles”. Chemistry–A European Journal. No. 18. 6243-6247.

[47] Wang Wan-Ying, Wu Jing-Yu, Liu Qing-Rong, Liu Xiu-Yan, *et al.* 2018. “Palladium/*N*-heterocyclic carbene (NHC)-catalyzed asymmetric [3+2] cycloaddition reaction of vinyl epoxides with allenic amides”. Organic Letters. No. 16. 4773-4776.

[48] Zhang Huihui, Gao Xing, Feng Jiang, Shi Wangyu, *et al.* 2020. “Palladium-catalyzed asymmetric [3+2] cycloaddition of vinyl-ethylene carbonates with 2-arylidene-1,3-indandiones: Synthesis of tetrahydrofuran-fused spirocyclic 1,3-indandiones”. European Journal of Organic Chemistry. No. 30. 4801-4804.

[49] Wang Junwei, Zhao Lei, Rong Quanjin, Lv Cheng, *et al.* 2020. “Asymmetric synthesis of 3,3′-tetrahydrofurylspiro-oxindoles via palladium-catalyzed [3+2] cycloadditions of methylene-indolin-ones with vinylethylene carbonates”. Organic Letters. No. 15. 5833-5838.

[50] Vetica Fabrizio, J. Bailey Stephen, Kumar Mukesh, Mahajan Suruchi, *et al.* 2020. “Palladium-catalyzed [3+2] cycloaddition of vinylaziridine and indane-1,3-diones: Diastereo-and enantioselective access to spiro-pyrrolidines”. Synthesis. No. 14. 2038-2044.

[51] Dong Lin, Qu Chuan-Hua, Huang Ji-Rong, Zhang Wei, *et al.* 2013. “Rhodium-catalyzed spirocyclic sultam synthesis by [3+2] annulation with cyclic *N*-sulfonyl ketimines and alkynes”. Chemistry–A European Journal. No. 49. 16537-16540.

[52] V. Pham Manh, Cramer Nicolai. 2016. “Enantioselective access to spirocyclic sultams by chiral Cp^x–rhodium(III)-catalyzed annulations”. Chemistry–A European Journal. No. 7. 2270-2273.

[53] Zhou Ming-Bo, Pi Rui, Hu Ming, Yang Yuan, *et al.* 2014. “Rhodium(III)-catalyzed [3+2] annulation of 5-Aryl-2,3-dihydro-1*H*-pyrroles with internal alkynes through $C(sp^2)$-H/alkene functionalization”. Angewandte Chemie. No. 42. 11520 -11523.

[54] V. Subba Reddy Basi, Reddy E. Pravardhan, Sridhar Balasubramanian, Rao Y. Jayaprakash. 2016. "Rhodium-catalyzed cycloaddition of carbonyl ylides for the synthesis of spiro[furo [2,3-*a*]xanthene-2,3′-indolin]-2′-one scaffolds". RSC Advances. No. 56. 50497-50499.

[55] Sharma Satyasheel, Oh Yongguk, Kumar Mishra Neeraj, De Umasankar, *et al.* 2017. "Rhodium-catalyzed [3+2] annulation of cyclic *N*-acyl ketimines with activated olefins: Anticancer activity of spiroisoindolinones". The Journal of Organic Chemistry. No. 7. 3359-3367.

[56] Jiang Yan-Hong, Sun Jing, Sun Qiu, Yan Chao-Guo. 2017. "Construction of spiro[indene-2,1′-pyrrolo[2,1-*a*]isoquinoline]s through a visible-light-catalyzed oxidative [3+2] cycloaddition reaction". Asian Journal of Organic Chemistry. No. 7. 862-866.

[57] Awata Atsuko, Arai Takayoshi. 2012. "Catalytic asymmetric *exo′*-selective [3+2] cycloaddition for constructing stereochemically diversified spiro[pyrrolidin-3,3′-oxindole]s". Chemistry–A European Journal. No. 27. 8278-8282.

[58] Wang Bin, Liang Man, Tang Jian, Deng Yuting, *et al.* 2016. "Gold/Lewis acid catalyzed cyclo isomerization/diastereoselective [3+2] cycloaddition cascade: Synthesis of diverse nitrogen-containing spiro heterocycles". Organic Letters. No. 18. 4614-4617.

[59] Arai Takayoshi, Ogawa Hiroki, Awata Atsuko, Sato Makoto, *et al.* 2015. "PyBidine–$Cu(OTf)_2$-Catalyzed asymmetric [3+2] cycloaddition with imino esters: Harmony of Cu–Lewis acid and imidazolidine-NH hydrogen bonding in concerto catalysis". Angewandte Chemie. No. 5. 1615-1619.

[60] Bhandari Sonal, P. Sakla Akash, Shankaraiah Nagula. 2020. "$FeCl_3$-Catalyzed [3+2] cycloaddition reaction: A mild synthetic approach to spirooxindolo-2-iminothiazolidine scaffolds". ChemistrySelect. No. 10. 2886-2891.

[61] Suresh Babu Adukamparai Rajukrishnan, Raghunathan Raghavachary. 2007. "$ZrOCl_2 \cdot 8H_2O$ mediated microwave induced [3+2] cycloaddition of azomethine ylides-a facile one-pot synthesis of novel dispiroheterocycles". Tetrahedron Letters. No. 2. 305-308.

[62] Nüske Hanno, Bräse Stefan, I. Kozhushkov Sergei, Noltemeyer Mathias, *et al.* 2002. "A new highly efficient three-component domino Heck–Diels–Alder reaction with bicyclopropylidene:

Rapid access to spiro[2.5]oct-4-ene derivatives". Chemistry–A European Journal. No. 10. 2350-2369.

[63] Shintani Ryo, Park Soyoung, Hayashi Tamio. 2007. "Palladium-catalyzed synthesis of spiro[2.4]heptanes: Ligand-dependent position control in the nucleophilic attack to a π-allylpalladium intermediate". Journal of the American Chemical Society. No. 48. 14866-14867.

[64] M. D'Souza Daniel, Kiel Alexander, Herten Dirk-Peter, Rominger Frank, J. J. Müller Thomas. 2008. "Synthesis, structure and emission properties of spirocyclic benzofuranones and dihydroindolones: A domino insertion–coupling–isomerization–diels–alder approach to rigid fluorophores". Chemistry-A European Journal. No. 2. 529-547.

[65] S. Ghotekar Ganesh, R. Shirsath Sachin, C. Shaikh Aslam, Muthukrishnan Murugan. 2020. "1,6-Conjugate addition initiated formal [4+2] annulation of *p*-quinone methides with sulfonyl allenols: A unique access to spiro[5.5]undeca-1,4-dien-3-one scaffolds". Chemical Communications. No. 37. 5022-5025.

[66] M. Trost Barry, Jiao Zhiwei. 2020. "Palladium-catalyzed enantioselective cycloaddition of carbonylogous 1,4-dipoles: Efficient access to chiral cyclohexanones". Journal of the American Chemical Society. No. 52. 21645-21650.

[67] Mao Biming, Liu Honglei, Yan Zhengyang, Xu Yi, *et al.* 2020. "Palladium-catalyzed asymmetric [4+2] cycloaddition of 2-methylidenetrimethylene carbonate with alkenes: Access to chiral tetrahydropyran-fused spirocyclic scaffolds". Angewandte Chemie. No. 28. 11412-11416.

[68] M. Trost Barry, Jiao Zhiwei, Liu Ying, Min Chang, Hung Chao-I. Joey. 2020. "Palladium-catalyzed enantioselective cycloadditions of aliphatic 1,4-dipoles: Access to chiral cyclohexanes and spiro[2.4]heptanes". Journal of the American Chemical Society. No. 43. 18628-18636.

[69] Hojo Daiki, Noguchi Keiichi, Hirano Masao, Tanaka Ken. 2008. "Enantioselective synthesis of spirocyclic benzopyranones by rhodium-catalyzed intermolecular [4+2] annulation". Angewandte Chemie. No. 31. 5904-5906.

[70] Barluenga Jos, Calleja Jonás, Mendoza Abraham, Rodríguez Félix, Francisco J. Fañanás. 2010. "Synthesis of polycyclic compounds by a cascade cycloisomerisation/Diels–Alder reaction". Chemistry–A European Journal. No. 24. 7110-7112.

[71] Sharma Pankaj, Praveen Kumar Niggula, Hari Krishna Namballa, Prasanna Daasi, *et al.* 2016. "Silver(I)-catalysed domino alkyne-annulation/Diels–Alder reaction: A mild synthetic approach to tetrahydrospiro[carbazole-4,3′-indoline] scaffolds". Organic Chemistry Frontiers. No. 11. 1503-1508.
[72] Wang Dan, Liu Shuai, Lan Xin-Chan, Paniagua Armando, *et al.* 2017. "Tunable dimerization and trimerization of β-alkynyl ketones via silver catalysis for accessing spiro and dispiro compounds containing 1*H*-isochromene". Advanced Synthesis & Catalysis. No. 18. 3186-3193.
[73] Suri Mrinaly, Liaquat Hussain Farhaz, Gogoi Chinu, Das Pankaj, Pahari Pallab. 2020. "Magnetically recoverable silica catalysed solvent-free domino Knoevenagel-hetero-Diels–Alder reaction to access divergent chromenones". Organic & Biomolecular Chemistry. No. 11. 2058-2062.
[74] Kong Ke, Moussa Ziad, Romo Daniel. 2005. "Studies toward a marine toxin immunogen: Enantioselective synthesis of the spirocyclic imine of (-)-gymnodimine". Organic Letters. No. 23. 5127-5130.
[75] Li Nai-Kai, Sun Bing-Bing, Chen Jun-Bo, Yang Hao-Di, *et al.* 2021. "Box-copper catalyzed asymmetric inverse-electron-demand oxa-hetero-Diels–Alder reaction for efficient synthesis of spiro pyranyl-oxindole derivatives". Organic Chemistry Frontiers. No. 9. 2009-2018.
[76] D. Lawlor Michael, W. Lee Thomas, L. Danheiser Rick. 2000. "Rhodium-catalyzed rearrangement of α-diazo thiol esters to thio-substituted ketenes. Application in the synthesis of cyclobutanones, cyclobutenones, and β-lactams". The Journal of Organic Chemistry. No. 14. 4375-4384.
[77] V. Subba Reddy Basi, Karthik Govindaraju, Rajasekaran Tamilselvan, Antony Aneesh, Sridhar Balasubramanian. 2012. "$Rh_2(OAc)_4$ Catalyzed substrate selective [4+2]/[2+2] cyclo-addition of acylketenes: A highly chemo-and regioselective synthesis of spiro(oxindolyl)oxazinones and β-lactams". Tetrahedron Letters. No. 19. 2396-2401.
[78] Xiao Meiling, Zhang Fuming, Du Zhe, Baochun Ma. 2018. "An efficient synthesis of 6-oxa-spiro[3.4]octan-1-one derivates through 3-diazochroman-4-one and alkene". Advanced Synthesis & Catalysis. No. 5. 911-916.
[79] Peng Shiyong, Gao Tao, Sun Shaofa, Peng Yanhong, *et al.* 2014. "Palladium-catalyzed [2+2+1] oxidative annulation of 4-hydroxy

coumarins with unactivated internal alkynes: Access to spirocyclopentadiene-chroman-2,4-dione complexes". Advanced Synthesis & Catalysis. No. 2-3319-324.

[80] Zuo Zhijun, Wang Hui, Fan Liangxin, Liu Jingjing. *et al.* 2017. "Modular assembly of spirocarbocyclic scaffolds through Pd^0-catalyzed intermolecular dearomatizing [2+2+1] annulation of bromonaphthols with aryl Iodides and alkynes". Angewandte Chemie. No. 10. 2811-2815.

[81] Li Mi-Zhuan, Tong Qi, Han Wen-Yong, Yang Si-Yi, *et al.* 2020. "Synthesis of chromone-containing polycyclic compounds via palladium-catalyzed [2+2+1] annulation". Organic & Bio-molecular Chemistry. No. 6.1112-1116.

[82] Wada Azusa, Noguchi Keiichi, Hirano Masao, Tanaka Ken. 2007. "Enantioselective synthesis of C_2-symmetric spiro-bipyridine ligands through cationic Rh(I)/modified-BINAP-catalyzed double [2+2+2] cycloaddition". Organic Letters. No. 7. 1295-1298.

[83] B. Suryawanshi Sharad, P. Dushing Mangesh, G. Gonnade Rajesh, V. Ramana Chepuri. 2010. "The isochroman-and 1,3-dihydroisobenzofuran-annulation on carbohydrate templates via [2+2+2] cyclotrimerization and synthesis of some tricyclic nucleosides". Tetrahedron. No. 32. 6085-6096.

[84] A. Golubev Artem, A. Smetanin Ilia, V. Agafonova Anastasiya, V. Rostovskii Nikolai, *et al.* 2019. "[2+1+1] Assembly of spiro-β-lactams by Rh(II)-catalyzed reaction of diazocarbonyl compounds with azirines/isoxazoles". Organic & Biomolecular Chemistry. No. 28. 6821-6830.

[85] Jiao Meng-Jie, Liu Dan, Hu Xiu-Qin, Xu Peng-Fei. 2019. "Photocatalytic decarboxylative [2+2+1] annulation of 1,6-enynes with *N*-hydroxyphthalimide esters for the synthesis of indene-containing polycyclic compounds". Organic Chemistry Frontiers. No. 23. 3834-3838.

[86] Göbel Angela, Imhof Wolfgang. 2001. "One-pot ruthenium catalyzed synthesis of spiro[pyrrolidin-2-one] derivatives by a [2+2+1] cycloaddition of ketimines, carbon monoxide and ethylene". Chemical Communications. No. 7. 593-594.

[87] Zhou Yuxuan, Ma Fanghui, Lu Ping, Wang Yanguang. 2019. "Preparation of spiro[imidazolidine-4,3′-indolin]-2′-imines via copper(I)-catalyzed formal [2+2+1] cycloaddition of 3-diazoindolin-2-imines and triazines". Organic & Biomolecular Chemistry. No. 39. 8849-8852.

[88] Liang Wu, Jiang Kun, Du Fei, Yang Jie, *et al.* 2020. "Iron-catalyzed, iminyl radical-triggered cascade 1,5-hydrogen atom transfer/(5+2) or (5+1) annulation: Oxime as a five-atom assembling unit". Angewandte Chemie International Edition. No. 43. 19222-19228.
[89] Lin En-Ze, Xu Yin, Ji Kegong, Ye Long-Wu. 2021. "Recent advances towards catalytic asymmetric Conia-ene-type reactions" . Chinese Chemical Letters. No. 3. 954-962.
[90] Schfer Christian, Miesch Michel, Miesch Laurence. 2012. "A silver-catalyzed spirocyclization of alkynyl silyl enol ethers". Chemistry–A European Journal. No. 26. 8028-8031.
[91] Li Jun, Lin Lili, Hu Bowen, Lian Xiangjin, *et al.* 2016. "Bimetallic gold(I)/chiral *N,N'*-dioxide nickel(II) asymmetric relay catalysis: Chemo-and enantioselective synthesis of spiroketals and spiroaminals". Angewandte Chemie International Edition. No. 20. 6075-6078.
[92] Raji Reddy Chada, Kumaraswamy Paridala. 2014. "Copper-promoted cascade reaction of active methylenes with MBH-acetates of acetylenic aldehydes to functionalized cyclopentenes". RSC Advances. No. 14. 7035-7040.
[93] Beltran Frédéric, Fabre Indira, Ciofini Ilaria, Miesch Laurence. 2017. "Direct spirocyclization from keto-sulfonamides: An approach to azaspiro compounds". Organic Letters. No. 19. 5042-5045.
[94] K. Corkey Britton, T. Heller Stephen, Wang Yi-Ming, Toste F. Dean. 2013. "Enantioselective cyclization of enamide-ynes and application to the synthesis of the kopsifoline core". Tetrahedron. No. 27-28. 5640-5646.
[95] Xu Chao, Deiana Luca, Afewerki Samson, Incerti-Pradillos Celia, *et al.* 2015. "The use of porous palladium(II)-polyimine in cooperatively-catalyzed highly enantioselective cascade transformations". Advanced Synthesis & Catalysis. No. 9. 2150-2156.
[96] Putatunda Salil, V. Alegre-Requena Juan, Meazza Marta, Franc Michael, *et al.* 2019. "Proline bulky substituents consecutively act as steric hindrances and directing groups in a Michael/Conia-ene cascade reaction under synergistic catalysis". Chemical Science. No. 14. 4107-4115.
[97] Das Arindam, Chang Hsu-Kai, Yang Cheng-Hua, Liu Rai-Shung. 2008. "Platinum-and gold-catalyzed hydrative carbocyclization of oxo diynes for one-pot synthesis of benzopyrones and bicyclic spiro ketones". Organic Letters. No. 18. 4061-4064.

[98] C. Hunter Arianne, Almutwalli Bilal, I. Bain Anae, Sharma Indrajeet. 2018. "Trapping rhodium carbenoids with amino-alkynes for the synthesis of diverse *N*-heterocycles". Tetrahedron. No. 38. 5451-5457.
[99] C. Hunter Arianne, Chinthapally Kiran, I. Bain Anae, C. Stevens Joseph, Sharma Indrajeet. 2019. "Rhodium/gold dual catalysis in carbene sp^2 C-H functionalization/Conia-ene cascade for the stereoselective synthesis of diverse spirocarbocycles". Advanced Synthesis & Catalysis. No. 12. 2951-2958.
[100] Hua Hui-Liang, He Yu-Tao, Qiu Yi-Feng, Li Ying-Xiu, *et al.* 2015. "Copper-catalyzed difunctionalization of activated alkynes by radical oxidation–tandem cyclization/dearomatization to synthesize 3-trifluoromethylspiro[4.5]trienones". Chemistry–A European Journal. No. 4. 1468-1473.
[101] Wu Xin-Xing, Shen Yi, Chen Wen-Long, Chen Si, *et al.* 2015. "Palladium-catalyzed dearomative cyclization by a norbornene-mediated sequence: A route to spiroindolenine derivatives". Chemical Communications. No. 94. 16798-16801.
[102] Hu Bei, Li Yuyuan, Dong Wuheng, Ren Kai, *et al.* 2016. "Visible light-induced intramolecular dearomative cyclization of α-bromo-*N*-benzyl-alkylamides: Efficient construction of 2-azaspiro [4.5]decanes". Chemical Communications. No. 18. 3709-3712.
[103] Dong Wuheng, Yuan Yao, Gao Xiaoshuang, Keranmu Miladili, *et al.* 2018. "Visible-light-lnduced lntermolecular dearomative cyclization of 2-bromo-1,3-dicarbonyl compounds and alkynes: Synthesis of spiro[4.5]deca-1,6,9-trien-8-ones". Organic Letters. No. 18. 5762-5765.
[104] Dong Wuheng, Yuan Yao, Xie Xiaomin, Zhang Zhaoguo. 2020. "Visible-light-driven dearomatization reaction toward the formation of spiro[4.5]deca-1,6,9-trien-8-ones". Organic Letters. No. 2. 528-532.
[105] Dong Wuheng, Yuan Yao, Gao Xiaoshuang, Keranmu Miladili, *et al.* 2019. "Visible-light-induced intermolecular dearomative cyclization of furans: Synthesis of 1-oxaspiro[4.4]nona-3,6-dien-2-one". The Journal of Organic Chemistry. No. 3. 1461-1467.
[106] Dong Wuheng, Yuan Yao, Liang Caiyun, Wu Feng, *et al.* 2021. "Photocatalytic radical *ortho*-dearomative cyclization: Access to spiro[4.5]deca-1,7,9-trien-6-ones". The Journal of Organic Chemistry. No. 5. 3697-3705.
[107] Magné Valentin, Blanchard Florent, Marinetti Angela, Voituriez Arnaud, Guinchard Xavier. 2016. "Synthesis of spiro[piperidine-

3,3′-oxindoles] via gold(I)-catalyzed dearomatization of *N*-propargyl-and *N*-homoallenyl-2-bromotryptamines". Advanced Synthesis & Catalysis. No. 21. 3355-3361.

[108] Zhang Yan, Zhang Junhua, Hu Boyue, Ji Mingming, *et al.* 2018. "Synthesis of difluoromethylated and phosphorated spiro[5.5] trienones via dearomative spirocyclization of biaryl ynones". Organic Letters. No. 10. 2988-2992.

[109] Sarkar Debayan, Rout Nilendri. 2019 "Ruthenium(VIII)-catalyzed *ipso*-dearomative spiro-etherification and spiro-amidation of phenols". Organic Letters. No. 11. 4132-4136.

[110] Wu Wen-Ting, Ding Lu, Zhang Liming, You Shu-Li. 2019. "Gold-catalyzed intramolecular dearomatization reactions of indoles for the synthesis of spiroindolenines and spiroindolines". Organic Letters. No. 4. 1233-1238.

[111] Dong Wei, Qi Lin, Song Jin-Yan, Chen Jia-Min, *et al.* 2020. "Direct synthesis of sulfonylated spiro[indole-3,3′-pyrrolidines] by silver-mediated sulfonylation of acrylamides coupled with indole dearomatization". Organic Letters. No. 5. 1830-1835.

[112] Xia Dong, Duan Xin-Fang. 2021. "Iron-catalyzed dearomatization of biaryl ynones with aldehydes via double C–H functionalization in eco-benign solvents: Highly atom-economical synthesis of acylated spiro[5.5]trienones". The Journal of Organic Chemistry. No. 21. 15263-15275.

[113] Xia Dong, Duan Xin-Fang. 2021. "Alkylative dearomatization by using an unactivated aryl nitro group as a leaving group: Access to diversified alkylated spiro[5.5]trienones". Organic Letters. No. 7. 2548-2552.

[114] Liao Xingrong, Zhou Fulin, Bin Zhengyang, Yang Yudong, You Jingsong. 2021. "Palladium-catalyzed cascade dearomative spirocyclization and C-H annulation of aromatic halides with alkynes". Organic Letters. No. 13. 5203-5207.

[115] P. Unsworth William, D. Cuthbertson James, J. K. Taylor Richard. 2013. "Total synthesis of spirobacillene A" . Organic Letters. No. 13. 3306-3309.

[116] Qian Peng-Cheng, Liu Yu, Song Ren-Jie, Xiang Jian-Nan, Li Jin-Heng. 2015. "Copper-catalyzed oxidative *ipso*-cyclization of *N*-(*p*-methoxyaryl)propiolamides with disulfides and water leading to 3-(arylthio)-1-azaspiro[4.5]deca-3,6,9-triene-2,8-diones". Synlett. No. 09. 1213-1216.

[117] Ouyang Xuan-Hui, Song Ren-Jie, Liu Bang, Li Jin-Heng. 2016. "Synthesis of 3-alkylspiro[4,5]trienones by copper-catalyzed

oxidative *ipso*-annulation of activated alkynes with unactivated alkanes". Chemical Communications. No. 12. 2573-2576.

[118] He Yicheng, Qiu Guanyinsheng. 2017. "$ZnBr_2$-Mediated oxidative spirobromocyclization of propiolamide for the synthesis of 3-bromo-1-azaspiro[4.5]deca-3,6,9-triene-2,8-dione". Organic & Biomolecular Chemistry. No. 16. 3485-3490.

[119] A. Nechaev Anton, Van Hecke Kristof, Zaman Manzoor, Kashtanov Stepan, *et al.* 2018. "Gold-catalyzed post-Ugi *ipso*-cyclization with switchable diastereoselectivity". The Journal of Organic Chemistry. No. 15. 8170-8182.

[120] Wang Yu-Chao, Liu Jin-Biao, Zhou Hongwei, Xie Wenlin, *et al.* 2019. "*ortho*-Hydroxylative *ipso*-cyclization of *N*-aryl-propiolamide" . The Journal of Organic Chemistry. No. 4. 1906-1914.

[121] Chen Pu, Xie Jun, Chen Zan, Xiong Bi-Quan, *et al.* 2021. "Visible-light-mediated nitrogen-centered radical strategy: Preparation of 3-acylated spiro[4,5]trienones". Advanced Synthesis & Catalysis. No. 18. 4440-4446.

[122] Xu Hui, Pan Zhengbing, Dai Lei, Mao Kaimin, Rong Liangce. 2019. "An efficient in situ reduction and cyclization reaction for synthesis of spiro compound derivatives in Fe–H_2O–AcOH medium from nitro compounds". Applied Organometallic Chemistry. No. 1. e4653.

[123] Guo Chenhao, Li Bin, Liu Huilai, Zhang Xiaopeng, *et al.* 2019. "Synthesis of fused or spiro polyheterocyclic compounds via the dehydrogenative annulation reactions of 2-arylindazoles with maleimides". Organic Letters. No. 18. 7189-7193.

[124] Ding Lu, Wu Wen-Ting, Zhang Liming, You Shu-Li. 2020. "Construction of spironaphthalenones via gold-catalyzed intramolecular dearomatization reaction of β-naphthol derivatives". Organic Letters. No. 15. 5861-5865.

[125] Wang Zhiming, Xu Xingzhu, Gu Zhanshou, Feng Wei. 2016. "Nazarov cyclization of 1,4-pentadien-3-ols: Preparation of cyclopenta[*b*]indoles and spiro[indene-1,4′-quinoline]s". Chemical Communications. No. 13. 2811-2814.

[126] Zheng Deping, Ma Zuodong, Gong Wanchun, Xie Zhixiang, Li Ying. 2010. "$FeCl_3$-Promoted one-step synthesis of spiro[4.5] decane derivatives". Synlett. No. 14. 2169-2173.

[127] M. Reddy L. Madhava, Reddy V. Veerabadra, Satteyyanaidu Vallabhareddy, K. Lakshmi Josyula, *et al.* 2018. "Tandem Prins cyclization for the synthesis of 1,8-dioxa-3-azaspiro[4.5]dec-2-ene derivatives". Tetrahedron Letters. No. 12. 1084-1086.

[128] Astruc Didier. 2005. "The metathesis reactions: From a historical perspective to recent developments". New Journal of Chemistry. No. 1. 42-56.
[129] Kress Steffen, Blechert Siegfried. 2012. "Asymmetric catalysts for stereocontrolled olefin metathesis reactions". Chemical Society Reviews. No. 12. 4389-4408.
[130] Kotha Ambasivarao, Lahiri Kakali. 2007. "Synthesis of diverse polycyclic compounds via catalytic metathesis". Synlett. No. 18. 2767-2784.
[131] C. Stewart Ian, Ung Thay, A. Pletnev Alexandre, M. Berlin Jacob, *et al.* 2007. "Highly efficient ruthenium catalysts for the formation of tetrasubstituted olefins via ring-closing metathesis". Organic Letters. No. 8. 1589-1592.
[132] Deraedt Christophe, d′Halluin Martin, Astruc Didier. 2013. "Metathesis reactions: Recent trends and challenges". European Journal of Inorganic Chemistry. No. 28. 4881-4908.
[133] W. Morzycki Jacek. 2011. "Application of olefin metathesis in the synthesis of steroids". Steroids. No. 10-11. 949-966.
[134] Monsigny Louis, Piątkowski Jakub, Trzybiński Damian, Woźniak Krzysztof, *et al.* 2021. "Activated Hoveyda-Grubbs olefin metathesis catalysts derived from a large scale produced pharmaceutical intermediate–sildenafil aldehyde". Advanced Synthesis & Catalysis, No. 19. 4590-4604.
[135] Požgan France, H. Dixneuf Pierre. 2007. "Recent applications of alkene metathesis for fine chemical and supramolecular system synthesis". In Metathesis Chemistry. 195-222.
[136] Nicolaou Kyriacos Costa, G. Bulger Paul, Sarlah David. 2005. "Metathesis reactions in total synthesis". Angewandte Chemie International Edition. No. 29. 4490-4527.
[137] Matsuda Takanori, Yamaguchi Yoshiyuki, Murakam Masahiro. 2008. "Synthesis of silole skeletons via metathesis reactions". Synlett. No. 04. 561-564.
[138] Kotha Sambasivarao, Rao Panguluri Nageswara, Ali Rashid. 2017. "Design and synthesis of spirocycles". European Journal of Organic Chemistry. No. 36. 5316-5342.
[139] Langer Peter, Albrecht Uwe. 2002. "Synthesis of spirocyclic butenolides by ring closing metathesis. Synlett. No. 11. 1841-1842.
[140] Kotha Sambasivarao, Manivannan Ethirajan. 2003. "Synthesis of spirocyclics via ring-closing metathesis". Arkivoc. 67-76.

[141] G. Csákÿ Aurelio, Medel Rocío, Murcia M. Carmen, Plumet Joaquín. 2005. "Stereoselective synthesis of tetrahydrofuran spiro-β-lactams by Ru-catalyzed metathesis of 7-oxabicyclo [2.2.1]heptenes". Helvetica Chimica Acta. No. 6. 1387-1396.

[142] Yu. Kuznetsov Nikolai, D. Kolomnikova Galina, N. Khrustalev Victor, G. Golovanov Denis, N. Bubnov Yuri. 2008. "The combination of diallylboration and ring-closing metathesis in the synthesis of spiro-β-amino alcohols and (±)-Cephalotaxine". European Journal of Organic Chemistry. 5647–5655.

[143] Nieczypor Piotr, C. Mol Johannes, B. Bespalova Natalia, N. Bubnov Yuri. 2004. "Synthesis of nitrogen-containing spiro compounds from lactams by allylboration and subsequent ring-closing metathesis". European Journal of Organic Chemistry. No. 4. 812-819.

[144] C. Majumdar Krishna, Islam Roushan, Rahaman Habibur, Roy Brindaban. 2006. "Synthesis of spiro-pyridopyridine analogues by Grubbs' catalyst mediated alkene and enyne metathesis reaction". Organic & Biomolecular Chemistry. No. 12. 2393-2398.

[145] Kersten Laura, H. Taylor Rachel, Felpin François-Xavier. 2009. "Synthesis of spiro-pyridones and spiro-quinolones by sequential palladium on carbon-catalyzed allylation and ring closing metathesis reactions". Tetrahedron Letters. No. 5. 506-508.

[146] Dhara Kalyan, Chandra Midya Ganesh, Dash Jyotirmayee. 2012. "A diversity-oriented approach to spirocyclic and fused hydantoins via olefin metathesis". The Journal of Organic Chemistry. No. 18. 8071-8082.

[147] R. Siddiqui Iqbal, Shamim Shayna, Rai Pragati, A. Waseem Malik, A. Abumhdi Afaf. 2013. "An $InBr_3$ catalyzed one-pot three-component synthesis of functionalized spirodihydrofuran oxindoles via intramolecular alkyne carbonyl metathesis". Tetrahedron Letters. No. 51. 6991-6994.

[148] Kotha Sambasivarao, Saifuddin Mohammad, Ali Rashid, Sreevani Gaddamedi. 2015. "Spiro annulation of cage polycycles via Grignard reaction and ring-closing metathesis as key steps". Beilstein Journal of Organic Chemistry. 11, No. 1. 1367-1372.

[149] Kotha Sambasivarao, Ali Rashid, Kumar Chinnam Ajay. 2014. "Diversity-oriented approach to spirocycles via ring-closing metathesis". Tetrahedron Letters. No. 32. 4492-4495.

[150] Kotha Sambasivarao, Ravikumar Ongolu, Sreevani Gaddamedi. 2016. "Design and synthesis of oxacycles from norbornene

derivatives via ring-opening metathesis and ring-rearrangement metathesis". Tetrahedron. No. 42. 6611-6615.

[151] Fábian Martin, Gonda Jozef, Dominika Jacková, Martinková Miroslava, *et al.* 2020. "Synthesis and *in vitro* cytotoxic evaluation of spiro-β-lactone-γ-lactam scaffolds". Tetrahedron. No. 18 131144.

5

SYNTHESIS OF SPIROCYCLES III: ORGANO-CATALYZED SYNTHESIS OF SPIRO COMPOUNDS

5.1. Introduction

The use of small organic molecules as catalysts has become increasingly important in recent years. These substances, known as organocatalysts, are less contaminating and less toxic than traditional metal catalysts. Many different organocatalysts have been found to be efficient in the synthesis of spiro compounds. Organocatalysts interact with molecules in various ways, from forming covalent bonds with substrates, such as prolines, to interacting via hydrogen bonds, such as (thio)ureas. This chapter does not aim to provide a comprehensive review of the literature, but rather highlights the role of organic catalysts in the synthesis of spiranes.

5.2. Alkylation reactions

In their study, Yoshida et al. aimed to synthesize spiroindanes **3** using an enantioselective organocatalytic approach via intramolecular Friedel–Crafts-type 1,4-addition [1]. To achieve this, they employed a primary amine based on cinchonidine as a catalyst and used water and *p*-phenol as accelerators. According to the proposed mechanism, the reaction between compounds **1** and **2** gives rise to imine **A**, which is subsequently activated by protonation through the deprotonation of the phenolic hydroxy group of **2** (**B**→**C**). The resulting intermediate **C**, undergoes intramolecular 1,4-addition to form spiro compound **D**, which upon hydrolysis releases the amino catalyst and yields the desired product **3** (see Scheme 5-1 for details). It is worth noting that this study represents the first successful application of an enantioselective organocatalytic method for the construction of spiroindanes.

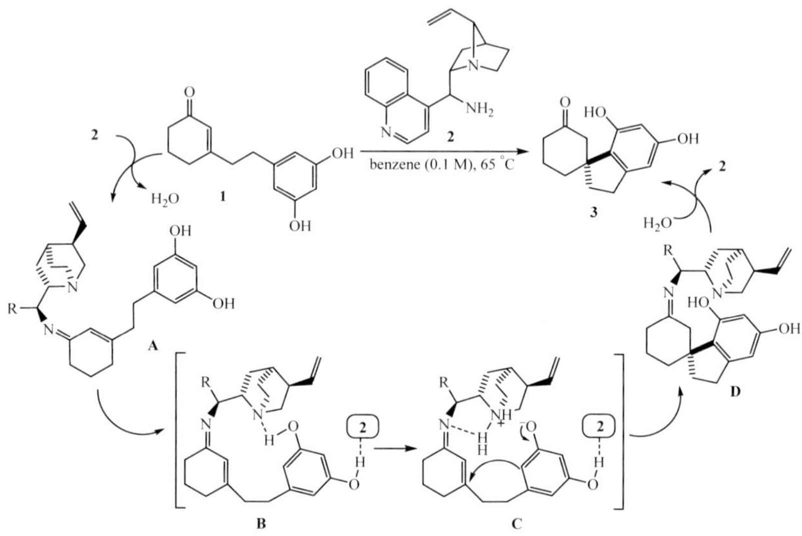

Scheme 5-1. Proposed mechanism for the construction of spiroindanes **3** by Friedel–Crafts-type 1,4-addition.

The Liu group developed an organocatalytic asymmetric synthesis method for spiro-bridged and spiro-fused heterocyclic compounds containing chromenes, indole, oxindole moieties, and three potential pharmacophores in a single molecule [2]. The synthesis is carried out in two stages:

i) In the first step, an asymmetric Michael addition is performed between compounds **4** and **5**, which results in the formation of a lactol intermediate **6** through iminium catalysis.

ii) In the second step, the lactol intermediate **6** undergoes acid-catalyzed intramolecular Friedel−Crafts alkylation by the nucleophilic attack of the indole moiety to an oxocarbenium ion, leading to the desired spiro-bridged heterocyclic products **7** (Scheme 5-2, path A).

Alternatively, when compounds **4** and **5** are reacted with **9**, chroman-2-ol **8** and 3-methoxy-3-indolyl-di-substituted oxindole **9** and **8**, respectively, are generated via the enamine-catalyzed C−C bond formation. The intermediate **10** obtained contains a lactol moiety, which is then treated with acid to provide the desired spiro-fused heterocyclic products **11** (Scheme 5-2, path B).

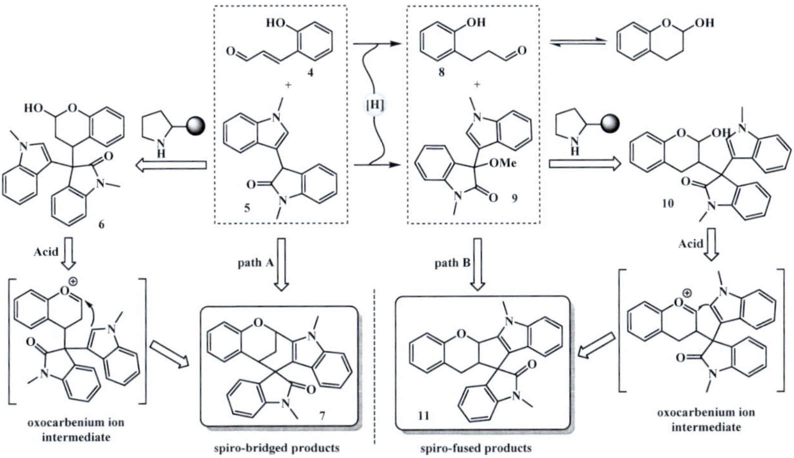

Scheme 5-2. Design of synthetic strategies.

Huang et al. developed a method for designing and synthesizing optoelectronic building blocks based on 9,9-diarylfluorenes through BF_3, OEt_2-mediated Friedel-Crafts reaction [3]. The reaction involved the use of 9-phenyl-fluoren-9-ol **12** and electron-rich aromatic substrates **13**, resulting in the formation of a new family of complex 9,9-diarylfluorenes **14** that were tethered with various functional substituents such as bulky spiro units, hole-transporting moieties, and fluorescent dyes (Scheme 5-3).

In 2007, Basavaiah et al. introduced an one-pot protocol for synthesizing indene-spiro-oxindoles through tandem Prins and intramolecular Friedel−Crafts (PFC) reactions. The $TiCl_4$-mediated reaction of 1,1-diarylethylenes **15** with isatin derivatives **16** results in the formation of two carbon−carbon bonds and thus provides 1*H*-indene-spiro-oxindoles **17** in moderate to excellent yields [4].

Scheme 5-4 shows a plausible mechanism for this transformation. According to this mechanism, 1 equiv of isatin **16** reacts with 1 equiv of diarylethylene **15** in the presence of $TiCl_4$ to afford the transient intermediate **T-A**, which can undergo intramolecular Friedel-Crafts reaction to yield intermediate **B**. The intermediate **B** then undergoes a Prins reaction with another molecule of diarylethylene **15** to furnish the desired 1*H*-indene-spiro-oxindole **17**. Alternatively, **T-A** might react with diarylethylene **15** to provide the first bisadduct **C**, which would then undergo a Prins reaction with another molecule of diarylethylene **15** to yield intermediate **D**. The latter would then undergo an intramolecular Friedel-Crafts reaction to provide the desired product **17** (path X).

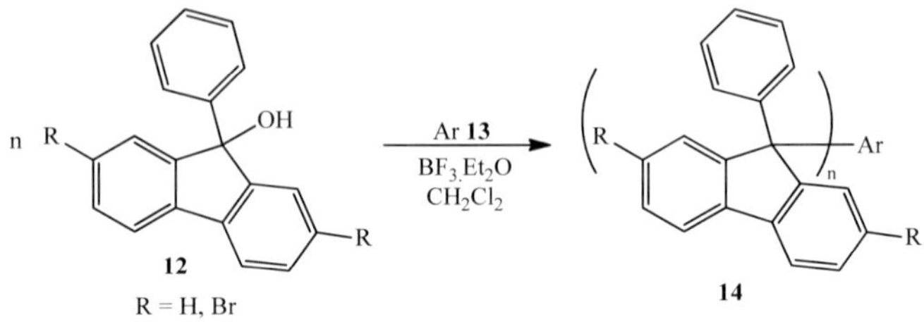

A few examples of 9,9-diarylfluorenes :

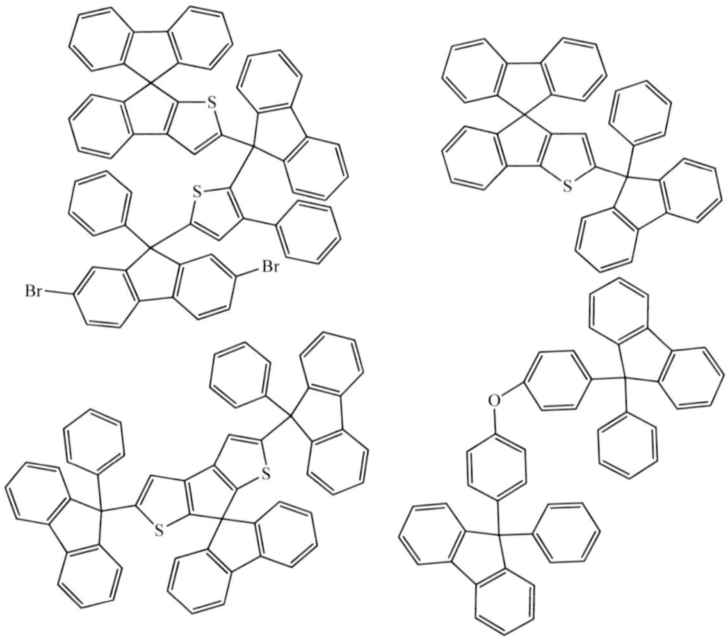

Scheme 5-3. The reaction of tertiary alcohol with different substrates.

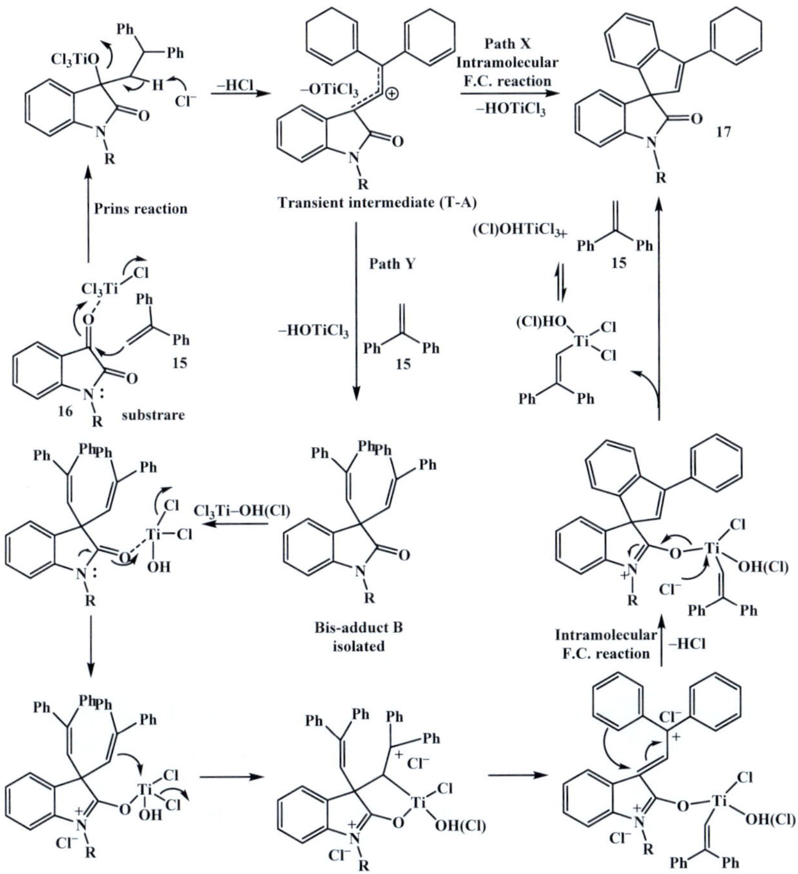

Scheme 5-4. Plausible proposed mechanism by Basavaiah et al.

Kobayashi et al. reported a novel method for constructing π-extended siloles, such as spiro-type silabifluorene **19**, through intramolecular sila-Friedel-Crafts reaction [5]. This reaction occurs under mild conditions and yields the desired product in relatively good yields, making it a versatile synthetic method. The basic strategy for developing the sila-Friedel–Crafts reaction is shown in Scheme 5-5.

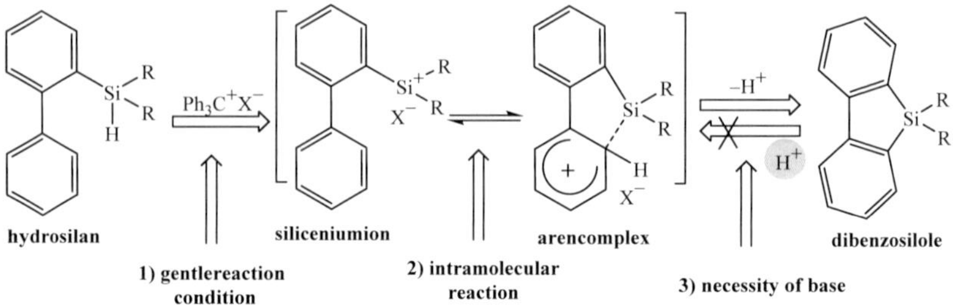

Scheme 5-5. The basic strategy for the development of the sila-Friedel–Crafts reaction.

The Kobayashi group hypothesized that the synthesis of spirobisilafluorene **19** could be achieved through the occurrence of two intramolecular sila-Friedel-Crafts reactions on the same silicon atom. To realize this, they utilized dihydrosilane **18** as a starting material in their reaction scheme (Scheme 5-6). Although the yield of the reaction was low (12%), the successful formation of the spirobiscilafluorene **19** highlights the potential of this method as a new strategy for multistep C-Si bond formation on the same silicon atom.

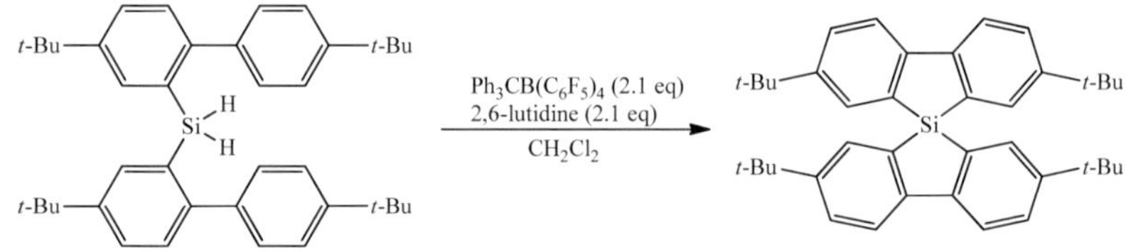

18 **19** (12%)

Scheme 5-6. Synthesis of spirobisilafluorene **19** by the sila-Friedel–Crafts reaction.

5.3. Pericyclic-type reactions

5.3.1. 1,3-DCRs

In 2009, the Gong group disclosed, for the first time, a chiral phosphoric acid (CPA)-organocatalyzed enantioselective 1,3-DC reaction of methyleneindolinones **20**, aldehydes **21**, and amino esters **22**. This reaction provided a wide range of spiro[pyrrolidin-3,3′-oxindole] derivatives **24-25** with high yield, unusual regiochemistry, and excellent stereoselectivities (up to 98% ee) (Scheme 5-7). Both the azomethine ylide and the methyleneindolinone **20** are hydrogen-bonded with the phosphoric acid **23**, which leads to the high enantio- and regioselectivity. This reaction demonstrates remarkable regioselectivity results from the favorable π-π stacking interaction between the oxo-indole ring and the conjugated esters [6].

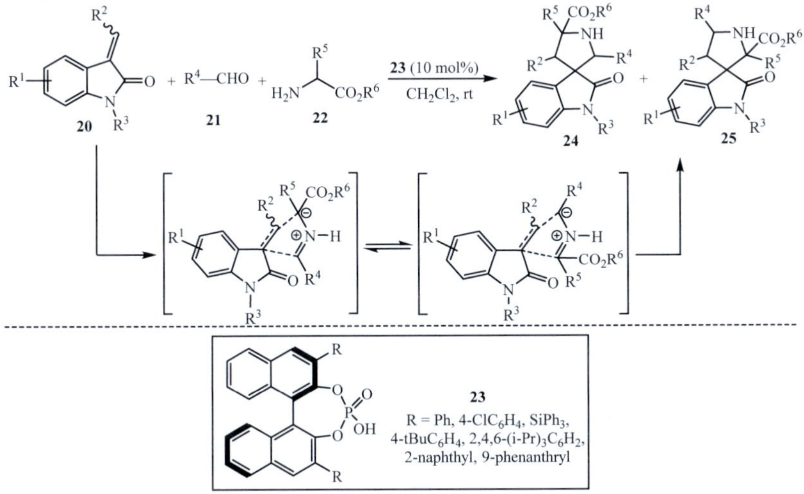

Scheme 5-7. Organocatalytic synthesis of spiro[pyrrolidin-3,3′-oxindole] derivatives **24-25**.

Gong et al. conducted an impressive total synthesis of two diastereomers of spirotryprostatin A using a chiral Brønsted acid as a key catalyst in a three-component asymmetric organocatalytic 1,3-DCR of electron-deficient olefins with azomethine ylides [7]. The reaction commenced with the 1,3-DC of prenyl aldehyde **26**, diethyl 2-amino-malonate **27**, and methyl 2-(4-methoxy-2-nitrophenyl)acrylate **28**, catalyzed by bisphosphoric acid **29**, resulting in **30** in 94% yield with >99:1 dr and >99% ee. The reduction of **30** with Zn/HOAc was then carried out, followed by protection of the resultant *N*-hydroxyoxindole with BnBr/TEA, cleavage of the N-O bond with SmI_2 in the presence of water in THF, and esterification with $SOCl_2$ in methanol to obtain **32** with a dr of 1.2:1. Treatment of **32** with the acid chloride derived from *N*-Boc-L-proline produced **33**. The proline moiety was then deprotected and subjected to subsequent lactamization, giving 18-epispirotryprostatin A and 9,18-bis-epispirotryprostatin A, respectively (Scheme 5-8).

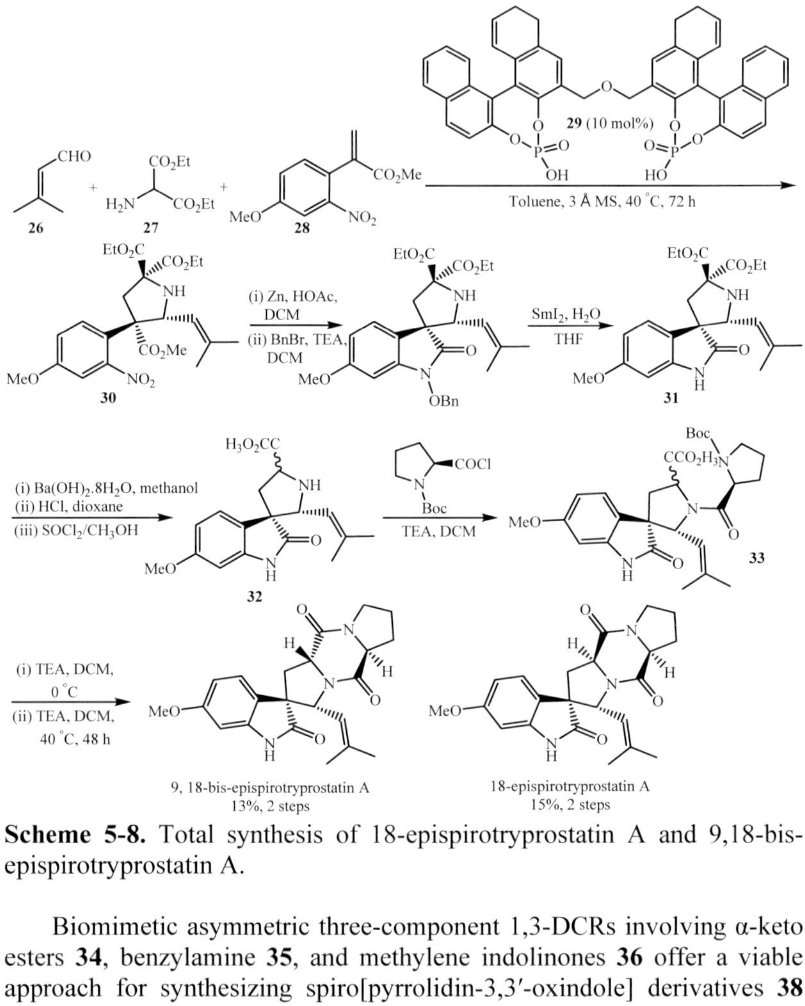

Scheme 5-8. Total synthesis of 18-epispirotryprostatin A and 9,18-bis-epispirotryprostatin A.

Biomimetic asymmetric three-component 1,3-DCRs involving α-keto esters **34**, benzylamine **35**, and methylene indolinones **36** offer a viable approach for synthesizing spiro[pyrrolidin-3,3′-oxindole] derivatives **38** (Scheme 5-9). Guo et al. [8] demonstrated that 3,3′-β-naphthyl phosphoric acid **37** (10 mol%) can catalyze this reaction in dichloromethane at room temperature for 2-3 days to produce spiro-oxindoles **38** with excellent yields and enantioselectivities.

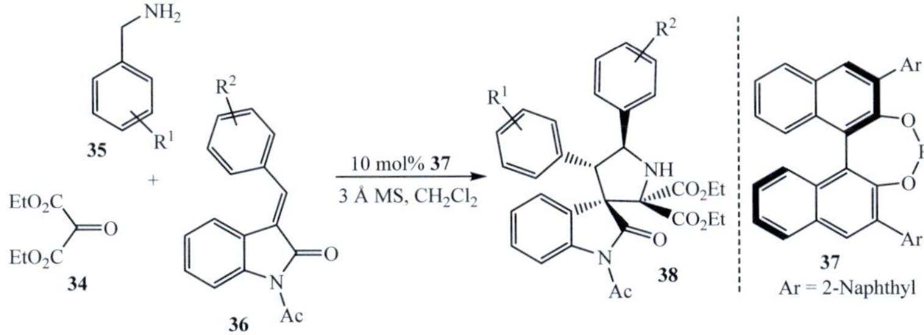

Scheme 5-9. Biomimetic asymmetric three-component 1,3-DC for the synthesis of spiro-oxindoles **38**.

In addition to the previously mentioned examples, a practical and straightforward method has been developed for synthesizing optically active highly functionalized spiro[γ-butyrolactone-pyrrolidin-3,3′-oxindole] tricyclic skeletons **40**, allowing excellent control over four contiguous stereocenters. This method utilizes a rosin-derived tertiary amine thiourea bifunctional catalyst **39** in an organocatalytic 1,3-DC reaction at room temperature (Scheme 5-10) [9].

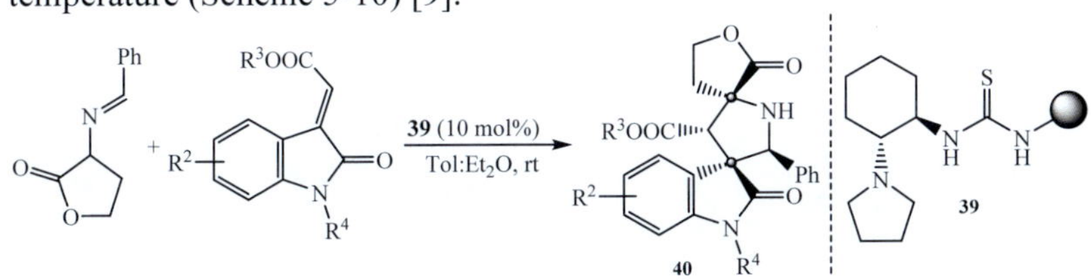

Scheme 5-10. Efficient construction of highly functionalized spiro[γ-butyrolactone-pyrrolidin-3,3′-oxindole]tricyclic skeletons **40**.

As illustrated in Scheme 5-11, Yang et al. reported the highly enantioselective synthesis of polycyclic spiro-oxindole scaffolds **43** with three contiguous stereocenters based on the prolinosulfonamide **42** catalyzed 1,3-DC of 2-cyclohexene-1-one **41** and azomethine ylides generated *in situ* from isatins and amino esters [10]. This catalytic system could effectively activate cyclohexenone **41** and enable the formation of hydrogen bonding between the catalyst and the dipole.

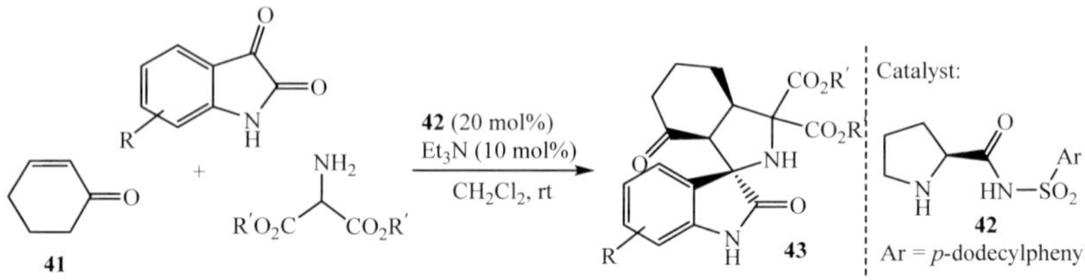

Scheme 5-11. Highly enantioselective construction of polycyclic spiro-oxindoles **43** by organocatalytic 1,3-DC.

Yan et al. reported a practical and efficient method for the synthesis of optically active 5′-CF_3 spiro[pyrrolidine-3,2′-oxindole] **47** in 2015 [11]. The reaction involved the use of 3 mol% (*R*)-2-(diphenyl((trimethylsilyl)oxy)methyl)pyrrolidine **46** as a catalyst and benzoic acid as an additive. The starting *N*-(2,2,2-trifluoroethyl) ketimines **44**, obtained from the condensation of trifluoroethylamine and isatins, were reacted with 2-enals **45** in a cycloaddition reaction to afford the spiro-fused product **47**. The reaction proceeded with excellent diastereoselectivity, stereoselectivity, and yield, affording the product with four contiguous stereogenic centers (Scheme 5-12).

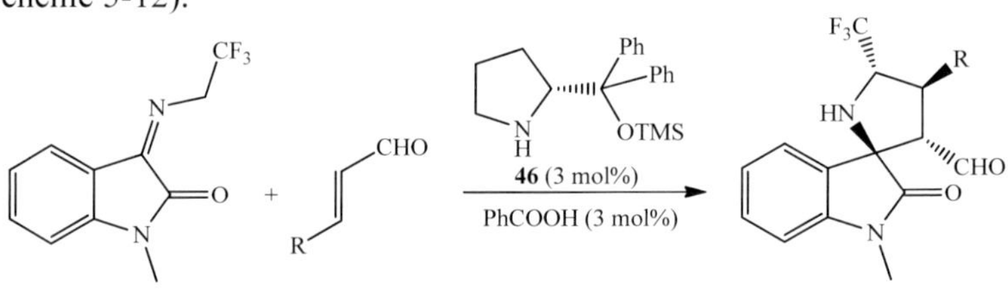

Scheme 5-12. The asymmetric synthesis of CF_3-containing spiro-fused product.

The proposed catalytic cycle for the formation of **47** is depicted in Scheme 5-13. Initially, the diphenylprolinol silyl ether **46** reacts with the 2-enals **45** to generate the iminium ion intermediate **A**. The bulky aryl groups effectively shield the *re*-face of the chiral iminium ion, leading to a diastereoselective *si*-facial cycloaddition on the oxindole-derived azomethine ylide, ultimately resulting in the formation of the cycloadduct iminium ion **B**. Subsequently, hydrolysis releases the product **47** from the catalytic cycle, and the catalyst is regenerated. It is worth noting that both steric and electronic effects influence the rate of these cycloaddition reactions. Overall, the reaction provided excellent diastereoselectivity, stereoselectivity, and yield.

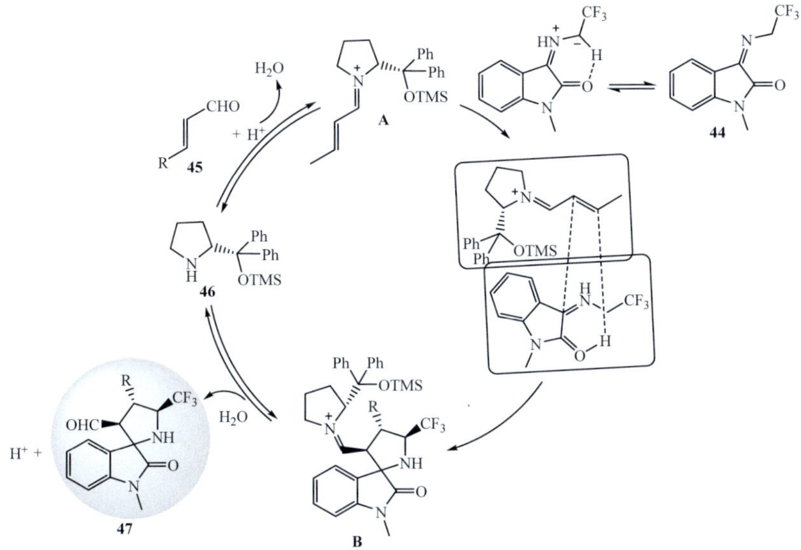

Scheme 5-13. Postulated catalytic cycle for the cycloaddition reaction.

In 2016, Shi et al. demonstrated the first organocatalytic asymmetric 1,3-DC reaction to create 3,3′-pyrrolidinyldispiro-oxindole frameworks **49** in high yields (up to 99%) and excellent diastereo- and enantioselectivities (up to >95:5 dr, 98% ee) (Scheme 5-14) [12]. The exceptional stereoselectivity observed in the reaction is largely attributed to the unique structure of the chiral bis-phosphoric acid (bis-PA) catalyst **48** and its powerful dual hydrogen-bonding activation mode towards the reactants. This reaction serves as an excellent example of catalytic asymmetric 1,3-DC's application in constructing enantioenriched bis-spiro-oxindole frameworks with structural complexity and rigidity.

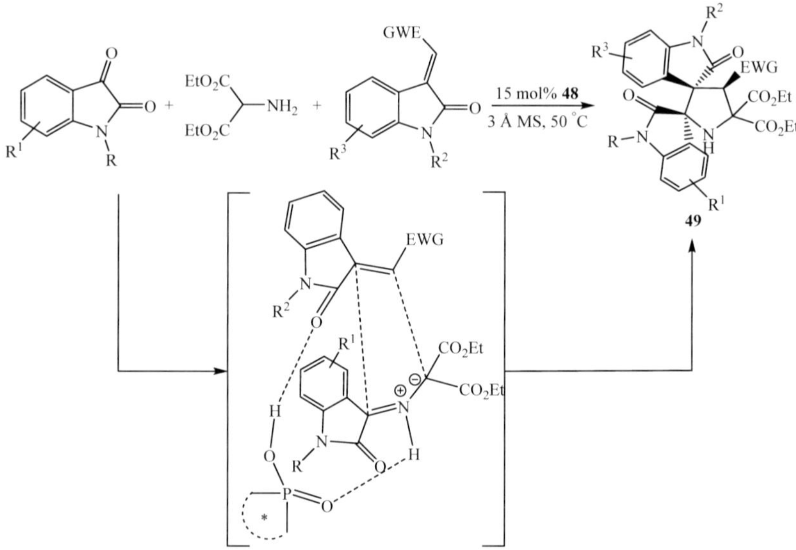

Scheme 5-14. Diastereo- and enantioselective construction of 3,3′-pyrrolidinyldispiro-oxindole framework **49**.

Simultaneously, Peng et al. reported the first catalytic asymmetric 1,3-DC of the Seyferth−Gilbert reagent (SGR) **51** with a series of isatylidene malononitriles **50** utilizing a cinchona alkaloid **52** as a bifunctional catalyst [13]. These reactions proceeded smoothly to generate the corresponding chiral spiro-phosphonyl pyrazolineoxindoles **53** in good yields with excellent enantioselectivities (Scheme 5-15).

According to the proposed mechanism, the cinchona alkaloid **52** acted as a bifunctional catalyst in which the hydroxyl group served as a Brønsted acid to activate the isatylidene malononitrile **51** through hydrogen bonding and the amine moiety acted as a base to activate the SGR **50** to attack the C3 position of the isatylidene malononitrile **51** from its *si*-face.

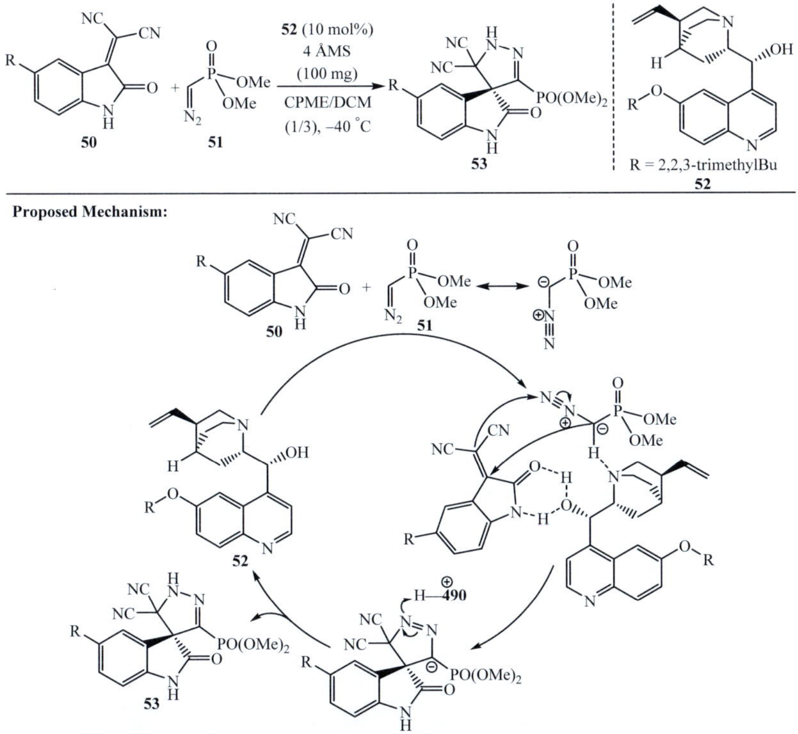

Scheme 5-15. One-pot construction of chiral spiro-pyrazolineoxindoles **53** and its proposed mechanism.

Among the selected examples, we can also highlight the synthesis of 5′-trifluoromethylspiro[pyrrolidin-3,2′-oxindole] derivatives **56** catalyzed by cinchona alkaloid-derived squaramide **55** [14]. By employing *N*-2,2,2-trifluoroethylisatin ketimines **54** and β-nitroalkenes as starting materials and utilizing cinchona alkaloid-derived squaramide **55** as the catalyst, the synthesis of 5′-trifluoromethylspiro[pyrrolidin-3,2′-oxindole]s **56** was achieved through an asymmetric 1,3-DCR under mild conditions (Scheme 5-16).

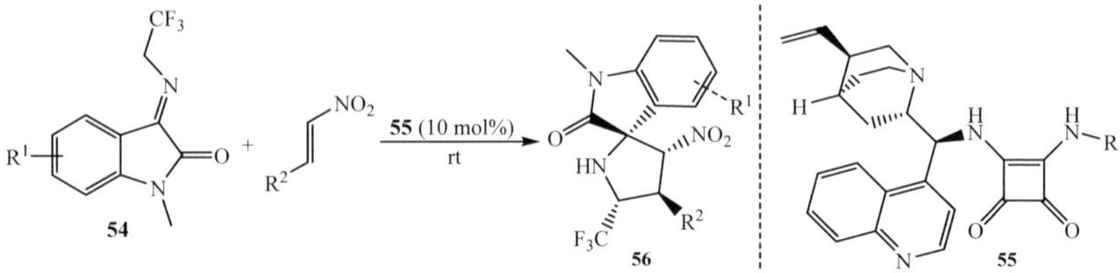

Scheme 5-16. The squaramide-catalyzed 1,3-DC.

Zhao et al. reported the 1,3-DCR of isatin-derived azomethine ylides with maleimides **57** using cinchona alkaloid-based squaramide **58** as an organocatalyst [15] (Scheme 5-17). The Chiral pyrrolidine-fused spiro-oxindoles **59** were obtained in 61–89% yields with >20:1 dr and 12 to >99% ee.

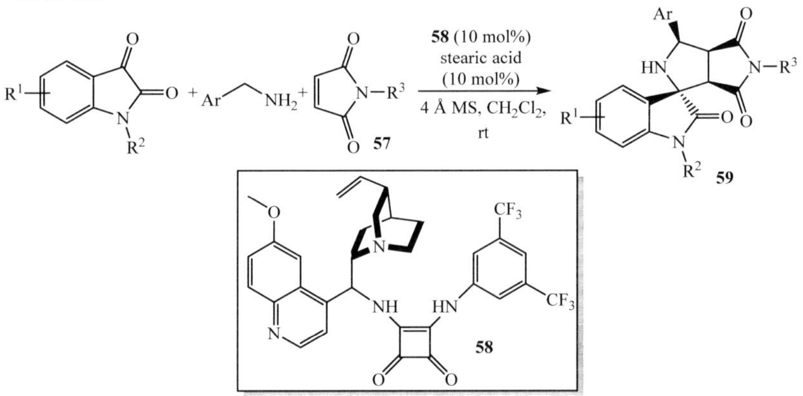

Scheme 5-17. Construction of pyrrolidine-fused spiro-oxindoles **59**.

In an elegant work, the Wang group introduced a catalytic asymmetric three-component 1,3-DC of 3-amino oxindoles **60** as azomethine ylide precursors with aldehydes and nitro-olefins under the catalysis of a CPA **61** (Scheme 5-18) [16].

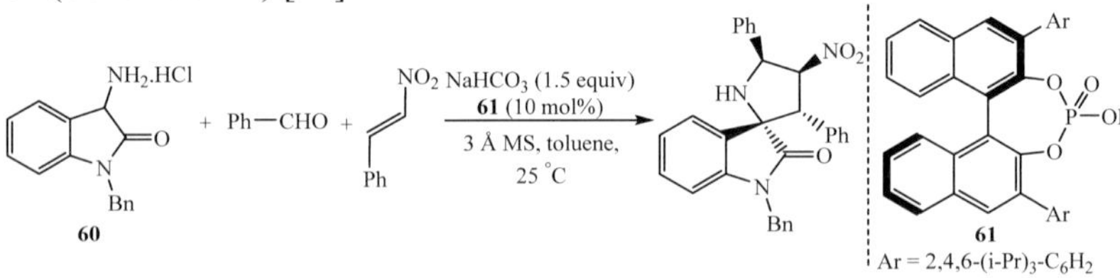

Scheme 5-18. CPA-Catalyzed 1,3-DC involving 3-amino oxindoles.

A few years later, in 2016, Yan et al. successfully synthesized functionalized spiro[indolin-3,2′-pyrrole]s **63** [17] through a one-pot, BF_3OEt_2-catalyzed 1,3-DCR. The reaction involved benzylamines **35**, isatins **16**, and dimethyl acetylenedicarboxylate **62** and resulted in moderate to good yields with high diastereoselectivity (Scheme 5-19).

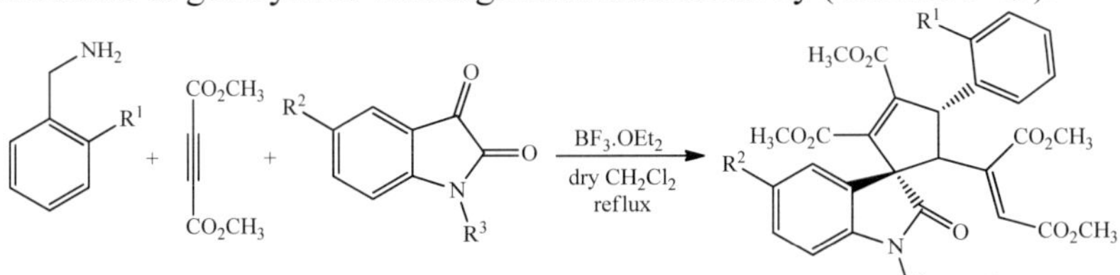

35 62 16 63

Scheme 5-19. Synthesis of the functionalized spiro[indolin-3,2′-pyrrole]s **63**.

Yan et al. [18] reported an umpolung/1,3-DCR sequence to synthesize 2′-trifluoro-methylated spiro-pyrrolidine-3,3′-oxindoles **65** (Scheme 5-20) in excellent yields, with high enantio-, and diastereoselectivities. This reaction is catalyzed by a quinine-derived squaramide **64** (1 mol%).

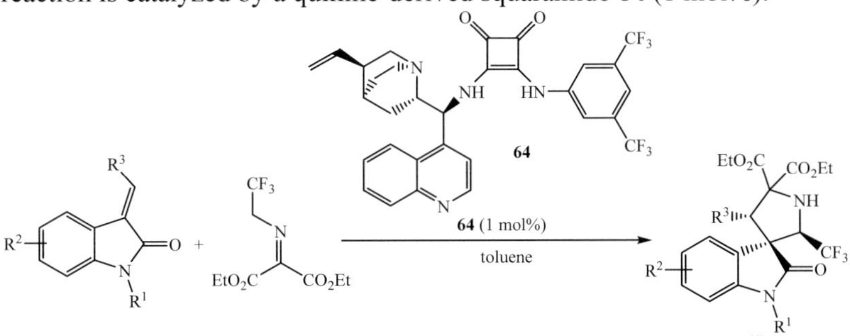

65

Scheme 5-20. Strategy for the synthesis of 2′-CF_3-spiro-pyrrolidine-3,3′-oxindoles **65**.

A CPA **68**-catalyzed 1,3-DC with isatin-derived imines **66**, aldehydes, and amino-ester **67** to synthesize spiro[imidazolidine-2,3′-oxindole] frameworks **69** in good yields, high diastereo- and enantioselectivities were first reported by Shi et al. (Scheme 5-21) [19].

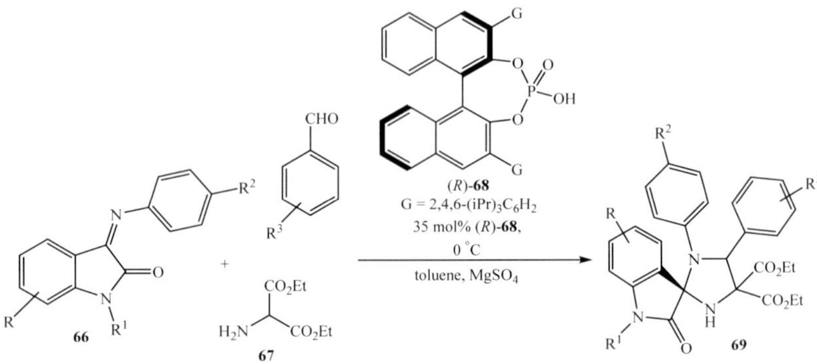

66 67 69

Scheme 5-21. Construction of spiro[imidazolidine-2,3'-oxindole] framework **69**.

Weng et al. used the organocatalytic 1,3-DCR to synthesize trifluoromethylated 3,3′-pyrrolidinyl-dispiro-oxindoles **73** [20]. CF_3-containing isatin-derived azomethine ylides **70** react with methylene indolinones **71** in the presence of bifunctional squaramide-tertiary amine organocatalysis **72**, yielding desired products **73** in excellent yields and stereocontrol (Scheme 5-22).

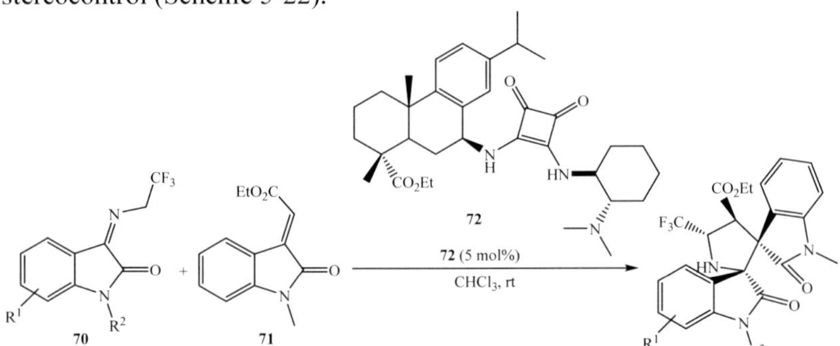

70 71 72 73

Scheme 5-22. Asymmetric synthesis of trifluoromethyl-substituted 3,3′-pyrrolidinyl-dispiro oxindoles **73**.

Moreover, Shang et al. [21] presented the synthesis of chiral spiro-[pyrrolidin-3,3′-oxindoles] **77** using a thiourea–quaternary ammonium **76** salts catalyzed asymmetric 1,3-DC starting from imino esters **74** and methyleneindolinones **75** under mild conditions (Scheme 5-23).

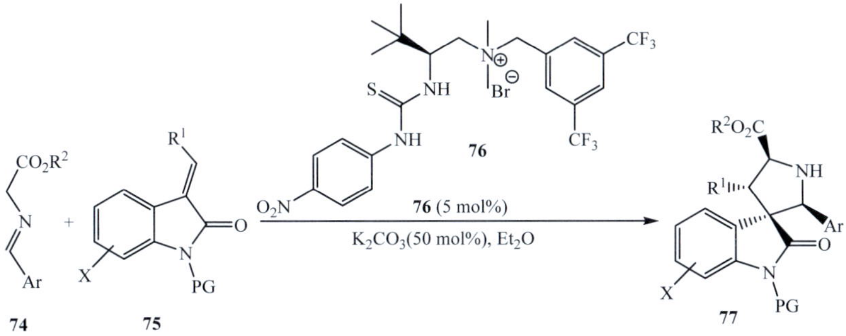

Scheme 5-23. Asymmetric 1,3-DC of imino esters **74** with methylene-indolinones **75**.

In 2018, Wang et al. reported the synthesis of spiro[pyrrolidine-2,3′-oxindole] products **80** via an asymmetric 1,3-DC reaction catalyzed by a 3,3′-(9-phenanthryl)-modified phosphoric acid **79** [22]. Starting with α,β-enones and *in situ* generated azomethine ylides from 3-amino oxindoles **78** and aldehydes, they achieved the desired products in high yields with excellent regio- and enantioselectivities (up to 99% yield, >20:1 dr, 99% ee) (Scheme 5-24).

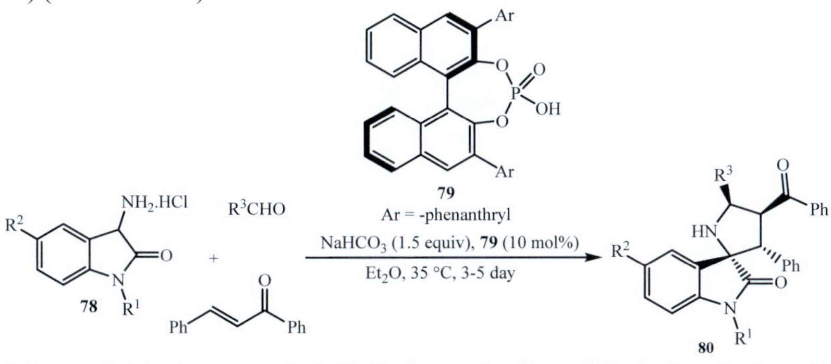

Scheme 5-24. Asymmetric 3,3′-(9-phenanthryl)-modified phosphoric acid **79** catalyzed 1,3-DCR.

In 2018, bispiro[γ-butyrolactone-pyrrolidin-4,4′-pyrazolone] scaffolds **84** were synthesized through a bifunctional squaramide **83** organo-catalyzed 1,3-DC using α-imino γ-lactones **81** and alkylidene pyrazolones **82** in Et_2O as depicted in Scheme 5-25 [23]. This efficient organocatalytic strategy affords facile access to a variety of highly functionalized drug-like compounds with two quaternary spirocenters.

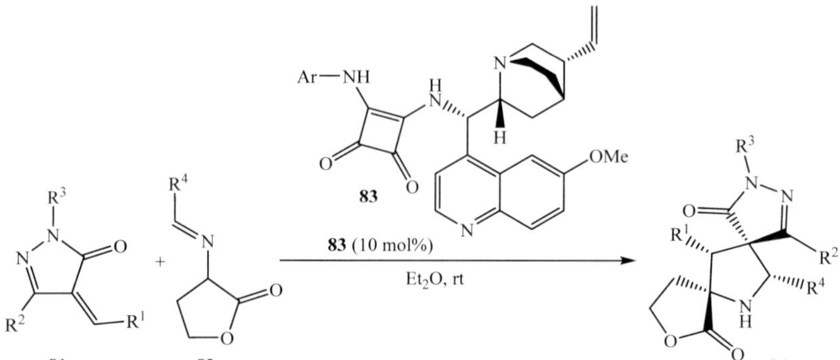

81 82 84

Scheme 5-25. Synthesis of bispiro[γ-butyrolactone-pyrrolidin-4,4′-pyrazolone] scaffolds **84**.

Zhao et al. developed a highly enantioselective 1,3-DCR for the synthesis of chiral spiro[benzofuran-2,3′-pyrrolidine] **87**. The reaction employed chiral thiourea-quaternary ammonium salts **86a-j** as catalysts, resulting in excellent enantioselectivity (Scheme 5-26).

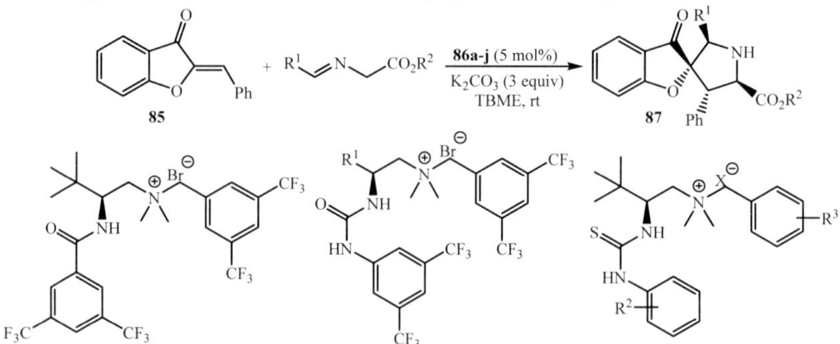

86a

86b: R^1 = t-Bu
86c: R^1 = 2-(s)-Bu
86d: R^1 = Bn

86e: $R_2 = R_3$ = 3,5-di-CF_3, X = Br
86f: R_2 = 4-NO_2, R_3 = 3,5-di-CF_3, X = Br
86g: R_2 = 4-F, R_3 = 3,5-di-CF_3, X = Br
86h: $R_2 = R_3$ = 4-NO_2, X = Br
86i: R_2 = 4-NO_2, R_3 = 3,5-di-CH_3O, X = Br
86j: R_2 = 4-NO_2, R_3 = 3,5-di-CF_3, X = Cl

Scheme 5-26. Synthesis of chiral spiro[benzofuran-2,3′-pyrrolidine] **87**.

The reaction involved the 1,3-DC of 2-arylidene-benzofuran-3(2*H*)-ones **85** with imino esters under mild conditions to afford the desired products in high yields (up to 99%) and with good enantioselectivities (up to 99% ee). The quaternary ammonium salts used in this reaction included amide, urea, and thiourea as the H-donor. Catalyst **86f** derived from L-isoleucine, which had higher hydrogen-bond strength and acidity due to the presence of thiourea, led to improved enantioselectivity of 98% ee along with an 11:1 dr.

Furthermore, a notable example in the field of 1,3-DCRs was recently reported by Yan et al. This study focused on the synthesis of spiro-pyrrolidine-linked oxindole and pyrazolone compounds **91**, which possessed four consecutive stereocenters and two vicinal spiro-quaternary chiral centers, demonstrating the remarkable synthetic capabilities in this area [25]. In the presence of a cinchonine-derived squaramide catalyst **89**, the cycloaddition of *N*-2,2,2-trifluoro-ethylisatin ketimines **54** with α,β-unsaturated pyrazolones **88** provided desired products **90** in excellent yields and stereoselectivities (Scheme 5-27).

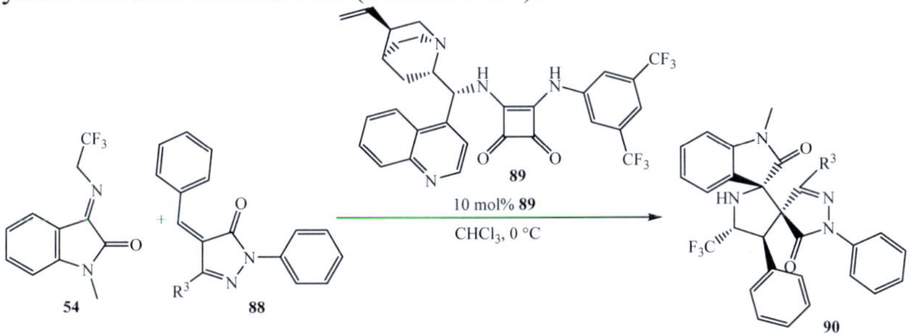

Scheme 5-27. Synthesis of spiro-pyrrolidine-linked oxindole and pyrazolone compounds **90**.

In the proposed mechanism by the authors, the reaction proceeds via the transition state **A**, where the squaramide catalyst **89** acts as a dual activation model. The tertiary amine of squaramide deprotonates ketamine and holds the resulting azomethine ylides through coordination, while the squaramide moiety binds and activates the α,β-unsaturated pyrazolone **88** through hydrogen bonding. The carbon anion then attacks the alkenyl from the *re*-face, and the addition of pyrazolone from the *si*-face to the *si*-face of isatin ketamine provides the corresponding product. The absolute configuration outcome of the reaction supports this proposed mechanism. (See Figure 5-1 for a visual representation of the mechanism)

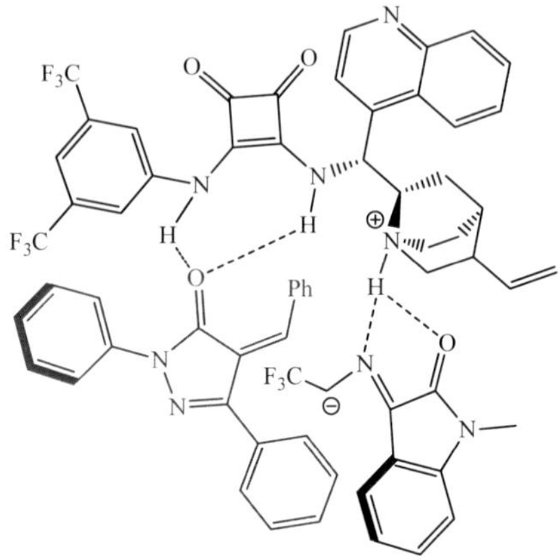

Fig. 5-1. Proposed transition state.

Lin et al. [26] recently reported the synthesis of spiro-oxindole-oxazolidines **93** by the cinchona derived bifunctional thiourea **92** catalyzed oxa-1,3-DCR between ketones and trifluoro-ethylamine-derived azomethine ylide **91** (Scheme 5-28). This protocol affords an efficient methodology for the facile synthesis of chiral CF_3-containing oxazolidines **93** in moderate to excellent yields, diastereoselectivities, and enantioselectivities. This oxazolidines **93** could be a facile conversion to CF_3-containing 1,2-amino alcohols **94** with vicinal stereogenic centers as a privileged structural motif in medicinal chemistry.

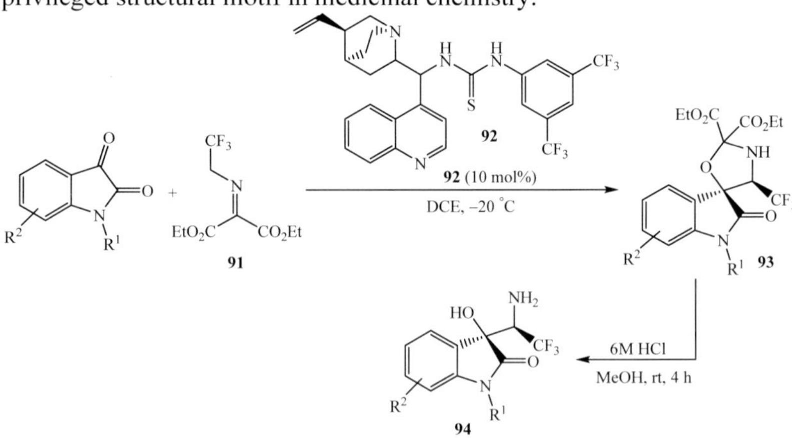

Scheme 5-28. Oxa-1,3-DC of trifluoroethylamine-derived azomethine ylide **91**.

5.3.2. [2+2] Cycloaddition reactions

In this context, chiral NHCs were found to be efficient catalysts for enantioselective [2+2] cycloaddition reaction of di-substituted ketenes **95** and isatins **96** to afford the corresponding spirocyclic oxindole-β-lactones **97** in good yields with good diastereoselectivities and excellent enantioselectivities (Scheme 5-29). Ye et al. investigated a series of NHCs derived from l-pyroglutamic acid **98**. NHCs possess a *tert*-butyl dimethylsilyl group, leading to high yields, good diastero- and excellent enantioselectivities. Interestingly, the reaction catalyzed by tetracyclic NHC **99** provided the desired products with opposite diastereoselectivity in low yield [27].

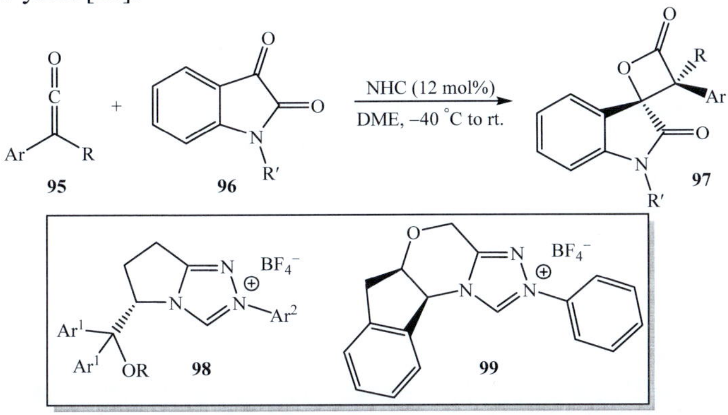

Scheme 5-29. Enantioselective [2+2] cycloaddition of isatins and ketenes.

In 2014, in their groundbreaking work, Wang et al. utilized the α,α-diphenylprolinol catalyst 101 to achieve the first-ever organocatalytic asymmetric synthesis of spirocyclobutyl oxindoles 102 through a [2+2] cycloaddition reaction [28]. This pioneering approach highlights the potential of organocatalysis in the construction of complex spirocyclic structures. Structurally complex spirocyclobutyl oxindoles **102**, which possess four contiguous stereocenters, including one spiro quaternary center, were formed in good yields (up to 83%) with excellent β,γ-regioselectivity (>19:1) and stereo control (up to >19:1 dr and 97% ee) from methyleneindolinones **100** and α,β-unsaturated aldehydes by using 20 mol% of catalyst **101**. The α,α-diphenylprolinol catalyst **101** combined H-bond directing and dienamine activation (Scheme 5-30).

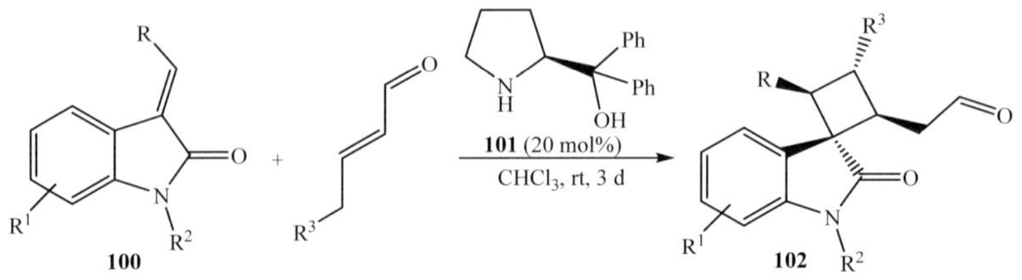

Scheme 5-30. Asymmetric synthesis of 3,3′-spiro-oxindoles fused with cyclobutanes **102**.

Silvani et al. also reported a DABCO-catalyzed diastereoselective [2+2] annulation reaction of isatin-derived ketimines **103** with allenoates **104** to synthesize chiral spiro-oxindole-based 4-methyleneazetidines **105** (Scheme 5-31), in moderate to excellent yields and good to excellent diastereoselectivies [29]. Based on the stereochemical outcome of the reaction, the authors propose that the nitrogen base DABCO acts as a nucleophile activator, producing a zwitterionic allylic carbanion from the allenoate reagent. Such an intermediate tolerates a γ-addition to the ketimine, followed by an intramolecular nucleophilic attack to generate the azetidine ring with regeneration of DABCO.

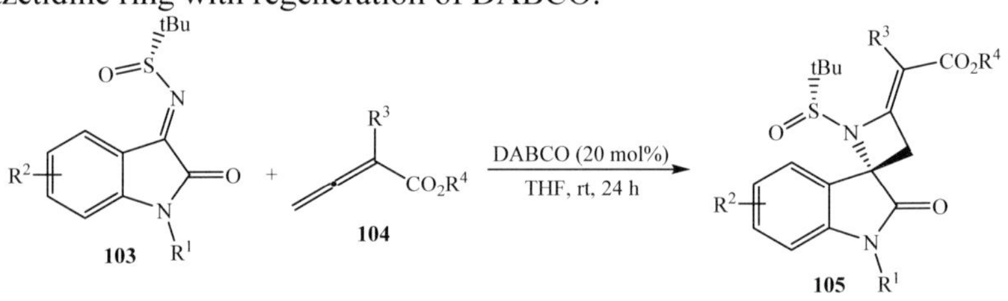

Scheme 5-31. Diastereoselective [2+2] annulation reaction of various isatin ketimines **103** with allenoates **104**, catalyzed by DABCO.

In 2017, Wang et al. elegantly introduced enantioselective synthesis of tetrahydropyridines/piperidines **109/110** through stepwise [4+2]/[2+2] cyclizations [30]. In the first step, a phosphine **108** catalyzed enantio-selective [4+2] annulation reaction between allene ketones **106** and 1-aza-dienes **107** led to tetrahydropyridines **109** in good yields and with excellent enantioselectivities. Subsequent exposure of tetrahydropyridines **109** to benzyne led to a [2+2] cycloaddition, creating optically enriched polycyclic piperidines **110** with a quaternary stereogenic center and a cyclobutene moiety (Scheme 5-32).

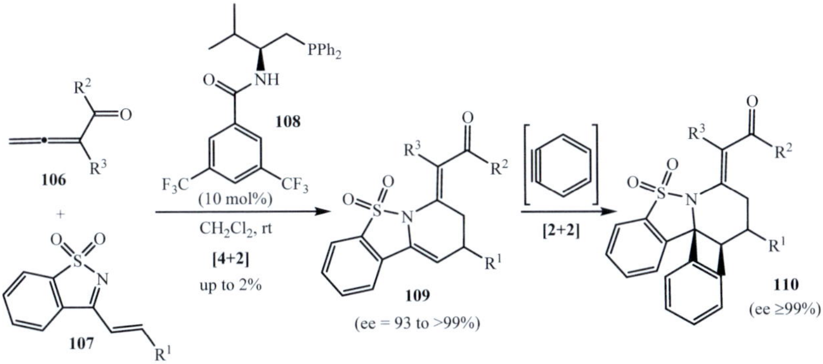

Scheme 5-32. Stepwise [4+2]/[2+2] cycloadditions for synthesizing tetrahydropyridines/piperidines **109/110**.

5.3.3. [3+2] Cycloaddition reactions

This section gives a brief overview of the interesting organocatalyzed [3+2] cycloaddition reactions for constructing spiro compounds.

In 2005, Ung et al. reported a new strategy for the synthesis of spiro-heterocyclic compounds **115-120** using the phosphine-catalyzed [3+2] cycloaddition of the 2-methylene γ-lactams **111** and **112** and the acrylate **113** with the ylides derived from the ethyl ester, the amide or the chiral camphorsultam of 2-butynoic acid **114a-c** (Scheme 5-33) [31].

Two years later, Cowen et al. utilized a protected multifunctional phosphine-containing α-amino acids **121a-f** as catalysts to generate cycloadducts **122α,γ** through a [3+2] cycloaddition of allenoate esters and enones (Scheme 5-34) [32].

Later, the Marinetti group studied the [Cy]-TMS-FerroPHANE **125** for the [3+2] annulation reactions between allenoates **123** and exocyclic enones **124**. **125** is a highly efficient nucleophilic promoter for enantioselective organocatalytic processes (Scheme 5-35) [33].

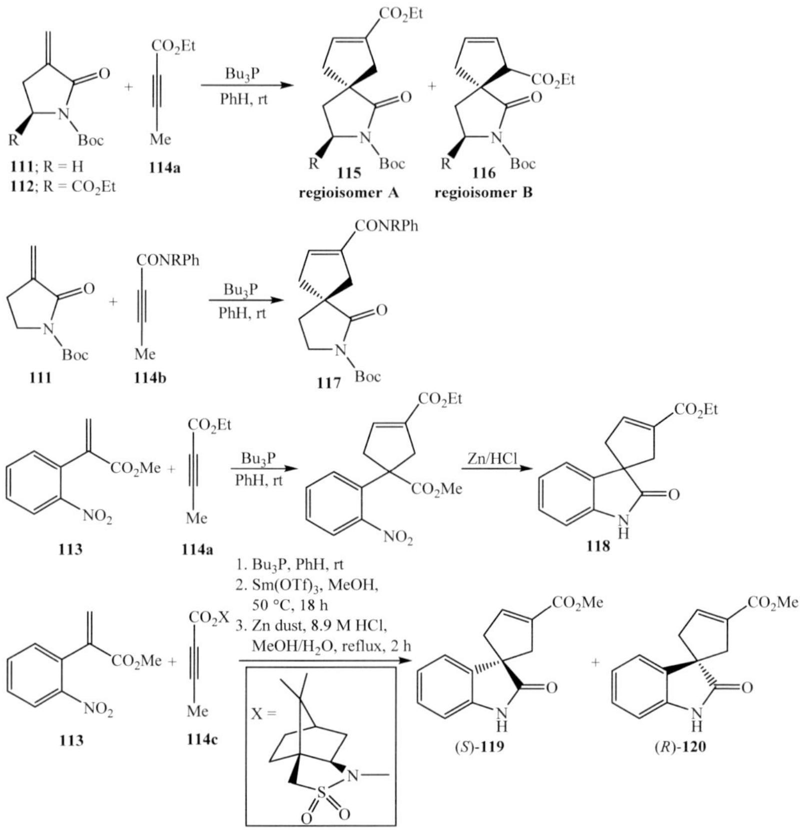

Scheme 5-33. Phosphine-catalyzed [3+2]-cycloaddition.

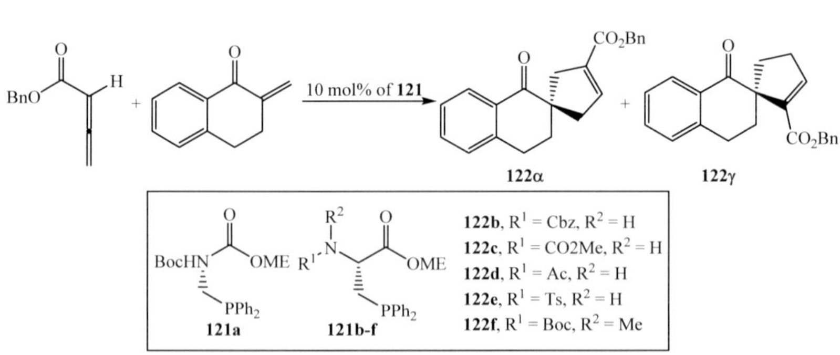

Scheme 5-34. Amino acid catalyst survey for [3+2]-cycloaddition.

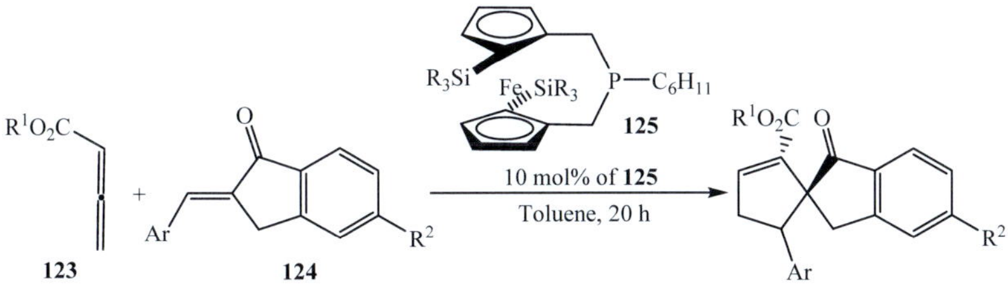

Scheme 5-35. Asymmetric [3+2] annulations on exocyclic enones promoted by FerroPHANE **125**.

In 2010, the same group also reported the first enantioselective approach for constructing spirocyclic oxindolic cyclopentanes **126/126′** based on a phosphine-mediated organocatalytic process (Scheme 5-36, Eq. 1) [34]. Additionally, they showed that phosphine catalyzed enantio-selective [3+2] cyclizations on 4-substituted 2,6-diarylidenecyclohexan-ones **127** and 2,4-diarylidenebicyclo[3.1.0]hexan-3-ones **128** take place with high diastereo- and enantioselectivity levels. The process gives spirocyclic compounds **129-130** with excellent stereochemical control of up to five stereogenic centers (Scheme 5-36, Eq. 2) [35].

Lian et al. disclosed an attractive phosphorus-containing Lewis base catalyzed [3+2] annulation of isatins with but-3-yn-2-one **131** to generate the corresponding spiro[furan-2,3-indoline]-2,4(5*H*)-diones **132** in good yields under mild conditions (Scheme 5-37) [36].

Concurrently, the chiral NHC bearing a proximal hydroxy group derived from L-pyroglutamic acid **133** was found to be an efficient catalyst for the [3+2] cycloaddition of enals and isatins to afford the corresponding spirocyclic oxindolo-γ-butyrolactones **134** in good yield with good diastereo- and enantioselectivities (Scheme 5-38) [37].

On the other hand, 3,3′-pyrrolidonyl spiro-oxindole skeletons **137** were synthesized through an efficient organocatalytic [3+2] cycloaddition reaction using α-isothiocyanato imides **135** as dienophiles and quinine **136** as a catalyst (Scheme 5-39).

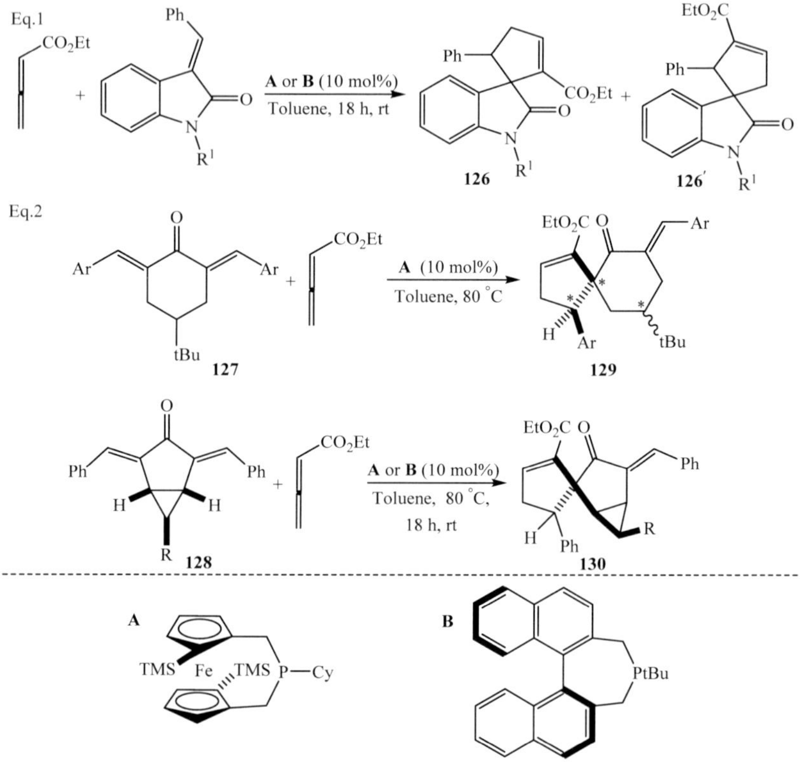

Scheme 5-36. Phosphine-promoted [3+2] annulation reactions.

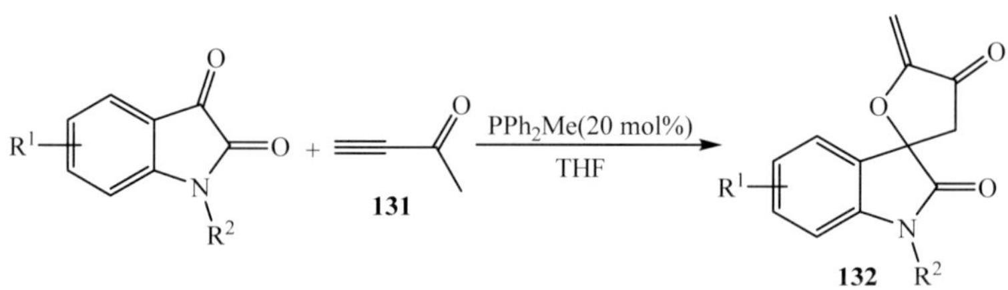

Scheme 5-37. The phosphorus-containing Lewis base catalyzed [3+2] annulation.

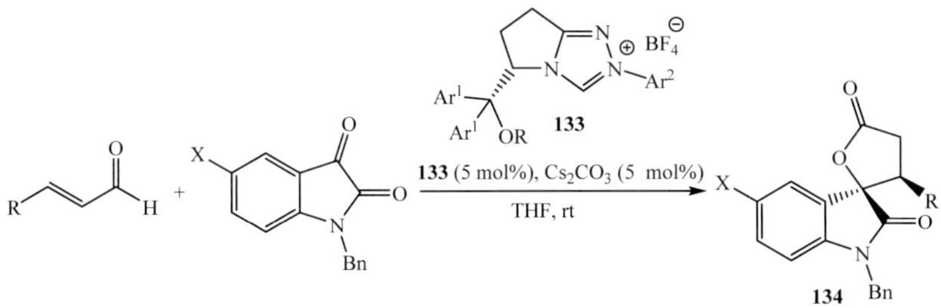

Scheme 5-38. Enantioselective NHC-catalyzed [3+2] annulation of enals and isatins.

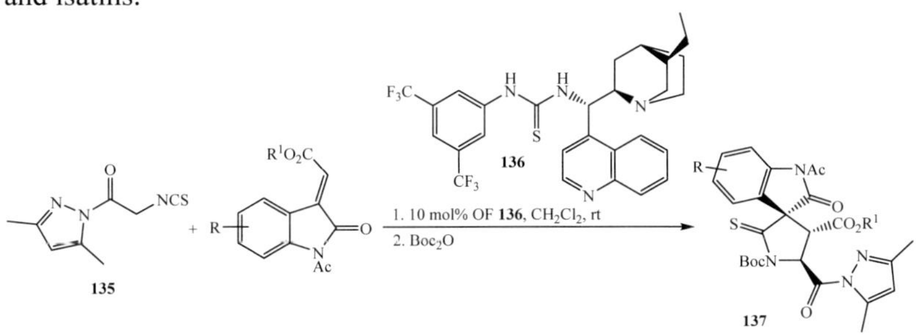

Scheme 5-39. Synthesis of 3,3′-pyrrolidonyl spiro-oxindole skeletons **137**.

The two substrates involved in the reaction are activated simultaneously by the catalyst, as shown in Figure 5-2 [38].

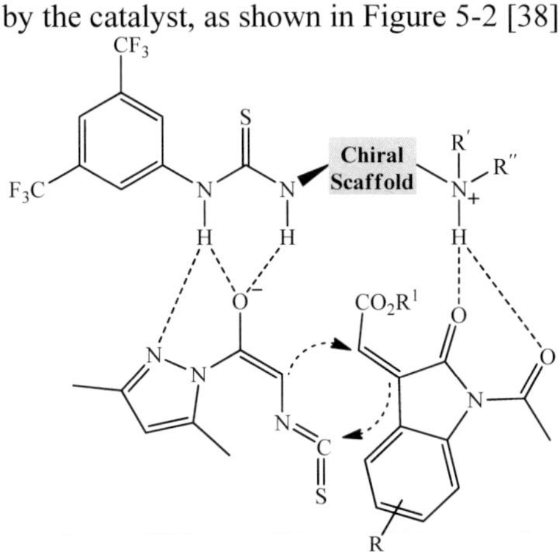

Fig. 5-2. Proposed mode of activation of substrates.

In 2012, Scheidt et al. utilized an NHC/Lewis acid strategy for the stereoselective synthesis of spiro-oxindole lactones **138** (Scheme 5-40). This strategy is one of the few common methods of enantioselective addition of NHC-generated homoenolate equivalents to ketones. It opens the possibility of including new, less reactive electrophilic partners with NHC–homoenolates using cooperative catalysis. The addition of lithium chloride as a Lewis acid with β-aryl substituted enals affords lactone products with high enantioselectivity (> 90%). Interestingly, lithium chloride was detrimental with enals bearing β-alkyl substituents, owing to the promotion of a competing pathway. In the absence of lithium chloride, a different triazolium pre-catalyst was used to produce the alkyl-substituted lactones in good yield and enantioselectivity [39].

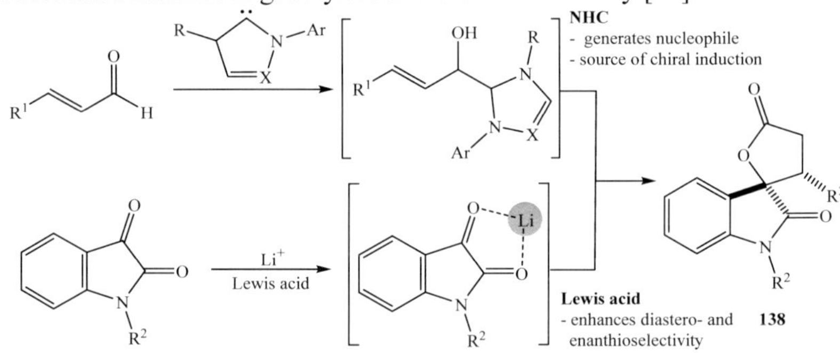

Scheme 5-40. Cooperative catalysis approach to spiro-oxindoles.

Asymmetric [3+2] annulation of 2-arylideneindane-1,3-diones **139** with Morita-Baylis-Hillman (MBH) carbonates **140** catalyzed by multifunctional thiourea-phosphines (**TP3**) was reported by the Shi group in the same year (Scheme 5-41) [40].

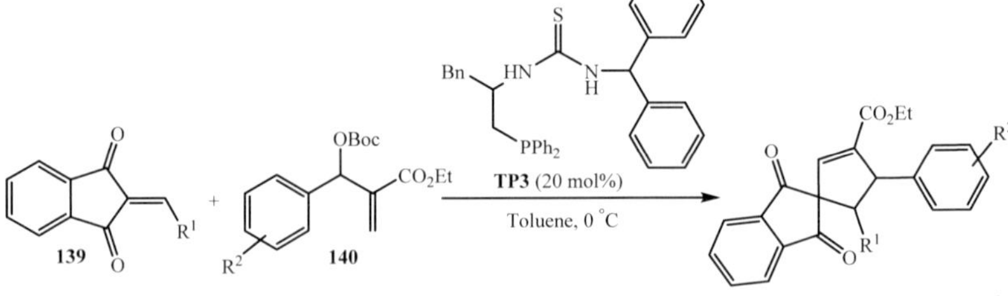

Scheme 5-41. Multifunctional thiourea-phosphines catalyzed asymmetric annulation.

Lian et al. developed a (4*S*,5*S*)-4,5-bis((diphenylphosphino)methyl)-2,2-dimethyl-1,3-dioxolane (DIOP) catalyzed highly enantioselective

[3+2] annulation of *N*-protected isatins with but-3-yn-2-one **131**, affording the corresponding functionalized spiro[furan-2,3′-indoline]-2′,4-(5*H*)-dione **141** containing a tertiary stereogenic center in moderate yields along with good to high enantioselectivities in Et_2O at −20 °C (Scheme 5-42) [41].

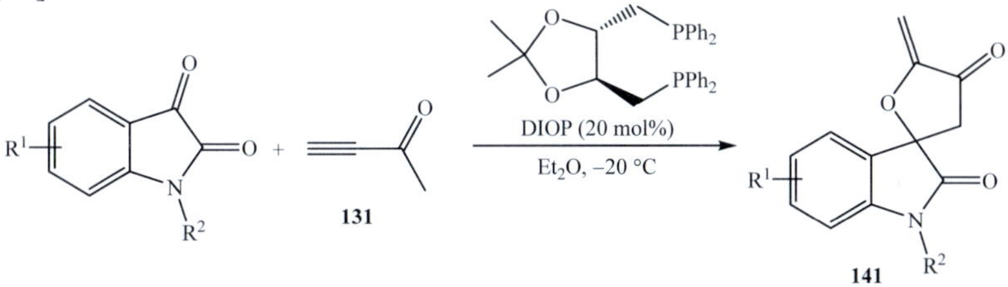

141

Scheme 5-42. DIOP-Catalyzed [3+2] annulation.

One year later, Xu et al. reported an efficient and powerful organocatalytic 1,3-H shift and [3+2] cycloaddition cased reaction [42]. This reaction gave the biologically important chiral spiro[pyrrolidin-3,2′-oxindole]s **143** reasonable yields and good stereoselectivity, employing easily available starting materials such as isatins benzylamines and nitroalkenes. This multicomponent cascade reaction is catalyzed by a bifunctional squaramide **142** (Scheme 5-43).

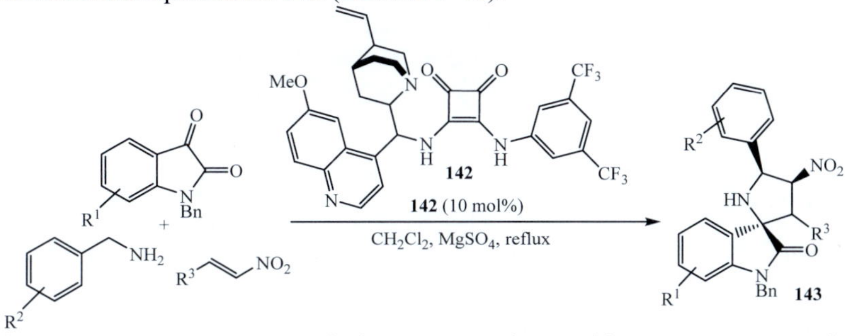

143

Scheme 5-43. Organocatalytic asymmetric multicomponent cascade reaction.

Shi et al. also explored the catalytic asymmetric [3+2] cycloaddition of (3-isothiocyanato)oxindoles **144** with azodicarboxylates **145** in the presence of $(DHQD)_2PHAL$ [43]. This process affords spiro-oxindoles **146** having two heterocycles at their C3-position in excellent yields with high enantioselectivities, as shown in Scheme 5-44.

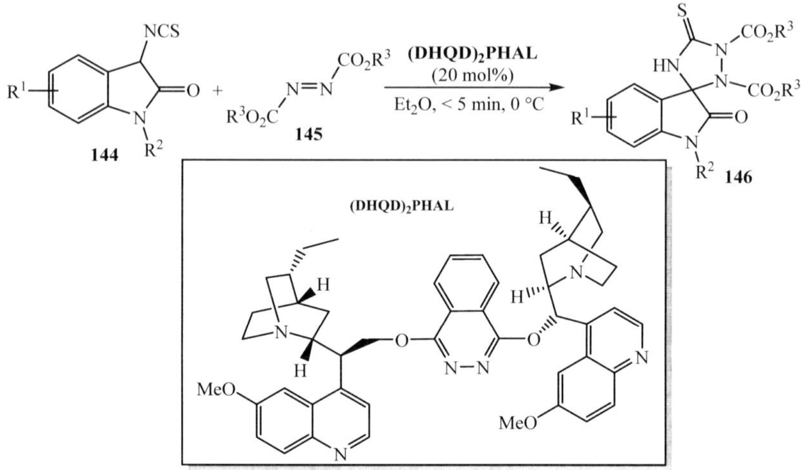

Scheme 5-44. (DHQD)$_2$PHAL-Catalyzed [3+2] cycloaddition.

Li et al. demonstrated the first example of the participation of an ester-carbonyl group in a carbonylative annulation reaction with homoenolate produced by the reaction of NHC with enals [44]. In this process, the NHC precursor **A**-catalyzed annulation of benzofuran-2,3-diones **147** and enals through homoenolate intermediates resulted in spiro-bis-lactones formation **148** and **149** in moderate to good yields with good regio-selectivity and high diastereoselectivity (Scheme 5-45).

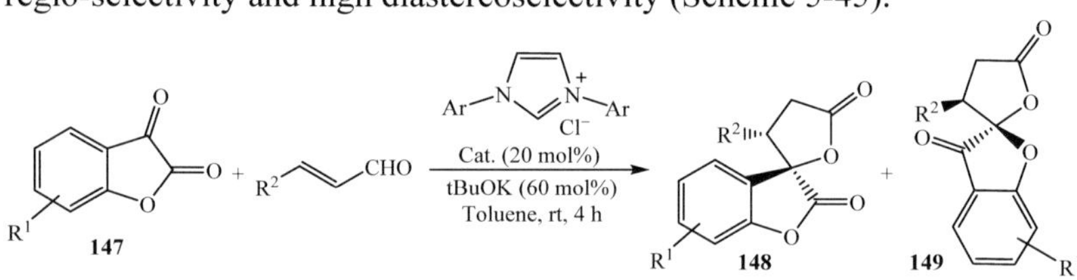

Scheme 5-45. Annulations of benzofuran-2,3-diones **147** and enals catalyzed by NHC.

On the other hand, chiral spiro-oxindole–butenolides **151** syntheses through asymmetric NHC-catalyzed formal [3+2] annulation of 3-bromoenals **150** and isatins reported by the Ma group in 2014 (Scheme 5-46). The resulting butenolides are of potential synthetic interest as they can be converted into various chiral spirocyclic scaffolds [45].

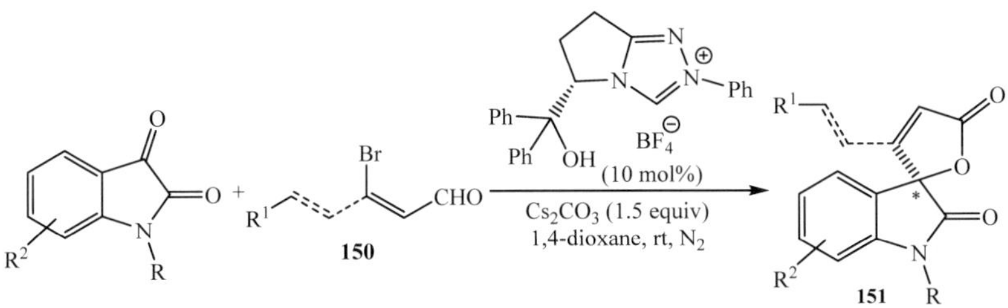

Scheme 5-46. NHC-Catalyzed formal [3+2] annulation of 3-bromoenals **150** and isatins.

Additionally, Glorius et al. developed a highly enantioselective NHC-catalyzed formal [3+2] annulation of α,β-unsaturated aldehydes with aza-aurones **152** or aurone **153** (Scheme 5-47). The protocol indicates a unique NHC-activation-based approach to access spiro heterocyclic derivatives **154-155** bearing a quaternary stereogenic center with high optical purity (up to 95% ee) [46].

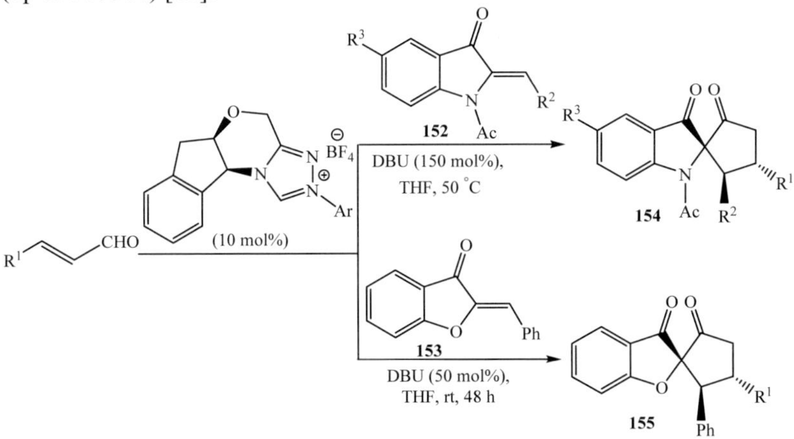

Scheme 5-47. NHC-Catalyzed formal [3+2] annulation of α,β-unsaturated aldehydes with aza-aurones **152** or aurone **153**.

As shown in Scheme 5-48, the most straightforward access to optically active spiro-oxindole oxazolines **158** through diastereo- and enantioselective [3+2] cycloaddition of isocyanoacetates **156** with isatins is the use of quinine-derived bifunctional amine-thiourea-bearing sulfonamide as multiple hydrogen-bonding donor catalysts **157** [47].

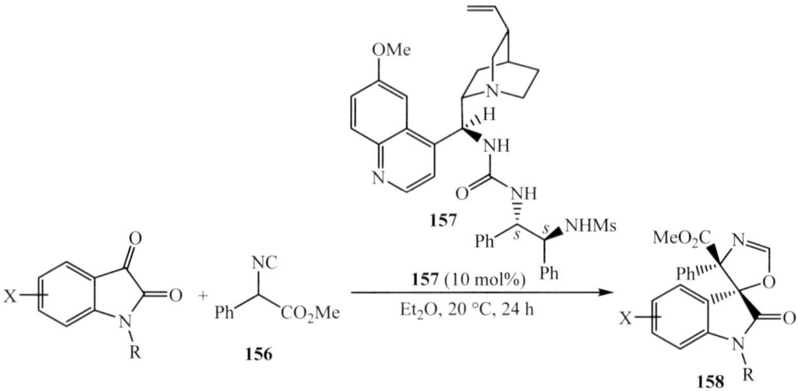

Scheme 5-48. Cinchona alkaloid-derived thiourea-catalyzed [3+2] cycloaddition reaction.

In 2015, the Cossío group reported the successful enantiodivergent synthesis of (5*R*) and (5*S*)-1,7-diazaspiro[4,4]nonan-6-ones **159** via sequential interrupted and completed [3+2] cycloadditions between metallated azomethine ylides and π−deficient alkenes (Scheme 5-49). The only source of chirality along the overall process is an enantiopure ferrocenyl pyrrolidine catalytic ligand [48].

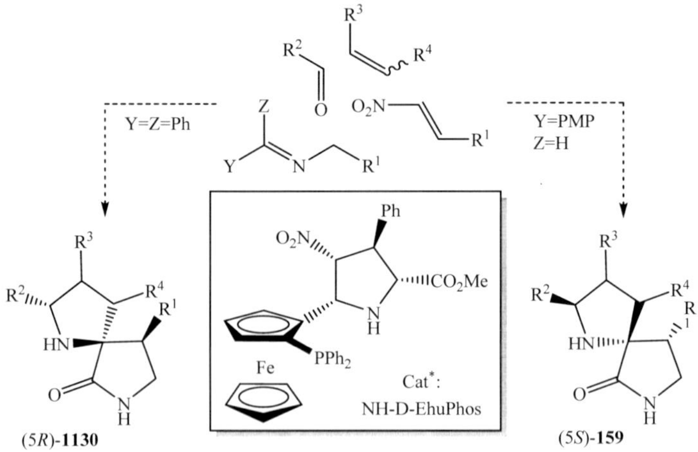

Scheme 5-49. Enantiodivergent synthesis of (5*R*) and (5*S*)-1,7-diazaspiro [4,4]nonan-6-ones **159**.

In the same year, Zhou et al. reported the first regioselective catalytic asymmetric [3+2] cycloaddition of benzofuranone-derived olefins with allenoates and substituted allenoates in the presence of a chiral phosphine **(*R*)-SITCP**, to afford different functionalized 3-spiro cyclopentene benzofuran-2-ones **160a,b** in good yields with high enantioselectivities under mild conditions (Scheme 5-50). Under the present catalytic system, γ-addition products **160a** and α-addition products **160b** can be formed in 48–99% yields with 87–99% ee values and 88:12 to >19:1 regioselectivity ratios and in 62–96% yields with 85–99% ee values and 84:16 to >19:1 regioselectivity ratios, respectively. The regioselective outcomes for this phosphine-catalyzed [3+2] cycloaddition reaction can be rationalized using DFT calculations [49].

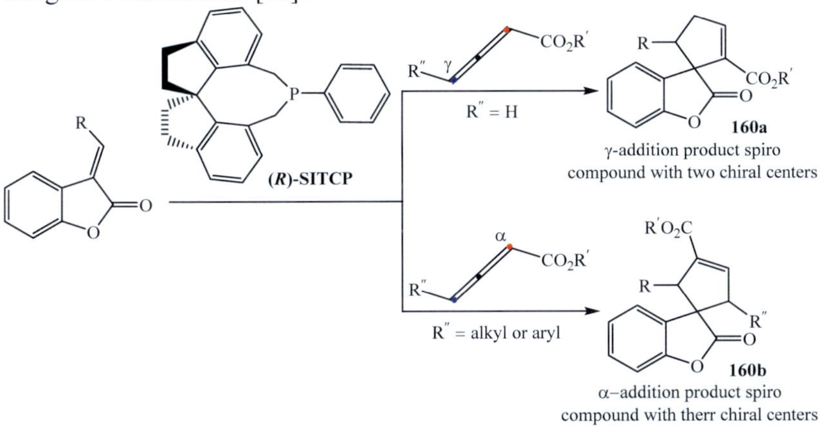

Scheme 5-50. Chiral phosphine-catalyzed cycloaddition reactions.

Yao et al. also developed an *in situ* NHC-catalyzed activation strategy to β-functionalize saturated carboxylic acid **162** [50]. This asymmetric [3+2] annulation could provide efficient access to spirocyclic oxindolo-γ-butyrolactones **162** from saturated carboxylic acid **161** and isatin in good yields with high to excellent enantioselectivities (Scheme 5-51). In Scheme 5-51, a plausible catalytic cycle is depicted. NHC is added to the ester substrate generated in situ from the saturated carboxylic acid **161** to afford NHC-bounded intermediate **A** in the proposed cycle. This intermediate undergoes deprotonation to give the enolate intermediate **B**, which converts into intermediate **C** through β-sp^3-H shift similar to the formation of homoenolate in the NHC-catalyzed reactions of carboxylic acids. Finally, intermediate **C** reacts with isatin derivatives through the nucleophilic addition giving the zwitterionic intermediate **E**. The release of the final cycloadduct **162** could then regenerate the catalyst.

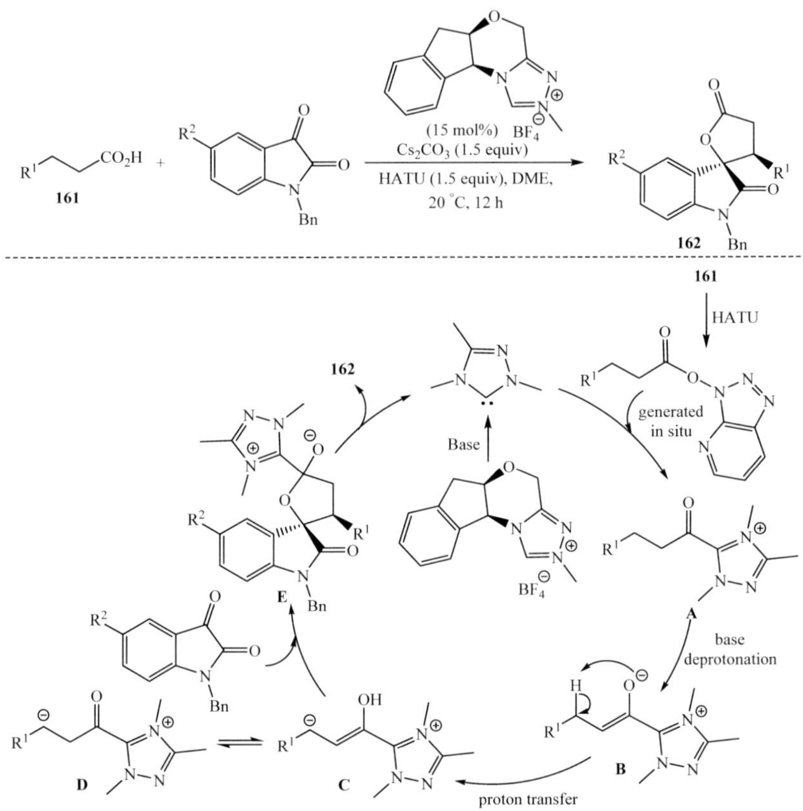

Scheme 5-51. An *in situ* NHC-catalyzed activation strategy and plausible catalytic cycle.

One year later, Xu et al. developed [51] a highly regioselective DBU-catalyzed [3+2] annulation of Morita-Baylis-Hillman carbonates **163** with isothiocyanates **164** to provide spirocyclic oxindole dihydrothiophene products **165** in moderate to high yields with excellent regioselectivities under simple conditions (Scheme 5-52).

Scheme 5-52. DBU-Catalyzed [3+2] annulation of Morita–Baylis–Hillman carbonates **163** with isothiocyanates **164**.

In 2016, Yuan et al. reported a highly diastereo- and enantioselective thiourea–tertiary amine **167** catalyzed [3+2] cycloaddition reaction rendering a series of spiro[pyrrolidin-3,2′-oxindole]s **168** bearing four consecutive stereocenters, including two vicinal spiroquaternary chiral centers, in excellent results (up to 99% yield, 20:1 dr, and 99% ee) for a variety of *N*-2,2,2-trifluoroethylisatin ketimines **54** and 3-alkenyl-5-arylfuran-2(3*H*)-ones **166** as represented in Scheme 5-53 [52].

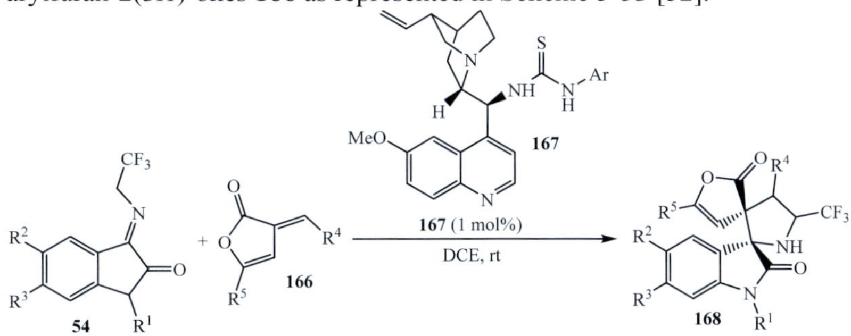

Scheme 5-53. Organocatalytic asymmetric [3+2] cycloaddition of *N*-2,2,2-trifluoroethylisatin ketimines **54**.

In search of developing useful new "click reactions", Ramachary et al. presented the organocatalytic azomethine imine-olefin [3+2] cycloaddition as a unique click reaction for the preparation of drug-like spiroindane-1,3-dione-pyrazolidinones **169** from simple precursors through an amino acid-catalysis (Scheme 5-54) [53].

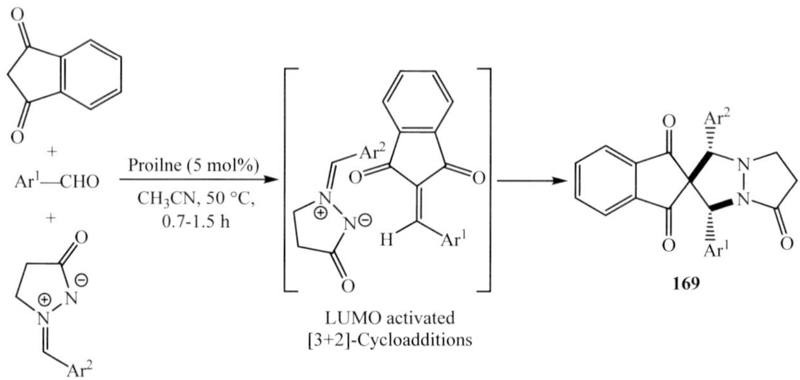

Scheme 5-54. Organocatalytic azomethine imine-olefin click reaction.

In the same year, Hui et al. presented another notable example of an NHC-catalyzed asymmetric formal [3+2] annulation reaction [54]. In this reaction, isatin *N*-Boc ketimines (Boc = *tert*-butyloxy-carbonyl) react with 3-bromoenals to the generation of spiro[indoline-3,2′-pyrrole] derivatives **170** with one chiral quaternary center in good yields (up to 81% yield) with excellent enantioselectivity (up to 99% ee) (Scheme 5-55).

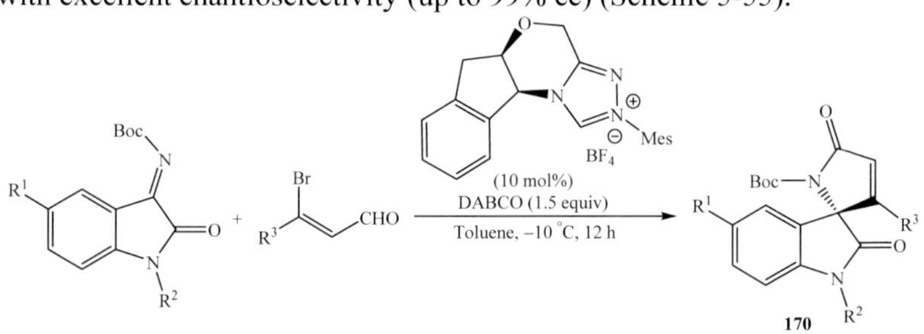

Scheme 5-55. Synthesis of spiro[indoline-3,2′-pyrrole] derivatives **170**.

Later, Yao et al. reported a notable study showcasing a novel intermolecular [3+2] annulation between *p*-QMs and vinyl cyclopropanes for the synthesis of spiro[4.5]deca-6,9-diene-8-ones **171**.The palladium and phosphine–thiourea cooperative catalysis system simultaneously actives *p*-QM and vinyl cyclopropane (Scheme 5-56) [55].

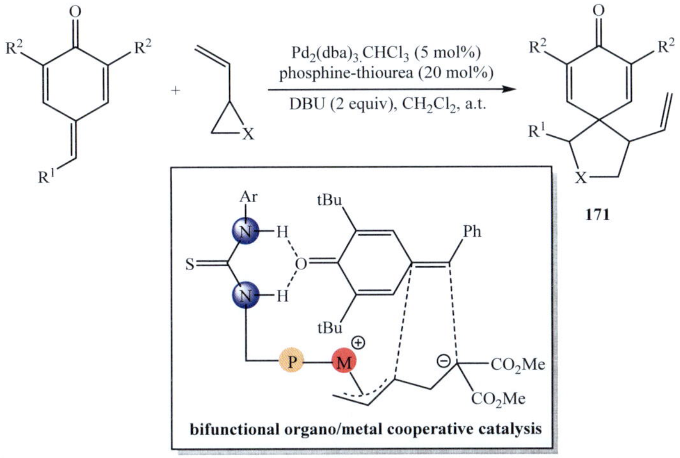

Scheme 5-56. Bifunctional organo/metal cooperatively catalyzed [3+2] annulation.

The enantioselective construction of a spirocyclic quaternary stereogenic carbon center at the C2 position of indole is a challenge in organic synthesis. To tackle the issue, Hu et al. developed a straightforward procedure for 2,2′-pyrrolidinyl-spiro-oxindole scaffolds **174** which successfully achieved by employing a rationally designed hydrogen-bonding network activation strategy [56]. As indicated in Scheme 5-57, the solution of bifunctional organocatalyst **173** (10 mol%), azaaruone (1.2 equiv), and *o*-hydroxy aromatic aldimine **172** (1.0 equiv) in dry DCE led to the desired products **174** in high yields with excellent stereoselectivities.

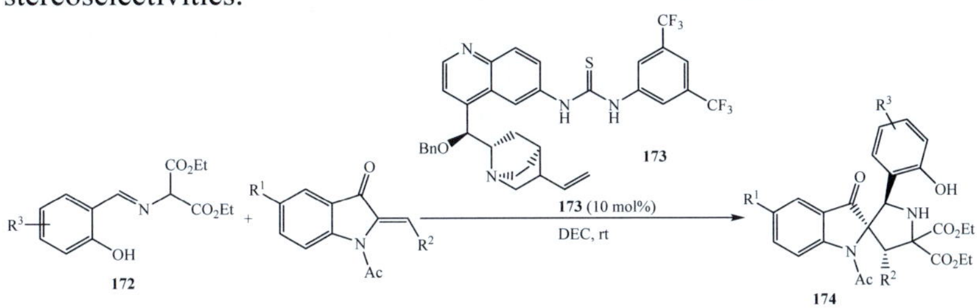

Scheme 5-57. Hydrogen-bonding network promoted [3+2] cycloaddition.

In addition to the previously mentioned examples of NHC-catalyzed [3+2] cycloaddition reactions, Yetra et al. reported the synthesis of spiro γ-butyrolactones **177** by the enantioselective formal [3+2] annulation catalyzed by the chiral triazolium salt **175** [57]. Under oxidative conditions using bisquinone oxidant **176**, the reaction between α,β-unsaturated aldehydes, and 3-hydroxy oxindoles resulted in the desired products **177** in moderate to good yield, enantioselectivity, and diastereoselectivity (Scheme 5-58). The reaction likely proceeds through the formation of NHC-bound α,β-unsaturated acylazolium intermediate from enals, which was intercepted by the dioxindoles in a formal [3+2] pathway to generate the spirocyclic compounds.

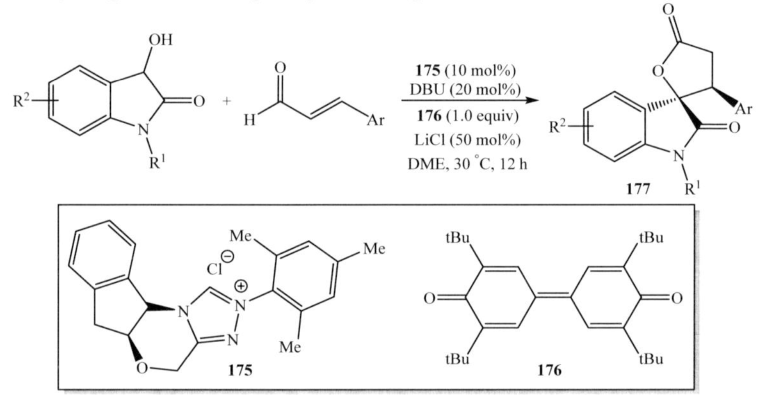

Scheme 5-58. Enantioselective synthesis of spiro γ-butyrolactones **177**.

The Wang group explored 3,3′-(9-phenanthryl)-modified phosphoric acid **179** catalyzed [3+2] cycloaddition of azomethine ylides generated *in situ* from 3-amino oxindoles and aldehydes, with α,β-ynones **178** to devise a straightforward route to spiro[dihydropyrrole-2,3′-oxindole] products **180** in high chemical yields with excellent stereoselectivities (Scheme 5-59) [58].

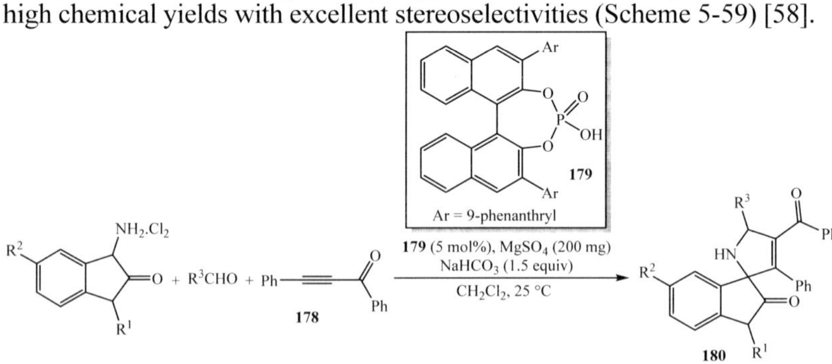

Scheme 5-59. 3,3′-(9-Phenanthryl)-modified phosphoric acid **179** catalyzed [3+2] cycloaddition.

In 2018, Lin et al. utilized a new type of all-carbon 1,3-dipole precursor, 3-homoacyl-coumarin **181**, to the stereoselective [3+2] cycloaddition with indandione alkylidenes **182** to produce a series of coumarin/indanedione-fused spirocyclopentanes **183** (Scheme 5-60).

Mechanistic studies demonstrated that the reaction followed a highly efficient concerted pathway, resulting in products with high stereoselectivities. Additionally, the step-wise path was ineffective under reaction conditions. The intermediate could only reverse to the starting materials that delivered the final product, making the concerted pathway even more efficient [59].

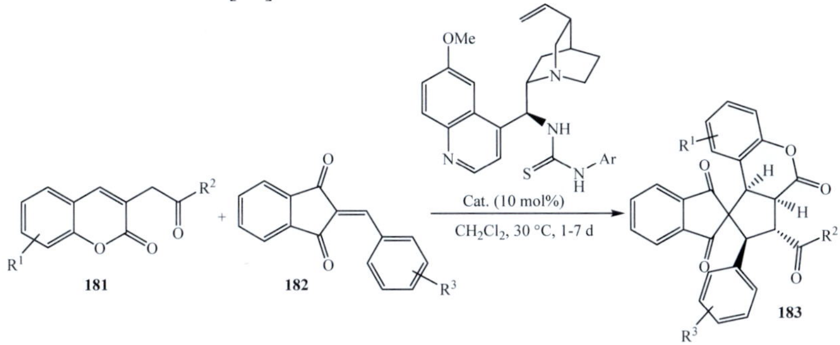

Scheme 5-60. Synthesis of coumarin/indanedione-fused spirocyclopentanes **183**.

One year later, Wu et al. developed a phosphine-catalyzed [3+2] cycloaddition reaction between *N*-2,2,2-trifluoroethylisatin ketimines **54** and γ-substituted allenoates **184**, leading to the synthesis of spiro[indoline-3,2′-pyrrole] scaffolds **185** (Scheme 5-61). The reactive *N*-2,2,2-trifluoroethylisatin ketimines **54** play a key role in the reaction as azomethine ylide precursors [60].

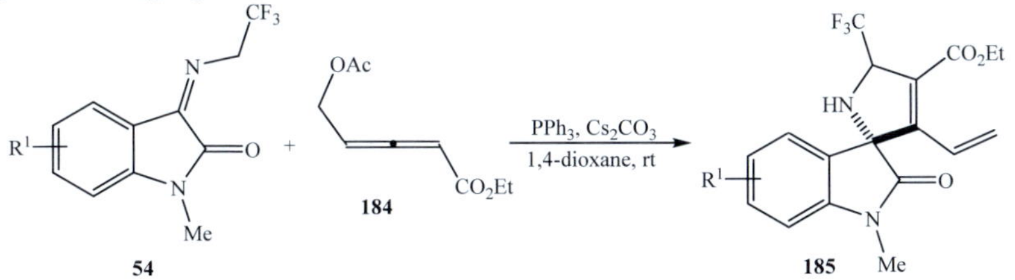

Scheme 5-61. Phosphine-catalyzed [3+2] cycloaddition of *N*-2,2,2-trifluoroethylisatin ketimines **54** with γ-substituted allenoates **186**.

Yan et al. reported the pioneering organocatalytic method for the asymmetric synthesis of CF_3-containing spiro-thiazolone-pyrrolidine compounds **189** (Scheme 5-62). In the presence of 1 mol% of the β-isocupreidine (cinchona alkaloid-derived) catalyst **188**, the [3+2] cycloaddition of trifluoroethylamine-derived ketimine **186** with 5-alkenyl thiazolones **187** furnished the compounds in excellent yields and stereoselectivities [61].

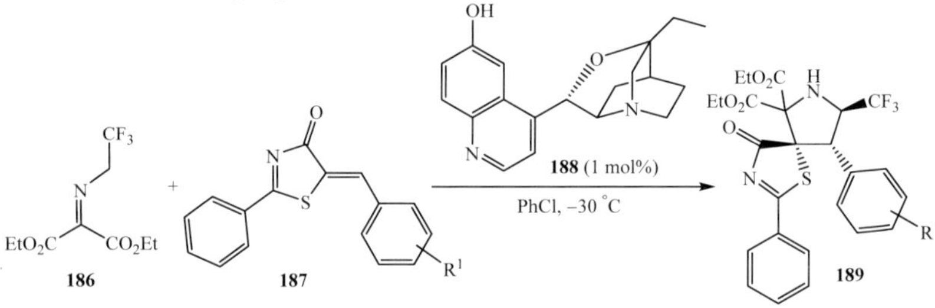

Scheme 5-62. The asymmetric construction of CF_3-containing spiro-thiazolone-pyrrolidine compounds **189**.

The Cui group performed a highly stereoselective [3+2] cycloaddition of isatin-derived MBH carbonates with β,γ-unsaturated α-keto esters for the construction of the chiral spiro-cyclopentene/cyclopentadiene-oxindole skeletons **190**, which were catalyzed by β-isocupreidine **188** (Scheme 5-63) [62].

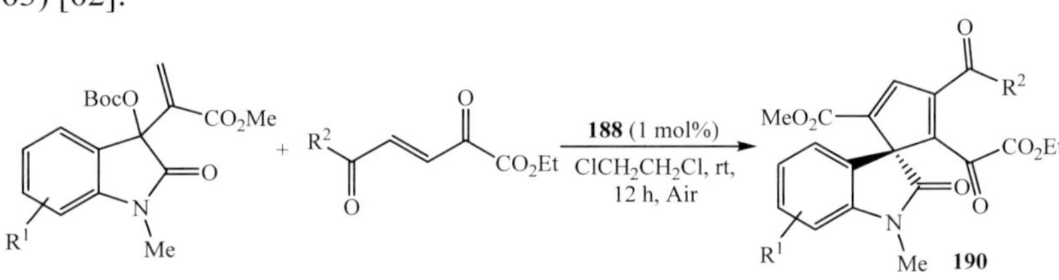

Scheme 5-63. Synthesis of chiral spiro-cyclopentene/cyclopentadiene-oxindoles **190**.

Recently, Zhong et al. investigated a ferrocene-derived bifunctional phosphine ($^*PR_3$) **192**-catalyzed enantioselective [3+2] annulation of γ-substituted allenoates **184** with pyrazoloneyldiene oxindoles **191** for the synthesis of spiro-cyclopentene-linked oxindole and pyrazolone compounds **193**. A possible mechanism and an activation model were proposed in Scheme 5-64 and Figure 5-3, respectively. Initially, the ferrocene-derived bifunctional phosphine **192** attacks to **184** to form the zwitterionic intermediate **Ia-Ib**, with the subsequent γ-addition to the more electron-deficient carbon of pyrazoloneyldiene oxindole **191** to give the intermediate **II**. Then, **II** was transformed to **III** through intramolecular Michael addition. Finally, a proton transfer followed along with the release of chiral phosphine catalyst yielded the [3+2] annulation product spirocyclopentene compound **193** combined with oxindole and pyrazolone [63].

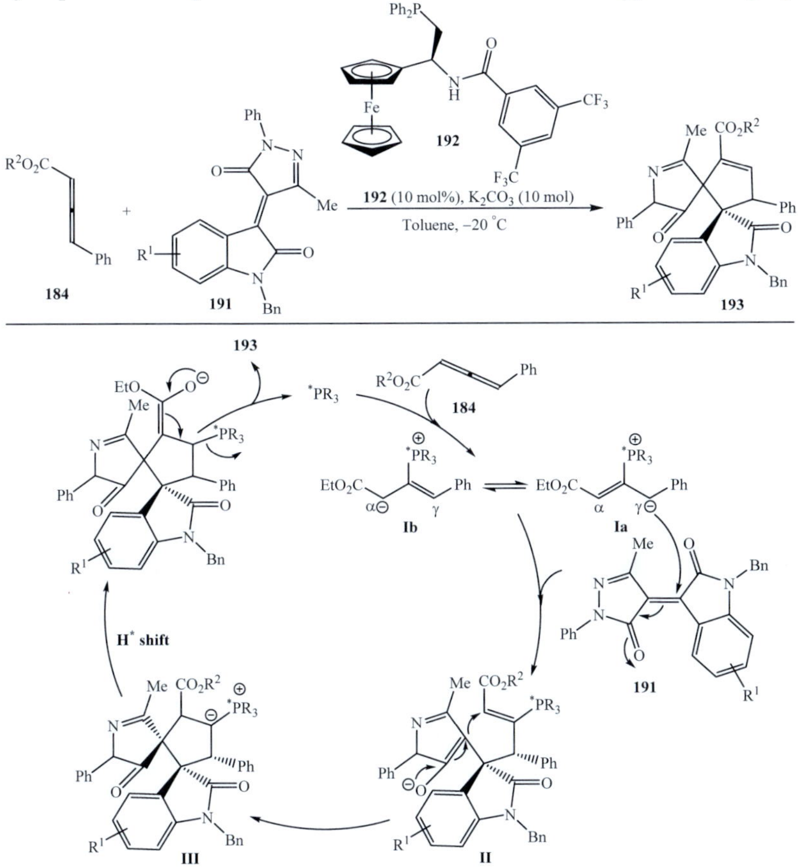

Scheme 5-64. Plausible mechanism.

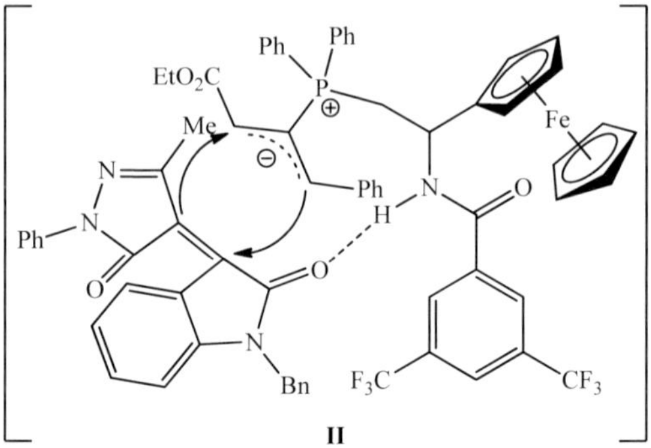

II

Fig. 5-3. Activation model.

5.3.4. Diels–Alder cycloaddition reactions

In 1999, Gelmi et al. explored the Diels–Alder cycloaddition reaction of 3-chloromethylene-2-indolones **194** with a series of dienes to synthesize spirocyclohexenindolones **195** (Scheme 5-65). The reaction proceeded in the presence of ethylaluminium dichloride with good diastereoselectivity and regioselectivity [64].

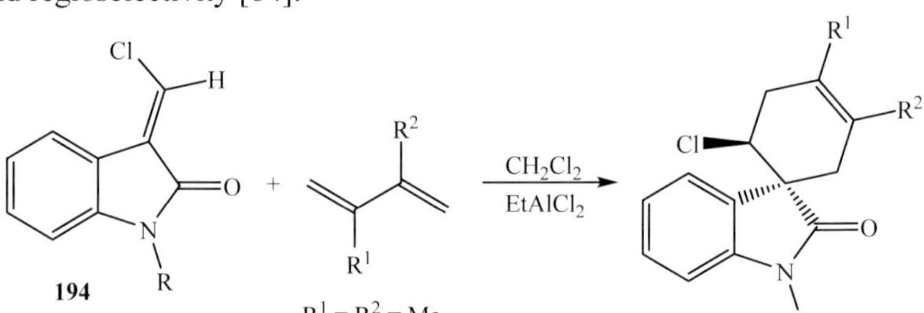

194 R = CO_2Et, H

$R^1 = R^2 = Me$
$R^1 = Me, R^2 = H$

195

Scheme 5-65. Synthesis of spirocyclohexenindolones **195**.

(-)-Gymnodimine **196** is a member of a unique class of potent marine toxins containing imines within a spirocyclic array. In 2000, Romo et al. reported the synthesis of the tetrahydrofuran fragment **197** and a strategy toward the spirocyclic imine fragment **198** of this family of toxins [65]. Key reactions involve a Diels−Alder reaction of α-methylene-γ-lactam **199** and a dienyne **200** to prepare spirocycle moiety **198**. The retrosynthetic analysis of spirocycle moiety **198** is shown in Scheme 5-66.

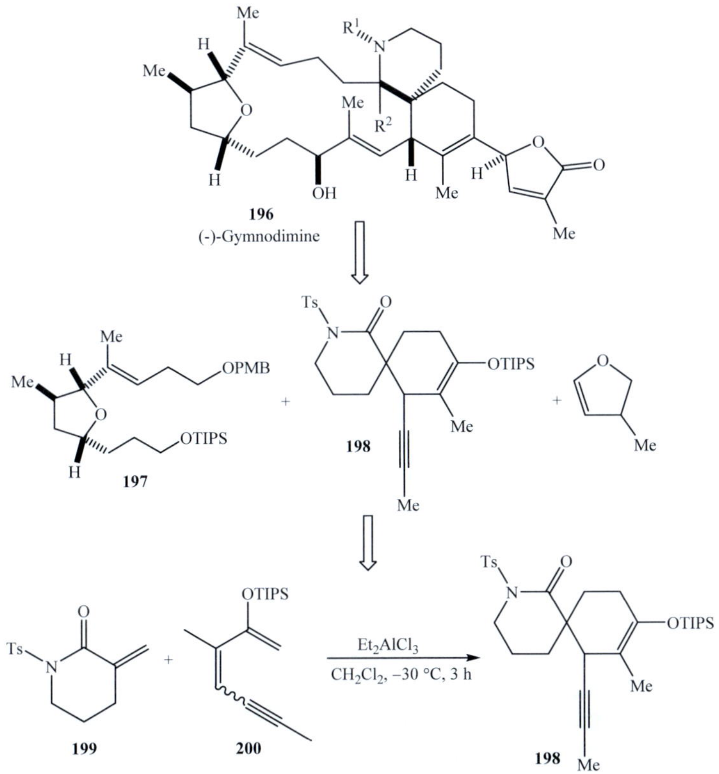

Scheme 5-66. Retrosynthetic analysis of spirocycle moiety **198**.

In the same year, a series of novel heterocyclic derivatives containing the spiro bicyclo[2.2.1]heptane system **203-204** were successfully prepared by *exo*-selective Diels–Alder reaction of exocyclic arylidenes **201-202** with cyclopentadiene (Scheme 5-67). The reaction was performed in a 5.0 M solution of lithium perchlorate in diethyl ether (LPDE) [66].

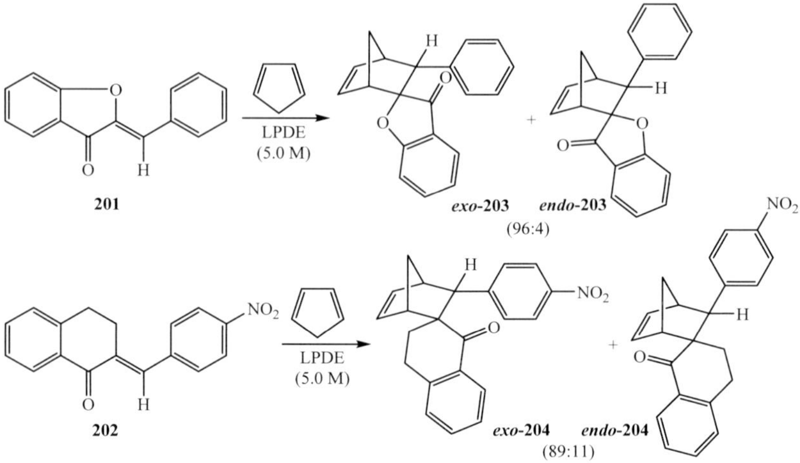

201

202

Scheme 5-67. *Exo*-selective Diels-Alder reaction in LPDE (5.0 M).

In 2003, Barbas et al. developed the first organocatalytic hetero-domino Knoevenagel/Diels–Alder/epimerization (K–DA-E) reactions of readily available precursors enones **205**, aryl-aldehydes and 1,3-indandione catalyzed by L-proline and pyrrolidine to afford highly substituted prochiral spiro[cyclohexane-1,2′-indan]-1′,3′,4-triones **206** in a highly diastereoselective fashion with excellent yields (Scheme 5-68) [67]. This domino's epimerization reaction, catalyzed by pyrrolidine, was significantly faster than proline catalysis.

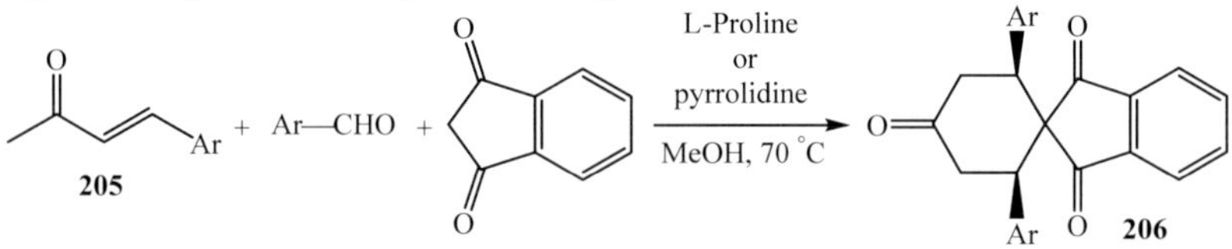

205

206

Scheme 5-68. The first organocatalytic hetero-domino K–DA-E reactions.

Later, this group also reported the first organocatalytic asymmetric domino Knoevenagel/Diels–Alder (K–DA) reaction catalyzed by L-proline to the diastereospecific and enantioselective construction of highly substituted spiro[5,5]undecane-1,5,9-triones **207** [68]. The three-component reaction of enone **205**, aldehyde and Meldrum's acid with a catalytic amount of L-proline in methanol at ambient temperature furnished the Diels–Alder product **207** as a single diastereomer in good yield. These products were accompanied by an unexpected symmetric spirocyclic ketone **208** in trace yield (Scheme 5-69).

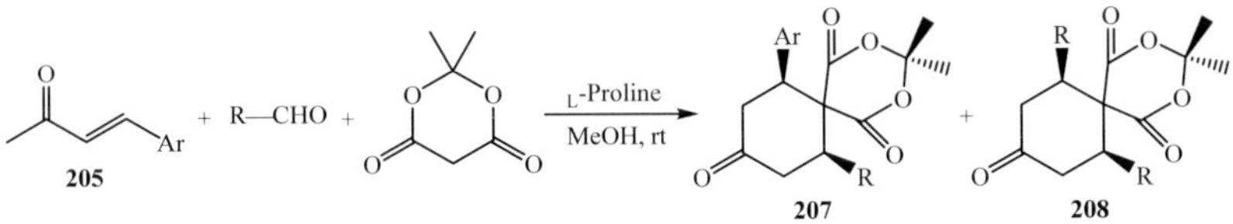

Scheme 5-69. The first organocatalytic domino K–DA reaction.

Matsuo et al. disclosed in 2008 a noteworthy intermolecular [4+2] cycloaddition catalyzed by boron trifluoride etherate. This reaction involved the coupling of 3-alkoxycyclobutanones **209a,b** with ketones, such as cyclohexanone and acetophenone, at the C2-C3 bond of the cyclobutanone ring (Scheme 5-70) [69]. In this cycloaddition, ketones were inserted into the less substituted C2-C3 bond of the cyclobutanone ring of **209a,b** and gave the corresponding spiro cycloadducts **210-211**.

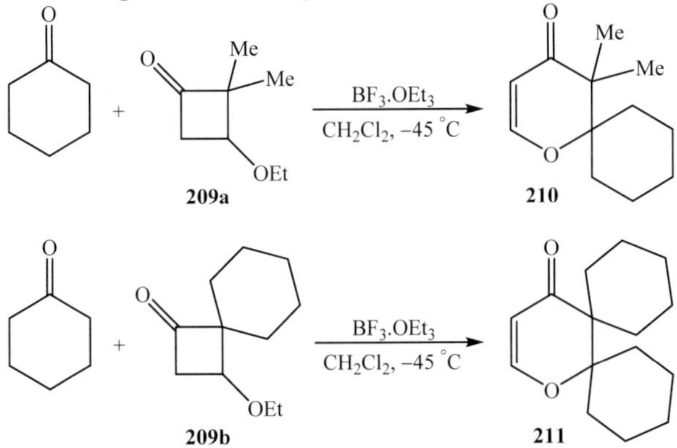

Scheme 5-70. Lewis acid-catalyzed intermolecular [4+2] cycloaddition of **209a,b**.

An efficient synthesis of azaspiro compounds having dissymmetric dienone cores was reported by Wada et al. in 2009 via *ipso*-iodocyclization of ethoxyethyl ether **212**, which act as oxygen nucleophiles, to alkynes [70]. They found that the combination of NIS and $BF_3{\cdot}OEt_2$ gave better results. A proposed mechanistic pathway based on experimental data is depicted in Scheme 5-71. First, the reaction of NIS and $BF_3{\cdot}OEt_2$ gives the active iodonium species **A**. The *anti*-attack of electrophile **A** and the electron-rich aromatic ring, followed by eliminating an oxonium ion from **C**, generates the spiro compound **E**. The oxonium ion could be trapped by imide salt **B** to produce the *N*,*O*-acetal **D**. Due to less hinder of ether and ester linkages (X = O) than those containing nitrogen, Diels–

Alder, dimerization of **E** occurs. The [4+2] cycloaddition of **F** proceeds in the sterically least hindered manner to generate dimer **G**.

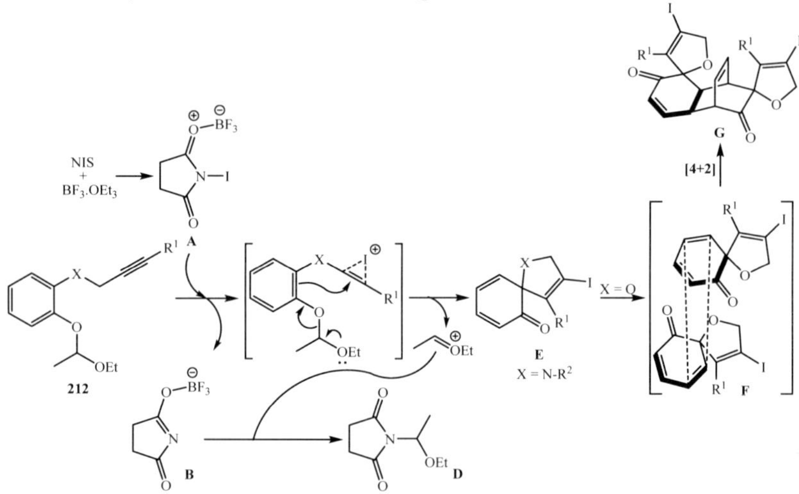

Scheme 5-71. Proposed mechanistic pathway.

One year later, Gong et al. used a Brønsted acid-Lewis base bifunctional organocatalyst **214** to catalyze [4+2] cycloaddition reaction of Nazarov reagents **213** and methyleneindolinones to synthesize spiro[4-cyclohexanone-1,3′-oxindoline]s **215** with excellent enantioselectivity (up to 98% ee) [71]. This reaction is performed through successive double Michael addition reactions in which the bifunctional chiral catalyst activates each reaction component by hydrogen bonding interactions (Scheme 5-72).

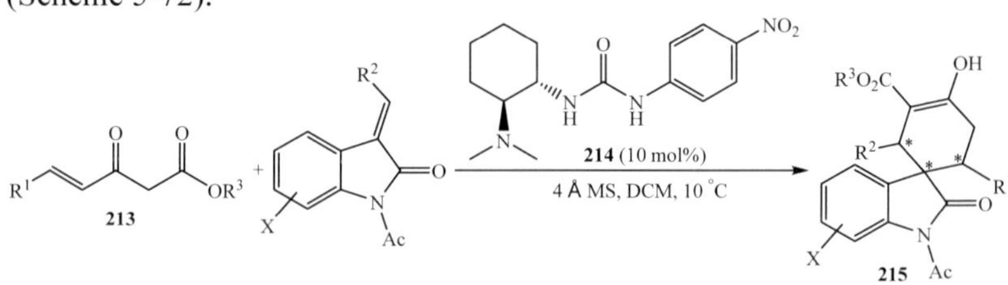

Scheme 5-72. The synthesis of spiro[4-cyclohexanone-1,3′-oxindoline]s **215**.

In an elegant work, the Melchiorr group introduced the first asymmetric catalytic Diels-Alder reaction of *in situ* generated heterocyclic *ortho*-quinodimethanes (*o*-QDMs) **216** with two distinct classes of olefinic dienophiles, nitro-olefins, and methyleneindolinones (Scheme 5-73). Asymmetric aminocatalysis uses chiral amines as catalysts and directs the pericyclic reactions with nitro-olefins and methyleneindolinones toward a highly stereoselective pathway [72].

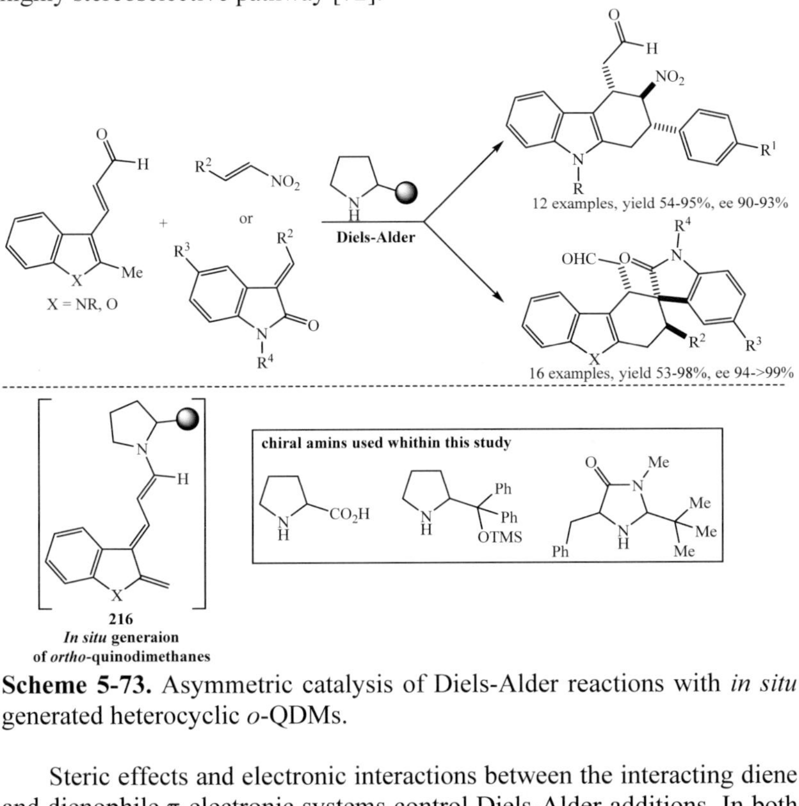

216
***In situ* generaion**
of *ortho*-quinodimethanes

Scheme 5-73. Asymmetric catalysis of Diels-Alder reactions with *in situ* generated heterocyclic *o*-QDMs.

Steric effects and electronic interactions between the interacting diene and dienophile π-electronic systems control Diels-Alder additions. In both systems (reaction with nitro-olefins and oxindole-derived dienophiles), the chiral catalyst selectively affords face shielding of the formed *o*-QDM intermediate, inducing the π-facial-selectivity and securing high ee. At the same time, the regioselectivity is directed by an optimal HOMO(*o*-QDM)-LUMO-(electron-deficient olefin) orbital overlap. The reaction with nitro-olefins proceeds with *exo*-selectivity relative to the nitro group, whereas the reaction with methyleneindolinones dienophiles proceeds by *endo*-procedure relative to the oxindole moiety.

Later on, Wang et al. suggested a highly efficient *in situ* generation/ activation strategy that enabled the enantioselective construction of densely functionalized azaspirocyclic skeletons **220** using a highly enantio -selective inverse-electron demand Diels-Alder reaction (IEDDAR) catalyzed by a bifunctional organocatalyst [73]. The rosin-derived bifunctional primary amine/thiourea **219** (10 mol%) in combination with AcOH (20 mol%) catalyzed the reaction of cyclic keto/enolate salts **217** with *N*-tosyl-2-methylenebut-3-enoates **218** to furnish the desired products **220** with excellent yields and diastereoselectivity (up to 99% yield, > 20:1 d.r., and > 99% ee) (Scheme 5-74). The authors proposed a possible model to explain the stereochemistry observed in these processes (Figure 5-4). The *N*-tosyl-2-methylenebut-3-enoate **218** is activated (lowering the LUMO energy) by the two thiourea hydrogen atoms through weak hydrogen bonds. In contrast, the cyclic keto/enolate salt **217** is activated (raising the HOMO energy) by the synchronous formation of ketiminium cation and protonation. The stereochemical control from the 1,2-diaminocyclohexane moiety and steric hindrance from the dehydroabietic amine moiety of the thiourea, high *re*-face, and *endo* selectivity would be applied to generate the desired chiral product **220**, which is consistent with the experimental results.

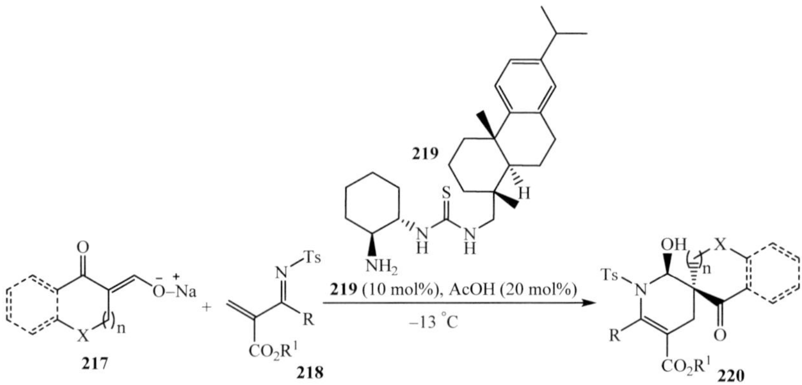

Scheme 5-74. Bifunctional organocatalytic strategy for IEDDAR.

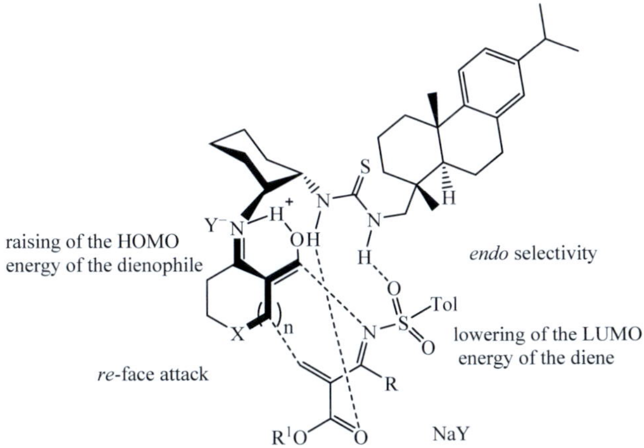

Fig. 5-4. Proposed transition state for the reaction.

In 2013, Cui et al. presented a highly diastereo- and enantioselective organocatalytic direct hetero-Diels-Alder (hDA) reactions of enones with isatins that gave access to substituted spiro-oxindole tetrahydropyranones **221** under mild conditions [74]. They found that reactions in the presence of amine **A** with acid **I**, amine **F** with acid **I**, and amine **A** with acid **J** in toluene gave good results (Scheme 5-75). Appropriate combinations of amine-acid and amine-acid-additive would allow *in situ* associations and interactions among the catalyst components and reactants for the desired diastereo- and enantioselective catalysis through the enamine formation.

Huang et al. presented a significant breakthrough in 2013 with the first phosphine-catalyzed [4+2] annulation of γ-substituted allenoates **184** with 2-arylidene-1*H*-indene-1,3(2*H*)-diones **222** [75]. Reactions were carried out at 60 °C in THF employing triphenylphosphine (10 mol%) to the construction of highly substituted spiro[4.5]dec-6-ene skeletons **223** in excellent yields and with complete regioselectivity and high diastereoselectivity (Scheme 5-76). In this annulation, allenoates act as a new type of 1,4-dipolar synthons.

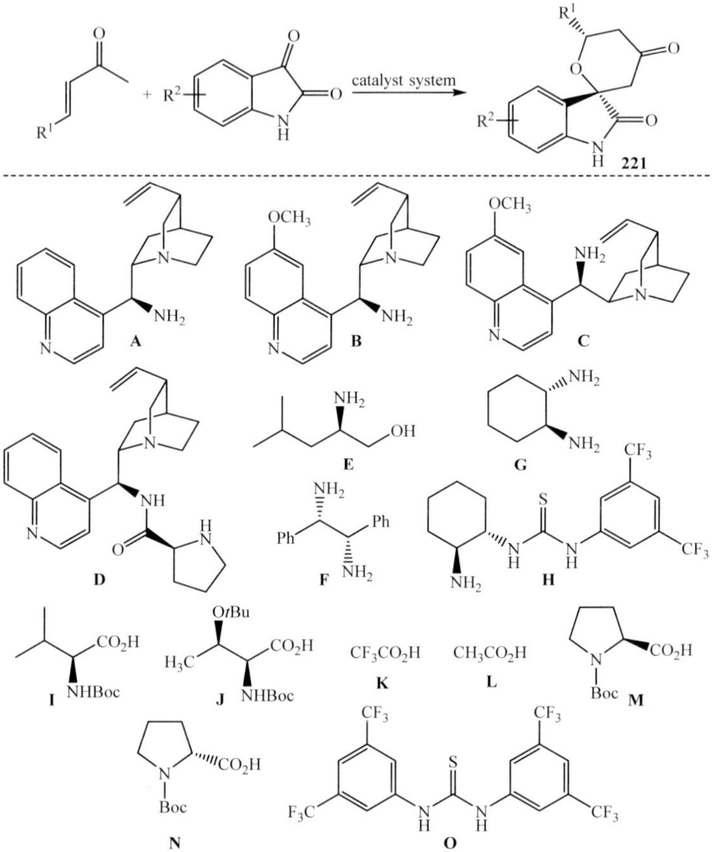

Scheme 5-75. Catalytic enantioselective formal hetero-Diels–Alder reactions.

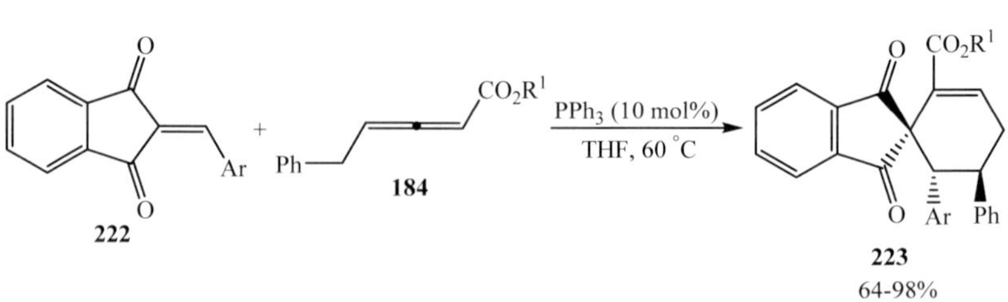

Scheme 5-76. Phosphine-catalyzed [4+2] annulation of γ-substituent allenoates **184**.

Almost simultaneously with the results of the Huang's research, Voituriez et al. disclosed another phosphine-catalyzed [4+2] cycloaddition of 3-arylideneoxindoles **224** and δ-aryl-substituted penta-2,3-dienoates **225** affording a great scope of 3,3′-spirocyclic oxindoles with functionalized six-membered rings **226** with good yields (Scheme 5-77) [76]. These reactions present excellent control of the relative stereochemistry of the three stereogenic centers.

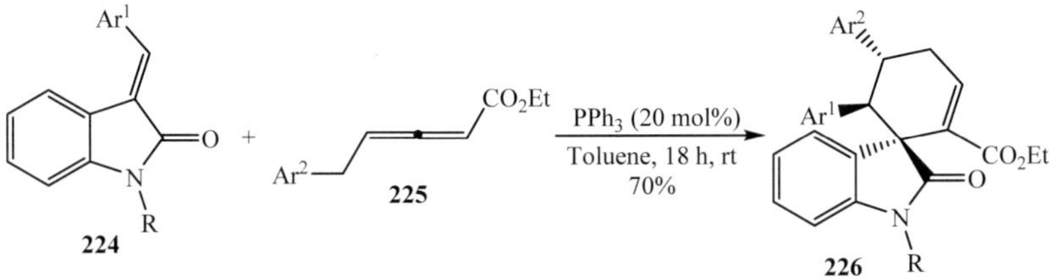

Scheme 5-77. Synthesis of 3,3′-spirocyclic oxindoles via phosphine-catalyzed [4+2] cycloaddition.

In addition, Shi et al. reported the first asymmetric [4+2] annulation of vinyl ketones with oxindole-derived α,β-unsaturated imines **227** that directly led to enantioselective synthesis of 2′,3′-dihydro-1′*H*-spiro [indoline-3,4′-pyridin]-2-ones **229** by utilizing a multifunctional thiourea-phosphine catalyst **228** derived from a natural amino acid (Scheme 5-78) [77].

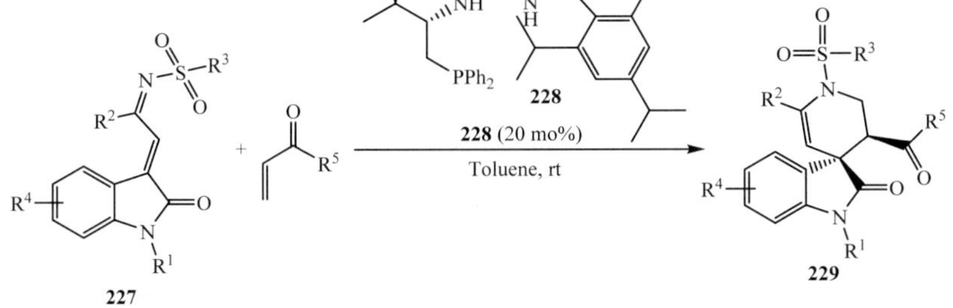

Scheme 5-78. Enantioselective synthesis of 2′,3′-dihydro-1′*H*-spiro [indoline-3,4′-pyridin]-2-ones **229**.

Li et al. applied the chiral amineocatalytic normal-electron-demand Diels-Alder reactions with HOMO-raised trienamines as a key step for the synthesis of spiro or fused frameworks [78]. In this approach, electron-deficient 1-aza-1,3-butadienes **230** containing a 1,2-benzoisothiazole-1,1-dioxide or 1,2,3-benzoxathiazine-2,2-dioxide motif operates as regio- and chemoselective dienophiles. The enantioenriched cycloadducts **231** could be efficiently transformed to spiro or fused framework **232-233** with high structural and stereogenic complexity by a sequential aza-benzoin reaction or other transformations (Scheme 5-79).

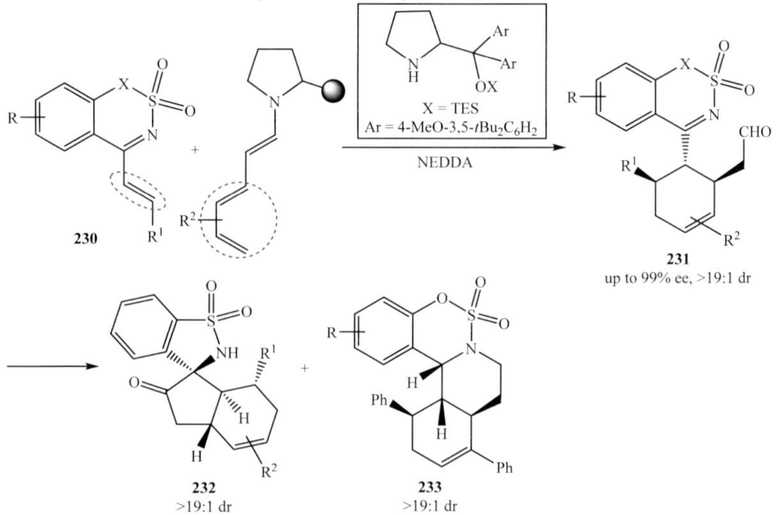

Scheme 5-79. 1-Azadienes as regio- and chemoselective dienophiles in aminocatalytic asymmetric Diels-Alder reaction.

Another aminocatalytic asymmetric Diels-Alder reaction was reported in 2013 (Scheme 5-80) [79]. Starting from phosphor-containing dienophiles **234** and 2,4-dienals upon catalysis of α,α-L-diphenylprolinol trimethylsilyl ether **235** and *o*-fluorobenzoic acid (OFBA) via the intermediacy of trienamine species (trienamine catalysis), the synthesis of spirocyclic oxindoles **236** was achieved in [4+2] cycloaddition in chloroform at room temperature.

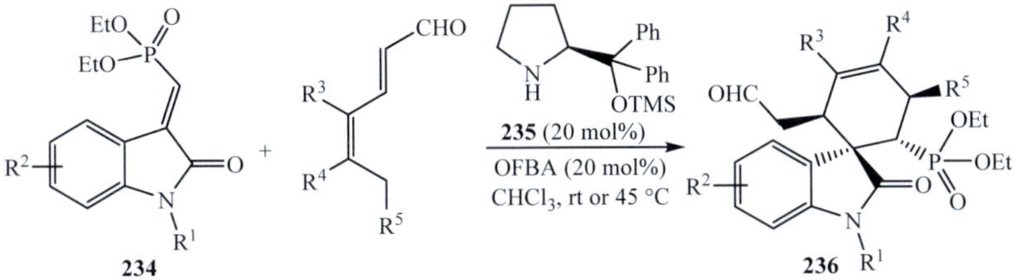

Scheme 5-80. Aminocatalytic asymmetric Diels-Alder reaction of phosphorus dienophiles **234**.

In other exciting examples, Ye et al. employed rhodanine/hydantoin derivatives **237** as dienophiles to develop an aminocatalyzed asymmetric Diels–Alder reaction for the synthesis of various spirocyclic compounds **238** with good yields (up to 98%) and excellent stereoselectivities (up to 99% ee and >19 : 1 dr) (Scheme 5-81) [80].

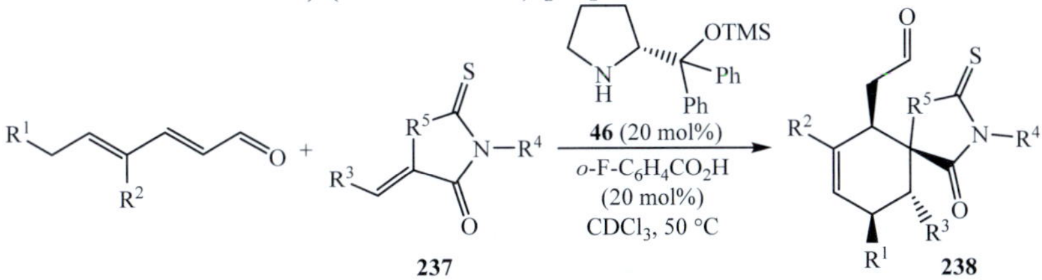

Scheme 5-81. Aminocatalyzed asymmetric Diels–Alder reaction of rhodanine/hydantoin derivatives **237**.

In 2014, an organocatalytic [4+2] annulation of 2-arylideneindane-1,3-diones **139** and glutaraldehyde **239** was developed to synthesize functionalized spirocyclohexane carbaldehydes **240** bearing one all-carbon quaternary stereogenic center [81]. The reaction proceeds through a Michael/Aldol sequence in good to high yields and with high levels of stereoselectivity (up to >95:5 dr and 99% ee) in the presence of chiral amine **235** (20 mol%) as catalyst and DIPEA (20 mol%) as an additive in ether at 0 °C (Scheme 5-82).

Li et al. reported an enantioselective construction of chiral spirocyclic benzofuran-2-ones **243** employing a trienamine activation strategy [82]. In this strategy, the Diels−Alder reaction of 3-olefinic benzofuran-2-ones **241** with polyenals **242** catalyzed by diphenylprolinolsilyl ether **235** gives the desired products in good yields with moderate diastereoselectivities and good enantioselectivities (Scheme 5-83).

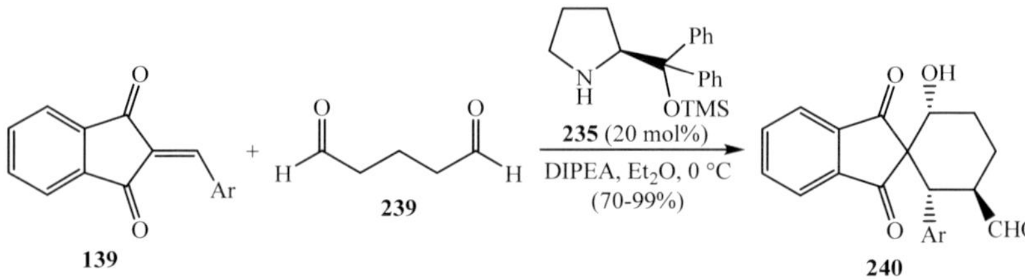

Scheme 5-82. Organocatalytic synthesis of functionalized spirocyclohexane carbaldehydes **240**.

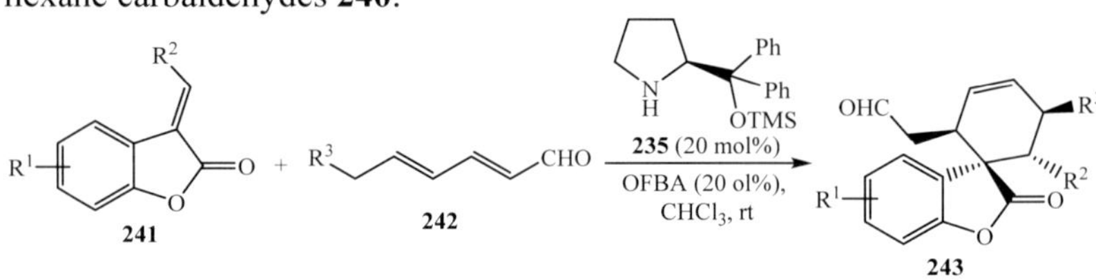

Scheme 5-83. Asymmetric Diels−Alder reaction of 3-olefinic benzofuran-2-ones **241**.

Access to optically pure spiro-oxindole 2,3-dihydropyran-4-ones **246** was reported by the Feng group in 2014 via the hetero-Diels–Alder reaction of Danishefsky's diene **244** with isatins catalyzed by a chiral *N*,*N'*-dioxide **245**/Mg(II) complex (0.1-0.5 mol%) (Scheme 5-84). The desired products **246** were obtained in excellent yields with excellent enantioselectivities (up to >99%ee) [83].

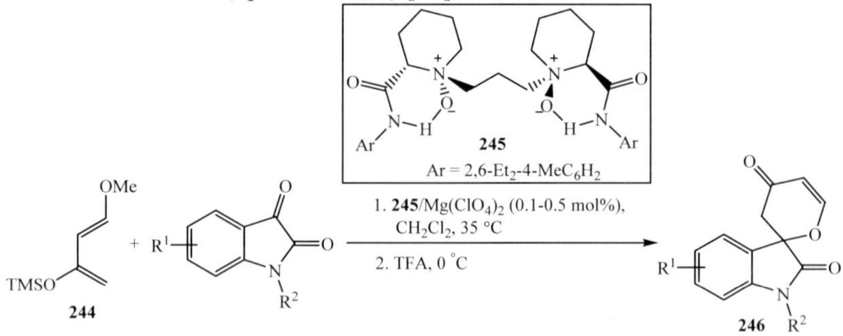

Scheme 5-84. Asymmetric hetero-Diels–Alder reaction of Danishefsky's diene **244**.

Zhang et al. explored amino-acid derived phosphine-catalyzed enantioselective 1,4-dipolar spiroannulation of cyclobutenones **247** and isatylidenemalononitrile **248**. Cyclobutenones **247**, as a new type of chiral

1,4-dipole four-carbon synthon, readily undergoes organophosphine-mediated C-C bond cleavage and asymmetric intermolecular 1,4-dipolar spiroannulation with isatylidene-malononitrile **248** in the presence of amino acid-derived chiral phosphine catalyst **249** to provide enantioenriched 3-spiro cyclohexenone 2-oxindoles **250** in good yield with up to 87% ee.

A reaction pathway for this chiral tertiary phosphine-catalyzed enantioselective 1,4-dipolar spiroannulation is represented in Scheme 5-85. The reaction is proceeded by nucleophilic 1,2-addition of the phosphine to cyclobutenone **247** to generate the zwitterionic intermediate **A**, followed by oxyanion-accelerated 4π ring-opening to form the corresponding 1,4-dipolar intermediate **B** (with two resonance forms), which undergoes Michael addition to **248** to generate intermediate **C** containing a carbanion. Finally, the carbanion undergoes nucleophilic attack on the acyl ion to furnish intermediate **D**, which, upon elimination of **249**, would give the spirocyclic product **250**. NaI operates as a counterion to stabilize the zwitterionic intermediates [84].

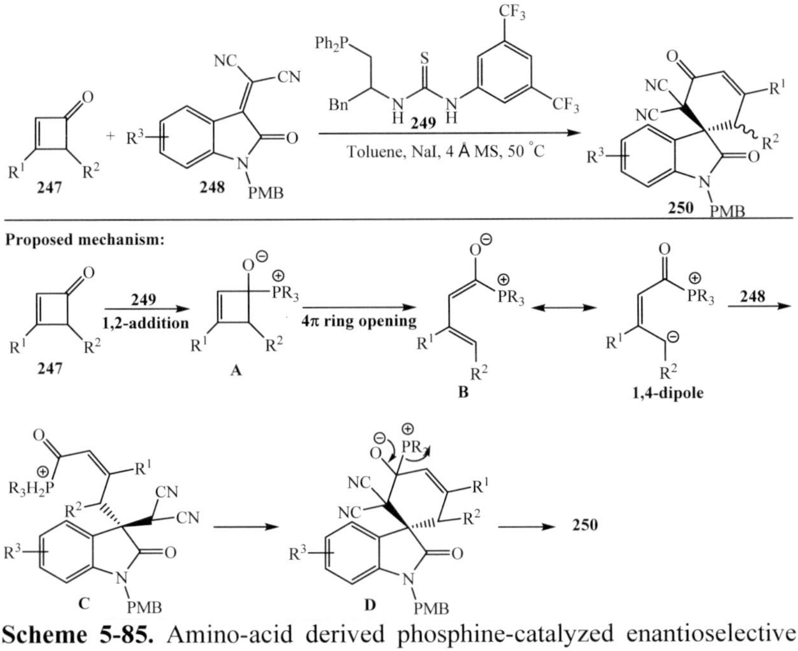

Scheme 5-85. Amino-acid derived phosphine-catalyzed enantioselective 1,4-dipolar spiroannulation and proposed mechanism.

In their proposed transition-state model, they believed that the hydrogen-bonding interactions between the thiourea moiety of the catalyst and the carbonyl group of the isatin moiety play a key role in asymmetric induction (Figure 5-5).

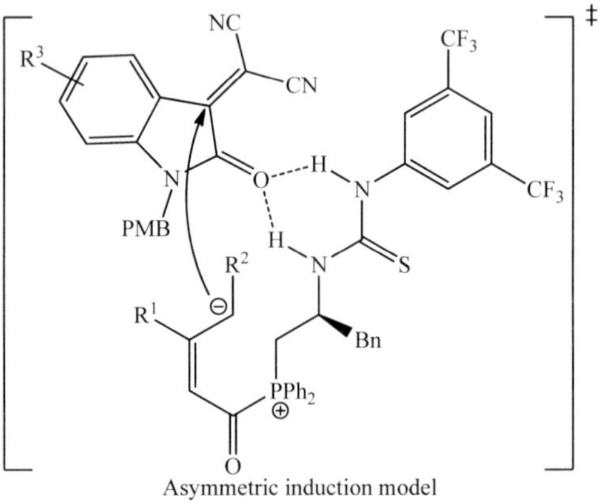

Fig. 5-5. Proposed transition-state model.

An efficient synthesis of chiral spiropyrazolones **251-254** was reported by Guo in 2015 via phosphine-catalyzed [4+2] cycloaddition of unsaturated pyrazolones with allenoates (Scheme 5-86). The desired products **251-254** were formed in moderate to excellent yields with moderate to excellent diastereoselectivities and enantioselectivities (89-95% ee) [85].

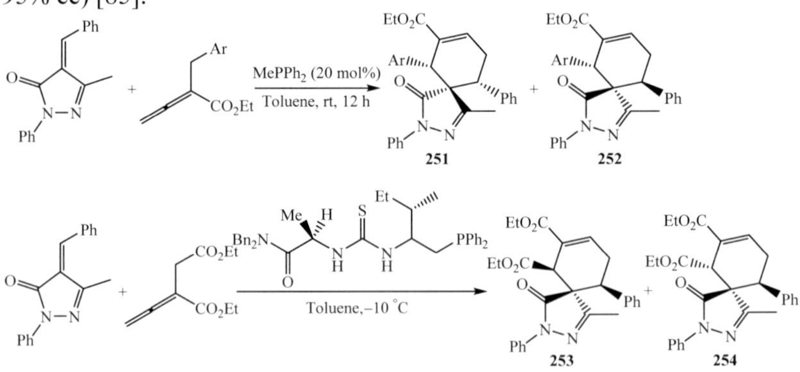

Scheme 5-86. Synthesis of chiral spiropyrazolones **251-254**.

Equally, chiral Brønsted acids were adopted in the asymmetric Diels−Alder reaction. The Shi group relied on the ability of phosphoric acid **256** to catalyze the reaction among 2-vinylindoles **255** and methyleneindolinones, which efficiently form the chiral spiro[tetrahydro carbazole-3,3′-oxindole] architecture **257** with one quaternary and three contiguous stereogenic centers in high yields (up to 99%) and excellent stereoselectivities (up to >95:5 dr, 97% ee) (Scheme 5-87) [86].

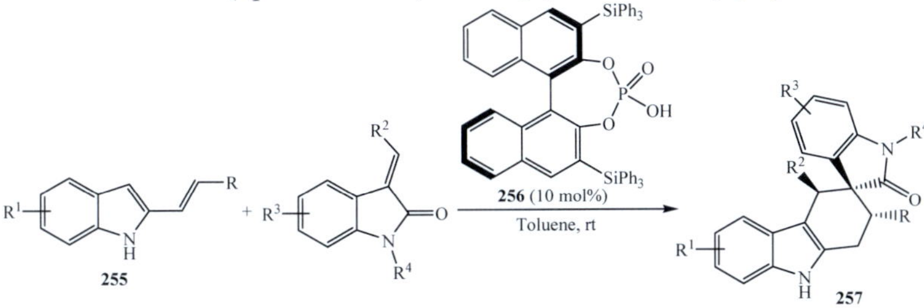

Scheme 5-87. Brønsted acid catalyzed asymmetric Diels−Alder reaction.

A highly efficient *N,N′*-dioxide (L-PiEt$_2$-Me)−Zn(II) complex catalytic system for the asymmetric Diels−Alder reaction of Brassard-type diene **258** with methyleneindolines Lin et al. developed **259** in 2015 [87]. This afforded the desired optically pure spiro[cyclohex[3]ene-1,3′-indoline]-1′-carboxylate-2, 2′-diones **260** containing three stereocenters in moderate yields with 98% to >99% ee in a stereospecific manner (Scheme 5-88).

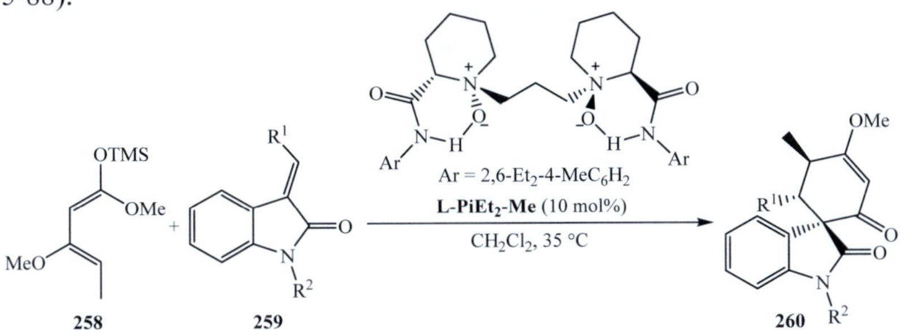

Scheme 5-88. Catalytic asymmetric Diels−Alder reaction of Brassard-type diene **258**.

To expand the synthetic applicability of the methodology, Ramachary et al. decided to utilize both *in situ*-generated 2-arylidene-1,3-indanediones as dienophiles and 2-aminobuta-1,3-dienes as the diene source [88]. They developed a three-component reaction between an arylidene-acetones **261**, 1,3-indandione, and aldehydes for the generation of nature-inspired bioactive spiro compounds **263**, catalyzed by the system 9-amino-9-

deoxyepiquinine **262** (10 mol%)/Brønsted acids (15 mol%) (Scheme 5-89). Different Brønsted acids (CH_3CO_2H, *o*-$FC_6H_4CO_2H$, *p*-$BrC_6H_4O_2$) were tested as co-catalysts, but only 2-fluorobenzoic acid (*o*-$FC_6H_4CO_2H$) gave promising results for *anti*-(+)-**263** formation in terms of yield, ee, and dr in toluene.

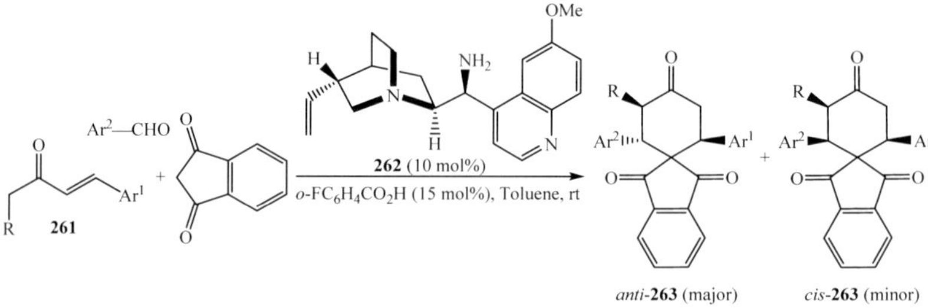

Scheme 5-89. Asymmetric synthesis of nature-inspired bioactive spiro compounds **263**.

In 2017, Llona-Minguez et al. reported a two-step diastereoselective methodology building on a multicomponent aza-Diels–Alder reaction to obtain diverse spiro-heterocyclic systems from commercial and pharmaceutically relevant starting materials under non-anhydrous mild conditions [89]. They have achieved the novel spiro-heterocyclic frameworks **264a,b** at the interphase between "drug-like" molecules and natural products using previously unexplored cyclic ketones, heterocyclic amines, and cyclopentadiene in the presence of Lewis acids (Scheme 5-90).

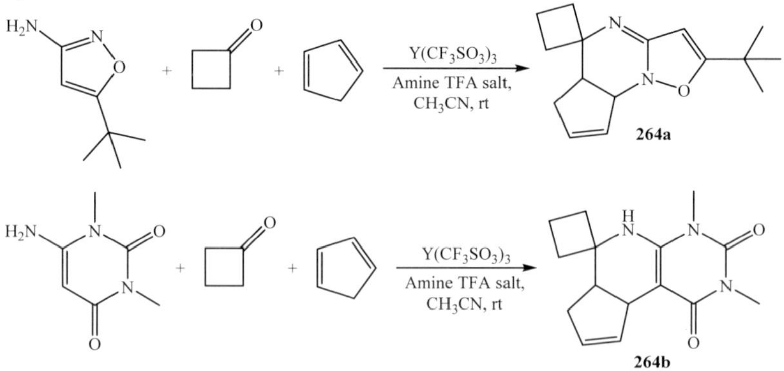

Scheme 5-90. Novel spirocyclic systems via multicomponent aza-Diels–Alder reaction.

Spirolides (PPXs, Figure 5-6) are a unique family of marine toxins and generated in 2019 by the dinoflagellates *Alexandrium ostenfeldii* and *Alexandrium peruvianum*.

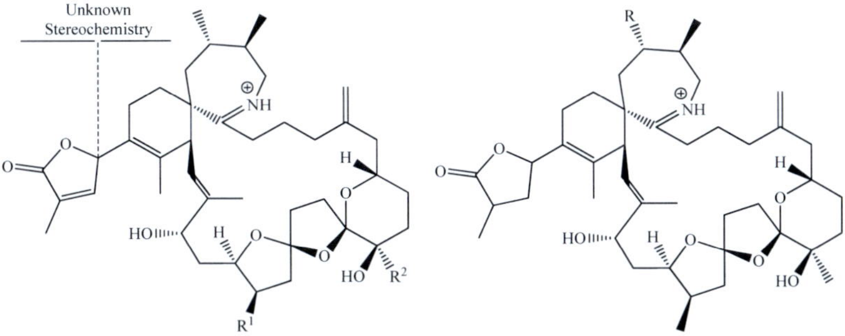

Spirolid (SPX) C: R^1 = Me; R^2 = Me
13-desMe SPX C: R^1 = H; R^2 = Me
13, 19-didesMe SPX C: R^1 = H; R^2 = H

SPX A: $\Delta^{2,3}$; R = H
SPX B: R = H
SPX D: R = Me

Fig. 5-6. The chemical structures of representative cyclic imine toxins.

Murata et al. achieved the efficient synthesis of spirocyclic imines **265**, which represent a key structural motif found in the marine toxin 13-desMe spirolide C, through a strategic sequence of reactions [90]. The synthesis involved the selective Diels-Alder reaction between a silatrane substituent **266** and a dienophile lactam **267**, followed by C-C bond formation and the facile formation of a γ-butenolide ring. This synthetic approach allows for the construction of complex spirocyclic architectures, providing access to valuable compounds with potential biological activities. The successful synthesis of spirocyclic imines **265** showcases the importance of strategic bond-forming reactions in the efficient assembly of intricate molecular structures. In the Diels-Alder reaction of lanthanides $Sc(OTf)_3$ and $Yb(OTf)_3$, their strong Lewis acidity led to the decomposition of both starting materials. The retrosynthetic analysis of spirocyclic imines **265** is displayed in Scheme 5-91.

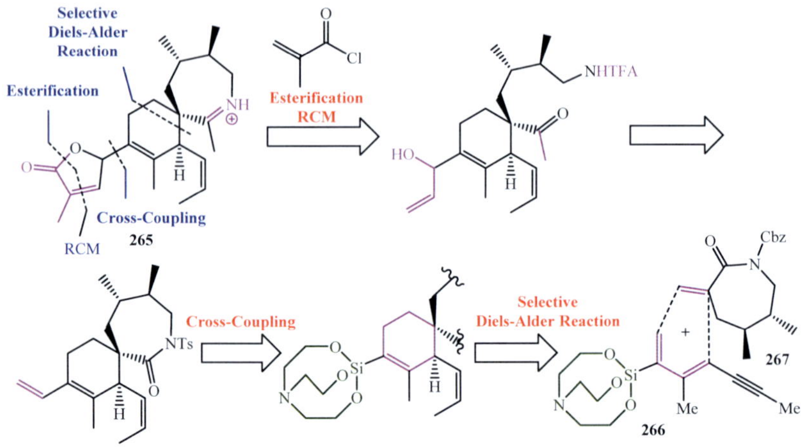

Scheme 5-91. Retrosynthetic analysis of spirocyclic imines **265**.

Furthermore, spirocyclohexane pyrazolone skeletons **271** can be prepared by a chiral NHC **270**-catalyzed [4+2] annulation of γ-chloroenals **268** and α-arylidene pyrazolinones **269** in the absence of expensive oxidants (Scheme 5-92).

A plausible mechanism for this reaction is depicted in Scheme 5-92. Firstly, a nucleophilic addition of NHC catalyst **270** to γ-chloroenal **268** and a 1,2-H migration to oxygen from carbon gives Breslow intermediate **I**, which subsequently undergoes tautomerisation and with a C−Cl bond cleavage to produce enolene **II**. Afterward, enone intermediate **III** generates through an enolene-enone rearrangement and a 1,5-H shift. Then, vinyl enolate **IV** forms by the base capture of the acidic γ-proton. Finally, the reaction finished with the cycloaddition of vinyl enolate **IV** with pyrazolone derivate **269**, generating the enantioenriched spiro-pyrazolones **271** and releasing the NHC catalyst **270** [91].

Remaining in the field of [4+2] cycloaddition reaction, Guo et al. [92] disclosed the phosphine-catalyzed cascade Michael addition/[4+2] cycloaddition reaction of tetrahydrobenzofuranone-derived allenoates **272** and 2-arylidene-1,3-indanediones, that afforded spirocyclic 1,3-indane-diones **273** in moderate to high yields with moderate to good diastereo-selectivities (Scheme 5-93).

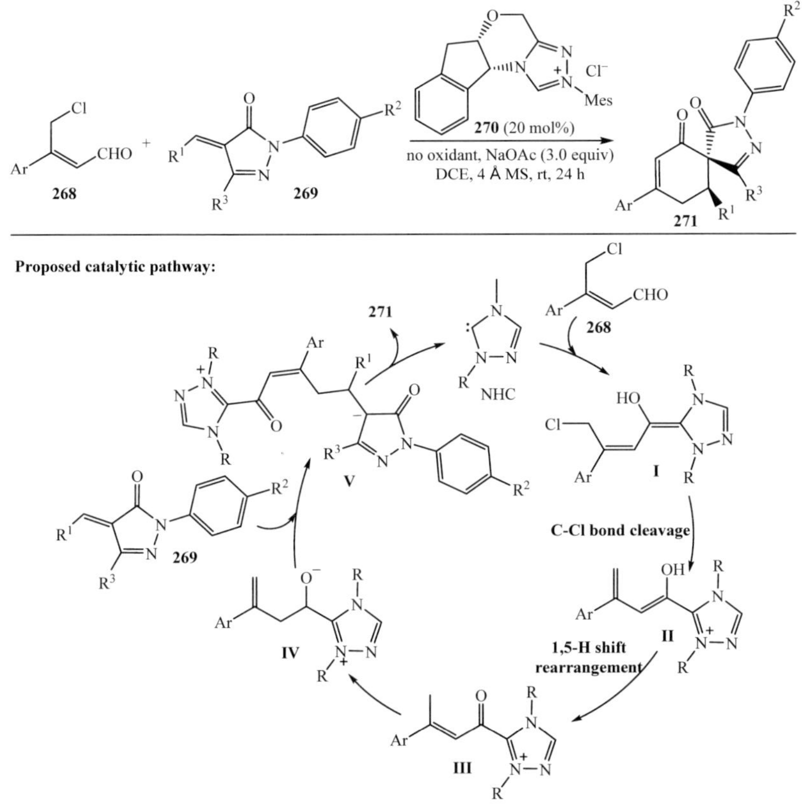

Scheme 5-92. NHC-Catalyzed asymmetric [4+2] annulation and proposed catalytic pathway.

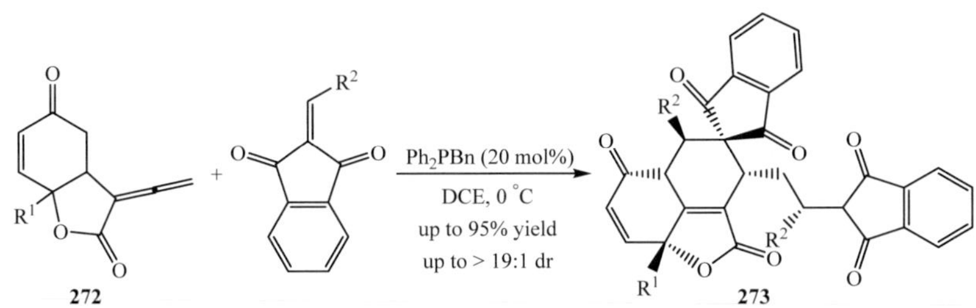

Scheme 5-93. Phosphine-catalyzed cascade Michael addition/[4+2] cycloaddition reaction of tetrahydro-benzofuranone-derived allenoates **272**.

Recently, diastereoselective synthesis of tetrahydrospiro[carbazole-1,3′-indoline]s **276** was accomplished in a highly effective one-pot reaction cascade that involved the sequential *in situ* generation of reactive dienophilic 3-alkenylindole, the Diels-Alder reaction, and the Lewis acid controlled diastereoisomerization process [93]. An $InBr_3$-catalyzed domino reaction of indoles, phenylacetylenes **274**, and various 3-methyleneoxindolines **275** provided an appropriate synthetic protocol for poly-substituted tetrahydrospiro[carbazole-1,3′-indoline]s **276** in good yields with excellent diastereoselectivity (Scheme 5-94).

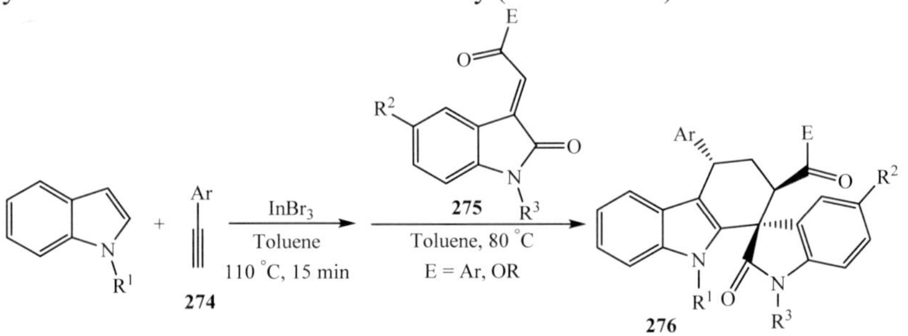

Scheme 5-94. $InBr_3$-Catalyzed the domino Diels−Alder reaction.

5.3.5. [3+3] Cycloaddition reactions

In the field of pericyclic reactions, extensive research has been conducted, uncovering numerous valuable insights. However, it is important to note that there are still many unexplored opportunities within this domain. For example, very few instances of the development of [3+3] cycloaddition reactions have been reported. Therefore, we expect a plethora of new methods to be developed to complement and improve existing methods.

The first [3+3] atom combination reaction was reported by Ghozlan et al. in 2009 [94]. In this reaction, as depicted in Scheme 5-95, the overall process involves a DBU-promoted Knoevenagel condensation of isatin derivatives **16** with active methylene **277** to afford 3-cyanomethylidene oxindoles **278**, which subsequently reacted with azaenamines **279** through a DBU-promoted Michael addition to furnish spirocyclic 2-oxindoles **280**.

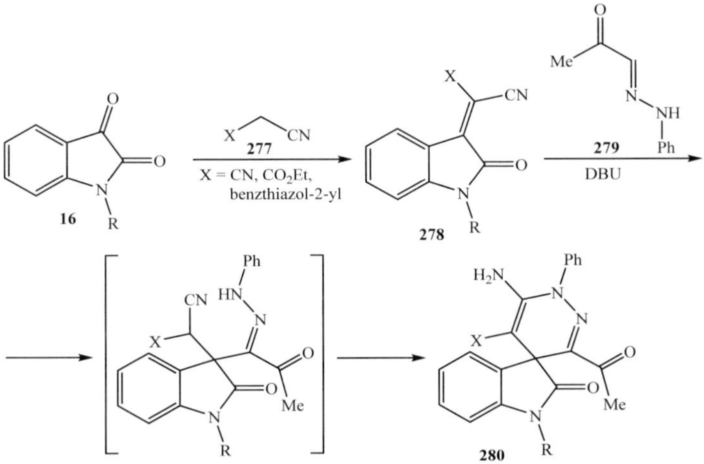

Scheme 5-95. DBU-Catalyzed spirocyclic 2-oxindoles **280**.

In 2012, Shi et al. reported the first [3+3] cycloaddition reaction between isatin ketonitrones **281** and cyclopropanes catalyzed by $Yb(NTf_2)_3$ wherein spiro[tetrahydro-1,2-oxazine]oxindoles **282** were synthesized in moderate to good yields with good regioselectivities (Scheme 5-96) [95].

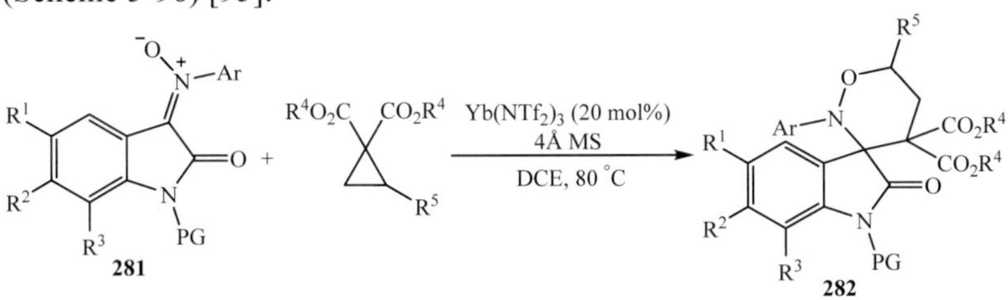

Scheme 5-96. $Yb(NTf_2)_3$-Catalyzed [3+3] cycloaddition between isatin ketonitrones **281** and cyclopropanes.

One year later, stereoselective synthesis of spirocyclohexadiene-pyrazolones **285** via organic base and/or hydrogen bonding assisted [3+3] annulation reaction was reported by the Wang group [96]. In this study, the DABCO-catalyzed [3+3] cycloaddition of α-arylidene-pyrazolinones **283** and 2-benzylidenemalononitriles **284** under mild reaction conditions lead to formation the spirocyclohexadiene-pyrazolones **285** in good yields with moderate to good diastereoselectivity (Scheme 5-97, Eq. 1). Additionally, an asymmetric version of this cycloaddition was also developed with the cinchona alkaloid derived squaramide-tertiary base catalyst **286**, which afforded optically active spirocyclohexadiene-pyrazolones **285** bearing an all-carbon quaternary chiral center in good yields with excellent enantioselectivities and moderate diastereo-selectivities (Scheme 5-97, Eq. 2).

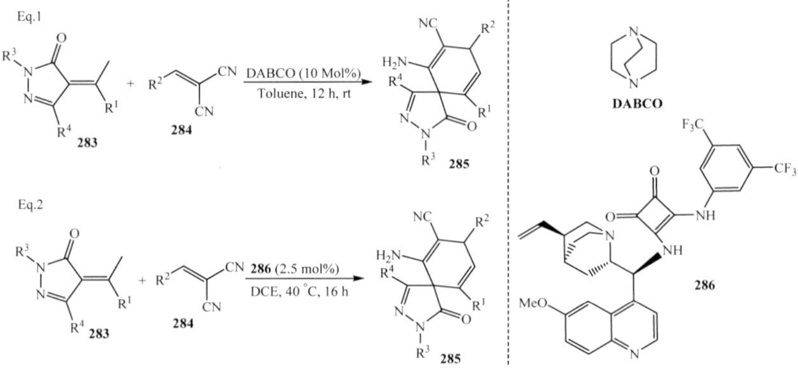

Scheme 5-97. Organic base and/or hydrogen bonding assisted [3+3] annulation reaction.

Bonne et al. utilized α,β-unsaturated acyl cyanides **287** as new bis-electrophiles for enantioselective organocatalyzed formal [3+3] spiro-annulation leading to the synthesis of enantioenriched azaspiro[4,5]-decanone ring systems **290** [97]. β-Ketoamide **288** as the bis-nucleophilic partner was reacted with α,β-unsaturated acyl cyanides **287** under hydrogen-bonding catalysis. Using the Takemoto's thiourea catalyst **289**, the domino-Michael-intramolecular nucleophilic substitution was efficient, and the desired product **290** was generated in 90% yield, albeit in low stereoselectivity (Scheme 5-98).

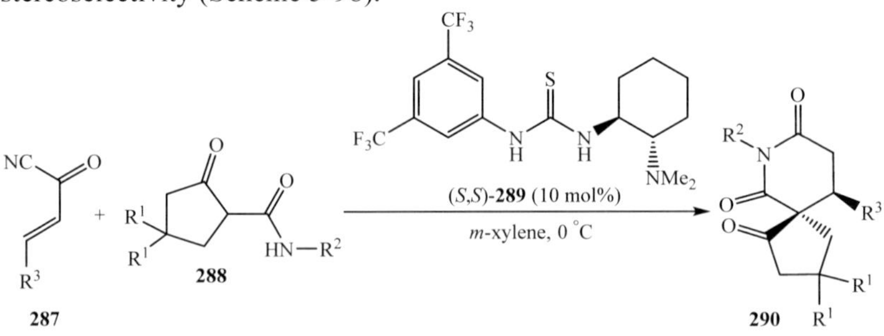

Scheme 5-98. Synthesis of enantioenriched azaspiro[4,5]decanone ring systems **290**.

In 2014, Shi et al. presented an elegant description of the first enantioselective organocatalytic [3+3] cycloaddition involving isatin-derived 3-indolyl methanols 291 [98]. Thus, CPA **292** with the bulky 9-phenanthrenyl group at the 3,3′-positions of the BINOL backbone efficiently activated the process between an aldehyde, an amino ester, and isatin-derived 3-indolylmethanols **291**, affording structurally diverse spiro[indoline-3,4′-pyridoindoles] **293** with one all-carbon quaternary stereogenic center in high yields and excellent enantioselectivities (up to 93% yield, > 99% ee) (Scheme 5-99).

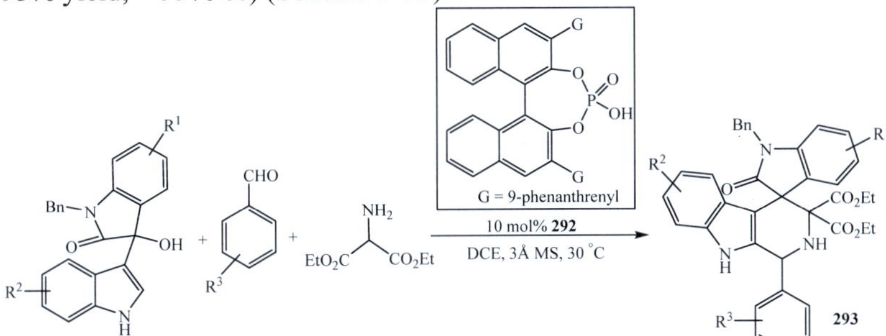

291

Scheme 5-99. The first enantioselective organocatalytic [3+3] cycloaddition of isatin-derived 3-indolylmethanols **291**.

Zhou et al. demonstrated the synthesis of spiro[thiopyranoindole-benzoisothiazole] scaffolds **297** through an enantioselective formal thio [3+3] spiroannulation reaction of indoline-2-thiones **294** and 1-azadienes **295** [99]. The reactions were carried out at -10 °C using 10 mol% of the quinine-derived bifunctional tertiary amine-thiourea catalyst **296** (Scheme 5-100). They provided the desired products **297** with spiro quaternary C-N/C-S stereogenic centers in high yields with good to excellent diastereo- and enantio-selectivities.

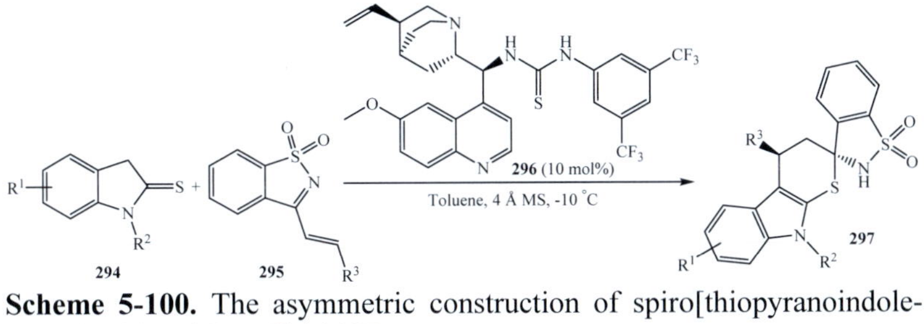

294 295

Scheme 5-100. The asymmetric construction of spiro[thiopyranoindole-benzoisothiazole] scaffold **297**.

Also, Chen et al. reported another example of an asymmetric formal [3+3] cycloaddition involving 1-azadienes **295**. This innovative transformation proceeded via a domino Michael addition–Mannich reaction, enabling the simultaneous formation of multiple bonds and stereocenters in a single step [100]. This report utilizes a variety of electron-deficient cyclic 1-azadienes **295** that undergo reaction with diversely structured aliphatic ketones by cascade enamine–enamine catalysis of a cinchona-based primary amine **298** as in the case of the example illustrated in Scheme 5-101. A variety of spirocyclic structures **299** were efficiently constructed in excellent diastereo- and enantioselectivity.

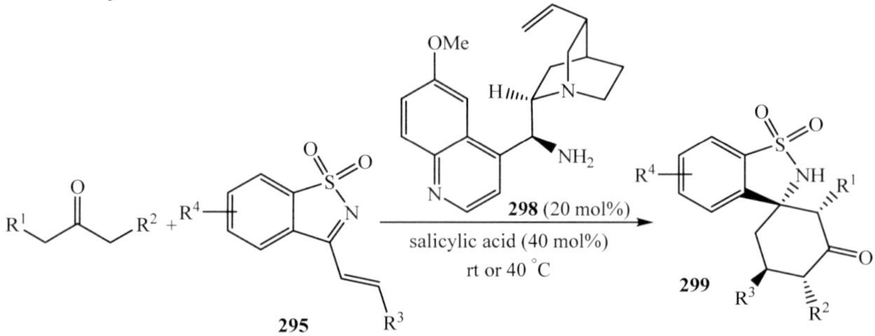

Scheme 5-101. Formal [3+3] cycloaddition reactions with structured aliphatic ketones and 1-azadiene **295**.

Yang et al. developed a chiral NHC-catalyzed [3+3] cycloaddition reaction of imines and oxindole-derived enals **300** for the production of spirocyclic oxindoles **303** using chiral *N*-mesityl-substituted triazolium salt **301** (20 mol%), oxidant **302** (1 equiv) and base 4-dimethylaminopyridine (DMAP) (50 mol%) (Scheme 5-102) [101].

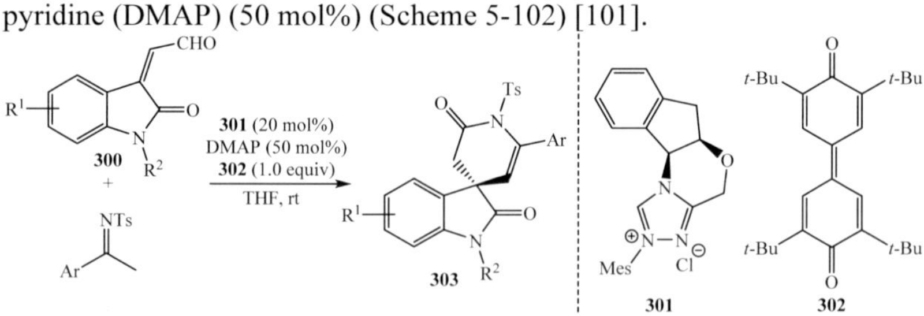

Scheme 5-102. *N*-Mesityl-substituted triazolium salt **301**-catalyzed [3+3] cycloaddition reaction.

Biju et al. demonstrated the enantioselective synthesis of pyrazolone-fused spirocyclohexadienones **305** by the reaction of α,β-unsaturated aldehydes with α-arylidene pyrazolinones **269** in the presence of carbene generated from the chiral triazolium salt **304** under oxidative conditions by using the quinone **302** (Scheme 5-103) [102]. This atom-economic [3+3] annulation reaction is involved a vinylogous Michael addition/spiroannulation/dehydrogenation cascade that allows the efficient preparation of spirocyclic compounds **305** with an all-carbon quaternary stereocenter in moderate to good yields and excellent ee values.

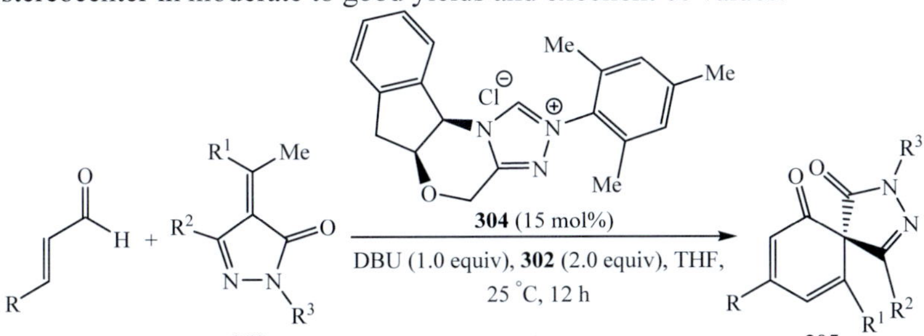

269 **305**

Scheme 5-103. Enantioselective synthesis of pyrazolone-fused spiro cyclohexadienones **305**.

Based on the proposed mechanism by the authors, the key to the success of the reaction is the cooperative NHC-catalyzed production of chiral α,β-unsaturated acyl azoliums **I** from enals, and base-mediated tandem production of dienolate/enolate intermediates **II** from pyrazolinones **269**, as shown in Scheme 5-104.

Lu et al. developed a cooperative NHC/Lewis acid-mediated [3+3] annulation of various isatin-derived 2-bromoenals **306** with 1,3-dicarbonyl compounds, affording the corresponding functionalized spiro-oxindole δ-lactones **308** in moderate to good yields [103]. The reaction was performed in 1,4-dioxane using Cs_2CO_3 (1.5 equiv) as a base in the presence of 15 mol% of pre-catalyst **307** and 1.1 equiv of LiCl (Scheme 5-105).

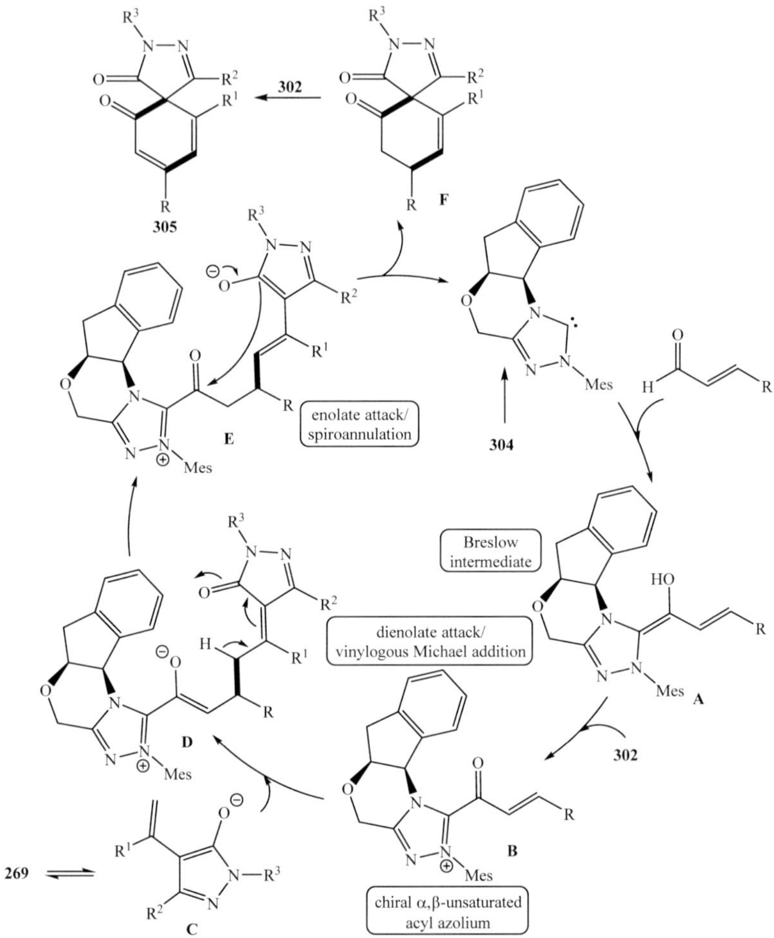

Scheme 5-104. Proposed mechanism of the reaction.

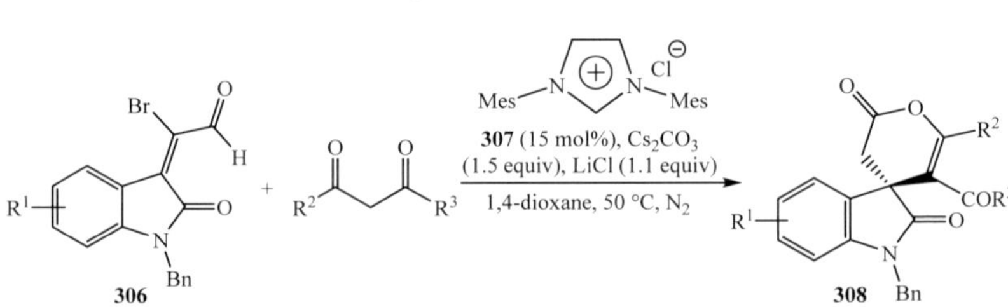

Scheme 5-105. Cooperative NHC/Lewis acid-mediated [3+3] annulation.

As an extension of this work, the same group also presented an *in situ* activation of isatin-derived α,β-unsaturated acids **309** for the generation of isatin-derived α,β-unsaturated acyl azoliums **II**, that undergoes a formal [3+3] annulation with 1,3-dicarbonyl compounds under the catalysis of **310** to furnish functionalized 3,4′-spiro-oxindole δ-lactones **311** (Scheme 5-106) [104]. The reactions were carried out at room temperature within 24 h using HATU (1.5 equiv) as a peptide coupling reagent (PCR) and Cs_2CO_3 (2.5 equiv) as a base in the presence of 20 mol% of pre-catalyst **310**.

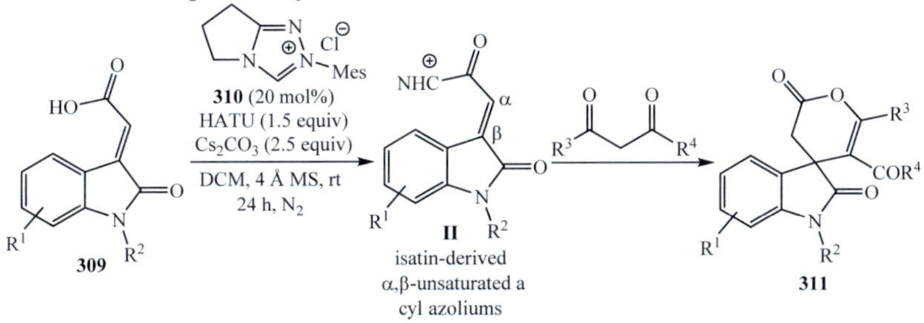

Scheme 5-106. NHC-Catalyzed [3+3] annulation of isatin-derived α,β-unsaturated acids **309**.

In 2017, Li et al. successfully disclosed a highly enantioselective [3+3] cycloaddition of α,β-unsaturated aldehydes with alkylidene pyrazolinones **312** via secondary amine **235** catalysis, in which both the β- and *ipso*-positions of the enal substrates as dielectrophilic sites are activated through an iminium ion pathway (Scheme 5-107). In all cases, spiropyrazolone skeletons **313** were obtained in outstanding yields with good diastereo- and excellent enantioselectivity [105].

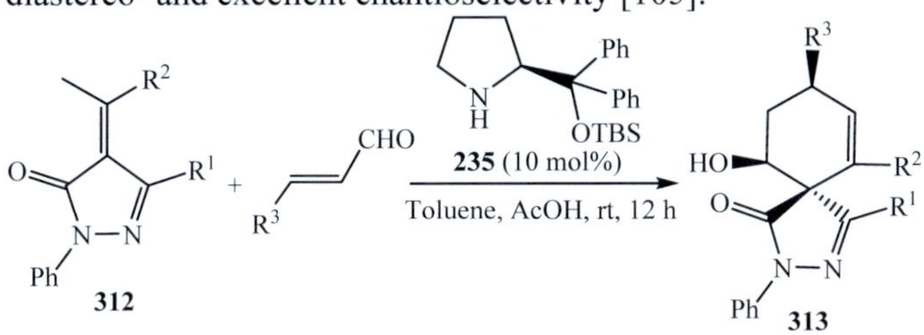

Scheme 5-107. The asymmetric construction of spiropyrazolone skeletons **313**.

Mei et al. also noted a CPA **315**-catalyzed asymmetric [3+3] cycloaddition of C3-substituted 2-indolylmethanols **314** with isatin-derived azomethine ylides, which provided chiral spiro-oxindoles **316** in considerable yields, moderate to good enantioselectivities and excellent diastereoselectivities (up to 85% yield, 96: 4 er, all >95: 5 dr) [106]. In this process, CPA played a crucial role in controlling diastereo- and enantioselectivities and in the generation of cyclic azomethine ylides from isatin and amino esters (Scheme 5-108).

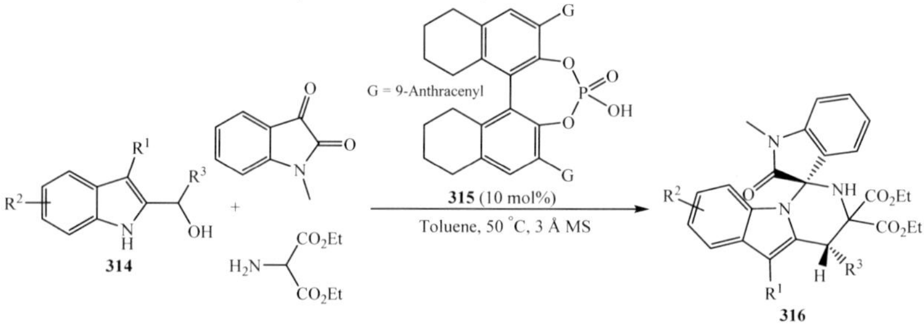

Scheme 5-108. CPA **315**-Catalyzed asymmetric [3+3] cycloaddition of C3-substituted 2-indolylmethanols **314**.

The synthesis of spiro-oxindoles **319** and dihydropyridinones **320a,b** was developed by Qi et al. [107] via the new strategy. In an NHC (**318**)-catalyzed [3+3] annulation reaction using enals or isatin-derived enals and 2-aminoacrylates **317** as nucleophiles under oxidative conditions in the presence of quinone **302**, the desired spiro-oxindoles **319** were obtained in good to excellent yields (Scheme 5-109). Modifying the standard base switched the carbon-carbon double bond formation from 5,6-positions to positions to produce 5,6-dihydropyridinones **320a** and 3,4-dihydropyridinones **320b**, respectively.

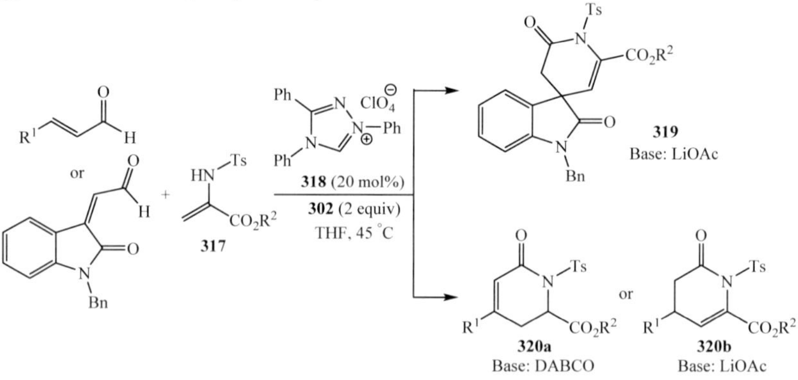

Scheme 5-109. Synthesis of spiro-oxindoles **319** and dihydropyridinones **320 (a and b)**.

In 2018, the Xu group achieved the asymmetric synthesis of functionalized spiro-piperidinones **324**, which featured two vicinal stereogenic centers, through an NHC-catalyzed asymmetric [3+3] spiroannulation reaction. This groundbreaking methodology involved the reaction of 2-bromoenal **321** with β-ketothioamide **322**, resulting in the formation of the spirocyclic scaffold (Scheme 5-110) [108].

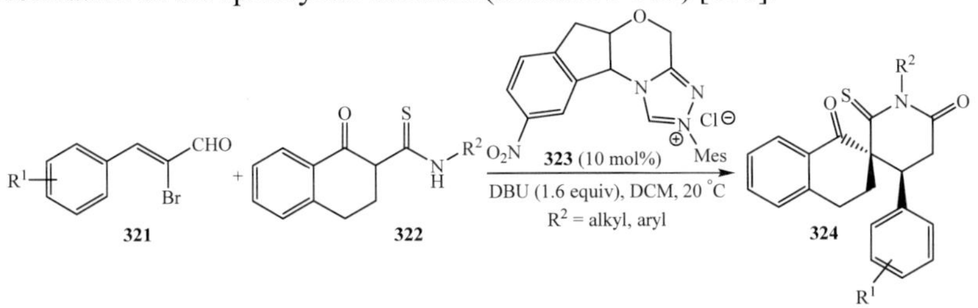

Scheme 5-110. Asymmetric synthesis of functionalized spiro-piperidinones **324**.

A plausible mechanistic cycle was suggested in Scheme 5-111. First, adding the NHC **323** to 2-bromoenal **321** gives Breslow intermediate **I**, which undergoes tautomerization and debromination to generate the key intermediate **II**. The resulting intermediate **II** undergoes Michael addition away from the hindered face generated by the aminoindanol phenyl framework to form intermediate **III**. After tautomerization, the intermediate **IV** undergoes H-migration and lactamization to provide the desired product **324** with the catalyst regeneration.

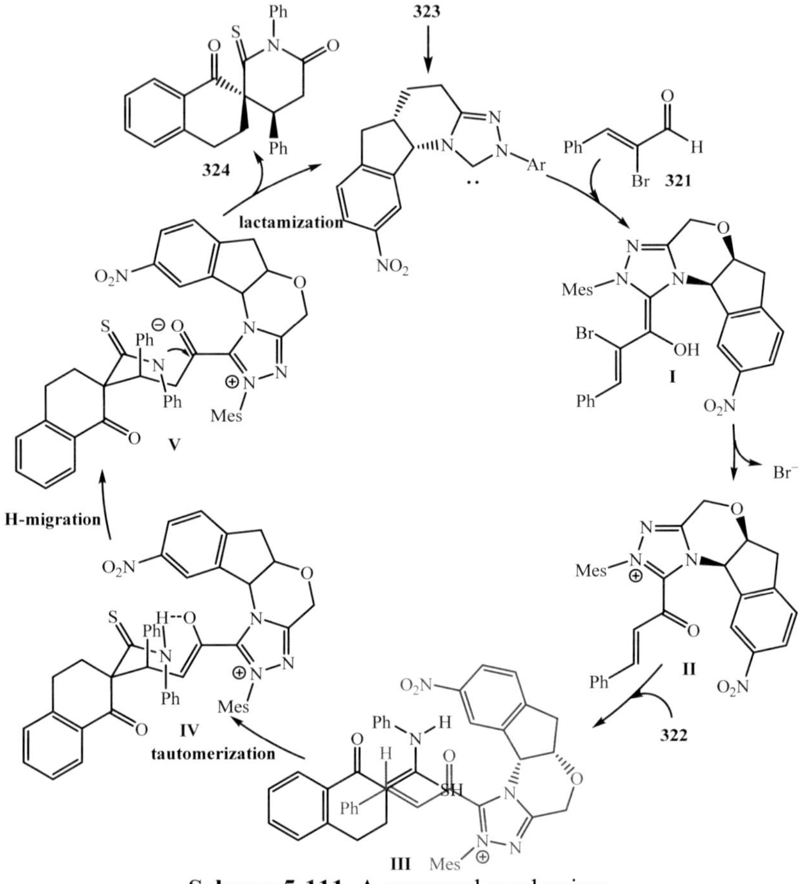

Scheme 5-111. A proposed mechanism.

Later on, a highly diastereoselective one-pot sequential [3+3] dipolar cycloaddition reaction of aldehyde or ketone, *N*-alkyl hydroxylamine **325**, and spirocyclopropyl oxindole **326** catalyzed by $Sc(OTf)_3$ was the key step on the synthesis of spirocyclic oxindole-tetrahydro-1,2-oxazines **327** (Scheme 5-112). The biological evaluation of selected oxindole-based spirocyclic tetrahydro-1,2-oxazines **327** disclosed that they exerted cytotoxic effects on human prostate cancer cells with the capacity to inhibit NF$_{\kappa}$B signaling in prostate cancer cells [109].

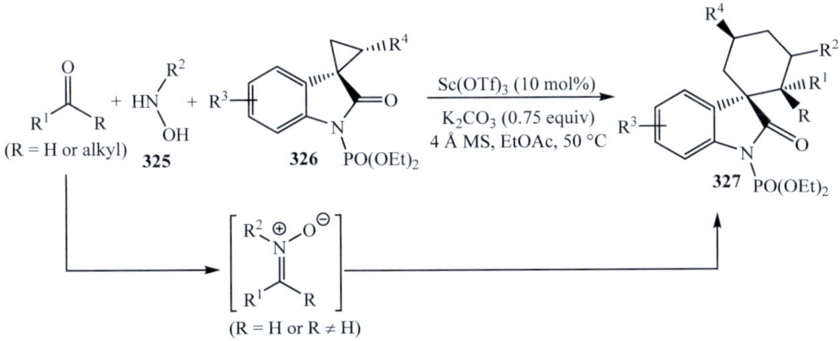

Scheme 5-112. One-pot sequential [3+3] dipolar cycloaddition.

In 2019, Wang et al. successfully developed a novel metal-free DBU-catalyzed [3+3] cycloaddition reaction through the C3 umpolung strategy of oxindole using the aqueous medium, which started from isatin derived cyclic imine 1,3-dipoles **328** with α,β-unsaturated aldehydes employing DBU as an organocatalyst furnished the desired spiro heterocyclic oxindoles **329** with azaquaternary centers in good yields and diastereo-selectivities (Scheme 5-113) [110].

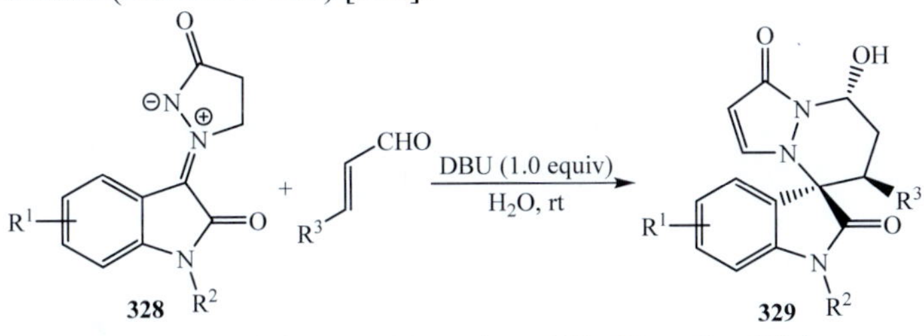

Scheme 5-113. Metal-free DBU-catalyzed [3+3] cycloaddition reaction.

Enders et al. suggested a convenient enantioselective route to new types of pentacyclic spiro-oxindoles through [3+3] annulation reactions of isatin-derived enals **330** and cyclic *N*-sulfonyl ketimines **331**, using NHC precatalyst **332**, quinone **302** as an oxidant and NaOAc as a base (Scheme 5-114). The protocol gives pentacyclic spiro-oxindoles **333** bearing a quaternary spiro stereocenter in good yields and good to high enantiomeric ratios (er) [111].

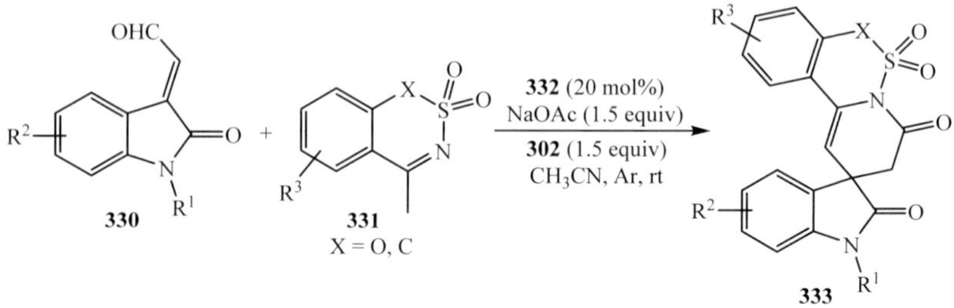

Scheme 5-114. NHC-Catalyzed asymmetric synthesis of pentacyclic spiro-oxindoles **333**.

Simultaneously, a series of biselectrophilic β,γ-unsaturated α-keto-esters derived from rhodanine **334** were used by Xu et al. to afford spirorhodanine-pyrans **336** from pyrazolones [112]. Under the catalysis of chiral squaramides **335**, the reaction led to the synthesis of chiral 2′-thioxo-5,6-dihydrospiro[pyrano[2,3-*c*]pyrazole-4,5′-thiazolidin]-4′-ones with excellent stereoselectivities **336** (Scheme 5-115).

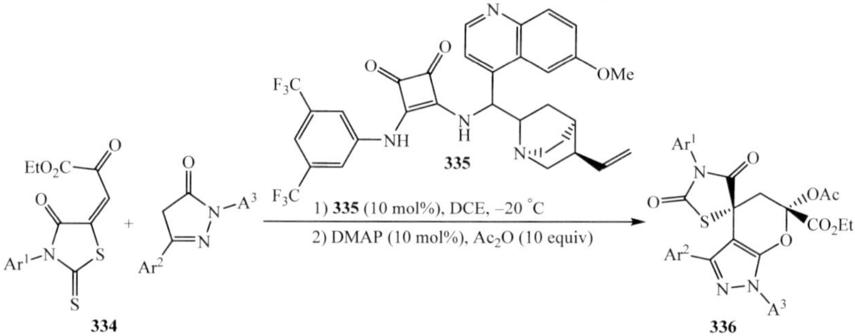

Scheme 5-115. Enantioselective synthesis of spirorhodanine-pyrans **336**.

One year later, the synthesis of highly substituted spiro *N*-heterocyclic oxindoles **338** was reported by Liu et al. [113]. The utilization of the chiral secondary amine as a catalyst **235** in the enantioselective C-*N*-*N* [3+3] annulation of isatin *N*,*N*'-cyclic azomethine imines **337** with α,β-unsaturated aldehydes provides the desired products **338** in good yields (up to 91%) and good to excellent enantioselectivities (up to >99% ee) (Scheme 5-116).

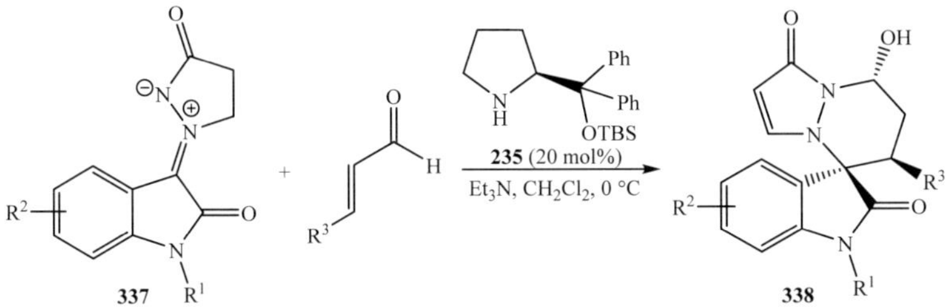

Scheme 5-116. Enantioselective C-*N*-*N* [3+3] annulation.

The formation of these oxindoles can be rationalized following the mechanism described in Scheme 5-117.

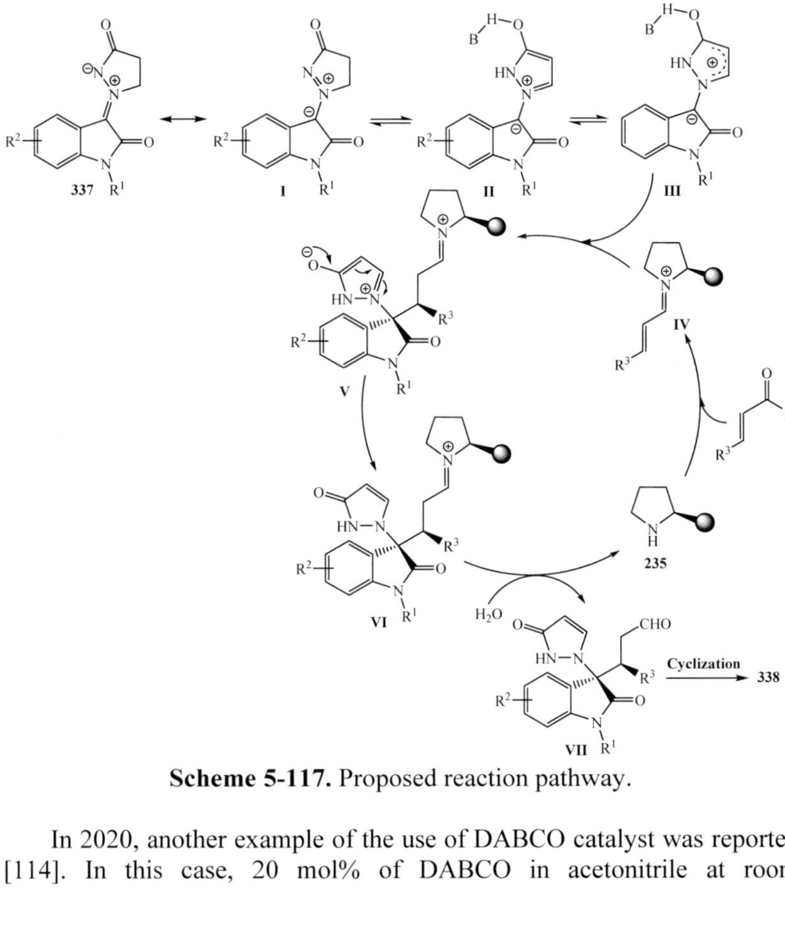

Scheme 5-117. Proposed reaction pathway.

In 2020, another example of the use of DABCO catalyst was reported [114]. In this case, 20 mol% of DABCO in acetonitrile at room temperature allowed the synthesis of the corresponding spirocyclohexene-oxindole **339** with excellent diastereoselectivity. The [3+3] annulation is found to occur through a vinylogous Michael-aldol cascade reaction (Scheme 5-118). Additionally, a bioactivity study of the compounds on mammalian sarcoma cells has reflected a cell growth inhibitory/anti-cancer property.

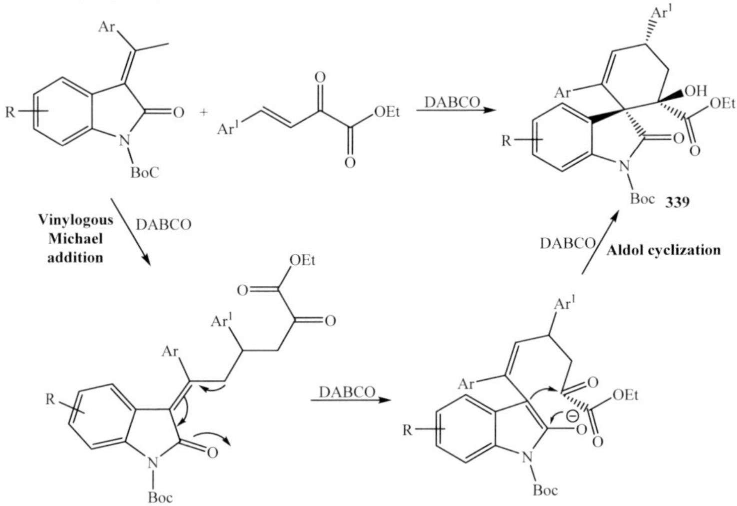

Scheme 5-118. Synthesis of spirocyclohexene-oxindole **339**.

Hu et al. presented the first example of asymmetric synthesis of optically active spirocyclic frameworks based on a catalytic propargylic cycloaddition strategy [115]. With this new method, various optically active spiro-oxindoles **343** bearing a spiro all-carbon quaternary stereocenter are readily accessed from 3-ethynyl-2-oxoindolin-3-yl acetates **340** and 1*H*-pyrazol-5(4*H*)-ones **341** in good to high yields and with up to 97% ee. The catalyst is a combination of $Cu(OTf)_2$ and chiral tridentate ketimine *P*,*N*,*N*-ligand **342** (Scheme 5-119).

The authors proposed the mechanism shown in Scheme 5-120. The reaction starts with the generation of a cationic alkyne−copper π-complex with **340**, which is deprotonated to furnish a copper acetylide complex **A**. Due to electronic and conjugated effects, the subsequent elimination of an acetyl group from **A** preferentially gives copper-allenylidene complex **C** instead of its propargylic cation tautomer **B**. Nucleophilic attack of the enolate of **341**′ at the C_γ-atom of **C** generates the corresponding Cu−acetylide complex **D**, which then undergoes 6-*endo*-dig cyclization to form Cu-alkenyl complex **E**. Finally, protonolysis of the Cu-alkenyl bond releases the spiro-oxindole product **343** and regenerates the copper catalyst.

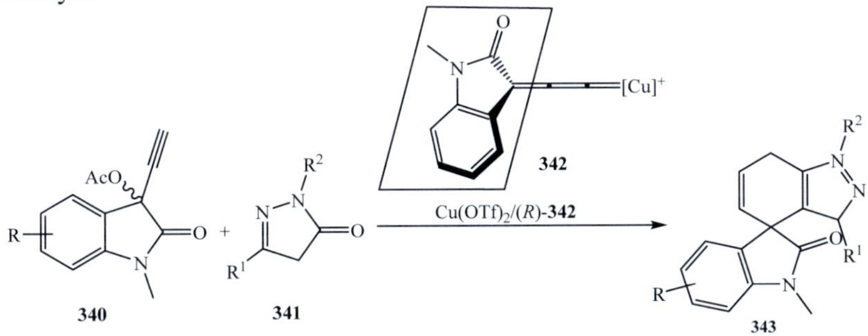

Scheme 5-119. Enantioselective Cu-catalyzed [3+3] cycloaddition.

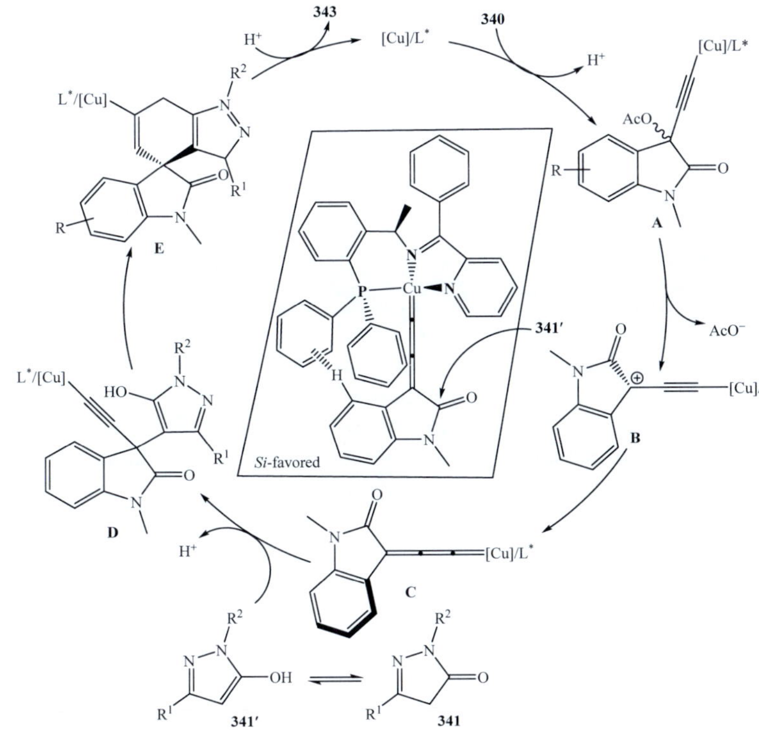

Scheme 5-120. Proposed Reaction Mechanism.

5.3.6. Miscellaneous reactions

In 2012, Mo et al. presented a Gimeracil **345**-catalyzed formal [4+1] annulation for synthesizing spirofluorenylpyrrolines **346** in good yields with high diastereoselectivity from easily available *N*-vinyl fluorenone nitrones **344** and allenoates (Scheme 5-121). In this annulation, allene serves as a C1 synthon. The catalyst might significantly promote enolation and control cyclization intermediates [116].

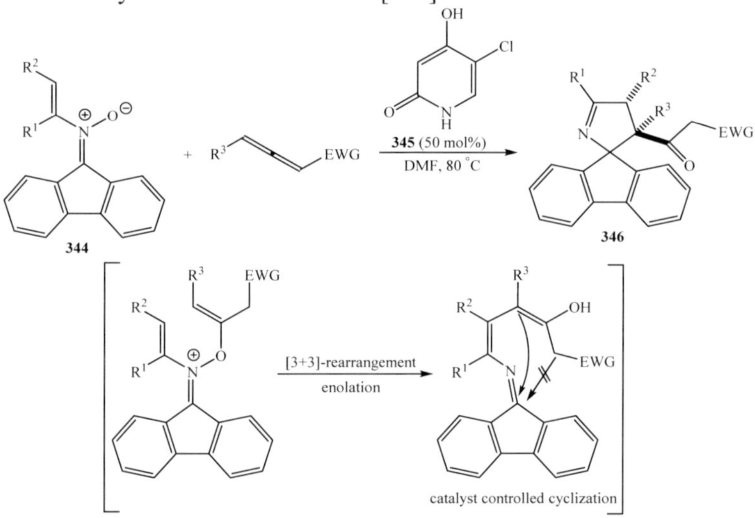

Scheme 5-121. Catalyst-controlled formal [4+1] annulation.

The Wang group also used the novel C1 synthons of 4-halo-pyrazolones **348** to synthesize enantiomerically enriched spiro-benzofuran pyrazolones **350** [117]. The synthesis of these compounds is achieved through chiral amine squaramide **349** catalyzed [4+1] annulation of *in situ* generated *ortho*-quinomethanes (*o*-QMs) **347** with 4-halo-pyrazolones **348** (Scheme 5-122).

Based on the experimental results, the authors suggested a plausible reaction mode for this asymmetric [4+1] cycloaddition (Scheme 5-123).

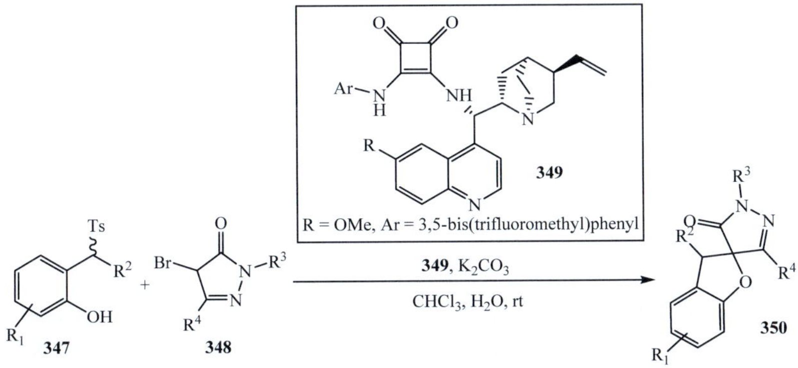

Scheme 5-122. Synthesis of enantiomerically enriched spiro-benzofuran pyrazolones **350**.

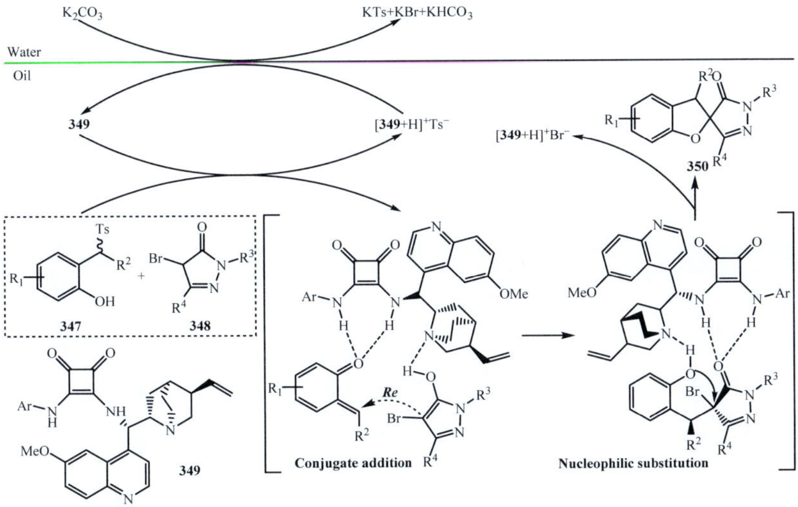

Scheme 5-123. Proposed mechanism of the [4+1] annulation.

In the organic phase, the basic nitrogen moiety of the chiral base **349** produced the benzylsulfones, which undergo the elimination of TsH (*p*-toluenesulfinic acid) and form the transient *o*-QMs **347** smoothly. Subsequently, the domino conjugate addition/nucleophilic substitution sequence was promoted by the dual-activating nature of the bifunctional catalyst, which synchronously activated the *o*-QMs **349** and 4-halo-pyrazolones **348** by forming multiple hydrogen bonds. In addition, the *re*-face approach of the 4-halo-pyrazolones **348** results in the formation of the major stereoisomer. In the aqueous phase, the neutralization reactions of K_2CO_3 with the ammonium salts of [**349**+H]$^+$Ts$^-$ and [**349**+H]$^+$Br$^-$ caused the regeneration of the catalyst **349**.

Another type of [4+1] annulation was reported by Lu et al. in 2014, in which α-substituted allenoate **351** was utilized as a C4 synthon under phosphine amide **352** catalysis, affording spiropyrazolones **353** [118]. Allenoate-derived MBH acetates **351** are reacted with pyrazolones to yield the corresponding spiro compounds **353** in good chemical yields and high enantioselectivities, as shown in Scheme 5-124. Scheme 5-124 displays the proposed mechanism for this reaction. The reaction is initiated by the nucleophilic addition of the phosphorus atom **352** over an allenoate **351** to form intermediate **A**. Subsequent elimination of an acetate group leads to **B**, which is most likely attacked at the γ-carbon position by the enolate derived from pyrazolone to generate phosphonium ylide **C**. The resulting ylide **C** undergoes proton transfer producing the intermediate **D**. This is followed by an intramolecular Michael addition and elimination of the phosphine catalyst **352** to provide the desired cycloadducts **353**.

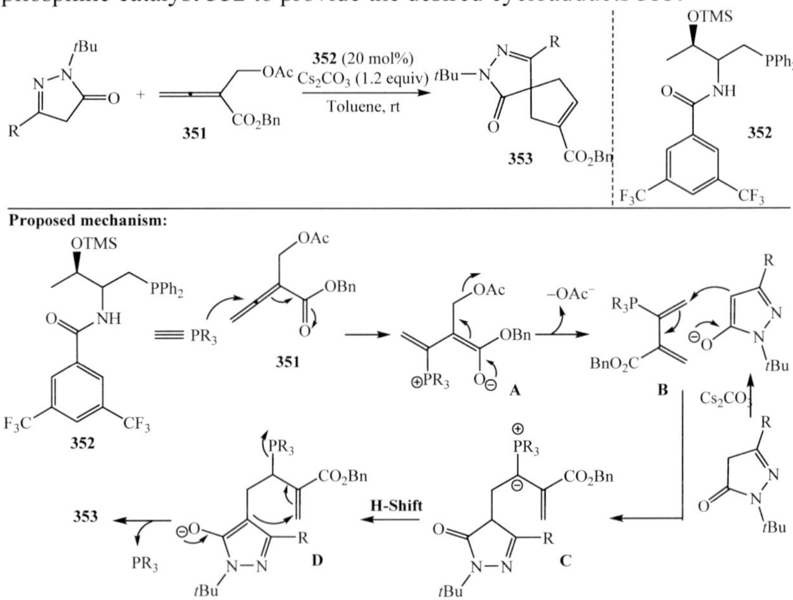

Scheme 5-124. Phosphine amid-catalyzed [4+1] annulation and proposed reaction mechanism.

Shi et al. reported in the same year an asymmetric [4+1] annulation of activated α,β-unsaturated ketones with MBH carbonates **354** catalyzed by phosphine [119]. In this case, 20 mol% of bifunctional chiral phosphines derived from an axially chiral binaphthyl scaffold **355** was used as a catalyst furnishing the corresponding spiro-oxindoles **356** with two adjacent quaternary stereocenters in good yields with high enantio-selectivities and moderate diastereoselectivities (Scheme 5-125).

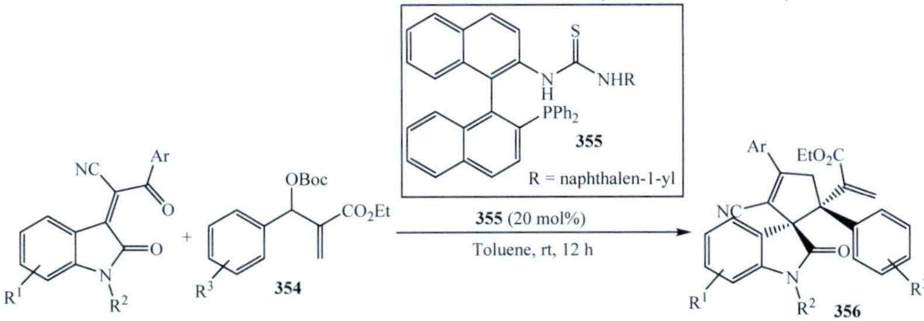

Scheme 5-125. Synthesis of spiro-oxindoles **356** containing two adjacent quaternary stereocenters.

In 2017, Zhou et al. exhibited the first realization of [4+1] annulation mode involving *N*-acyldiazenes **357** to construct five-membered heterocycles [120]. [Tris(dimethylamino)phosphine]-mediated deoxy-genative [4+1] cycloaddition reaction of *N*-acyldiazenes **357** with α-di-carbonyl compounds such as isatins, α-keto esters, and α-diketones afforded the corresponding spiro-oxindole-1,3,4-oxadiazoles **358** in good to excellent yields (Scheme 5-126).

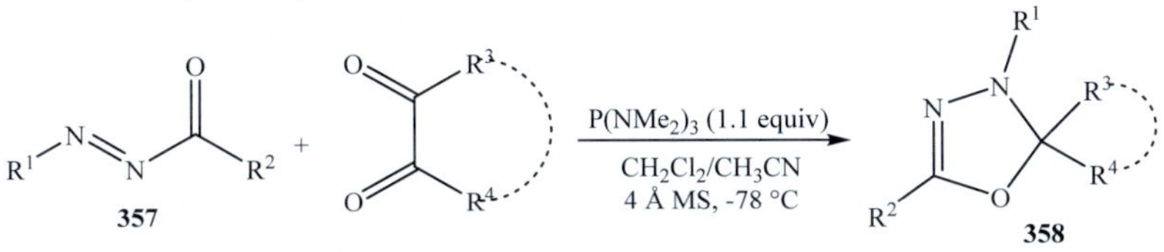

Scheme 5-126. $P(NMe_2)_3$-Mediated deoxygenative [4+1] cycloaddition reaction of *N*-acyldiazenes **357**.

Likewise, an asymmetric [4+1] annulation of 3-diazooxindoles/4-diazooxisoquinolines **359a,b** with *p*-QMs, catalyzed via a CPA **360** has been reported by Cui et al. in 2020 (Scheme 5-127) [121].

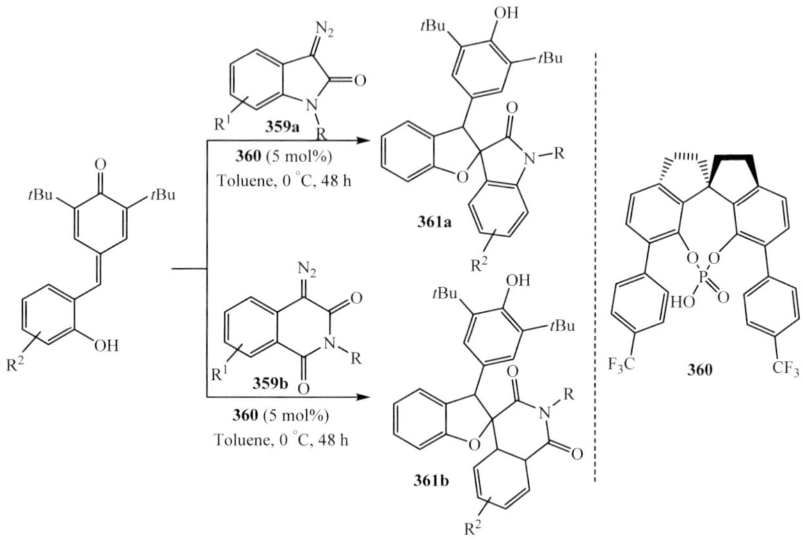

Scheme 5-127. CPA-Catalyzed [4+1] annulation.

A plausible reaction mechanism is illustrated in Scheme 5-128. In this process, the *ortho*-hydroxyphenyl-substituted *p*-QMs are isomerized to the *o*-QMs through a proton transfer step. Meanwhile, the heterocyclic diazos **359a,b** tend to exist in an ionic state. According to these key points, a double activation mode for controlling the reaction stereochemistry is suggested, where the *re*-face of activated *o*-QMs enabled by the chiral hydrogen-bond induction from the catalyst is initially attacked by the hindered heterocyclic diazos **359a,b** from its *si*-face in a conjugated 1,4-addition. Finally, the intramolecular oxygen attack on the C3/C4 position of the heterocyclic diazos **359a,b** with a release of nitrogen completed the cyclization process, giving the desired enantioenriched spiro-heterocyclic dihydrobenzofurans **361a,b**.

An efficient strategy for the diversity-oriented synthesis of spiropentadiene pyrazolones **364** and 1*H*-oxepino[2,3-*c*]pyrazoles **365** was disclosed by Lin et al. [122]. The reaction of α,β,γ,δ-unsaturated pyrazolones **362** and acyl chloride catalyzed by PBu_3 in the presence of Et_3N furnished the spiropentadiene pyrazolones in high yields by an unprecedented phospha-1,6-addition/*O*-acylation/δ-C-acylation/cyclization /Wittig reaction sequence (Scheme 5-129). The mechanistic studies disclosed that forming the betaines **363** is the key step to generating the products through an intramolecular Wittig reaction or an unusual δ-*C*-acylation/cyclization/Wittig reaction.

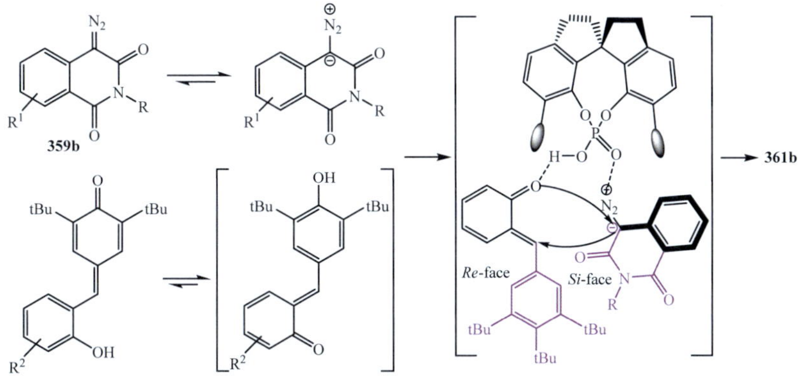

Scheme 5-128. Proposed pathway.

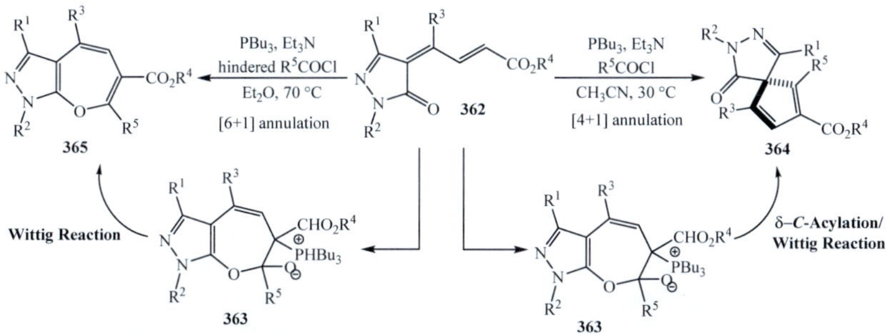

Scheme 5-129. Diversity-oriented synthesis of **364** and **365**.

Recently, Tang et al. [123] employed allenic ketones **366** dielectrophilic C4 synthons and oxindoles as a nucleophilic C1 reaction partners to construct the 3-spirocyclopentene-2-oxindoles **368** with quaternary stereogenic centers via a phosphine **367**-catalyzed enantio-selective [4+1] annulation (Scheme 5-130).

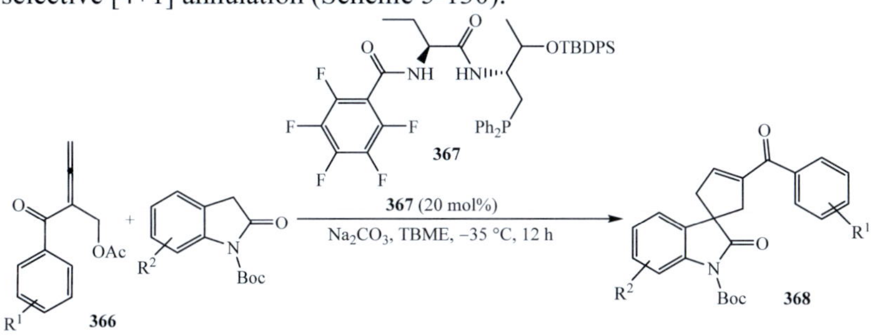

Scheme 5-130. Synthesis of 3-spirocyclopentene-2-oxindoles **368**.

Enantioselective construction of multi-functionalized spiro-cyclopentane oxindoles **371** bearings α,α-di-substituted α-amino-β-keto esters through [4+1] annulation of oxindoles and α-imine-β-oxo-γ,δ-unsaturated esters **369** in the presence of 1 mol% of a bifunctional squaramide catalyst **370** reported by the Sun group (Scheme 5-131) [124].

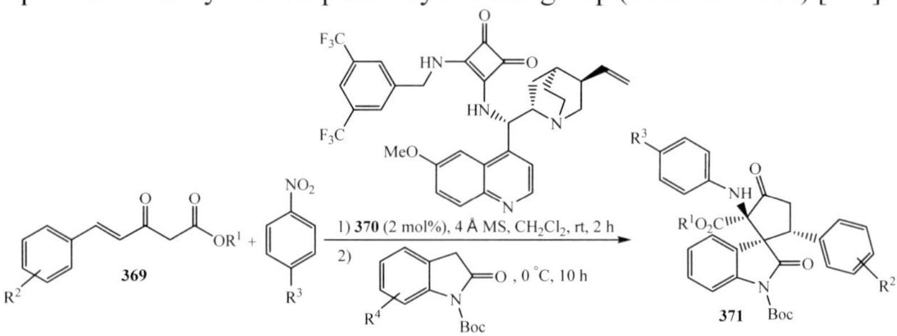

Scheme 5-131. Enantioselective construction of multi-functionalized spirocyclopentane oxindoles **371**.

Xu et al. successfully developed a tandem organocatalytic strategy for synthesizing spirocycle-pentanone pyrazolones **374** bearing three contiguous stereocenters and two quaternary carbons with good stereoselectivities under mild conditions from simple 4-unsubstituted pyrazolones **372**, as represented in Scheme 5-132. Starting from α,β-unsaturated keto esters, nitrosobenzenes, and pyrazolones using 20 mol% bifunctional catalyst **373**, the synthesis of the desired products **374** was achieved in good yields with high enantio-selectivities. Using pyrazolones as efficient C1 synthons and involving a polarity reversal process not only overcome the energy barrier of the dearomatization process but also avoided nucleophilic addition of the hydroxy group in the enol form tautomer [125].

In 2012, Ji et al. [126] applied cascade [5+1] Michael/Michael addition reactions catalyzed by a combination of the easily available 9-amino-9-deoxy-*epi*quinine **378** with *N*-Boc-*D*-phenyl glycine (Boc-*D*-Phg-OH) to the construction of enantiomerically enriched spiro[cyclohexanone-oxindole]s **379/379′** and spiro[cyclohexanone-pyrazolone]s **380/380′** from divinyl ketones **375** and either *N*-unprotected oxindoles **376** or *N*-phenyl-protected pyrazolones **377** (Scheme 5-133).

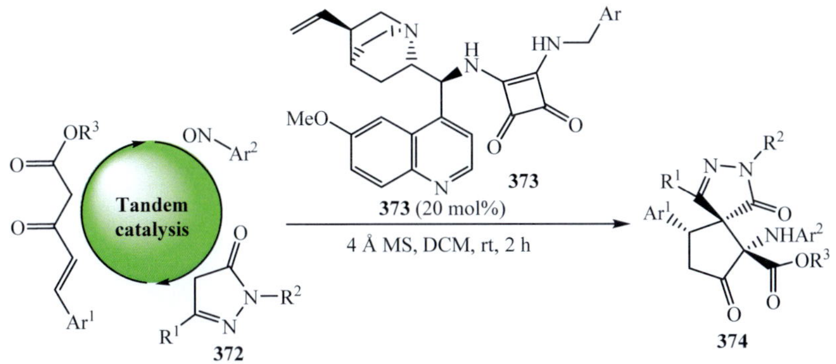

Scheme 5-132. A tandem catalytic approach to highly functionalized spiropyrazolones **374**.

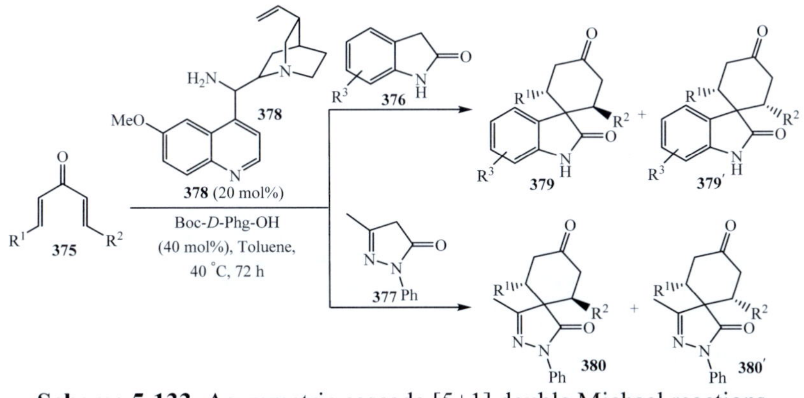

Scheme 5-133. Asymmetric cascade [5+1] double Michael reactions.

A year later, another cascade protocol was developed by Roy et al. [127]. As shown in Scheme 5-134, the reaction of indane 1,3-diones/ oxindoles/coumaranone as the dinucleophilic components with (*E*)-5-nitro-6-aryl-hex-5-ene-2-one **380** as electrophiles in the presence of DABCO afforded functionalized six-membered spirocyclic compounds **381-382** in reasonable-to-good yields (30-84%) and high level of stereoselectivities (up to >95:5 dr).

The enantioselective synthesis of stereochemically and structurally diverse spirocyclic oxindoles by [5+2] cycloaddition of chiral crotylsilanes **383** bearing primary alcohol was elegantly reported by Zhang et al. in 2009 [128]. The complexity of the spiro-oxindoles **384a,b** can be enhanced by using different combinations of functionalized silyl alcohols or substituted isatin reaction partners (Scheme 5-135).

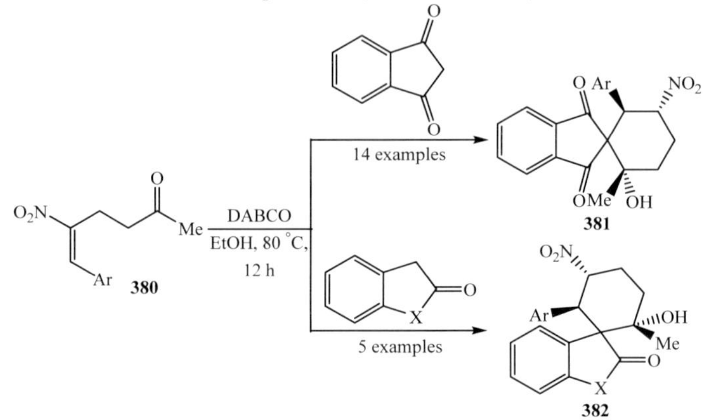

Scheme 5-134. Diastereoselective cascade synthesis of functionalized six-membered spirocyclic indane-1,3-diones/oxindoles **381-382**.

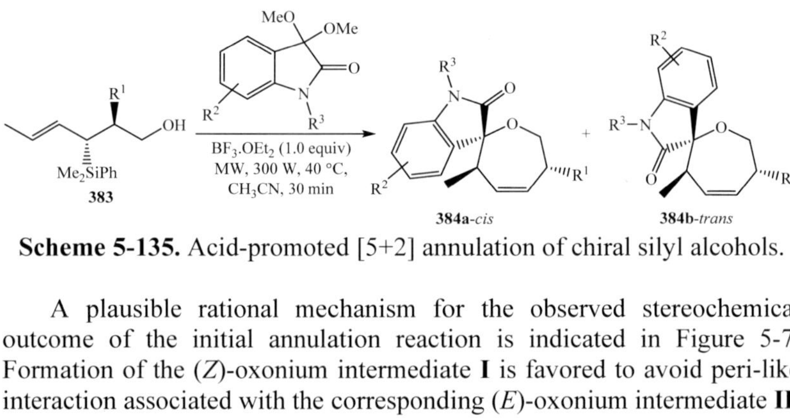

Scheme 5-135. Acid-promoted [5+2] annulation of chiral silyl alcohols.

A plausible rational mechanism for the observed stereochemical outcome of the initial annulation reaction is indicated in Figure 5-7. Formation of the (*Z*)-oxonium intermediate **I** is favored to avoid peri-like interaction associated with the corresponding (*E*)-oxonium intermediate **II**. Designating the aryl ring as the larger steric subscriber orients the carbonyl pseudoaxial. Therefore, transition state **I** is preferred and results in the formation of *cis*-**384a** as a major diastereomer under kinetic conditions.

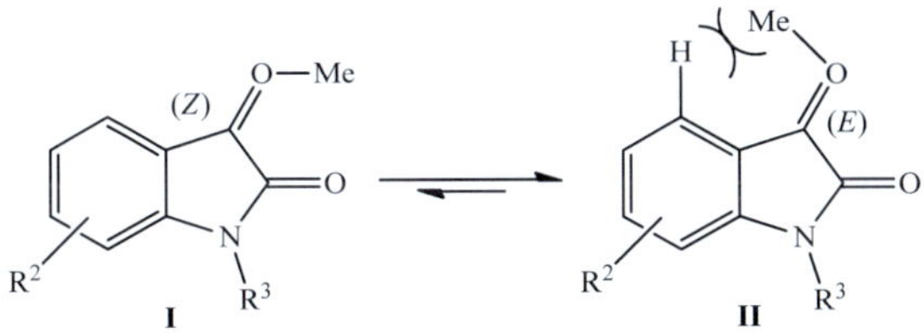

Fig. 5-7. Proposed transition state.

The Chen group developed an efficient α,α-L-diphenylprolinol trimethylsilyl ether **235**-catalyzed [2+2+2] annulation strategy rendering spirocyclic oxindoles **385** with versatile molecular complexity (up to six contiguous stereogenic centers, high diastereo- and enantioselectivities) for a variety of α,β-unsaturated aldehydes, aliphatic aldehyde, and electron-deficient olefinic oxindoles as represented in Scheme 5-136. The amine-catalyzed stereoselective Michael addition of aliphatic aldehydes to electron-deficient olefinic oxindole motifs afforded chiral C3 components, which were further combined with diverse electrophiles (activated olefins or imines) to furnish the desired products [129].

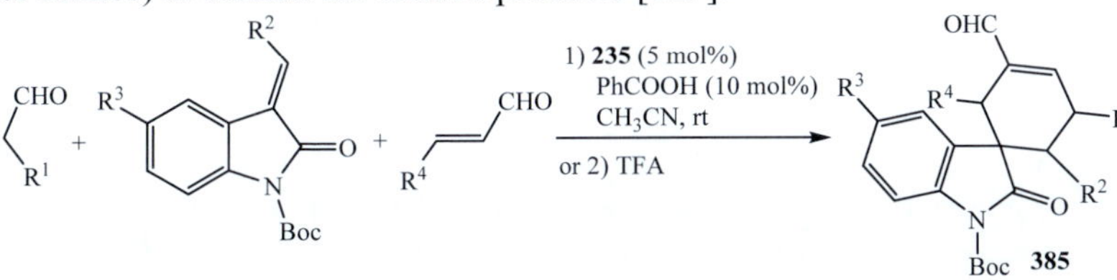

Scheme 5-136. A **235**-catalyzed [2+2+2] annulation strategy.

An interesting organocatalytic [2+2+2] annulation reaction between 2-arylideneindane-1,3-diones **139** and aldehydes was reported by Kuan et al. in 2013 [130]. The reaction proceeds via a unique Michael/Michael/Aldol reaction that requires 2 equiv of the 2-arylideneindane-1,3-dione **139**. The dispirocyclohexanes **386** were formed in reasonable-to-good chemical yields and with high stereoselectivities (>95:5 d.r. and up to 99% ee) using a catalytic amount of α,α-L-diphenylprolinol trimethylsilyl ether **235** (5 mol%) and DABCO (20 mol%) in DMF at -20 °C (Scheme 5-137).

In 2015, Zhang et al. introduced a Rauhut–Currier-initiated organo-cascade reaction (RC-reaction) between 2-arylideneindan-1,3-diones **139** and activated alkenes in the presence of DABCO to prepare functionalized dispirocyclohexanes **387** with a wide range of substituents in satisfactory to excellent chemical yields (58–99%) with high chemo- and diastereoselectivities (>95:5 dr) [131]. The Rauhut–Currier reaction, also called the vinylogous Morita–Baylis–Hillman reaction, is a reaction of latent enolates with a second Michael acceptor (Scheme 5-138).

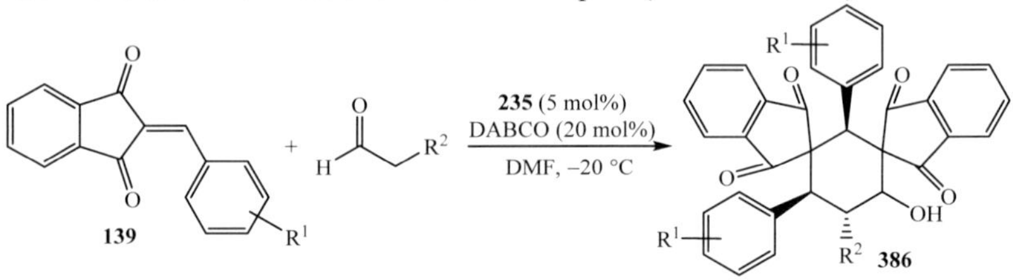

Scheme 5-137. Synthesis of fully substituted dispirocyclohexanes **386**.

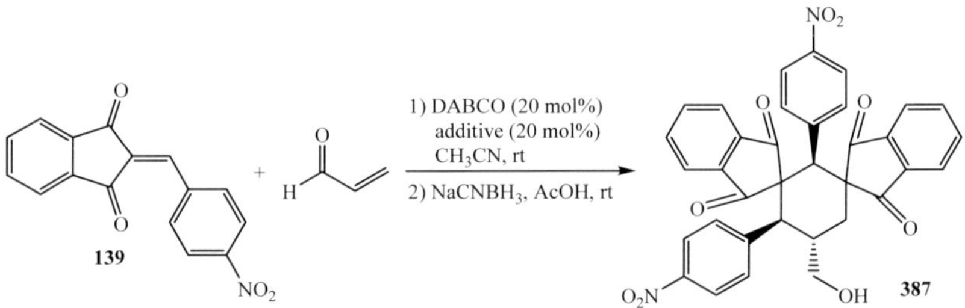

Scheme 5-138. A Rauhut–Currier-initiated organocascade reaction.

Tan et al. employed lactols **388** in a one-pot organo dual catalytic system-induced [2+2+2] annulation reaction for the asymmetric construction of six-membered carbocycles [132]. The enamine-based stereoselective Michael addition of lactols **388** to electron-deficient olefinic oxindole motifs afforded chiral C4 components, which were further combined with triethylamine catalyzed Michael/Henry sequential reactions providing spirocyclic oxindoles **389** with excellent enantioselectivities as a single diastereoisomer (Scheme 5-139).

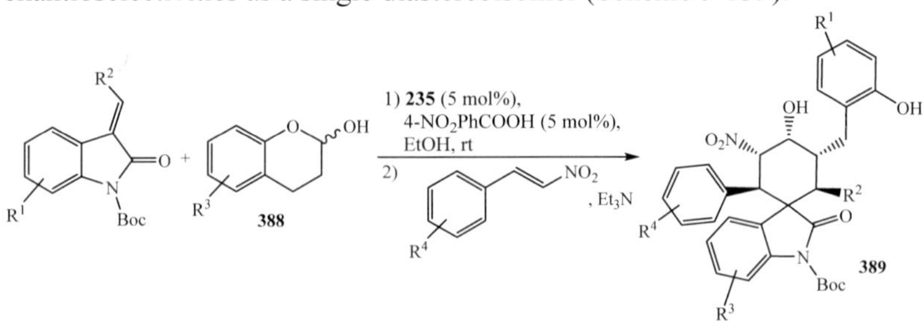

Scheme 5-139. A one-pot organo/organo dual catalytic system-induced [2+2+2] annulation reaction.

5.4. Cyclization reactions

Various aspects of cyclization reactions were discussed in chapters 3 and 4, and we will reveal a vision of the new organocatalyst contributions to the readers in this chapter.

5.4.1. Organocatalyzed Prins cyclization reactions

The Prins reaction is the acid-catalyzed electrophilic addition of alkenes or alkynes to aldehydes or ketones, which is applied widely in the fragrance industry [133]. In this context, we want to show those Prins reactions that lead to spiranes.

Organofluorines are key substrates in medicinal, agrochemical, and material sciences. In 2008, the Fuchigami group reported an efficient synthesis of 4-fluorinated heterocycle **390** through Prins cyclization of homoallylic alcohol with a cyclic ketone in ionic liquid hydrogen fluoride salt (Et_4NF_5HF). However, the protons of Et_4NF_5HF have a poor ability to activate the carbonyl groups (Scheme 5-140) [134].

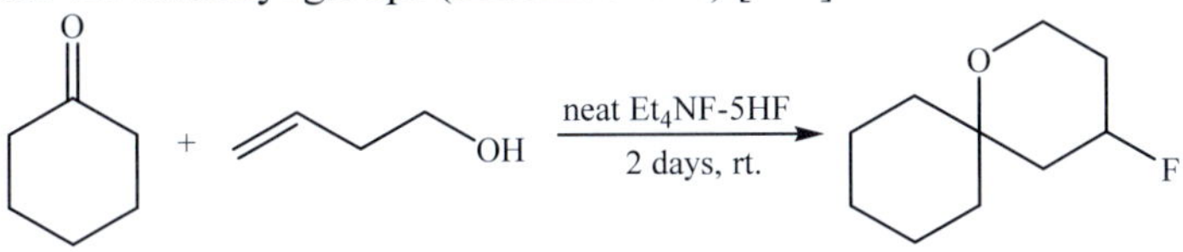

390

Scheme 5-140. Prins cyclization of cyclic ketone with homoallylic alcohol in Et_4NF_5HF.

After this pioneering work, Castaldi et al. synthesized enantiopure spirocyclic oxindole pyran and oxepene **392-394** frameworks using highly stereoselective Prins cyclizations of both homoallylic and *bis*-homoallylic alcohols and isatin ketals **391** (Scheme 5-141) [135].

The Lewis acid-catalyzed Prins-pinacol annulation of alkene diols **395** with a wide range of aliphatic or aromatic aldehydes and ketones was reported by Min et al. in 2009 [136]. This reaction led to the stereoselective synthesis of 2-oxaspiro[*m*,*n*]alkanes **396** in high yields. The author suggested that alkene diol **395** condenses with aldehyde, giving the corresponding oxocarbenium intermediate under acidic conditions, which undergoes Prins cyclization followed by quick pinacol rearrangement to afford 3-substituted 2-oxaspiro compound **396** (Scheme 5-142). The same group previously used this protocol to construct the oxatricyclic ring system.

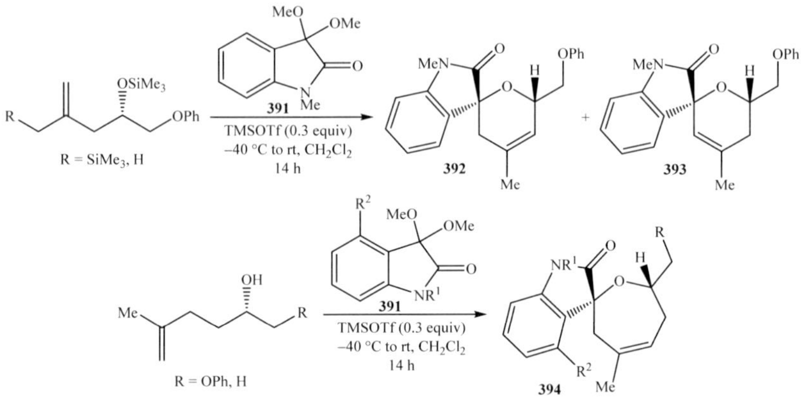

Scheme 5-141. Synthesis of spirocyclic oxindole pyran and oxepene **392-394**.

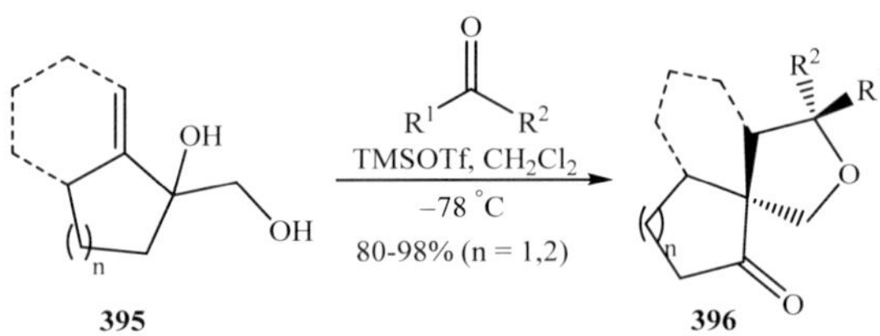

Scheme 5-142. Stereocontrolled synthesis of oxaspirobicycles via Prins-Pinacol Annulation.

In addition, a Prins cascade strategy allowed the construction of spiro[tetrahydropyran-3,3′-oxindole] derivatives **398/398′** in good yields with excellent selectivity, as reported in a work of 2014 [137]. Hence, a variety of aldehydes were coupled with 4-hydroxy-*N*-methyl-2-methylene-*N*-phenylbutanamide **397** in the presence of $BF_3{\cdot}OEt_2$ under ambient conditions to generate the corresponding spiro-oxindoles **398/398′** (Scheme 5-143).

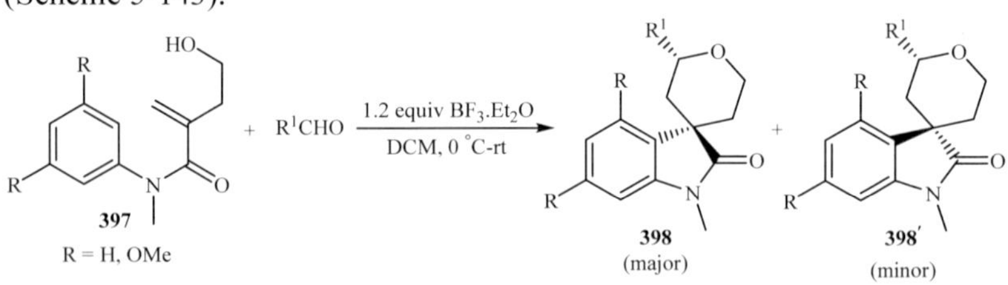

Scheme 5-143. Stereoselective synthesis of spiro[tetrahydropyran-3,3′-oxindole] derivatives **398/398′**.

One year later, a study by Damera et al. reported the highly diastereoselective synthesis of 1-methyl-3′,4′,5′,6′-tetrahydrospiro [indoline-3,2′-pyran]-2-ones **399** employing trimethylsilyl triflate (TMSOTf) as a catalyst (Scheme 5-144) [138].

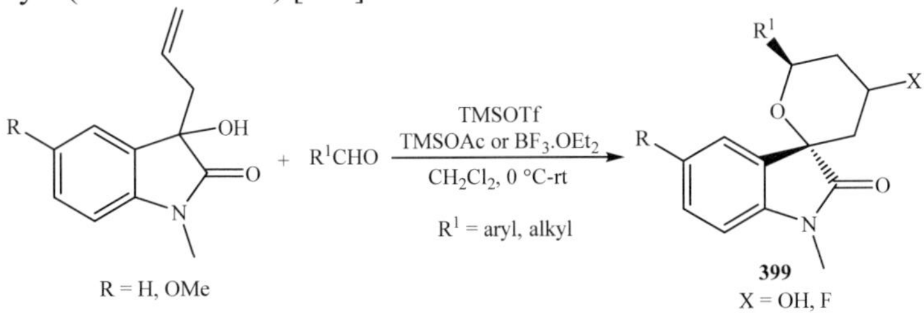

Scheme 5-144. TMSOTf-Promoted Prins cyclization.

Likewise, a cascade cyclization of a 3-[(3-methylbut-2-enylamino) methyl]but-3-en-1-ol **400** or 3-[(4-methylpent-3-enylamino)methyl]but-3-en-1-ol **401** with aldehydes was developed by Reddy et l. [139]. They synthesized spiropyrrolidine and spiropiperidines **402-405** in the presence of TMSOTf at 0 °C under mild conditions, following their developed procedure as shown in Scheme 5-145.

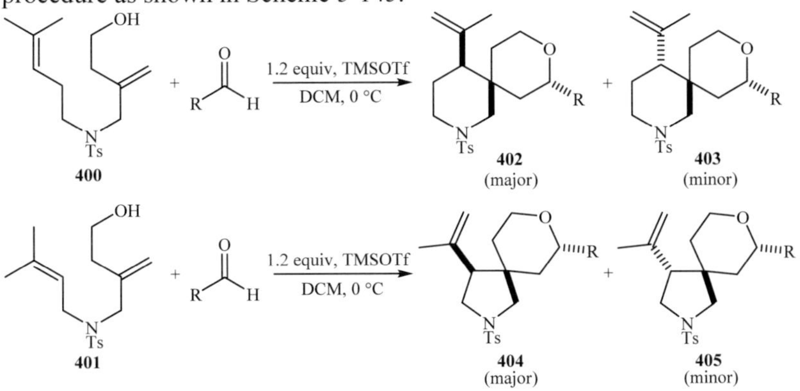

Scheme 5-145. Preparation of spiropyrrolidine and spiropiperidines **402-405**.

Later on, the Reddy group reported a methodology based on an intramolecular Prins cascade process that enabled the production of hexahydro-1*H*-spiro[isoquinoline-4,4′-pyran] scaffolds **407** in a highly stereoselective manner [140]. The reaction was carried out by the condensation of 3-((benzylamino)methyl)but-3-en-1-ol **406** with aldehydes using $BF_3 \cdot OEt_2$ (Scheme 5-146).

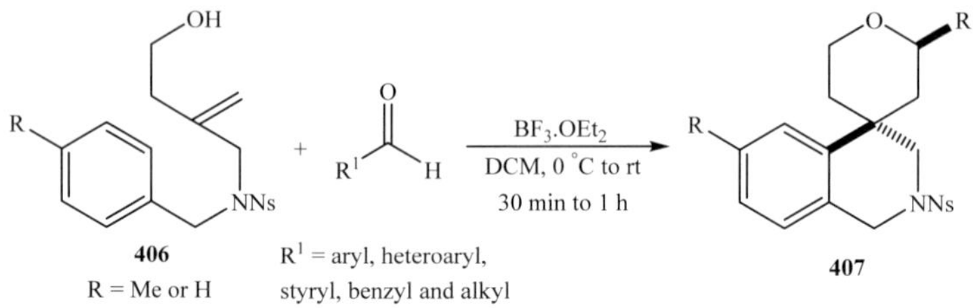

Scheme 5-146. Stereoselective synthesis of hexahydro-1*H*-spiro[isoquinoline-4,4′-pyran]s **407**.

The same group also reported the stereoselective synthesis of biologically spiro-β-tetralones **408** using the Prins cyclization terminated by a pinacol 1,2-shift (Scheme 5-147) [141]. This method uses trimethylsilyl triflate (10 mol%) in dichloromethane under mild conditions.

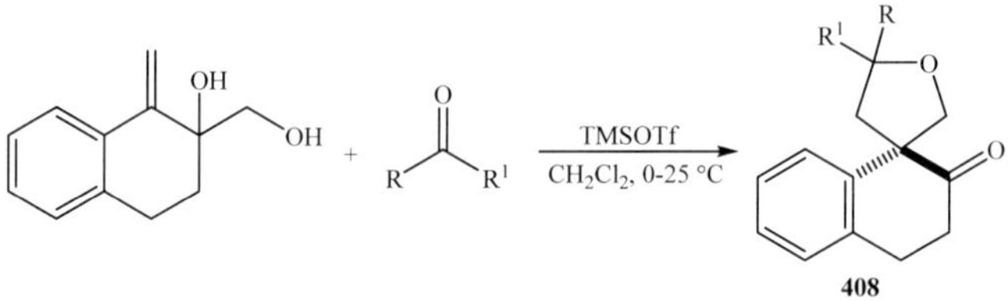

Scheme 5-147. Synthesis of spiro-β-tetralones **408**.

The first report on synthesizing eight-membered spirodiazocane scaffolds **409/409′**, which are less accessible due to ring strain but more relevant to drug discovery, was presented by the Reddy group in 2020 [142]. Using a tandem Prins cyclization, the desired products **409/409′** were obtained in high yields with excellent diastereoselectivity under catalysis with a Lewis acid, as indicated in Scheme 5-148.

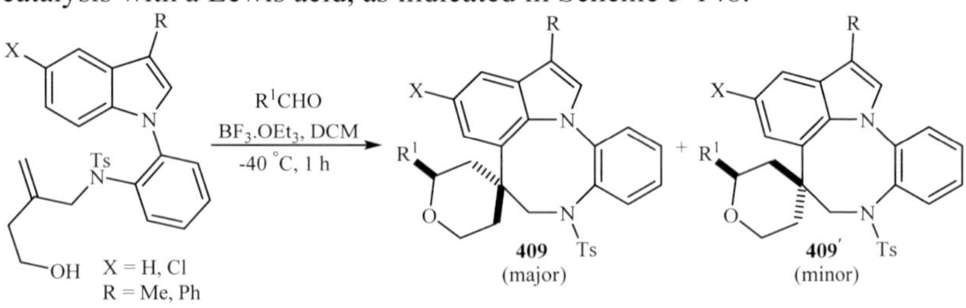

Scheme 5-148. Synthesis of eight-membered spirodiazocane scaffolds **409/409′**.

An elegant example of cyclization reactions was devised in 2021 with the development of an unprecedented synthetic strategy based on umpolung allylation/aza-Prins cyclization of *N*-2,2,2-trifluoroethylisatin ketimines **54** for the syntheses of spiro[indoline-3,2′-piperidin]-2-ones **410** and 5′,6′-dihydro-1′*H*-spiro[indoline-3,2′-pyridin]-2-ones **411** in a one-pot process [143]. Reactions proceeded smoothly with allyl bromide in the presence of 1,8-diazabicyclo[5.4.0]undec-7-ene, H_2O, and trimethylsilyl bromide (Scheme 5-149).

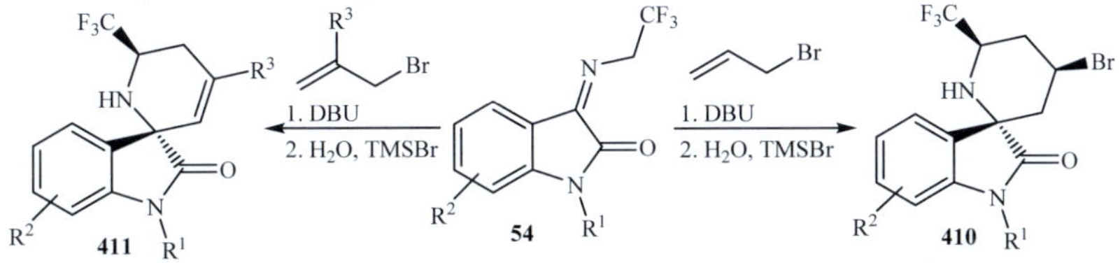

Scheme 5-149. One-pot process of umpolung allylation/aza-Prins cyclization.

5.4.2. Other organocatalyzed cyclization reactions

In this section, different types of organocatalysis applied in various cyclization reactions will be discussed briefly, as well as other methods that improve the efficiency of these reactions.

In 2018, Taylor et al. used a dearomatizing spirocyclization procedure based on a one-pot, double alkylation process to obtain 3,3′-spirocyclic indolenines and indolines **412-413**, which were obtained in high yields [144]. Unfunctionalized indoles with electrophilic dihalides in the presence of *t*-BuOK/triethylborane (BEt_3) produced corresponding spiro compounds, as shown in Scheme 5-150.

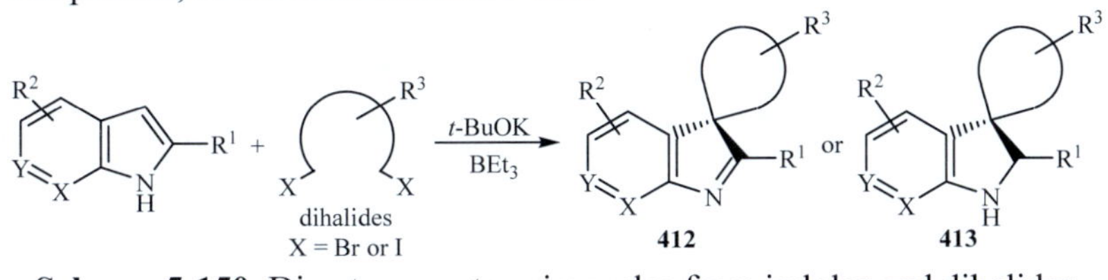

Scheme 5-150. Direct access to spirocycles from indoles and dihalides.

Jui et al. showed that the dearomative cyclization of linear amides to complex spirocyclic butyrolactams **415** is possible by photoredox catalysis via a reductive radical−polar crossover mechanism (Scheme 5-151). This method employs a simple organic catalyst/reductant pair to give products in a highly flexible manner for substitution. The reaction is accomplished by using the Zeitler organic photocatalyst 3DPAFIPN (5 mol%), blue light, Hünig's base **414** (3 equiv), and a mixture of water and acetonitrile as a solvent [145].

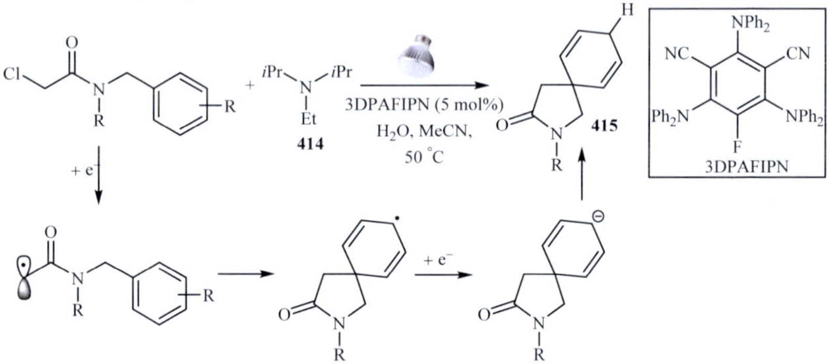

Scheme 5-151. Dearomative cyclization of linear amides.

Moreover, diverse 3-sulfonylspiro[4,5]trienones **417** were accessed by the visible-light-mediated *ipso*-carbosulfonylation of alkynes **416** with sulfonyl chlorides [146]. The target products **417** were formed in excellent yields by using Eosin Y (2 mol%) as a photocatalyst, Na_2CO_3 (2 equiv) as a base, and H_2O (4 equiv), at 100 °C in acetonitrile under an Ar atmosphere and 5 W blue LED light source for 20 hours. In this conversion, the O atom in the newly formed carbonyl is derived from H_2O and features a broad substrate scope. The plausible reaction mechanism is shown in Scheme 5-152.

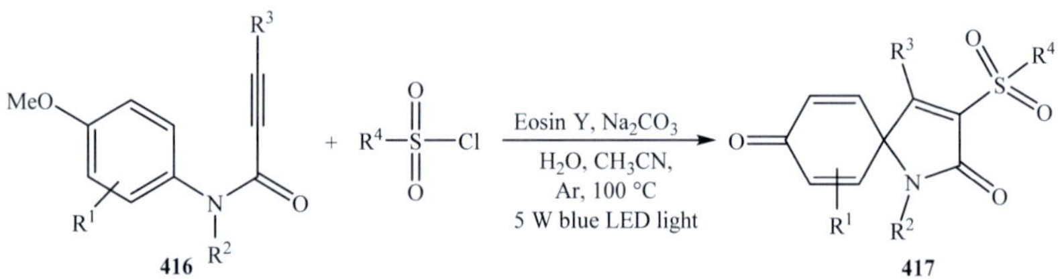

Scheme 5-152. Visible-light-mediated *ipso*-carbosulfonylation of alkynes.

The same group reported a year later similar reaction for *N*-aryl-propiolamide **416**, aromatic amines, and DABCO·$(SO_2)_2$ (as a sulfonyl source to access sulfone), employing Eosin Y (10 mol%) as a photocatalyst (Scheme 5-153) [147]. As shown in Scheme 5-153, according to a plausible mechanism, this sulfonylation/*ipso*-cyclization reaction proceeds through a radical process and undergoes arylsulfonyl radical generation, sulfonylation, and *ipso*-cyclization. Aromatic amines, acting as an aryl source, undergoes diazotization and decomposition to form aryl radicals. The *in situ*-generated sulfonyl radicals are trapped by *N*-aryl propiolamides **416** to generate the sulfonated spiro[4,5]trienones **418**.

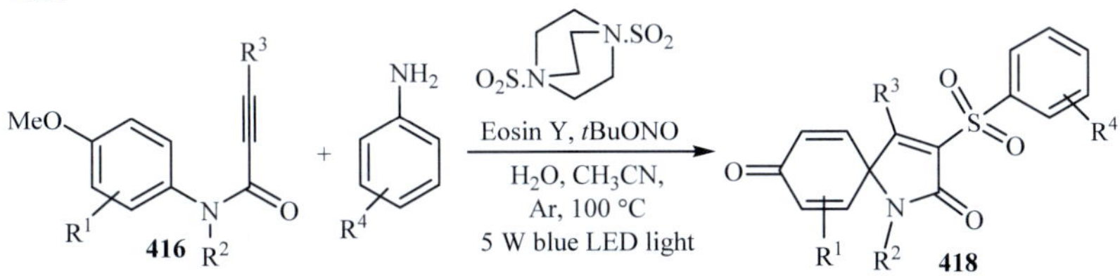

Scheme 5-153. Visible-light-promoted one-pot synthesis of sulfonated spiro[4,5]trienones **418**.

A metal-free photocatalytic thiocyanate radical addition/*ipso*-cyclization/oxidation cascade reaction starting from *N*-phenyl propynamides and ammonium thiocyanate was presented by Chen et al. for the synthesis of thiocyanato-containing azaspirotrienediones **419** (Scheme 5-154). The organic dye acridinium perchlorate was employed as a photocatalyst [148].

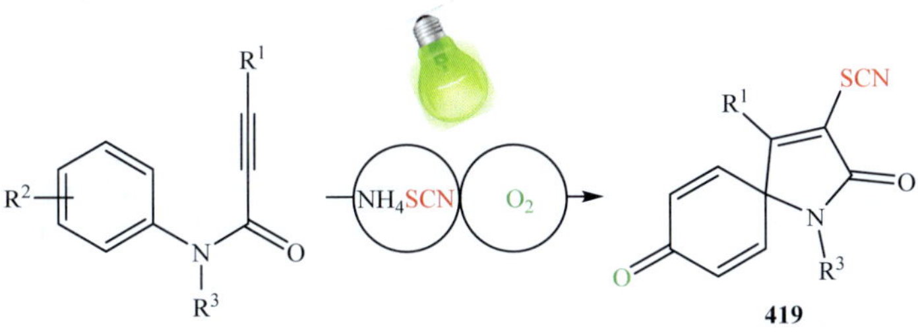

Scheme 5-154. Visible-light-mediated selective thiocyanation/*ipso*-cyclization/oxidation cascade.

Huang et al. found that the construction of spiro[fluorene-9,9′-xanthene] (SFX) building blocks **421** for the preparation of blue-light-emitting materials, which exhibit high thermal stability, is possible by a thermodynamic-controlled process under excessive acid catalyst (Scheme 5-155) [149].

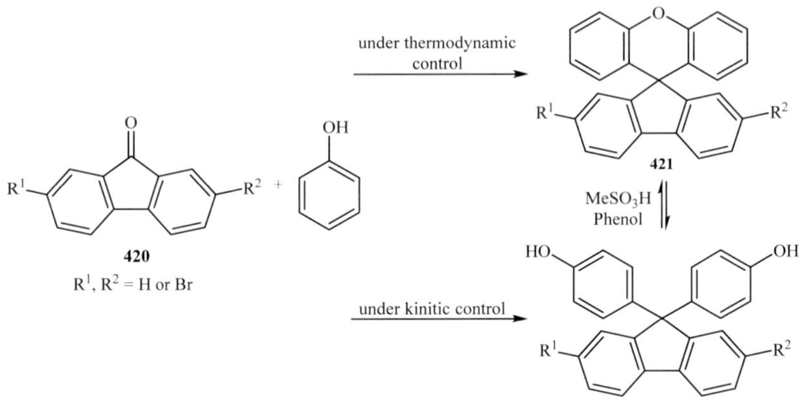

Scheme 5-155. Synthesize spiro[fluorene-9,9′-xanthene] building blocks **421**.

The key step involves a thermodynamically controlled cyclization reaction. Scheme 5-156 displays a plausible reaction pathway of the ring-closure process in the one-pot method. Firstly, an electrophilic attack of protonated fluorenone **420** to the *o*- or *p*-position of phenol and the loss of equivalent water gives a carbon cation, which subsequently attacks the *o*- or *p*-position of another phenol to generate diphenol. Later, intramolecular cyclization through the loss of H_2O provides the SFX with product **421** in the presence of an acid catalyst if diphenol is the *o*,*o*-substituted intermediate. The equilibrium should be hypothesized between the *o*,*o*-substituted intermediates and the *o*,*p*-substituted or *p*,*p*-substituted intermediates. The equilibrium proceeds forward when the ether bond is formed with the loss of H_2O because the last ring-closure step is irreversible. An important factor for this reaction is the amount of $MeSO_3H$ acid employed because forming the ether bond requires an excessive acid catalyst, which causes the equilibrium movement.

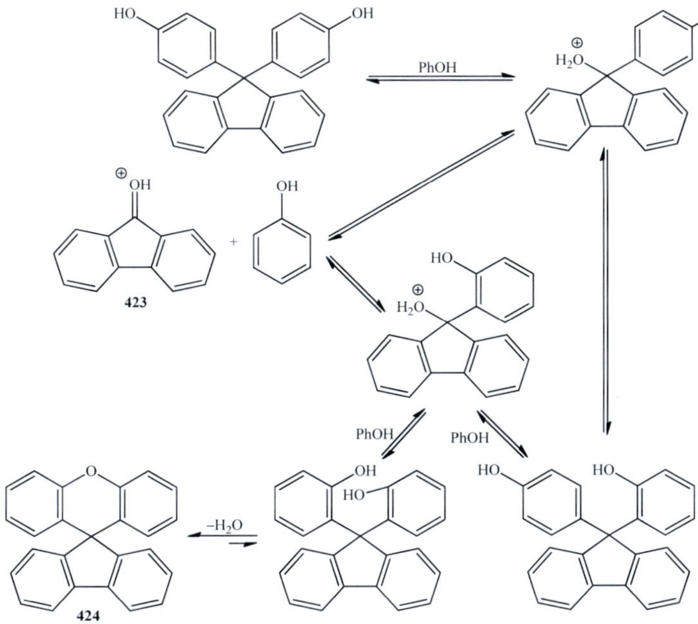

Scheme 5-156. Plausible mechanism.

The Du group reported a divergent synthesis of iodinated quinoline-2-ones and spiro[4,5]trienones **423-424** by using PIFA as both a nonmetal oxidant and an iodination reagent in the presence of $BF_3 \cdot Et_2O$ (Scheme 5-157) [150].

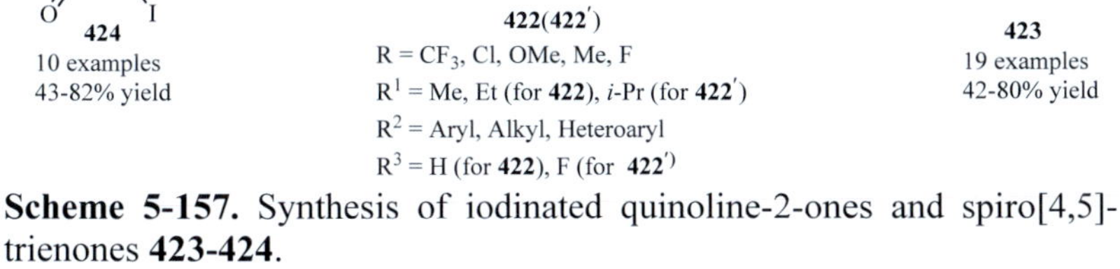

Scheme 5-157. Synthesis of iodinated quinoline-2-ones and spiro[4,5]-trienones **423-424**.

Depending on the substituent pattern of *N*-arylpropynamide **422**, the reaction proceeds in two different ways (path **A** and **B**), as depicted in Scheme 5-158. When the *N*-arylpropynamide **422** possesses a *para*-fluorine on the aniline ring, the spiro compound **424** is formed through an exclusive defluorination process; otherwise, the product was quinoline-2-one **423**.

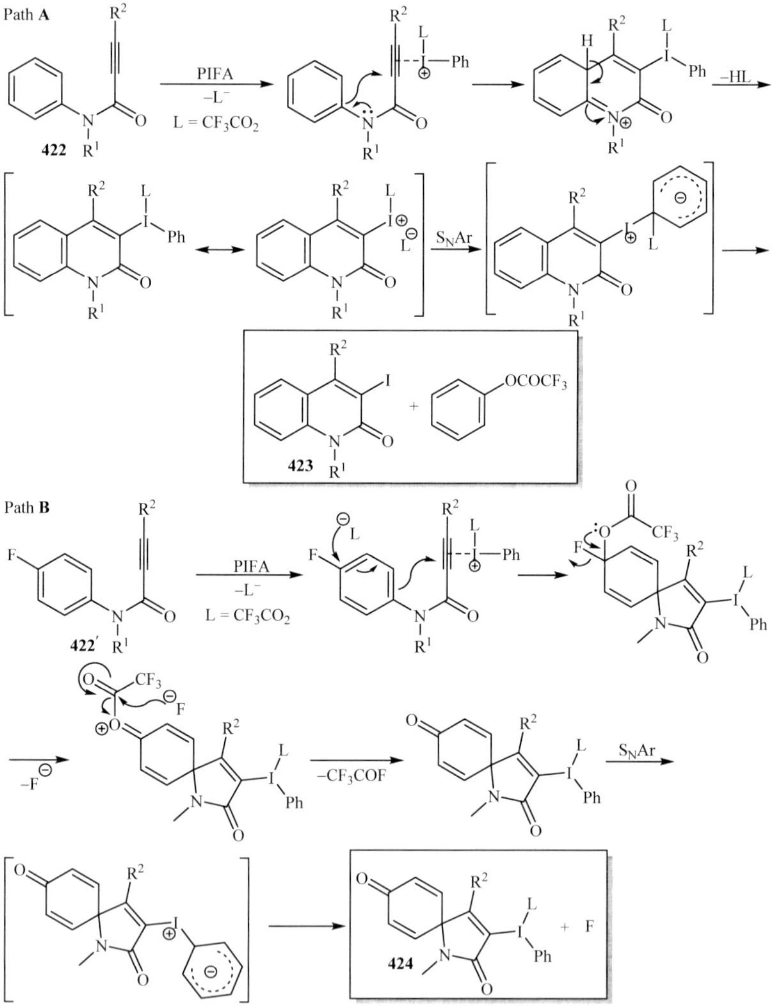

Scheme 5-158. Proposed mechanistic pathways.

Likewise, 3-phosphoryl/trifluoromethyl/thiocyanato azaspiro[4.5]-trienones **428-429** were successfully prepared by a 4CzIPN-photo-catalyzed metal-free cascade cyclization [151]. These photocatalytic strategies proceed via radical-initiated cascade annulation of *N*-arylpropiol-amides **425** with phosphorus reagents **426**, 1-trifluoromethyl-1,2-benziodoxol-3(1*H*)-one (Togni's reagent II) **427** or NH_4SCN (Scheme 5-159).

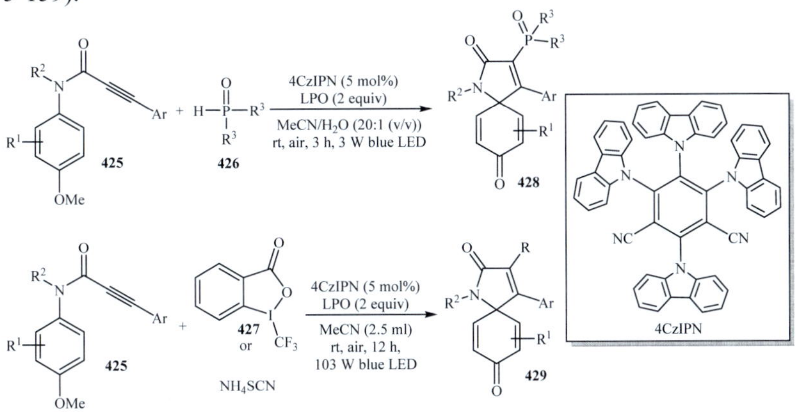

Scheme 5-159. Synthesis of azaspiro[4.5]trienones **431-432.**

In 2015, Tu et al. reported the first direct example of synthesizing asymmetrically cyclopentanones with four stereocenters using Nazarov cyclization [152]. In this work, a series of chiral spiro[4.4]nonane-1,6-diones **430** with up to 96% yield and 97% ee was synthesized through a novel organocatalytic asymmetric tandem Nazarov cyclization/ semipinacol rearrangement reaction using "unactivated" substrates (Scheme 5-160).

Scheme 5-160. Organocatalytic asymmetric tandem Nazarov cyclization/ semipinacol rearrangement.

Recently, Kalesse et al. investigated the reactivity of different trienones **431**, differentiated by the side chain length and functionalities, under Lewis and Brønsted acids catalysis (Scheme 5-161). These reactions led to the formation of the different cyclization products and carbon backbones that originated either from a conjugate Prins cyclization or an interrupted Nazarov cyclization. This reaction results in the stereoselective construction of quaternary spirocenters **432**, with a β-hydroxyl group acting as an excellent stereo-directing element. The authors observed a recurring skeleton, representing a newly interrupted Nazarov reaction, in which the carbocationic intermediate is trapped by an ene-type cyclization [153].

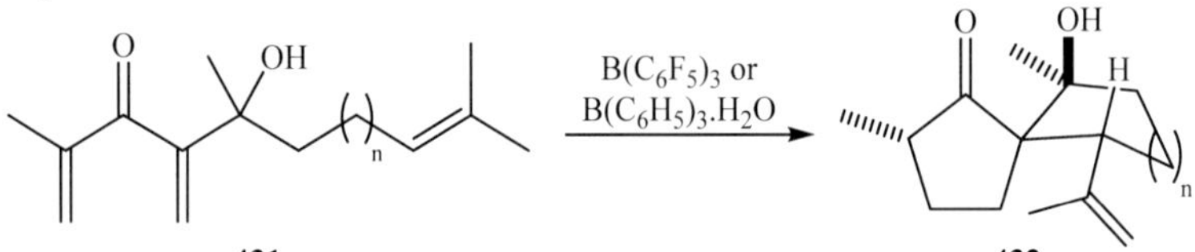

431 **432**

Scheme 5-161. A Nazarov-ene tandem reaction for the stereoselective construction of spiro compounds **432**.

As previously mentioned in Section 2.4.3, the Pictet–Spengler reaction is a chemical reaction that involves the condensation and ring closure between a β-arylethylamine and a carbonyl compound.

In this field, an asymmetric Michael-hemiketalization-oxa-Pictet-Spengler cyclization was successfully applied by Qiao et al. in the synthesis of chiral bridged and spiro heterocyclic skeletons with one spiro stereogenic carbon center and two bridgehead carbon centers [154]. In this developed procedure, the reaction of 3-pyrrolyloxindoles **433** with (*E*)-2-hydroxyaryl-2-oxobut-3-enoates **434** using cooperative catalysts of a Takemoto thiourea catalyst **435** and a triflimide (Tf_2NH) gave bridged and spiro heterocyclic skeletons **436** (Scheme 5-162).

Also, a series of enantioenriched 5′,11′-dihydrospiro-[indoline-3,6′-indolo[3,2-*c*]quinolin]-2-ones **439** bearing quaternary stereogenic centers prepared following BINOL-derived phosphoric acid **438**-catalyzed Pictet-Spengler reactions of 2-(1*H*-indolyl)anilines **437** and isatins by the condensation/cyclization process (Scheme 5-163). The desired products were obtained with excellent yields and up to >99% ee [155].

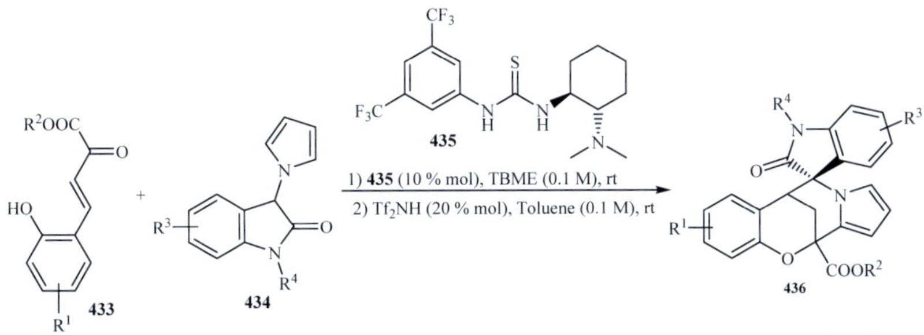

433 434 435 436

Scheme 5-162. Organocatalyzed asymmetric Michael–hemiketalization–oxaPictet–Spengler cyclization.

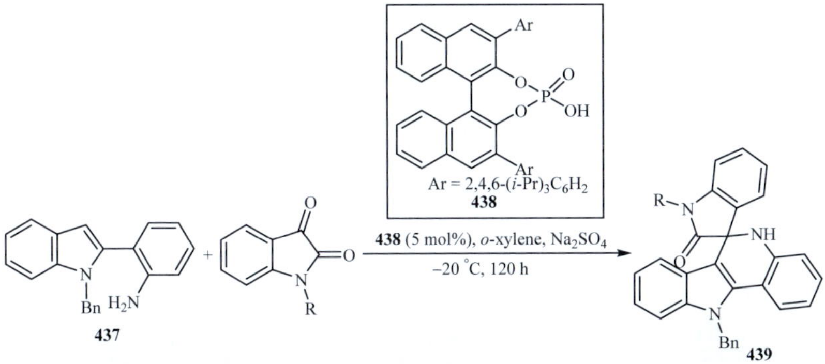

438 437 439

Scheme 5-163. CPA-Catalyzed Pictet–Spengler reactions.

The authors explained the observed absolute stereochemistry of final products based on the plausible transition state depicted in Figure 5-8. In the presence of CPA, aniline reacts with isatin to generate ketimine through dehydration. The CPA activates the C=N bond through hydrogen-bonding interactions, and the subsequent cyclization occurs. The triisopropyl phenyl groups at the 3,3′-positions of the catalyst shield the ketimine, and nucleophilic attack preferentially occurs at the *re*-face to generate the *R*-configured adduct **439**.

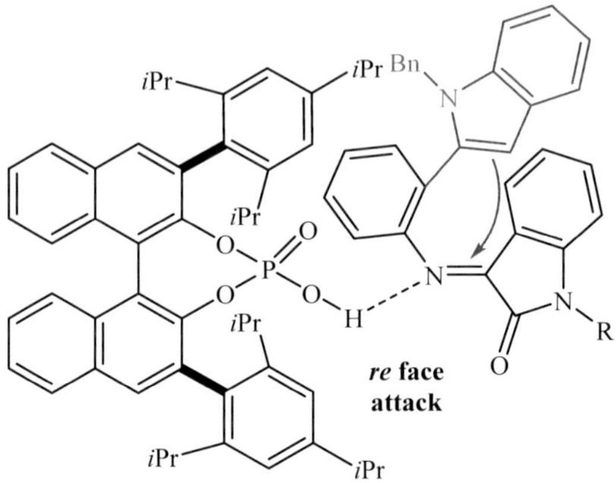

Fig. 5-8. Plausible transition state.

References

[1] Yoshida Keisuke, Itatsu Yukihiro, Fujino Yuta, Inoue Hiroki, Takao Ken-ichi. 2016. "Enantioselective organocatalytic construction of spiroindane derivatives by intramolecular Friedel -Crafts-type 1,4-addition". Angewandte Chemie International Edition. No. 23. 6734-6738.

[2] You Zhi-Hao, Chen Ying-Han, Tang Yu, Liu Yan-Kai. 2018. "Organocatalytic asymmetric synthesis of spiro-bridged and spiro -fused heterocyclic compounds containing chromane, indole, and oxindole moieties". Organic Letters. No. 21. 6682-6686.

[3] Xie Ling-Hai, Hou Xiao-Ya, Hua Yu-Ran, Tang Chao. *et al.* 2006. "Facile synthesis of complicated 9,9-diarylfluorenes based on $BF_3.Et_2O$-mediated Friedel-Crafts reaction". Organic Letters. No. 17. 3701-3704.

[4] Basavaiah Deevi, Ramesh Reddy Kanumuri. "Simple and one-pot protocol for synthesis of indene-spiro-oxindoles involving tandem Prins and Friedel-Crafts reactions". Organic Letters. No. 1. 57-60.

[5] Furukawa Shunsuke, Kobayashi Junji, Kawashima Takayuki. 2010. "Application of the sila-Friedel-Crafts reaction to the synthesis of π-extended silole derivatives and their properties". Dalton Transactions. No. 39. 9329-9336.

[6] Chen Xiao-Hua, Wei Qiang, Luo Shi-Wei, Xiao Han, Gong Liu-Zhu. 2009. "Organocatalytic synthesis of spiro[pyrrolidin-3,3′-oxindoles] with high enantiopurity and structural diversity". Journal of the American Chemical Society. No. 38. 13819-13825.

[7] Cheng Mou-Nuo, Wang Hao, Gong Liu-Zhu. 2011. "Asymmetic organocatalytic 1,3-dipolar cycloaddition of azomethine ylide to methyl 2-(2-nitrophenyl)acrylate for the synthesis of diastereo-isomers of spirotryprostatin A". Organic Letters. No. 9. 2418-2421.

[8] Guo Chang, Song Jin, Gong Liu-Zhu. 2013. "Biomimetic asymmetric 1,3-dioplar cycloaddition: Amino acid precursors in biosynthesis serve as latent azomethine ylides". Organic Letters. No. 11. 2676-2679.

[9] Wang Long, Shi Xiao-Mei, Dong Wei-Ping, Zhua Li-Ping, Wang Rui. 2013. "Efficient construction of highly functionalized spiro-[γ-butyrolactone-pyrrolidin-3,3′-oxindole]tricyclic skeletons via an organocatalytic 1,3-dipolar cycloaddition". Chemical Communications. No. 33. 3458-3460.

[10] Xiao Jun-An, Liu Qi, Ren Ji-Wei, Liu Jian, *et al.* 2014. "Highly enantioselective construction of polycyclic spiro-oxindoles by organocatalytic 1,3-dipolar cycloaddition of 2-cyclohexenone catalyzed by proline-sulfonamide". European Journal of Organic Chemistry. No. 26. 5700-5704.

[11] Ma Mingxia, Zhu Yuanyuan, Sun Quantao, Li Xiaoyuan, *et al.* 2015. "The asymmetric synthesis of CF_3-containing spiro [pyrrolidin-3,2′-oxindole] through the organocatalytic 1,3-dipolar cycloaddition reaction". Chemical Communications. No. 42. 8789-8792.

[12] Dai Wei, Jiang Xiao-Li, Wu Qiong, Shi Feng, Tu Shu-Jiang. 2015. "Diastereo-and enantioselective construction of 3,3′-pyrrolidinyldispiro-oxindole framework via catalytic asymmetric 1,3-dipolar cycloadditions" . The Journal of Organic Chemistry. No. 11. 5737-5744.

[13] Du Taiping, Du Fei, Ning Yanqiang, Peng Yungui. 2015. "Organocatalytic enantioselective 1,3-dipolar cycloadditions between Seyferth–Gilbert reagent and isatylidene malononitriles: Synthesis of chiral spiro-phosphonylpyrazoline-oxindoles". Organic Letters. No. 5. 1308-1311.

[14] Sun Quantao, Li Xiaoyuan, Su Jinhuan, Zhao Long, *et al.* 2015. "The squaramide-catalyzed 1,3-dipolar cycloaddition of nitro-alkenes with *N*-2,2,2-trifluoroethylisatin ketimines: An approach for the synthesis of 5′-trifluoromethyl-spiro[pyrrolidin-3,2′-oxindoles]". Advanced Synthesis & Catalysis. No. 14-15. 3187-3196.

[15] Zhao Hong-Wu, Yang Zhao, Meng Wei, Tian Ting, *et al.* 2015. "Diastereo-and enantioselective synthesis of chiral pyrrolidine -fused spiro-oxindoles via organocatalytic [3+2] 1,3-dipolar cyclo-addition of azomethine ylides with maleimides". Advanced Synthesis & Catalysis. No. 11. 2492-2502.

[16] Zhu Guodong, Wang Baomin, Bao Xiaoze, Zhang Huanrui, *et al.* 2015. "Catalytic asymmetric construction of spiro[pyrrolidine-2,3′-oxindole] scaffolds through chiral phosphoric acid-catalyzed 1,3-dipolar cycloaddition involving 3-aminooxindoles". Chemical Communications. No. 85. 15510-15513.

[17] Wu Ping, Gao Hong, Sun Jing, Yan Chao-Guo. 2017. "1,3-Dipolar cycloaddition reaction for diastereo-selective synthesis of functionalized dihydrospiro[indoline-3,2′-pyrroles]". Chinese Chemical Letters. No. 2. 329-332.

[18] Su Jinhuan, Ma Zelin, Li Xiaoyuan, Lin Li, *et al.* 2016. "Asymmetric synthesis of 2′-trifluoromethylated spiro-pyrrolidine-3,3′-oxindoles via squaramide-catalyzed umpolung and 1,3-dipolar cycloaddition". Advanced Synthesis & Catalysis. No. 23. 3777-3785.

[19] Wang Yue-Ming, Zhang Hong-Hao, Li Can, Fana Tao, Shi Feng. 2016. "Catalytic asymmetric chemo-selective 1,3-dipolar cycloadditions of an azomethine ylide with isatin-derived imines: Diastereo- and enantioselective construction of a spiro [imidazolidine-2,3′-oxindole] framework". Chemical Communications. No. 9. 1804-1807.

[20] Huang Wei-Jie, Chen Qing, Lin Ning, Long Xian-Wen, *et al.* 2017. "Asymmetric synthesis of trifluoromethyl-substituted 3,3′-pyrrolidinyl-dispiro-oxindoles through organocatalytic 1,3-dipolar cycloaddition reactions". Organic Chemistry Frontiers. No. 3. 472-482.

[21] Zhang Jia-Xing, Wang Hong-Yu, Jin Qiao-Wen, Zheng Chang-Wu, *et al.* 2016. "Thiourea–quaternary ammonium salt catalyzed asymmetric 1,3-dipolar cycloaddition of imino esters to construct spiro[pyrrolidin-3,3′-oxindoles]". Organic Letters. No. 19. 4774-4777.

[22] Zhu Guodong, Wei Qian, Chen Hongbo, Zhang Yanpeng, *et al.* 2017. "Asymmetric [3+2] cycloaddition of 3-amino oxindole-based azomethine ylides and α,β-enones with divergent diastereocontrol on the spiro[pyrrolidine-oxindoles]". Organic Letters. No. 7. 1862-1865.

[23] Chen Nannan, Zhu Liping, Gan Lu, Liu Zhenquan, *et al.* 2018. "Asymmetric synthesis of bispiro[γ-butyrolactone-pyrrolidin-4,4′-pyrazolone] scaffolds containing two quaternary spirocenters via an organocatalytic 1,3-dipolar Cycloaddition". European Journal of Organic Chemistry. No. 23. 2939-2943.

[24] Du Ting, Li Zhaokun, Zheng Changwu, Fang Guosheng, *et al.* 2018. "Highly enantioselective 1,3-dipolar cycloaddition of imino esters with benzofuranone derivatives catalyzed by thiourea-quaternary ammonium salt". Tetrahedron. No. 52. 7485-7494.

[25] Wang Cui, Wen Dongwa, Chen Hui, Deng Yabo, *et al.* 2019. "The catalytic asymmetric synthesis of CF_3-containing spiro-oxindole-pyrrolidine-pyrazolone compounds through squaramide -catalyzed 1,3-dipolar cycloaddition". Organic & Biomolecular Chemistry. No. 22. 5514-5519.

[26] Zhu Wen-Run, Su Qiong, Lin Ning, Chen Qing, *et al.* 2020. "Organocatalytic synthesis of chiral CF_3-containing oxazolidines and 1,2-amino alcohols: Asymmetric oxa-1,3-dipolar cycloaddition of trifluoroethylamine-derived azomethine ylide". Organic Chemistry Frontiers. No. 21. 3452-3458.
[27] Wang Xiao-Na, Zhang Yan-Yong, Ye Song. 2010. "Enantioselective synthesis of spirocyclic oxindole-β-lactones via *N*-heterocyclic carbene-catalyzed cycloaddition of ketenes and isatins". Advanced Synthesis & Catalysis. No. 11-12. 1892-1895.
[28] Qi Liang-Wen, Yang Yu, Gui Yong-Yuan, Zhang Yong, *et al.* 2014. "Asymmetric synthesis of 3,3′-spiro-oxindoles fused with cyclobutanes through organocatalytic formal [2+2] cycloadditions under H-bond-directing dienamine activation". Organic Letters. No. 24. 6436-6439.
[29] Rainoldi Giulia, Faltracco Matteo, Lo Presti Leonardo, Silvani Alessandra, Lesma GiordaNo. 2016. "Highly diastereo-selective entry into chiral spiro-oxindole-based 4-methyleneazetidines in formal [2+2] annulation reaction". Chemical Communications. No. 77. 11575-11578.
[30] Wang Zhen, Xu Huacheng, Su Qin, Hu Ping, *et al.* 2017. "Enantioselective synthesis of tetrahydropyridines/piperidines via stepwise [4+ 2]/[2+ 2] cyclizations". Organic Letters. No. 12. 3111-3114.
[31] R. Yong Sarah, C. Williams Morwenna, G. Pyne Stephen, T. Ung Alison, *et al.* 2005. "Synthesis of 2-azaspiro[4.4]nonan-1-ones via phosphine-catalysed [3+2]-cycloadditions" . Tetrahedron. No. 34. 8120-8129.
[32] J. Cowen Bryan, J. Miller Scott. 2007. "Enantioselective [3+2]-cycloadditions catalyzed by a protected, multifunctional phosphine-containing α-amino acid". Journal of the American Chemical Society. No. 36. 10988-10989.
[33] Voituriez Arnaud, Panossian Armen, Fleury-Brgeot Nicolas, Retailleau Pascal, Marinetti Angela. 2009. "Synthesis of chiral 2-phospha[3]ferrocenophanes and their behaviour as organocatalysts in [3+2] cyclization reactions". Advanced Synthesis & Catalysis. No. 11-12. 1968-1976.
[34] Voituriez Arnaud, Pinto Nathalie, Neel Mathilde, Retailleau Pascal, Marinetti Angela. 2010. "An organocatalytic [3+2] cyclisation strategy for the highly enantioselective synthesis of spiro-oxindoles". Chemistry-A European Journal. No. 42. 12541-12544.

[35] Pinto Nathalie, Retailleau Pascal, Voituriez Arnaud, Marinetti Angela. 2011. "Organocatalytic enantioselective desymmetrization of cyclic enones via phosphine promoted [3+2] annulations". Chemical Communications. No. 3. 1015-1017.

[36] Lian Zhong, Shi Min. 2012. "Nitrogen-and phosphorus -containing Lewis base catalyzed [4+2] and [3+2] annulation reactions of isatins with but-3-yn-2-one". European Journal of Organic Chemistry. 581-586.

[37] Sun Li-Hui, Shen Li-Tao, Ye Song. 2011. "Highly diastereo-and enantioselective NHC-catalyzed [3+2] annulation of enals and isatins". Chemical Communications. No. 36. 10136-10138.

[38] Tan Bin, Zeng Xiaofei, Wen Yi Leong Wendy, Shi Zugui, *et al.* 2012. "Core structure-based design of organocatalytic [3+2] -cycloaddition reactions: Highly efficient and stereo-controlled syntheses of 3,3′-pyrrolidonyl spiro-oxindoles". Chemistry-A European Journal. No. 1. 63-67.

[39] Dugal-Tessier Julien, A. OBryan Elizabeth, B. H. Schroeder Thomas, T. Cohen Daniel, A. Scheidt Karl. 2012. "An *N*-heterocyclic carbene/Lewis acid strategy for the stereoselective synthesis of spiro-oxindole lactones". Angewandte Chemie International Edition. No. 20. 4963-4967.

[40] Hu Fangle, Wei Yin, Shi Min. 2012. "Enantioselective synthesis of spirocyclic cyclopentenes: Asymmetric [3+2] annulation of 2-arylideneindane-1,3-diones with MBH carbonates derivatives catalyzed by multifunctional thiourea–phosphines". Tetrahedron. No. 38. 7911-7919.

[41] Lian Zhong, Shi Min. 2012. "Asymmetric [3+2] annulation of *N*-protected isatins with but-3-yn-2-one catalyzed by DIOP: Facile creation of enantioenriched spiro[furan-2,3′-indoline]-2′,4(5*H*)-dione". Organic & Biomolecular Chemistry. No. 40. 8048-8050.

[42] Tian Li, Hu Xiu-Qin, Li Yun-Han, Xu Peng-Fei. 2013. "Organocatalytic asymmetric multicomponent cascade reaction via 1,3-proton shift and [3+2] cycloaddition: An efficient strategy for the synthesis of oxindole derivatives". Chemical Communications. No. 65. 7213-7215.

[43] Jiang Yu, Pei Cheng-Kui, Du Dan, Li Xiao-Ge, *et al.* 2013. "Enantioselective synthesis of spiro-oxindoles: Asymmetric [3+2] cycloaddition of (3-isothiocyanato)oxindoles with azodicarboxy- lates". European Journal of Organic Chemistry. No. 35. 7895-7901.

[44] Wang Ze-Dong, Wang Feng, Li Xin, Cheng Jin-Pei. 2013. "*N*-Heterocyclic carbene catalyzed annulation of benzofuran-2,3-diones and enals: A concise synthesis of spiro-bis-lactone". Organic & Biomolecular Chemistry. No. 34. 5634-5641.
[45] Zheng Chenguang, Yao Weijun, Zhang Yucheng, Ma Cheng. 2014. "Chiral spiro-oxindole–butenolide synthesis through asymmetric *N*-heterocyclic carbene-catalyzed formal (3+2) annulation of 3-bromoenals and isatins". Organic Letters. No. 19. 5028-5031.
[46] Guo Chang, Schedler Michael, G. Daniliuc Constantin, Glorius Frank. 2014. "*N*-Heterocyclic carbene catalyzed formal [3+2] annulation reaction of enals: An efficient enantioselective access to spiro-heterocycles". Angewandte Chemie International Edition. No. 38. 10232-10236.
[47] Zhao Mei-Xin, Zhou Hao, Tang Wen-Hao, Qu Wei-Song, Shi Min. 2013. "Cinchona alkaloid-derived thiourea-catalyzed diastereo-and enantioselective [3+2] cycloaddition reaction of isocyanoacetates to isatins: A facile access to optically active spiro-oxindole oxazolines". Advanced Synthesis & Catalysis. No. 7. 1277-1283.
[48] Conde Egoitz, Rivilla Ivan, Larumbe Amaia, P. Cossío Fernando. 2015. "Enantiodivergent synthesis of bis-spiropyrrolidines via sequential interrupted and completed (3+2) cycloadditions". The Journal of Organic Chemistry. No. 23. 11755-11767.
[49] Wang De, Wang Guo-Peng, Sun Yao-Liang, Zhu Shou-Fei, *et al.* 2015. "Chiral phosphine-catalyzed tunable cycloaddition reactions of allenoates with benzofuranone-derived olefins for a highly regio-, diastereo- and enantioselective synthesis of spiro-benzo-furanones". Chemical Science. No. 12. 7319-7325.
[50] Xie Yuanwei, Yu Chenxia, Li Tuanjie, Tu Shujiang, Yao Changsheng. 2015. "An NHC-catalyzed *in situ* activation strategy to β-functionalize saturated carboxylic acid: An enantioselective formal [3+2] annulation for spirocyclic oxindolo-γ-butyro lactones". Chemistry-A European Journal. No. 14. 5355-5359.
[51] Zhao Yuan-Yuan, Zhao Shuai, Xie Ji-Kang, Hu Xiu-Qin, Xu Peng-Fei. 2016. "Synthesis of spirocyclic oxindole dihydro-thiophenes by DBU-catalyzed [3+2] annulation of Morita–Baylis–Hillman carbonates with isothiocyanates". The Journal of Organic Chemistry. No. 21. 10532-10537.
[52] Wang Zhen-Hua, Wu Zhi-Jun, Yue Deng-Feng, Hu Wen-Fei, *et al.* 2016. "Organocatalytic asymmetric [3+2] cycloaddition of *N*-

2,2,2-trifluoroethylisatin ketimines with 3-alkenyl-5-arylfuran-2(3*H*)-ones". Chemical Communications. No. 78. 11708-11711.

[53] B. Ramachary Dhevalapally, Reddy T. Prabhakar, Kumar A. Suresh. 2016. "Organocatalytic azomethine imine-olefin click reaction: high-yielding stereoselective synthesis of spiroindane-1,3-dione-pyrazolidinones". Organic & Biomolecular Chemistry. No. 27. 6517-6522.

[54] Wang Chengyuan, Zhu Shiya, Wang Guanjun, Li Zhuopeng, Hui Xin-Ping. 2016. "Enantioselective synthesis of spiro[indoline-3,2′-pyrroles] through *N*-heterocyclic-carbene-catalyzed formal [3+2] annulation". European Journal of Organic Chemistry. No. 34. 5653-5658.

[55] Yuan Zhenbo, Wei Weiwei, Lin Aijun, Yao Hequan. 2016. "Bifunctional organo/metal cooperatively catalyzed [3+2] annulation of *para*-quinone methides with vinylcyclopropanes: Approach to spiro[4.5]deca-6,9-diene-8-ones". Organic Letters. No. 14. 3370-3373.

[56] Zhang Liang-Jie, Wang Yao, Hu Xiu-Qin, Xu Peng-Fei. 2016. "Hydrogen-bonding network promoted [3+2] cycloaddition: Asymmetric catalytic construction of spiro-pseudoindoxyl derivatives". Chemistry–An Asian Journal. No. 6. 834-838.

[57] Mukherjee Subrata, Joseph Sumi, Bhunia Anup, G. Gonnade Rajesh, *et al.* 2017. "Enantioselective synthesis of spiro γ-butyrolactones by *N*-heterocyclic carbene (NHC)-catalyzed formal [3+2] annulation of enals with 3-hydroxy oxindoles". Organic & Biomolecular Chemistry. No. 9. 2013-2019.

[58] Zhu Guodong, Wu Shiqi, Bao Xiaoze, Cui Longchen, *et al.* 2017. "Asymmetric [3+2] cycloaddition of 3-amino oxindole-based azomethine ylides with α,β-ynones: A straightforward approach to spiro-oxindoles incorporating 2,5-dihydropyrroles and pyrroles". Chemical Communications. No. 34. 4714-4717.

[59] Chen Yi-Ru, Reddy Ganapuram Madhusudhan, Hsieh Kai-Hong, Chen Kai-Han, *et al.* 2018. "3-Homoacyl coumarin: An all carbon 1,3-dipole for enantioselective concerted (3+2) cycloaddition". Chemical Communications. No. 90. 12702-12705.

[60] Wu Xiao-Yun, Gao Yu-Ning, Shi Min. 2019. "Phosphine-catalyzed [3+2] annulation of *N*-2,2,2-trifluoroethyl-isatin ketimines with γ-substituted allenoates: Synthesis of spiro [indoline-3,2′-pyrrole]". European Journal of Organic Chemistry. No. 7. 1620-1626.

[61] Li Boyu, Liu Jikun, Gao Fengyun, Sun Mengmeng, *et al.* 2019. "The asymmetric construction of CF_3-containing spiro-thiazolone-pyrrolidine compounds via [3+2] cycloaddition". Organic & Biomolecular Chemistry. No. 11. 2892-2895.
[62] Chen Yu, Cui Bao-Dong, Bai Mei, Han Wen-Yong, *et al.* 2019. "Synthesis of chiral spirocyclopentene/cyclopentadiene-oxindoles through an asymmetric [3+2] cycloaddition of isatin-derived MBH carbonates and β,γ-unsaturated α-keto esters". Tetrahedron. No. 21. 2971-2979.
[63] Luo Wenjun, Shao BingXuan, Li Jingyi, Xiao Xiao, *et al.* 2020. "Construction of spiro-oxindole-fused spiropyrazolones containing contiguous three stereogenic centres via [3+2] annulation utilizing a ferrocene derived bifunctional phosphine catalyst". Organic Chemistry Frontiers. No. 8. 1016-1021.
[64] Maria Beccalli Egle, Clerici Francesca, Luisa Gelmi Maria. 1999. "6-Chloro-spirocyclohexenindol-2-ones: An unusual ring transformation to ethyl 2-(cyclohexa-1,4-dienyl)phenylcarbamates". Tetrahedron. No. 28. 8579-8586.
[65] Yang Ju, T. Cohn Stephen, Romo Daniel. 2000. "Studies toward (-)-gymnodimine: Concise routes to the spirocyclic and tetrahydrofuran moieties". Organic Letters. No. 6. 763-766.
[66] Shanmugasundaram Muthian, Raghunathan Raghavachary. 2000. "High, exoselective Diels–Alder reaction in 5.0 M Lithium perchlorate in diethyl ether medium: Efficient synthesis of novel heterocyclic derivatives containing a spirobicyclo[2.2.1]heptane system". Tetrahedron. No. 29. 5241-5245.
[67] B. Ramachary Dhevalapally, S. Chowdari Naidu, F. Barbas III Carlos. 2003. "The first organocatalytic hetero-domino Knoevenagel-Diels-Alder-epimerization reactions: Diastereoselective synthesis of highly substituted spiro[cyclohexane-1,2′-indan]-1′,3′,4-triones". Synlett. No. 12. 1910-1914.
[68] B. Ramachary Dhevalapally, S. Chowdari Naidu, F. Barbas III Carlos. 2003. "Organocatalytic asymmetric domino Knoevenagel /Diels-Alder reactions: A bioorganic approach to the diastereospecific and enantioselective construction of highly substituted spiro[5,5]undecane-1,5,9-triones". Angewandte Chemie International Edition. No. 35. 4233-4237.
[69] Matsuo Jun-ichi, Sasaki Shun, Tanaka Hiroyuki, Ishibashi Hiroyuki. 2008. "Lewis acid-catalyzed intermolecular [4+2] cyclo-addition of 3-alkoxycyclobutanones to aldehydes and

ketones". Journal of the American Chemical Society. No. 35. 11600-11601.

[70] Okitsu Takashi, Nakazawa Daisuke, Kobayashi Akihiro, Mizohata Masahiro, *et al.* 2010. "*ipso*-Iodocyclization of ethoxy ethyl ethers to alkynes at the ortho-position: An efficient synthesis of functionalized spiro compounds". Synlett. No. 02. 203-206.

[71] Wei Qiang, Gong Liu-Zhu. 2010. "Organocatalytic asymmetric formal [4+2] cycloaddition for the synthesis of spiro[4-cyclo hexanone-1,3′-oxindoline] derivatives in high optical purity". Organic Letters. No. 5. 1008-1011.

[72] Liu Yankai, Nappi Manuel, Arceo Elena, Vera Silvia, Melchiorre Paolo. 2011. "Asymmetric catalysis of Diels–Alder reactions with *in situ* generated heterocyclic *ortho*-quinodimethanes". Journal of the American Chemical Society. No. 38. 15212-15218.

[73] Jiang Xianxing, Shi Xiaomei, Wang Shoulei, Sun Tao, *et al.* 2012. "Bifunctional organocatalytic strategy for inverse-electron -demand Diels–Alder reactions: Highly efficient in situ substrate generation and activation to construct azaspirocyclic skeletons". Angewandte Chemie. No. 9. 2126-2129.

[74] Cui Hai-Lei, Tanaka Fujie. 2013. "Catalytic enantioselective formal hetero-Diels–Alder reactions of enones with isatins to give spiro-oxindole tetrahydropyranones". Chemistry-A European Journal. No. 20. 6213-6216.

[75] Li Erqing, Huang You, Liang Ling, Xie Peizhong. 2013. "Phosphine-catalyzed [4+2] annulation of γ-substituent allenoates: Facile access to functionalized spirocyclic skeletons". Organic Letters. No. 12. 3138-3141.

[76] Gicquel Maxime, Gomez Catherine, Retailleau Pascal, Voituriez Arnaud, Marinetti Angela. 2013. "Synthesis of 3,3′-spirocyclic oxindoles via phosphine catalyzed [4+2] cyclizations". Organic Letters. No. 15. 4002-4005.

[77] Zhang Xiao-Nan, Chen Gen-Qiang, Dong Xiang, Wei Yin, Shi Min. 2013. "Phosphine-catalyzed asymmetric [4+2] annulation of vinyl ketones with oxindole-derived α,β-unsaturated imines: Enantioselective syntheses of 2′,3′-dihydro-1′*H*-spiro[indoline -3,4′-pyridin]-2-ones". Advanced Synthesis & Catalysis. No. 17. 3351-3357.

[78] Ma Chao, Gu Jing, Teng Bin, Zhou Qing-Qing, *et al.* 2013. "1-Azadienes as regio- and chemoselective dienophiles in amino-

catalytic asymmetric Diels–Alder reaction". Organic Letters. No. 24. 6206-6209.

[79] Zhou Qing-Qing, Yuan Xin, Xiao You-Cai, Dong Lin, Chen Ying-Chun. 2013. "Aminocatalytic asymmetric Diels–Alder reaction of phosphorus dienophiles and 2,4-dienals". Tetrahedron. No. 48. 10369-10374.

[80] Zhu Kailong, Huang Huicai, Wu Wenbin, Wei Yuan, Ye Jinxing. 2013. "Aminocatalyzed asymmetric Diels–Alder reaction of 2,4-dienals and rhodanine/hydantoin derivatives". Chemical Communications. No. 21. 2157-2159.

[81] Anwar Shaik, Li Shao Ming, Chen Kwunmin. 2014. "Organocatalytic synthesis of substituted spirocyclohexane carbaldehydes via [4+2] annulation strategy between 2-arylideneindane-1,3-diones and glutaraldehyde". Organic Letters. No. 11. 2993-2995.

[82] Li Xin, Lin Mao-Hui, Han Yu, Wang Feng, Cheng Jin-Pei. 2014. "Asymmetric Diels-Alder reaction of 3-olefinic benzofuran-2-ones and polyenals: Construction of chiral spirocyclic benzofuran -2-ones". Organic Letters. No. 1. 114-117.

[83] Zheng Jianfeng, Lin Lili, Fu Kai, Zhang Yulong, *et al.* 2014. "Asymmetric hetero-Diels–Alder reaction of danishefsky's diene with α-ketoesters and isatins catalyzed by a chiral *N*,*N*′-dioxide/ Mg(II) complex". Chemistry-A European Journal. No. 44. 14493-14498.

[84] Li Yangyan, Su Xiao, Zhou Wei, Li Wenbo, Zhang Junliang. 2015. "Amino acid derived phosphine-catalyzed enantioselective 1,4-dipolar spiroannulation of cyclobutenones and isatylidene-malononitrile". Chemistry-A European Journal. No. 11. 4224-4228.

[85] Yang Wenjun, Zhang Yunpeng, Qiu Shuxian, Zhao Chuanqi, *et al.* 2015. "Phosphine-catalyzed [4+2] cycloaddition of unsaturated pyrazolones with allenoates: A concise approach toward spiropyrazolones". RSC Advances. No. 77. 62343-62347.

[86] Wang Yang, Tu Man-Su, Yin Lei, Sun Meng, Shi Feng. 2015. "Brønsted acid catalyzed asymmetric Diels–Alder reactions: Stereoselective construction of spiro[tetrahydrocarbazole-3,3′-oxindole] framework". The Journal of Organic Chemistry. No. 6. 3223-3232.

[87] Zheng Jianfeng, Lin Lili, Fu Kai, Zheng Haifeng, *et al.* 2015. "Synthesis of optically pure spiro[cyclohexane-oxindoline] derivatives via catalytic asymmetric Diels–Alder reaction of

brassard-type diene with methyleneindolines". The Journal of Organic Chemistry. No. 17. 8836-8842.
[88] B. Ramachary Dhevalapally, M. Krishna Patoju. 2016. "Asymmetric synthesis of nature-inspired bioactive spiro compounds through organocatalytic Diels–Alder reactions". Asian Journal of Organic Chemistry. No. 6. 729-734.
[89] Llona-Minguez Sabin, Throup Adam, Steiner Emilie, Lightowler Molly, *et al.* 2017. "Novel spirocyclic systems via multi-component aza-Diels–Alder reaction". Organic & Biomolecular Chemistry. No. 37. 7758-7764.
[90] Minamino Kou, Murata Michio, Tsuchikawa Hiroshi. 2019. "Synthesis of 7,6-spirocyclic imine with butenolide ring provides evidence for the relative configuration of marine toxin 13-desMe spirolide C". Organic Letters. No. 22. 8970-8975.
[91] Zhao Chengtao, Shi Kuangxi, He Gao, Gu Qingxiu, *et al.* 2019. "NHC-Catalyzed asymmetric formal [4+2] annulation to construct spirocyclohexane pyrazolone skeletons". Organic Letters. No. 19. 7943-7947.
[92] Shi Wangyu, Mao Biming, Xu Jiaqing, Wang Qijun, *et al.* 2020. "Phosphine-catalyzed cascade Michael addition/[4+2] cyclo-addition reaction of allenoates and 2-Arylidene-1,3-indane-diones". Organic Letters. No. 7. 2675-2680.
[93] Wang Daqian, Sun Jing, Liu Ru-Zhang, Wang Yang, Yan Chao-Guo. 2021. "Diastereo-selective synthesis of tetrahydrospiro [carbazole-1,3′-indolines] via an $InBr_3$-catalyzed domino Diels-Alder reaction". The Journal of Organic Chemistry. No. 8. 5616-5629.
[94] A. Abdelhamid Ismail, H. Mohamed Mona, M. Abdelmoniem Amr, A.S. Ghozlan Said. 2009. "DBU-Catalyzed, facile and efficient method for synthesis of spirocyclic 2-oxindole derivatives with incorporated 6-amino-4*H*-pyridazines and fused derivatives via [3+3] atom combination". Tetrahedron. No. 48. 10069-10073.
[95] Yang Hai-Bin, Shi Min. 2012. "$Yb(NTf_2)_3$-Catalyzed [3+3] cycloaddition between isatin ketonitrones and cyclopropanes to construct novel spiro[tetrahydro-1,2-oxazine]oxindoles". Organic & Biomolecular Chemistry. No. 41. 8236-8243.
[96] Sun Bing-Bing, Zhang Jun-Qi, Chen Jun-Bo, Fan Wei-Tai, *et al.* 2019. "Stereoselective synthesis of spirocyclohexadiene-pyrazol-ones via organic base and/or hydrogen bonding assisted [3+3]

annulation reactions". Organic Chemistry Frontiers. No. 11. 1842-1857.

[97] Goudedranche Sbastien, Bugaut Xavier, Constantieux Thierry, Bonne Damien, Rodriguez Jean. 2014. "α,β-Unsaturated acyl cyanides as new bis-electrophiles for enantioselective organo-catalyzed formal [3+3] spiroannulation". Chemistry–A European Journal. No. 2. 410-415.

[98] Shi Feng, Zhu Ren-Yi, Dai Wei, Wang Cong-Shuai, Tu Shu-Jiang. 2014. "Catalytic asymmetric formal [3+3] cycloaddition of an azomethine ylide with 3-indolylmethanol: Enantioselective construction of a six-membered piperidine framework". Chemistry-A European Journal. No. 9. 2597-2604.

[99] Chen Xiang, Zhang Jun-Qi, Yin Shao-Jie, Li Hai-Yan, *et al.* 2015. "Asymmetric construction of spiro[thiopyranoindole-benzoisothiazole] scaffold via a formal [3+3] spiroannulation". Organic Letters. No. 17. 4188-4191.

[100] He Xiao-Long, Xiao You-Cai, Du Wei, Chen Ying-Chun. 2015. "Enantioselective formal [3+3] cycloadditions of ketones and cyclic 1-azadienes by cascade enamine–enamine catalysis". Chemistry-A European Journal. No. 8. 3443-3448.

[101] Xie Danbo, Yang Limin, Lin Youqiang, Zhang Zhiming, *et al.* 2015. "Rapid access to spirocylic oxindoles: Application of asymmetric *N*-heterocyclic carbene-catalyzed [3+3] cycloaddition of imines to oxindole-derived enals". Organic Letters. No. 10. 2318-2321.

[102] Reddy Yetra Santhivardhana, Mondal Santigopal, Mukherjee Subrata, G. Gonnade Rajesh, T. Biju Akkattu. 2016. "Enantioselective synthesis of spirocyclohexadienones by NHC-catalyzed formal [3+3] annulation reaction of enals". Angewandte Chemie. No. 1 276-280.

[103] Xu Junyu, Zhang Weiguo, Liu Yishan, Zhu Suzhen, *et al.* 2016. "Formal [3+3] annulation of isatin-derived 2-bromoenals with 1,3-dicarbonyl compounds enabled by Lewis acid/*N*-heterocyclic carbene cooperative catalysis". RSC Advances. No. 22. 18601-18606.

[104] Zhang Weiguo, Xu Junyu, Cao Jing, Fang Chao, *et al.* 2017. "*N*-Heterocyclic carbene-mediated formal [3+3] annulation of isatin-derived α,β-unsaturated acids: Access to functionalized 3,4′-spiro-oxindole δ-lactones". Tetrahedron. No. 23. 3249-3254.

[105] Leng Hai-Jun, Li Qing-Zhu, Zeng Rong, Dai Qing-Song, *et al.* 2018. "Asymmetric construction of spiropyrazolone skeletons via

amine-catalyzed [3+3] annulation". Advanced Synthesis & Catalysis. No. 2. 229-234.
[106] Li Can, Lu Han, Sun Xiao-Xue, Mei Guang-Jian, Shi Feng. 2017. "Diastereo-and enantioselective construction of spiro-oxindole scaffolds through a catalytic asymmetric [3+3] cycloaddition". Organic & Biomolecular Chemistry. No. 22. 4794-4797.
[107] Zhao Liang-Liang, Li Xing-Shuo, Cao Li-Li, Zhang Rui, *et al.* 2017. "Access to dihydropyridinones and spiro-oxindoles: Application of *N*-heterocyclic carbene-catalyzed [3+3] annulation of enals and oxindole-derived enals with 2-aminoacrylates". Chemical Communications. No. 44. 5985-5988.
[108] Lu Hong, Tan Chang-Yin, Zhang Huan-Xin, Zhang Jia-Lu, *et al.* 2018. "Participation of β-ketothioamides in *N*-heterocyclic carbene-catalyzed [3+3] spiroannulation: Asymmetric synthesis of functionalized spiro-piperidinone derivatives". The Journal of Organic Chemistry. No. 24. 15245-15255.
[109] Xu Peng-Wei, Chen Chen, Liu Jia-Kuan, Song Yu-Ting, *et al.* 2018. "One-pot sequential [3+3] dipolar cycloaddition of aldehyde or ketone and hydroxylamine with spirocyclopropyl oxindole". The Journal of Organic Chemistry. No. 20. 12763-12774.
[110] Wang Zhan-Yong, Yang Ting, Chen Rongxiang, Ma Xueji, *et al.* 2020. "1,3-Dipolar cycloaddition of isatin *N,N′*-cyclic azomethine imines with α,β-unsaturated aldehydes catalyzed by DBU in water". RSC Advances. No. 41. 24288-24292.
[111] Liua Qiang, Chen Xiang-Yu, Lia Sun, Rissanen Kari, Enders Dieter. 2019. "*N*-Heterocyclic carbene catalyzed asymmetric synthesis of pentacyclic spiro-oxindoles via [3+3] annulations of isatin-derived enals and cyclic *N*-sulfonyl ketimines". Advanced Synthesis & Catalysis. No. 9. 1991-1994.
[112] Ji Dong-Sheng, Luo Yong-Chun, Hu Xiu-Qin, Xu Peng-Fei. 2020. "Enantioselective synthesis of spirorhodanine-pyrans via organocatalytic [3+3] annulation reactions between pyrazolones and rhodanine-derived ketoesters". Organic Letters. No. 3. 1028-1033.
[113] Gu Boqi, Wu Shuxiao, Xu Hui, Yang Wulin, *et al.* 2021. "Organocatalytic asymmetric [3+3] annulation of isatin *N,N′*-cyclic azomethine imines with enals: Efficient approach to functionalized spiro *N*-heterocyclic oxindoles". Chinese Chemical Letters. No. 2. 672-675.

[114] K. Jaiswal Manish, Singh Bhuvnesh, De Shreemoyee, Singh Neetu. 2020. "Stereoselective formal [3+3] annulation of 3-alkylidene-2-oxindoles with β,γ-unsaturated α-ketoesters". Organic & Biomolecular Chemistry. No. 48. 9852-9862.

[115] Xu You-Wei, Li Ling, Hu Xiang-Ping. 2020. "Enantioselective copper-catalyzed [3+3] cycloaddition of tertiary propargylic esters with 1*H*-pyrazol-5(4*H*)-ones toward optically active spiro-oxindoles". Organic Letters. No. 24. 9534-9538.

[116] Wei Cui, Zhang Jin-Qi, Zhang Jia-Jie, Liang Cui, Mo Dong-Liang. 2020. "Catalyst-controlled formal [4+1] annulation of *N*-vinyl fluorenone nitrones and allenoates to prepare spirofluorenyl pyrrolines". Organic Chemistry Frontiers. No. 12. 1520-1526.

[117] Chu Ming-Ming, Qi Suo-Suo, Wang Yi-Feng, Wang Biao, *et al.* 2019. "Organocatalytic asymmetric [4+1] annulation of *in situ* generated *ortho*-quinomethanes with 4-halopyrazolones: Straightforward access to chiral spiro-benzofuran pyrazolones". Organic Chemistry Frontiers. No. 12. 1977-1982.

[118] Han Xiaoyu, Yao Weijun, Wang Tianli, Ren Tan Yong, *et al.* 2014. "Asymmetric synthesis of spiropyrazolones through phosphine-catalyzed [4+1] annulation". Angewandte Chemie International Edition. No. 22. 5643-5647.

[119] Hu Fang-Le, Wei Yin, Shi Min. 2014. "Phosphine-catalyzed asymmetric [4+1] annulation of activated α,β-unsaturated ketones with Morita–Baylis–Hillman carbonates: Enantioselective synthesis of spiro-oxindoles containing two adjacent quaternary stereocenters". Chemical Communications. No. 64. 8912-8914.

[120] Zhou Rong, Han Ling, Zhang Honghui, Liu Rongfang, Li Ruifeng. 2017. "A deoxygenative [4+1] annulation involving *N*-acyldiazenes for an efficient synthesis of 2,2,5-trisubstituted 1,3,4-oxadiazole derivatives". Advanced Synthesis & Catalysis. No. 22. 3977-3982.

[121] Wu You-Cai, Cui Bao-Dong, Long Yan, Han Wen-Yong, *et al.* 2021. "Chiral phosphoric acid catalyzed (4+1) annulation of 3-diazooxindoles/4-diazooxisoquinolines with *para*-quinone methides to access chiral spiro[dihydrobenzofuran-2,3′-oxindoles/2,4′-oxisoquinolines]". Advanced Synthesis & Catalysis. No. 6. 1702-1713.

[122] V. Khairnar Pankaj, Wu Chi-Yi, Lin Yi-Fang, Edukondalu Athukuri, *et al.* 2020. "Diversity-oriented synthesis of spiropentadiene pyrazolones and 1*H*-oxepino[2,3-*c*]pyrazoles

from doubly conjugated pyrazolones via intramolecular Wittig reaction". Organic Letters. No. 12. 4760-4765.
[123] Tang Xiaodong, Ni Huanzhen, Lu Yixin. 2021. "Phosphine-catalyzed enantioselective [4+1] annulation of oxindoles with allenic ketones for the construction of spirocyclopentene oxindoles". Organic Chemistry Frontiers. No. 16. 4485-4489.
[124] Wang Chuan-Chuan, Huang Jian, Li Xin-Hao, Kramer Søren, *et al.* 2018. "Asymmetric organocatalytic [4+1] annulations: Enantioselective construction of multifunctionalized spirocyclo pentane oxindoles bearing α,α-disubstituted α-amino-β-keto esters". Organic Letters. No. 10. 2888-2891.
[125] Tan Chang-Yin, Lu Hong, Zhang Jia-Lu, Liu Jin-Yu, Xu Peng-Fei. 2019. "Asymmetric organocatalytic [4+1] annulations involving a polarity reversal process: A tandem catalytic approach to highly functionalized spiropyrazolone derivatives". The Journal of Organic Chemistry. No. 2. 594-602.
[126] Wu Bin, Chen Jian, Li Mei-Qiu, Zhang Jin-Xin, *et al.* 2012. "Highly enantioselective synthesis of siro[cyclohexanone-oxindoles] and spiro[cyclohexanone-pyrazolones] by asymmetric cascade [5+1] double michael reactions". European Journal of Organic Chemistry. 1318-1327.
[127] Roy Suparna, Amireddy Mamatha, Chen Kwunmin. 2013. "Organocatalytic formal [5+1] annulation: Diastereoselective cascade synthesis of functionalized six-membered spirocyclic indane-1,3-diones/oxindoles via Michael–aldol reaction". Tetrahedron. No. 41. 8751-8757.
[128] Zhang Yun, S. Panek James. 2009. "Stereocontrolled synthesis of spiro-oxindoles through lewis acid-promoted [5+2]-annulation of chiral silyl alcohols". Organic Letters. No. 15. 3366-3369.
[129] Jiang Kun, Jia Zhi-Jun, Chen Shi, Wu Li, Chen Ying-Chun. 2010. "Organocatalytic tandem reaction to construct six-membered spirocyclic oxindoles with multiple chiral centres through a formal [2+2+2] annulation". Chemistry–A European Journal. No. 9. 2852-2856.
[130] Kuan Hsuan-Hao, Chien Chung-Han, Chen Kwunmin. 2013. "Synthesis of fully substituted dispirocyclohexanes by organo-catalytic [2+2+2] annulation strategy between 2-arylideneindane-1,3-diones and aldehydes". Organic Letters. No. 11. 2880-2883.
[131] Zhang Yi-Ya, Gurubrahamam Ramani, Chen Kwunmin. 2015. "Rauhut–Currier-initiated organocascade reaction: Synthesis of substituted dispirocyclohexanes through a [2+2+2] strategy

between 2-arylideneindan-1,3-diones and activated alkenes". Advanced Synthesis & Catalysis. No. 11. 2457-2463.

[132] Li Zhi-Long, Liu Chao, Tan Rui, Tong Zhi-Ping, Liu Yan-Kai. 2016. "Organocatalytic, asymmetric [2+2+2] annulation to construct six-membered spirocyclic oxindoles with six continuous stereogenic centers". Catalysts. No. 5. 65.

[133] Doro Franco, Akeroyd Niels, Schiet Franc, Narula Anubhav. 2019. "The Prins reaction in the fragrance industry: 100th anniversary (1919–2019)". Angewandte Chemie. No. 22. 7248-7253.

[134] Kishi Yuichiro, Nagura Hirokatsu, Inagi Shinsuke, Fuchigami Toshio. 2008. "Facile and highly efficient synthesis of fluorinated heterocycles via Prins cyclization in ionic liquid hydrogen fluoride salts". Chemical Communications. No. 33.3876-3878.

[135] Castaldi M. Paola, M. Troast Dawn, A. Porco Jr John. 2009. "Stereoselective synthesis of spirocyclic oxindoles via Prins cyclizations". Organic Letters. No. 15. 3362-3365.

[136] N Chavre Satish, Reddy Ullapu Punna, Min Sun-Joon, Kyun Lee Jae, *et al.* 2009. "Stereocontrolled synthesis of oxaspirobicycles via Prins-Pinacol annulation". Organic Letters. No. 17. 3834-3837.

[137] V. Subba Reddy Basi, Swathi Devabhaktuni, Swain Manisha, Pal Bhadra Manika, *et al.* 2014. "Stereoselective synthesis of spiro[tetrahydropyran-3,3′-oxindole] derivatives employing Prins cascade strategy". Organic Letters. No. 24. 6267-6269.

[138] Damera Krishna, Yu Bingchen, Wang Binghe. 2015. "Stereoselective synthesis of 1-methyl-3′,4′,5′,6′-tetrahydrospiro [indoline-3,2′-pyran]-2-one derivatives via Prins cyclization". The Journal of Organic Chemistry. No. 11. 5457-5463.

[139] V. Subba Reddy Basi, Yarlagadda Suresh, Reddy C. Ravikumar, Reddy M. Rajashekhar, *et al.* 2015. "Tandem Prins strategy for the synthesis of spiropyrrolidine and spiropiperidine derivatives". European Journal of Organic Chemistry. No. 14. 3076-3085.

[140] V. Subba Reddy Basi, Medaboina Durgaprasad, Sridhar Balasubramanian. 2015. "Stereoselective synthesis of hexahydro-1*H*-spiro[isoquinoline-4,4′-pyran] scaffolds through an intramolecular Prins cascade process". The Journal of Organic Chemistry. No. 1. 653-660.

[141] V. Subba Reddy Basi, Raju N. Prudhvi, Mohan Reddy B. Jagan, Sridhar Balasubramanian. 2016. "Sequential Prins/Pinacol

strategy for the stereoselective synthesis of spiro-β-tetralones". Asian Journal of Organic Chemistry. No. 3. 411-416.

[142] Rapelli Chandrashekhar, Sridhar Balasubramanian, V. Subba Reddy Basi. 2020. "Tandem Prins cyclization for the synthesis of indole fused spiro-1,4-diazocane scaffolds". Organic & Biomolecular Chemistry. No. 34. 6710-6715.

[143] Jang Woo Cheol, Jung Myeongjin, Min Ko Haye. 2021. "Synthesis of six-membered spiro azacyclic oxindole derivatives via a one-pot process of umpolung allylation/aza-Prins cyclization". Organic Letters. No. 4. 1510-1515.

[144] T. R. Liddon John, A. Rossi-Ashton James, J. K. Taylor Richard, P. Unsworth William. 2018. "Dearomatizing spiroannulation reagents: Direct access to spirocycles from indoles and dihalides". Organic Letters. No. 11. 3349-3353.

[145] A. McDaniel Kelly, R. Blood Anna, C. Smith Gavin, T. Jui Nathan. 2021. "Dearomatization of unactivated arenes via catalytic hydroalkylation". ACS Catalysis. No. 9. 4968-4972.

[146] Liu Yu, Wang Qiao-Lin, Xiong Bi-Quan, Zhang Pan-Liang, *et al.* 2018. "Visible-light-mediated *ipso*-carbosulfonylation of alkynes: Synthesis of 3-sulfonylspiro[4,5]trienones from propiolamides and sulfonyl chlorides under transition-metal-free conditions". Synlett. No. 18. 2396-2403.

[147] Liu Yu, Wang Qiao-Lin, Chen Zan, Zhou Quan, *et al.* 2019. "Visible-light promoted one-pot synthesis of sulfonated spiro-[4,5]trienones from propiolamides, anilines and sulfur dioxide under transition metal-free conditions". Chemical Communications. No. 81. 12212-12215.

[148] Chen Yuan, Chen Yu-Jue, Guan Zhi, He Yan-Hong. 2019. "Visible-light-mediated selective thiocyanation/*ipso*-cyclization/oxidation cascade for the synthesis of thiocyanato-containing azaspirotrienediones". Tetrahedron. No. 51. 130763.

[149] Xie Ling-Hai, Liu Feng, Tang Chao, Hou Xiao-Ya, *et al.* 2006. "Unexpected one-pot method to synthesize spiro[fluorene-9,9′-xanthene] building blocks for blue-light-emitting materials". Organic Letters. No. 13. 2787-2790.

[150] Zhou Ying, Zhang Xiang, Zhang Yong, Ruan Linxin, *et al.* 2017. "Iodocyclization of *N*-arylpropynamides mediated by hypervalent iodine reagent: Divergent synthesis of iodinated quinolin-2-ones and spiro[4,5]trienones". Organic Letters. No. 1. 150-153.

[151] Zeng Fan-Lin, Chen Xiao-Lan, Sun Kai, Zhu Hu-Lin, *et al.* 2021. "Visible-light-induced metal-free cascade cyclization of *N*-aryl-

propiolamides to 3-phosphorylated, trifluoromethylated and thiocyanated azaspiro[4.5]trienones". Organic Chemistry Frontiers. No. 4. 760-766.

[152] Yang Bin-Miao, Cai Pei-Jun, Tu Yong-Qiang, Yu Zhi-Xiang, *et al.* 2015. "Organocatalytic asymmetric tandem Nazarov cyclization/semipinacol rearrangement: Rapid construction of chiral spiro[4.4]nonane-1,6-diones". Journal of the American Chemical Society. No. 26. 8344-8347.

[153] Etling Christoph, Tedesco Giada, Kalesse Markus. 2021. "A Nazarov-ene tandem reaction for the stereoselective construction of spiro compounds". Chemistry-A European Journal. No. 36. 9257-9262.

[154] Fan Wei-Tai, Li Nai-Kai, Xu Lumei, Qiao Chunhua, Wang Xing-Wang. 2017. "Organo-catalyzed asymmetric Michael–hemiketalization–oxa-Pictet–Spengler cyclization for bridged and spiro heterocyclic skeletons: Oxocarbenium ion as a key intermediate". Organic Letters. No. 24. 6626-6629.

[155] Wang Xin-Wei, Li Xiang, Chen Mu-Wang, Wu Bo, Zhou Yong-Gui. 2021. "Chiral phosphoric acid-catalyzed Pictet–Spengler reactions for synthesis of 5′,11′-dihydrospiro[indoline-3,6′-indolo-[3,2-*c*]qui-nolin]-2-ones containing quaternary stereo-centers". The Journal of Organic Chemistry. No. 9. 6897-6906.

6

APPLICATION OF SPIROCYCLIC FRAMEWORKS

6.1. Introduction

Spirocyclic compounds are unique in their twisted structure, where two or more rings are connected at the same atom in a noncoplanar fashion, as first reported by von Baeyer in 1900. The book covers the latest research and technical developments in spiro compound synthesis and their potential medicinal properties. It serves as a valuable resource for anyone interested in understanding the importance of spirocyclic compounds and their applications.

6.2. Spiro compounds as drugs

Spiro heterocyclic compounds possess unique properties, such as spiroconjugation, spiro-hyperconjugation, and an anomeric effect, which give them considerable biological activities. The special space configurations of these compounds match the spatial structures of macromolecules in organisms, making them suitable as pharmacophore units in the construction of various drugs, including anti-neoplastic, anti-depressant, ant-iviral, anti-bacterial, and anti-anxiety drugs [1]. Therefore, we conducted a comprehensive review of the literature and technical developments of spiro compounds to provide an in-depth understanding of their synthesis and medicinal properties.

6.2.1. Spiro compounds as anti-oxidant drugs

Ashok et al. achieved successful synthesis of a series of innovative chromane scaffolds that include spirochromanones **3a-i** and **4a-i**. This synthesis was accomplished through the utilization of 2-hydroxy acetophenone **1a-b** and cyclic alkanones **2a-i**, employing microwave irradiation. The reaction demonstrated high efficiency, delivering the desired products in good yields (Scheme 6-1) [2]. The synthesized compounds were evaluated for their anti-oxidant activities and compared with standard drugs.. Among all the synthesized compounds, **3d** (6.09), **3g** (5.32), **3h** (2.03), **4a** (1.17), **4b** (0.50), **4c** (6.59), **4d** (7.86), **4e** (6.85), **4f**

(4.82), and **4g** (6.59) showed higher anti-oxidant activity comparable to that of ascorbic acid (IC_{50} 8.64 μM).

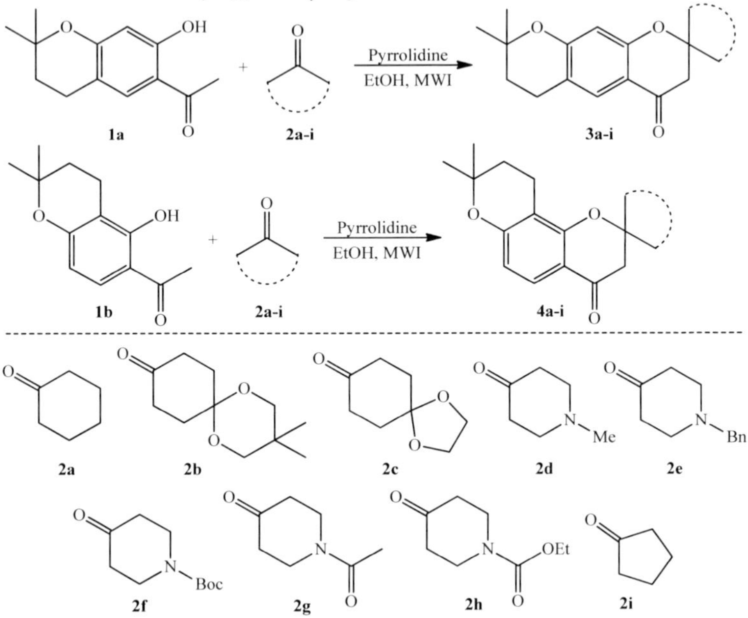

Scheme 6-1. Synthesis of spirochromanones **3a–i** and **4a–i**.

Furthermore, the same research group reported a highly efficient and cost-effective synthesis of novel indolin-2-one fused spirochromanone conjugates **6a-p**. This was achieved using Kabbe condensation of substituted 3-[2-(5-acetyl-2,4-dihydroxyphenyl)-2-oxoethylidene]-5-substituted-indolin-2-ones **5a-b** with various cycloalkanones in the ionic liquid [bmim]Cl·$FeCl_3$ under microwave irradiation (Scheme 6-2) [3]. The synthesized compounds **6a-p** were evaluated for their anti-oxidant activity, with compounds **6b** and **6d** exhibiting the highest activity. The IC_{50} values for compounds **6b** and **6d** were found to be 1.25 and 1.74 μM, respectively, which is comparable to that of ascorbic acid (IC_{50} 8.64 μM), a standard anti-oxidant agent.

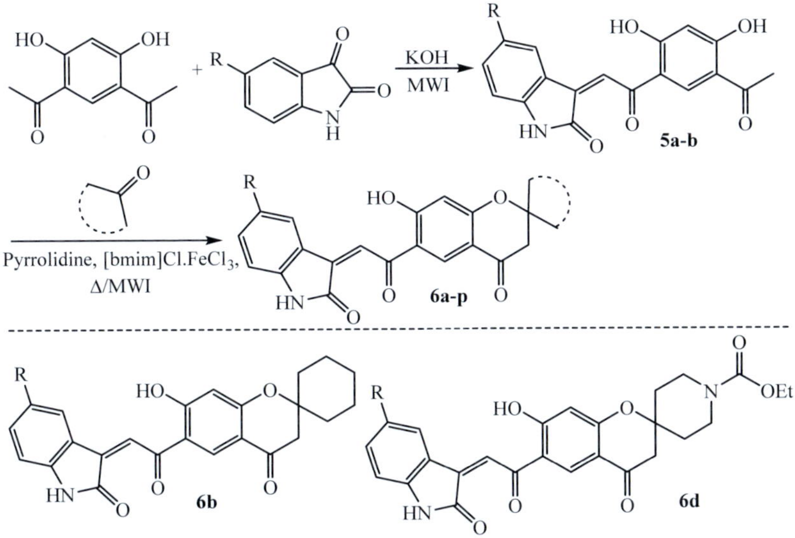

Scheme 6-2. Synthesis of indolin-2-one fused spirochromanone conjugates **6a-p**.

The Rajendran group [4] reported on a high atom-economical, one-pot synthesis of biologically active novel spiro-indenoquinoxaline pyrrolizines **7a-f** and **8a-f**. These compounds accumulate three pharmocophoric cores, namely heterocyclic quinoline scaffold, pyrrolizines, and indeno-quinoxaline, in a single molecular framework. The synthesis was achieved through a four-component reaction between ninhydrin, *o*-phenylene-diamine, L-proline, and quinolinyl chalcones in methanol via [3+2] cycloaddition (Scheme 6-3). The newly synthesized compounds were evaluated for in vitro anti-oxidant activity using DPPH, nitric oxide, and superoxide radicals. Notably, all the compounds exhibited excellent anti-oxidant activity compared to the standard molecule, i.e., BHT.

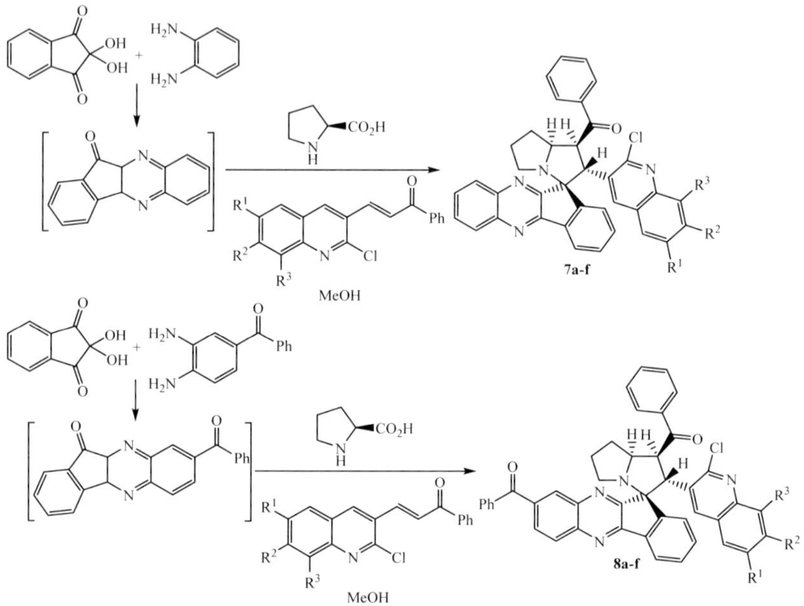

Scheme 6-3. Synthesis of biologically active novel spiro-indeno-quinoxaline pyrrolizines **7a-f** and **8a-f**.

6.2.2. Spiro compounds as anti-diabetic drugs

In 2008, Somsák et al. introduced glucopyranosylidene-spiro-oxa thiazoles **9a-d** as a novel class of compounds with nanomolar inhibitory activity against glycogen phosphorylase. These compounds were identified as potential anti-diabetic agents [5]. The synthetic route to the glucopyrano -sylidene-spiro-oxathiazoles **9a-d** is schematically represented in Scheme 6-4. As mentioned before, the GP inhibitor candidates were evaluated for their inhibitory activity against the RMGPb enzyme, and the resulting inhibitor constants (Ki) are shown in Scheme 6-4. The research team discovered that incorporating a rigid spiro-bicyclic structure, along with the introduction of a large hydrophobic aromatic moiety in an optimal orientation for optimal interaction with the enzyme, significantly enhances the biological activity of the molecules. Accordingly, the 2-naphthyl substituted **9d** is currently the most potent glucose-based inhibitor of GP, with inhibition observed in the nanomolar range.

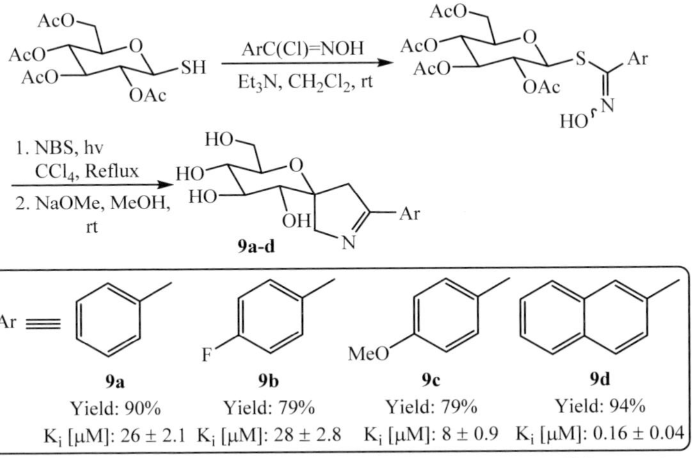

Scheme 6-4. Synthetic route to the glucopyranosylidene-spiro-oxathiazoles **9a-d**.

In another study, Narayanan et al. developed an efficient method for synthesizing regio- and stereocontrolled di-spiropyrrolidines **10** of thiazolidinedione and rhodanine which are found to expose attractive anti-diabetic properties to male Wistar rats (Scheme 6-5) [6].

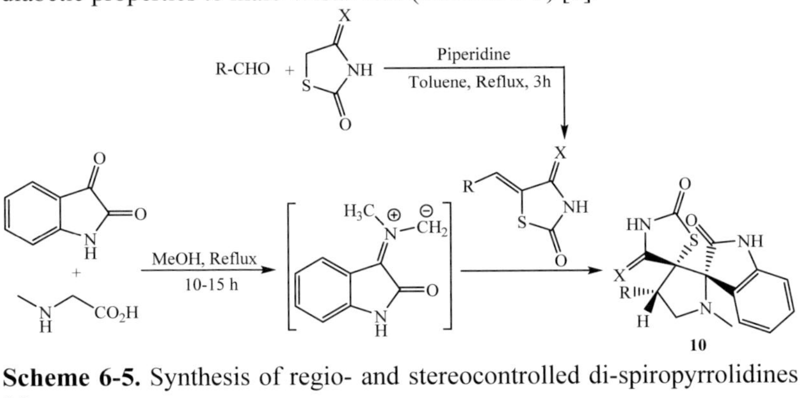

Scheme 6-5. Synthesis of regio- and stereocontrolled di-spiropyrrolidines **10**.

Askri et al. [7] proposed a three-component 1,3-DCR of stabilized azomethine ylide, which was generated *in situ* by the condensation of methyl glycine ester with isatin derivatives, to various arylidene rhodanines, producing a range of novel rhodanine-fused spiro[pyrrolidin-2,3′-oxindole]s **11a-s** (Scheme 6-6). Evaluation of these derivatives in vitro against the α-amylase enzyme revealed that among the series **11**, derivatives **11g**, **11k**, and **11s** exhibited the highest anti-diabetic activity. The structure-activity relationship suggested the influence of the substituents' nature and electronic effect on the isatin moiety. Notably, the phenyl part of (*Z*)-5-arylidine-2-thioxothiazolidin-4-ones had a significant impact on the inhibitory activity. Additionally, substituted spiro-pyrrolidines exhibited greater inhibitory activity compared to their unsubstituted counterparts.

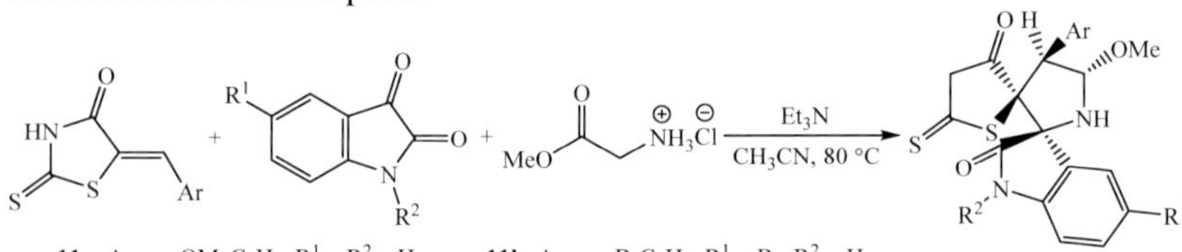

11a-s

11a: Ar = *p*-OMeC$_6$H$_4$, R^1 = R^2 = H
11b: Ar = *p*-OMeC$_6$H$_4$, R^1 = H, R^2 = Bn
11c: Ar = *p*-OMeC$_6$H$_4$, R^1 = Br, R^2 = H
11d: Ar = *p*-OMeC$_6$H$_4$, R^1 = Cl, R^2 = H
11e: Ar = C$_6$H$_5$, R^1 = H, R^2 = H
11f: Ar = C$_6$H$_5$, R^1 = Cl, R^2 = H
11g: Ar = C$_6$H$_5$, R^1 = Br, R^2 = H
11h: Ar = C$_6$H$_5$, R^1 = H, R^2 = Bn
11i: Ar = *p*-OMeC$_6$H$_4$, R^1 = Br, R^2 = H
11j: Ar = *p*-BrC$_6$H$_4$, R^1 = Cl, R^2 = H
11k: Ar = *p*-BrC$_6$H$_4$, R^1 = Br, R^2 = H
11l: Ar = *p*-BrC$_6$H$_4$, R^1 = NO$_2$, R^2 = H
11m: Ar = *p*-FC$_6$H$_4$, R^1 = Br, R^2 = H
11n: Ar = *p*-FC$_6$H$_4$, R^1 = H, R^2 = Bn
11o: Ar = *p*-FC$_6$H$_4$, R^1 = Cl, R^2 = H
11p: Ar = *p*-ClC$_6$H$_4$, R^1 = H, R^2 = H
11q: Ar = *p*-ClC$_6$H$_4$, R^1 = H, R^2 = Bn
11r: Ar = *p*-ClC$_6$H$_4$, R^1 = Cl, R^2 = H
11s: Ar = *p*-ClC$_6$H$_4$, R^1 = Br, R^2 = H

Scheme 6-6. Synthesis of rhodanine-fused spiro[pyrrolidin-2,3′-oxindole]s **11a-s**.

6.2.3. Spiro compounds as anti-Alzheimer drugs

Spiro dioxolanes such as (**12**: ethyl 2-methyl-1,3-dioxa-8-azaspiro-[4.5]decane-8-carboxylate) and oxathiolanes such as (**13**: 2′,2′-dimethyl-4-azaspiro[bicyclo[2.2.2]octane-2,5′-[1,3]oxathio-lane]) were patented for the treatment of Alzheimer's disease.

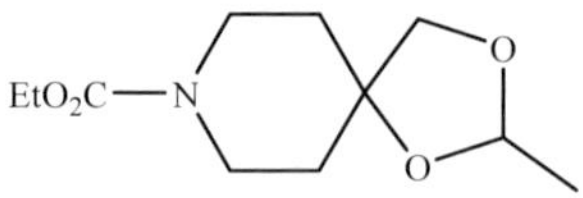

12

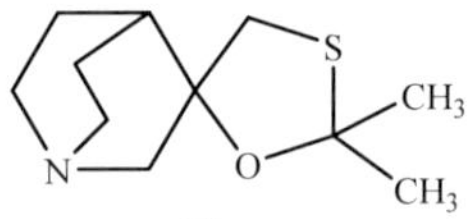

13

Quiroga et al. [8] designed the synthesis of fused spiropyrazolo[1,5-*c*]quinazolines **14a-h** as potential cholinesterase inhibitors. They evaluated their inhibitory activity against AChE and butyryl- cholinesterase (BuChE) and used molecular docking and molecular mechanics with generalised Born and surface area solvation (MM/GBSA) calculations to analyze the protein-ligand interactions for further rational design of more potent inhibitors (Scheme 6-7). The most potent compound exhibited a moderate

AChE inhibitory activity (**14f**: IC_{50} = 84 μM), and molecular docking studies showed that its binding mode shares common characteristics with the galantamine/donepezil-AChE complexes. Free binding energy (ΔG) calculations agreed with the experimental biological activity values. Theoretical results suggested that halogen bond interactions may be involved in the differential potency of these compounds and provide a new starting point for designing novel pyrazolo- [1,5-*c*]quinazolines as anti-Alzheimer agents.

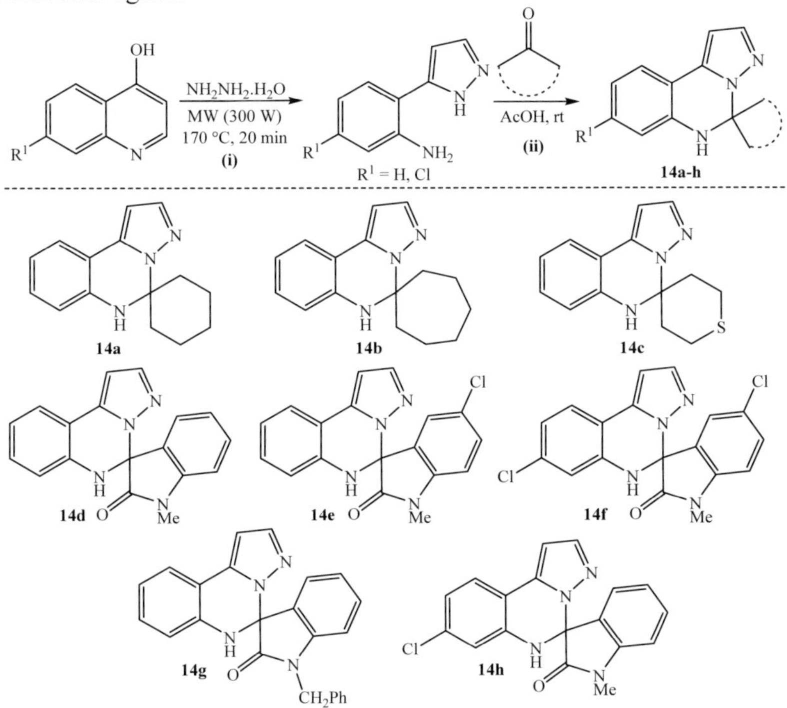

Scheme 6-7. Synthesis of fused spiro-pyrazolo[1,5-*c*]quinazolines **14a-h**.

The Hasaninejad group conducted a convenient synthesis of substituted spiroindenoquinoxalines **15a-l** at mild and green conditions [9]. The multicomponent reaction of substituted phenylene diamines, ninhydrin, malononitrile, and *N,N'*-substituted-2-nitroethane-1,1-diamines generated the target compounds, as shown in Scheme 6-8. The ability of twelve new spiroindenoquinoxalines **15a-l** in inhibiting acetyl and butyrylcholinesterases cholinesterases was evaluated both *in vitro* and *in* *silico*. All compounds disclosed moderate-level activity against both acetyl and butyryl cholinesterases.

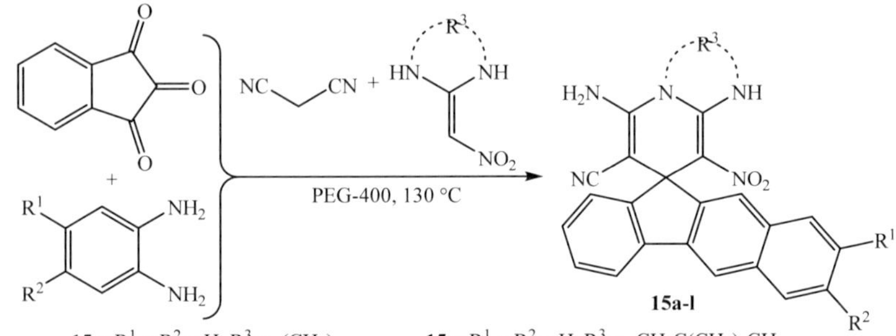

15a; $R^1 = R^2 = H$, $R^3 = -(CH_2)_2-$ **15g;** $R^1 = R^2 = H$, $R^3 = -CH_2C(CH_3)_2CH_2-$
15b; $R^1 = R^2 = Me$, $R^3 = -(CH_2)_2-$ **15h;** $R^1 = R^2 = Me$, $R^3 = -CH_2C(CH_3)_2CH_2-$
15c; $R^1 = NO_2$, $R^2 = H$, $R^3 = -(CH_2)_2-$ **15i:** $R^1 = NO_2$, $R^2 = H$, $R^3 = -CH_2C(CH_3)_2CH_2-$
15d; $R^1 = R^2 = H$, $R^3 = -(CH_2)_3-$ **15j;** $R^1 = R^2 = H$, $R^3 = 2PhCH_2-$
15e; $R^1 = R^2 = Me$, $R^3 = -(CH_2)_3-$ **15k;** $R^1 = R^2 = Me$, $R^3 = -(CH_2)_2-$
15f; $R^1 = NO_2$, $R^2 = H$, $R^3 = -(CH_2)_3-$ **15l;** $R^1 = R^2 = NO_2$, $R^3 = -(CH_2)_2-$

Scheme 6-8. Synthesis of substituted spiroindenoquinoxalines **15a-l**.

6.2.4. Spiro compounds as anti-bacterial drugs

In 2010, Chen et al. reported the synthesis of a series of C-12 pyrazolinyl spiro ketolides **16a-d** [10], which replaced the natural C-12 methyl group in the clarithromycin core with different pyrazolinyl spiros through chemical synthesis (Scheme 6-9). The anti-bacterial activities of the compounds with various alkane and ester groups at pyrazolinyl spiros were evaluated against erythromycin-susceptible and erythromycin-resistant bacteria, with particular emphasis on respiratory pathogens. The results showed that all derivatives had better anti-bacterial activities than erythromycin A and clarithromycin against *Staphylococcus aureus* strains, and almost equivalent bioactivities against *Streptococcus pneumonia* and *Haemophilus influenza* strains. Among them, compounds (**16c** and **16d**) with ester substituents exhibited better anti-bacterial activities than those with alkyl substituents (**16a** and **16b**).

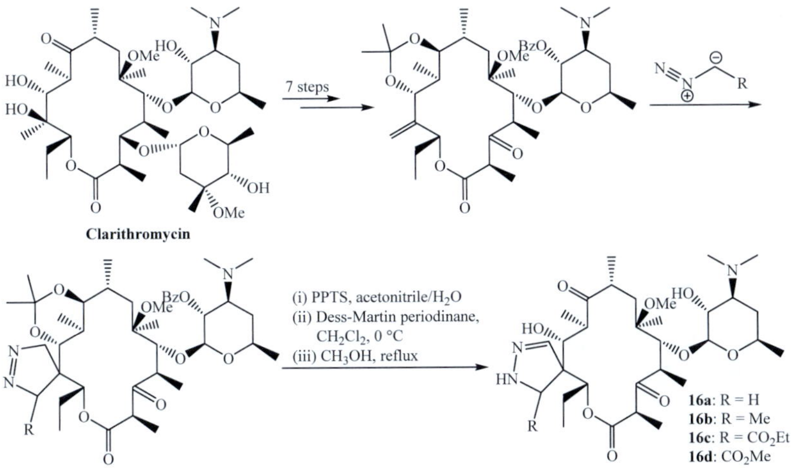

Scheme 6-9. Synthesis of C-12 pyrazolinyl spiro ketolides **16a-d**.

Spiro-oxindoles **17a-j** were synthesized by a three-multicomponent reaction involving 1-methyl quinoline-2,4(1*H*,3*H*)-dione, isatins, and malononitrile/cyanoacetic ester in the task-specific ionic liquid [DBU][Ac] (Scheme 6-10) [11]. The desired products **17a-j** were evaluated for anti-bacterial activity against gram-positive and gram-negative bacteria (*Staphylococcus aureus*, *Escherichia coli*, *Bacillus cereus*, *Bacillus subtilis*, *Salmonella typhimurium*, and *Klebsiella pneumonia*). The four compounds, **17b**, **17e**, **17f**, **17g**, and **17j**, represented a moderate activity against all the strains compared to the standard employed.

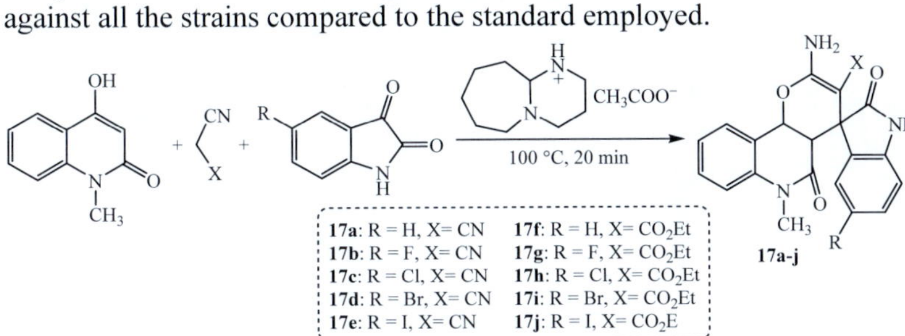

Scheme 6-10. Synthesis of spiro-oxindoles **17a-j**.

Recently, two spiro compounds **18** and **19** (Figure 6-1) were isolated from the acetone extract of *Artemisia Pallens* Wall roots using bioassay-guided isolation. The anti-bacterial activity of the acetone extract, fractions, and isolated compounds was evaluated against two gram-negative bacterial strains (*Escherichia coli* and *Pseudomonas aeruginosa*) and two gram-positive bacterial strains (*Staphylococcus aureus* and *Bacillus subtilis*). Compound **19** exhibited the best spectrum of activity, with IC_{50} and MIC values ranging from 2.48-3.08 and 12.78-21.77 μM, respectively, for all four bacterial strains tested. Compound **18** showed weaker activity, with IC_{50} and MIC values ranging from 2.57-3.69 and 38.17-80.57 mM, respectively. However, both compounds were found to be inactive against *Mycobacterium tuberculosis* [12].

Artemisia pallens

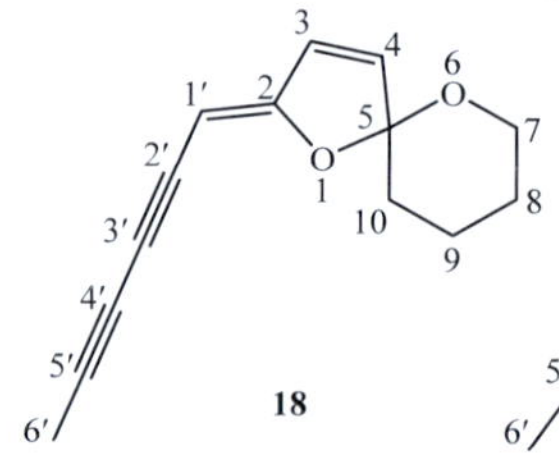

18

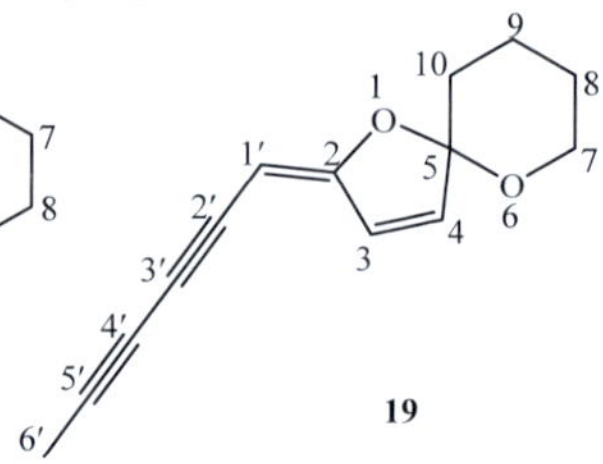

19

Fig. 6-1. Structures of compounds **18** and **19**.

6.2.5. Spiro compounds as anti-platelet drugs

Peng et al. investigated the EtOAc extract of L. Japonicus in 2015 and discovered six new bis-spirolabdane diterpenoids, **20-25**, along with four known analogs, **26-29** (Figure 6-2) [13]. These compounds were evaluated for their inhibitory activity against abnormal platelet aggregation induced by adenosine diphosphate. Interestingly, only the (13*R*)-configured bis-spirolabdane diterpenoids exhibited significant activity against platelet aggregation.

L. japonicus

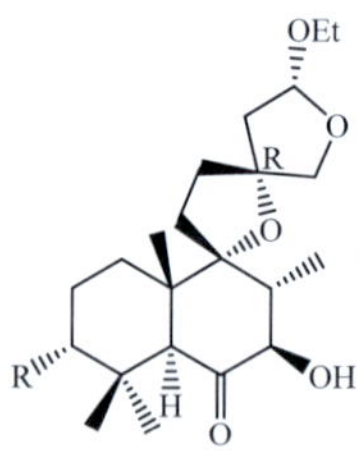

20 R = OAc
21 R = H

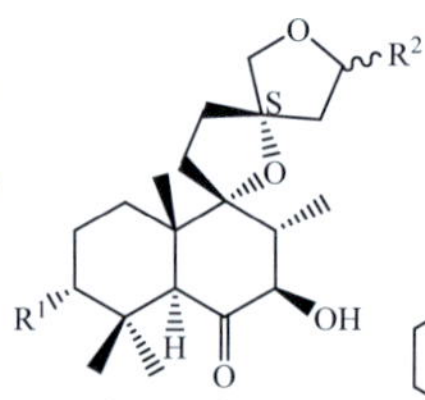

22 R^1 = H, R^2 = α-OEt
23 R^1 = H, R^2 = β-OMe
24 R^1 = OAc, R^2 = β-OMe
26 R^1 = OAc, R^2 = α-OEt
27 R^1 = OAc, R^2 = β-OEt
28 R^1 = OAc, R^2 = α-OH
29 R^1 = OAc, R^2 = β-OH

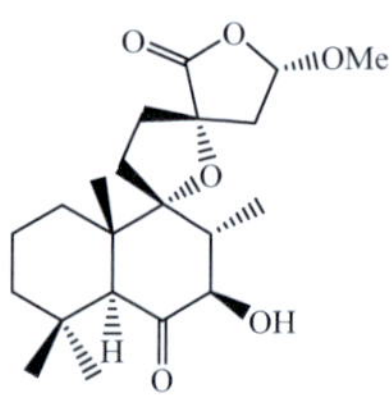

25

Fig. 6-2. Chemical structures of compounds **20-29**.

Anti-platelet drugs are widely used in the treatment and prevention of acute coronary syndromes and recurrent events. However, many patients exhibit resistance to current anti-platelet therapy. To address this issue, researchers have explored 5-HT_{2A} receptor antagonists as an alternative anti-platelet therapy.

Czopek et al. conducted a study in which they synthesized and examined the affinity of two compounds, featuring a 4-phenyl cyclohexane-5-spiro- or 5-methyl-5-phenyl-hydantoin core linked to different aryl piperazine moieties (compounds **30-43**), towards the 5-HT_{2A} receptor (Figure 6-3) [14].

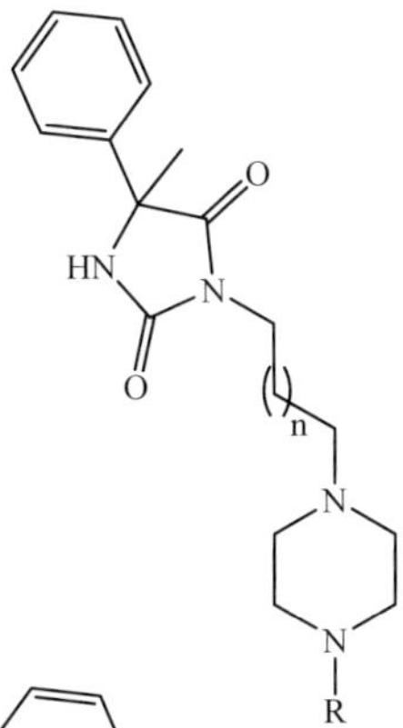

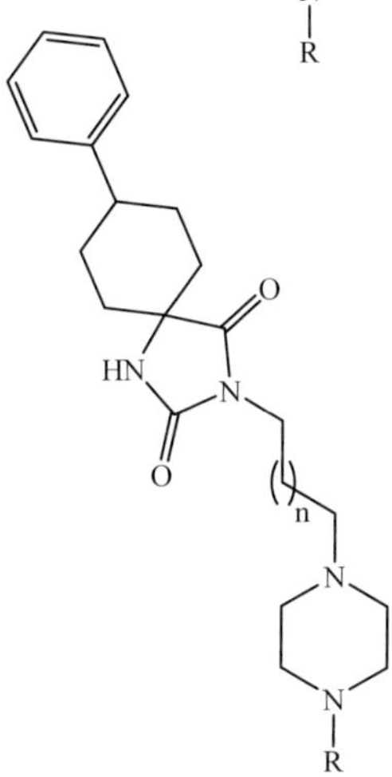

Compound	*n*	R	Ki[nM] ± SEM
30	0	–CO4FPh	nd
31	0	–CH(Ph)$_2$	45.0 ± 43.9
32	1	–2FPh	516.0 ± 52.0
33	1	–3FPh	340.0 ± 57.0
34	1	–4FPh	48.0 ± 4.9
35	1	–CO4FPh	nd
36	1	–CO4FPh	nd
37	0	–CH(Ph)$_2$	nd
38	1	–2FPh	220.0 ± 19.6
39	1	–3FPh	297.0 ± 19.9
40	1	–4FPh	21.0 ± 1.7
41	1	–CO4FPh	2584.0 ± 302.0
42	1	–Ph	20.0 ± 4.8
43	1	–3ClPh	24.0 ± 2.5
Mianserin			1.2 ± 0.2

Fig. 6-3. Chemical structures of compounds **30-43** and their radioligand binding data and reference drug (mianserin) for serotonin 5-HT_{2A} receptors.

Functional bioassays showed that most of the synthesized compounds had moderate affinity and antagonistic activity for rat and human serotonin 5-HT_{2A} receptors. Among them, compound **42**, with the chemical name 8-phenyl-3-(3-(4-phenylpiperazin-1-yl)propyl)-1,3-diazaspiro[4.5]-decane-2, 4-dione, was found to be the most potent inhibitor of collagen-stimulated aggregation of whole rat blood, with an IC_{50} value of 27.3 μM, which is comparable to sarpogrelate (IC_{50} = 66.8 μM) and ketanserin (IC_{50} = 32.1 μM).

The tested compounds were also evaluated for their inhibitory effect on 5-HT-amplified and ADP- or collagen-induced platelet aggregation. The 5-HT-dependent aggregation studies showed that compounds **32** and **42** were the most active. In conclusion, 5-HT_{2A} antagonists have the potential to effectively inhibit platelet aggregation and represent an exciting target for the development of novel anti-platelet agents with an alternative mechanism of action.

6.2.6. Spiro compounds as anti-cancer drugs

Das et al. reported the synthesis of several fluoro-substituted spiro-isoxazoline ethers and lactones **44a-z** and **45a-k** by electrophilic fluorine-mediated decorative spirocyclization (Scheme 6-11) [15]. The *in vitro* biological tests of synthesized compounds were explored for anti-viral activity against human cytomegalovirus (HCMV) and cytotoxicity against glioblastomas (GBM6) and triple-negative breast cancer (MDA MB 231). Interestingly, compounds **4d** and **4n** illustrated a considerable activity against HCMV (IC_{50} = 10 μM), while **44l** and **45f** showed the highest cytotoxicity with IC_{50} = 36 to 80 μM.

Hamme et al. explored 1,3-DC some synthetic intermediates for spiro-isoxazolines **46** via intramolecular cyclization of either alcohol or carboxylic acid precursors (Scheme 6-12) [16]. The biochemical anti-proliferative activity was investigated *in vitro* by using two breast cancer cell lines (MCF-7 and MDA-MB-231) and two prostate cancer cell lines (PC-3 and DU-145) using the MTT viability assay, and the IC_{50} values were achieved. Spiro-isoxazolines possessing a *p*-chloro or an *o*-dichloro aromatic substituent at the 3-position of the isoxazoline exhibited significant anti-tumor activities in all four cell lines with IC_{50} values ranging from 43 μM to 56 μM.

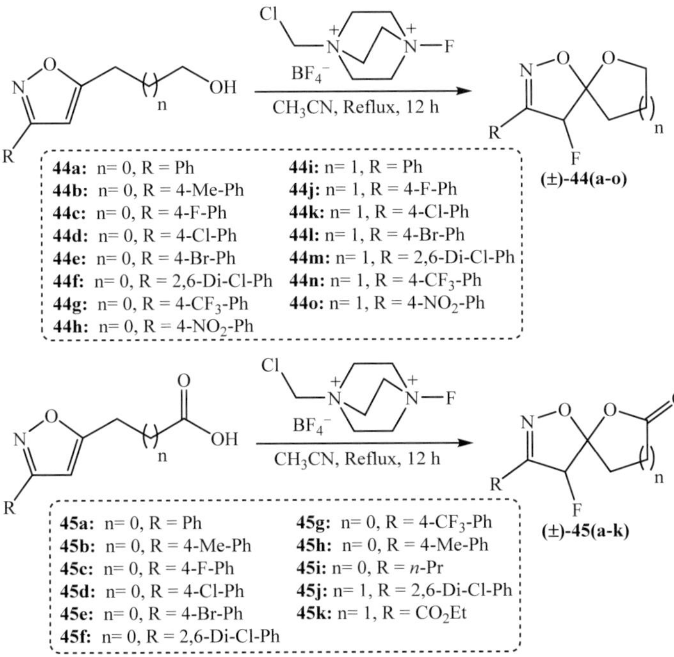

Scheme 6-11. Synthesis of fluoro-substituted spiro-isoxazoline ether and lactones **44a-z** and **45a-k**.

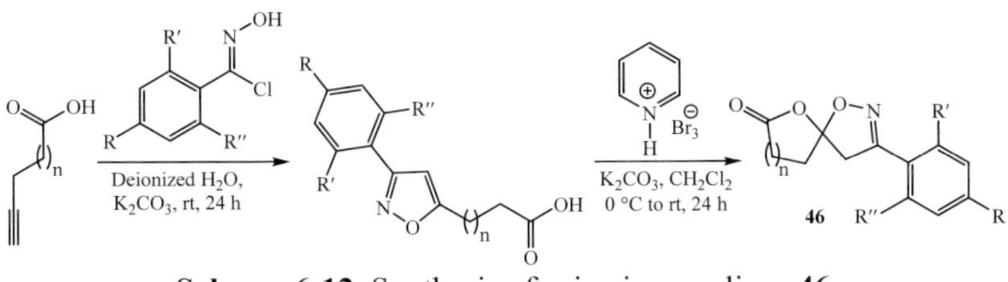

Scheme 6-12. Synthesis of spiro-isoxazolines **46**.

Recently, Hati et al. introduced a library of compounds possessing spiro[pyrrolidine-3,3′-oxindole] motifs **47** that displayed unusual inhibitory activity against the proliferation of MCF-7 breast cancer cells [17]. As indicated in Scheme 6-13, the process involved a one-pot Pictet Spengler-oxidative ring contraction of tryptamine to the desired scaffolds. It was performed in 1:1 THF and water with catalytic trifluoroacetic acid and stoichiometric *N*-bromosuccinimide as an oxidant. Phenotypic profiling depicted that these molecules induce apoptotic cell death in MCF-7 cells.

Target deconvolution with the most potent compound, **47k** from the library, using chemical proteomics exhibited histone deacetylase 2 (HDAC2) and prohibitin 2 as the potential cellular binding partners. Molecular docking of **47k** with HDAC2 gives insights pertinent to putative binding interactions.

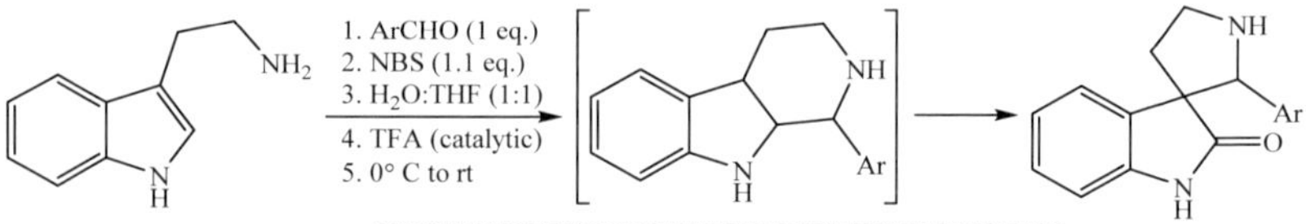

Tryptamine **47a-n**

47a: Ar = 3-CF_3-Ph
47b: Ar = 3,5-Di-CH_3-Ph
47c: Ar = 3,5-Di-F-Ph
47d: Ar = 3-F-Ph
47e: Ar = 3-OCH_3-Ph
47f: Ar = 3,4-Di-OCH_3-Ph
47g: Ar = 4-OCH_3-Ph
47h: Ar = 2,4,6-Tri-OCH_3-Ph
47i: Ar = 2-F-4-OCH_3-Ph
47j: Ar = 3-CH_3-4-OCH_3-Ph
47k: Ar = Ph
47l: Ar = 2,5-Di-CF_3-Ph
47m: Ar = 2-Thiophen
47n: Ar = 4-Pyridine

Scheme 6-13. Synthesis of a library of spiro[pyrrolidine-3,3′-oxindole] compounds **47a-n**.

6.2.7. Spiro compounds as anti-microbial drugs

Safari et al. worked on the multicomponent synthesis of a new library of spiroindeno-pyridazine-4*H*-pyrans **48a-j** starting from cyanoaceto-hydrazide, ninhydrin, malononitrile, and various cyclic CH-acids in refluxing EtOH (Scheme 6-14) [18]. The desired products were probed *in vitro* for anti-bacterial effects on *Escherichia coli* and *Staphylococcus aureus*. Moreover, the cytotoxic activity of the spiropyrans on non-small-cell lung cancer cells (A549 cells), a breast epithelial cancer cell line (MCF-7), human malignant melanoma cells (A375), prostate cancer cells (PC3 cells, LNCaP cells) and normal cells HDF (human dermal fibroblast) was evaluated. They found that compounds **48a**, **48b**, **48f**, **48g,** and **48i** have the best MIC against *Staphylococcus aureus* and compound **48a** disclosed the most potent activity against A549, A375, and LNCaP tumor cells. In addition, DAPI staining of the A549 cancer cell lines treated with **48a** was related to the death of A549 cells.

The occurrence of apoptosis in A549 cells was ultimately confirmed through the application of the RT-PCR (reverse transcription and polymerase chain reaction) method. This confirmation was based on the observed up-regulation of Bax expression and the down-regulation of Bcl-2 expression, indicating the involvement of the mitochondrial pathway of apoptosis.

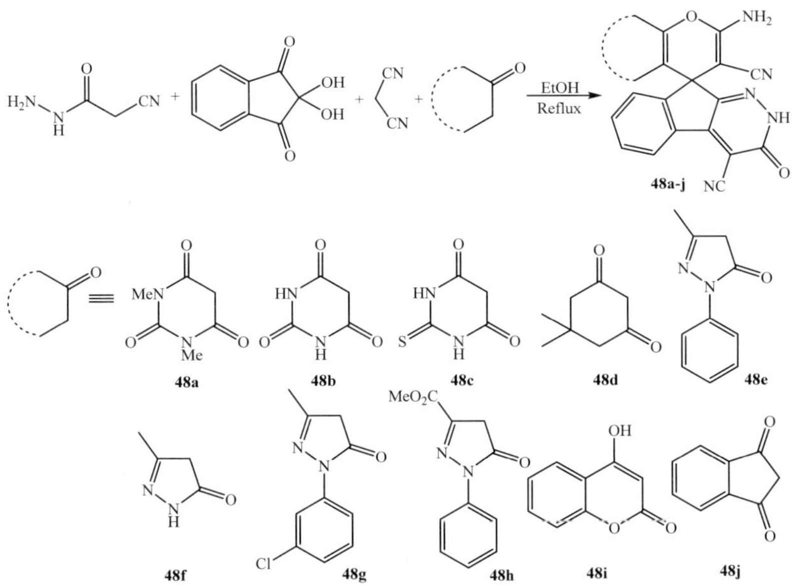

Scheme 6-14. Synthesis of spiroindenopyridazine-4*H*-pyrans **48a-j**.

6.2.8. Spiro compounds as anti-inflammatory drugs

A new series of functionalized spiropyridine **49a-i** was synthesized through an effective MCR strategy using an eco-friendly solvent, ethanol (Scheme 6-15) [19]. The anti-inflammatory activity of protein denaturation and RBC membrane stabilization methods were studied. Among the tested derivatives, the anti-inflammatory activity of compounds **49g**, **49d,** and **49i** was relatively equal to the standard diclofenac, and that of compounds **49a**, **49b**, **49h,** and **49f** was indicated to be moderate.

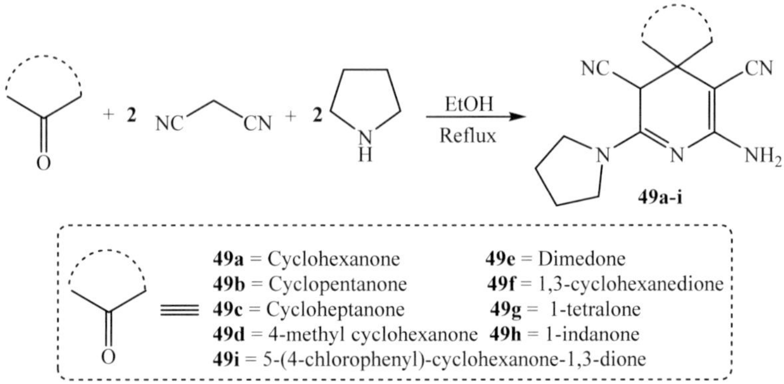

Scheme 6-15. Synthesis of functionalized spiropyridine **49a-i**.

Ashok et al. reported the selective synthesis of monospiro chromanones **50a–c** from which a series of spirofurochromanones **51a-i** and **52a-u** were synthesized following the microwave irradiation method, and their anti-inflammatory and anti-oxidant activities were ascertained (Scheme 6-16) [20]. The reaction protocol needs cheap starting materials and is performed under mild conditions. Compounds **51a**, **51b**, **51c**, **51d**, **51e**, **52g**, **52h**, **52j**, **52l**, **52n,** and **52q** were found to be promising with IC_{50} values ranging from 77.11 to 116.10 µM/mL showing that they are potent anti-inflammatory agents. The binding mode of the synthesized compounds with protein active site was predicted using *in silico* docking. All the compounds **51a–i** and **52a–u** were docked well into the binding pocket of the target protein at all five sites. The compounds exhibited good docking at site I and site V.

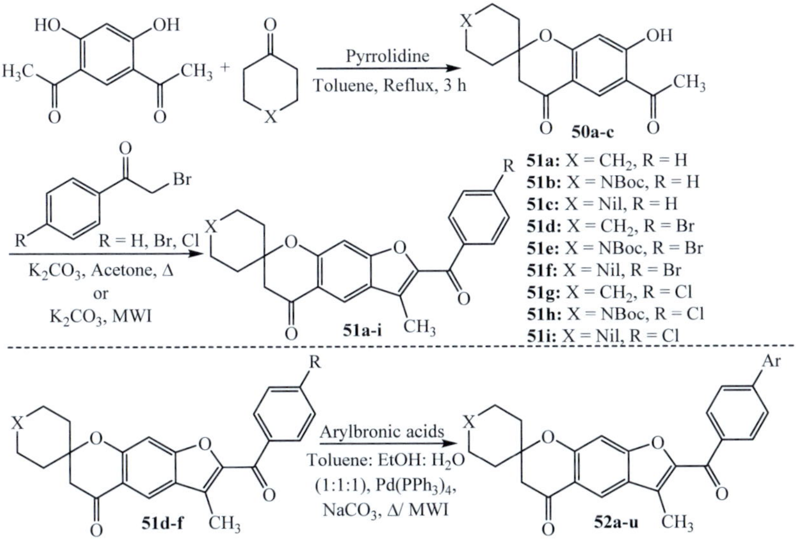

52a: X = CH_2, Ar = Ph **52h:** X = NBoc, Ar = Ph **52o:** X = Nil,
Ar = Ph
52b: X = CH_2, Ar = 4-CHO-C_6H_4 **52i:** X = NBoc, Ar = 4-CHO-C_6H_4 **52p:** X = Nil,
Ar = 4-CHO-C_6H_4
52c: X = CH_2, Ar = 4-Cl-C_6H_4 **52j:** X = NBoc, 4-Cl-C_6H_4 **52q:** X = Nil,
4-Cl-C_6H_4
52d: X = CH_2, Ar = 4-CH_3-C_6H_4 **52k:** X = NBoc, Ar =$_4$-CH_3-C_6H_4 **52r:** X = Nil,
Ar =$_4$-CH_3-C_6H_4
52e: X = CH_2, Ar = 1-Naphthyl **52l:** X = NBoc, Ar = 1-Naphthyl **52s:** X = Nil,
Ar = 1-Naphthyl
52f: X = CH_2, Ar = 3,5-diCl-C_6H_4 **52m:** X = NBoc, Ar = 3,5-diCl-C_6H_4 **52t:** X = Nil,
Ar = 3,5-diCl-C_6H_4
52g: X = CH_2, Ar = 4-F-C_6H_4 **52n:** X = NBoc, Ar =$_4$-F-C_6H_4 **52u:** X = Nil,
Ar =$_4$-F-C_6H_4

Scheme 6-16. Synthesis of spirofurochromanones **51a-i** and **52a-u**.

6.2.9. Spiro compounds as anti-pyretic drugs

Spiro-oxindole-type frameworks exhibit significant bioactivities, making them of great importance in various fields. Wang et al. reported a highly efficient and convenient strategy for synthesizing these frameworks, providing an elegant example of their synthesis [21]. This strategy affords straightforward access to unique optically active spiro[oxazoline-3,3′-oxindole]s **53a-n** with high levels of enantio- and diastereoselectivity (up to >99% ee and >99:1 dr) through the organocatalyzed asymmetric synthesis of spirocyclic thiocarbamates via an aldol reaction (Scheme 6-17). They investigated the biological activities of several spiro oxazolines on fever by intracerebroventricular injection of lipopolysaccharide (a component of the outer membrane of gram-negative bacteria) using a model of acute neuroinflammation in mice. Interestingly, the lipopoly-saccharide-induced fever was considerably decreased by the co-injection of several analogues, including **53f** (20 nmol, $p < 0.05$) and **53i** (20 nmol, $p < 0.05$); **53f** (20 nmol) or **53i** (20 nmol).

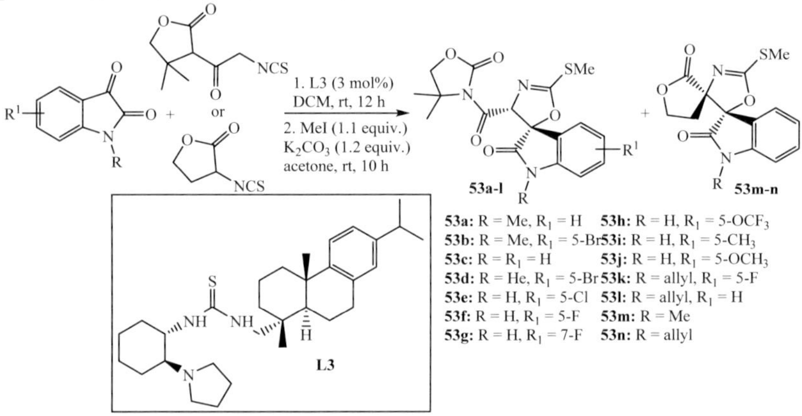

Scheme 6-17. Synthesis of biologically active chiral spiro-oxazolines **53a-n**.

6.2.10. Spiro compounds as anti-bacterial drugs

Khan et al. reported the synthesis of novel steroidal (6*R*)-spiro-1,3,4-thiadiazolines **54a-c** by cyclizing steroidal thiosemicarbazones (Scheme 6-18). The anti-bacterial activity of these compounds was evaluated by the disc diffusion method against the culture of bacteria. The biological behavior of these compounds demonstrated that chloro- and acetoxy-substituents on the 3β-position of the steroidal thiadiazoline ring promoted the anti-bacterial activity. Among them, compounds **54a** and **54b** displayed better anti-bacterial activity than the respective drug Amoxicillin. The results demonstrated that steroidal (6*R*)-spiro-1,3,4-thiadiazolines are new leads in anti-bacterial chemotherapy [22].

In 2017, Bishnoi et al. successfully prepared a series of novel spiro derivatives incorporating pyrido[1,2-*a*]pyrimidine **55a-g** and quinoline **56a-g** moieties [23].The synthetic route for synthesizing spiro derivatives containing pyrido-pyrimidine **55a-g** and quinoline **56a-g** moieties is shown in Scheme 6-19. Their anti-microbial activity against a panel of various bacterial strains and unicellular and multicellular pathogenic fungal strains was measured, and the result revealed that some of the title compounds illustrated excellent activities against reference drugs. The spiro compounds with different substituents, **55a**, **55c,** and **56e**, indicated promising *in vitro* anti-microbial activity.

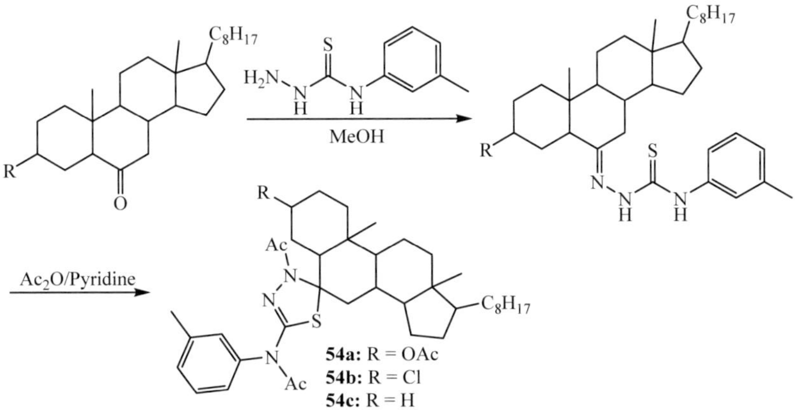

Scheme 6-18. Schematic diagram showing the synthesis of compounds **54a-c**.

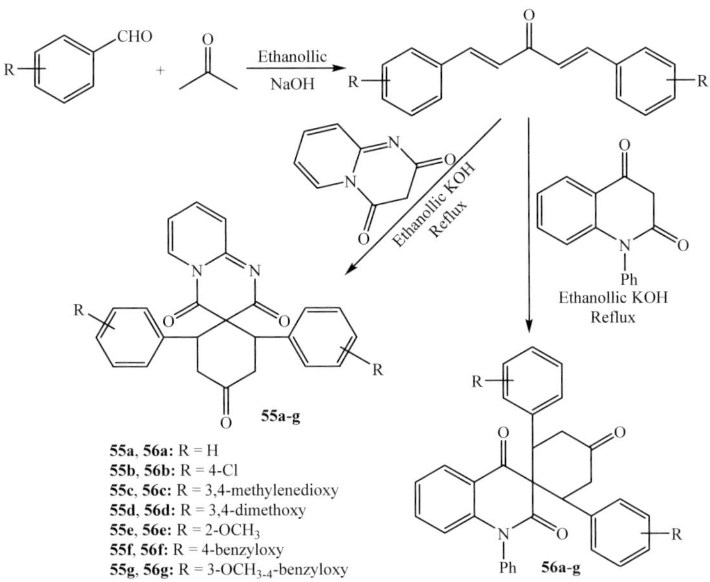

Scheme 6-19. Synthetic route for synthesizing spiro derivatives containing pyrido-pyrimidine **55a-g** and quinoline **56a-g** moieties.

6.2.11. Spiro compounds as anti-malarial drugs

In 2005, Vennerstrom et al. introduced spiro- and dispiro-1,2,4-trioxolanes **57-61** as a new class of synthetic anti-malarial peroxides. They discovered an achiral dispiro-1,2,4-trioxolane **59**, in which the trioxolane is flanked by a spiroadamantane and spirocyclohexane as a novel anti-malarial lead (Figure 6-4) [24]. Nonperoxidic 1,3-dioxolane isosteres (**62-63**) of **59** were inactive, as were trioxolanes without the spiro-adamantane. The trioxolanes were markedly less effective in a standard oral suspension formulation than in a solubilizing formulation. They were more active when administered subcutaneously than orally, and exhibited significant biopharmaceutical responsibilities. Nonetheless, despite their limited oral bioavailability, the more lipophilic trioxolanes generally had better oral activity than their more polar counterparts.

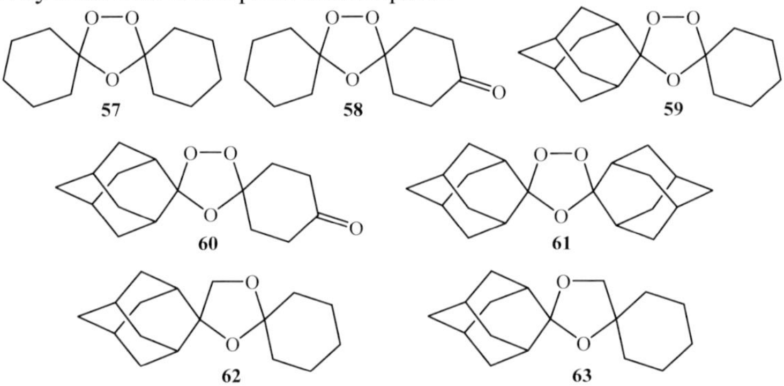

57 58 59 60 61 62 63

Fig. 6-4. Trioxolanes **57-61** and nonperoxidic isosteres **62-63**.

Jarrahpour et al. synthesized several new mono-, bis-spiro- and dispiro-β-lactams **64-67** using a [2+2] cycloaddition reaction with imines derived from 9*H*-fluorene-9-one and a ketene derived from 9*H*-xanthene-9-carboxylic acid or phenoxy acetic acid (Scheme 6-20) [25]. The biological activity of these monocyclic β-lactams was successfully explored against *Plasmodium falciparum* K14 resistant strain with excellent EC_{50} values up to 5 μM.

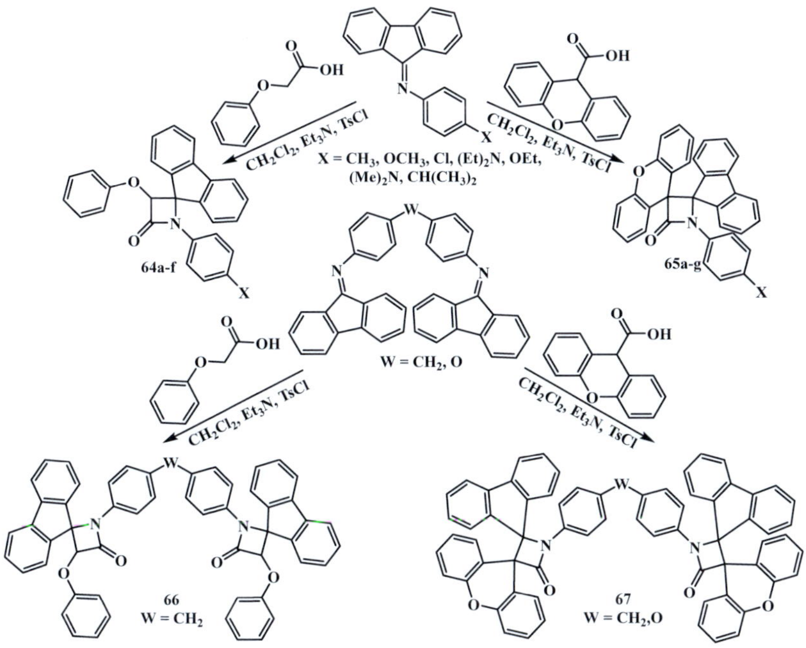

Scheme 6-20. Synthesis of mono-, bis-spiro- and dispiro-β-lactams **64-67**.

6.2.12. Spiro compounds as anti-fungal drugs

The Singh group developed a microwave-assisted method to synthesize a series of spiro indoline-based heterocycles **68** at atmospheric pressure in an open vessel (Scheme 6-21) [26]. The synthesized compounds have been investigated *in vitro* for anti-fungal activity against two fungal strains (i.e., *Aspergillus niger* and *Aspergillus flavus*). All compounds have exhibited notable activity against these pathogens.

The alkyne-azide 1,3-dipolar cycloaddition reaction of 3-*O*-propargyl-3-azido-D-glucofuranose (**IXd**) provided adduct **69** that was applied as intermediate, by judicious manipulation of the 1,2- and 5,6-*O*-isopropylidene groups, in the preparation of a library of spiro-iminosugars **70**, **71**, **72a–g**, and **73a–g** linked with the 1,2,3-triazole-fused morpholine (Figure 6-5), as reported by Dhavale et al. [27].

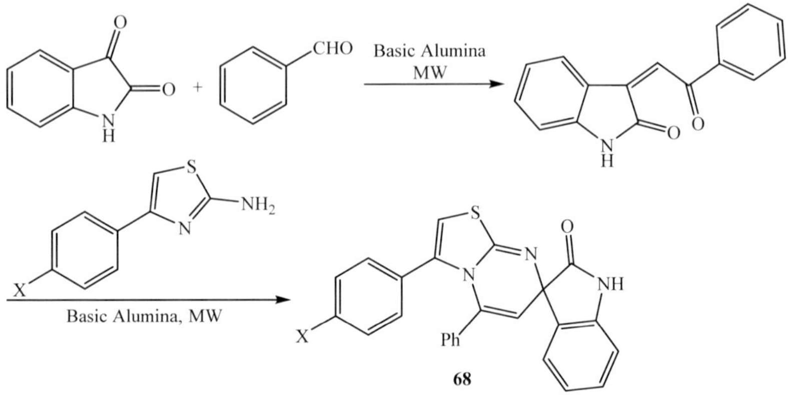

X = H, Me, Et, OH, OMe, Cl, NO_2

Scheme 6-21. Synthesis of spiro indoline-based heterocycles **68**.

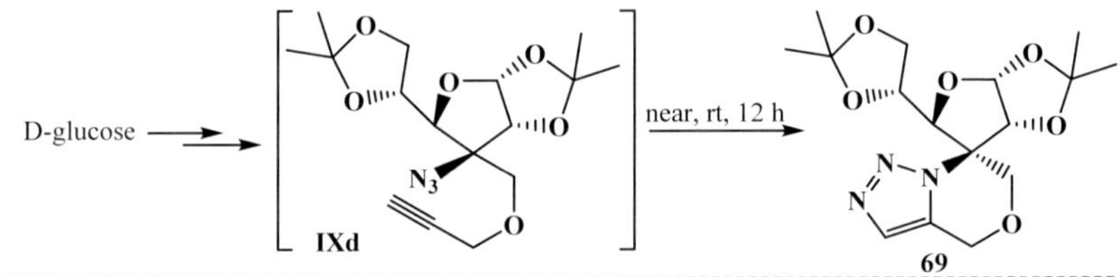

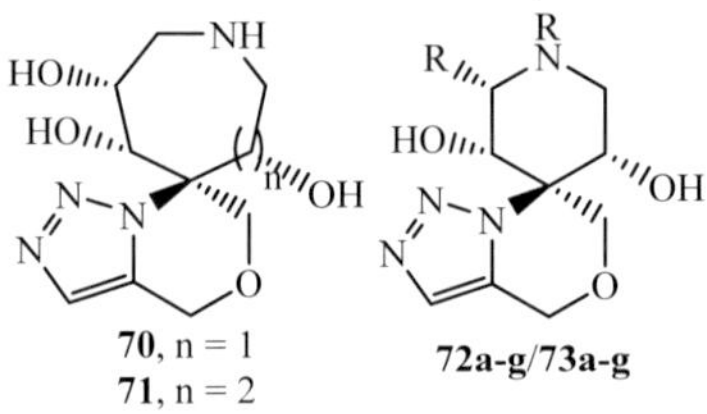

72a: R = R′ = H
72b: R = H, R′ = ethyl
72c: R = H, R′ = buthyl
72d: R = H, R′ = hexyl
72e: R = H, R′ = octhyl
72f: R = H, R′ = decyl
72g: R = H, R′ = dodecyl

73a: R = CH_2OH, R′ = H
73b: R = CH_2OH, R′ = ethyl
73c: R = CH_2OH, R′ = buthyl
73d: R = CH_2OH, R′ = hexyl
73e: R = CH_2OH, R′ = octhyl
73f: R = CH_2OH, R′ = decyl
73g: R = CH_2OH, R′ = dodecyl

Fig. 6-5. Chemical structures of spiro-iminosugars **70**, **71**, **72a–g**, and **73a–g** are linked with the 1,2,3-triazole-fused morpholine.

The presence of two pharmacophores, such as iminosugar and triazole, indicated dual biological activity. Also, the synergetic effect of two pharmacophores allows the augmentation of the glycosidase inhibitory and anti-fungal activity. Among them, the spiroiminosugars **72a** and **73a** were found to be more potent inhibitors of α-glucosidase than miglitol and showed high anti-fungal activity against *C. Albicans* than amphotericin B. The *N*-alkyl derivatives demonstrated enhancement of α-glucosidase inhibition and improvement of anti-fungal activity compared to their respective parent iminosugar.

6.2.13. Spiro compounds as anti-mycobacterial drugs

In 2008, Perumal et al. introduced the stereoselective synthesis of spiro-piperidine-4-ones **74a-m**, **75a-m,** and **76a-m** by a 1,3-DC of azomethine ylides generated *in situ* from isatin and α-amino acids (Scheme 6-22). The synthesized compounds were probed for their *in vitro* and *in vivo* activity against MTB, MDR-TB, and *Myco-bacterium smegmatis* (MC2).

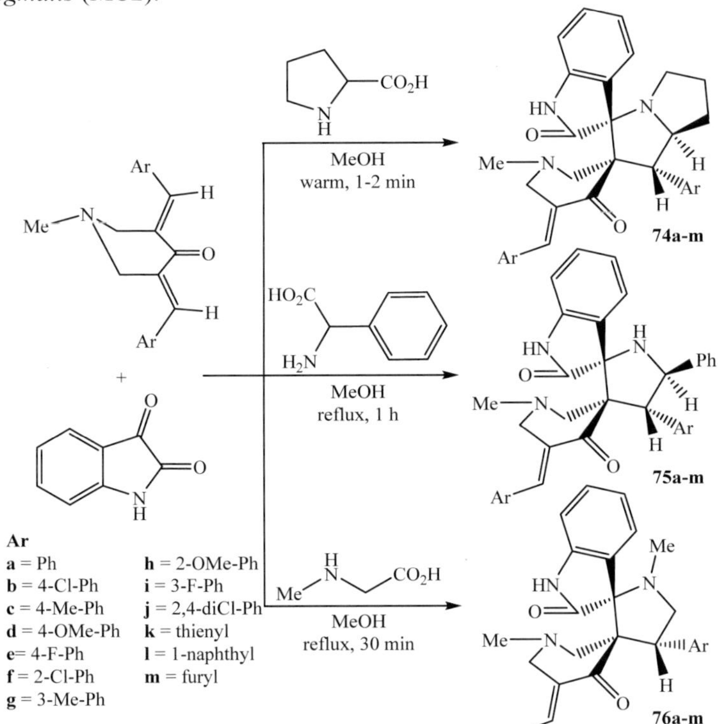

Scheme 6-22. Synthesis of spiro-piperidine-4-ones **74a-m**, **75a-m,** and **76a-m**.

Compound 4-(4-fluorophenyl)-5-phenylpyrrolo(spiro[2.3″]oxindole)-spiro[3.3′]-1′-methyl-5′-(4-fluorophenylmethylidene)piperidine-4′-one **75e** was found to be the most active *in vitro* with a MIC value of 0.07 μM against MTB. It was 5.1 and 67.2 times as potent as Isoniazid and

Ciprofloxacin, respectively. *In vivo*, compound **75e** reduced the bacterial load in lung and spleen tissues with 1.30 and 3.73-log 10 protections, respectively. It was considered promising in reducing bacterial count in lung and spleen tissues [28].

In addition, Muthusubramanian et al. reported the three-component 1,3-DC of azomethine ylides generated *in situ* from ninhydrin and sarcosine with a series of 1-aryl-1*H*-pyrrole-2,5-diones under microwave irradiation conditions to produce dihydro-2′*H*-spiro[indene-2,1′-pyrrolo-[3,4-*c*]pyrrole]-tetraones **77a-n** (Scheme 6-23) [29]. The compounds were evaluated for their anti-mycobacterial and AChE inhibition activities. Compound **77b** (IC_{50} 1.30 μM) has been found to show the twelve-fold anti-mycobacterial activity compared to cycloserine, and it is thirty-seven times more active than pyrimethamine. Compound **77h** displays maximum AChE inhibitory activity with an IC_{50} value of 0.78 ± 0.01 μmol/L.

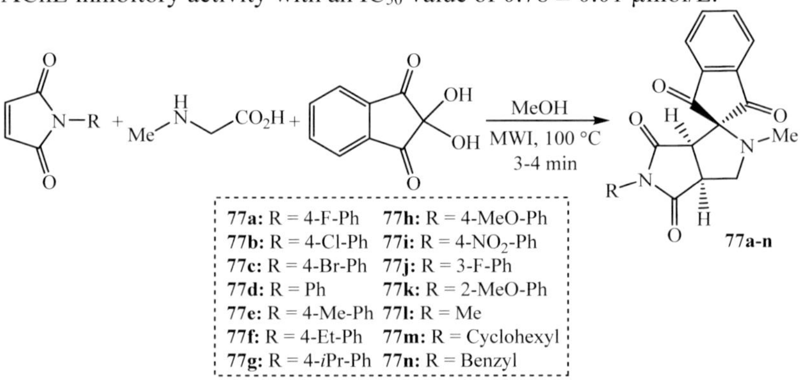

Scheme 6-23. Synthesis of dihydro-2′*H*-spiro[indene-2,1′-pyrrolo-[3,4-*c*]pyrrole]-tetraones **77a-n**.

6.2.14. Spiro compounds as anti-biotic drugs

In this context, a strain of *Streptomyces nodosus* (NPS007994) isolated from marine sediment collected in Scripps Canyon, La Jolla, California, was found to generate lajollamycin **78**, a nitro-tetraene spiro-α-lactone-γ- lactam anti-biotic (Figure 6-6) [30]. The structure was determined by a complete analysis of spectroscopic data and the comparison with known anti-biotics oxazolomycin **79**, 16-methyl oxazolomycin **80**, and triedimycin B **81**. Lajollamycin **78** revealed anti-microbial activity against both drug-sensitive and -resistant gram-positive bacteria and inhibited the growth of B16-F10 tumor cells *in vitro*.

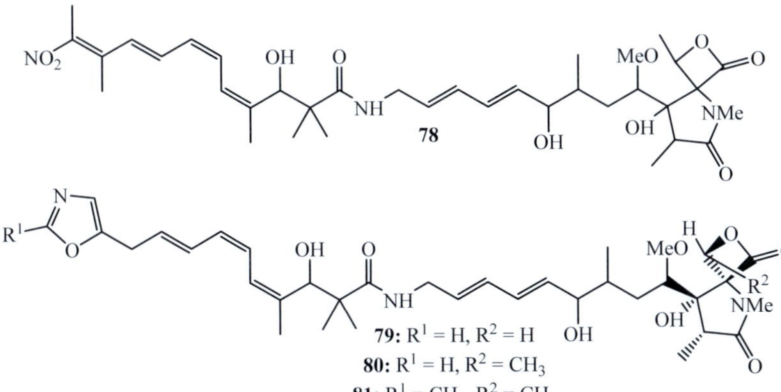

79: $R^1 = H$, $R^2 = H$
80: $R^1 = H$, $R^2 = CH_3$
81: $R^1 = CH_3$, $R^2 = CH_3$

Fig. 6-6. Nitro-tetraene spiro-α-lactone-γ-lactam anti-biotic **78**.

Armeniaspirols **82** are potent anti-biotics with an unusual spiro[4.4]non-8-ene moiety. Muller et al. defined the cloning and functional analysis of the armeniaspirol biosynthetic gene cluster [31]. Gene inactivation studies and subsequent isolation of previously unknown biosynthetic intermediates shed light on intriguing biosynthetic details. Scheme 6-24 indicated the proposed biosynthesis scheme for armeniaspirols **82**. Interestingly, the elimination of Ams15, encoding a protein bearing a flavin-binding domain, causes the depletion of several non-spiro intermediates with varying numbers of chlorine substitutions on the pyrrole moiety. The trichloropyrrole and dichloropyrrole species were transformed by Streptomyces albus expressing Ams15 into dichlorinated and monochlorinated spiro derivatives. Additionally, *in vitro* transformation of these non-spiro intermediates into des-*N*-methyl spiro intermediates by the cell lysate of the same recombinant strain asserted Ams15 to be responsible for spiro formation through oxidative dehalogenation.

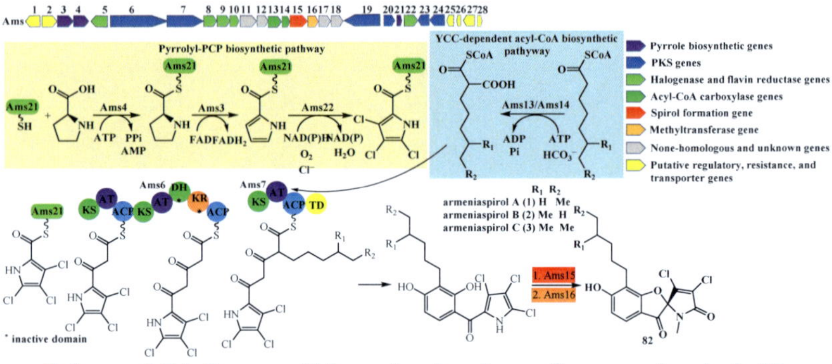

Scheme 6-24. Proposed biosynthesis scheme for armeniaspirols **82**.

6.2.15. Spiro compounds as anti-convulsant drugs

Stephani et al. [32] synthesized thirteen *N*-1′,*N*-3′-di-substituted-2′*H*,3*H*,5′*H*-spiro-(2-benzo-furan-1,4′-imidazolidine)-2′,3,5′-triones **83a-m** and their anti-convulsant effects determined by standard pentylenetetrazol (scPTZ test) and maximum electroshock seizure (MES test) models in mice (Scheme 6-25). Most of the compounds illustrated the ability to protect against pentylenetetrazol-induced convulsions.

Compound **83l** (the *N*-1′-*p*-nitrophenyl, *N*-3′-ethyl) in the *N*-1′-aryl, *N*-3′-alkyl di-substituted series displayed maximum activity with ED_{50} of 41.8 mg/kg in scPTZ convulsion model.

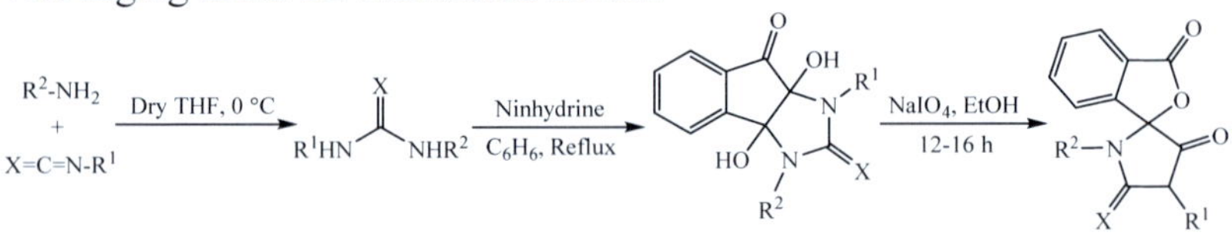

83a-m

83a: R^1 = Me; R^2 = Et; X = O
83b: R^1 = R^2 = *n*-C_3H_7; X = O
83c: R^1 = R^2 = *n*-C_4H_9; X = O
83d: R^1 = R^2 = *i*-C_3H_7; X = O
83e: R^1 = R^2 = cyclo-C_6H_{11}; X = O
83f: R^1 = C_6H_5; R^2 = Me; X = O
83g: R^1 = C_6H_5; R^2 = Me; X = S
83h: R^1 = *p*-F-C_6H_4; R^2 = Me; X = O
83i: R^1 = *p*-CF_3-C_6H_4; R^2 = Et; X = O
83j: R^1 = C_6H_5; R^2 = Et; X = O
83k: R^1 = *p*-F-C_6H_4; R^2 = Et; X = O
83l: R^1 = *p*-NO_2-C_6H_4; R^2 = Et; X = O
83m: R^1 = *p*-OMe-C_6H_4; R^2 = Et; X = O

Scheme 6-25. Synthesis of **83a-m** by Stephani et al.

Two series of spiro[*b*]benzothiophene-3,2′-[1,3]thiazolidine]-2,4′-diones **85a-e** and **86a-e** were successfully prepared by the reaction of benzothiophene-2,3-dione with certain arylamines followed by cyclization of the resulting intermediate **84** with thioglycolic or this lactic acids, as demonstrated in Scheme 6-26 [33]. The final compounds were screened for their anti-convulsant activity. Among the synthesized compounds, the naphthyl derivatives **85e** and **86e** were the most effective in prolonging phenobarbitone-induced sleeping time and protecting against strychnine-induced seizures.

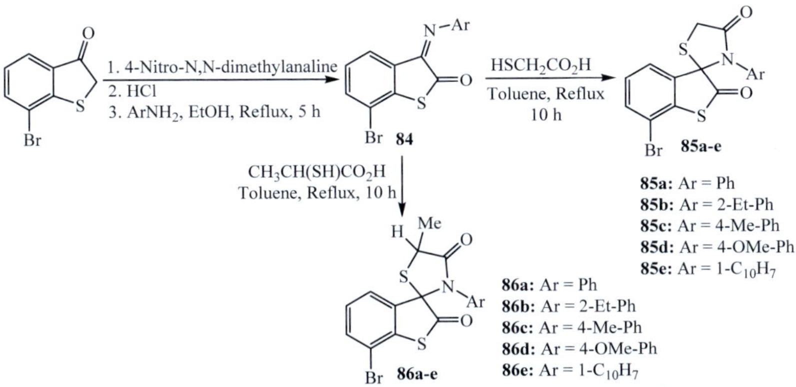

Scheme 6-26. Synthesis and anti-convulsant activity of spiro[*b*]benzothiophene-3,2′-[1,3]thiazolidine]-2,4′-diones **85a-e** and **86a-e**.

6.2.16. Spiro compounds as anti-acetylcholinesterase drugs

Niknam et al. prepared the spiro[indoline-3,9-xanthene]-trione **87** and hydroxyspiro[indoline-3,9-xanthene]-trione **88** and tested them for the *in vitro* anti-bacterial activity against *Staphylococcus aureus* and *Escherichia coli* (Figure 6-7). The anti-cholinesterase activities for six substitutes of hydroxy derivatives **88a-f** were also evaluated through molecular docking. All concentrations of compounds dispensed a dose-dependent anti-bacterial activity. The docking results exhibited that all compounds are more constant than galantamine. Compound **88a** shows the minimum binding energy in AChE and BChE enzymes [34].

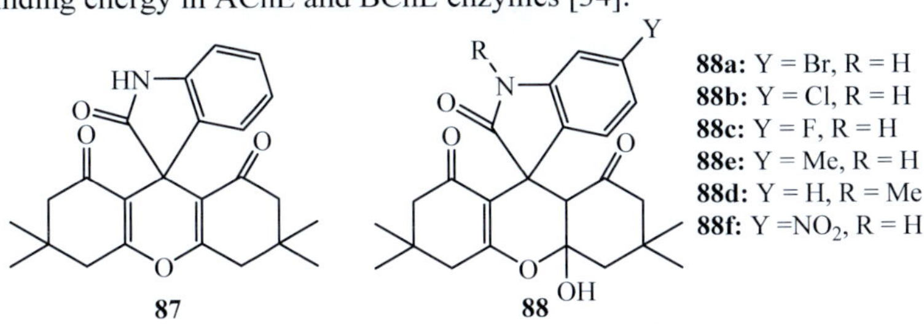

Fig. 6-7. Structures of spiro[indoline-3,9-xanthene]-trione **87** and hydroxy spiro[indoline-3,9-xanthene]-trione **88a-f**.

Lycium europaeum L. (Solanaceae) is a spiny shrub of the Solanaceae family, known in Algeria as 'Awsaj' and applied as a food and herbal drug. In the phytochemical study of the alkaloid extract of roots and leaves, Benamar et al. successfully isolated and characterized one alkaloid (harmine **91**) and four sesquiterpenes (C-1′ epimer of (2*R*,5*S*,10*R*)-2-(1′,2′-dihydroxy-1′-methylethyl)-6-10-dimethylspiro[4.5]dec-6-en-8-one **89**, C-1′ epimer of 2-(1′2′-di-hydroxy-1′-methylethyl)-6,10-di-methylspiro[4.5] dec-6,9-dien-8-one **90**, (β)-dehydrovomifoliol **92**, vomifoliol **93**) (Figure 6-8). The isolated compounds have demonstrated anti-oxidant and AChE inhibitory activities in the bioautography assays. The ethanolic and alkaloid extracts of roots and leaves were also probed, and, mainly, the alkaloid extracts were the most active ones as anti-oxidant and AChE inhibitors [35].

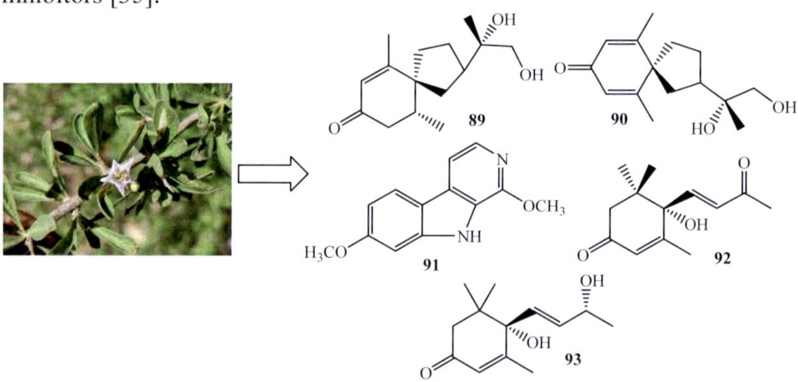

Fig. 6-8. Alkaloids from roots and leaves of *Lycium europaeum* L.

6.2.17. Spiro compounds as anti-tumor drugs

Tumor cells have specific features involving angiogenesis induction, cell-cycle dysregulation, and immune destruction evasion. Inducing a T helper type 2 (Th2) immune response, tumor cells may favor immune tolerance within the tumor, allowing cancer growth to progress. Hence, Sobral et al. [36] prepared the novel spiro-acridine (*E*)-5′-oxo-1′-((3,4,5-trimethoxybenzylidene)amino)-1′,5′-dihydro-10*H*-spiro[acridine-9,2′-pyrrole]-4′-carbonitrile (**AMTAC-17**) and tested it for anti-tumor effects (Figure 6-9). Toxicity assessment was done in mice after acute treatment (2000 mg/kg, intraperitoneally, i.p.). The Ehrlich ascites carcinoma model was applied to study the anti-tumor activity of **AMTAC-17** (12.5, 25, or 50 mg/kg,i.p.) after seven days of treatment. Effects on the cell cycle, angiogenesis, and inflammatory responses were evaluated. LD_{50} (lethal dose 50%) was calculated to be higher than 5000 mg/kg. **AMTAC-17** decreased the Ehrlich tumor's total viable cancer cells count and peritumoral micro-vessels density and caused an enhancement in the sub-G1 peak. Besides, there was an increase in Th1 cytokine profile levels (IL-1β, TNF-α, and IL-12). Therefore, the spiro-acridine compound **AMTAC-17** exhibits low toxicity, and its *in vivo* anti-tumor effect involves modulation of the immune system to a cytotoxic Th1 profile and a devaluation of tumor angiogenesis.

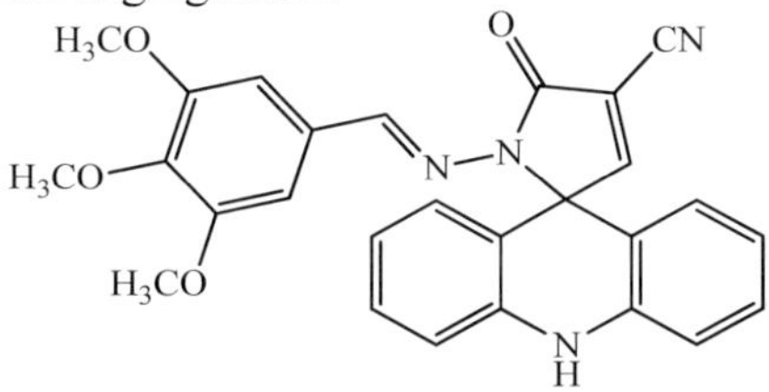

Fig. 6-9. Structure of spiro-acridine compound **AMTAC-17**.

The Wang group developed an efficient method that allows the synthesis of spiro-tetrahydrothiopyran oxindoles **94** in moderate to good yields (51–78%) with excellent diastereoselectivities (>20: 1 dr) using a proline-catalyzed Michael–Aldol cascade reaction (Scheme 6-27) [37]. Interestingly, the oxindoles **94** disclosed moderate to good *in vitro* anti-tumor activities and were confirmed as p53-MDM2 inhibitors, which indicated promising lead compounds for anti-tumor drug discovery.

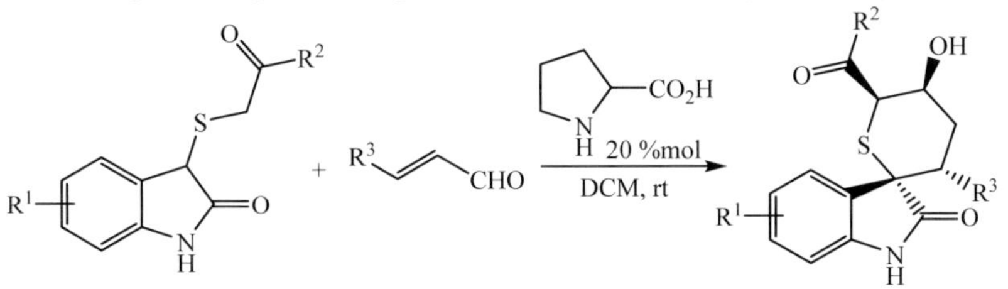

94

Scheme 6-27. Synthesis of spiro-tetrahydrothiopyran oxindoles **94**.

6.2.18. Spiro compounds as anti-viral drugs

In this field, Chu et al. synthesized a novel spiro-carbocyclic adenosine **95** and studied it as a potential anti-viral agent (Figure 6-10) [38]. The synthesized compound **95** illustrated considerable anti-viral activity against HCV with EC_{50} values of 0.273 and 0.368 μM in genotypes 1A and 1B, respectively.

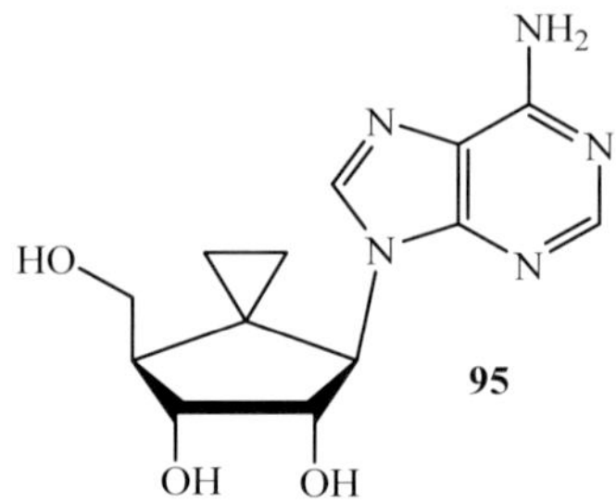

95

Fig. 6-10. Spiro-carbocyclic adenosine **95**.

Pestaloxazine A **96** is an alkaloid with an unprecedented symmetric spiro[oxazinane-piperazinedione] skeleton that was isolated from a *Pestalotiopsis species* fungus derived from a soft coral (Figure 6-11) [39]. (+)-Pestaloxazine A showed potent anti-viral activity against EV71 with an IC_{50} value of 14.2 ± 1.3 μM, which was stronger than that of the positive control Ribavirin (IC_{50} = 256.1 ± 15.1 μM).

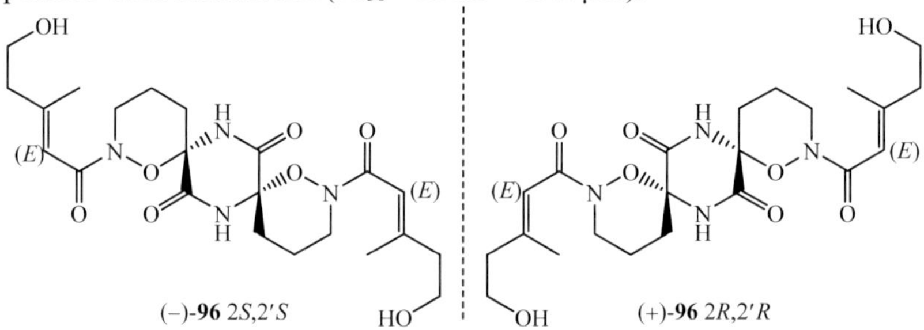

(−)-**96** 2*S*,2′*S* (+)-**96** 2*R*,2′*R*

Fig. 6-11. Structures of (-)-**96** (2*S*, 2′*S*) and (+)-**96** (2*R*, 2′*R*).

6.2.19. Spiro compounds as anti-tubercular drugs

The Perumal group explored a 1,3-DC between azomethine ylide generated *in situ* from substituted isatin and 1,3-thiazolane-4-carboxylic acid with 1-methyl-3,5-bis[(*E*)-arylmethylidene]tetrahydro-4(1*H*)-pyridinones to devise a straightforward route to thirty-eight spiro-pyrrolothiazoles **97-99** chemo-, regio- and stereoselectively in quantitative yields, as depicted in Scheme 6-28 [40]. Among them, spiro[5.3″]-5″-nitrooxindole-spiro-[6.3′]-1′-methyl-5′-(2,4-di-chlorophenylmethylidene)-tetrahydro-4′-(1*H*)-pyridinone-7-(2,4-dichlorophenyl)tetra-hydro-1*H*-pyrrolo[1,2-*c*][1,3] thiazole **99k** was found to be the most active with an MIC of 0.6 μM against *Mycobacterium tuberculosis* $H_{37}Rv$ (MTB and MDR-TB).

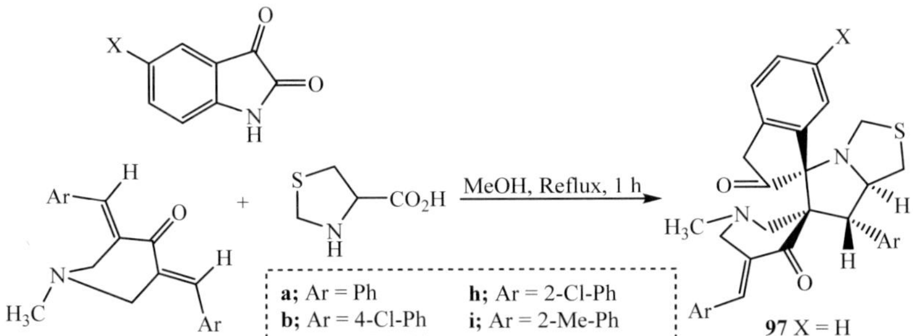

a; Ar = Ph
b; Ar = 4-Cl-Ph
c; Ar = 4-Me-Ph
d; Ar = 4-OMe-Ph
e; Ar = 4-F-Ph
f; Ar = 4-Br-Ph
g; Ar = 4-iPr-Ph
h; Ar = 2-Cl-Ph
i; Ar = 2-Me-Ph
j; Ar = 2-OMe-Ph
k; Ar = 2,4-diCl-Ph
l; Ar = 2,4-$(OMe)_2$Ph
m; Ar = 2-Thienyl

97 X = H
98 X= Cl
99 X = NO_2

Scheme 6-28. Synthesis of spiro-pyrrolothiazoles **97-99**.

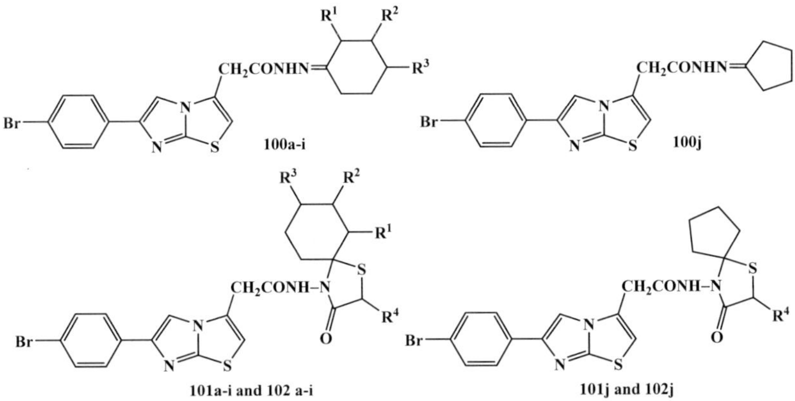

100a: R^1 = Me, R^2 = R^3 = H
100b: R^1 = H, R^2 = Me, R^3 = H
100c: R^1 = H, R^2 = H, R^3 = Me
100d: R^1 = H, R^2 = H, R^3 = Et
100e: R^1 = H, R^2 = H, R^3 = Pr
100f: R^1 = H, R^2 = H, R^3 = *t*Bu
100g: R^1 = H, R^2 = H, R^3 = Ph
100h: R^1 = H, R^2 = H, R^3 = 4-OH-Ph
100i: R^1 = H, R^2 = R^3 = H
101a: R^1 = Me, R^2 = R^3 = R^4 =H
101b: R^1 = H, R^2 = Me, R^3 = R^4 = H
101c: R^1 = H, R^2 = H, R^3 = Me, R^4 = H
101d: R^1 = H, R^2 = H, R^3 = Et, R^4 = H
101e: R^1 = H, R^2 = H, R^3 = Pr, R^4 = H
101f: R^1 = H, R^2 = H, R^3 = *t*Bu, R^4 = H
101g: R^1 = H, R^2 = H, R^3 = Ph, R^4 = H
101h: R^1 = H, R^2 = H, R^3 = 4-OH-Ph, R^4 = H
101i: R^1 = H, R^2 = R^3 = R^4 = H
101j: R^4 = H
102a: R^1 = Me, R^2 = R^3 = H, R^4 =H
102b: R^1 = H, R^2 = Me, R^3 = R^4 = Me
102c: R^1 = H, R^2 = H, R^3 = Me, R^4 = Me
102d: R^1 = H, R^2 = H, R^3 = Et, R^4 = Me
102e: R^1 = H, R^2 = H, R^3 = Pr, R^4 = Me
102f: R^1 = H, R^2 = H, R^3 = *t*Bu, R^4 = Me
102g: R^1 = H, R^2 = H, R^3 = Ph, R^4 = Me
102h: R^1 = H, R^2 = H, R^3 = 4-OH-Ph, R^4 = Me
102i: R^1 = H, R^2 = R^3 = R^4 = Me
102j: R^4 = Me

Scheme 6-29. Synthesis of imidazo[2,1-*b*]thiazoles **100-102**.

6.2.20. Spiro compounds as anti-liposomal drugs

Interestingly, Dincel et al. developed a series of novel imidazo[2,1-*b*]-thiazole moiety bearing oxo-hydrazone and spirocondensed-thiazolidines **100-102** and studied their anti-oxidant activity (Scheme 6-29) by three methods: inhibition of $FeCl_3$/ascorbate system-induced lipid peroxidation of lecithin liposome (anti-LPO), scavenging activity against ABTS radical, and ferric reducing anti-oxidant power (FRAP) activity. **100h**, **101h**, and **102h** revealed the highest anti-LPO and ABTS radical removal activity. Also, in FRAP analysis, **100i** and **100a** disclosed the best activity [41].

6.2.21. Spiro compounds as anti-depressant drugs

Pawlowski et al. presented the synthesis of 5-(cyclo)alkyl-5-phenyl- and 5-spiro imidazol-idine-2,4-diones with an aryl piperazinyl propyl moiety **103a-l** and their *in vitro* and *in vivo* pharmacological properties and molecular characteristics (Scheme 6-30) [42]. The synthesized compounds displayed a high affinity for 5-HT_{1A} (**103b-k**) and 5-HT_{2A} (**103g**, **103i**, **101j**, **103l**) receptors and diversified pharmacological profiles. Compounds **103f**, **103i,** and **103k** demonstrated antagonistic, partial agonistic, and agonistic activity, respectively, toward the 5-HT_{1A} receptor, and they were explored as potential anti-depressants and/or anxiolytics. Interestingly, compound **103k**, a pre- and postsynaptic 5-HT_{1A} receptor agonist, generated an anti-depressant-like effect, which was more pronounced than imipramine in the forced swim test in mice without affecting locomotor activity. In addition, compound **103k** brought about an anxiolytic-like weak impact in the four-plate test in mice. Molecular docking studies of compound **103k** to the homology model of the 5-HT_{1A} receptor exhibited that a 3′,4′-dihydro-2′*H*-spiro-[imidazolidine-4,1′-naphthalene]-2,5-dione moiety played a significant role in stabilizing the ligand-receptor complex.

In another study, Hassan et al. indicated that the reaction of 3-(dicyano-methylene)-2-indolone in a solution of ethanol/piperidine with 4-substituted thiosemicarbazides gives the derivatives of 5′-(substituted amino)-3′*H*–spiro(indoline-3,2′-[1,3,4]thiadiazol-2-one **104a-e** (Scheme 6-31). The prepared compounds were studied for anti-depressant activity using the tail suspension behavioral despair test and anti-convulsant activity against pentylenetetrazol-induced seizures in mice. All the compounds induced substantial anti-depressant activity compared with imipramine at a 10 mg/kg dose level. Compounds **104a**, **104b**, and **104d** illustrated the most protective class of the tested compounds against clonic seizures induced by intraperitoneal injection of PTZ, the effect of which was nearly close to that of phenobarbital sodium at a dose level of 30 mg/kg [43].

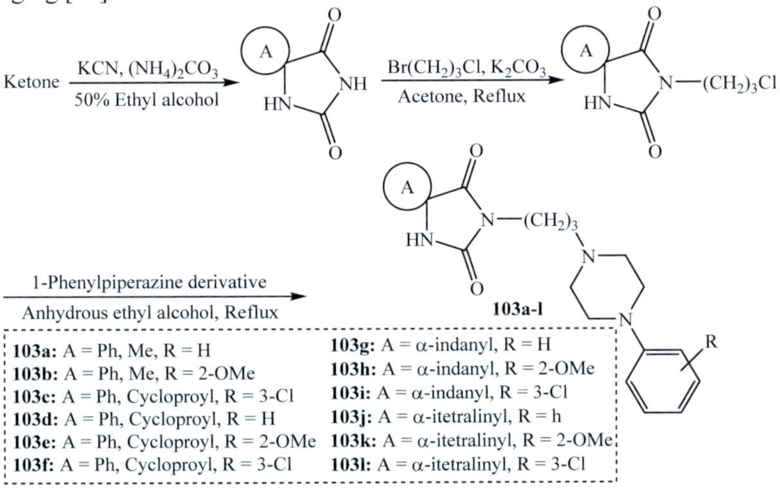

Scheme 6-30. Synthesis of 5-(cyclo)alkyl-5-phenyl- and 5-spiroimidazolidine-2,4-diones **103a-l**.

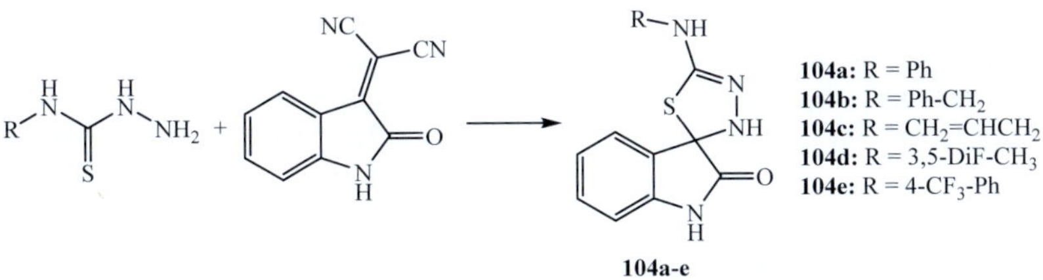

Scheme 6-31. Synthesis of spiro(indolone-3,20-[1,3,4]thiadiazol)-2-ones **104a-e**.

6.2.22. Spiro compounds as anti-septic drugs

Tumor necrosis factor-alpha (TNF-α) is a prominent proinflammatory cytokine that plays a crucial role in regulating inflammatory and immune responses, as well as processes such as proliferation and apoptosis.The human TNF-α gene is expressed in a variety of cell types (e.g., T and B lymphocytes and macrophages) and regulates further cytokine induction, especially of IL-1β and IL-6 in many human diseases involving cancer, immune disorders, and inflammatory diseases like septic shock, rheumatoid arthritis, and Crohn's disease.

In a search for compounds inhibiting the inducible TNF-α promoter activity in T cells, Erkel et al. presented a new spiro compound, designated as oxaspirodion (Figure 6-12), isolated from fermentations of the ascomycete *Chaetomium subspirale*. Oxaspirodion inhibited TNF-α promoter-driven luciferase reporter gene expression with an IC_{50} value of 2.5 μg/ml (10 μM) in TPA/ionomycin-stimulated Jurkat T cells. Explores the compound's mode of action and discloses that inhibition of the phosphorylation of the ERK1/2 kinases causes the inhibition of the TNF-α promoter activity. Moreover, oxaspirodion inhibited the activation of the transcription factor NF-$_{k}$B, which is involved in the inducible expression of many proinflammatory genes [44].

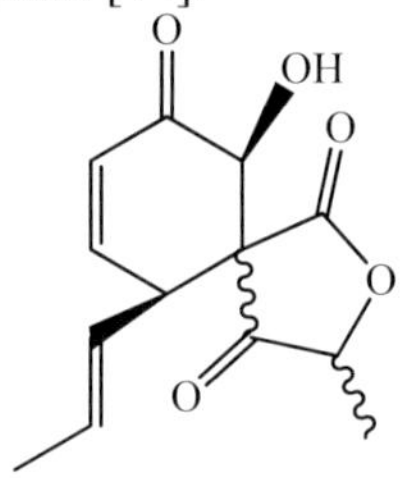

Fig. 6-12. Structure of oxaspirodion.

6.2.23. Spiro compounds as anti-photoaging drugs

Extracts of *Dioscorea composita* or *Dioscorea villosa* were used as supplemental health foods containing large amounts of the plant steroid, diosgenin, etc. The efficacy of diosgenine against skin was proved with the *in vitro* human 3D skin equivalent model (Figure 6-13), and demonstrated the enhanced DNA synthesis, bromodeoxyuridine uptake and intracellular cAMP level in adult human keratinocytes.

These results express that for restoration of keratinocyte proliferation in aged skin, diosenin has a potential effect [45].

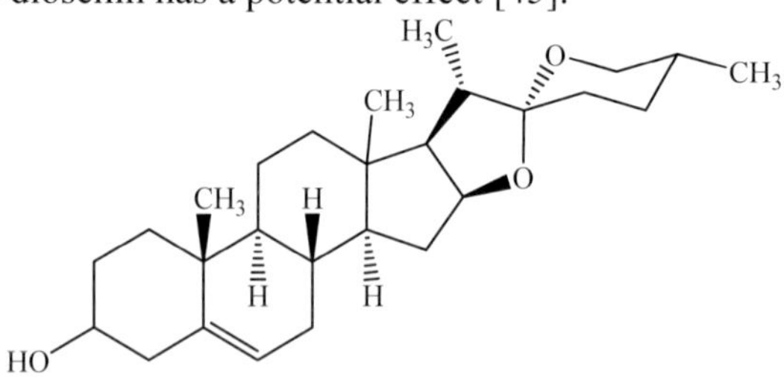

Fig. 6-13. Structure of diosgenin.

6.2.24. Spiro compounds as anti-acne drugs

7α-Acetylthio-3-oxo-17α-pregn-4-ene-21,17-carbolactone (Aldactone, spironolactone, Figure 6-14), was first synthsized in the late 1950s as a potassium-sparing diuretic for the treatment of hypertension and condestive heart failure which has been proved to be effective in treating acne, as well [46, 47].

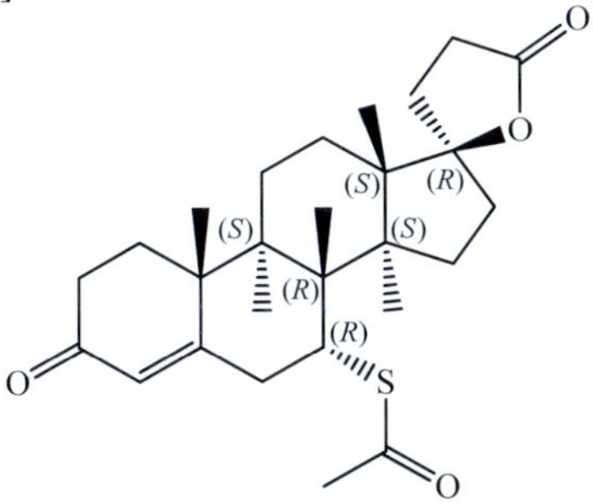

Fig. 6-14. Structure of spironolactone.

6.2.25. Spiro compounds as anti-HIV drugs

In 2016, a protocol for the rapid access to spiro cyclopropyl oxindoles **105a-p** was reported by a domino Michael/intramolecular nucleophilic substitution pathway with variously substituted vinyl selenones and enolizable oxindoles in aqueous sodium hydroxide solution (Scheme 6-32) [48]. Some of the synthesized compounds were chosen for biological evaluation. Compound **105m** exhibited selective anti-HIV-1 activity thanks to its ability to inhibit the reverse transcriptase.

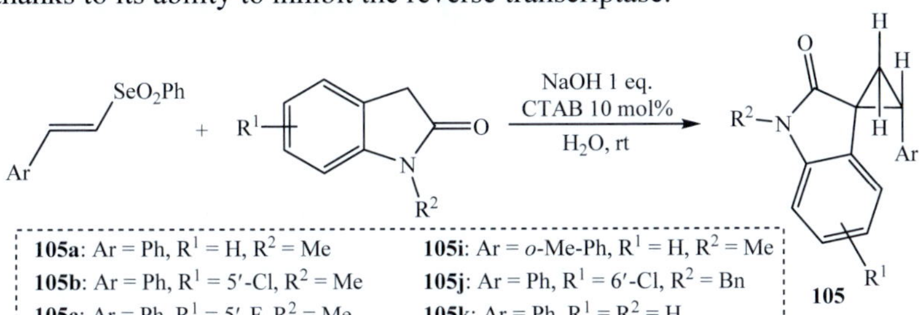

105

105a: Ar = Ph, R^1 = H, R^2 = Me
105b: Ar = Ph, R^1 = 5′-Cl, R^2 = Me
105c: Ar = Ph, R^1 = 5′-F, R^2 = Me
105d: Ar = Ph, R^1 = 6′-Cl, R^2 = Me
105e: Ar = Ph, R^1 = 5′-Me, R^2 = Me
105f: Ar = Ph, R^1 = 5′-NO_2, R^2 = Me
105g: Ar = *p*-Cl-Ph, R^1 = H, R^2 = Me
105h: Ar = *p*-Me-Ph, R^1 = H, R^2 = Me
105i: Ar = *o*-Me-Ph, R^1 = H, R^2 = Me
105j: Ar = Ph, R^1 = 6′-Cl, R^2 = Bn
105k: Ar = Ph, R^1 = R^2 = H
105l: Ar = Ph, R^1 = 5′-Cl, R^2 = H
105m: Ar = Ph, R^1 = 5′-F, R^2 = H
105n: Ar = Ph, R^1 = 5′-Me, R^2 = H
105o: Ar = Ph, R^1 = 6′-Br, R^2 = H
105p: Ar = *p*-Me-Ph, R^1 = H, R^2 = H

Scheme 6-32. Spirocyclopropyl oxindoles **105a-p** endowed with anti-HIV RT activity.

Taveira et al. presented new spiro-β-lactam derivatives with potent (nM) activity against HIV and Plasmodium and no activity against bacteria and yeast (Figure 6-15). **BSS-730A** is the best-performing molecule of the series that inhibited both HIV-1 and HIV-2 replication with an IC_{50} of 13 ± 9.59 nM and P. berghei hepatic infection with an IC_{50} of 0.55 ± 0.14 μM with a clear impact on parasite development. **BSS-730A** was also active against the Erythrocytic stages of *Plasmodium falciparum*, with an estimated IC_{50} of 0.43 ± 0.04 μM. Time-of-addition studies revealed that **BSS-730A** potentially affects all stages of the HIV replicative cycle, suggesting a complex mechanism of action. **BSS-730A** was active against multidrug-resistant HIV isolates, with a median 2.4-fold higher IC_{50} relative to control isolates. **BSS-730A** was equally active against R5, and X4 HIV isolates and represented strong synergism with the entry inhibitor AMD3100. **BSS-730A** is a promising candidate for development as a potential therapeutic and/or prophylactic agent against HIV and Plasmodium [49].

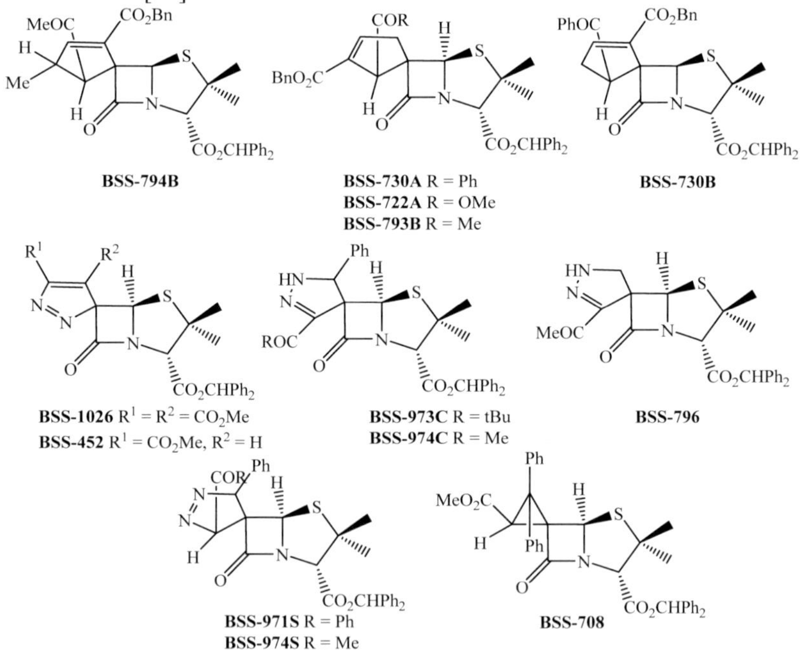

Fig. 6-15. Structures of spiro-β-lactams biologically evaluated.

6.2.26. Spiro compounds as anti-coagulant drugs

In this context, two new phthalides, angesinenolides A and B (**106** and **107**), were isolated from the roots of *Angelica Sinensis,* and their structures were determined using HRMS, NMR, and X-ray crystallographic data (Figure 6-16). Also, their melting points were determined at 154-155 and 139-140 °C, respectively. Angesinenolide A **106** is the first example of a phthalide trimer presumably established via two [2+2] cycloaddition reactions. Angesinenolide B **107** is a unique dimeric phthalide with a peroxy bridge between C-3a and C-6. Both phthalides were investigated for *in vitro* anti-coagulation activities. Angesinenolide A **106** decreased the level of fibrinogen (FIB). Angesinenolide B **107** remarkably increased thrombin time and activated partial thromboplastin time, as well as markedly reduced the content of FIB [50].

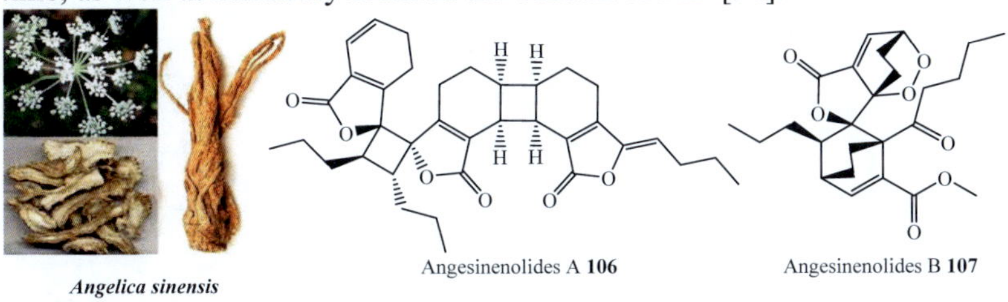

Fig. 6-16. Chemical structures of angesinenolides A and B (**106** and **107**).

Mchiri et al. [51] recently used a three-component cascade reaction to synthesize a new series of spiropyrrolizidines **108-110** with wide structural complexity and diversity through an unprecedented isomerization with moderate to excellent yield (Scheme 6-33). The structures of prepared spiro[pyrrolizidin-2,3-oxindole]s **108-110** were characterized by NMR spectroscopic and elemental analyses. The synthesized compounds were probed *in vitro* to study their anti-microbial activity against different species of bacteria and fungi and their anti-coagulant activity. Their oxindoles' anti-microbial and anti-coagulant activities are moderate to strong compared to standard drugs.

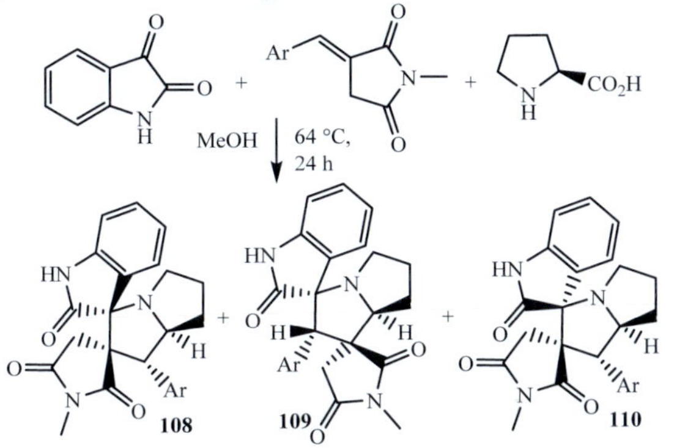

Scheme 6-33. Spiro[pyrrolizidin-2,3-oxindole]s **108-110**.

6.2.27. Spiro compounds as anti-allergic drugs

Spirograterpene A (**111**), a spiro-tetracyclic diterpene, was isolated from the deep-sea-derived fungus *Penicillium granulatum* MCCC 3A00475, together with two biosynthetically related cyclopianes, conidiogenone I (**112**) and conidiogenone C (**113**) (Figure 6-17). Extensive spectroscopic analyses determined the relative configuration of **111**, and the absolute structure was elucidated by the modified Mosher's method. It is the second example of a diterpene featuring a 5/5/5/5 spirocyclic carbon skeleton. Also, this compound demonstrated modest anti-allergic activity [52].

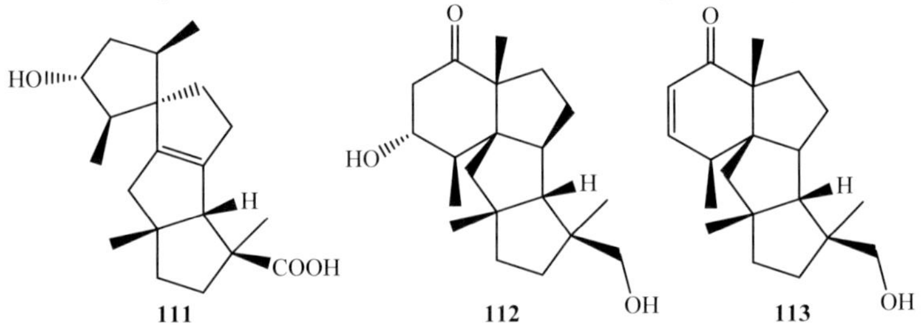

Fig. 6-17. Spirograterpene A **111** together with two biosynthetically related cyclopianes **112** and **113**.

6.2.28. Spiro compounds as anti-proliferative drugs

Nuclear factor kappa-light-chain-enhancer of activated B cells (NF-κB) is a protein complex that controls transcription of DNA, cytokine production, and cell survival, affecting most aspects of cellular physiology. Its deregulation is associated with inflammatory diseases and cancer. Bouhenna et al. conducted a study to investigate the cytotoxic and NF-κB inhibition potential of spiro(lactone-cyclohexanone) compounds **114-119** in two human leukemia cell lines (U937 and K562) [53]. The structures of spiro(lactone-cyclohexanone)s **114-119** are shown in Figure 6-18. The anti-proliferative effects of these compounds were estimated using trypan blue staining on human K562 and U937 cell lines. The study evaluated their involvement in NF-κB regulation using a luciferase reporter gene assay, and Caspase-3/7 activities were assessed to analyze apoptosis induction. Both spiro(coumarin-cyclohexanone) **114** and spiro(6-methyllactone cyclohexanone) **119** reduced cancer cell viability and proliferation. Compound **114** inhibited TNF-α-induced NF-κB activation in a dose-dependent manner and induced caspase-dependent apoptosis in both leukemia cell lines. These findings suggest that compounds **114** and **119** have the potential to be used as anti-cancer agents. Additionally, compound **114** demonstrated NF-κB inhibition activity in leukemia cancer cells.

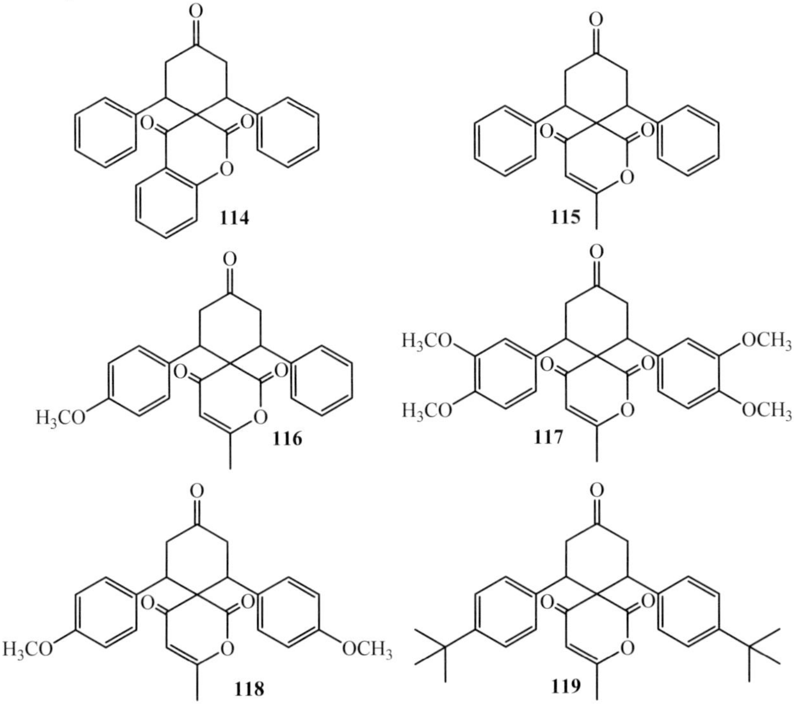

Fig. 6-18. Chemical structures of spiro(lactone-cyclohexanone)s **114-119**.

In a separate study, 3-spiro[3-azabicyclo[3.1.0]hexane]oxindole frameworks **120a-c** were evaluated as potential anti-tumor agents (Scheme 6-34) [54]. *In silico* ADMET (absorption, distribution, metabolism, excretion, and toxicity) analysis was conducted on compounds **120a-c**, which showed promising anti-proliferative activity. Compounds **120a** and **120c** exhibited cytotoxicity against 3T3-SV40 cells compared to 3T3 cells. The tested compounds caused considerable cell-cycle perturbation, leading to a higher accumulation of cells in the G0/G1 phase, which was consistent with DNA cytometry studies. Confocal microscopy revealed that treatment with compounds **120a** and **120c** caused actin filaments to disappear in 82-97% of 3T3 cells, and granular actin was distributed diffusely in the cytoplasm. The number of 3T3-SV40 cells with stress fibers increased to 7-30% compared to 2% in the control group. Furthermore, treatment with compounds **120a** and **120c** significantly reduced the number of transformed 3T3-SV40 cells with filopodium-like membrane protrusions, from 86% in control cells to 6-18% after treatment, indicating a reduction in cell motility. Therefore, the investigated compounds **120a** and **120c** have a cytostatic effect, which reduces the number of filopodium-like membrane protrusions, making them potential candidates as anti-tumor agents.

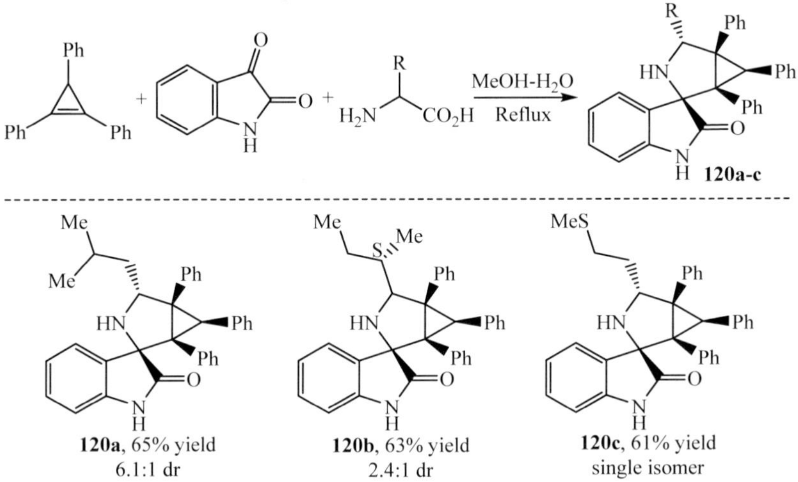

Scheme 6-34. Synthesis of 3-spiro[3-azabicyclo[3.1.0]hexane]oxindole frameworks **120a-c**.

6.2.29. Spiro compounds as anti-arrhythmic drugs

In 2015, Yao et al. isolated four spiroalkaloids **121-124**, including a new compound Shensongine A (**121**), from the anti-arrhythmic TCM formula Shensong Yangxin capsule (SSYX). The Shensong Yangxin capsule is a traditional Chinese medicine that has been used widely to treat cardiac arrhythmia involving 12 Chinese materia medica such as ginseng radix et rhizoma, Corni Fructus, Salviae miltiorrhizae radix et rhizoma, Taxilli herba, Ziziphi spinosae semen, Paeoniae radix rubra, etc. The chemical structures of compounds **121-124** are indicated in Figure 6-19. Compounds **121** and **123** exposed cardiovascular activities by shortened APD in rat myocardial cells. These compounds were possibly formed from precursors in different composed herbal medicines while processing the TCM formula [55].

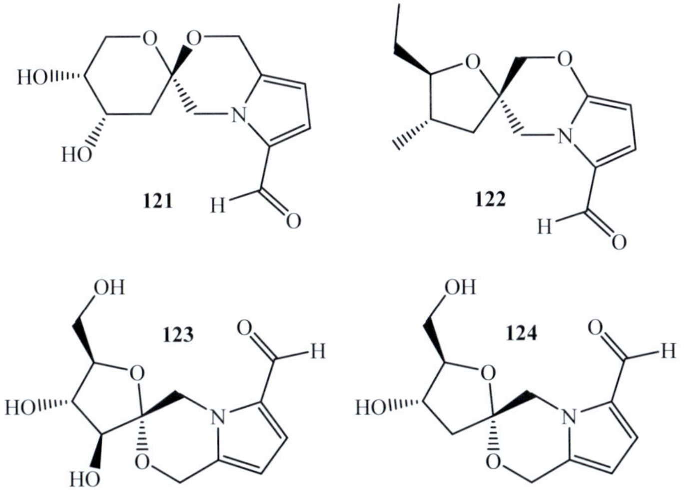

Fig. 6-19. Chemical structures of compounds **121-124**.

Compound **125**, which is a new spirocyclic benzopyran activator of the mitochondrial ATP-dependent potassium channels (mitoKATP), has been studied by the Gonca group for its potential to prevent ischemia/reperfusion (I/R)-induced ventricular arrhythmias (Figure 6-20) [56]. They hypothesized that combining the activation of mitoKATP with the inhibition of sarcolemmal ATP-dependent potassium channels (sarcKATP) may result in a stronger anti-arrhythmic effect. In anesthetized rats, myocardial ischemia was induced by ligating the left main coronary artery, followed by reperfusion. The results showed that compound **125** significantly reduced arrhythmia scores and the total length of arrhythmias at a dose of 10 mg/kg, while it was inactive at a dose of 3 mg/kg. Pre-treatment with **5-HD**, a selective mitoKATP blocker, suppressed the anti-arrhythmic effect of compound **125**. Both diazoxide, a selective mitoKATP opener, and **HMR 1098**, a selective sarcKATP blocker, considerably reduced the total length of arrhythmias. However, the combination of neither diazoxide nor compound **125** with **HMR 1098** demonstrated any additional remedial benefit. These results suggest that compound **125** may have a dose-dependent anti-arrhythmic effect that is more pronounced than the anti-arrhythmic effect of diazoxide, and that its anti-arrhythmic effect may depend on mitoKATP activation.

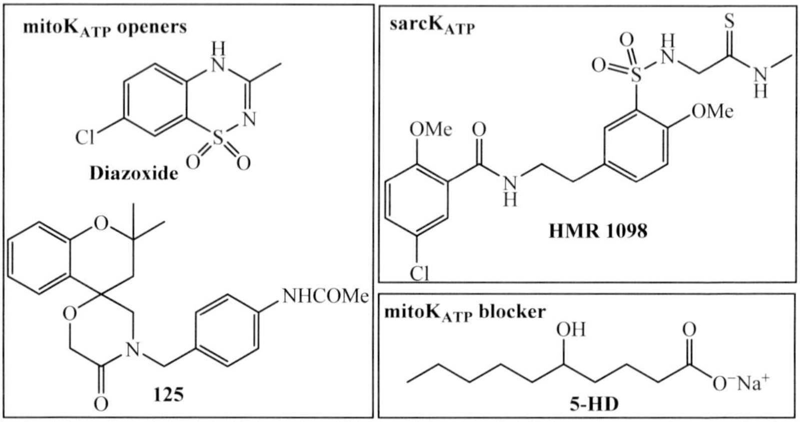

Fig. 6-20. Chemical structures of diazoxide, **125**, **HMR 1098**, and **5-HD**.

6.2.30. Spiro compounds as anti-psychotic drugs

Czopek et al. synthesized a series of novel spirohydantoin derivatives **126a-o** with a 4-aryl piperazinylbutyl moiety and different substituents, and evaluated their binding affinity for serotonin 5-HT_{1A}, 5-HT_{2A}, 5-HT_7, and dopamine D_2 receptors (Scheme 6-35). Four compounds were selected for further binding affinity assays on dopamine D_1, D_3, D_4, and 5-HT_{2C}, 5-HT_6, and adrenergic α_1 and α_{2C} receptors, which are implicated in various CNS diseases such as schizophrenia, anxiety, and/or depression. The most promising compound, **126f**, exhibited a unique multireceptor profile as a 5-HT_{1A} receptor partial agonist and a D_2/5-HT_{2A} receptor antagonist. The compound displayed specific anti-psychotic-like activity in an animal model of schizophrenia without causing side effects, catalepsy, and inactive sedation doses. Additionally, in a mouse model of depression, compound **126f** behaved as a potential anti-depressant, unlike the anti-psychotic reference drug. It is worth noting that in spirohydantoin derivatives, this combination of pharmacological activities-anti-psychotic and anti-depressant-is distinct and could be advantageous in the treatment of schizophrenia [57].

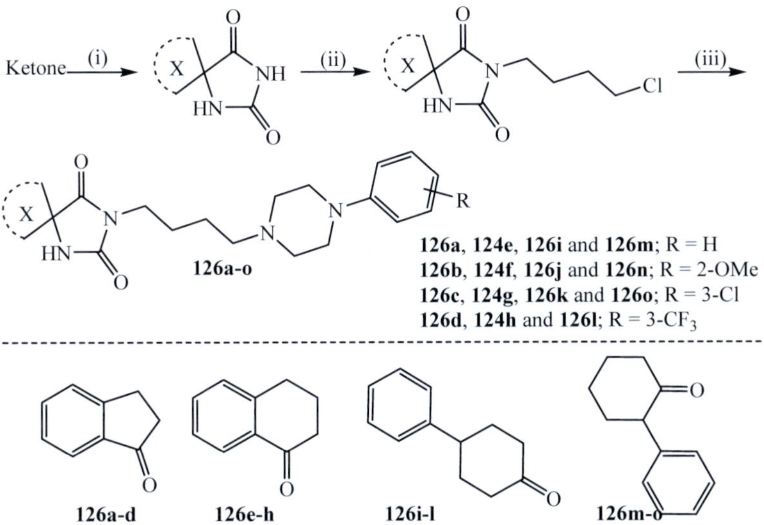

126a-o

126a, **124e**, **126i** and **126m**; R = H
126b, **124f**, **126j** and **126n**; R = 2-OMe
126c, **124g**, **126k** and **126o**; R = 3-Cl
126d, **124h** and **126l**; R = 3-CF_3

126a-d **126e-h** **126i-l** **126m-o**

Scheme 6-35. Synthetic pathways of compounds **126a-o**. Reagents and conditions: (i) KCN, $(NH_4)_2CO_3$, 50% ethyl alcohol, more than 20 h, 56 °C; (ii) 1-bromo-4-chlorobutane, K_2CO_3, KJ, acetone, reflux 7-20 h; (iii) 4-substituted piperazines, K_2CO_3, KJ, acetonitrile (**126a-l**) or 2-methoxy ethanol (compound **126m-o**), reflux 20–40 h.

6.2.31. Spiro compounds as anti-leishmanial drugs

Serrano-Martín et al. developed a simple synthesis method for spiro dihydroquinoline-oxindoles **127-129** via $BF_3.OEt_2$-catalyzed imino Diels-Alder reaction between ketimine-isatins and *trans*-isoeugenol (Scheme 6-36) [58]. Ten spiro-oxindoless were tested against the lifecycle of Leishmania braziliensis parasites that cause cutaneous leishmaniasis in South America. Among the compounds tested, 8′-ethyl-4′-(4-hydroxy-3-methoxyphenyl)-3′-methyl-3′,4′-dihydro-1′*H*-spiro[indoline-3,2′-quinolin] -2-one (**127d**) showed significant growth inhibition of promastigotes without affecting the viability of mammalian cells. It also reduced the number of intracellular amastigotes of L. braziliensis. The spiro compound **127d** disrupted the regulatory volume decrease (RVD) and affected the sterol biosynthetic pathway at the level of squalene epoxidase (SE) enzyme. The results suggest that the spiroannulation between quinoline and oxindole scaffolds can enhance anti-leishmanial activity and may facilitate the development of potent quinoline-oxindole hybrids against L. braziliensis, the primary agent of cutaneous leishmaniasis in South America.

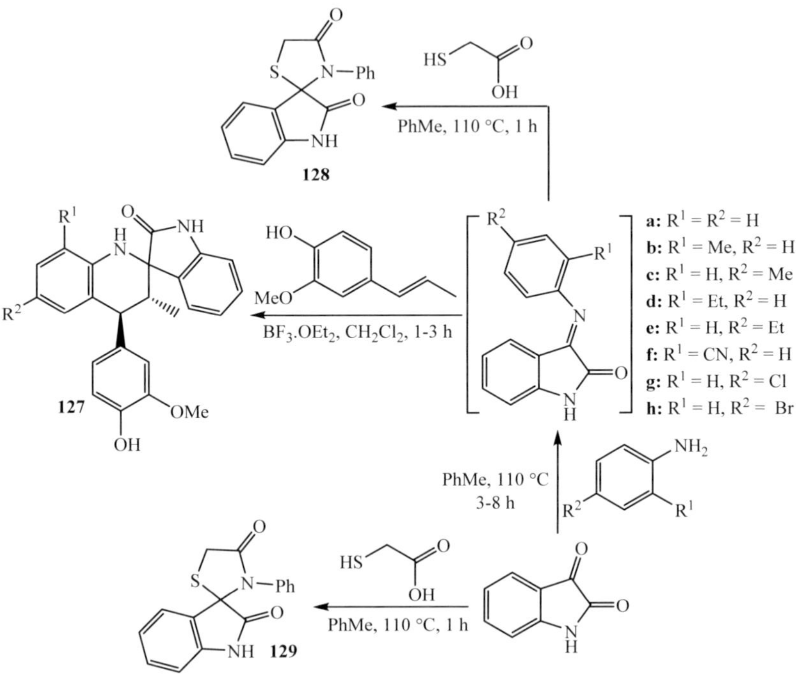

Scheme 6-36. Synthesis of spiro dihydroquinoline-oxindoles **127-129**.

Olímpio de Moura et al. investigated the anti-leishmania activity of spiroacridine compounds **(AMTACs) 130a-i** against both the promastigote and amastigote forms of the best compounds in leishmania infantum. They suggested an *in silico* study with possible selective therapeutic targets for L. infantum [59]. The substituted dimethyl-amine compound (**AMTAC 130f**) illustrated the best leishmanicidal activity *in vitro* and was found to interact with TryR - and LdTopoI (Figure 6-21).

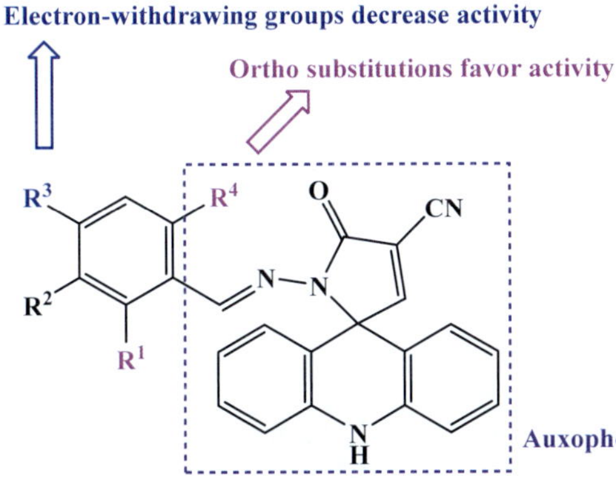

AMTAC	R^1	R^1	R^1	R^1
130a	H	H	H	H
130b	H	H	Cl	H
130c	H	H	F	H
130d	Cl	Cl	Cl	H
130e	Cl	Cl	H	Cl
130f	H	H	NMe_2	H
130g	NO_2	H	H	H
130h	H	H	NO_2	H
130i	H	OEt	OH	H

Fig. 6-21. Chemical structure of **AMTACs**.

6.2.32. Spiro compounds as anti-hypertensive drugs

Spironolactone and Irbesartan are the most currently used drugs as anti-hypertensive drugs, also spironolactone has been an aldosterone antagonist used to treat high blood pressure since the 1960s. Spironolactone decreases morbidity and mortality in severe heart failure and effectively lowers blood pressure in hypertensive patients without hyperaldosteronism (Figure 6-22) [60, 61].

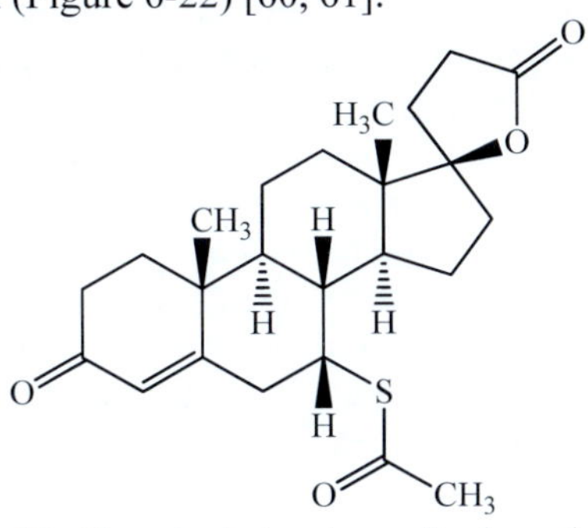

Fig. 6-22. Chemical structure of spironolactone.

In addition, Irbesartan is a non-peptide angiotensin II receptor antagonist applied to treat hypertension, heart diseases, heart strokes, diabetic neuropathy, and congestive heart diseases. Irbesartan is currently accessible in the market as an anti-hypertensive drug under the brand name Avapro that operate as a specific and competitive angiotensin I receptor antagonist, thereby blocking the actions of endogenous ligand angiotensin II and, in turn, preventing the rise in blood pressure (Figure 6-23) [62, 63].

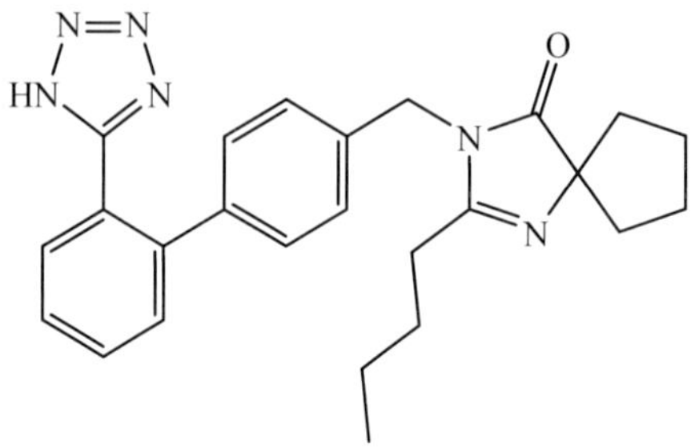

Fig. 6-23. Chemical structure of Irbesartan.

6.2.33. Spiro compounds as anti-hyperglycemic drugs

Glycogen phosphorylase (GP) is an enzyme responsible for the depolymerization of glycogen into glucose, which affects glucose levels in the bloodstream. Praly et al. developed a series of D-gluco-pyranosylidene -spiro-isoxazolines **131a-s** via regio- and stereoselective 1,3-DC of nitrile oxides generated *in situ* by treating the corresponding oximes with bleach (Scheme 6-37) [64]. The resulting *O*-unprotected spiro-isoxazolines **131a-s** were evaluated as GP inhibitors and displayed IC_{50} values ranging from 1 to 800 mM. Selected inhibitors were further examined *in vitro* using rat and human hepatocytes and demonstrated considerable inhibitory properties in primary cell culture. In tests on human hepatocytes, the tetra-*O*-acetylated spiro-isoxazoline bearing a 2-naphthyl residue displayed a much lower IC_{50} value (2.5 μM) than the *O*-unprotected analogue (19.95 μM). The most promising compounds were evaluated in the Zucker *fa*/*fa* rat model in acute and sub-chronic assays and reduced hepatic glucose production, which is known to be elevated in type 2 diabetes. This indicates that glucose-based spiro-isoxazolines have potential as anti-hyperglycemic agents for the treatment of type 2 diabetes.

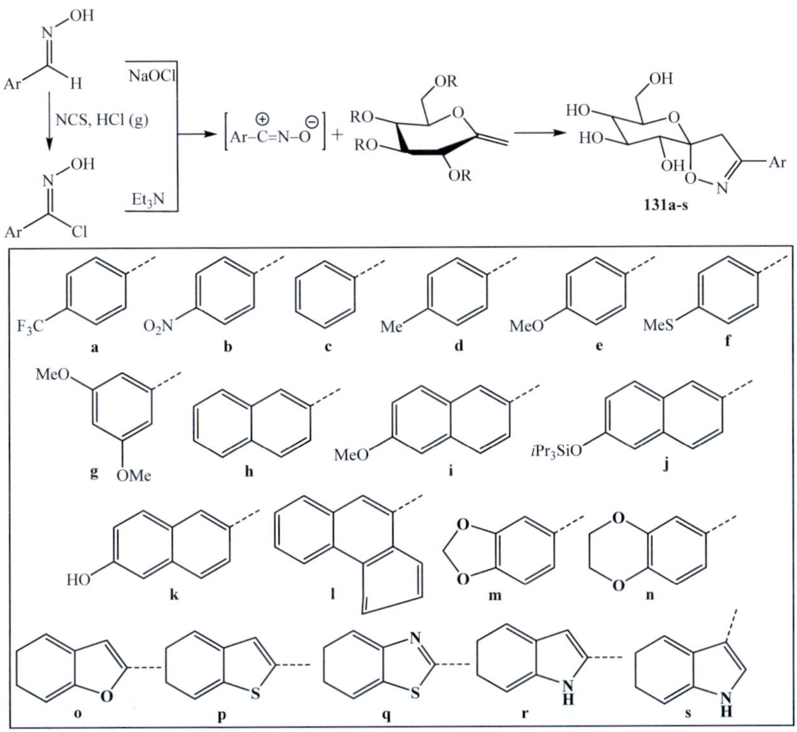

Scheme 6-37. Synthesis of spiro-isoxazolines by nitrile oxides cycloaddition to *exo*-glucans.

In 2020, the same group presented spiro-oxathiazoles **132a-l** as potent GP inhibitors [65]. The desired products were prepared through 1,3-DC of an aryl nitrile oxide to a glucono-thionolactone, as indicated in Scheme 6-38. The 2-naphthyl substituted glucose-based spiro-oxathiazole **132h**, recognized as one of the most potent GP inhibitors (Ki = 160 nM against RMGPb), could be generated on the gram scale from this procedure. Further evaluation *in vitro* using rat and human hepatocytes disclosed that compound **132h** is a druggable anti-hyperglycemic compound performing slightly better than 1,4-dideoxy-1,4-imino-D-arabinitol (DAB) applied as a positive control.

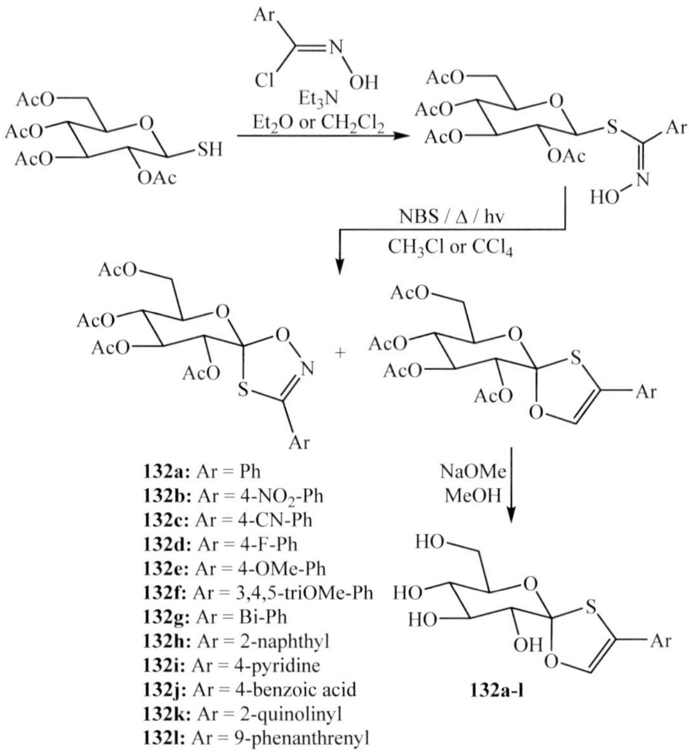

Scheme 6-38. Synthesis of spiro-oxathiazoles **132a-l**.

6.2.34. Spiro compounds as anti-parasitic drugs

Trypanothione reductase (TR) is a crucial enzyme responsible for the reduction of trypanothione, which is a dithiol anti-oxidant that safeguards Trypanosomatid parasites from oxidative stress caused by the host defense systems. Bresciani et al. discovered and validated a range of spiro-containing derivatives as inhibitors of *Trypanosoma brucei* TR (TbTR) through small-molecule screening, utilizing an optimized luminescence assay to measure TbTR activity *in vitro*. The hit series, discovered by high throughput screening, was demonstrated to bind TbTR reversibly and compete with the trypanothione (TS2) substrate. Compound **133**, a representative hit compound of this new class of TR inhibitors, has shown the ability to kill T. brucei parasites in vitro. The X-ray structure of the TbTR/compound **133** complex identified a new pocket in the TS2 binding site where compound **133** binds. The general structures of the hit series and the representative molecule **133** with TR inhibition potency are shown in Scheme 6-39. This new inhibitor is specific to TbTR, and no activity was detected against structurally similar human glutathione reductase (hGR). The central spiro scaffold is considered convenient for active brain compounds in humans, making it an attractive starting point for developing future treatments for the central nervous system stage of *Trypanosoma brucei* infections [66].

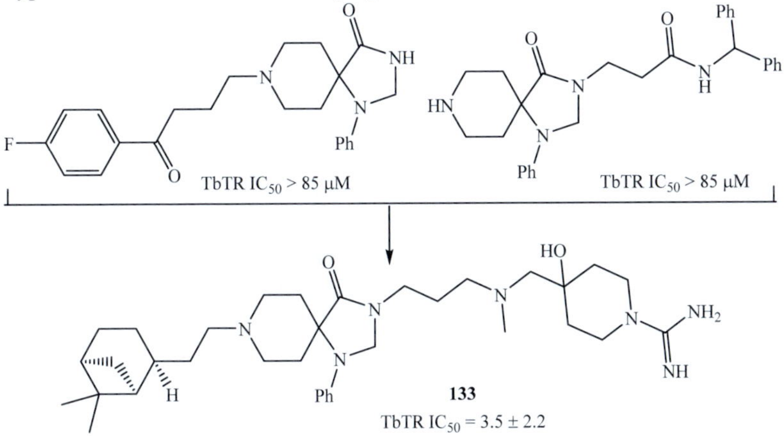

Scheme 6-39. General structures of hit series and representative molecule with TR inhibition potency.

6.2.35. Spiro compounds as anti-metabolic drugs

6-[5-(Cyclopropylmethyl)-4,5-dihydro-1′*H*,3*H*-spiro[1,5-benzoxazepine-2,4′-piperidin]-1′-yl]-*N*-(2-hydroxy-2-pyridin-3-ylethyl)pyridazine-3-carboxamide **134** and 2-(5-(3-(trifluoro methyl)benzyl)-4,5-dihydro-3*H*-spiro[benzo[*b*][1,4]oxazepine-2,4'-piperidin]-1'-yl)thiazole-5-carboxylic acid **135** are inhibitors of stearoyl-CoA desaturase-1 (SCD1), a therapeutic target for metabolic disorders and other indications (Figure 6-24). Compound **134** inhibited human SCD1 with an IC_{50} of 10 nM. Compound **135** is potent against mSCD1 (IC_{50} = 8 nM) with a good PK profile and discloses a moderate decrease of the plasma desaturation index 4 hours after oral dosing at 30 mg/kg [67, 68].

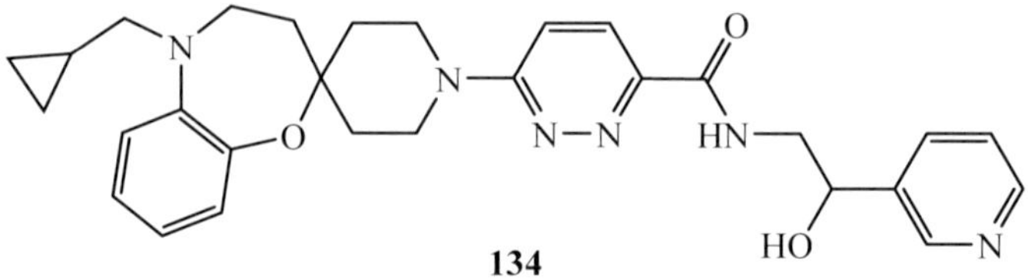

134

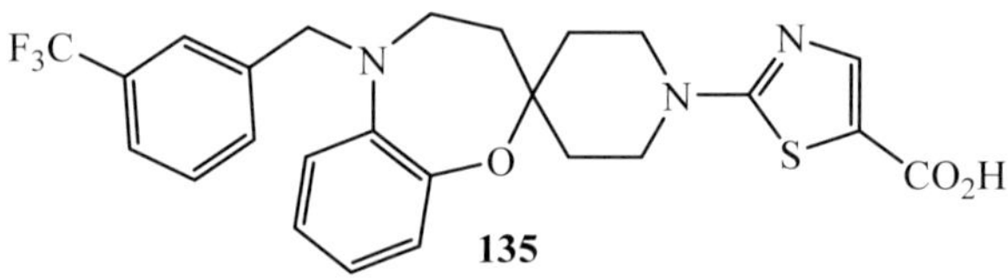

135

Fig. 6-24. Structure of spirocyclic scaffolds **134** and **135**.

6.2.36. Spiro compounds as anti-influenza drugs

In 2013, Kim et al. developed a two-step synthetic procedure to synthesize various substituted 3*H*,3′*H*-spiro[benzofuran-2,1′-isobenzofuran]-3,3′-diones **136-164** from readily available starting materials (Scheme 6-40) [69]. The synthesized compounds were screened against influenza virus type A, specifically A/Taiwan/1/86 (H1N1), A/Hong Kong/8/68 (H3N2), and type B viruses, including B/Panama/45/90, B/Taiwan/2/62, B/Lee/40, B/Brisbane/60/2008. Out of the 31 compounds screened, some showed good activity (selective index values >10) against these influenza viruses, especially type B. The most potent compound, **138**, exhibited activity in the 3.0-16.1 mM range with a selectivity index value between 30 and 166 against these type B viruses, which was comparable to the anti-viral agent favipiravir. Additionally, compound **138** was found to be inactive against other enveloped viruses such as HIV and HSV, suggesting its specificity for influenza viruses.

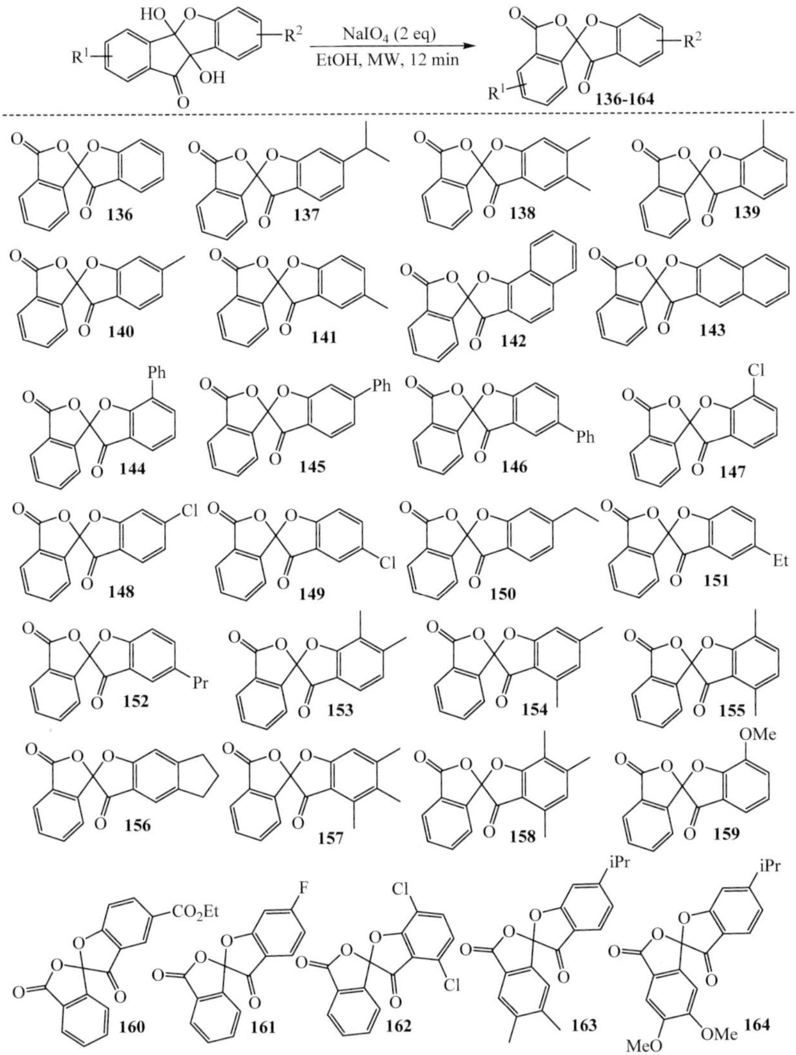

Scheme 6-40. Synthesis of various substituted 3*H*,3′*H*-spiro[benzofuran-2,1′-isobenzofuran]-3,3′-diones **136-164**.

In another study, a new series of *N*-(3-oxo-1-thia-4-azaspiro [4.5] decan-4-yl)carboxamides **165a-l** was synthesized by condensation f 2-methyl furan-3-carbohydrazide, appropriate carbonyl compounds, and sulfanyl acids and tested as anti-viral agents (Scheme 6-41) [70]. The IR,

^{1}H NMR, ^{13}C NMR, mass spectrometry, and elemental analysis confirmed the structures of new molecules. Six analogs were demonstrated to be active against the influenza A/H3N2 virus, the two most potent analogs, **165i** and **165j**, having an EC_{50} value of about 1 μM. These findings assist in defining the SAR of spirothiazolidinone-based inhibitors of the influenza virus membrane fusion process.

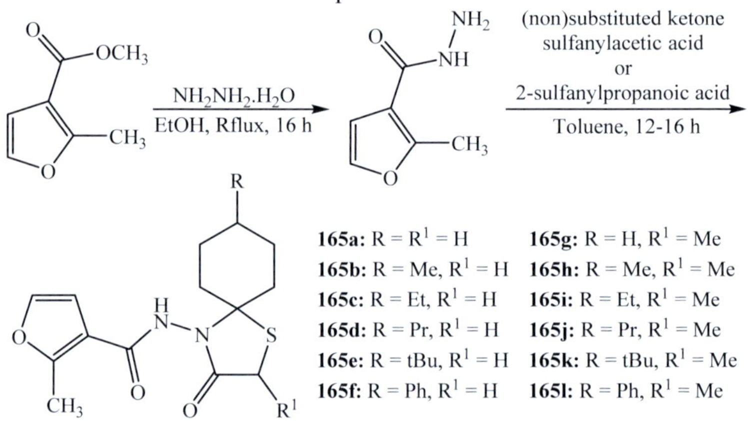

165a-l

Scheme 6-41. *N*-(3-Oxo-1-thia-4-azaspiro[4.5]decan-4-yl)carboxamides **165a-l**.

6.2.37. Spiro compounds as anti-hepatic drugs

In this context, noteworthy spiro[4.6.0] structures have been reported in the literature. For instance, spiro-lanostane triterpenoids **166a-c**, which were isolated from *Ganoderma applanatum*, were found to exhibit anti-hepatic fibrosis activities (Figure 6-25) [71].

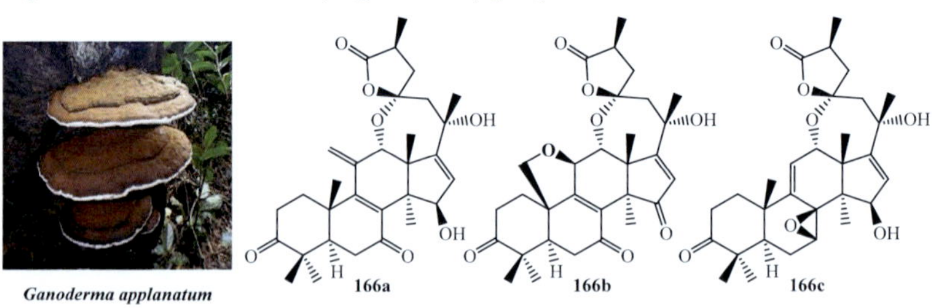

Fig. 6-25. Structure of spiro-lanostane triterpenoids **166a-c** isolated from *Ganoderma applanatum*.

6.2.38. Spiro compounds as anti-fibrotic drugs

A novel meroterpenoid dimer, namely Applanatumin A, was isolated from the fungus *Ganoderma applanatum* (Figure 6-26). Applanatumin A possesses a new hexacyclic skeleton containing a spiro[benzofuran-2,1′-cyclopentane] motif, and its structure and absolute configuration were characterized based on spectroscopic and computational data. Applanatumin A exposes potent anti-fibrotic activity in TGF-β1-induced human renal proximal tubular cells [72].

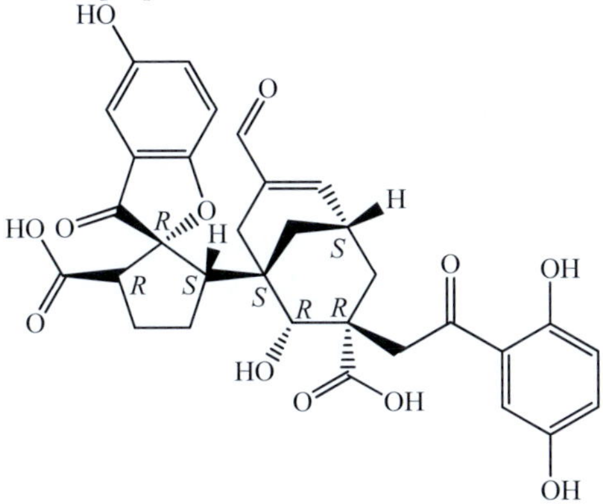

Fig. 6-26. Applanatumin A.

6.2.39. Spiro compounds as anti-asthmatic drugs

The Kubota group reported the synthesis and pharmacological properties of (±)-*N*-[2-(3,4-dichlorophenyl)butyl]-*N*-methylbenzenamide (YM-35375) as a potential new lead compound for anti-asthmatic drugs [73]. Through analysis of the structural requirements of compound SR48968, a potent NK receptor antagonist, it was determined that the 4-phenyl group of the piperidine is essential for binding with the NK_2 receptor and occupies an equatorial position (Figure 6-27). Further energy calculations of various substituted 4-phenyl piperidines showed that spiro-[isobenzofuran-1(3*H*),4′-piperidine] possesses a conformationally restricted equatorial phenyl group. Compound YM-35375 contains this spiro-substituted piperidine, which binds to the NK_2 receptor with an IC_{50} value of 84 nM and the NK_1 receptor with an IC_{50} value of 710 nM. *In vivo* studies revealed that YM-35375 demonstrated more potent inhibitory activity (ID_{50} 41 μg/kg) against [β-Ala8]-NKA(4-10)-induced broncho-constriction in guinea pigs than (±)-SR48968 (ID_{50} 68 μg/kg). These results suggest that YM-35375 has potential to serve as a lead compound for NK_2 receptor antagonists or NK_1-NK_2 dual antagonists in the development of anti-asthmatic drugs.

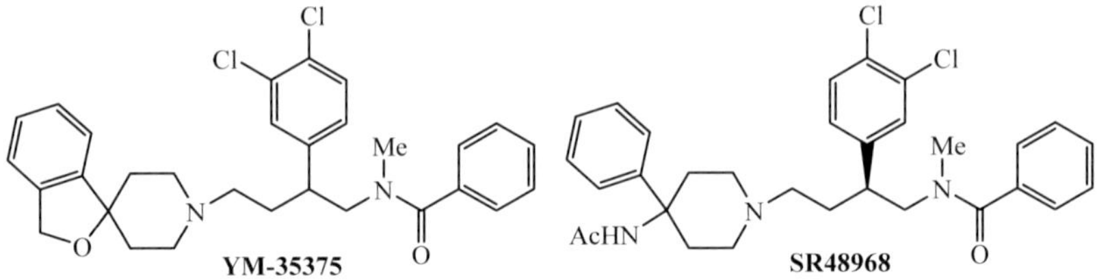

Fig. 6-27. Chemical structures of **YM-35375** and **SR48968**.

6.2.40. Spiro compounds as anti-thrombotic drugs

Preclinical anti-thrombotic efficacy and bleeding models have represented that $P2Y_1$ antagonists are efficacious as anti-platelet agents and may present a safety advantage over $P2Y_{12}$ antagonists regarding decreased bleeding liabilities. This way, spiropiperidine-based compound **167a-d** from BMS was reported as a potent, orally available $P2Y_1$ antagonist and displayed a robust anti-thrombotic effect in rat thrombosis and hemostasis models (Figure 6-28) [74].

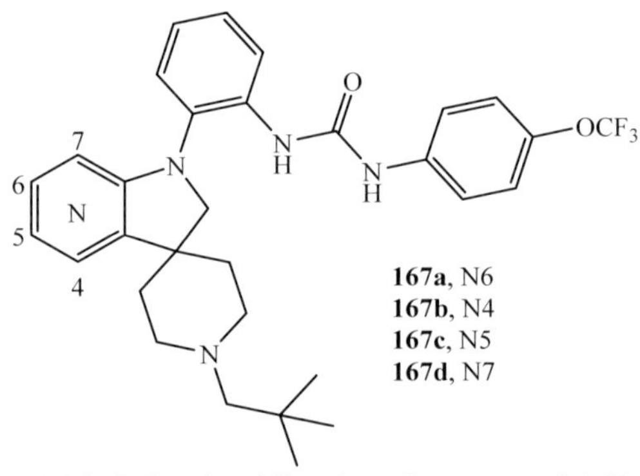

167a, N6
167b, N4
167c, N5
167d, N7

Fig. 6-28. Spiropiperidine-based compound **167a-d**.

6.2.41. Spiro compounds as anti-trypanosoma drugs

Human African Trypanosomiasis (HAT), commonly known as sleeping sickness, is a significant health problem in many African countries, with thousands of new cases reported annually. Chemotherapy, the primary form of control against HAT, has been recently characterized by issues related to toxicity and drug resistance. Morinda Lucida Benth (Rubiaciae), also referred to as brimstone tree, a widely distributed medicinal plant in Africa, has been shown to possess anti-protozoal activities. In a study by Shoyama et al., molucidin **168** (Figure 6-29), a tetracyclic iridoid, was isolated from the $CHCl_3$ fraction of *Morinda*

Lucida leaves using bioassay-guided fractionation and purification [75]. Its structure was determined by ^{1}H and ^{13}C NMR spectroscopy, including HMQC, HMBC, H–H COSY, and NOESY, revealing a tetracyclic iridoid skeleton with a defined absolute configuration. Molucidin **168** exhibited potent anti-trypanosomal activity with an IC_{50} value of 1.27 μM. The compound demonstrated selectivity towards normal fibroblasts with a selectivity index (SI) greater than 4.73, indicating its potential as a lead compound for further development of anti-HAT drugs.

Morinda lucida

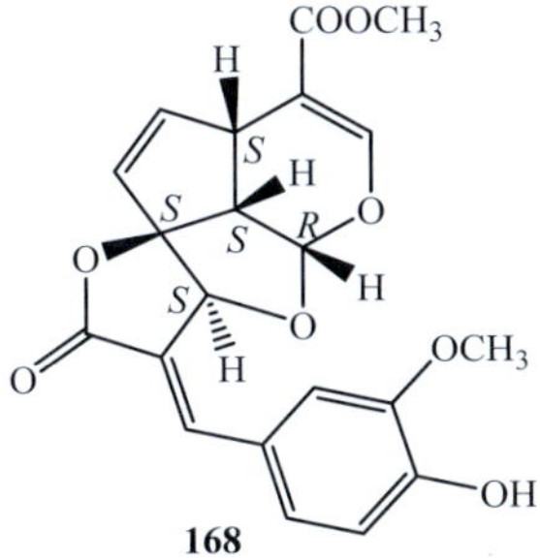

168

Fig. 6-29. Structure of Molucidin **168**.

6.2.42. Spiro compounds as anti-arthritic drugs

Yucca schidigera is a medicinal plant that may have practical effects in the debarment and treatment of arthritis. Active components of *Yucca* contain steroidal saponins and polyphenolics such as resveratrol and yuccaols. Saponins **169-176** may have anti-arthritic effects associated with their anti-protozoal activity (Figure 6-30). *Yucca* polyphenolics may have various roles in anti-arthritic activity. Folk medicine and anecdotal reports express that whole yucca plant powder aids in the debarment and treatment of arthritis [76].

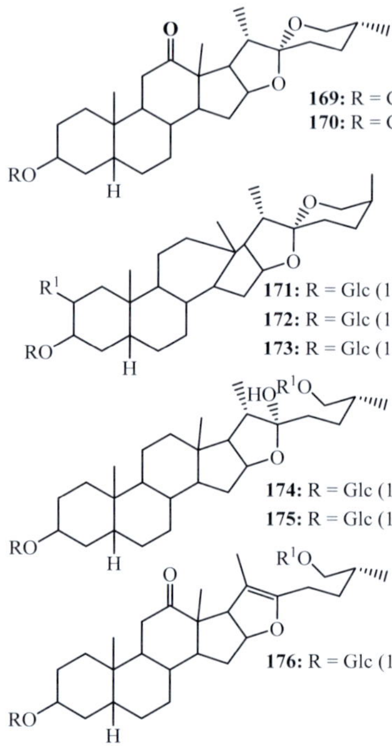

169: R = Glc (1-2) [Xyl (1-3)] Glc
170: R = Glc (1-2) [Glc (1-3)] Glc

171: R = Glc (1-2) Xyl R[1] = H
172: R = Glc (1-2) [Xyl (1-3)] Glc R[1] = H
173: R = Glc (1-2) Gal R[1] = OH

174: R = Glc (1-2) Glc R[1] = Glc
175: R = Glc (1-2) [Xyl (1-3)] Glc R[1] = Glc

176: R = Glc (1-2) [Xyl (1-3)] Glc R[1] = Glc

Yucca schidigera

Fig. 6-30. Structures of saponins identified in *Yucca schidigera* bark. Compounds **169–172** represent monodesmosidic and **174–175** bidesmosidic structures.

6.2.43. Spiro compounds as anti-nociceptive drugs

Yang et al. presented a series of pyrrolo-benzo-1,4-diazine analogs **177** (Figure 6-31), which displayed significant inhibitory activity against $Na_v1.7$, a voltage-gated sodium channel. These compounds demonstrated moderate selectivity over $Na_v1.5$, an integral membrane protein and tetrodotoxin-resistant voltage-gated sodium channel subunit. Compound **178** indicated oral anti-nociceptive efficacy at 100 mpk in both the rat CFA and SNL pain models with an EC_{50} value of 75 μM [77]. In addition, compound **179** demonstrated anti-nociceptive efficacy in the rat Complete Freund's Adjuvant (CFA) pain model at 100 mpk oral dosing [78].

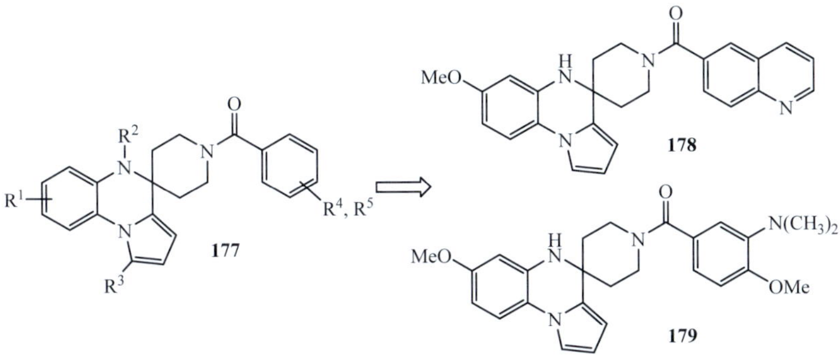

Fig. 6-31. Pyrrolo-benzo-1,4-diazine analogs **177**.

6.2.44. Spiro compounds as anti-spermatogenic drugs

19-Norspiroxenone (2′,3′α-tetrahydrofuran-2′-spiro-17-[estr-4-en-3-one]) **180** is a potent anti-estrogen with no known estrogenic action. It produces a striking arrest of spermatogenesis in the early stages of meiosis in the adult male rat with a complete block of fertility (Figure 6-32) [79].

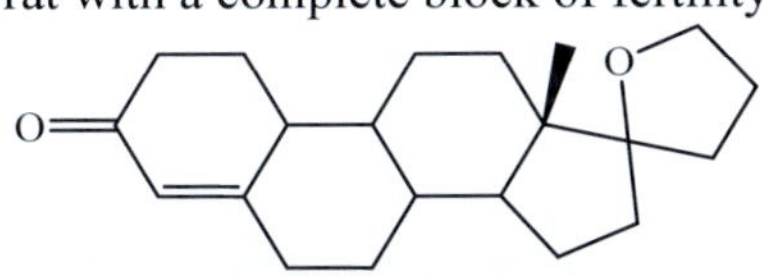

180

Fig. 6-32. Chemical structure of **180**.

6.2.45. Spiro compounds as anti-fertility drugs

The control of pregnancy has been achieved through the use of various natural and synthetic substances, which can interrupt pregnancy at different stages. One of the most significant scientific achievements of the 20th century was the development of hormonal contraceptives, including steroidal drugs like estradiol and progesterone. In recent decades, spiro compounds have emerged as promising anti-fertility agents.

Joshi et al. synthesized spiro[3,3′-pyrozol]-2-ones **181** and spiro[3*H*-indole-3,2-thiazolidine]-2,4(1*H*)-diones **182**, which showed anti-implantation activity of 10 and 30 mg/kg, respectively, in pregnant rats. Additionally, 1,4-dioxospiro[4,5]decanes **183** exhibited good anti-fertility activity [80].

Yuehchukene (YCK) **184** is a bis-indole alkaloid with weak estrogenic activity, first isolated from the root of *Murrsya paniculsta* (L) Jack. Biochemical studies have shown that YCK can attenuate estrogenic action. It has also been found to possess considerable anti-implantation activity in rats, making it a potential anti-fertility agent. The spiro analogues of YCK, such as **185**, exhibit potent anti-implantation activity in rats and are considered potential fertility regulatory agents. The presence of a bulky group at C-7 in YCK restricts the free rotation of the C-6 indole and encourages it to adopt the required conformation for biological activity (Figure 6-33) [81].

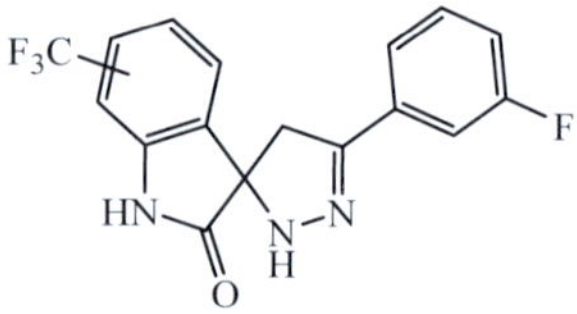

181

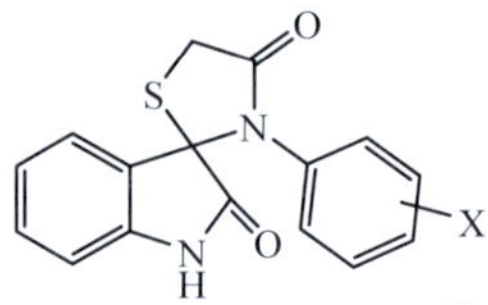

182

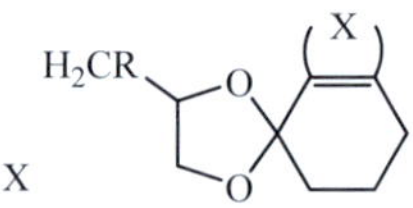

183

R = H, Halogen, OH, Alkoxy, NH_2
X = 5 or 6 membered ring

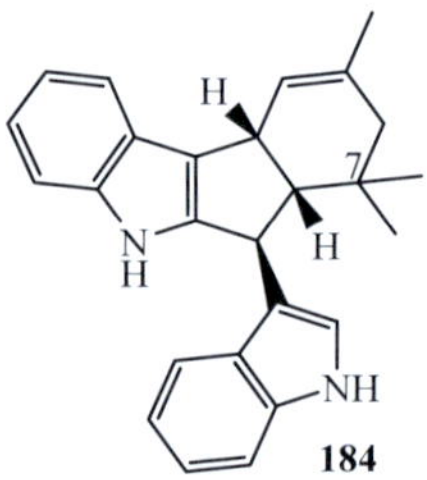

184

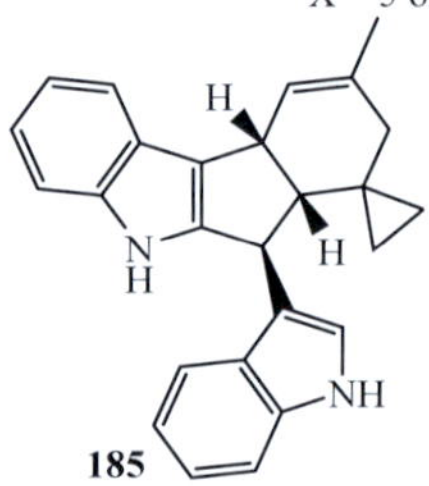

185

Fig. 6-33. Anti-fertility compounds.

6.2.46. Spiro compounds as anti-feedant drugs

Leucosceptroid O **186**, an unusual sesterterpenoid possessing a spiro α,β-unsaturated γ-lactone moiety, was isolated from the flowers of a large woody Labiatae, *Leucosceptrum canum* (Figure 6-34). NMR, MS, and single-crystal X-ray diffraction (with copper radiation) analyses assessed its structure containing absolute stereochemistry. The apparent anti-feedant activity of **186** against the generalist plant-feeding insect Helicoverpa armigera indicated that it is a new defensive sesterterpenoid of *Leucosceptrum canum* [82].

Leucosceptrum canum

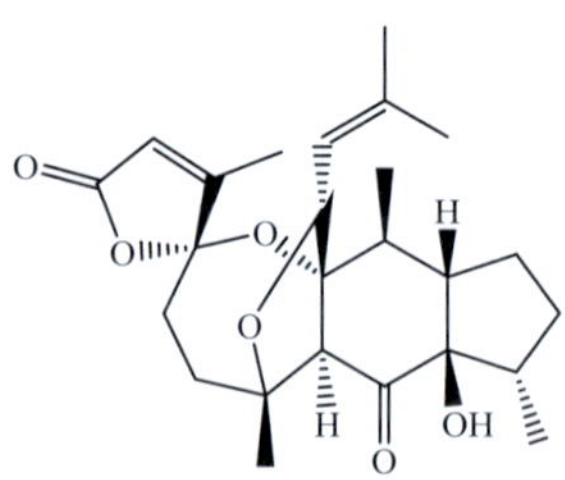

186

Fig. 6-34. Leucosceptroid O **186**.

Pedonin **187**, a ring-D cleaved spiro tetranortriterpenoid insect anti-feedant, has been isolated from the roots of *Harrisonia abyssinica* (Figure 6-35) [83].

Harrisonia abyssinica

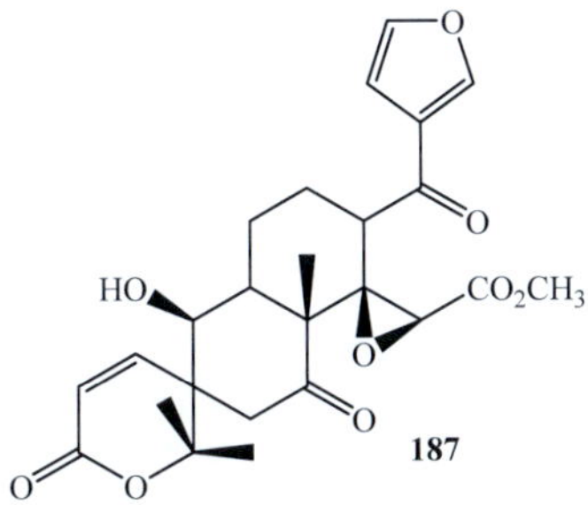

Fig. 6-35. Pedonin **187**.

The anti-feedant activity of a series of spiro enol ether analogs **188a-r** was investigated using larvae of *Pieris rapae* (Figure 6-36). Notably, compounds **188r** and **188l** exhibited higher anti-feedant activity compared to the other analogs in the study.

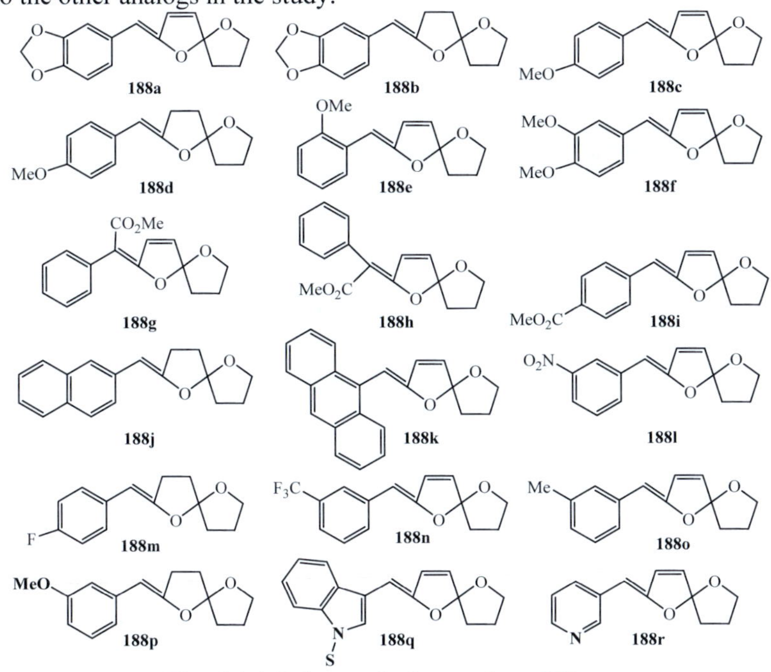

Fig. 6-36. Spiro enol ether analogs **188a-r**.

In the non-choice test, AFC5O values within 24 h of compounds **188r** and **188l** against 3rd instar larvae of *P. rapae* were 226.93 pg/mL and 370.00 pg/mL, and that in a choice test against 4th larvae were 280.54 μg/mL and 398.88 μg/mL, respectively. Compound **188r** could elongate the egg's hatch time and diminish the hemolymph and protein content in the hemolymph of 4th instar larvae. Compound **188r** could protect tested leaves and control larvae of *Pieris rapae* effectively [84].

6.2.47. Spiro compounds as anti-dermatitis drugs

Dermatologic disorders such as atopic dermatitis have complex and multifactorial causes, including genetic and environmental factors, with aberrant immunological reactions being a leading etiology. Immuno-suppressive agents such as topical steroids are commonly used to treat these disorders, but they can cause skin thinning. To address this issue, Noji et al. developed a series of JAK inhibitors with novel three-dimensional spiro motifs that exhibit excellent physicochemical properties and anti-dermatitis efficacy in animal models. Among the compounds tested, **JTE-189** (Figure 6-37) has shown impressive results and is well-tolerated in human clinical trials. As a result, **JTE-189** has been approved and prescribed in Japan as the first drug among Janus kinase inhibitors for the treatment of atopic dermatitis [85].

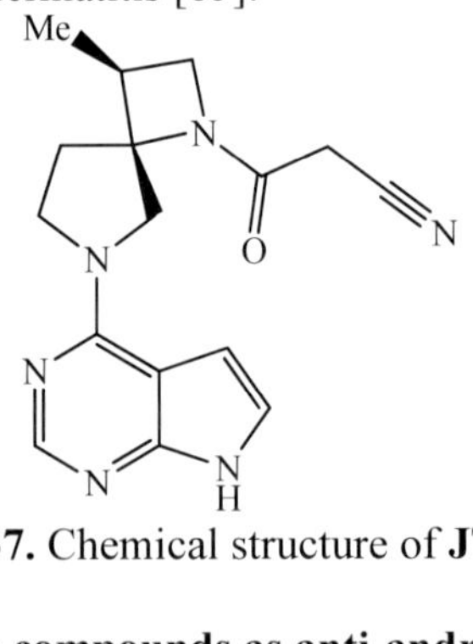

Fig. 6-37. Chemical structure of **JTE-189**.

6.2.48. Spiro compounds as anti-androgenic drugs

Bonne et al. conducted a study to compare the anti-androgenic effects of spironolactone with those of two potent anti-androgenic steroids, cyproterone acetate and R 2956, on rat prostate *in vivo* and *in vitro*. They found that spironolactone, like these compounds, inhibits the formation of the cytoplasmic specific androstane receptor complex and consequently, nuclei. Of its dethioacetylated metabolites, only canrenone exhibits the same *in vitro* action, while canrenoate-K does not [86].

Spiro steroidal-6,3-1-pyrazolines **190** and **191** exhibit valuable pharmacological properties such as anti-androgenic, anti-inflammatory, and anti-oxidant activities (Figure 6-38) [87].

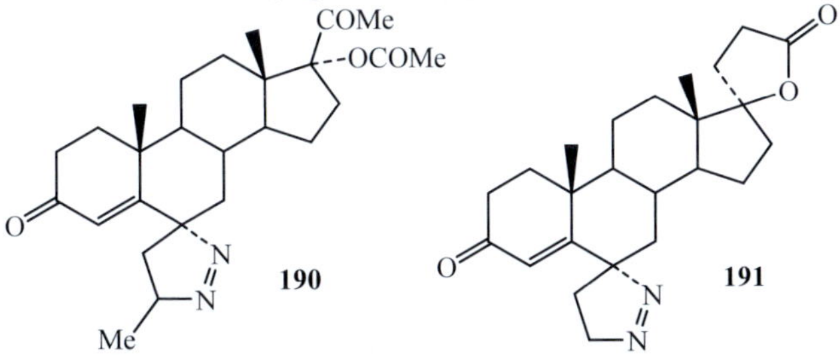

Fig. 6-38. Spiro steroidal-6,3-1-pyrazolines **190** and **191**.

6.2.49. Spiro compounds as anti-mineralocorticoid drugs

Clinically available active drugs against Epstein–Barr virus (EBV) and other human herpesviruses are limited to those targeting viral DNA replication. To identify compounds directed against other steps in the viral life cycle, Verma et al. probed the drugs active against the EBV SM protein, which is essential for infectious virus production. SM has a highly gene-specific mode of action and preferentially improves the expression of several late lytic cycle EBV genes. They showed that spironolactone, a mineralocorticoid receptor antagonist approved for clinical use, inhibits SM function and infectious EBV production. Expression of EBV viral capsid antigen is highly SM dependent. Spironolactone inhibits viral capsid antigen synthesis and capsid formation, blocking EBV virion generation at a step further to viral DNA replication. Additionally, spironolactone inhibits the expression of other SM-dependent genes essential for infectious virion formation. They further showed that molecules structurally related to spironolactone with similar anti-mineralocorticoid blocking activity do not inhibit EBV production [88].

6.2.50. Spiro compounds as anti-neoplastic drugs

The synthesis of fluoro-substituted dispiro[3*H*-indole-3,2′-pyrrolidine-3′,3″-piperidine] **192** was reported by Hall et al. in 2015 [89]. The authors reported the synthetic route to obtain this compound, as shown in Scheme 6-42. Single crystal X-ray studies of the regioselectively synthesized fluoro-substituted dispiro-oxindole **192** revealed that the structure belongs to the monoclinic space group $P2_1/n$, with four molecules in the unit cell. The authors also conducted computational studies using AM1, PM3, and

DFT methods to evaluate the structure. Compound **192** exhibited high potency against the HeLa cell line.

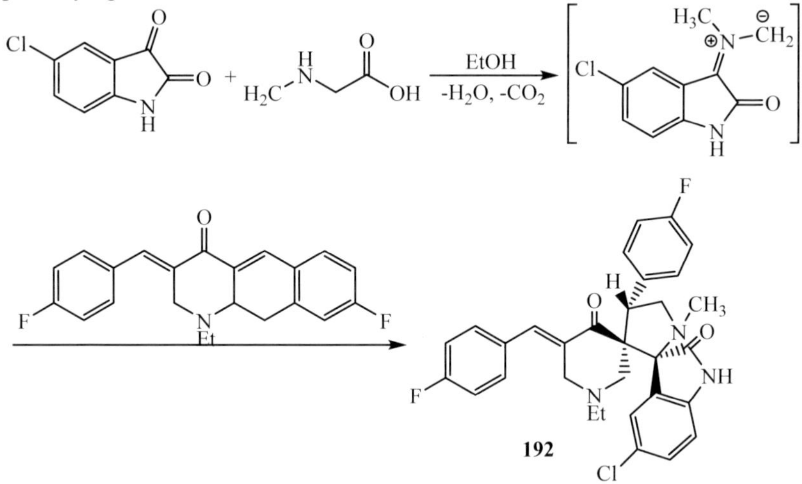

Scheme 6-42. Synthesis of fuoro-substituted dispiro[3*H*-indole-3,2′-pyrrolidine-3′,3″-piperidine] **192**.

As exemplified in Scheme 6-43, dibenzo-1-azaaspiro[4.5]decanes **193** were obtained in an enantiomerically pure form from *tert*-butane sulfinamide, 2-bromobenzyl bromide, and tetralone-type ketones [90]. The benzo-1-azaspiro[4.5]-decanes **193** were effective on models of leukemias with diverse adaptations to evade cytotoxicity, depending on the pattern of substitutions on the molecular scaffold. The thiochromane compounds were the most promising, whereas adding bromine or fluorine caused collateral sensitivity to the multidrug resistance phenotype (MDR) cells. The obtained results highlight the azaspiro motif as a substantial lead for developing new anti-neoplastic drugs targeted at chemo-refractory neoplasias, a sought-after profile for managing cancer.

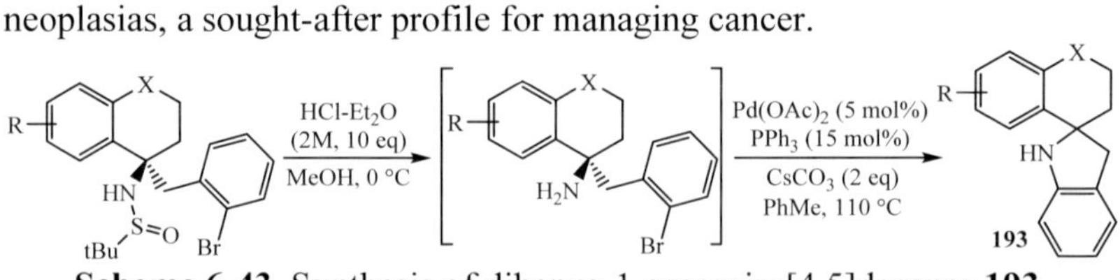

Scheme 6-43. Synthesis of dibenzo-1-azaaspiro[4.5]decanes **193**.

6.2.51. Spiro compounds as anti-histamine drugs

Arya et al. [91] developed an efficient strategy using Brønsted acidic ionic liquid containing nitrogen-based organic cations 1-methyl imidazolium and 1-butyl-3-methylimidazolium and inorganic anions such as BF_4^-, PF_6^-, and $PTSA^-$, as catalysts and reaction medium for synthesis of fluorinated spiro[3*H*-indole-3,2′-tetrahydro-1,3-thiazine]-2,4′(1*H*)-diones (**194a-f**), fluorinated spiro[3*H*-indole-3,2′-thiazolidine]-2,4′(1*H*)-diones (**195a-f**) in 90–97% yield by one-pot environmentally benign microwave-induced technique (Scheme 6-44). The synthesized compounds inhibit the contractions induced by histamine on guinea pig ileum. The measurement of pA_2 values showed that the reported compounds revealed H1-antagonism (the H1-receptor is a key element in the pathophysiology of allergic responses).

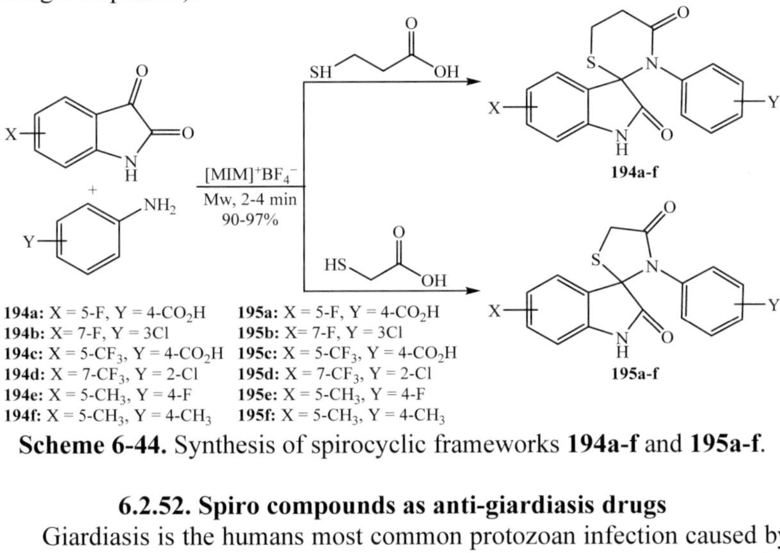

Scheme 6-44. Synthesis of spirocyclic frameworks **194a-f** and **195a-f**.

6.2.52. Spiro compounds as anti-giardiasis drugs

Giardiasis is the humans most common protozoan infection caused by the protozoan parasite *Giardia intestinalis* [92]. Lesyk et al. prepared a series of *rel*-(6′R,7′R)-3′,7′-dihydro-2*H*,2′*H*,5*H*-spiro[pyrolidin-3,6′-thiopyrano[2,3-*d*]thiazol]-2,2′,5-triones **196a-d** and studied as potential anti-trypanosomal agents (Scheme 6-45). *Trypanosoma* is a genus of kinetoplastids (Trypanoso-matidae, a monophyletic group of unicellular parasitic flagellate protozoa). The most active analog, **196b**, inhibited *Trypanosoma brucei brucei* and *Trypanosoma brucei gambiense* with an IC_{50} of 0.26 and 0.42 μM, respectively [93].

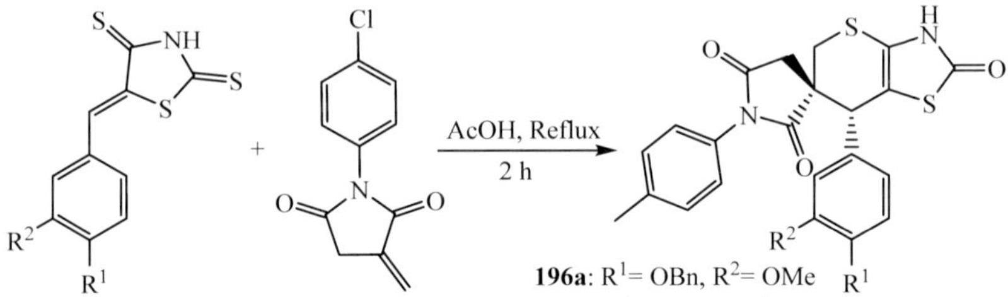

196a: R^1= OBn, R^2= OMe
196b: R^1= F_2CHO, R^2= OMe
196c: R^1= NMe_2, R^2= H
196d: R^1= NEt_2, R^2= H

Scheme 6-45. (6′R,7′R)-3′,7′-Dihydro-2*H*,2′*H*,5*H*-spiro[pyrolidin-3,6′-thio pyrano[2,3-*d*]thiazol]-2,2′,5-triones **196a-d**.

In another study, oxiranes **197** and **198** were synthesized respectively by the reaction of α-lapachone and tetrachloro-*p*-quinone with diazomethane in a mixture of ether and ethanol (Scheme 6-46). The synthsized compounds displayed the highest anti-plasmodial activity and low cytotoxicity against human fibroblasts (MCR-5 cell line). The active compounds could represent a good prototype for an anti-malarial lead molecule [94].

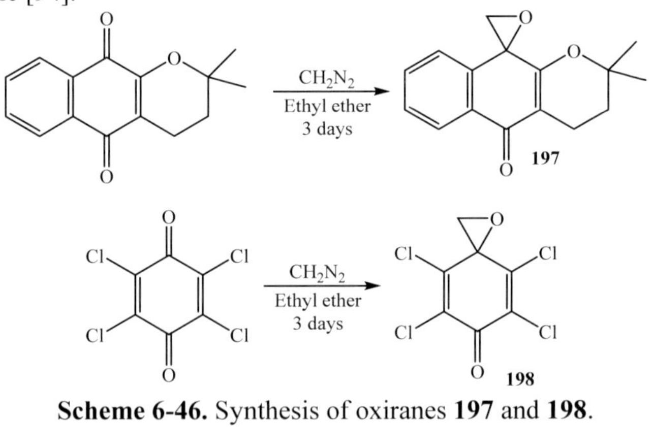

Scheme 6-46. Synthesis of oxiranes **197** and **198**.

6.2.53. Spiro compounds as anti-emetic drugs

Neurokinin NK1 receptors possess various physiological and potential therapeutic effects in animal models and humans. A. Duffy's studies illustrated that Rolapitant (SCH 619734, (5*S*)-8(*S*)-[[1(*R*)-[3,5 bis(trifluoro methyl)phenyl]ethoxy]methyl]-8-phenyl-1,7-diazaspiro[4,5]decan-2-one) **199** (Figure 6-39) is a selective, bioavailable, central nervous system

(CNS) penetrant neurokinin NK1 receptor antagonist that exhibits behavioral effects in animals models of emesis. *In vitro* studies demonstrated that Rolapitant has a high affinity for the human NK1 receptor of 0.66 nM and high selectivity over the human NK2 and NK3 subtypes of >1000-fold, as well as a preferential relationship for human, guinea pig, gerbil, and monkey NK1 receptors over the rat, mouse, and rabbit. Rolapitant is a functionally competitive antagonist, estimated by calcium efflux, with a calculated Kb of 0.17 nM. Rolapitant reversed NK1 agonist-induced foot tapping in rats following intravenous and oral administration for up to 24 hours at a minimal effective dose (MED) of 0.1 mg/kg. Rolapitant was active at 0.1 and 1 mg/kg in both critical and delayed emesis models in ferrets, respectively, consistent with clinical data for other NK1 antagonists. The clinical efficacy of anti-emetics is highly correlated with effectiveness in the ferret emesis model, suggesting that Rolapitant is an effective clinical candidate for this indication [95].

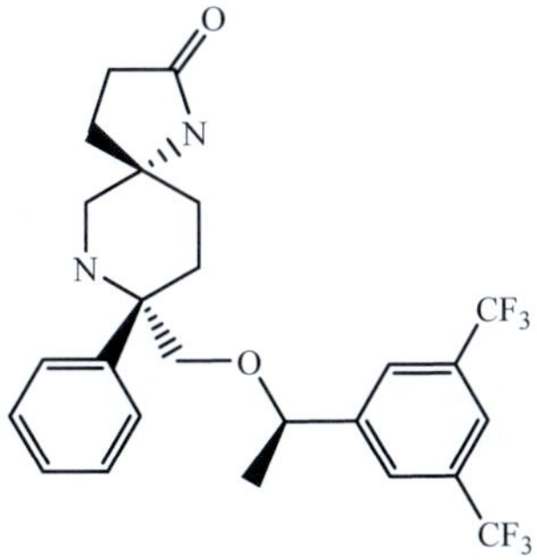

Fig. 6-39. Chemical structure of Rolapitant (5*S*)-8(*S*)-[[1(*R*)-[3,5-bis(trifluoromethyl)phenyl]ethoxy] methyl]-8-phenyl-1,7-diazaspiro[4,5]decan-2-one) **199**.

6.2.54. Spiro compounds as anti-diuretic drugs

Vasopressin is involved in most cases of sustained low sodium concentration in the blood (hyponatremia). Hence, using specific blockers of vasopressin receptors is a logical approach in treating patients with the syndrome of inopportune anti-diuretic hormone secretion (SIADH). The vasopressin V_2-receptor antagonist **200** (satavaptan) was established by Sanofi Aventis for the treatment of hyponatremia as well as cirrhosis-related ascites (Scheme 6-47). Treatment of patients with SIADH and mild or moderate hyponatremia with satavaptan, a new oral V_2-receptor antagonists (V_2RA), corrects SNa with a good safety profile. This effect was approved for the first time on a long-term basis [96, 97].

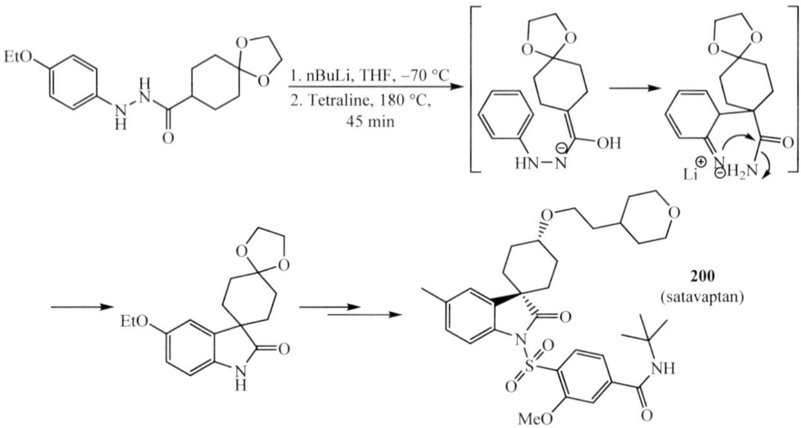

Scheme 6-47. Synthesis of satavaptan **200**.

6.2.55. Spiro compounds as anti-Parkinsonian drugs

Parkinson's disease is a neurodegenerative disorder caused by progressive loss of midbrain dopaminegic neurons. The loss of neurons is strongly correlated with aging and aging-associated disorders. In 2019, Abdel-Magid introduced indazolyl-spiro[2.3]hexane-carbonitriles **201** as leucine-rich repeat kinase 2 (LRRK2) inhibitors (Figure 6-40). The prepared compounds are potent inhibitors of LRRK2 kinase. They may effectively treat or prevent diseases in which the LRRK2 kinase is involved, such as Parkinson's disease and possibly other neuro-degenerative diseases and disorders [98].

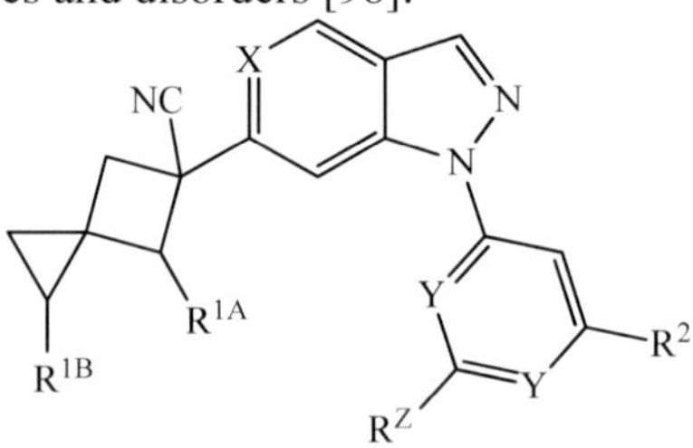

Fig. 6-40. Indazolyl-spiro[2.3]hexane-carbonitriles **201** as LRRK2 inhibitors.

In another study, cell viability tests and mitochondrial function were accomplished to consider the effect of new spiro-pyrazoles **202a,b,c,d,f,g,h,i** and **203a,d,e,i** generated from aldehydes and 3-amino-1-phenyl-2-pyrazolin-5-one, on neuroprotection in an *in vitro* model of dopaminergic cell death induced by 1-methyl-4-phenylpyridinium (MPP^+)

(Scheme 6-48). The percentages of neuroprotection by derivatives were found between 21.26% and 52.67% at chosen concentrations (10-50 μM), with compound **203d** exerting the best neuroprotective effect [99].

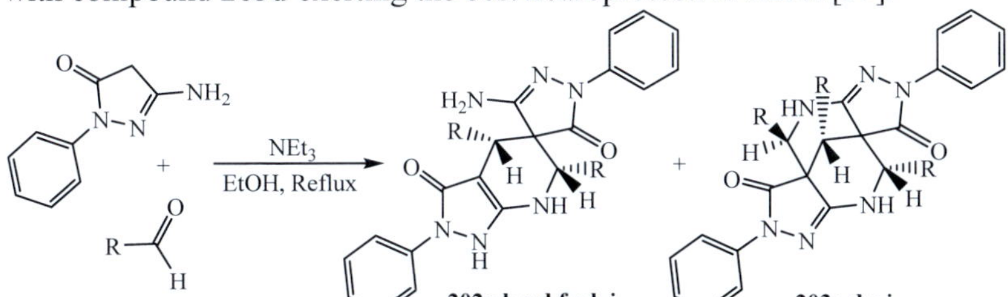

202a,b,c,d,f,g,h,i **203a,d,e,i,**

202a: R = Ph	**202f:** R = 4-Me-Ph	**203a:** R = Ph
202b: R = 2-NO_2-Ph	**202g:** R = 4-OMe-Ph	**203d:** R = Thiophene
202c: R = 3-Br-Ph	**202h:** R = 3-NO_2-Ph	**203e:** R = 4-Cl-Ph
202d: R = Thiophene	**202i:** R = Furan	**203i:** R = Furan

Scheme 6-48. Synthesis of spiro-pyrazoles **202a,b,c,d,f,g,h,i** and **203a,d,e,i**.

6.2.56. Spiro compounds as anti-cough drugs

A series of hydantoin compounds **204a-g** (Scheme 6-49, Eq. 1) and **205a,b** (Scheme 6-49, Eq. 2) was prepared with 3-bromophenylacetic acid and 3,4-dimethylbromobenzene as starting material, 7-bromo-2-tetralone and 2-bromo-5,6,8,9-tetrahydro-7-benzocycloheptenone as intermediate and Ullmann reaction, Suzuki reaction and Bucherer-Berg's reaction as key steps. The chemical structures of the key intermediate and target compounds were characterized by ^1H NMR, ^{13}C NMR, IR, and MS. A bioactivity study indicated that target compounds **204a**, **204c**, **204d**, **204e,** and **205b** had a remarkable anti-tussive effect on the ammonia-induced cough of mice [100].

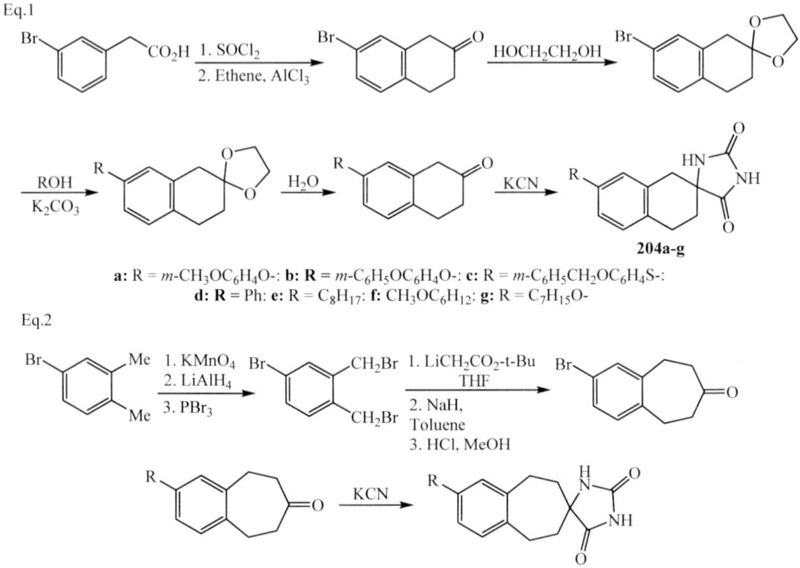

a: R = *m*-$CH_3OC_6H_4O$-: **b: R** = *m*-$C_6H_5OC_6H_4O$-: **c:** R = *m*-$C_6H_5CH_2OC_6H_4S$-:
d: R = Ph: **e:** R = C_8H_{17}: **f:** $CH_3OC_6H_{12}$: **g:** R = $C_7H_{15}O$-

205a: R = C_8H_{17}: **205b:** R = C_4H_9

Scheme 6-49. Hydantoin compounds **204a-g** and **205a,b**.

Twelve known compounds were isolated from the roots of *Stemona sessilifolia* and their structures confirmed by one- and two-D NMR spectra and other spectroscopic studies. The main alkaloidal constituents, stemospironine **206** and maistemonine **207** revealed an anti-tussive activity in a citric acid-induced guinea pig cough model following peripheral administration; stemonamine **208** had anti-tussive activity following i.c.v. administration (Figure 6-41) [101].

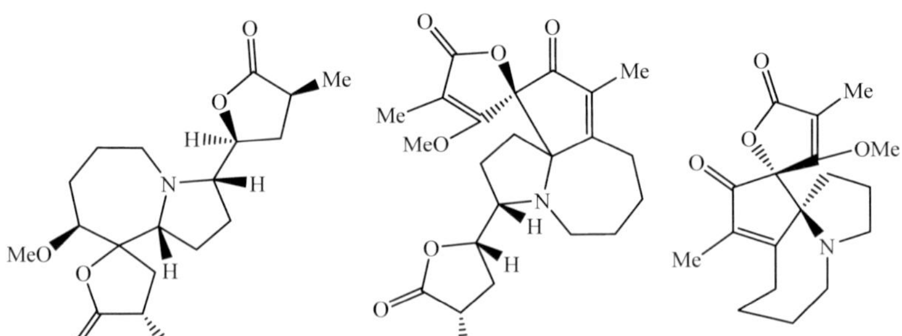

206 **207** **208**

Fig. 6-41. Alkaloids possess anti-tussive activity.

6.2.57. Spiro compounds as anti-retroviral drugs

Classical anti-retroviral therapy involves drugs from two different classes, Reverse Transcriptase Inhibitors and Protease Inhibitors. Reverse transcriptase inhibitors inhibit the process of reverse transcription at stage 4 of HIV replication (the life cycle of HIV is: 1. viral attachment, 2. viral fusion, 3. uncoating, 4. reverse transcription, 5. nuclear import, 6. integration, 7. transcription, 8. nuclear export, 9. translation, 10. assembly, 11. budding, 12. release, and 13. maturation). Protease inhibitors such as Saquinavir, Ritonavir, and Indinavir inhibit the budding process and thus block stage 10 and prorogate the generation of mature HIV [102]. Aplaviroc (APL; GW873140) **209** is a chemokine receptor 5 (CCR5) antagonist that was established for the treatment of patients with HIV infection/Division of Acquired Immunodeficiency Syndrome (AIDS) (Figure 6-42) [103].

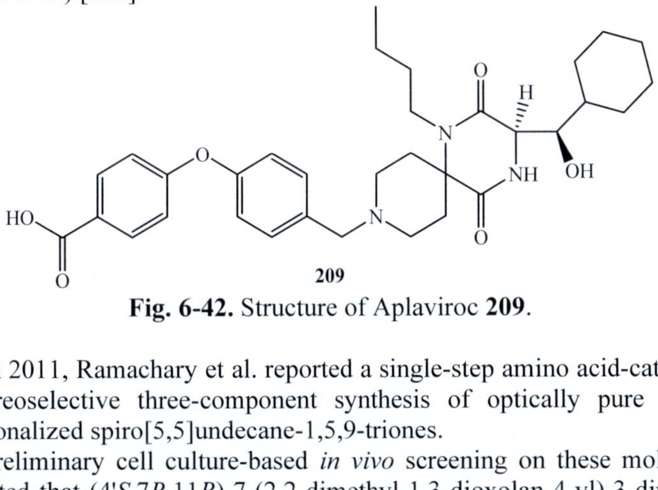

209

Fig. 6-42. Structure of Aplaviroc **209**.

In 2011, Ramachary et al. reported a single-step amino acid-catalyzed diastereoselective three-component synthesis of optically pure highly functionalized spiro[5,5]undecane-1,5,9-triones.

Preliminary cell culture-based *in vivo* screening on these molecules exhibited that (4'*S*,7*R*,11*R*)-7-(2,2-dimethyl-1,3-dioxolan-4-yl)-3-dimethyl -11-phenyl-2,4-dioxaspiro[5.5]undecane-1,5,9-trione **210** and (4'*S*,7*R*,11*S*) -7-(2,2-Dimethyl-[1,3]dioxolan-4-yl)-11-furan-2-yl-3,3-dimethyl-2,4-dioxaspiro[5.5]undecane-1,5,9-trione **211** are better lead compounds for HIV-1 treatment than the known anti-retroviral drug azidothymidine **212** (Figure 6-43) [104].

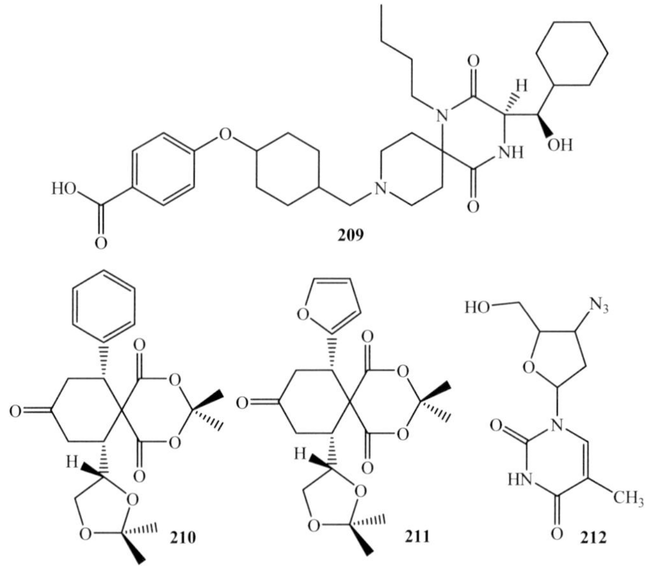

Fig. 6-43. Structure of compounds **209-212**.

6.2.58. Spiro compounds as anti-phlogistic drugs

Anti-phlogistic agents are compounds, either derived from plants or synthesized chemically, that can reduce inflammation. In order to target inflammation, it is important to inhibit both pro-inflammatory enzymes, 5-lipoxygenase, and cyclooxygenase-2, and spiro compounds have been synthesized as potential anti-phlogistic agents. Two new spiro[5.5] undecanes, 3-(hydroxymethyl)-7-(methoxy-methyl)-3,11-dimethyl-9-oxo-spiro[5.5]undec-4-en-10-methyl-butanoate **213** and 4-ethoxy-11,11-di-methyl-7-methylene-8-(propionyloxy)-spiro[5.5]undec-2-en-10^4,10^6-di-hydroxy-tetra-hydro-2*H*-pyran-10-carboxylate **214**, were isolated from *Gracilaria salicornia* seaweed and were characterized by various spectroscopic methods. Both compounds demonstrated potential anti-inflammatory properties by inhibiting 5-lipoxygenase (5-LOX) with IC_{50} values < 2.80 µM. Interestingly, their selectivity indices were significantly greater than Ibuprofen (0.89) ($p < 0.05$), which can be attributed to the selective anti-inflammatory potencies of the spiro[5.5]undecanes against inducible cyclooxygenase-2 rather than constitutive cyclooxygenase-1. Spiro[5.5]undec-2-en-10^4,10^6-dihydroxytetrahydro-2*H*-10-carboxylate analog **214** also exhibited radical scavenging potential against oxidants, 2,2-diphenyl-1-picrylhydrazyl and 2,2'-azino-bis-3-ethylbenzothiozoline-6-sulfonic acid, with IC_{50} values < 1.25 µM, which was greater than the commercial standard, α-tocopherol (IC_{50} 1.42-1.79 μM). Furthermore, the compound showed significant hydrogen-bonding interactions and binding affinity with 5-LOX, further corroborating its potential anti-inflammatory activity [105].

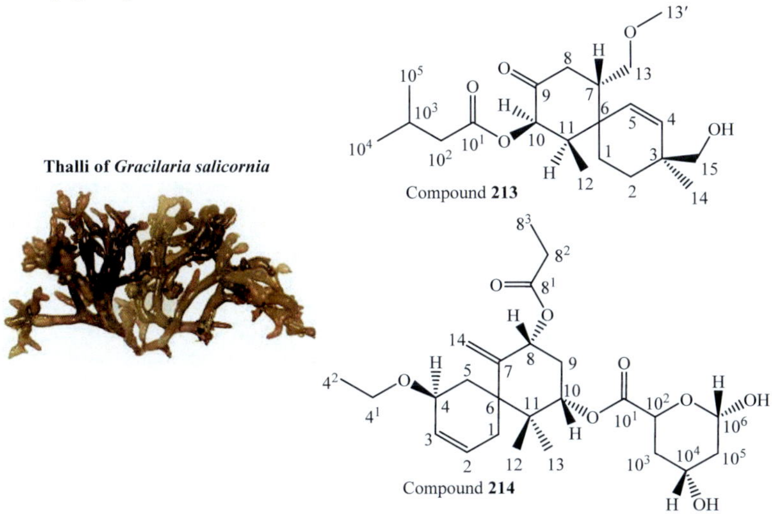

Fig. 6-44. Structures of spiro[5.5]undecane compound **213** and **214** isolated from the seaweed *Gracilaria salicornia*. The representative image of the thalli of *Gracilaria Salicornia* was indicated.

6.2.59. Spiro compounds as anti-metabolite drugs

An anti-metabolite is a class of drugs that disrupts normal cellular metabolic processes by acting as structural analogs of metabolites [106]. Anti-metabolites, such as anti-folates that interfere with the use of folic acid, can competitively inhibit enzymes in metabolic pathways and have toxic effects on cells. These compounds are commonly used in cancer chemotherapy to interfere with DNA production, cell division, and tumor growth, as cancer cells spend more time dividing than normal cells [107]. Anti-metabolites are used to treat a variety of cancers, including leukemia, breast cancer, ovarian cancer, gastrointestinal tract cancer, and others.

A recent study by Barakat investigated the *in vitro* anti-proliferative effects of tetrahydro-1*H*-spiro[indoline-3,5′-pyrrolo[1,2-*c*]thiazol]-2-one hybrids **215a-n** (Figure 6-45) on various cancer cell lines. The spiro-oxindole analog **215g** demonstrated a broad anti-proliferative activity against HepG2 (liver cancer), MCF-7 (breast cancer), and HCT-116 (colon cancer) cell lines compared to cisplatin. The study highlights the potential of these synthesized compounds as effective anti-cancer agents for liver, breast, and colorectal cancers [108].

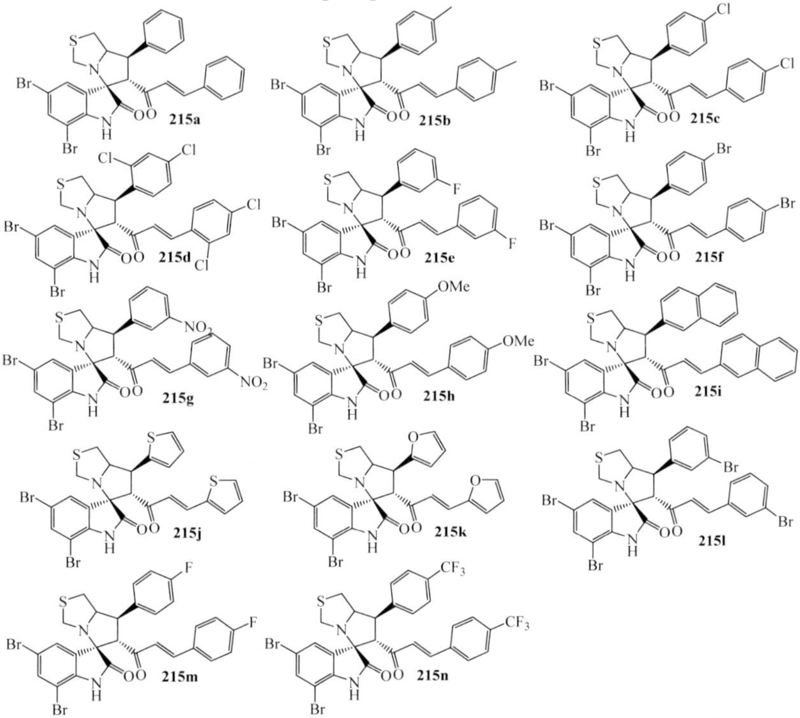

Fig. 6-45. Tetrahydro-1*H*-spiro[indoline-3,5′-pyrrolo[1,2-*c*]thiazol]-2-ones **215a-n**.

6.3. Organic optoelectronic devices

Traditional electronic and optoelectronic devices are typically fabricated on rigid and fragile substrates that lack deformability [109]. Optoelectronic devices, whether based on inorganic or organic materials, are electronic and optical devices that emit, guide, modulate, and receive light. Advanced optoelectronic devices, such as light-emitting diodes (LEDs), LASERs, solar cells, and photodetectors, have been extensively studied for various applications in energy, communication, and healthcare [110]. Recently, devices using organic materials as active elements, such as organic light-emitting diodes (OLEDs), organic photovoltaic devices (OPVs), organic field-effect transistors (OFETs), and organic photorefractive devices, have garnered significant attention for their potential technological applications and fundamental science. Organic materials offer advantages such as light weight, potentially low cost, and capability for thin-film, large-area, and flexible device fabrication.

Organic semiconductors (OSCs) are often used as charge transporting materials (CTMs) in these devices. In contrast to inorganic semiconductors, organic semiconductors are essentially electrical insulators. Charge transport is an essential operation process in these devices and requires CTMs, which are often based on π-electron systems [111].

OSCs offer advantages over their inorganic counterparts in terms of their ability to be deposited by many methods, including vapor deposition and solution-based processes such as spin-coating and printing techniques. These thin films can be deposited at room temperature, and the construction costs are notably lower than those of the corresponding inorganic optoelectronic devices. The optoelectronic devices can also be constructed on low-cost, flexible substrates, paving the way for all-plastic electronics and transparent devices.

However, controlling the reproducibility of device performance and device lifetime are important challenges in organic optoelectronics devices, posing challenges for materials scientists and device engineers. This issue is considered in industrial research and at an academic level in universities and institutional laboratories worldwide. Understanding the morphology and microscopic organization of the applied materials in single-, poly-, and liquid crystals and amorphous glasses is crucial for controlling and optimizing the corresponding optoelectronic device performance. The amorphous glassy state is particularly desirable for active materials in various optoelectronic applications for several reasons. Firstly, achieving the amorphous state is relatively simple compared to single-crystal growth. Secondly, the resulting thin films are free from grain boundaries, leading to spatial homogeneity of the transport properties. Thirdly, the isotropic stable glassy state has desirable optical properties. Polycrystalline materials lead to a large amount of scattering, while single crystals cause birefringence, which is not always desirable. Therefore, when manufacturing amorphous films, high stability of the glassy state is required to prevent device degradation in applications.

A significant physical parameter indicating the stability of the amorphous state of a given material is the glass transition temperature (T_g). Heating an amorphous film above this temperature causes molecular movement to increase rapidly, leading to transition into the crystalline state. Thus, designing molecules with low molecular weight is essential to obtain high T_g [112].

Molecular CTMs are superior to charge transport polymers such as poly(*p*-phenylene vinylene) (PPV) in terms of carrier mobilities. However, they often suffer from poor thermal stability. To confer higher stability to the chemical structure of low-molecular-weight compounds, different synthetic strategies have been pursued.

The Salbeck group introduced the concept of linking two aryl amine moieties through a spiro center, 9,9′-spirobifluorene (SBF), which introduces a 90-degree angle between them. This raises the thermal stability of the amorphous state, significantly increasing the glass transition temperature [113].

In this book, we briefly summarize the attempts at functionalizing and applying compounds based on spiro linkage in OLEDs, solar cells, and organic transistors.

6.3.1. Organic light-emitting diodes (OLEDs)

In 2000, Steuber et al. prepared two series of white LEDs using spiro-linked low molecular weight structures with high glass transition temperature [114]. In devices of type A and B, 2,2′,7,7′-tetrakis-(diphenylamino)-9,9′-spirobifluorene (**spiro-TAD**) and 4,4′,4′4″-tris(*N*-(3-methylphenyl)-*N*-phenylamino)-triphenylamine (***m*-MTDATA**) is used as the hole transport layer (HTL), respectively (Table 6.1 and Figure 6-46).

Table 6.1. Layer setup of samples with doped emitter layers [a].

Type	Hole Transport Layer	Emitter Layer	Electron Transport Layer
A1	spiro-TAD, 75 nm	spiro-Octo:rubrene, 35 nm	Alq_3, 30 nm
A2	spiro-TAD, 75 nm	spiro-4Φ:rubrene, 35 nm	Alq_3, 30 nm
B	*m*-MTDATA, 60 nm	α-NPD:rubrene, 20 nm	Alq_3, 60 nm
Anode Layer: Indium Tin Oxide 100 nm			
Cathode Layer $Mg_{0.9}Ag_{0.1}$ 150 nm / Ag 150 nm			

[a] In devices of type B, *m*-MTDATA is used as the HTL, and *N*,*N*′-di (naphthalene-1-yl)-*N*,*N*′-diphenyl-benzidine (α-NPD) is used as the EML host material. In all devices, rubrene is used as the emitter dopant. Alq_3= tris(8-hydroxyquinolinolato)aluminum(III).

In all devices, 5,6,11,12-tetraphenylnaphthacene (**rubrene**) is used as the emitter dopant. They achieved a luminous efficiency of 2.4 cd/A, and the maximum luminance was as high as 11800 cd/m^2 at a voltage of 15 V for devices based on **spiro-Octo** (type A1). An efficiency of 1.6 cd/A and maximum luminance of 4100 cd/m^2 was achieved for devices based on **spiro-4Φ** (type A2). The higher efficiency and luminance acquired in devices of type A1 can be rationalized because of the blue emission spectra. They found that these devices also operate at ambient temperatures well above 100 °C due to the high morphological stability of the spiro-linked compounds for HTL and the emission of blue light.

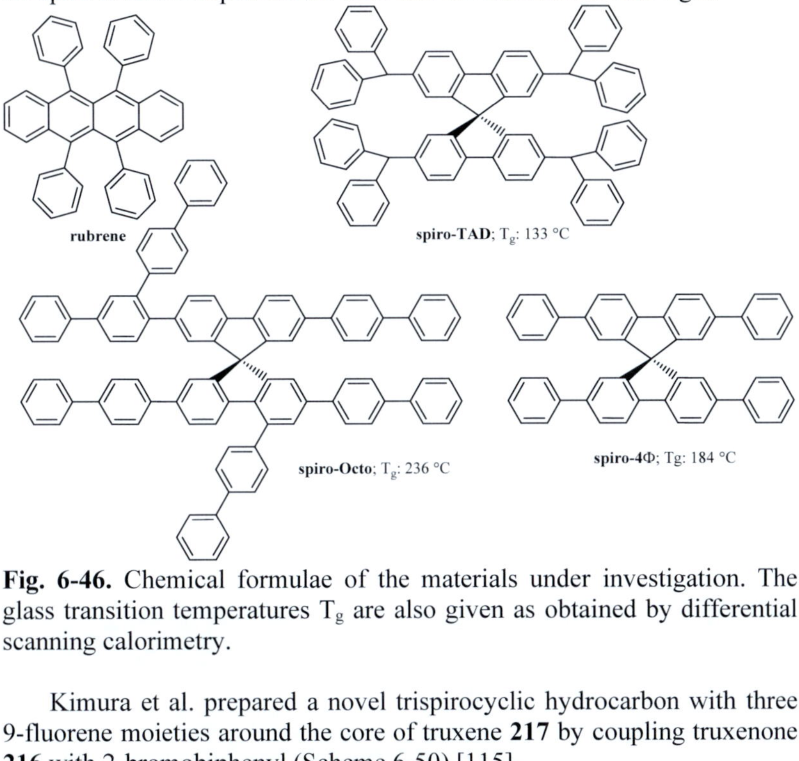

Fig. 6-46. Chemical formulae of the materials under investigation. The glass transition temperatures T_g are also given as obtained by differential scanning calorimetry.

Kimura et al. prepared a novel trispirocyclic hydrocarbon with three 9-fluorene moieties around the core of truxene **217** by coupling truxenone **216** with 2-bromobiphenyl (Scheme 6-50) [115].

Then, a triphenylamine bearing six diphenylamino groups at the 2- and 7-positions of 9-fluorene moiety **218** (Figure 6-47) was designed to achieve high thermal stability as an improved material of the *N,N'*-diphenyl-*N,N'*-bis(3-methyl-phenyl)-1,1′-biphenyl-4,4′-diamine (TPD) type Hole-Transporting Molecule (HTM) for the application of **217** to an impressive HTM in the OLED. The hexakis(diphenylamino) substituted trispirocyclic **218** was easily synthesized using 4,4′-bis(diphenyl amino)-2-bromobiphenyl and exhibited a high T_g of 170 °C. It leads to the stable cation radical upon electro-oxidation in solution and amorphous thin films in solid. It affects the formation of its stable cation radical upon electro-oxidation in solution and amorphous thin films in solid. Their molecular design allows **218** to be applied as a thermally durable HTM in combination with Alq_3 at high temperatures. The electroluminescence (EL) characteristics of **218** as HTM demonstrated to be like or better than the representative TPD. Therefore, trispirocyclic amine **218** can be used for advanced full-color EL displays.

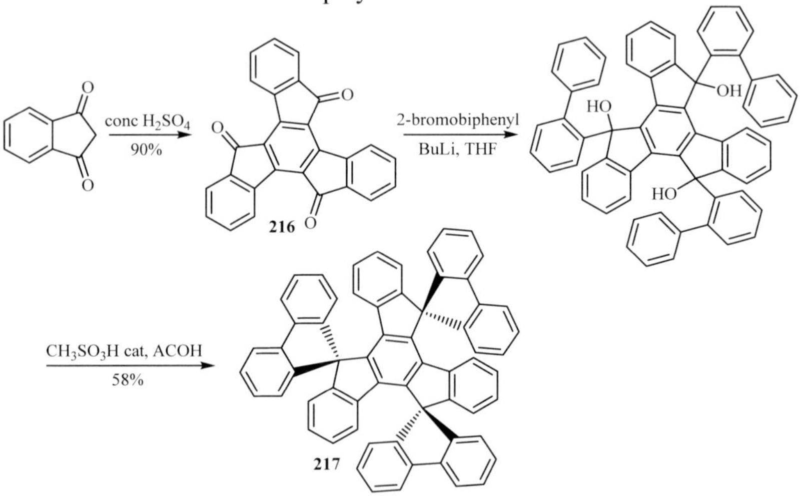

Scheme 6-50. Preparation of threefold spirobifluorene **217**.

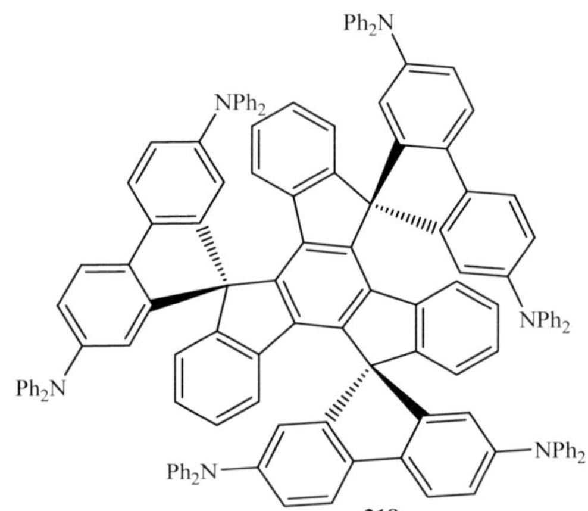

218

Fig. 6-47. The chemical structures of hexakis(diphenylamino) substituted trispirocyclic **218**.

Huang et al. accomplished the successful synthesis of ter(9,9′-spiro bifluorene)s analog 2′,7′-bis(9,9′-spirobifluoren-2-yl)spiro[[8*H*]indeno[2,1-*b*]-thiophene-8,9′-fluorene] (**BSBF–SITF**), a regioselective blue light-emitting material, through Suzuki coupling reactions. The synthetic pathway for this achievement is outlined in Scheme 6-51. **BSBF–SITF** has revealed excellent thermal stability and luminescent properties in the spirobifluorene system, which shows that **BSBF–SITF** will be promising for manufacturing complicated optoelectrical systems [116].

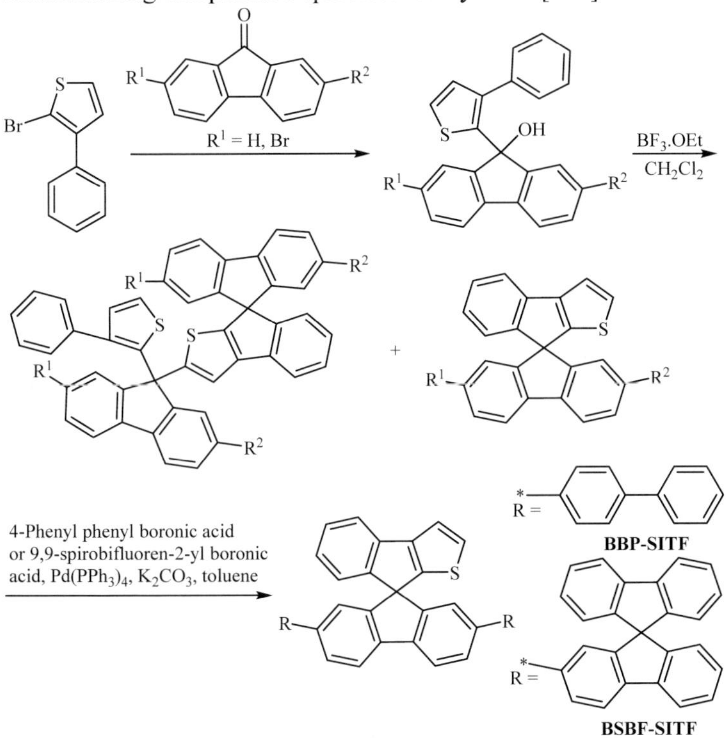

Scheme 6-51. Preparation of ter(9,9′-spirobifluorene)s analog **BSBF–SITF**.

In 2008, the Bo group developed a novel route to spiro-bridged ladder-type polymers, which displays excellent thermal and color stability and is free of low-energy ketonic defects [117]. The synthetic pathway leading to the spiro-bridged ladder-type poly(*p*-phenylene)s (**LPPP**) is represented in Scheme 6-52. The obtained polymer discloses excellent thermal stability, undergoing less than 5% weight loss up to 435 °C and loss of about 50% of its weight at about 700 °C. No distinct glass transition was observed from 25 to 350 °C in its differential scanning calorimetry curves of the second heating and cooling runs (10 °C/min). In addition, the obtained polymer exhibited well-resolved spectra with the absorption and emission maxima at 462 and 464 nm, respectively. A minimal Stokes shift of 2 nm reflects that such spiro-bridged **LPPP** is exceptionally rigid. No low-energy green emission was observed after the film was annealed at 110 °C for 24 h in air. Spiro-bridged ladder-type polymer is the first **LPPP** that renders excellent thermal and optical stability.

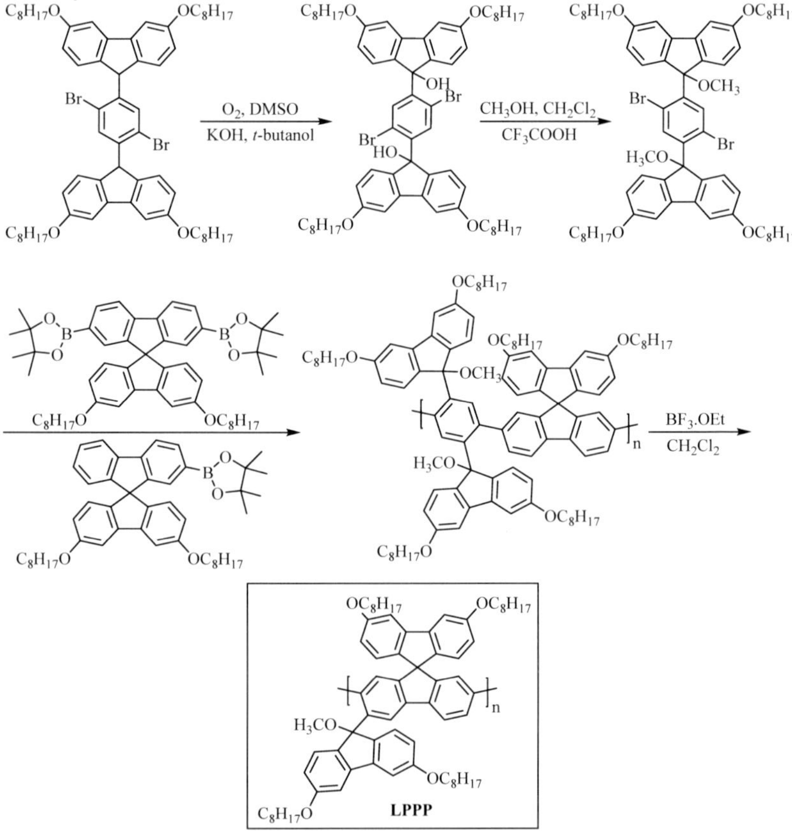

Scheme 6-52. Synthetic route to spiro-bridged **LPPP**.

The Liu group investigated the synthesis of a novel ladder-type molecule with carbazole core and spiro linkage, the first ladder-type hexaphenylene with bent conformation, using the *ortho*-linked spirobifluorene (Scheme 6-53) [118]. The judicious design eliminates the possibility of forming geometrical isomers by using the *ortho*-linked spirobifluorene, which are usually encountered in the synthesis of typical ladder-type molecules. As a blue emitter, the bent ladder-type hexaphenylene (**BLHPC**) indicates good thermal and color stability. The incorporation of the central carbazole core significantly decreases the ionization potential, and therefore the OLED device configurations could be simplified without using PEDOT:PSS as the hole-injection layer. A simple light-emitting device fabricated from **BLHPC** discloses a maximum current efficiency of 1.46 cd/A and maximum luminance of 505 cd/m^2.

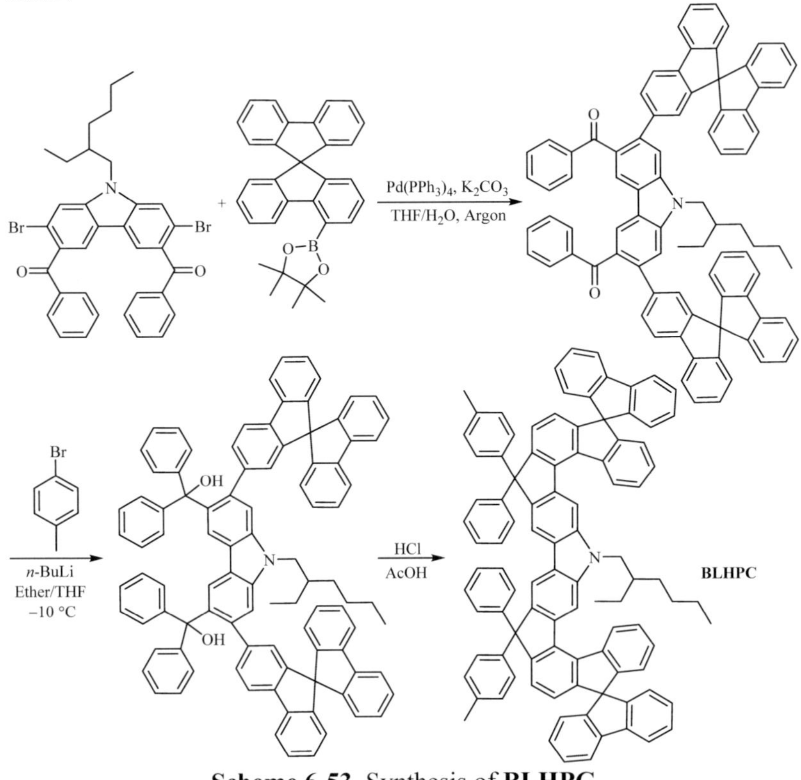

Scheme 6-53. Synthesis of **BLHPC**.

The synthesis of a new blue fluorophore dispiroxanthene-indenofluorene (**DSX-IF**) possessing an indenofluorenyl core spiro-linked to two xanthene units were reported by Rault-Berthelot et al. in 2010 (Scheme 6-54) [119]. **DSX-IF** includes good morphological and color stability upon heating, has a high quantum yield, and may be easily polymerized via anodic oxidation. Small molecule organic light-emitting diodes (SMOLEDs), using this promising new dixanthene as a blue emissive layer, reach a high brightness of ca. 3800 $Cd.m^{-2}$ with a maximum luminous efficiency of 1 $Cd.A^{-1}$, which can be further facilitated by adjusting the nature and the thicknesses of different layers.

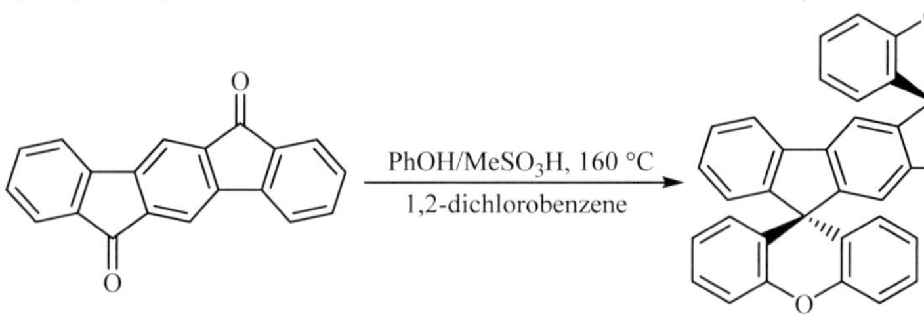

Scheme 6-54. Synthesis of **DSX-IF**.

A year later, Poriel et al. reported the incorporation of spiroxanthene units within blue-emitting oligophenylene frameworks [120]. Dispiro xanthene-indenofluorene (**DSX-IF**) and dispiroxanthene-ladderpenta phenylene (**DSX-LPP**) were designed and synthesized through short and efficient synthetic approaches (Scheme 6-55). Since the central π-conjugation is expanded going from **DSX-IF** to **DSX-LPP**, they observed a gap contraction from 3.64 eV for **DSX-IF** to 3.25 eV for **DSX-LPP**. Compared with their all-carbon analogs, **DSF-LPP** and **DSF-IF**, xanthene derivatives **DSX-LPP** and **DSX-IF** exhibit enhanced thermal stability while maintaining similar optical (high quantum yield, blue fluorescence, and very small Stokes shift) and electrochemical properties. Similarly, the nonoptimized double-layer **DSX-LPP**-based device reaches an attractive brightness of approximately 1400 $Cd.m^{-2}$ with a maximum luminous efficiency of 0.35 $Cd.A^{-1}$.

In addition, the spiro connection of xanthene units with either **DSX-IF** or **DSX-LPP** compounds afforded new generations of violet and blue fluorophores based on a "3π-2 spiro" molecular architecture. Fouassier et al. in 2015 investigated the ability of **DSX-LPP** as a photoinitiator (PI) in the presence of an iodonium salt for the cationic ring-opening polymerization (CP) of a diepoxide and the free-radical polymerization (FRP) of an acrylate under violet LED exposure (385 and 405 nm). Good-to-excellent rates of polymerization and final conversions were achieved in this work. Adding a poly(ionic liquid) improves the cationic polymerization profiles [121].

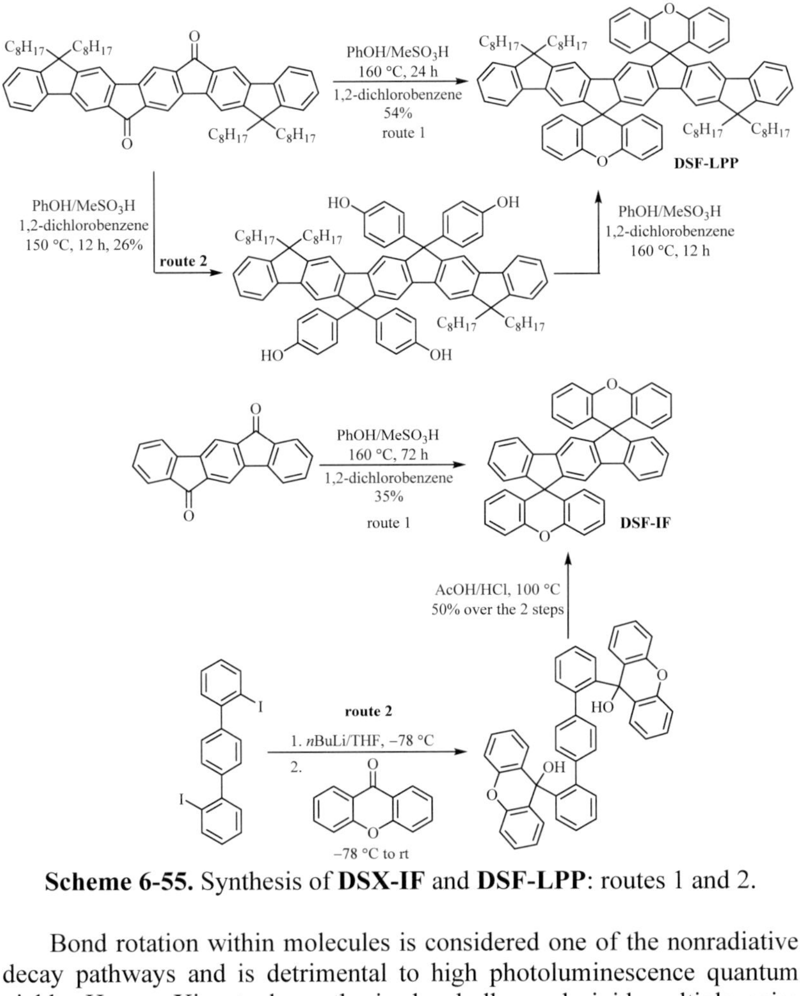

Scheme 6-55. Synthesis of **DSX-IF** and **DSF-LPP**: routes 1 and 2.

Bond rotation within molecules is considered one of the nonradiative decay pathways and is detrimental to high photoluminescence quantum yields. Hence, Xia et al. synthesized a bulky and rigid multiple spiro bifluorene-based blue emitters (**spiro-F**) with a high photoluminescence quantum yield (Scheme 6-56) [122]. The rigid structure with spiro-carbon linkages demonstrates benefits in thermal stability and solution process ability, respectively. In this work, high-quality amorphous films can be achieved, whereas avoiding keto defects. Due to fluorenes introducing a steric effect, no intermolecular π-π interactions are observed in the single crystals. **Spiro-F** can operate as an active component of blue OLEDs and as a host and blue emitter of white OLEDs. The WOLED presents a low turn-on voltage of 3.5 V and high current efficiency of 3.6 cd A^{-1}, together with CIE (0.29, 0.33) coordinates.

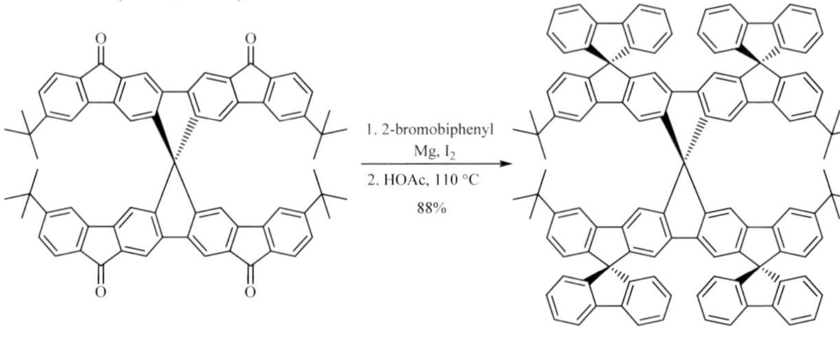

Scheme 6-56. Synthesis of **spiro-F**.

In another study, new fluorescence blue host materials based on spiro[benzotetraphenefluorene] (**SBTF**) derivatives, 3-(1-naphthyl)-1′-naphthylspiro[benzo[*ij*]tetraphene-7,9′-fluorene] **219**, 3-(2-naphthyl)-1′-naphthylspiro[benzo[*ij*]tetraphene-7,9′-fluorene] **220**, and 3-[2-(6-phenyl) naphthyl]-1′-naphthylspiro[benzo[*ij*]tetraphene-7,9′-fluorene] **221** were successfully synthesized through multi-step Suzuki coupling reactions and used to fabricate blue OLEDs (Scheme 6-57) [123]. Introducing various aromatic groups into the **SBTF** core reduces the band gap and determines the color purity and luminescence efficiency. Substitution of the aromatic group on the 3-position of 10-naphthylspiro[benzo[*ij*]tetraphene-7,9′-fluorene] (**NSBTF (219-221)**) core contorted the conjugated structure little and enhanced T_g for good thermal stability. The EL emission spectra of the devices were about 472 nm. A device obtained from three materials doped with *p*-bis(*p*-*N*,*N*-diphenyl-aminostyryl)benzene (**DSA-Ph**) revealed color purity of 0.148 and 0.239, a luminance efficiency of 7.91 cd/A, and an external quantum efficiency (EQE) > 4.75% at 5 V. A device obtained from three materials doped with **DSA-Ph** showed color purity of 0.148 and 0.239, a luminance efficiency of 7.91 cd/A, and an EQE > 4.75% at 5 V. According to these characteristics, these blue light-emitting materials have the adequate potential for fluorescent OLED applications.

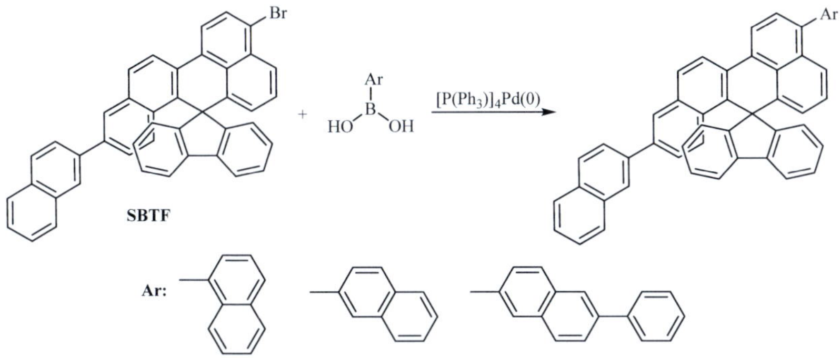

Scheme 6-57. Preparation of three NSBTF derivatives **219-221**.

In the same year, a study conducted by researchers resulted in the successful synthesis of four high triplet organic semiconductors based on the Donor-Spiro-Acceptor (D-Spiro-A) design. The study extensively examined the physicochemical and photophysical properties of these compounds and explored their application in phosphorescent organic light-emitting diodes (PhOLEDs) [124]. Scheme 6-58 shows the synthetic route to prepare organic semiconductors (**STX-TXO$_2$**, **SPA-TXO$_2$**, **STX-DAF,** and **SPA-DAF**) based on a Donor-Spiro-Acceptor (D-Spiro-A) molecular design. This chemical design is based on a π-conjugation disruption developed by an insulating spiro bridge between a hole transporting unit, which is phenylacridine (PA) or thioxanthene (TX) moiety, and an electron transporting unit that is dioxothioxanthene (TXO_2) or diazafluorene (DAF) moiety. This molecular design allows (i) a spatial separation of HOMO and LUMO retaining a high ET, (ii) HOMO/LUMO levels of the constituting building blocks adapted to efficient charge injections, and (iii) good physical properties, a key feature for device stability and performance. These molecules have been incorporated as hosts in sky blue PhOLEDs with EQE at 10 mA/cm^2 ranging from 2% to more than 10%. The best performance was obtained with **SPA-DAF** and **SPA-TXO$_2$**, highlighting the significance of the selected donor-acceptor combination on the device performance.

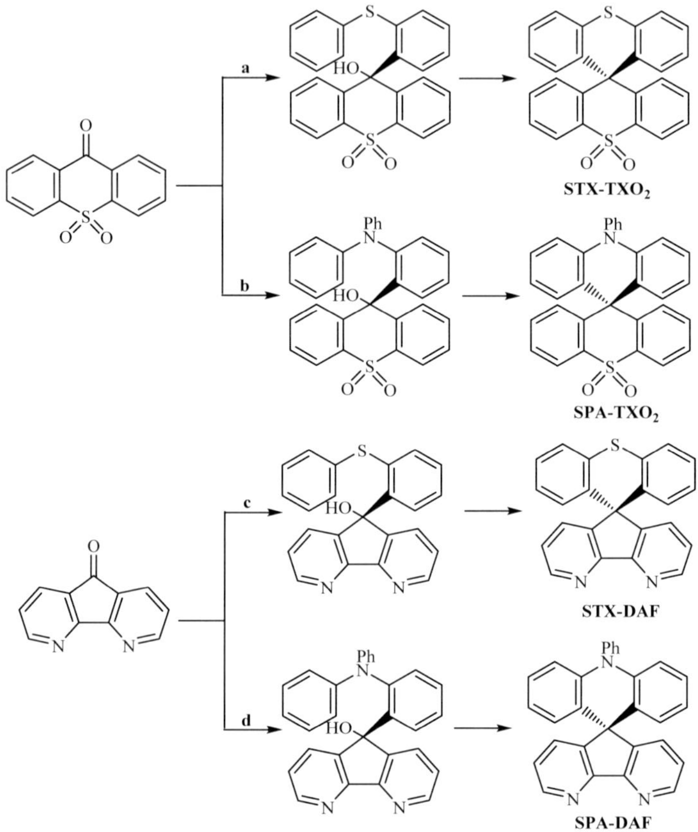

Scheme 6-58. Synthesis of STX-TXO$_2$, SPA-TXO$_2$, STX-DAF, and SPA-DAF **a**: (2-iodophenyl)(phenyl)sulfane, *n*-BuLi, THF from -80 to 20 °C then HCl, AcOH reflux 88%, **b**: 2-bromo-*N*,*N*-diphenylamine, *n*-BuLi, THF from -80 to 20 °C then HCl, AcOH reflux 78%. **c**: (2-iodophenyl)-(phenyl)sulfane, *n*-BuLi, THF from -80 to 20 °C, then MsOH, 1,2-dichloro -benzene reflux 39%, **d**: 2-bromo-*N*,*N*-diphenylaniline, *n*-BuLi, THF from -80 to 20 °C, then MsOH, 1,2-dichlorobenzene, reflux 47%.

Moreover, the Rault-Berthelot group reported the synthesis, structural, electrochemical, and photophysical properties of new dihydroindeno[2,1-*b*]fluorene-based semiconductors, i.e., dispiro[thioxanthene-9,5′-indeno-[2,1-*b*]fluorene-7′,9″-thioxanthene] **223** and dispiro[dioxothio-xanthene-9,5′-indeno[2,1-*b*]fluorene-7′,9″-thioxanthene] **224** [125]. These properties were compared with pure hydrocarbon dispiro[fluorene-9,5′-indeno[2,1-*b*]-fluorene-7′,9″-fluorene] **222**. Scheme 6-59 indicates the synthetic route to synthesizing the target molecules **222-224**.

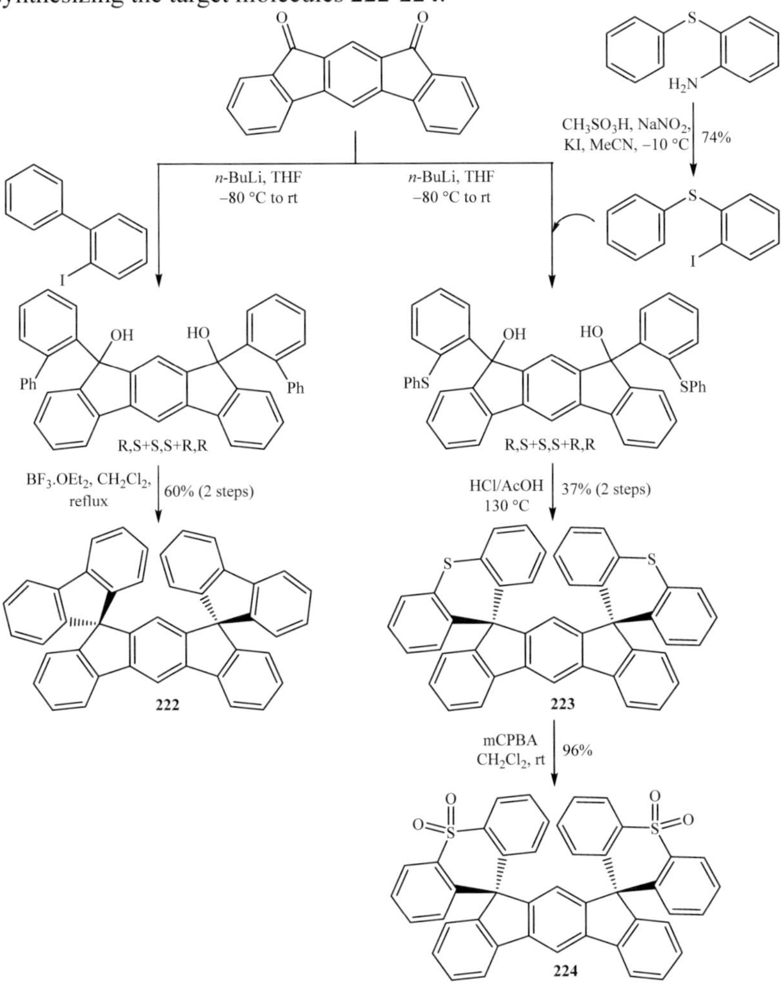

Scheme 6-59. Synthesis of the target molecules **222-224**.

The incorporation of the sulfur atom within the bridged units leads to tuning the HOMO and LUMO energy levels, retaining nevertheless the high E_T of the dihydroindeno[2,1-*b*]fluorenyl core ([2,1-*b*]IF), a key feature for further device applications. This control of the properties allowed incorporating **224** as a host material in blue and green PhOLED with EQE of 12.8% (at 1 mA/cm^2 with corresponding current efficiency (CE) and power efficiency (PE) of 48.6 cd/A and 29.5 lm/W) for the green device and EQE of 4.8% (at 1 mA/cm^2 with corresponding CE and PE of 14.8 cd/A and 8.2 lm/W) for the sky-blue device.

Two novel spiro-based host materials, namely 3-(9,9′-spirobi[fluoren]-6-yl)-9-phenyl-9*H*-carbazole (**SF3Cz1**) and 9-(3-(9,9′-spirobi[fluoren]-6-yl)phenyl)-9*H*-carbazole (**SF3Cz2**) were also designed and fabricated in 2016 (Scheme 6-60) [126]. Due to the meta-linkage of the spiro-bifluorene backbone, both **SF3Cz1** and **SF3Cz2** bear triplet energies over 2.70 eV, stating they could serve as suitable hosts for blue and even white PhOLEDs. The prepared bis(4,6-(difluorophenyl)-pyridinato-*N*,C′) picolinate (FIrpic) based PhOLEDs hosted by **SF3Cz1** and **SF3Cz2** disclosed excellent performance with maximum EQEs of 18.1% and 19.7%, respectively. Two-color warm white PhOLEDs fabricated using **SF3Cz1** and **SF3Cz2** as hosts also obtained high EQEs and extremely low-efficiency roll-offs.

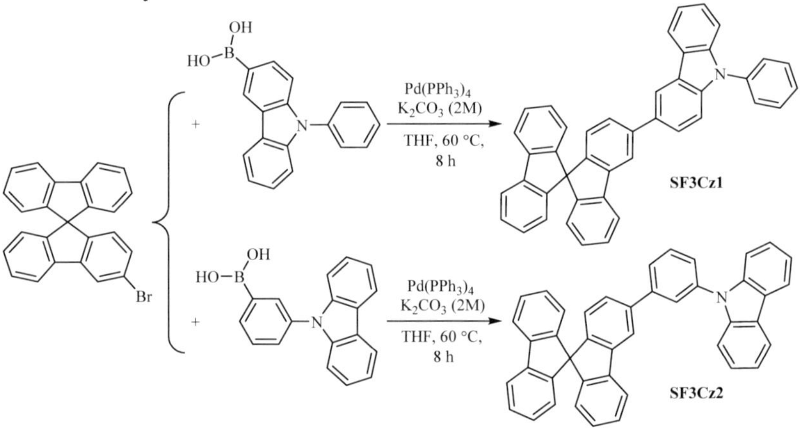

Scheme 6-60. Synthetic routes of **SF3Cz1** and **SF3Cz2**.

Similarly, Yam reported an efficient synthesis of a series of bifunctional heterocyclic spiro compounds **225-228**, which proved to be valuable for constructing both organic photovoltaic (OPV) devices and OLEDs. The synthetic route for these compounds is summarized in Scheme 6-61 [127]. Taking advantage of their excellent hole-transporting properties, highly efficient small molecular OPV devices based on these heterocyclic compounds as donors with low dopant concentrations were prepared; mainly, a high open-circuit voltage (V_{OC}) of up to 1.10 V and a power conversion efficiency (PCEs) of up to 5.12% was realized. Moreover, most of these heterocyclic spiro derivatives are highly emissive in solutions with photoluminescence quantum yields of up to 0.91, and high-performance deep-blue-emitting OLEDs were achieved. Such devices represent a stable deep blue emission with the Commission Internationale de L'Eclairage (CIE) γ-coordinate of (0.16, 0.04) and high EQEs of up to 4.7%, which is one of the best values among the described OLEDs with $CIE_{\gamma} < 0.08$.

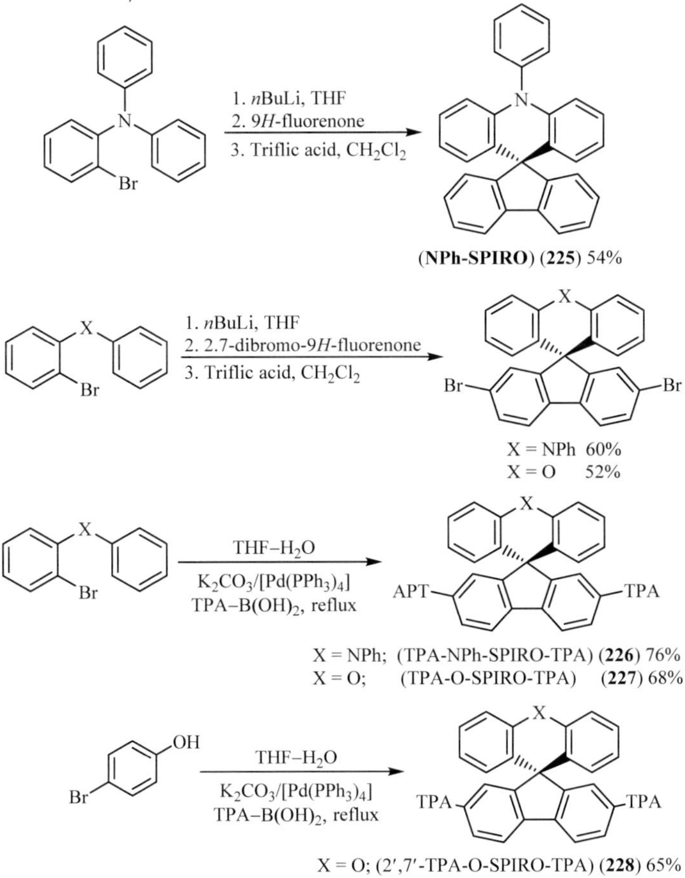

Scheme 6-61. Synthetic scheme for compounds **225-228**.

Typical π-π stacking and aggregation caused by quenching could be repressed in the film-state by the spiro conformation molecular design in the OLED field. Hence, a bipolar deep-blue emitting material, namely 1-

(4-(*tert*-butyl)phenyl)-2-(4-(10-phenyl-10*H*-spiro[acridine-9,9′-fluoren]-2-yl) phenyl)-1*H*-phenanthro[9,10-*d*]imidazole (**SAF-BPI**), was designed and prepared by Li et al. which involves spiro-acridine-fluorene (**SAF**) as a donor part and phenanthroimidazole (**BPI**) as an acceptor part [128]. The molecular structure and synthetic routes were shown in Scheme 6-62. Owing to the rigid **SAF** skeleton, this compound exposes high thermal stability with a glass transition temperature (T_g) of 198 °C. The compound displays bipolar transporting characteristics indicated by the single-carrier devices. The non-doped OLEDs based on the **SAF-BPI** as the emitting layer exhibits maximum emission at 448 nm, maximum luminance of 2122 cd m^{-2}, maximum CE of 3.97 cd A^{-1}, maximum PE of 2.08 lm W^{-1}. The chromaticity coordinate is stable at (0.15, 0.10) at the voltage of 7-11 V. The device exhibits a slow efficiency roll-off with CE 3.35 and 2.85 cd A^{-1} at 100 and 1000 cd m^{-2}, respectively.

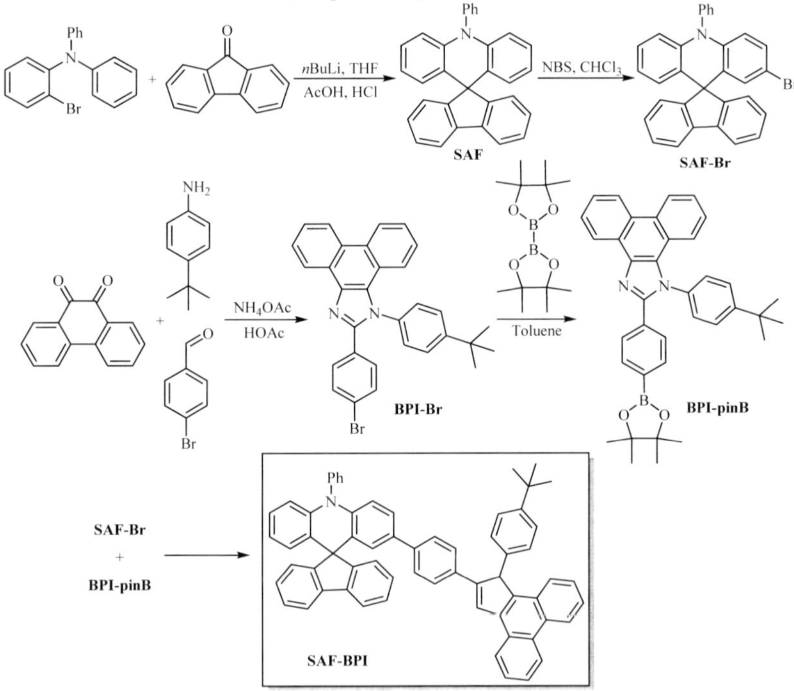

Scheme 6-62. Molecular structure and synthetic routes of **SAF-BPI**.

Novel carbazolyl-substituted spiro[acridine-9,9′-fluorene] derivatives (**SAF-CzB**, **SAF-2CzB**, and **F-2CzB**) as deep-blue emitting materials were also synthesized by the same group (Scheme 6-63) [129]. Two compounds **SAF-CzB** and **SAF-2CzB** have a rigid **SAF** central core. The counterpart **F-2CzB**, which contains 9,9-dimethyl fluorescent as the core, was also prepared for comparative purposes. **SAF-2CzB** shows the best thermal stability (Td of 534 °C), hole-transporting property, and a convenient HOMO energy level (−5.44 eV). The presence of bulky spiro-annulated triphenylamine moiety at the C9 position of the fluorene can intercept the fluorenone defect problem. Its HOMO and LUMO were separated on the side of the **SAF** group, providing individual holes and electron transporting channels. Both **SAF-2CzB-** and **F-2CzB**-based devices show good color stability with a CIE coordinate of (0.16, 0.11) at 7–12 V.

The two main techniques, i.e., backbone modification and position engineering, can cooperate in developing a rigid scaffold to realize spatial donor/acceptor intramolecular interaction, which limits the donor/acceptor distance.

In an elegant work, the Jiang group systematically researched the electron-rich spiro backbones, which could greatly assist the carrier injection/transport. The new derivation positions (C1, C3, and C4 at fluorene moiety) can remarkably affect the energy levels of triplet excitons, which is crucial for host materials in OLEDs, especially for the blue ones [130]. The **SAF** unit is the first example to improve the performance of blue and violet fluorescent emitters, which have attained the theoretical upper limit of fluorescent OLEDs (FOLEDs). For PhOLEDs, the spiro-type compounds were mainly applied as host materials. The substitution positions of the 9,9′-spirobifluorene (**SBF**) unit were thus thoroughly studied to furnish the host materials with high triplet energy. The **SAF** unit was also modified to form ambipolar host materials to balance hole/electron injection for better white PhOLEDs. They found that the orthogonal arrangement in the spiro structure was essential for ambipolar material design.

Besides, the spiro-type compounds can represent thermally activated delayed fluorescence (**TADF**) properties by dividing HOMO and LUMO in an unconjugated motif if the donor and acceptor are orthogonally located at two halves of the spiro system. Hence, TADF materials based on the spiro compounds (DiSAAF and DiSFAQ) were also introduced. They demonstrated that the donor/acceptor (D/A) spatial distance and conformation are critical in molecular design, and highly efficient devices based on these new **TADF** emitters were also obtained.

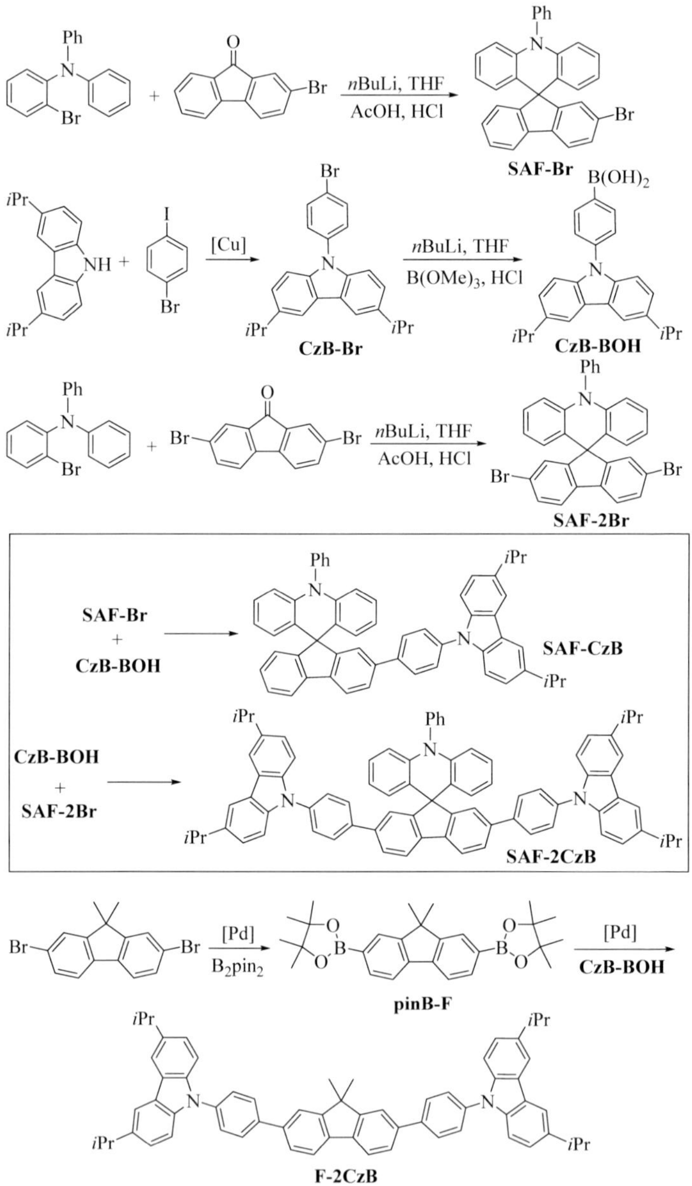

Scheme 6-63. The molecular structure and synthetic routes of **SAF-CzB**, **SAF-2CzB**, and **F-2CzB**.

6.3.2. Solar cells (SCs)

Another application of compounds based on spiro linkage is their use to construct solar cells (SCs). Renewable energies, particularly solar energy, have attracted much attention as a sustainable and environmentally friendly approach to replacing fossil fuels. Extensive attempts have been invested to develop highly efficient solar energy conversion technologies, and the most prospective process is converting solar energy into electricity. The emerging photovoltaic industry has been overgrowing in recent years. Silicon-based and inorganic SCs were initially developed, reaching high levels of efficiency. The cost of solar power from state-of-the-art inorganic silicon technology still notably exceeds that of the electrical grid, thus prohibiting the large-scale application of silicon-based solar energy. Recent reductions in the manufacturing expenses of mainstream silicon SC technology assure the future large-scale use of photovoltaics, with a current prognosticate anticipating photovoltaics will contribute nearly a third of new electricity production capacity worldwide between now and 2030. As in microelectronics, silicon has a compound of strengths that has made it difficult to displace as the favored photovoltaic material. Opportunities exist for technologies that promise either significantly higher energy conversion efficiencies or significantly lower processing expenses. A new generation of mixed organic-inorganic halide perovskites presents tantalizing prospects on both fronts.

Perovskite solar cells (PSCs) generally consist of electrodes, charge transport layers (CTLs), and a perovskite layer sandwiched between two electrodes. Electron and the hole transport layers are inserted into each interface to assist charge transfer, transport, and collection. Since the first report on a long-term durable solid-state PSC with a PCE of 9.7% in 2012, a PCE as high as 19.3% was shown in 2014, and a certified PCE of 17.9% was indicated in 2014. Such high photovoltaic performance is attributed to optically high absorption characteristics and balanced charge transport properties with long diffusion lengths [131-134].

Generally, a PSC device involves five fundamental layers: a cathode, an HTL, a perovskite layer, an electron transporting layer (ETL), and an anode [135]. Schematic illustrations of the principle operation in PSCs and the typical device configurations of PSCs are shown in Figures 6-48a,b. Among different layers, the HTL, where the extraction of photogenerated holes from the perovskite occurs so that these positive charges can be transported to the cathode, plays a key role in enhancing the PCE and reaching high stability of both regular (n-i-p) and inverted (p-i-n) architectures. Indeed, this HTL is crucial for improving the stability of the device because it can preserve the perovskite layer from the influence of moisture/oxygen through the electrode or HTL itself. This is the major bottleneck on the way to the commercialization of PSCs.

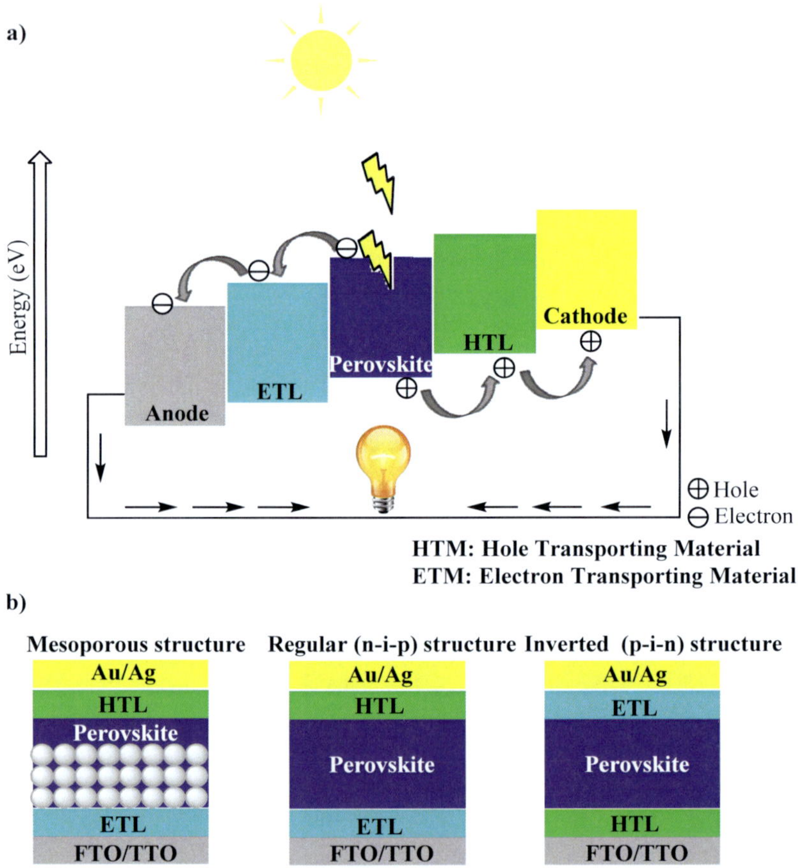

Fig. 6-48. a) Schematic illustration of the main operation in PSCs; b) the typical device configurations of PSCs.

Moreover, a higher PCE can be reached with an HTL owning to its role in suppressing charge recombination and blocking electron movement. In addition, the HTLs importance in inverted layouts is important because the growth of perovskite on top of the HTL layer governs the device performance, penetrating the perovskite thin film formation. Furthermore, the efficiency of PSCs using organic HTMs, involving polymers, organometallic compounds, and small molecules is the highest among all HTLs. Organic HTMs also allow for achieving high stability in PSCs. Small molecular HTMs can be categorized into three main groups: (a) linear, (b) star-shaped, and (c) spiro-orthogonal structures, as illustrated in Figure 6-49.

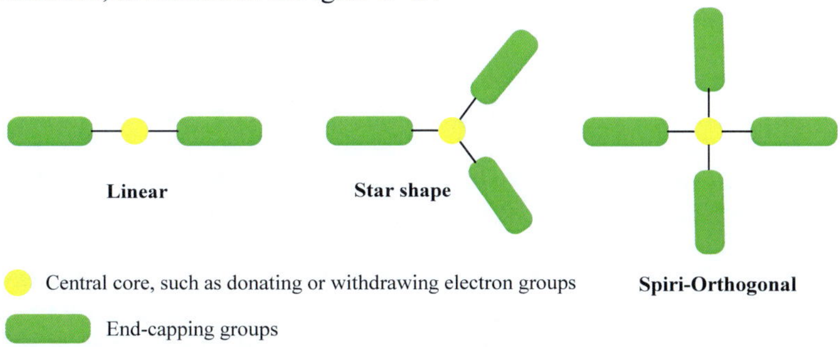

Fig. 6-49. Scheme of three types of molecular geometries regularly designed for small molecular HTMs of PSCs.

The most widely applied small molecular HTM is 2,2′,7,7′-tetrakis-(*N*,*N*′-di-*p*-methoxy-phenylamine)-9,9′-spirobifluorene), which was known better as **spiro-OMeTAD** (Figure 6-50) 2,2′,7,7′-tetrakis-(*N*,*N*′-di-*p*-methoxy-phenylamine)-9,9′-spirobifluorene), also known as **spiro-OMeTAD**, is a widely used small molecular HTM (Figure 6-50). However, it is a tremendously expensive material with mediocre hole carrier mobility. To ensure the wide-scale application of PSC-based technologies, alternative HTMs are being presented [136].

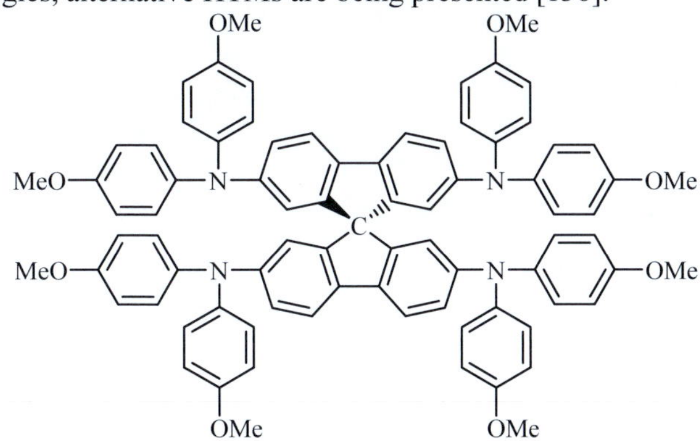

Fig. 6-50. The molecular structure of **spiro-OMeTAD**.

In this area, Gao et al. reported in 2012 a cost-effective spiro-type HTM 4,4′,4″,4‴-(2*H*,2′*H*,4*H*,4′*H*-3,3′-spiro-bi[thieno[3,4-*b*][1,4]dioxepine]-6,6′,8,8′-tetrayl)tetrakis(*N*,*N*-bis(4-methoxy-phenyl)aniline) (**PST1**) through facile synthetic route [137]. The chemical structure and the synthetic way of **PST1** are displayed in Scheme 6-64. A parallel X-ray crystallography study on single crystals of **PST1** and **spiro-OMeTAD** showed completely different types of spiro configuration and found multiple CH/π, O-C, and π-π intermolecular short contacts. This is the first report on the crystal structure of **spiro-OMeTAD** to demonstrate that dopant is critical due to the lack of π−π intermolecular stacking. The PCS employing **PST1** as HTM layer depicted an overall conversion efficiency of 13.44%, comparable with that achieved using the **spiro-OMeTAD** under the same fabrication conditions. This can be partially ascribed to the increased open circuit potential (V_{oc}) with **PST1** as the HTM layer, which has a lower HOMO level than **spiro-OMeTAD**. Only an intense decrease in PCE was observed when the cobalt dopant was removed from **PST1** leading to a PCE of 12.7%. This is considerably in contrast to the performance drop to 9.57% in the device built with **spiro-OMeTAD** under the same condition.

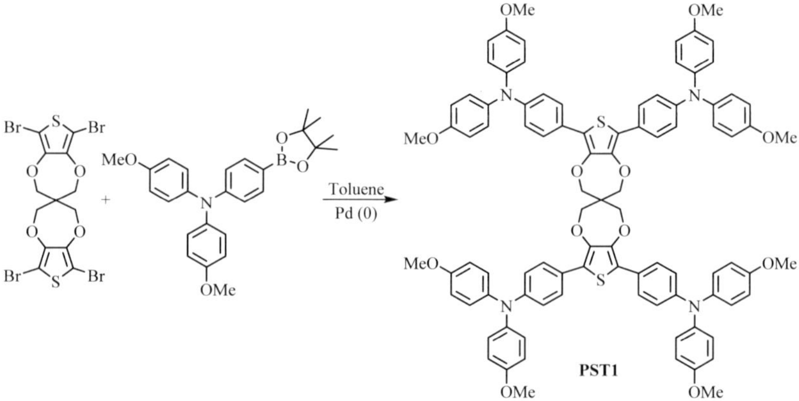

Scheme 6-64. The chemical structure and the synthetic route of **PST1**.

Chemical doping is often applied to raise the electric conductivity of the conjugated molecule as HTM for application in optoelectronics. However, chemical dopants can improve ion migration in the electrical field, which deteriorates the device efficiency and increases the fabrication cost. Hence, a small molecule HTM, named **SAF-OMe**, is fabricated and implemented, which combines star subunits of poly(triarylamine) (PTAA) with **spiro-OMeTAD** (Scheme 6-65) [138]. By incorporating the *N*-benzene-*N*-structure, which is the main unit of PTAA, into the spiro-core, **SAF-OMe** exhibits a better hole-transporting ability than PTAA and thermal stability superior to **spiro-OMeTAD**. This concept allows us to partially enhance the weak points of both **spiro-OMeTAD** and PTAA while fully utilizing both advantages in **SAF-OMe**. As expected, the PCE based on **SAF-OMe** is notably facilitated with the best PCE of 16.73% and 12.39% with and without dopant. Moreover, the device based on **SAF-OMe** displays much-improved stability compared to the **spiro-OMeTAD**-based one, mainly resulting from the better hole mobility of **SAF-OMe**.

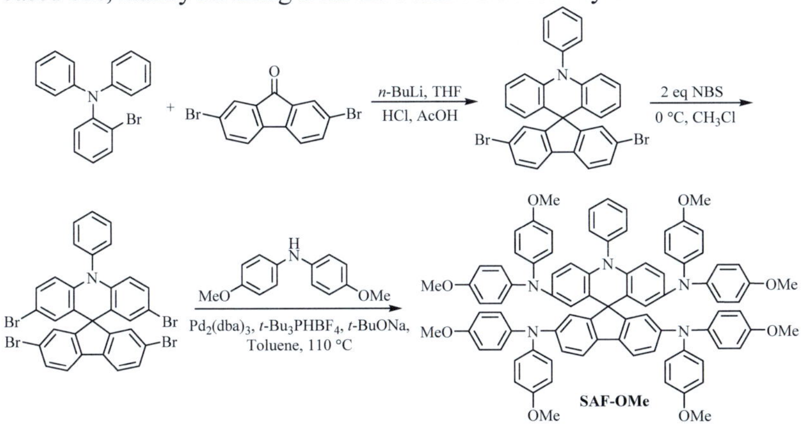

Scheme 6-65. Synthesis of **SAF-OMe**.

In an elegant work, Qi et al. studied the effects of environmental H_2O vapor with relative humidity (RH%) of 90% at room temperature (RT) of 25 ºC that is generated by letting a stream of dry N_2 gas flow through an H_2O bubbler, high purity dry O_2 gas (>99.5%), and ambient air (RH 50%, RT 25 ºC) on electrical properties of hole-only devices of LiTFSI-doped **spiro-MeOTAD** films, using mercury drop electrode measurements [139]. They also explored chemical and electronic properties using X-ray photoelectron spectroscopy (XPS) and ultraviolet photoemission spectroscopy (UPS). They found that H_2O vapor exposure caused an irreversible enhancement of LiTFSI-doped HTL conductivity. XPS indicated that this enhancement was mainly due to LiTFSI redistribution across the HTL and was almost independent of the used bias. On the other hand, O_2 exposure is caused by a reversible enhancement under used bias in the devices. This enhancement was due to doping, which was affirmed by UPS, specifically after long O_2 exposure times (39 h). The mercury drop electrode measurements on air-exposed samples showed a more than four-folds increase in conductivity compared to about two orders of magnitude in each film exposed to H_2O vapor and O_2. Such a combined effect is believed to be the major reason for the raised cell performance when PSCs are exposed to ambient air for some time (typically several hours).

Likewise, four different fluorene-dithiophene-based HTMs (**229-232**) have been synthesized by the Nazeeruddin group through a facile route and were successfully applied in the construction of formamidinium lead bromide PSCs (Scheme 6-66) [140]. Electrochemical measurements, combined with theoretical studies, disclosed a higher oxidation potential for **229-232**, which rationalizes the higher V_{oc} achieved using **229** compared to its reference counterpart, i.e., **spiro-OMeTAD**. The photovoltaic performance generally revealed good reproducibility, and current−voltage curves showed negligible hysteresis. Time-resolved photoluminescence (TRPL) decay measurements revealed that the injection of holes from the valence band of $FAPbBr_3$ perovskite into the HOMO of HTM molecules majorly followed the oxidation potentials of the latter.

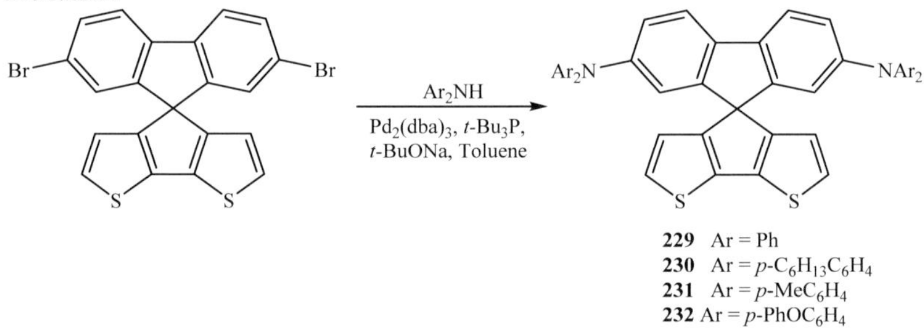

Scheme 6-66. General conditions for the synthesis of **229-232**.

Four heterocyclic spiro-types HTMs, **SFX-TPAM**, **SFX-TPA**, **SFT-TPAM,** and **SFT-TPA**, can also be prepared through a low-cost facile route with high yields for PSC applications, as shown in Figure 6-51 [141]. The structure-physicochemical property relationship of these four molecules was well-perused, and their application as HTLs in the inverted PSC was investigated. The results indicated that when sulfur was introduced onto the spiro-core (substitution with oxygen), it decreased the HOMO and LUMO energy levels of the molecules due to the weak interaction between the 3p orbital of a sulfur atom and the π-orbital of the phenyl ring. On the other hand, adding electron-donating alkyl groups will elevate the HOMO level (such as **SFX-TPAM/SFT-TPAM** pair). Consequently, only **SFX-** with the HOMO level compatible with $MAPbI_3$ perovskite performs good photovoltaic properties. Inverted PSC based on dopant-free **SFX-TPAM** HTL obtains a high PCE of 10.23%, which is higher than that (8.17%) of the cell using dopant-free **spiro-OMeTAD** as an HTL. The better performance of the **SFX-TPAM** HTL is due to the high mobility, therefore has extracted holes more efficiently than dopant-free **spiro-OMeTAD**.

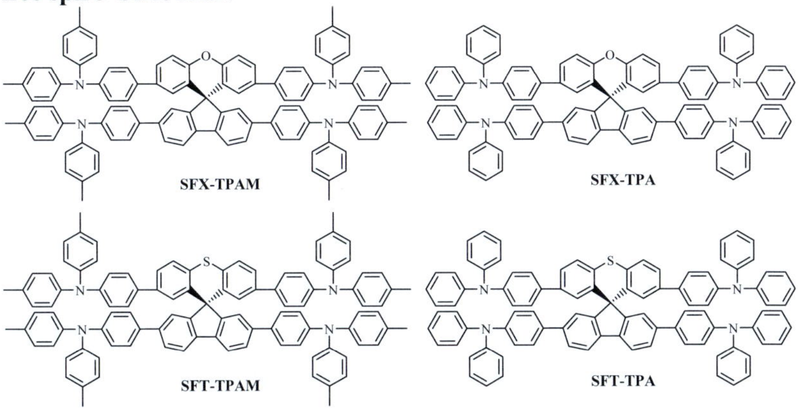

Fig. 6-51. The chemical structures of four new spiro-based HTMs.

In 2019, Paek et al. successfully afforded four new spirobisacridine (**SBA**)-based HTMs with various alkyl substituents attached to the in-plane N atoms of the spiro-connected acridines [142]. Scheme 6-67 demonstrates the synthetic approach for obtaining the **SBA**-based family of HTMs. The **SBA** cores were endowed with di-*p*-methoxy phenylamine (DMPA) entities to mimic **spiro-MeOTAD** structurally and to study how the introduction of a heteroatom and alkyl chains into the molecular scaffold of HTM affects the device performance and long-term stability. The **SBA** HTMs used in PSCs achieved efficiencies on par with the benchmark **spiro-MeOTAD** and insignificant current density-voltage. The results reveal that even minor differences in chemical structure can remarkably change the rates of hole extraction from perovskite, which, however, only weakly influences the device efficiency but strongly affects the long-term stability and evolution of devices over time. The best results were obtained using **MeSBA-DMPA** with the shortest C1-substituent, and the elongation of the side arms result the slow deterioration of device metrics and poorer stability.

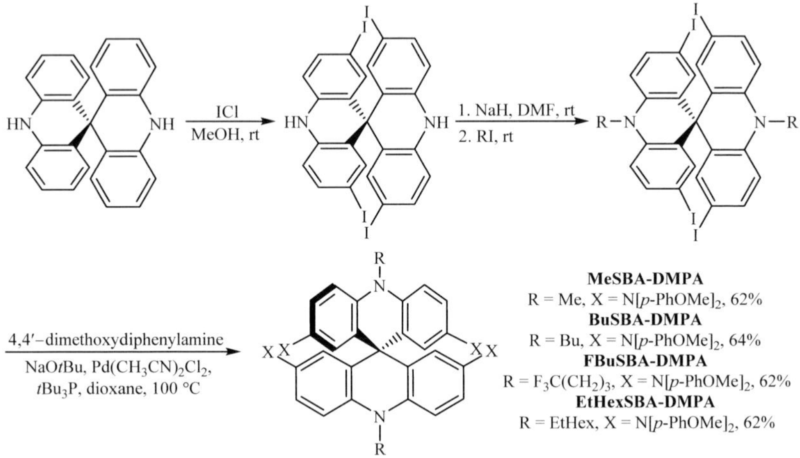

Scheme 6-67. Synthetic procedure for the **SBA**-Based HTMs.

To further cure the performance of **spiro-OMeTAD**-based PSCs, the Lewis base can be introduced into the structure of **spiro-OMeTAD** advisedly, which can preserve the advantages of **spiro-OMeTAD** while raising the functionality of Lewis base in passivating the surface of perovskite. Hence, two spiro-type HTMs, spiro-CNOMeTAD (**spiro-CN**) and spiro-PS-OMe-TAD (**spiro-PS**), were synthesized by Gao et al. in 2020 (Figure 6-52) [143]. The dicyano group of **spiro-CN** as an intense electron-withdrawing center can not only diminution the highest occupied molecular orbit (HOMO) energy level but also facilitate the passivation of the perovskite layer. Meanwhile, **spiro-PS** with the thiocarbonyl group can also passivate the defect in perovskite through the binding interaction with under-coordinated Pb^{2+} cations. Consequently, **spiro-CN,** with the best surface passivation effect on the perovskite layer, has the excellent film-forming ability, improved hole mobility, and more appropriate HOMO interfacial band alignment with perovskite.

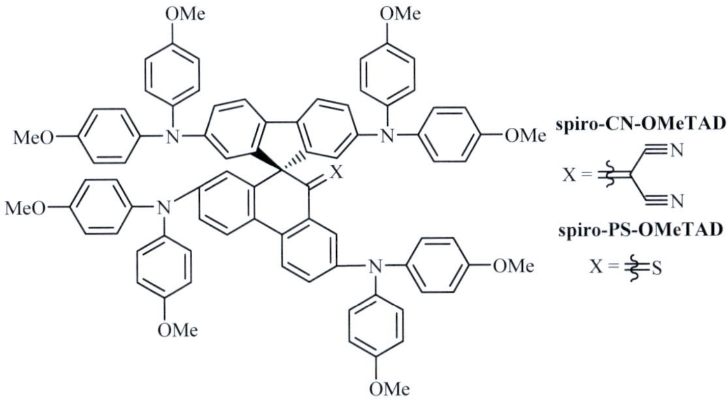

Fig. 6-52. The chemical structures of **spiro-CN** and **spiro-PS**.

At the same time, C_{2v}-symmetric spiro-configured and cyclopenta-dithiophene-based **HTM-1** has been tailor-made by a multistep synthesis and successfully applied as a hole conductor in highly efficient and stable PSCs (Scheme 6-68). **HTM-1** represented higher hole mobility, higher thermal, and better moisture stability than the well-known **spiro-MeOTAD**. In the light of these properties, PSCs based on **HTM-1** disclosed an impressive PCE of 21.0%, one of the highest efficiencies for organic HTMs reported to date. Moreover, PSCs using **HTM-1** revealed negligible hysteresis and excellent reproducibility. Importantly, **HTM-1**-based devices exposed excellent long-term operational stability after aging over 400 h [144].

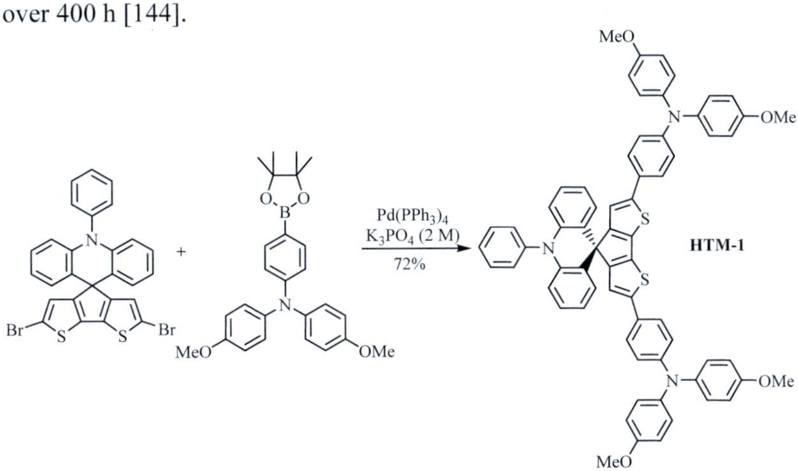

Scheme 6-68. Preparation of **HTM-1.**

Three new HTMs **234-236**, were synthesized by Guo et al. via inserting benzene, naphthalene, and anthracene groups between the **SFX** core and diphenylamine (DPA) group of the mother molecule **233**, respectively (Scheme 6-69) [145].

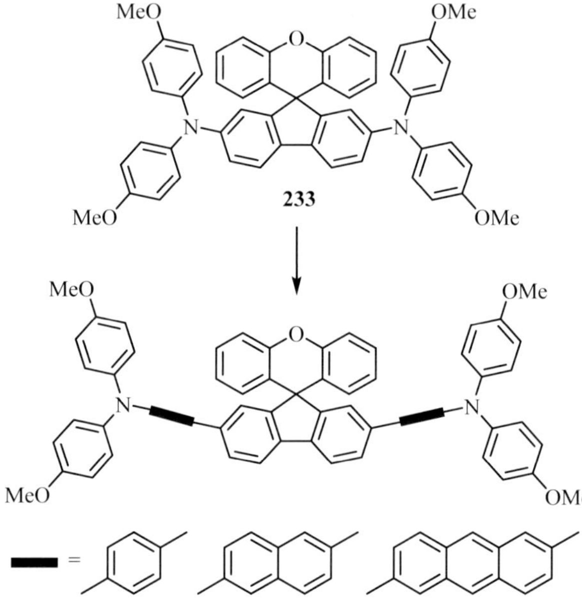

234 **235** **236**

Scheme 6-69. Sketch structures of studied molecules.

Theoretical calculations disclose that the intramolecular charge transfer (ICT) states are formed in the newly designed molecules because of the introduction of conjugated groups, which develops a self-doping effect to obtain efficient hole mobility without any external dopant. The outstanding hole mobility of **234** and **236** as dopant-free HTMs in PSCs was forestalled computationally and verified experimentally, which cases from the generated ICT states and good molecular planarity. The fact that hole mobility of **235** is lower than that of **233** is mainly due to the severe distortion of its molecular structure, thus preventing intermolecular stacking. The combined computational and experimental study illustrated that the PSC device based on **234** as a dopant-free HTM achieves a champion PCE of 16.29% under one sun illumination, much higher than that of the PSC device with **233** as a dopant-free HTM (14.64%). The present study indicates introducing the conjugated groups to produce charge-separated states which are efficient and feasible for the formation of dopant-free HTMs in PSCs.

Shafei et al. developed a simple synthetic approach for preparation of the spiro fluorene-based dopant-free HTM denoted as **spiro-IA** from inexpensive starting materials with high yield (Scheme 6-70) [146]. The unit expense of **spiro-IA** can be 1/9 times as low as that of the conventional **spiro-OMeTAD**. **Spiro-IA** indicates good solubility in different organic solvents, e.g., $CHCl_3$, acetone, EtOH, and DMF, and disclosed favorable charge-transport ability and greater photocurrent density compared to **spiro-OMeTAD**. The UV absorption/emission spectra of **spiro-IA** (λ_{max} = 430 nm, E_{max} = 601 nm) are red-shifted compared to those of **spiro-OMeTAD** (λ_{max} = 388 nm, E_{max} = 414 nm) with larger stoke shift values which assists to repress the loss of incident photons absorbed by the HTM and is more beneficial for improving the performance of PSCs. Optical and electrochemical studies exhibit that **spiro-IA** fulfilled the basic requirements of the hole transfer and electron regeneration process in the constructed devices. PSCs fabricated (surface area = 1.02 cm^2) with dopant-free **spiro-IA** obtained a maximum PCE of 15.66% (short-circuit current density (J_{SC}) = 22.14 mA cm^{-2}, open-circuit voltages (V_{OC}) = 1.042 V, FF = 0.679%), which was comparable to that of the most commonly employed Li-doped **spiro-OMeTAD** (PCE = 15.93%, J_{SC} = 20.37 mA cm^{-2}, V_{OC} = 1.057 V, FF = 0.74%) and exceeded that of the dopant-free **spiro-OMeTAD** (PCE = 9.34%). Besides, the PSCs based on dopant-free **spiro-IA** achieved outstanding long-term stability, and good conductivity ($\sigma = 2.104 \times 10^{-4}$ S cm^{-1}) compared to those found on **spiro-OMeTAD** ($\sigma = 9.00 \times 10^{-8}$ S cm^{-1}). DFT studies were conducted using Gaussian 09 at the B3LYP/6-31G (d/p) level to explore their electron cloud delocalization in HOMO/LUMO levels. These results demonstrated that **spiro-IA** could be a promising candidate for low-cost PSC technology and has a great chance to supersede the expensive **spiro-OMeTAD**.

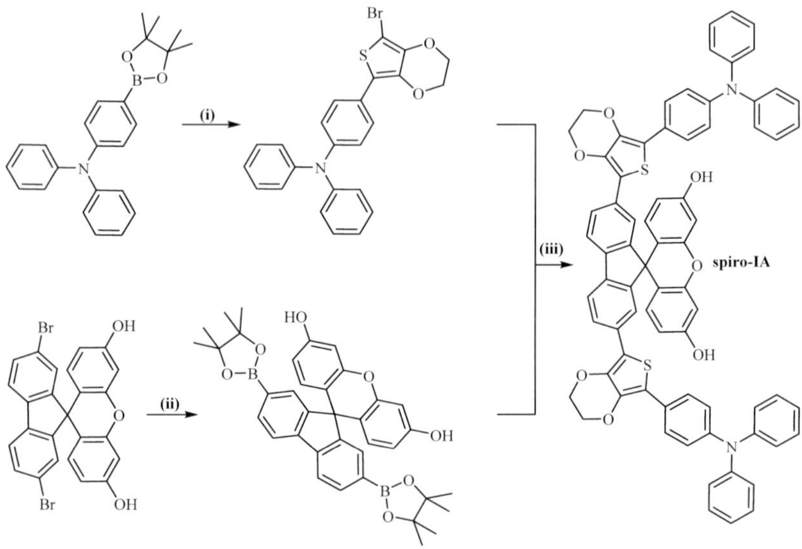

(i) 2,5-dibromo-3,4-ethylenedioxythiophene, K_2CO_3, Aliquat-336, $Pd(PPh_3)_4$, toluene–EtOH–H_2O/90 °C,
(ii) ACOK, $Pd(dppf)_2Cl_2$, DMF, 80–90 °C, bis(pinacolato)diboron, and
(iii) Aliquat-336, K_2CO_3, $Pd(PPh_3)_4$, toluene–EtOH–H_2O/90 °C.

Scheme 6-70. The synthetic routes to **Spiro-IA**.

In 2021, four new small molecules A–D–A type (acceptor–donor-acceptor) exposed by **237-240** based on spirofluorene xanthene (**SFX**) and its central entity attached with other acceptor groups through thiophene unit was investigated with density functional theory utilizing four different functionals B3LYP, CAM-B3LYP, MPW1PW91 and WB97XD at 6-31G (d, p) as a basis set (Figure 6-53) [147]. Designed molecules display better optoelectronic properties when compared with **SFX** because of their electron-withdrawing acceptor moieties. Compound **239** shows maximum absorption (λ_{max} = 675 nm) in the visible region and a small energy gap (2.81 eV) between HOMO and LUMO. Open-circuit voltages (V_{oc}) of R (**SFX**) and designed molecules **237-240** are 1.61 V, 1.41 V, 1.82 V, 1.65 V, and 1.78 V, respectively. **238** revealed the highest open-circuit voltage (1.82 V). Further, transition density matrix (TDM) calculation represents remarkable dissociation of charges in all cases from **237** to **240** in interrelation with **SFX**, explaining the higher density of *J*sc for all molecules from **237** to **240** encouraging properties of optoelectronics as 3D non-fullerene based small molecule acceptors.

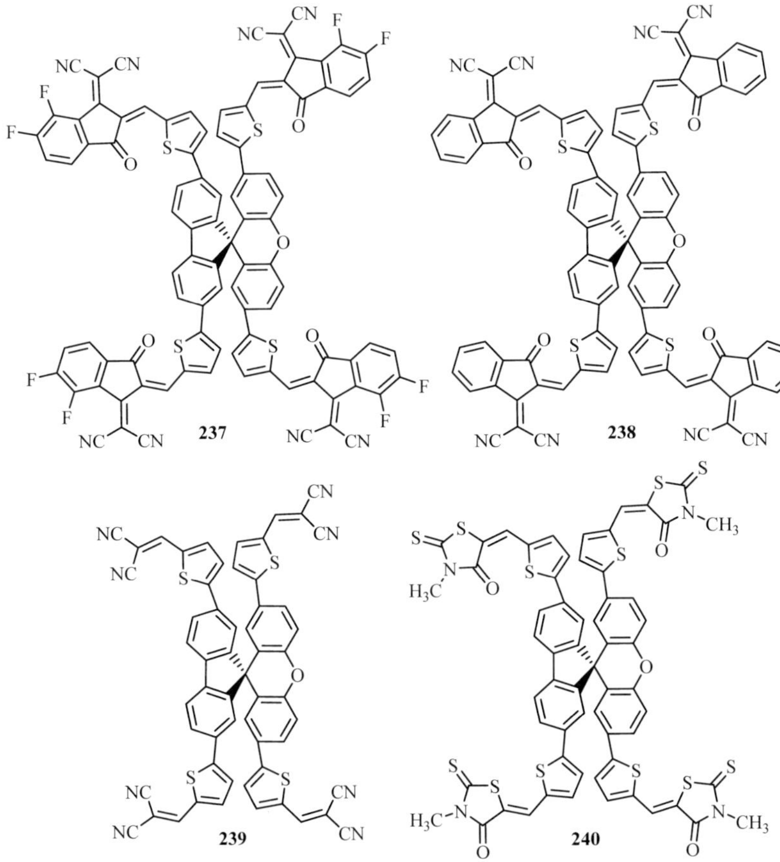

Fig. 6-53. The chemical structures of **237-240**.

The Wang group synthesized the three asymmetric **SFX**-based HTMs with *N*-(*p*-methoxy-phenyl)-*N*′-(9,9′-dimethylfluoren-2-yl)amino (**FPA**) substituents by a facile process, which were named **SFX-F**, **SFX-FM**, and **SFX-FP** (Figure 6-54) [148]. Compared with two branched **SFX-F**, four branched HTMs **SFX-FM** and **SFX-FP** have a slightly deeper HOMO energy level, higher T_g, and smoother film surface morphology.

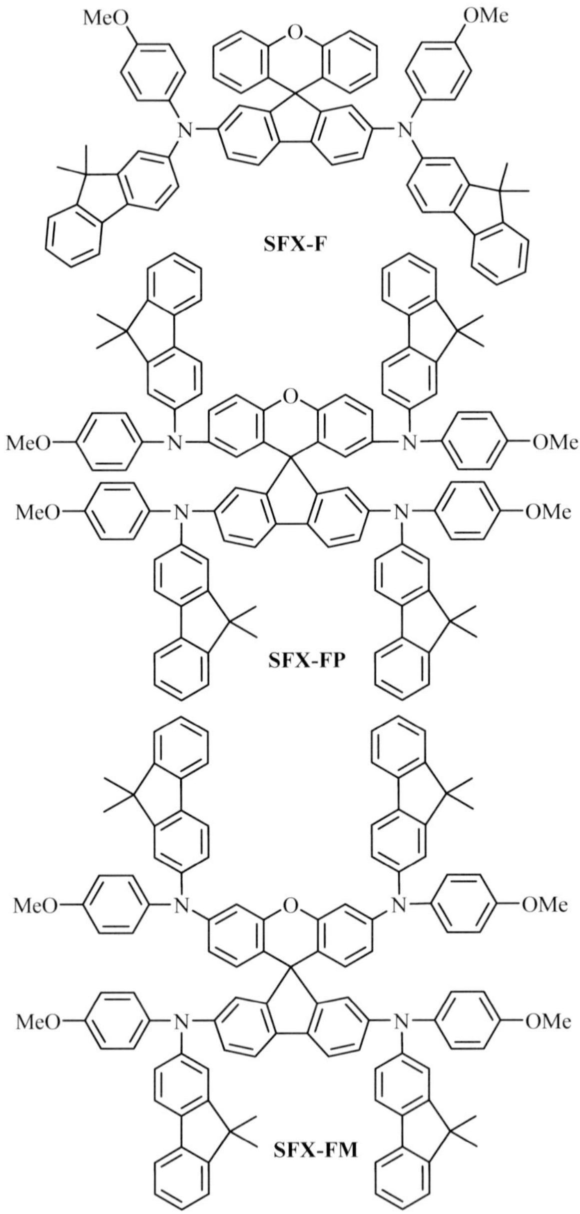

Fig. 6-54. Structure of **SFX-F**, **SFX-FM,** and **SFX-FP**.

Authors found that the T_g of HTM with substitution at the *meta*-position of the compound (**SFX-FM**) is 14 °C higher than the *para*-substitution one (**SFX-FP**), and 46 °C higher than the unsubstituted one (**SFX-F**). Due to the suitable HOMO energy level and higher hole mobility, the PSC constructed with **SFX-FM** achieved a PCE value of 17.29%, which is higher than those of the analogous devices provided with **SFX-F** (14.54%), **SFX-FP** (15.45%) and the reference one with **spiro-OMeTAD** (15.14%) as HTMs.

Dye-sensitized solar cells (DSCs) are another type of SCs that belong to the thin-film SCs that have been under extensive research for more than two decades due to their low cost, simple preparation methodology, low toxicity, and ease of production [149]. In general, a DSC contains a nanocrystalline titanium dioxide (TiO_2) electrode modified with a dye fabricated on a transparent conducting oxide (TCO), a platinum (Pt) counter electrode, and an electrolyte solution with a dissolved iodide ion/triiodide (I^-/I_3^-) ion redox couple between the electrodes [150]. A schematic representation of the DSC is indicated in Figure 6-55a [151]. In the cell, the dye monolayer, after absorbing light, injects electrons into the mesoporous TiO_2, followed by dye regeneration. Electrolyte-based DSCs have revealed up to 13% PCE. In solid-state dye solar cells, the liquid electrolyte is replaced by a solid hole-transporting material, as demonstrated in Figure 6-55b [152].

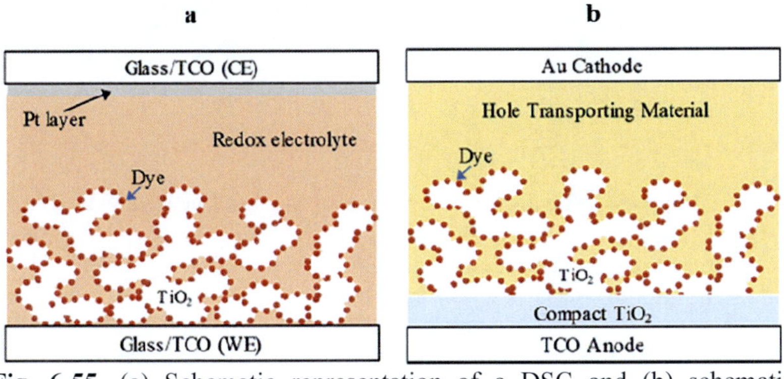

Fig. 6-55. (a) Schematic representation of a DSC and (b) schematic illustration of a solid-state DSC.

DSCs differ from traditional semiconductor devices in that they separate the function of light absorption from charge carrier transport. Dye sensitizer absorbs the incident sunlight and exploits the light energy to induce a vectorial electron transfer reaction. Thus, DSCs have the following benefits compared with Si-based photovoltaics:

1) It is not sensitive to the defects in semiconductors, such as defects in Si,

2) The semiconductor-electrolyte interface (SEI) is easy to form, and it is expense effective for production, and

3) It is possible to realize the direct energy transfer from photons to chemical energy [153]. After some brief notes on DSC, we want to disclose some selected examples in this field.

In 2005, Schmidt-Mende et al. reported very efficient solid-state SC with the amorphous organic HTM **spiro-OMeTAD** as a hole conductor and, for the first time, an organic metal-free indoline dyes sensitizer [154]. Devices sensitizer for solid-state dye display record efficiency of over 4% for solid-state dye-sensitized SC. Therefore, the indoline dye is the first organic dye that is a real alternative to previously applied ruthenium dyes. Figure 6-56 shows the chemical structure of the indoline dye **241**. The higher extinction coefficient converts it into an ideal dye for solid-state dye-sensitized SC. This dye indicates light-harvesting properties superior to those of ruthenium dyes such as N719 when used as a sensitizer of mesoporous TiO_2 layers. Furthermore, the simple and cheap synthesis makes this dye extremely interesting for industrial production.

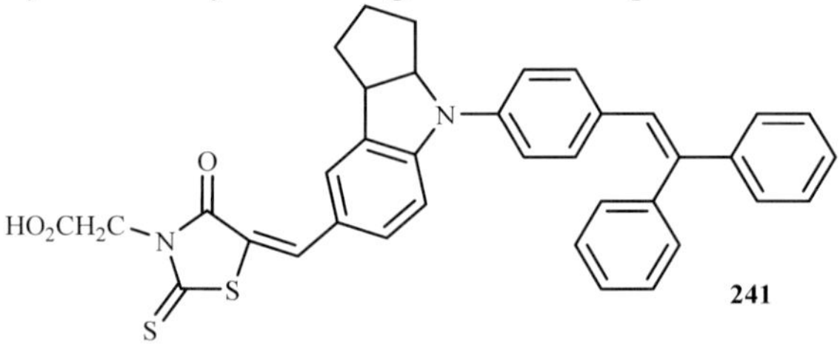

241

Fig. 6-56. The chemical structure of the indoline dye **241**.

One of the primary factors that restrict the efficiency of these cells is the high rate of back reaction of electrons with the oxidized **spiro-MeOTAD**. This reaction occurs not only at the interface with the nanocrystalline TiO_2 but also through the conducting glass substrate. For this reason, the compact blocking layers are essential to obtain the working cells. This significant loss mechanism could be decreased by blending the **spiro-MeOTAD** matrix with a combination of *t*-Bu pyridine (*t*-BP) and $Li[CF_3SO_2]_2N$, yielding devices with a remarkably improved device performance [155].

To understand the effect of *t*-BP on the electron transfer at the interfaces between the **spiro-MeOTAD** and the TiO_2 (both compact and nanocrystalline), the interaction of *t*-BP with the **spiro-MeOTAD** radical cation was investigated by Howie et al. in 2007 [156]. They found that *t*-BP raises the open-circuit voltage of DSC that possesses **spiro-MEOTAD** as a hole conductor. Spectroscopic evidence shows that *t*-BP lowers the doping density of the **spiro-MEOTAD**. *t*-BP increases the blocking ability of the TiO_2 compact layer to more negative potentials, increasing V_{oc} and hence device efficiency. A further increase in V_{oc} may be achieved by engineering the interface between the TiO_2 blocking layer and the hole transport layer to extend the film-blocking behavior to more negative potentials. The EIS and current-voltage data indicate that the surface of the blocking layer may be inhomogeneous. It has non-ideal diode characteristics and contains a high surface state density.

The Boschloo group introduced photoinduced absorption (PIA) spectroscopy to systematically study dye regeneration and pore filling in solid-state DSC [157]. Without a redox mediator, the absorption spectra of electrons in TiO_2 and oxidized dye can be measured. In the presence of a redox mediator, the spectrum of the oxidized redox mediator is observed instead of the spectrum of the oxidized dye. If the redox mediator cannot regenerate the dye or is not in contact with every dye molecule, then the spectrum of the oxidized dye would still be observed in the mediator's presence. Therefore, if, under certain conditions, pores remain unfilled by **spiro-MeOTAD**, signals for the oxidized dye should be seen in the PIA spectrum. The same group applied two different TiO_2 film thicknesses and a variety of **spiro-MeOTAD** concentrations for the spin-coating solutions to study the effect on dye regeneration systematically. ID28 is a perylene dye (Figure 6-57), which performed well in the liquid cell. Perylenes are promising dyes for DSC applications due to their stability, but to obtain high efficiencies in the solid state, thicker films are required for better light harvesting. **242** binds to TiO_2 via ring-opening of the anhydride group. It has a bis((*tert*-octyl)phenyl)amine group at the other end, similar to the methoxyphenyl amine groups on **spiro-MeOTAD**. Therefore, a good contact between **242** and **spiro-MeOTAD** is expected, which should lead to good wetting of the dye molecules by the hole conductor.

The onset of oxidation of a solid film of **spiro-MeOTAD** was found to be 0.15 V versus Fc/Fc^{+}, and extinction coefficients of **spiro-MeOTAD**$^{+}$ were found to be 33000 M^{-1} cm^{-1} at 507 nm and 8500 M^{-1} cm^{-1} at 690 nm. Electrons in TiO_2 films were displayed to alter the ground-state absorption spectra of **242** attached to TiO_2. PIA measurements revealed an excellent contact between **242** and **spiro-MeOTAD** for different **spiro-MeOTAD** concentrations for both 2- and 6-μm thick TiO_2 films. They discuss the possibility of estimating the quality of pore filling from the positions of absorption peaks. Results demonstrated that with a **spiro-MeOTAD** concentration of 300 mg/mL in chlorobenzene, a uniform distribution of **spiro-MeOTAD** in the pores of the 6-μm thick TiO_2 film could be obtained.

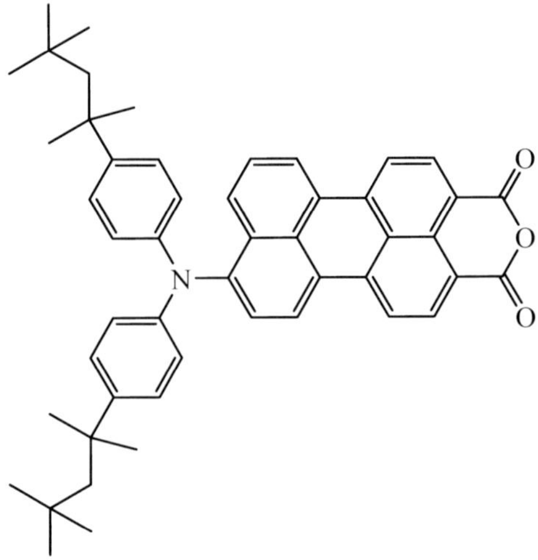

Fig. 6-57. The chemical structure of the perylene dye (**242**).

Later in 2009, the McGehee group quantified for the first time the pore filling of **spiro-OMeTAD** in mesoporous TiO_2 films using XPS depth profiling and UV–Vis absorption spectroscopy [158]. It was demonstrated that **spiro-OMeTAD** can interpenetrate the entire depth of the film, and its concentration is constant throughout the film. They determine that in a 2.5-mm-thick film, the volume of the pores is 60–65% filled. The pores become less filled when thicker films are employed. Such filling fraction is much higher than the solution concentration because the excess solution on top of the film can act as a reservoir during the spin coating process. Lastly, they showed that by using a lower spin coating speed and higher **spiro-OMeTAD** solution concentration, the filling fraction and, consequently, the device's efficiency can be increased.

In 2011, continuous efforts by the McGehee group on this topic led to improve the efficiency and photovoltaic characteristics of solid-state dye-sensitized solar cells (SSDSCs) improve with the pore filling fraction (PFF) of the HTM, **spiro-OMeTAD** [159]. The mechanisms through which photovoltaic characteristics improved were investigated with transient absorption spectroscopy and transient photovoltage/photocurrent measurements. Results show that as **the spiro-OMeTAD** PFF is increased from 26% to 65%, there is a higher hole injection efficiency from dye cations to **spiro-OMeTAD** because more dye molecules are covered with **spiro-OMeTAD**, an order-of-magnitude slower recombination rate because holes can diffuse further away from the dye/HTM interface, and a 50% higher ambipolar diffusion coefficient due to an improved percolation network. Device simulations have shown that if 100% PFF could be obtained for thicker devices, the efficiency of SSDSCs using a conventional ruthenium dye would increase by 25% beyond its current value.

Continuing the research on SSDSCs, Chu et al. directed their attention towards commercially available p-type dopants, specifically emphasizing those that are cost-effective and can be synthesized on a large scale [160]. They found that incorporating the strong electron acceptor perfluoro-tetracyano quinodimethane (**F4TCNQ**) into the solid HTMs **spiro-MeOTAD** develops a **spiro-MeOTAD/ F4TCNQ** polaron charge-transfer complex (Figure 6-58).

Careful exploration reveals that the formation of the polaron charge-transfer complex not only facilitates the conductivity of HTMs but also prevents charge recombination across the heterojunction interface, i.e., photo anode/ HTMs and/or counter electrode/HTMs. The performance of SSDSCs is substantially improved by using the organic dye **243**. At AM1.5 illumination, the short circuit current densities J_{sc} increase from 8.29 mA cm^{-2} (w/o **F4TCNQ**) to 10.95 mA (w/**F4TCNQ**), accompanied by a 20% increase in the overall power conversion efficiency, η, from 4.55 to 5.44%.

In contrast to organic SCs, where oxygen leads to the degradation of devices, SSDSCs using the hole conductor **spiro-MeOTAD** require oxygen during fabrication and storage. Device performance has been indicated to decrease during storage under argon or vacuum [161]. Snaith et al. attributed this effect to a shunting at the metal back contact for SnO_2-based SSDSC [162]. In elegant work, Cappel determined the concentrations of oxidized **spiro-MeOTAD** within devices under different operating and storage conditions by UV-vis spectroscopy. Relative concentrations of **spiro-MeOTAD**$^+$ were greater than 10% after illumination for standard SSDSCs, where no chemical dopant had been applied in the SC construction, but oxygen and lithium ions were present. They offer that oxidized **spiro-MeOTAD** is created by oxygen reduction at the TiO_2 surface during cell illumination. Furthermore, they explored the effect of light soaking under different conditions and associated changes in **spiro-MeOTAD**$^+$ concentration on the SC measurements. Their findings give insights into photochemical reactions occurring within

SSDSCs and provide guidelines for doping levels in device construction in the absence of oxygen (Figure 6-59).

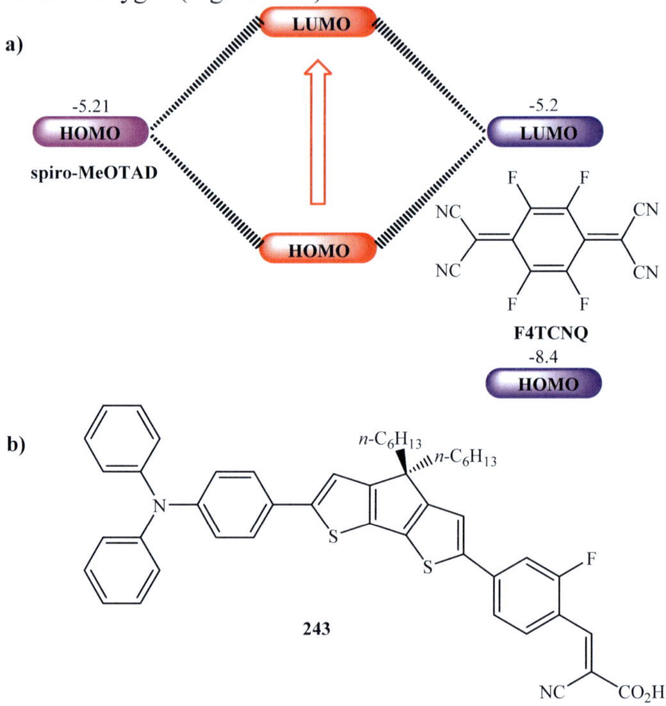

Fig. 6-58. (a) The structural drawing of **spiro-MeOTAD**, **F4TCNQ**, and the occurrence of charge transfer interaction. (b) The structure of dye **243**.

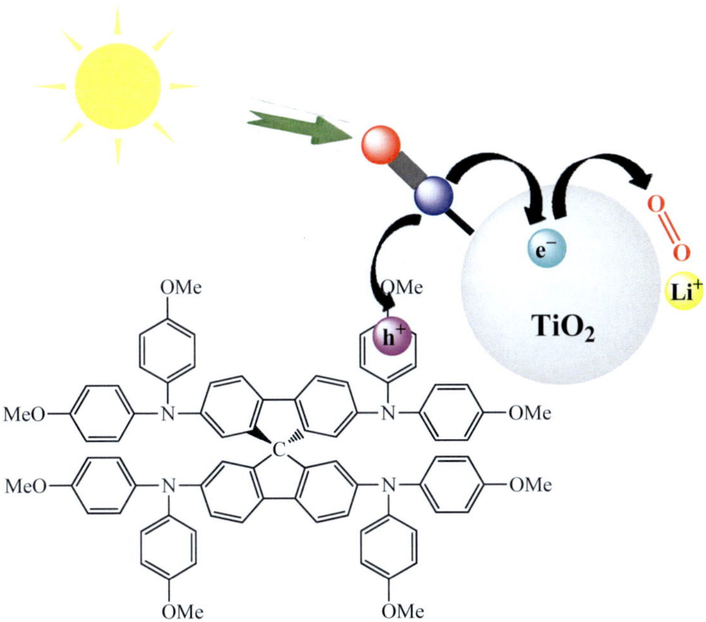

Fig. 6-59. Oxygen-induced doping of **spiro-MeOTAD** in SSDSCs.

To improve the stability and processability of DSCs to allow efficient roll-to-roll manufacturing, extensive research is being invested in studying solid and quasi-solid alternatives for the liquid electrolyte. Endeavors have focused on organic and inorganic semiconductors and conducting polymers to meet all the requirements for a good HTM. In organic semiconductors, the p-type **spiro-MeOTAD** remains the material of choice for high PCEs. Electrochemical impedance spectroscopy (EIS) has become a powerful tool to characterize the internal electrical processes occurring within devices such as DSSCs. In 2013, the internal transport and recombination parameters of SSDSCs using the amorphous organic semiconductor **spiro-MeOTAD** as an HTM were studied by Moehl et al. using ESI [163]. Devices were constructed using flat and nanostructured TiO_2 and compared to systems using nanostructured ZrO_2 to differentiate between the transports processes within the different components of the SSDSC. The effect of chemically p-doping the HTM on its transport was perused, and its temperature dependence was studied and analyzed using the Arrhenius equation. Using this approach, the activation energy of the hole hopping transport within the undoped **spiro-MeOTAD** film was determined to be 0.34 (0.02 and 0.40 (0.02 eV for the mesoporous TiO_2 and ZrO_2 systems, respectively.

Spiro-OMeTAD relies on an uncontrolled oxidative process to reach appreciable conductivity. Lin et al. expressed using a dicationic salt of **spiro-OMeTAD**, named **spiro(TFSI)$_2$**, as a facile means of controllably raising the conductivity of **spiro-OMeTAD** to 10^{-3} S cm^{-1} without relying on oxidation in the air [164]. **Spiro(TFSI)$_2$** was prepared by oxidizing **spiro-OMeTAD** with silver bis-(trifluoromethanesulfonyl) imide (AgTFSI) and was easily isolated in high yield (93%) through precipitation in diethyl ether, as shown in Scheme 6-71.

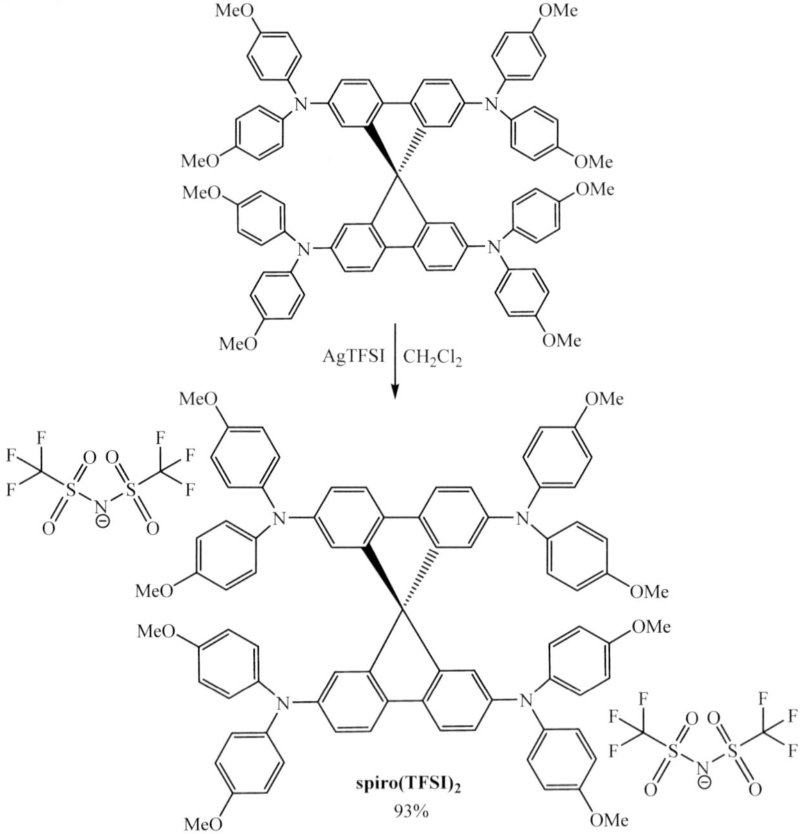

Scheme 6-71. Synthesis of **spiro(TFSI)$_2$**.

Spiro(TFSI)$_2$ enables the first demonstration of SSDSCs constructed and operated with the complete elimination of oxygen after deposition of the sensitizer with higher and more reproducible device performance. Using **spiro(TFSI)$_2$** will assist in facilitating the commercialization of these technologies by enabling sealing in a dry, oxygen-free environment to eschew potential degradation of the absorbers and organic materials by moisture and photo-oxidation.

Bruder et al. reported that the growth of large crystalline domains in the hole-transporting layer is one of the reasons for SC degradation at elevated temperatures [165]. Crystallization of the HTM inside the pores or on the interface affects the contact between the light absorber and the hole transport. The symmetrical **spiro-OMeTAD** is prone to crystallization, mainly if processed from solution. The presence of residue solvent and liquid additives, such as 4-*tert*-butyl pyridine accelerates this process. They operate as plasticizers, remarkably decreasing the glass transition temperature and overall morphological stability of the amorphous state. One of the possible solutions to the problem is the introduction of local asymmetry into the **spiro-OMeTAD** molecule. This tactic represses crystal growth in the HTM layer and considerably elevates SC resistance to thermal stress and the overall lifetime of the device. For this purpose, 3 new **spiro-OMeTAD** analogs (**HTM244**–**HTM246**) have been synthesized and evaluated. Molecules of **HTM244**–**HTM246** have a varying degree of asymmetry introduced into them (Figure 6-60). Devices with **HTM244**, a **spiro-OMeTAD** analog with four other strategically well-placed methyl groups, exhibited impressively improved lifetime at elevated temperatures and retained ~90% of their initial efficiency after 1000 hours at 60 °C, while SCs with **spiro-OMeTAD** destroyed rapidly under identical conditions.

In 2013, Pozzi prepared a new donor/acceptor (D-A) spiro dye (**SCPDT1**) featuring two bithiophene units connected through an sp^3-hybridized carbon atom by a multistep synthetic sequence involving the convenient assembly of the spiro system under mild catalytic conditions [166]. Scheme 6-72 shows the synthetic approach for obtaining the **SCPDT1**. The photocurrent spectrum of DSSCs incorporating **SCPDT1** covers the spectral region ranging from 350 to 700 nm and attains a wide maximum of ~80% in the 420-560 nm range. Power conversion efficiencies of up to 6.02% were achieved.

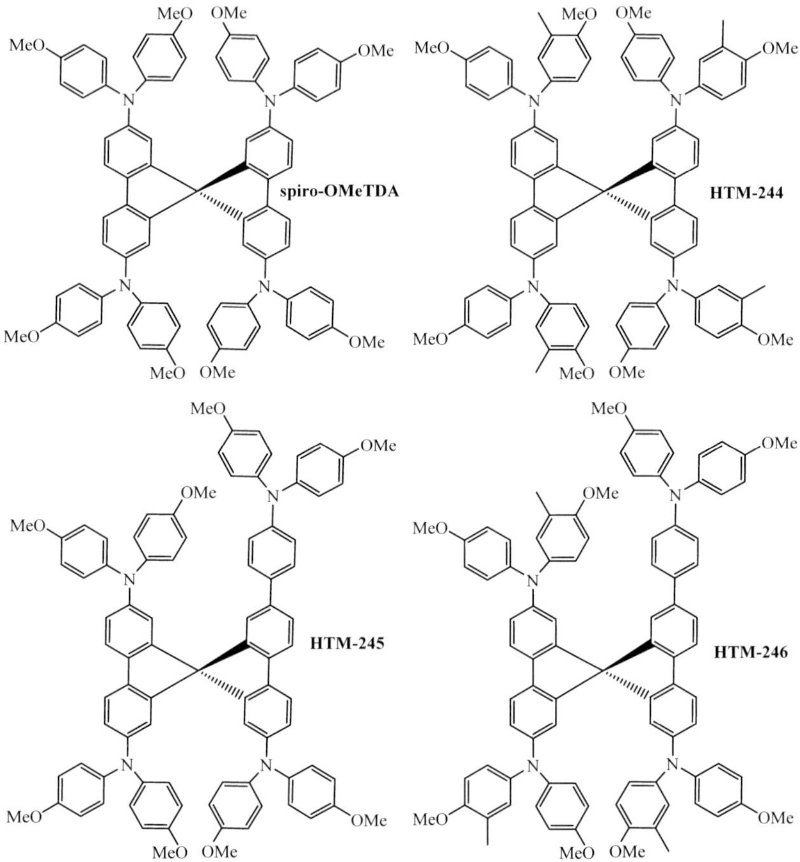

Fig. 6-60. Structures of the synthesized **spiro-OMeTAD** analogs.

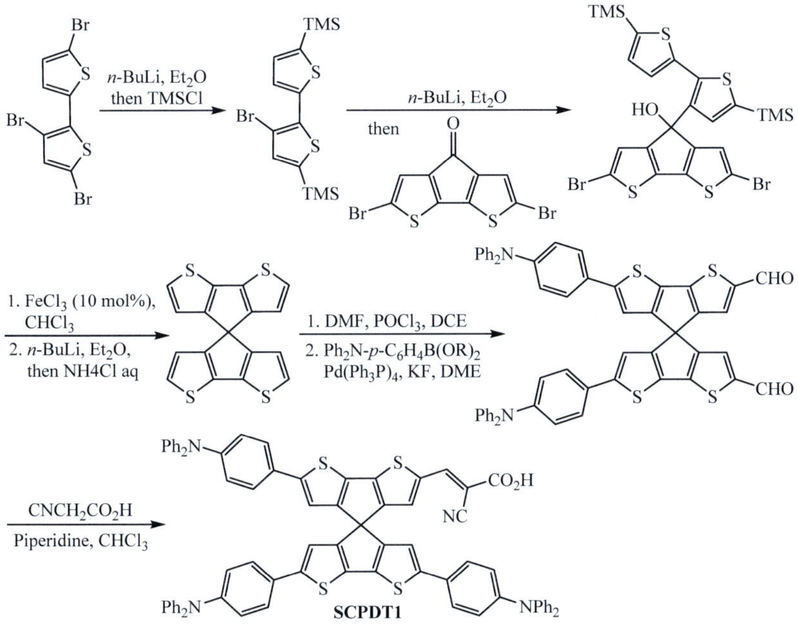

Scheme 6-72. Synthesis of **SCPDT1**.

Inopportune orientation and self-assembly of dyes on the TiO_2 surface are the fundamental weaknesses at the dye-TiO_2 interface that affect both charge injection and dye regeneration processes. Hence, a series of homo spiro-dimeric donor-π-spacer-acceptor (D-π-A) dyes based on bithiophene. Terthiophene spacers with the two-anchoring groups containing dyes **247-249** were synthesized by the Nithyanandhan group via the Pd-catalyzed successive Suzuki coupling in between different donor moiety with different formylated spiro-bipropylenedioxythiophene (**spiro-BiProDOT**) in 2019 (Figure 6-61) [167].

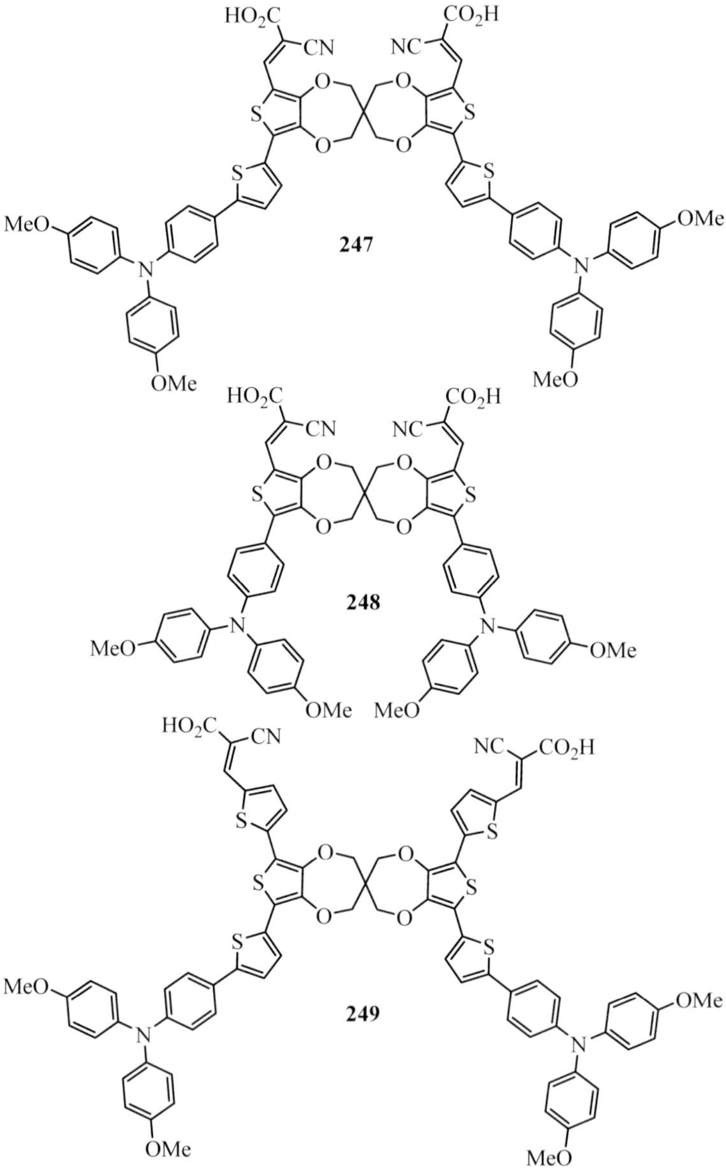

Fig. 6-61. Molecular structures of **247-249** spiro dyes.

The position of the branching center in **247** and **248** was systematically varied, where the spiro unit was placed near and away from the anchoring carboxylic acid unit, respectively. The dimeric spiro-dye **248** disclosed device performance, which is 1.6-fold higher than its structural isomeric analog **247**, of 3.9% with a V_{oc} and J_{sc} of 0.593 V and 9.09 mA/cm^2, respectively, whereas for the isomeric analog **247** a device performance of 2.45% (V_{oc} of 0.568 V and J_{sc} of 6.25 mA/cm^2) under resembled 1 Sun (100 mW/cm^2) condition. The improved V_{oc} for the dye **248** compared to **247** dye was attributed to the dipole moment applied by the dyes on the TiO_2 surface. Moreover, the maximum PCE of 4.16% (V_{oc} 0.589 V, J_{sc} 9.79 mA/cm^2, respectively) was apperceived when an additional thiophene unit was inserted between the donor and π-spacer in dye **249**. The shifting of the position of the branching **spiro-BiProDOT** π-spacer makes an impact on the device performance by synergistically enhancing both V_{oc} and J_{sc}.

6.3.3. Field-effect transistor (FET)

A field-effect transistor (FET) is a device that uses an electric field to control the flow of current in a semiconductor material, and it has a wide range of applications in integrated circuits and flexible electronics [168]. The OFET is a key component in modern organic electronics. Its main advantages over a silicon transistor include the low cost of organic materials, the ability to generate organic elements through low-cost and eco-friendly printing technologies, and their universal qualities such as low weight, flexibility, transparency, long shelf life, and reliability [169]. A variety of conjugated small molecules, polymers, and room-temperature liquid crystals are used as active channel semiconductors in OFET devices. The selection of these materials depends on the highest occupied molecular orbital (HOMO) and lowest unoccupied molecular orbital (LUMO) energy levels of these molecules and their proximity to the work function of metals. The difference between HOMO and LUMO serves as the bandgap of OSCs, as determined from cyclic voltammetry data [170]. OFETs have been the subject of much research in the fields of chemistry, physics, materials science, and microelectronics since their first discovery based on OSCs in the 1980s. From a technological perspective, OFETs are considered a key component of organic integrated circuits, with potential applications in flexible smart cards, low-cost radio frequency identification (RFID) tags, and organic active matrix displays. From a scientific perspective, OFETs are also useful for testing novel OSCs and investigating organic electronics. The charge carrier mobility and the ON/OFF current ratio are the key material parameters that control the

OFET characteristics, with a high ON/OFF ratio being particularly important for applications in digital circuits and pixel switches for electronic ink and liquid crystal displays [171].

To improve the ON/OFF ratio in organic FETs, many efforts have been made to improve device design and post-treatment procedures. In recent years, there has been increasing interest in studying the effect of an external magnetic field on OS devices. Scientists have constructed at least three different types of devices based on a diode structure consisting of two electrodes: spin-valves (with electrodes fabricated from ferromagnetic materials with various coercive fields), organic magneto-resistance (OMAR) devices (with electrodes fabricated from non-magnetic materials), and asymmetric devices (with electrodes fabricated from ferromagnetic and non-magnetic materials, respectively) [172]. In this context, OFETs based on spiro compounds are currently being studied.

OSCs are attractive for making large-area and flexible organic electronic devices at a meager cost. The reliability of organic electronic devices depends on the morphological stability of the corresponding materials, which can be quantified by the material glass-transition temperature (T_g). One important class of functional materials used in OLEDs and OFETs is HTMs. One well-known example is *N*,*N*′-diphenyl-*N*,*N*′-bis(1-naphthyl)-1,1′-biphenyl-4,4′-diamine (α-NPB), which has a T_g of 95 °C and ionization potential of 5.1 eV. However, the T_g of α-NPB is not high enough to exclude thin films of the material from crystallizing. In 1996, Salbeck presented the spiro concept to enhance the morphological stability of functional organic glasses while retaining their functionality. Joining two charge-transport moieties (for instance, two identical α-NPB molecules) through a spiro-carbon center has the effect of increasing the T_g because the raised steric demand of the resulting spiro-linked compound, 2,2′,7,7′-tetra(*N*-phenyl-1-naphthylamine)-9,9′-spirobifluorene (spiro-α-NPB), effectively hinders crystallization. The T_g of the resulting spiro-α-NPB is 147 °C. The perpendicular arrangement of the two α-NPB moieties in spiro-α-NPB is attended to give only relatively weak electronic interaction between the molecular halves. In general, spiro-linked compounds are known to be glass-forming materials with high T_g and good morphological stability, which makes them well-suited for organic devices.

In an elegant work in 2006, a comparison between the hole-transport properties in thin films of spiro-linked triarylamine compounds and their corresponding parent compounds was performed. The materials employed in this study are indicated in Figure 6-62: **spiro-TAD**, 2,2′,7,7′-tetrakis(*N*,*N*′-di-*p*-methyl-phenyl-amino)-9,9′-spirobifluorene (**spiro TTB**),

2,2′,7,7′-tetra(*m*-tolyl-phenylamino)-9,9′-spirobi-fluorene (**spiro-TPD**), 2,2′,7,7′-tetra(*N*-phenyl-1-naphthylamine)-9,9′-spirobifluorene (**spiro-*α*-NPB**), *N*,*N*,*N*′,*N*′-tetraphenyl benzidine (**TAD**), *N*,*N*,*N*′,*N*′-tetrakis(4-methylphenyl)benzidine (**TTB**), *N*,*N*′-bis(3-methyl phenyl)(1,1′-biphenyl)-4,4′-diamine (**TPD**), and *N*,*N*′-diphenyl-*N*,*N*′-bis(1-naphthyl)-1,1′-biphenyl-4,4′-diamine (α-**NPB**).

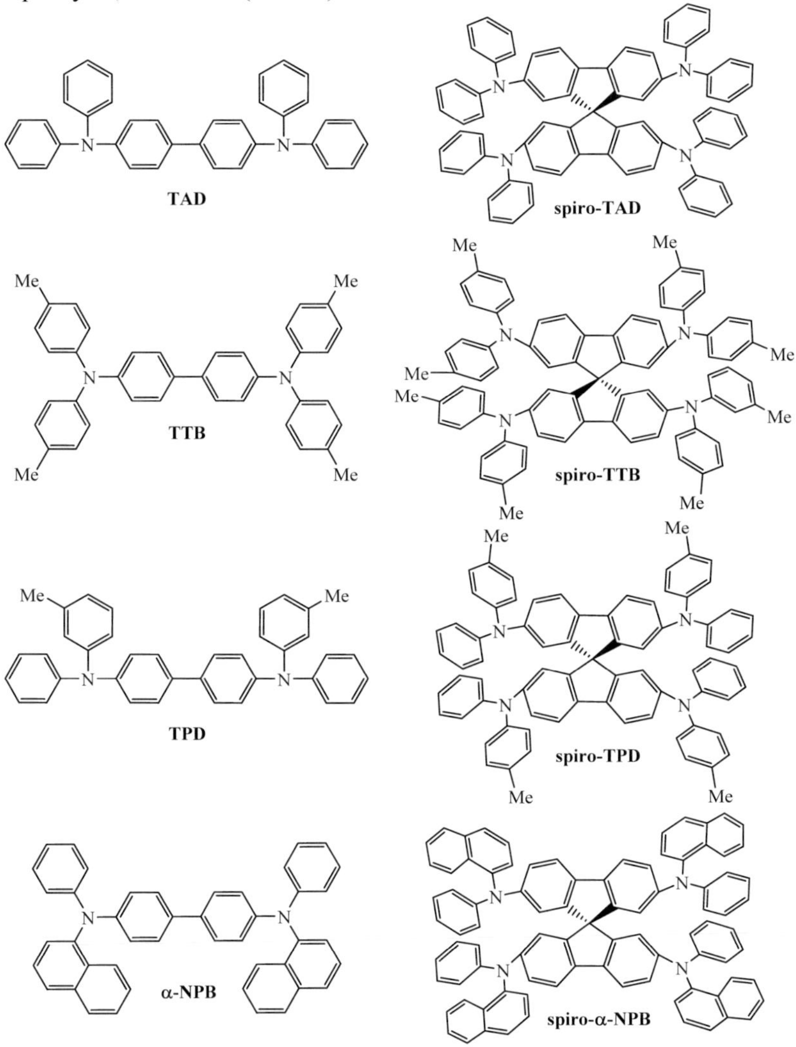

Fig. 6-62. The chemical structures of materials used in this study.

OSCs have emerged as a promising alternative for producing large-area and flexible organic electronic devices at a low cost. The reliability of these devices heavily depends on the morphological stability of the corresponding materials, which can be quantified by their glass-transition temperature (T_g). One crucial class of functional materials utilized in OLEDs and OFETs is. For instance, *N*,*N*′-diphenyl-*N*,*N*′-bis(1-naphthyl)-1,1′-biphenyl-4,4′-diamine (α-NPB) is a well-known example of HTMs, with a T_g of 95 °C and ionization potential of 5.1 eV. However, the T_g of α-NPB is not high enough to prevent thin films of the material from crystallizing.

To address this issue, in 1996, Salbeck proposed the spiro concept to enhance the morphological stability of functional organic glasses while retaining their functionality. By linking two charge-transport moieties (e.g., two identical α-NPB molecules) through a spiro-carbon center, the resulting spiro-linked compound, 2,2′,7,7′-tetra(*N*-phenyl-1-naphthyl amine)-9,9′-spiro-bifluorene (spiro-α-NPB), has its Tg increased to 147 °C due to the heightened steric demand, effectively hindering crystallization. The perpendicular arrangement of the two α-NPB moieties in spiro-α-NPB is known to give only relatively weak electronic interaction between the molecular halves. Spiro-linked compounds are generally known as glass-forming materials with high T_g and good morphological stability, making them well-suited for organic devices.

In a study conducted in 2006, a comparison between the hole-transport properties in thin films of spiro-linked triarylamine compounds and their corresponding parent compounds was performed. The hole mobilities in new thin films of spiro-linked compounds were found to be of the same magnitude as those of the corresponding parent compounds. However, the transistor operation of the parent compounds disappeared after the sample was stored in an ambient atmosphere for several days due to the thin films' lack of morphological stability. In contrast, the hole mobilities of the spiro-linked compounds did not change remarkably after the samples were stored in an ambient atmosphere for up to nine months. The temperature-dependence measurements revealed that the Arrhenius model better fits the experimental data, with significant differences in the fit parameters for the spiro-linked compounds compared to their parent compounds. Thus, the spiro concept, designed to enhance morphological stability, significantly influences the charge-transport properties of the resulting compound.

They found that the hole mobilities in new thin films of spiro-linked compounds are of the same magnitude as those of the corresponding parent compounds. However, the transistor operation of the parent compounds disappeared after the sample was stored in an ambient atmosphere for several days because of the thin films' lack of morphological stability. In contrast, the hole mobilities of the spiro-linked compounds did not change remarkably after the samples were stored in an ambient atmosphere for up to nine months. From the temperature-dependence measurements, it has been found that the Arrhenius model better fits the experimental data. There are substantial differences in the fit parameters for the spiro-linked compounds compared with their parent compounds. Therefore, the spiro concept, designed to enhance morphological stability, influences the charge-transport properties of the resulting compound [173].

Moreover, Ohshita et al. prepared spirobi(dithienosilole)s **sDTS250-252** as the novel–conjugated compounds, and their optical and FET properties were studied. **sDTS250-252** were readily achieved, as shown in Scheme 6-73. They exhibited absorption maxima at 358–368 nm, a little red-shifted from those of the corresponding non-spiro type dithienosiloles, indicating that spiro-conjugation operates in these molecules to an extent. The field-effect hole mobility of spirobi[bis(trimethylsilyl)dithienosilole] **sDTS250** was determined to be 1.4 × 10^{-6} cm^2/Vs [174].

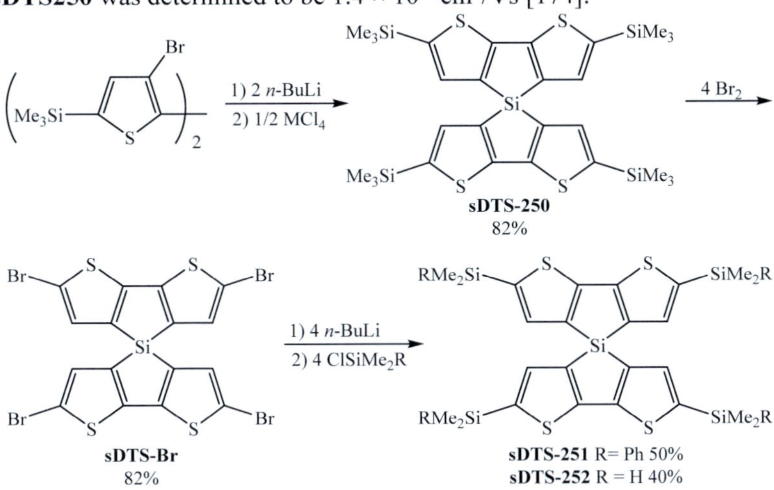

Scheme 6-73. Preparation of **sDTS250-252**.

In OSCs, low magnetic fields can influence the spin correlation between quasi-particles, resulting in significant magneto-resistive effects at room temperature. This OMAR is highly interesting for addressing fundamental questions about spin transport and manipulation in organic materials, as well as for developing weightless, flexible, and multifunctional magneto-optoelectronic applications. Investigating OMAR in OFETs is of high significance, as this three-terminal device geometry allows for the verification of the charge carrier sign in the conducting channel and the determination of the corresponding mobility.

In 2012, the Salbeck group reported the first magneto-resistive FETs that are sensitive to ultra-small magnetic fields (<5 mT) at room temperature without any additional irradiation. The magneto-sensitivity of their devices is obtained by blending the strong acceptor molecule 1,4,5,8,9,11-hexaazatriphenylene hexacarbonitrile (**HAT-CN**) with the electron donor 2,2′,7,7′-tetrakis-(*N*,*N*-dip-methylphenylamino)-9,9′-spiro-bifluorene (**spiro-TTB**), leading to a high density of intermolecular radical pair (**IRP**) states with magnetic-field-dependent formation, dissociation, and interaction rates (Figure 6-63). By changing the blending composition, the amount of IRP states can be tuned, and the magneto response of the devices can also be regulated. Using donor-acceptor blends as active thin films in OSC devices is a powerful tool for constructing magneto-resistive devices without ferromagnetic materials [175].

Stutzmann et al. introduced the concept of the vertical organic field-effect transistor (VOFET) as an approach to reduce the geometric channel length of transistors to a few tens of nanometers, requiring minimal structuring efforts. VOFETs redirect charge transport from the lateral to the vertical direction, akin to traditional thin-film transistors. This redirection enables high current densities and switching frequencies, which are essential for flexible electronic applications like flexible RFID tags and active-matrix displays. In 2018, Sawatzki et al. analyzed the working principle of VOFETs based on simulations and experiments (Figure 6-64). They found that VOFETs operate, in principle, as OFETs, with the only distinction being the short channel length and unique channel geometry. Charge carriers are accumulated at the semiconductor dielectric interface and extracted at the drain, similar to lateral OFETs. The size and shape of the channel area are determined by the gate-source voltage, drain-source voltage, and the OSCs lateral and vertical mobility.

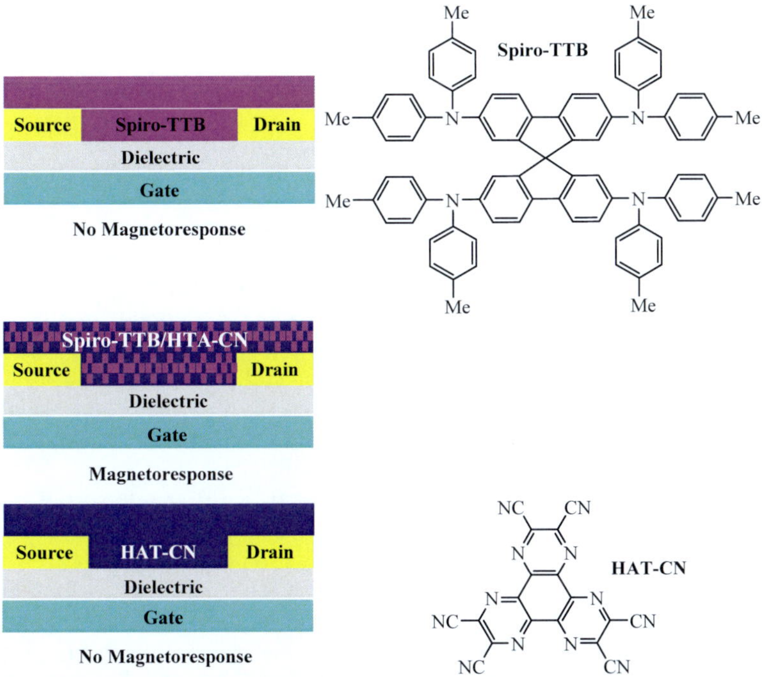

Fig. 6-63. Schematic representation of their experimental schemes. On the left and right sides, nonmagneto-resistive FETs based on p-type **Spiro-TTB** and n-type **HAT-CN**, as well as the corresponding chemical structures, are indicated. Devices with a mixture of **Spiro-TTB** and **HAT-CN** exhibit a significant response to an external magnetic field of 3.2 mT (indicated in the middle).

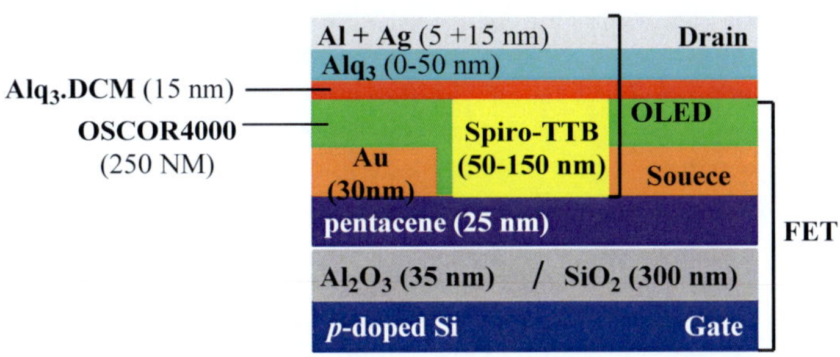

Fig. 6-64. Sample structure and stack of experimental (V)OLET devices. (Some of the samples are made with doped **Spiro-TTB**, and some are undoped.)

Sawatzki et al. used drift-diffusion simulations and a light-emitting transistor to measure the dimension of the vertical channel's independence of the bias conditions. They found good agreement between simulations and experiments regarding the dependencies of the channel size on the gate and drain-source voltages. The area of light emission in the (V)OLETs grows with the overall thickness of the OLED stack and the charge-carrier imbalance at the light-emitting interface. They realized an upper limit for the required distance between adjacent VOFETs of 4 μm from experiments. The simulations forecast a minimal channel depth of under 1 μm for pentacene-based VOFETs of this design. This required distance is even lower when driving the device in the saturation regime.

Sawatzki et al. also found that having a more favorable mobility configuration, such as higher vertical than lateral mobility, would further diminish the channel size. Therefore, there is no applicable limit for the integration of vertical organic transistors to be expected within the currently applied alignment and patterning limits.

Sawatzki et al. used a light emission method for charge-carrier path detection, which could be used in other vertical design stacks, such as organic permeable-base transistors or VOFETs with nanostructured electrodes. Their findings from experiments and simulations contradict previous reports analyzing charge transport in vertical organic transistors by simulations. They found strong evidence that the main path of current flow in the vertical direction is close to the source insulator. However, the high density of grid points required for stringent modeling might be a potential reason for the discrepancy [176].

The injection of charges in OFETs is a crucial factor that affects the subthreshold swing and operating voltage of the devices. To enhance charge carrier injection, various strategies have been proposed to achieve low contact resistance (R_c). One approach is to modify the work-function of the metal electrodes by using self-assembled monolayers (SAMs), which can shift the vacuum level by forming dipole layers. SAMs commonly applied to Au and Ag electrodes include alkane terminals (which decrease work-function) and terminals with halogen atoms (which increase work-function). For example, 1*H*,1*H*,2*H*,2*H*-perfluorodecanethiol can increase the work-function of Au by 0.6 to 5.5 eV, whereas hexadecane thiol can decrease it by 0.8 eV.

Another approach is to insert a doped layer at the interface between the electrode and active OSC film. The strong band bending in the doped layer leads to charge injection via tunneling. For example, hexafluorotetracyanonaphthoquinodimethane (**F6TCNNQ**) is a typical p-type dopant that has an electron affinity of ≈5.6 eV, which can effectively inject charges into OSCs with an ionization energy (IE) lower than 5.6 eV. $FeCl_3$ is another widely used dopant that can significantly reduce Rc, from 200 × 103 to 8.8 × 103 Ω cm, by inserting it into the contact interface of the Au electrode and 2,7-dioctyl[1]benzothieno[3,2-*b*][1]benzothiophene (**C8-BTBT**) film. The improved charge injection is attributed to a decrease in the depletion layer thickness at the contact interface and the occupation of trap states in the access region.

Metal-oxide electrodes, such as molybdenum oxide (MoO_3), are also applied as hole injection layers in organic electronics due to their high work function of 6.7–7.0 eV. Recently, Marszalek et al. demonstrated a significantly improved charge-carrier injection from high work-function metal-oxide electrodes in OFETs for amorphous OSCs by using organic interlayers (IL) with a high IE [177]. The effect of the IL on hole injection for p-type OFETs was investigated by making bottom-gate and top-contact (BGTC) transistors based on heavily doped silicon wafers with a 300 nm SiO_2 dielectric layer (Figure 6-65). MoO_3 showed limited injection into amorphous **spiro-TAD** and tris(4-carbazoyl-9-ylphenyl)amine (**TCTA**) active layers, resulting in high contact resistance and threshold voltage.

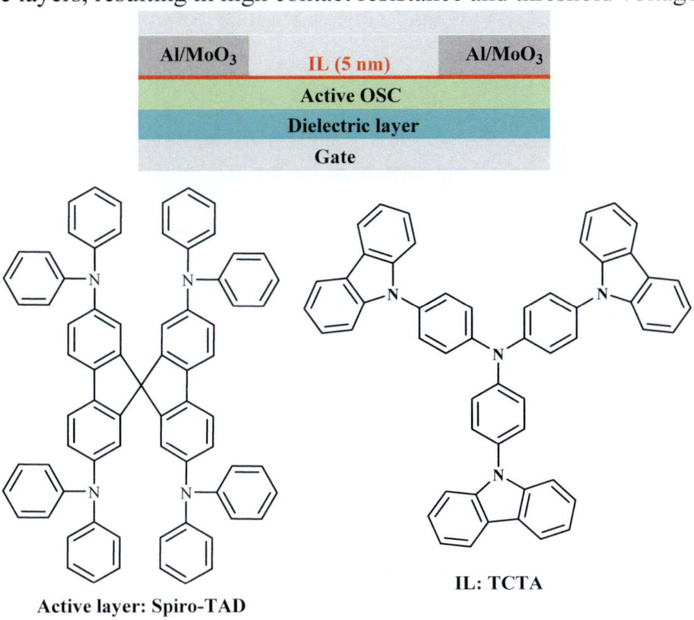

Fig. 6-65. Schematic BGTC transistor configuration and the molecule structures of **spiro-TAD** and **TCTA**.

However, by inserting a few nanometers thick IL with high IE between the MoO_3 electrode and the amorphous OSC films, the **spiro-TAD** and **TCTA** OFETs demonstrated a considerable improvement in hole current, subthreshold swing, and adequate charge carrier mobility due to the decreased contact resistance. Nonetheless, the effect of the IL decreases for discontinuous interlayers formed on distinct grain domains, as in the case of polycrystalline **C8-BTBT** films. These results suggest that using smooth interlayers with IE higher than the semiconductor is a general approach to raising the hole injection into amorphous OSCs with high IE in OFETs.

High dielectric constants are crucial for enhancing the performance of various types of OSC devices. This includes improving the piezoelectric, pyroelectric, and ferroelectric effects and photoelectric conversion efficiency in OPVs, carrier mobility in OFETs, and charge density in charge-trapping memories. A high dielectric constant with a value of $k \geq 10$ is particularly important for achieving band-like carrier transport and Wannier-Mott-like excitons in high-performance inorganic semi-conductors. However, it is challenging to enhance *k* values solely through carbon elements due to the small electronic cloud volume, which results in low polarization efficiency.

Carbon nanostructures, such as organic nanopolymers, have excellent dielectric properties and play a critical role in wave absorption materials. Ling et al. designed organic nanopolymers to facilitate the diffusion of dipole polarization by effectively overcoming the limitations of molecular nanoscale structures [178]. They created a state-of-the-art ultra-rigid spiro-nanopolymer to investigate the effect of covalent nanoscale ordering on ultra-low energy disorder and high dielectric constant.

The researchers prepared nanopolyspirogrids (NPSGs) via simple cross-scale Friedel-Crafts polygridization of A_2B_2-type nanomonomers (Scheme 6-74). Spiro-polygridization effectively suppressed conformational entropy to achieve recording rigidity with a persistence length (l_p) of approximately 41 nm. The high dielectric constant ($k = 8.43$) of NPSG was obtained by locking the spiropolygridization effect, which enhances dipole polarization. Doping NPSG into a polystyrene-based dielectric layer increased the field-effect carrier mobility from 0.20 to 0.90 $cm^2V^{-1}s^{-1}$ in pentacene OFET devices. These results exceeded all amorphous π-conjugated polymer semiconductors reported in the literature.

This polygridization strategy is a powerful cross-scale chemistry to modulate the nature of organic nanopolymers hierarchically, opening new opportunities to improve the physical optoelectronic performance and the recording performance of semiconducting devices. Utilizing this type of molecular integration technology (MIT), organic nanopolymer semiconductors could become potential candidates for fourth-generation semiconductors with high-performance, multifunctionality, intelligence, and ubiquity to meet the requirements of flexible electronics, nanoelectronics, and organic intelligence.

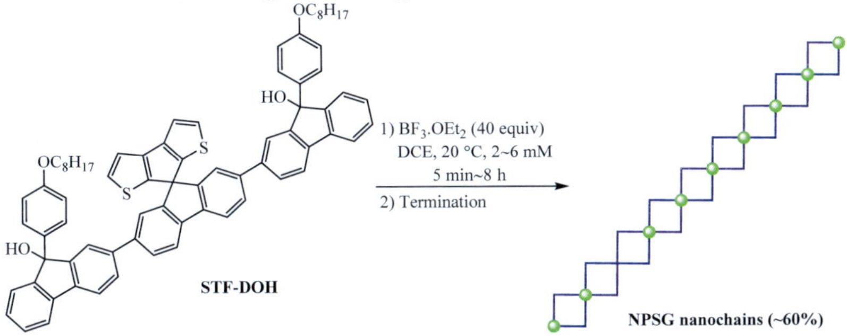

Scheme 6-74. Synthesis of **NPSG**.

6.4. Application of spiro compounds as stimuli-chromism of photoswitches in smart polymers

Polymers that respond sharply to little changes in physical or chemical conditions with relatively large phase or property changes are Smart polymers. Chemical and physical properties result in different names, e.g., 'stimuli-responsive' polymers, 'smart' polymers, 'environmental-sensitive' polymers, or 'intelligent' polymers. The stimulus can be physical (ionic strength, temperature, radiation), chemical (specific ions, pH, chemical agents), or biochemical (enzyme, substrates, ligands). It can also be internal, external, single, or associated with two or more stimuli [179]. Stimuli-chromism is a phenomenon in smart materials related to color change by an external stimulant, such as light, pH, temperature, mechanical stress, polarity, electrical potential, and metal ion coordination [180].

Spiropyran, spirooxazine, diarylethene, oxazine, and oxazolidine are some of the most important photoswitches that show different stimuli-chromisms depending on the induced stimuli, such as photochromism, mechanochromism, acidochromism, ionochromism, thermochromism, piezochromism, hydrochromism, electrochromism, and solvatochromism [181, 182].

6.4.1. Spiro compounds as photochromic materials

Compared with the other stimulants, light is a clean, fast, and efficient stimulus with the ability to induce from outside the responsive system. Moreover, light can control smart behavior spatially, temporally, quickly, and safely [180].

The spiropyran (**SP**) class was the first prominent family of light-responsive colorants whose photochromism was investigated systematically. The literature contains reports of an enormous number of photochromic **SP**-derived substances. They range from 'small molecule' dyes to high molecular weight polymers and biomolecules. The general structure of **SP** dye is indicated in Figure 6-66. SP is isomerized from the spiro(non-polar) to merocyanine (high polar, zwitterionic) isomer (**MC**) upon UV irradiation, and reisomerization to the initial state could be obtained under visible light irradiation.

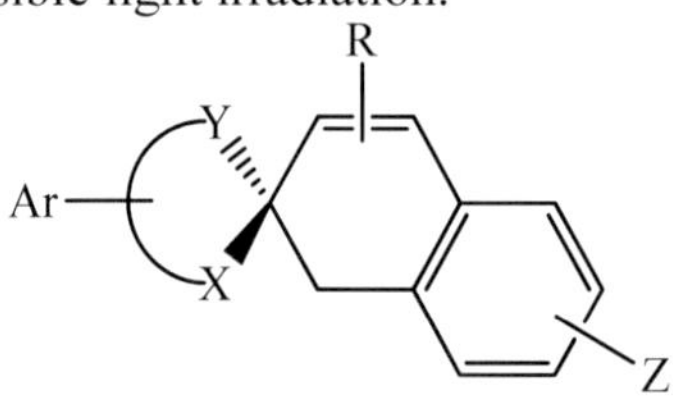

X = N, O, S, etc.
Y = $C(Me)_2$, S, etc.
Ar = H, Benzo, Naphtho, etc.
R = H, Alkyl, Aryl
Z = H, OR, Halogen, NO_2, Benzo, etc.

Fig. 6-66. The general structure of **SP** dye.

The **SP** dye class is known for displaying photochromism, specifically T-type photochromism, in which the dye changes color in one direction upon exposure to light and in the opposite direction upon exposure to thermal energy. The color change is typically positive, with the dye transitioning from a colorless or weakly colored state to a highly colored state upon exposure to light. This process is reversible, with the dye fading back to its original state in the absence of light. The rate of thermal reversion slows with decreasing temperature [183, 184].

SPs have been used extensively in the development of anti-counterfeiting materials because they display both photochromism and fluorescence emission upon UV irradiation [181]. Roghani-Mamaqani et al. developed high-security photochromic and fluorescent anti-counter-feiting inks using SP-modified latex nanoparticles [185]. The functionalized latex nanoparticles were prepared using semi-continuous emulsifier-free emulsion polymerization and then modified with hydroxylated spiropyran (SPOH) and carboxylated spiropyran (SPCOOH) chromophores using physical and chemical methods. The nanoparticles had various morphologies depending on the concentration and type of functional groups.

The researchers applied epoxy-functionalized latex nanoparticles containing SP to prepare high-security photochromic and fluorescent anti-counterfeiting inks for printing on different cellulosic security documents. The printed marks and fingerprints on the documents exhibited photochromism and red fluorescence emission upon UV irradiation with high intensity and brightness.

The same group also fabricated flexible and invisible high-security anti-counterfeiting inks by chemically incorporating SP into copolymer latex nanoparticles [186]. The inks had various polarities and glass transition temperatures and displayed photochromism and fluorescence emission upon UV irradiation. The nanoparticles had a suitable coating and stability on the surface of the cellulosic substrate, and the photochromic and fluorescence properties were a function of the polarity and flexibility of the polymer chains. The invisible high-security ink was loaded onto stamps with different marks and applied to print-marking other security documents.

In a third study, the researchers presented a novel strategy for encrypting cellulosic papers and optically authenticating confidential documents using ecofriendly multicolor photoluminescent inks. These inks incorporated coumarin (**CUOH**) and SPs (**SPOH** and **NSPOH**) into functional latex nanoparticles and exhibited ultra-bright fluorescence emissions with different colors and intensities depending on the polarity of the polymer matrix and its functional groups (Figure 6-67) [187]. The nanoparticles had various sizes and morphologies, and their stability and printability on cellulosic substrates were verified. The photoluminescent latex nanoparticles were loaded onto stamps with different patterns for security marking cellulosic papers, and the printed marks revealed ultra-bright fluorescence emissions under UV light, which could act as an optical sensor for detecting originality and counterfeiting in security documents.

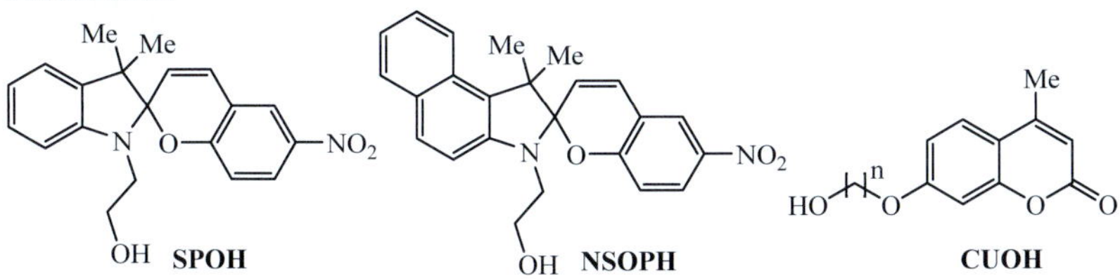

Fig. 6-67. Structure of coumarin (**CUOH**) and spiropyrans (**SPOH** and **NSPOH**).

One of the primary aims of nanotechnology is to create three-dimensional structures at the nanoscale. Modern fabrication systems, such as scanning electron beam lithography, have enabled the production of features with nanometer resolution. Optical interference lithography is an attractive technique for rapidly and inexpensively patterning three-dimensional features in polymer photoresists. Ullal et al. developed a superresolution lithography system that combines photochromic switching with thiol-Michael addition chemistry (Figure 6-68) [188]. Small molecule studies using ethyl maleimide and spirothiopyran (**STP**) showed that they have the necessary properties for superresolution lithography. A functional photoresist was prepared by copolymerizing a spirothiopyran-functionalized methacrylate monomer with a PEGylated methacrylate monomer. This copolymer was crosslinked with a multifunctional maleimide under UV irradiation, and the green light was demonstrated to inhibit cross-linking with excellent spatial and temporal control. Photokinetic simulations for 1D interference lithography using a 2 W 532 nm laser show that the spirothiopyran-maleimide system can produce 45 nm lines over an area several hundred microns in size.

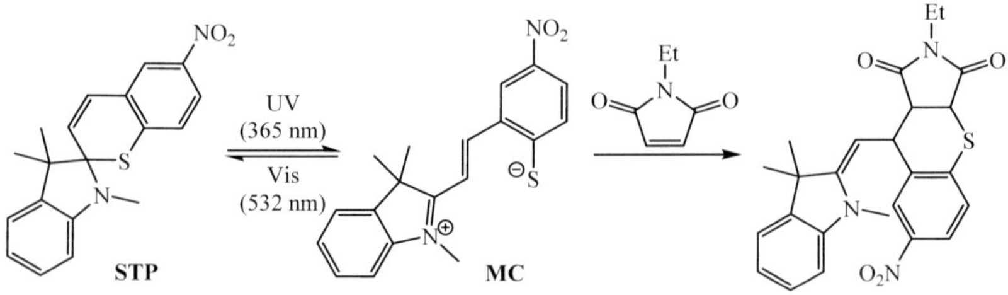

Fig. 6-68. The super-resolution lithography writing scheme using spirothiopyran.

In addition, a German-Australian group introduced a one-component photoresist that contains a copolymer of methyl methacrylate and **STP**-functionalized methacrylate with acetophenone as a suitable solvent (Figure 6-69) [189]. Upon two-photon absorption at 820 nm, the **STP** moieties transit into their open **MC** forms, which allowes the interaction between **SP** and **MC** moieties from the same and neighboring polymer chains, resulting in efficient physical entanglement of the copolymer chains. Using this photoswitchable system, it was possible to fabricate freestanding 3D structures by direct laser writing (DLW). They presented a reversible depletion effect that can be applied in stimulated-emission depletion (STED)-inspired DLW, employing a 640 nm cw depletion laser. By utilizing a half-space phase mask for the depletion beam, the linewidth could be decreased by almost a factor of 2, down to 31 nm, compared to that of diffraction-limited DLW in the same resist. Line gratings also showed the capability of the mechanism to support the subdiffraction resolution.

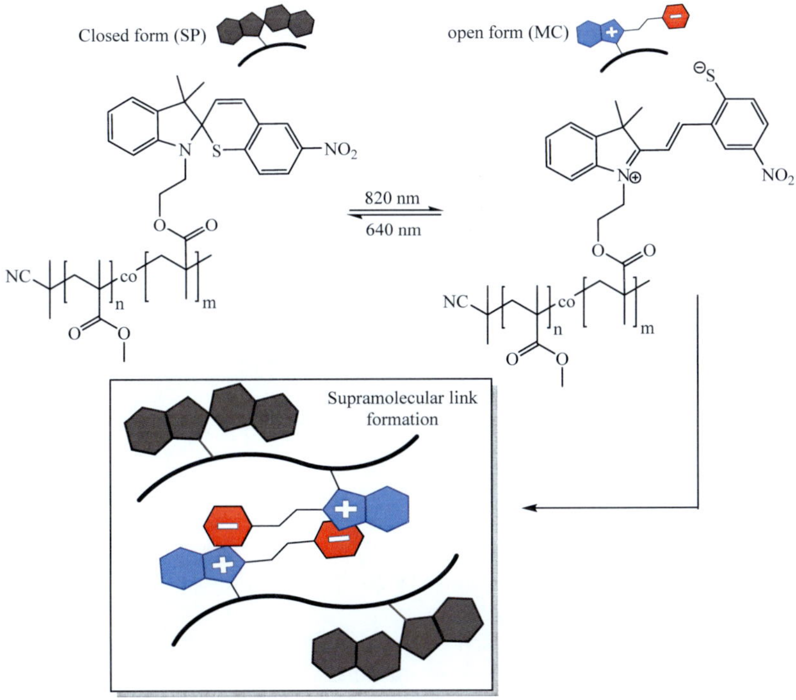

Fig. 6-69. The spirothiopyran-bearing copolymer in the closed-form (**SP**) can transit into its open merocyanine form (**MC**) by two-photon absorption at 820 nm. The reverse photoisomerization (**MC** to **SP**) occurs upon irradiation at 640 nm. The **MC** form can develop supramolecular links with other **SP** or **MC** moieties.

6.4.2. Spiro compounds as electrochromic materials

Electrochromism refers to the reversible change in optical properties that can occur when a material is electrochemically oxidized or reduced, and it has gained significant attention in both academic and commercial fields. Typically, electrochromic materials change between a transparent state and a colored state, which allows for the darkening of windows with just the flick of a switch. This principle has already been applied in electrochromic car rearview mirrors, which automatically darken at night when a driver is dazzled by the headlights of a car behind [190].

One noteworthy example of an electrochromic material is poly-(spiroBiPrDOT) **253**, which was introduced by Reynolds et al. in 2002 (Scheme 6-75). This polymer displays three color states and a luminance change of 30% at intermediate potentials. Since the change in luminance takes place around the $E_{1/2}$ of the polymer, this can result in electrochromic devices with improved stability and longer lifetimes. Luminance investigations showed that the polymer was initially cathodically coloring, indicating a dark red film in the neutral state, and subsequently anodically coloring, indicating a dark blue color in the oxidized form [191].

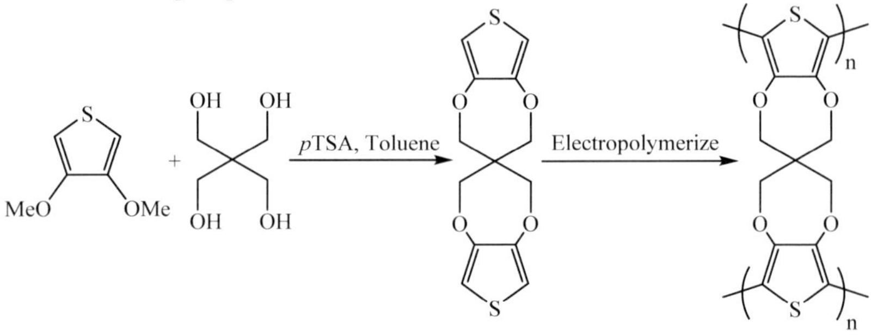

253

Scheme 6-75. Synthesis of poly(spiroBiPrDOT) **253**.

Fluorescent switches based on spirocyclic zwitterionic Meisenheimer (**SZMC**) complexes are stimuli-responsive organic molecules with applications in various areas. To develop their functionality, Guirado et al. presented novel switching mechanisms for these systems (Figure 6-70): (a) acid and redox-triggered formation of an additional protonation state with distinct optical properties, and (b) solvent-induced fluorescence modulation. These new features, which allow multi-stimuli and multistate operation of **SZMC** switches, can be applied in the construction of smart organic materials: wide-range pH optical probes, electrochromic and electrofluorochromic films, and polymer-based fluorescent detectors of organic liquids [192].

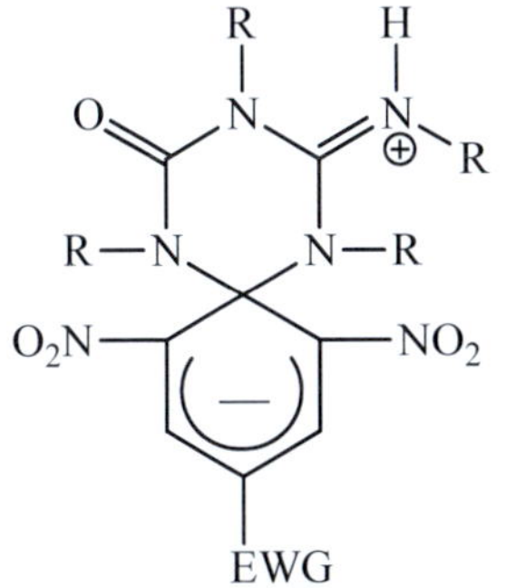

Fig. 6-70. SZMC complexes.

Usluer et al. synthesized (2,7-bis(carbazol-9-yl)-9,9′-spiro[cyclo dodecane-1,9′-fluorene]) (**SFC**) through Ullmann coupling and confirmed its structure using FT-IR, ^{1}H, and ^{13}C NMR, and MALDI-ToF [193]. The synthesis of the **SFC** monomer is shown schematically in Scheme 6-76. The **SFC** monomer was then electrochemically polymerized onto a transparent ITO/glass surface to produce a very stable **poly-SFC** with a high contrast ratio (DT% = 58% at 800 nm). Spectroelectrochemical studies showed that the **poly-SFC** film can generate different colors because it has multiple electroactive moieties scanned cathodically between 0 and 2.0 V. An electrochromic devices (ECD) was also assembled from **poly-SFC** in a sandwich cell with an ITO/**poly-SFC**//gel electrolyte//PEDOT/ITO configuration. Both the oxidation and reduction response time of this device were found to be about 1 s. The ECD sandwich cell exhibits reversible electrochemical oxidation and strong color changes between transparent and dark blue upon applying a potential between –1.0 and +2.0 V, with a contrast ratio of 36%. The optical activity of the device is retained by 96.4% after 1000 cycles. Therefore, **poly-SFC** is considered a stable electrochromic material that maintains its redox behavior and stability even at high scan rates and oxidation levels compared to other electrochromic polymers. Based on the observed results, particularly the transparency values at neutral and oxidized states, **poly-SFC** appears to be a useful electrochromic material for the fabrication of smart windows.

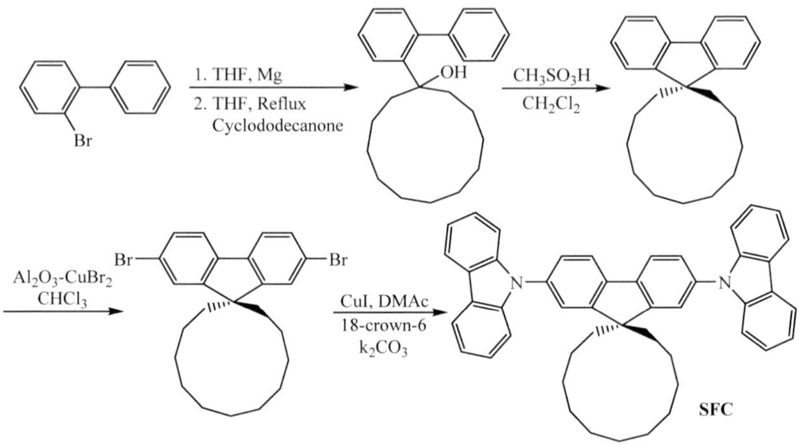

Scheme 6-76. Schematic representation of the synthesis of **SFC** monomer.

6.4.3. Spiro compounds as piezoelectric materials

In recent decades, there has been a growing interest in piezochromic materials (PCMs) due to their potential in various applications and the need for understanding their mechanism of action and molecular design. These stimuli-responsive materials exhibit significant changes in emission color upon grinding, pressing, or stretching, and are ideal for pressure sensors, optical recording, and memory applications. However, there are only a few reports on PCMs that exhibit photoluminescence (PL) spectra covering the deep red/near-deep-red to infrared (IR) regions with clear color differences [194, 195].

The first report on PCMs was by Wilson et al. in 1975, where they evaluated the effects of pressure on the chromism of PS and PMMA films containing SP derivatives (**6-N-BIPS** and **BMN-BIPS**) (Figure 6-71). They observed pressure-induced coloration independent of time, rate of pressure application, or irradiation in the pressure region of 30-140 kbars. The evidence supports a direct thermal transition between the uncolored and colored ground states. However, in the pressure range of 8-40 kbars, they found that pressure significantly decreased the efficiency of the photochromic process. This process occurs through a triplet state of the uncolored molecule, and with increasing pressure, the energy of a triplet state increases to the excited singlet state, decreasing the probability of intersystem crossing and, therefore, the formation of the colored species. Isomerization of the SP to MC was observed for pressures above 30 kbars, but not below 10 kbars [196].

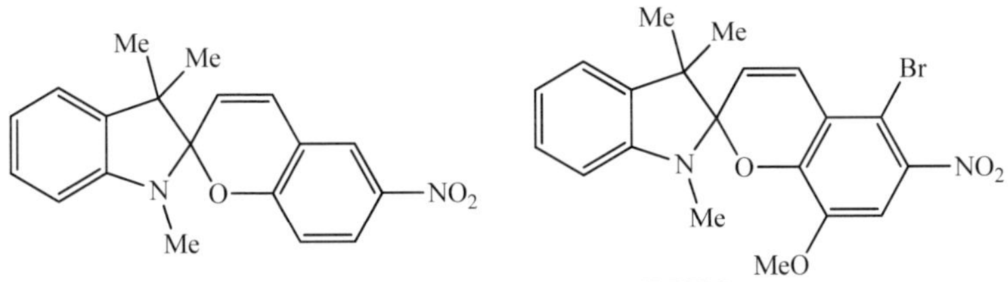

6-N-BIPS **BMN-BIPS**

Fig. 6-71. Structure of the SPs used by Wilson and Drickamer.

In 2004, Favaro et al. represented that spirooxazine is photochromic and piezochromic in the pure microcrystalline solid phase and indicated acidichromism when kept in contact with a hydrogen-donating solid phase [197]. The photoproduced merocyanine (**MC**) thermally reconverts to the original spiro isomer (Scheme 6-77). Piezochromism was detected using a mechanical force on the solid powder, using either a hydraulic press or manual crushing. After discontinuing pressure, biexponential bleaching kinetics were observed. **MC** micro-spot formation inside the crystalline phase explains the biexponential nature of the fading processes.

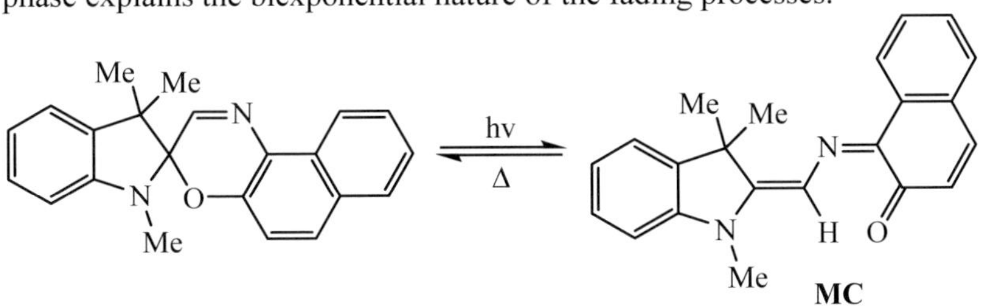

Scheme 6-77. Reconvert **MC** into the original spiro isomer.

Funasako et al. developed multi-stimuli-responsive polymer films that incorporate cationic SP into a Nafion membrane (Figure 6-72). The resulting film displays thermochromism under ambient light, meaning it exhibits reversible color changes only with visible light and heat, without the need for UV light irradiation. Additionally, the film shows photochromism and significant piezochromism. These characteristic behaviors arise from the stabilization of the zwitterionic merocyanine cation in the polar environment of Nafion. The film is advantageous because it can be easily made from Nafion membranes through cation exchange and is flexible, robust, and transparent [198].

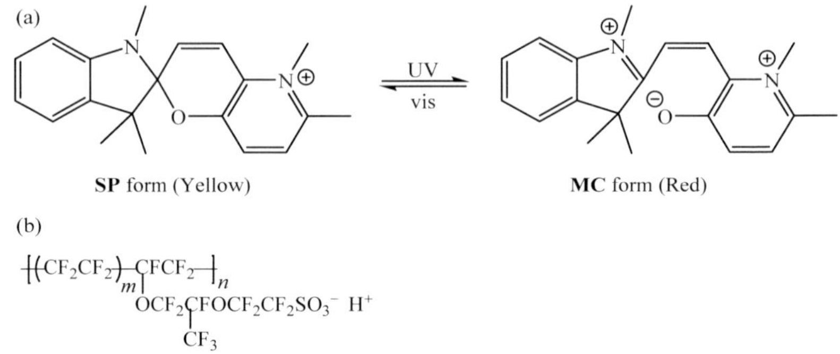

Fig. 6-72. (a) Photoisomerization of **SP** cation (b) The structural formula of Nafion.

Moreover, the Ma group synthesized a novel multi-colored switching mechanochromic materials based on an SP and pyrene (**SP–Pyr**) that responded to anisotropic grinding and isotropic compression (Figure 6-73). When applying 15 GPa pressure on **SP–Pyr**, the emission exhibited a continuous red-shift with irreversibility mainly due to the formation of the isomers caused by the broken C-O bond under high pressure. This molecule demonstrates excellent potential for applications in pressure sensors, pressure calibration, mechanoprobes, and security inks [199].

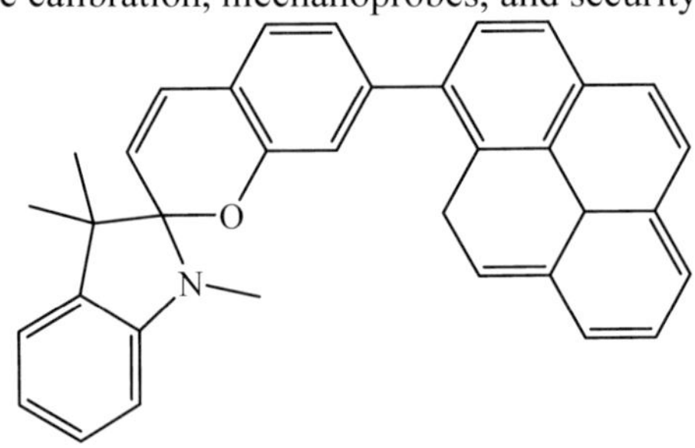

Fig. 6-73. Structure of **SP-Pyr**.

Two luminescent materials comprised of tetraphenylethene (**TPE**) plus spirobifluorene **(TPE)$_2$PF** and **TPE(SF)$_2$** were successfully prepared by a Chinese group (Figure 6-74). **(TPE)$_2$PF** represents significant reversible piezochromic luminescence which indicates blue light (445 nm) in the crystalline state and green light (503 nm) in the amorphous state. Results illustrate that these adducts combine the merits of high solid-state emission efficiency, stimuli-responsive luminescence, and efficient electrolumine-scence, which enables them to find an array of high-tech applications in mechano-sensors, optoelectronic devices, etc. [200].

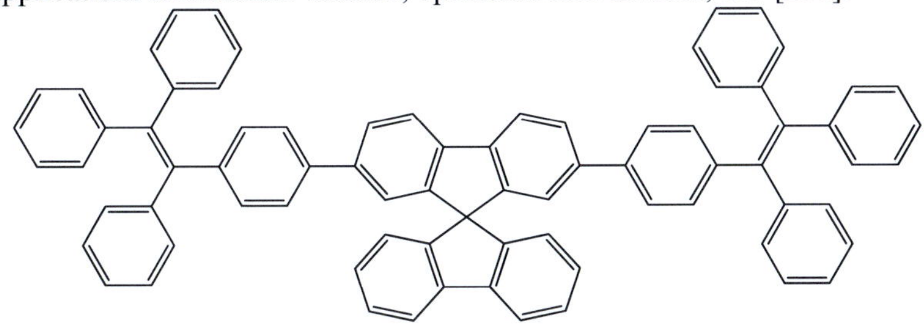

$(TPE)_2PF$

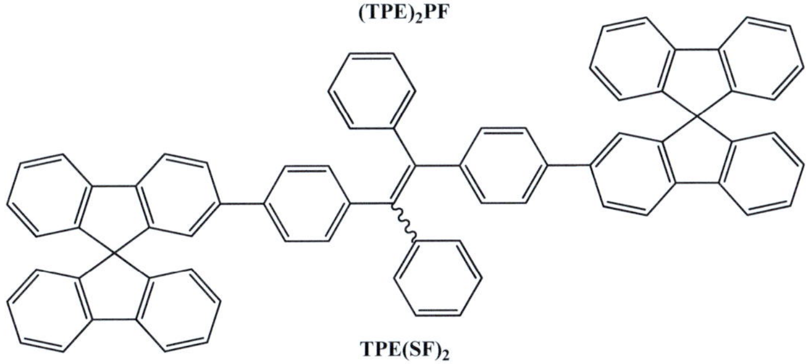

$TPE(SF)_2$

Fig. 6-74. Tetraphenylethene (**TPE**) plus spirobifluorene **$(TPE)_2PF$** and **$TPE(SF)_2$**.

6.5. Application of spiro compounds in the synthesis of organic compounds

Spiro compounds have been applied in the synthesis of diverse types of heterocyclic and carbocyclic compounds and are assumed to be a valuable building block in organic synthesis. This section aims to review the advances in using spiro compounds as ligands or catalysts in synthesizing various organic compounds.

6.5.1. Spiro compounds as organocatalysts

In the last decades, chiral organocatalysts with a spiro skeleton as an environmentally friendly methodology for asymmetric synthesis because of their unique structural properties and high asymmetric induction efficiency [201]. In this chapter, some remarkable examples of reactions performed by using chiral spiro-type organocatalysts will be discussed.

In 2011, Sasai et al. introduced a unique spiro-type Brønsted acid-Lewis base organocatalyst (*S*)-**254** for the enantioselective aza-Morita-Baylis–Hillman (aza-MBH) reaction [202]. The new bifunctional spiro organocatalyst improved aza-MBH reactions to yield adducts **255** with up to 98% ee, as shown in Scheme 6-78.

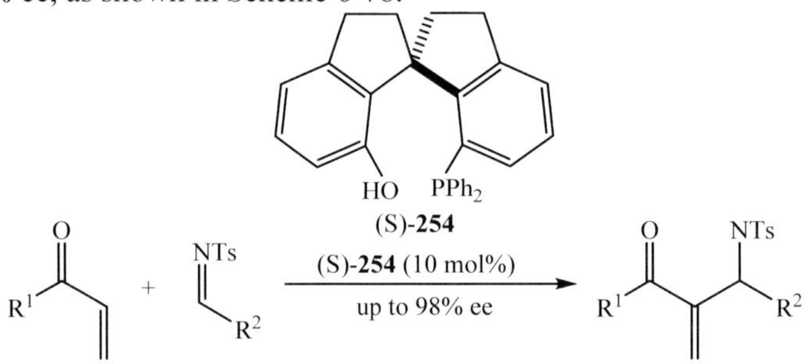

255

Scheme 6-78. Spiro-type acid-based organocatalyst (*S*)-**254** for the aza-MBH reaction.

They also reported the enantioselective organocatalytic synthesis of tetrahydropyridines **256a-l** bearing a chiral tetra-substituted carbon stereogenic center via [4+2] cycloaddition of saccharin-derived ketimines and α-methyl allenoate (Scheme 6-79). The spiro-type monoaryl phosphine catalyst, (*R*)-**SITCP**, promoted cycloaddition reactions to provide the corresponding six-membered *N*-heterocycles in high yields and excellent regioselectivities with up to 93% ee [203].

The reaction starts with the addition of the phosphine catalyst to the α-methyl allenoate, followed by the kinetically favored γ-addition of the monoaryl spiro-type phosphine to the ketimine and intramolecular cyclization, as indicated in Scheme 6-80. To avoid steric interactions between the R substituent of cyclic ketimines and the phenyl substituent in the catalyst, the reaction with (*R*)-**SITCP** would favor the formation the (*R*)-configured products **256a–i** or the (*S*)-configuration products **256j** and **256k**.

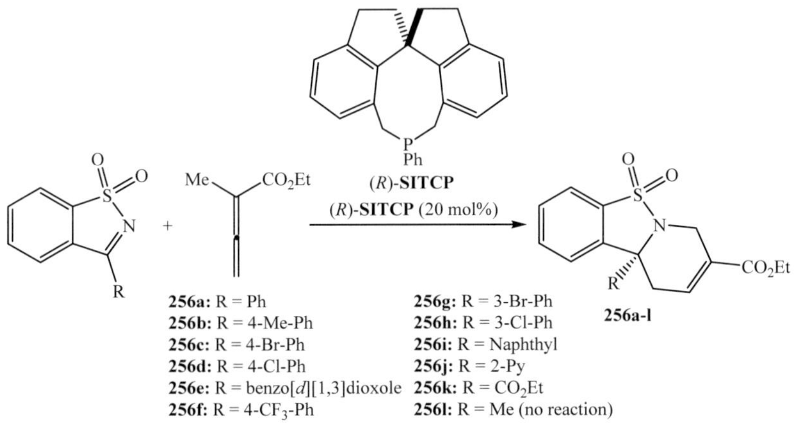

Scheme 6-79. Formal [4+2] cycloaddition of cyclic ketimine with α-methyl allenoate.

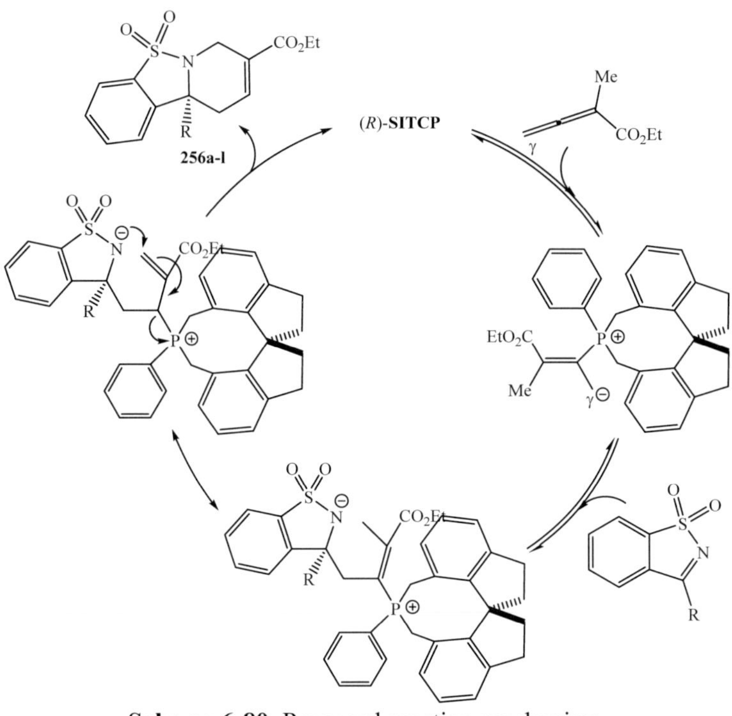

Scheme 6-80. Proposed reaction mechanism.

Another exciting example reported by Shi et al. consists of (*R*)-**SITCP** catalyzed highly regio-, diastereo- and enantioselective [3+2] cycloaddition of alkylidene azlactones with various allenic esters (Scheme 6-81). The corresponding functionalized spirocyclic products **257** were obtained in moderate to excellent yields under mild conditions. Authors suggested that the azlactone may attack the zwitterionic intermediate from the *re*-face to form the products due to the steric interaction [204].

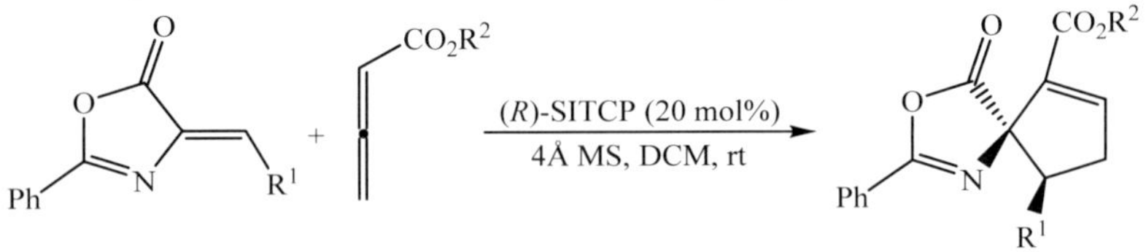

Scheme 6-81. (*R*)-**SITCP** catalyzed [3+2] cycloaddition.

SPINOL-derived chiral Brønsted acid catalyst was applied by You et al. in an asymmetric synthesis of α-tetra-substituted (3-indolyl)(diaryl) methanamines **258** via aza-Friedel-Crafts reaction of cyclic α-diaryl *N*-acyl imines with indoles (Scheme 6-82). The desired products were obtained in excellent yields and enantioselectivities (up to 98% yield, up to >99:1 e.r.) [205].

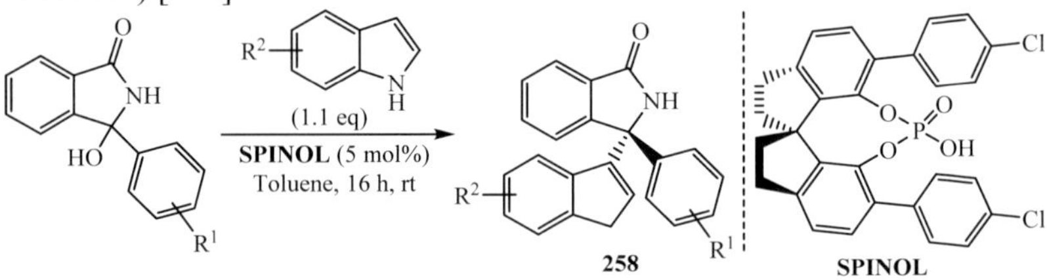

Scheme 6-82. Synthesis of α-tetra-substituted (3-indolyl)(diaryl)methanamines **258**.

In addition, Chung et al. reported that spiro phosphepine **(*S*)-259** catalyzes the synthesis of tetrahydrofurans **260**, tetrahydropyrans **261**, and dihydrobenzopyrans **262** with high efficiency (Scheme 6-83) [206].

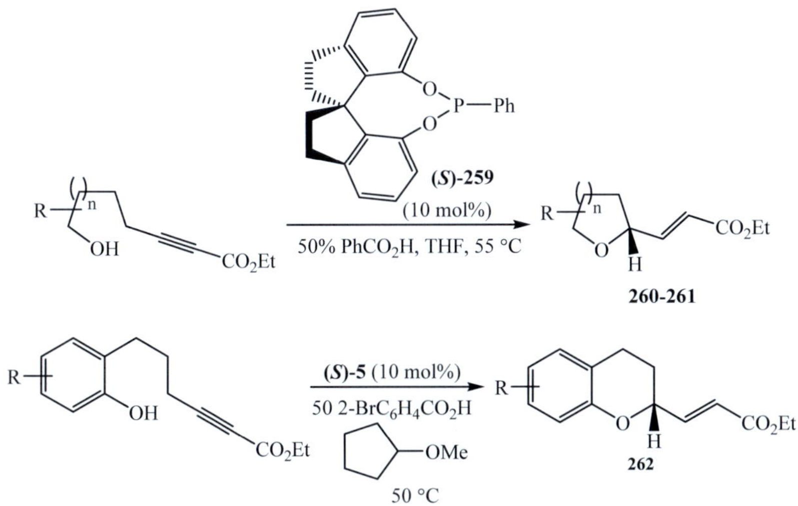

Scheme 6-83. Catalytic enantioselective synthesis of compounds **260-262**.

Tang et al. presented a catalytic asymmetric version of the intramolecular ylide annulation that gives high ee values and diastereoselectivities [207]. In this process, the spirobiindane-based chiral phosphine **(*S*)-DMM-SITCP** was used as the organocatalyst which exhibited the excellent activity. Both optically active benzobicyclo[4.3.0] compounds **263** and **263′** with three contiguous stereogenic centers could be achieved as major products selectively under neutral and mild conditions just by choice of an additive (Scheme 6-84).

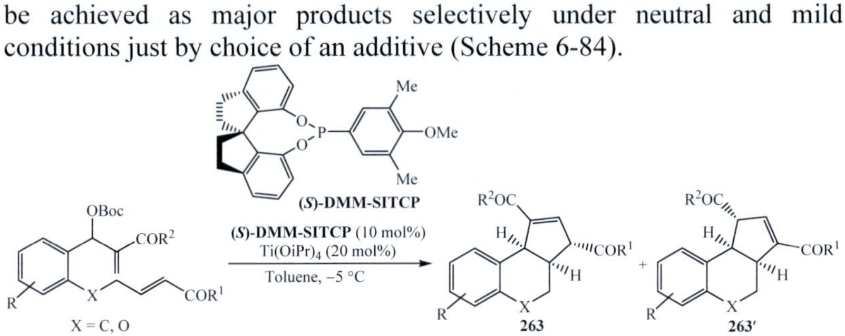

Scheme 6-84. Synthesis of benzobicyclo[4.3.0] compounds **263** and **263′**.

Zhou et al. reported the enantioselective synthesis of pharmaceutically essential spirobarbiturate-cyclohexenes **264** using asymmetric [4+2] annulation of barbiturate-derived alkenes with allenoates (Scheme 6-85) [208]. This reaction is catalyzed by a spirocyclic chiral phosphine **(*R*)-**

DTBM-SITCP and proceeds with high yields and high diastereo- and enantioselectivities.

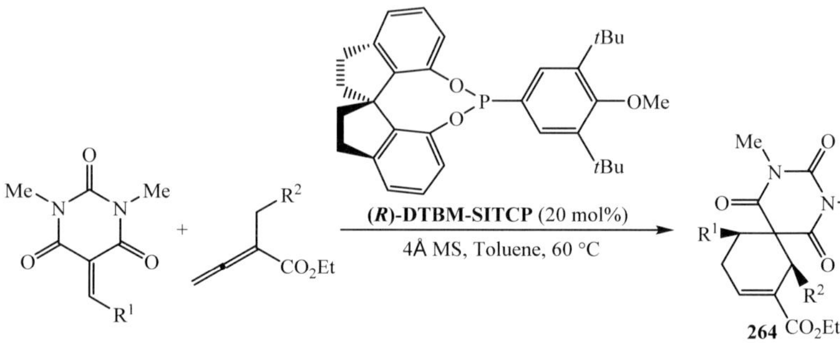

Scheme 6.85. Enantioselective synthesis of spirobarbiturate-cyclohexenes **264**.

6.5.2. Spiro compounds as chiral ligand

Developing enantiomerically pure chiral ligands is a significant challenge in the development of asymmetric catalysts. Chiral spiranes with axial chirality are of particular interest due to their rigid chiral environments with C_2 symmetry by the perpendicular arrangement of two planes [209]. Since the discovery of the first spiro compound in 1900, a significant number of exciting spiro compounds have been reported as chiral ligands. In this section, we will extensively discuss these compounds.

In 2001, Sasai et al. reported the first catalytic asymmetric Wacker-type cyclization for rapid access to dihydropyrans **265a-c** [210]. They used a catalyst system based on spiro bis(isoxazoline) (*M*,*S*,*S*)-H-SPRIX and $Pd(OOCCF_3)_2$ to promote the asymmetric Wacker-type cyclization of alkenyl alcohols in the presence of *p*-benzoquinone in CH_2Cl_2, giving the 6-*endo* cyclized product **265a-c** in high yield (Scheme 6-86).

Dupont et al. synthesized (±)-1,6-bisoximo[4.4]nonane **267** by the reaction of hydroxylamine hydrochloride and (±)-spiro[4.4]-1,6-nonane-dione **266** that can act as a helical chiral bridging ligand for the preparation of bimetallic compounds (Scheme 6-87) [211].

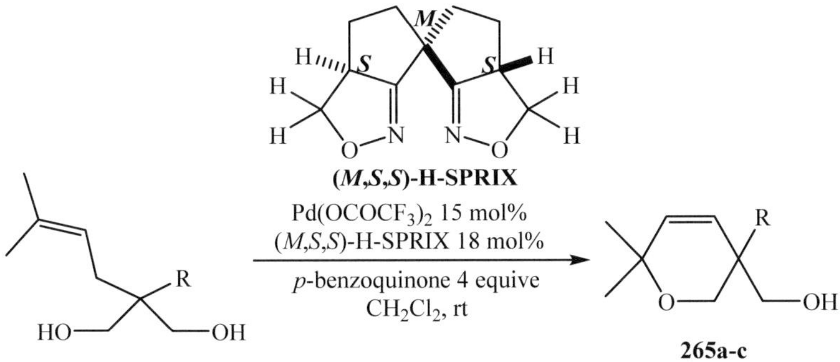

Scheme 6-86. Catalytic asymmetric Wacker-type cyclization.

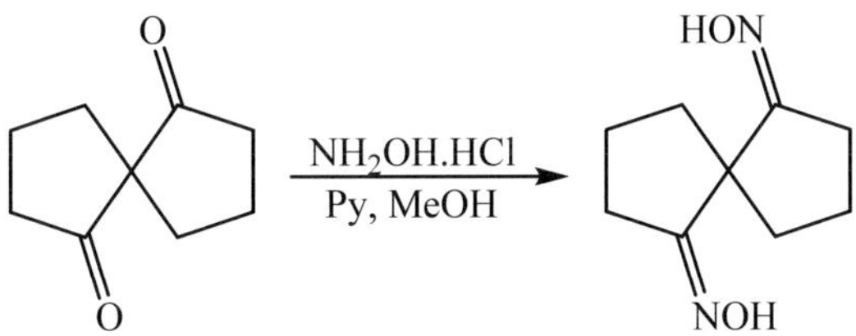

(±)-**266** (±)-**267**

Scheme 6-87. Synthesis of (±)-1,6-bisoximo[4.4]nonane **267**.

Zhou et al. also reported the synthesis of monodentate chiral spiro phosphoramidite ligands **268a-d** and **269** in good yields from enantiomerically pure 1,1′-spirobiindane-7,7′-diol (Figure 6-75). The synthesized ligands were highly efficient in the Cu-catalyzed conjugate addition of Et_2Zn to enones with up to 98% ee [212].

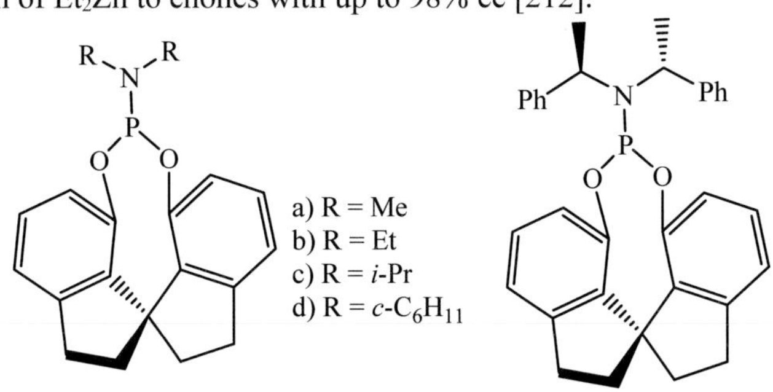

(*R*)-**268** (*R*,*R*,*R*)-**269**

Fig. 6-75. Structure of ligands used in Cu-catalyzed conjugate addition of Et_2Zn to enones.

Likewise, bis(oxazoline) ligand **271** bearing a spiro skeleton, as a new class of bidentate ligand, is developed via double ring-closing metathesis (RCM) of tetraene **270** and subsequent oxazoline ring formation using *N*-bromosuccinimide, as shown in Scheme 6-88. The Cu-spiro bis(oxazoline) complexes prepared from **271** and $Cu(OTf)_2$ or $Cu(OAc)_2$ act efficiently as chiral catalysts for promoting carbonyl-ene reactions or Henry reactions with good enantiocontrol [213].

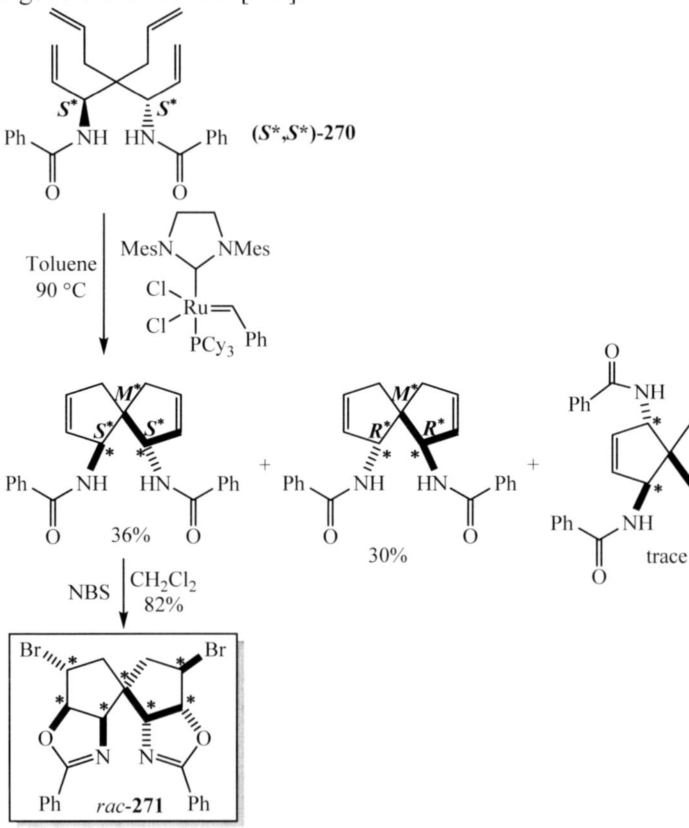

Scheme 6-88. Synthesis of spiro bis(oxazoline) **271**.

Zhou et al. presented a highly efficient Rh-catalyzed asymmetric addition of arylboronic acids to aldehydes using chiral spiro phosphite ligands (*S*)-**272** that gives diarylmethanols **273** in excellent yields and good enantiomeric excesses (Scheme 6-89).

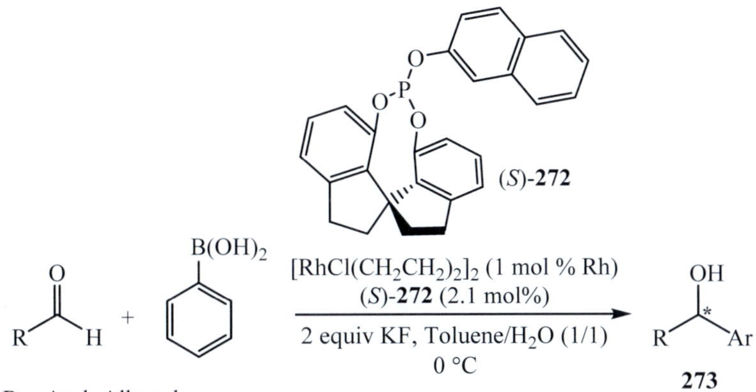

R = Aryl, Alkenyl

Scheme 6-89. Highly efficient Rh-catalyzed asymmetric addition of arylboronic acids to aldehydes.

A stereorecognition model was suggested in Figure 6-76 to rationalize the products' absolute configuration in adding arylboronic acids to aldehydes. It can be found from the model that the transfer of the phenyl group to the *re*-face of aldehyde is much more favorable [214].

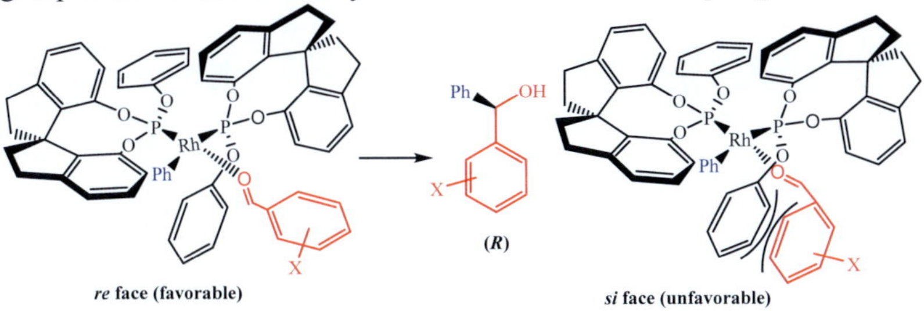

Fig. 6-76. Stereorecognition model for the Rh-catalyzed asymmetric addition of arylboronic acids to aldehydes.

Sasai et al. synthesized the novel chiral hybrid ligands **274** and **274′** containing an unsymmetrical spiro[4.5]decane skeleton with isoxazole and isoxazoline as the two coordinating units. The Pd complexes of these ligands were found to be effective in activating olefins towards enantioselective tandem cyclization of a dialkenyl alcohol, affording the cyclized products **275-277** in up to 97% ee (Schemes 6-90 and 6.91) [215]

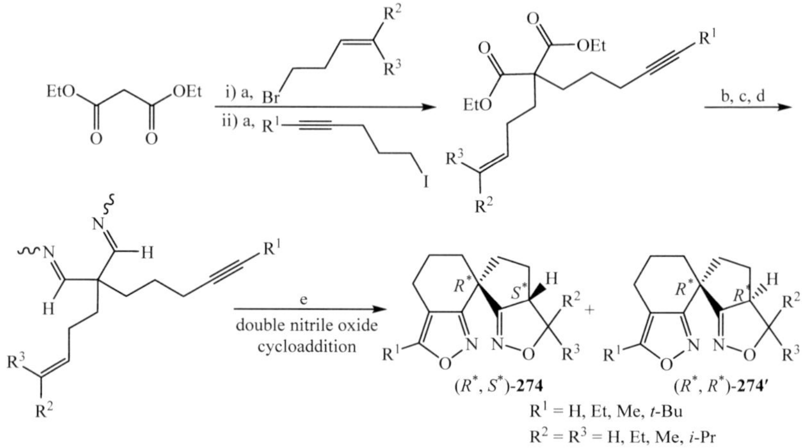

Scheme 6-90. Synthesis of spiro (isoxazole–isoxazoline) ligands **274** and **274′**. Reagents and conditions: (a) NaH (1 equiv), THF, 0 °C then rt; (b) $LiAlH_4$, THF, 0 °C then rt; (c) Swern oxidation. $(COCl)_2$, CH_2Cl_2, DMSO, 78 °C, Et_3N; (d) $NH_2OH.HCl$, Py, 0 °C then rt; (e) aq NaOCl, CH_2Cl_2, 0 °C then rt.

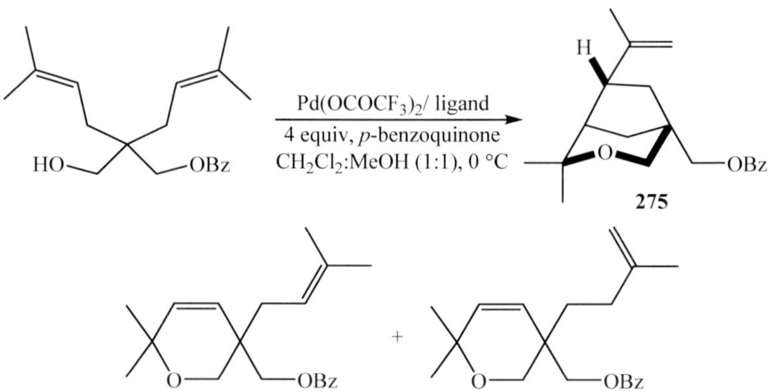

276 **277**

Scheme 6-91. Pd-Catalyzed tandem cyclization of dialkenyl alcohol.

The Zhou group [216] reported the first highly enantioselective catalytic to insert *R*-diazoesters into the N-H bonds in the presence of the copper complex of chiral spiro bisoxazoline ligand **278** as catalysts, affording *R*-amino acids **279** in high yields with excellent enantioselectivities (up to 98% ee) (Scheme 6-92). Non-coordinating ion $BARF^-$ improved the enantioselectivity of the reaction.

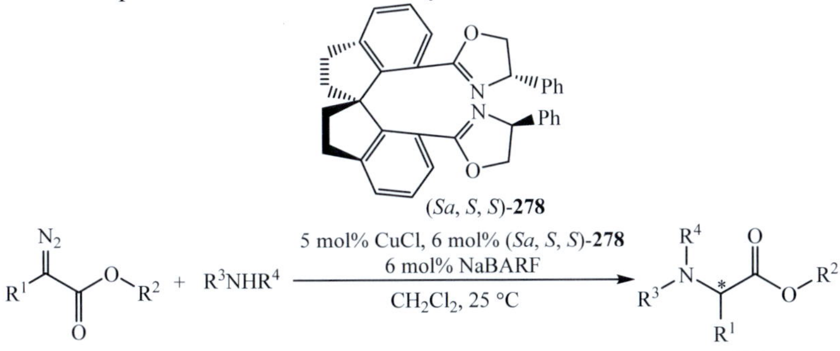

Scheme 6-92. Catalytic asymmetric N-H insertion of amines with diazoesters.

Zhou et al. reported the synthesis of chiral spiro-diphosphine ligands **SDP** with a 1,1′-spirobiindane backbone and **SFDP** with a 9,9′-spirobifluorene backbone, as well as spiro monophosphorus ligands, including phosphoramidites, phosphites, phosphonites, and phospholane, with a 1,1′-spirobiindane backbone, in good yields from enantiomerically pure 1,1′-spirobiindane-7,7′-diol and 9,9′-spirobifluoren-1,1′-diol (Figure 6-77). They discovered that the Ru complexes of chiral spiro diphosphine ligands are highly effective catalysts for asymmetric hydrogenations of ketones, α-arylaldehydes, and α,β-unsaturated acids. Furthermore, the Rh complexes of chiral spiro monophosphorus ligands are highly enantio-selective in the asymmetric hydrogenation of α- and β-dehydroamino acids, α-arylethenyl acetamides, and non-*N*-acyl enamines. The spiro monophosphorus ligands are also highly efficient in the Rh-catalyzed asymmetric addition of arylboronic acids to aldehydes and *N*-tosylarylimines, Pd-catalyzed asymmetric allylation of aldehydes with allylic alcohols, Cu-catalyzed asymmetric ring-opening reactions with Grignard reagents, and Ni-catalyzed asymmetric hydrovinylation of styrene derivatives with ethylene [217].

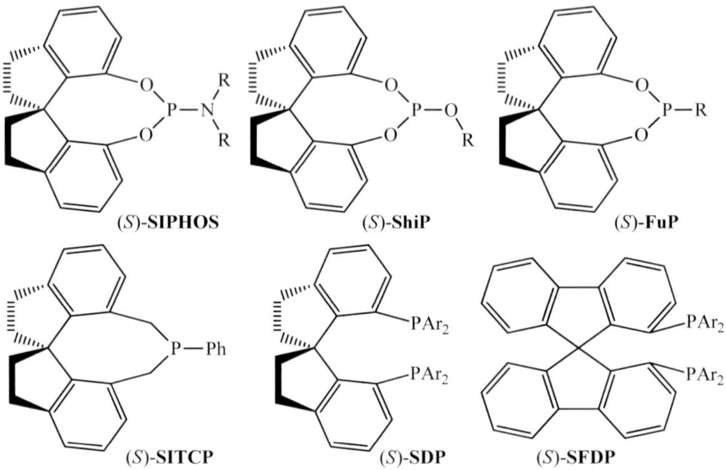

Fig. 6-77. Chiral phosphorus ligands with spiro backbones.

Ma et al. prepared a new chiral spiro-bisoxazoline ligand, i.e., β-naphthylmethyl-substituted spiro-BOX [(*Ra*,*S*,*S*)-**280**]. They successfully used it in the Pd-catalyzed enantioselective cyclization reaction of simple allenes with *o*-aminoiodobenzenes, which led to highly optically active 3-alkylideneindolines **281** in good yields with excellent enantiomeric excesses (Scheme 6-93) [218].

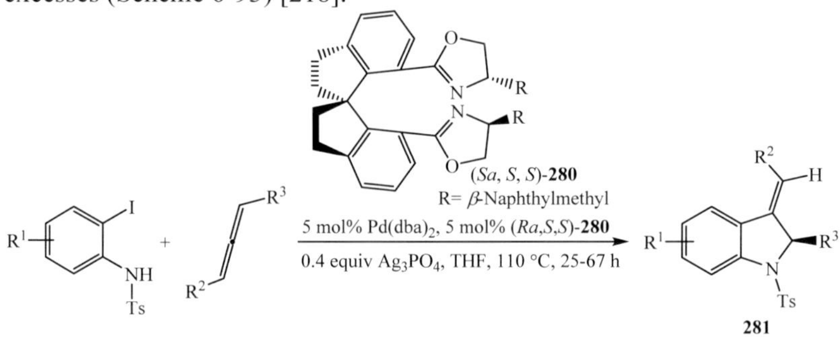

Scheme 6-93. Pd-Catalyzed cyclization reaction of allenes with *o*-amino iodobenzenes.

In 2009, a novel optically active 2-(oxazolinyl)pyridine ligand (Spymox) (*R*)-**282** containing a spiro binaphthyl backbone (Scheme 6-94, Eq. 1) was also synthesized and successfully applied in Pd-catalyzed asymmetric allylic alkylations to provide the corresponding alkylated products **283** with 99% ee (Scheme 6-94, Eq. 2) [219].

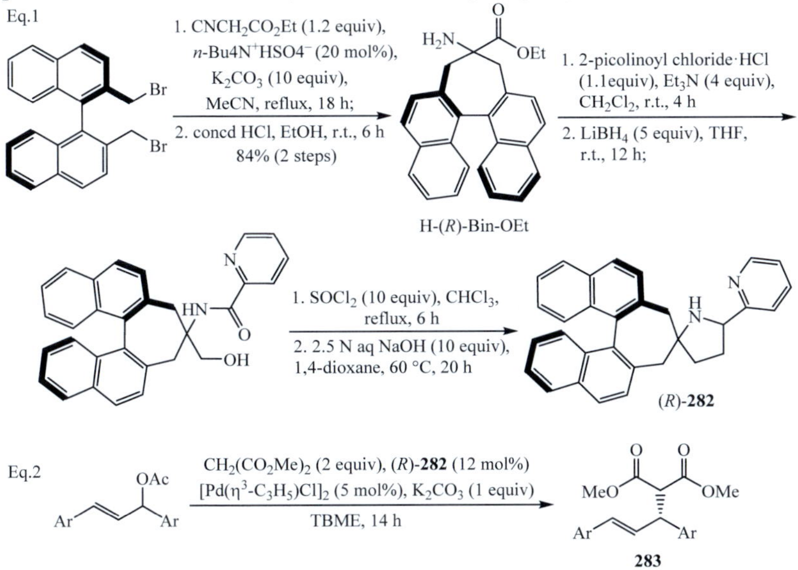

Scheme 6-94. Synthesis of active 2-(oxazolinyl)pyridine ligand (*R*)-**282** and its application.

The Zhou group suggested in 2010 that the Ir/chiral spiro phosphino-oxazoline (Ir-SIPHOX) cationic complexes **284a-f** (Figure 6-78) are highly efficient catalysts for the enantioselective hydrogenation of α-aryloxy- and α-alkoxy-substituted α,β-unsaturated carboxylic acids, providing a variety of optically active α-oxyfunctionalized carboxylic acids in extremely high enantioselectivities under mild reaction conditions.

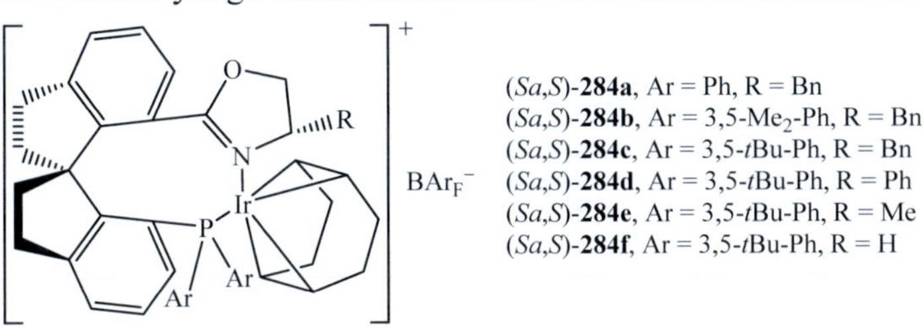

Fig. 6-78. Ir-SIPHOX cationic complexes **284a-f**.

Among them, the hydrogenation of α-benzyloxy-substituted α-unsaturated acids afforded an appropriate methodology for synthesizing chiral α-hydroxy acids. Based on the coordination model of the unsaturated acids with the Ir metal center, a mechanism involving a catalytic cycle between Ir(I) and Ir(III) was proposed (Scheme 6-95) [220].

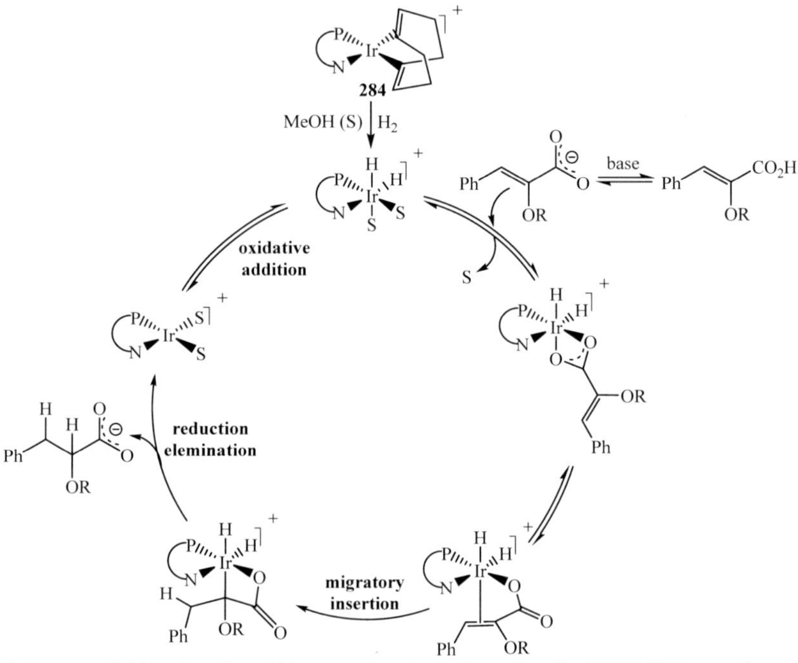

Scheme 6-95. A plausible mechanism for the Ir-SIPHOXs-catalyzed hydrogenation of α,β-unsaturated acids.

Moreover, α-chloro-α-fluoro-β-keto **286** phosphonate and α-chloro-α-fluoro-β-keto **287** ester compounds prepared by highly enantioselective *gem*-chlorofluorination of active methylene compounds using a Cu(II) complex of a chiral spiro pyridyl monooxazoline ligand (SPYMOX) **(*R*)-285**. The resulting dihalo-β-keto esters **287** were converted into various α-fluoro-α-heteroatom-substituted carbonyl compounds **288-290** through nucleophilic substitution without loss of optical purity. A fully protected β-amino acid **291** with a *gem*-chlorofluoromethylene function was also prepared (Scheme 6-96) [221].

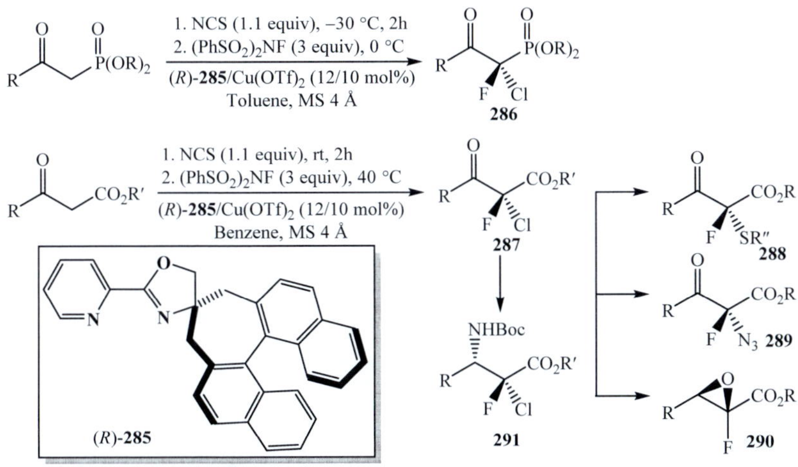

Scheme 6-96. (*R*)-**285**/Cu(OTf)$_2$-Catalyzed *gem*-chlorofluorination of active methylene compounds.

Likewise, the Cu(II)/SPYMOX complex also catalyzed the highly enantioselective fluorination of α-alkyl-β-keto esters (Scheme 6-97) [222]. The fluorination proceeded in a highly enantioselective manner when cyclic and acyclic substrates were used for the reaction.

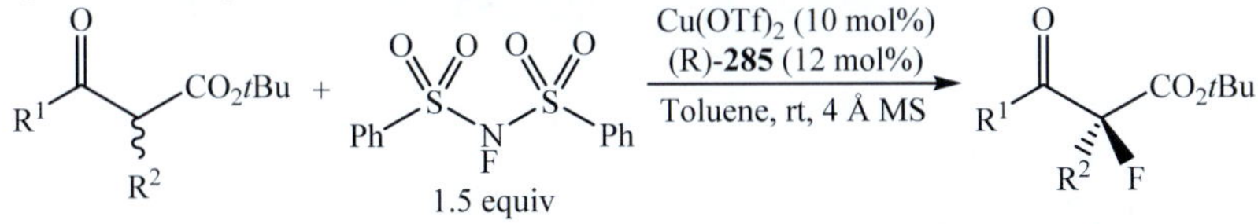

Scheme 6-97. Enantioselective α-fluorination of β-keto esters.

In 2012, Ding et al. reported the highly efficient enantioselective chlorination of cyclic β-keto esters with *N*-chlorosuccinimide (NCS) catalyzed by chiral Cu(II) and Zn(II) complexes of a spiro-2,2′-bischromane based bisoxazoline (SPANbox) ligands (*S*,*R*,*S*)-**292a-h** and (*S*,*S*,*S*)-**292a-g** (Scheme 6-98) [223]. The corresponding α-chloro-β-keto esters were obtained in excellent yields in 5–30 min with ee values up to 97%. The Cu(II) triflate and Zn(II) triflate complexes of a single SPANbox ligand exhibited complementary results in terms of the enantioselection, providing both antipodes of the chlorinated product enantiomers with good to excellent optical purities.

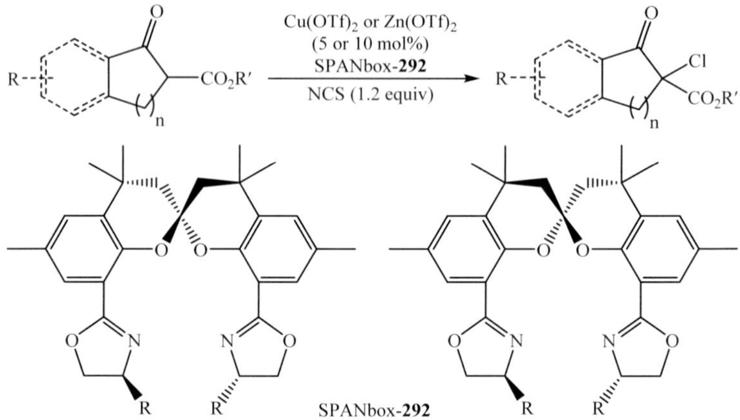

(*S*,*R*,*S*)-**292a-h** (*S*,*S*,*S*)-**292a-g**

R = Ph, (*S*,*R*,*S*)-**292a** and (*S*,*S*,*S*)-**292a**
R = *i*-Pr, (*S*,*R*,*S*)-**292b** and (*S*,*S*,*S*)-**292b**
R = *s*-Bu, (*S*,*R*,*S*)-**292c** and (*S*,*S*,*S*)-**292c**
R = *t*-Bu, (*S*,*R*,*S*)-**292d** and (*S*,*S*,*S*)-**292d**
R = Bn, (*S*,*R*,*S*)-**292e** and (*S*,*S*,*S*)-**292e**
R = 1-Np, (*S*,*R*,*S*)-**292f** and (*S*,*S*,*S*)-**292f**
R = 2-Np, (*S*,*R*,*S*)-**292g** and (*S*,*S*,*S*)-**292g**
R = In, (*S*,*R*,*S*)-**292h**

Scheme 6-98. $Cu(OTf)_2$- or $Zn(OTf)_2$/SPANbox-**292** catalyzed asymmetric chlorination of β-oxo esters.

Xie et al. developed a highly efficient asymmetric hydrogenation of β-aryl β-ketoesters catalyzed by chiral Ir catalysts bearing spiro pyridine-aminophosphine (SpiroPAP) ligands **(*R*)-293a-e** (Scheme 6-99). The reaction allows a readily accessible method for the synthesis of β-hydroxy esters in excellent enantioselectivity (up to 99.8% ee) and extremely high turnover numbers (up to 1230000) that render the highest level of efficiency in the asymmetric hydrogenation of β-ketoesters [224].

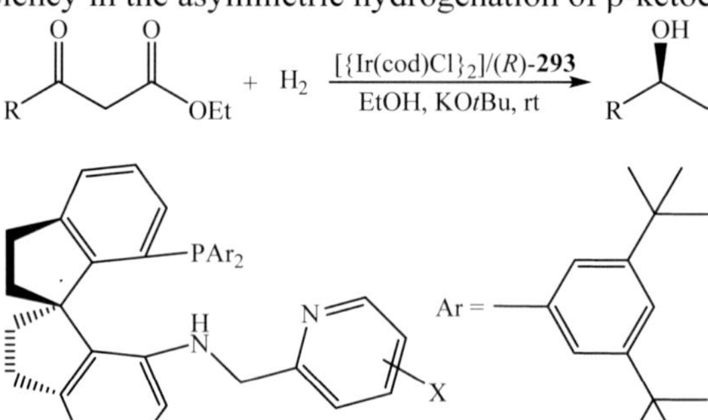

(*R*)-**293**

(*R*)-**293a**, X = H
(*R*)-**293b**, X = 3-Me
(*R*)-**293c**, X = 4-*t*Bu
(*R*)-**293d**, X = 6-Me
(*R*)-**293e**, X = 6-Et

Scheme 6-99. Asymmetric hydrogenation of β-ketoesters with Ir/Spiro-PAP complexes.

Later, they showed that the introduction of a 1,3-dithiane moiety into the chiral tridentate spiro P-N-S ligand (SpiroSAP) ***(R)*-294a-k** led to a highly efficient chiral Ir catalyst for asymmetric hydrogenation of β-alkyl-β-ketoesters, affording chiral β-alkyl-β-hydroxyesters with excellent enantioselectivities (95–99.9% ee) and turnover numbers of up to 355000 [225]. Scheme 6-100 shows the synthetic route to prepare SpiroSAP ligands (*R*)-**294a-k**.

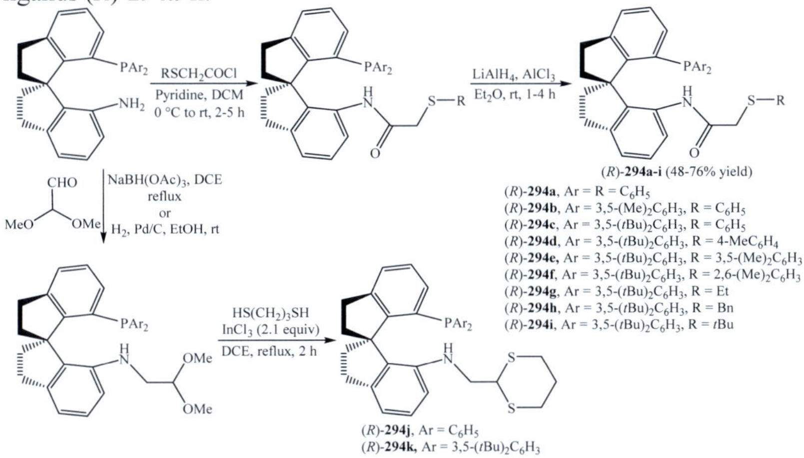

Scheme 6-100. The synthesis of SpiroSAP ligands (*R*)-**294a-k**.

A new type of tridentate chiral spiro aminophosphine-oxazoline ligands (SpiroOAP) **295a-e** was also synthesized by the same group, which are highly efficient for the asymmetric hydrogenation of α-keto amides, giving a variety of synthetically helpful α-hydroxy amides with excellent enantioselectivity (up to 98% ee) and turnover numbers (up to 10,000) (Scheme 6-101) [226].

As demonstrated in Scheme 6-102, Ziegler et al. synthesized four spiro-fused D-*fructo*- and D-*psico*-configurated oxazoline ligands **296-299** via a straightforward synthetic route. The key steps in their synthesis were a Ritter-type condensation reaction of partially benzyl-protected D-*fructose* and D-*psicose* derivatives with thiocyanic acid to furnish the corresponding 1,3-oxazolidine-2-thiones, and the cross-coupling of the latter under modified Suzuki–Miyaura conditions.

The synthesized ligands **296-299** exhibited active pre-catalysts for the asymmetric allylic alkylation of 1,3-diphenylallyl acetate with dimethyl malonate (Scheme 6-103). The D-*fructo*-configurated ligands **296** and **298** presented the (*R*)-enantiomer with up to 93% ee, whereas the D-*psico*-configurated ligands **297** and **299** provided the (*S*)-enantiomer in a somewhat lower selectivity (up to 59% ee). The stereochemical outcome of the reaction could be explained by a proposed transition state of the allylic substitution [227].

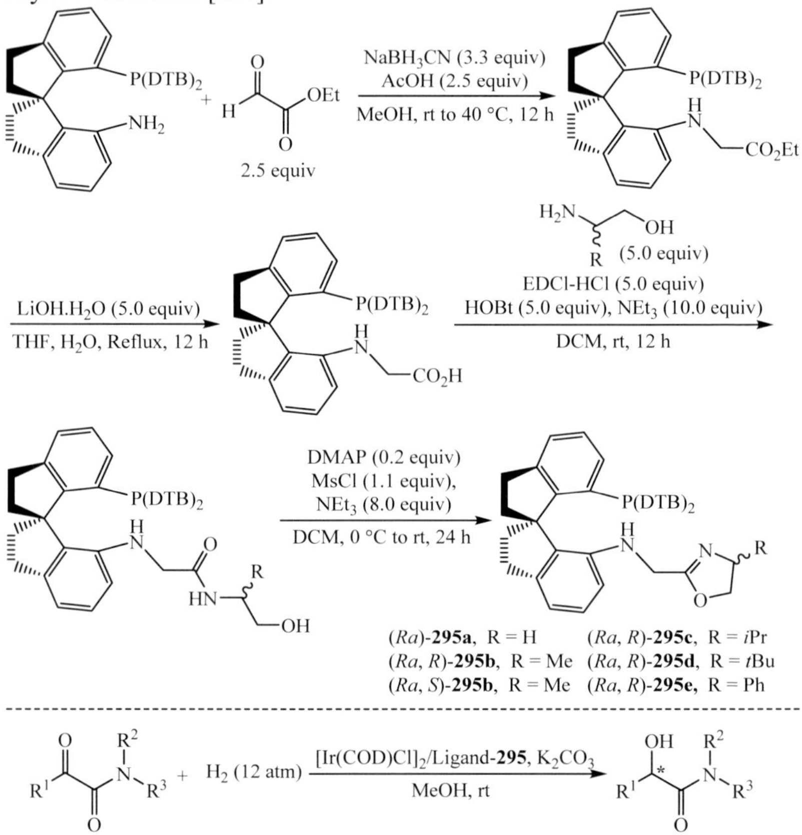

Scheme 6-101. The synthesis of SpiroOAP ligands (*R*)-**295a-e**.

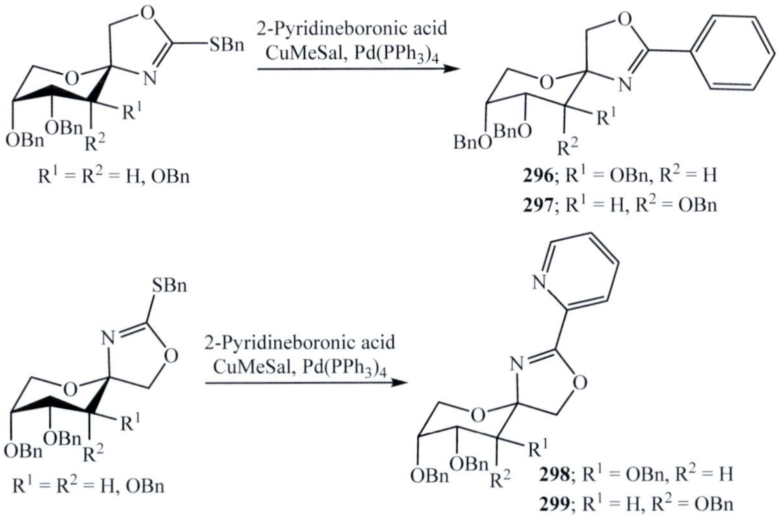

Scheme 6-102. Synthesis of spiro-fused D-*fructo*- and D-*psico*-configurated oxazoline ligands **296-299**.

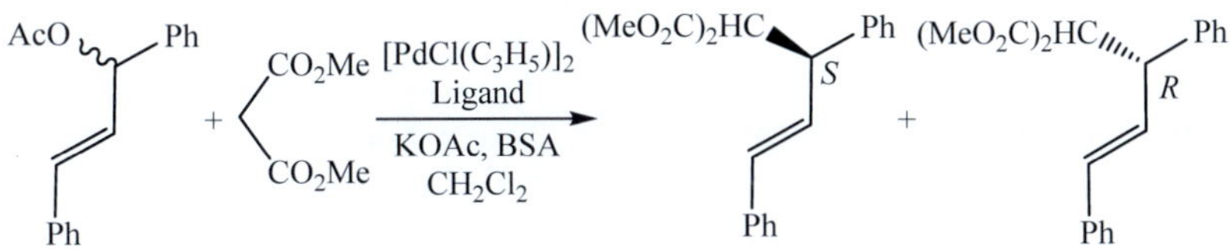

Scheme 6-103. Pd-Catalyzed allylic substitution.

Recently, the Zhou group reported a highly efficient synthesis of chiral sulfides **301** by the chiral spiro phosphoramide **300** which catalyze the sulfa-Michael addition/enantioselective protonation of exocyclic enones (Scheme 6-104). The precise chiral control in the protonation of the enol intermediate ensured high enantioselectivity. The reaction features high activity (yields up to 99%, turnover numbers up to 8400) and high enantioselectivity (up to 97% ee) with a broad substrate scope [228].

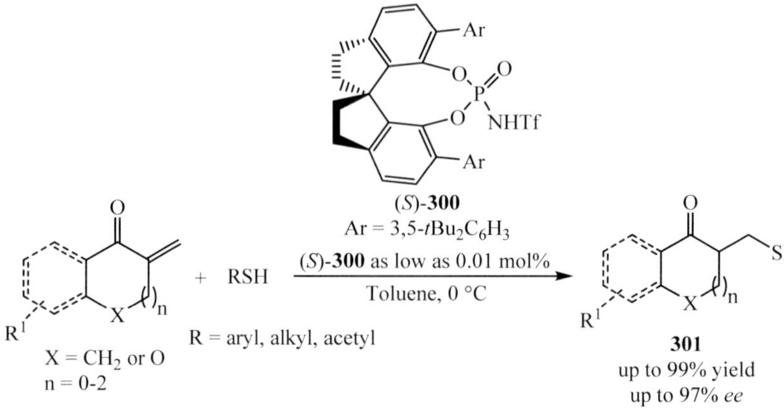

Scheme 6-104. Highly efficient synthesis of chiral sulfides **301**.

The discovery of appropriate ligands to simultaneously modulate the reactivity and effectively control the stereoselectivity is a central topic in the field of enantioselective C-H activation. Hence, Lan et al. developed the synthesis of axially chiral biaryls by Pd-catalyzed atroposelective C-H olefination. A novel chiral spiro phosphoric acid, STRIP (*R*)-**302**, was presented as a superior ligand for this transformation (Scheme 6-105). A broad range of axially chiral quinolines **303** was achieved in good yields with excellent enantioselectivities (up to 98% ee) [229].

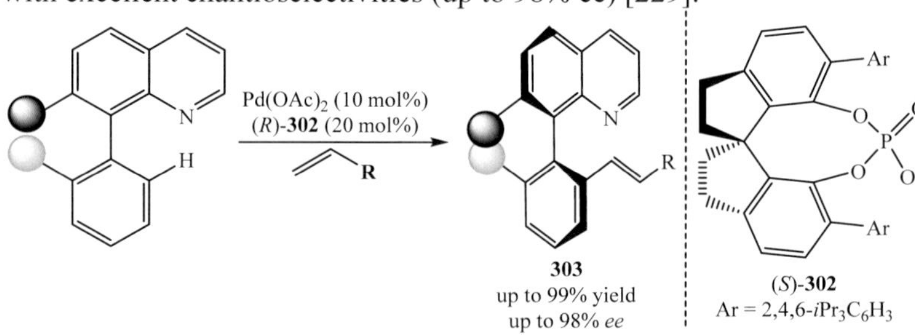

Scheme 6-105. Enantioselective synthesis of chiral quinolines **303**.

6.5.3. Spiro compounds as chiral PTC

Phase-transfer catalysis (PTC) was first reported by Starks et al. in 1971 as a technique that utilizes tetraalkylammonium or phosphonium salts to catalyze reactions between two substances in different immiscible phases. PTC has since become a popular synthetic tool in both laboratory and industrial processes due to its mild reaction conditions, simple experimental procedures, use of safe and inexpensive reagents and solvents, and the possibility of using organocatalysts (catalysts without metals) and air/moisture tolerant organocatalysts. Chiral nonracemic onium salts and crown ethers are commonly used as effective PTCs, especially for enantioselective carbon-carbon or carbon-heteroatom bond-forming reactions under mild biphasic conditions [230-232].

In 1999, the Maruoka group developed a new class of C_2-symmetric PTCs based on a structurally rigid, quaternary spiro-ammonium salt **304a-c** derived from commercially available chiral binaphthol. These chiral binaphthol-based PTCs, known as Maruoka catalysts, have been shown to be highly efficient in the large-scale preparation of biologically active molecules and have become an important tool in asymmetric synthesis of α-amino acids (Scheme 6-106) [233].

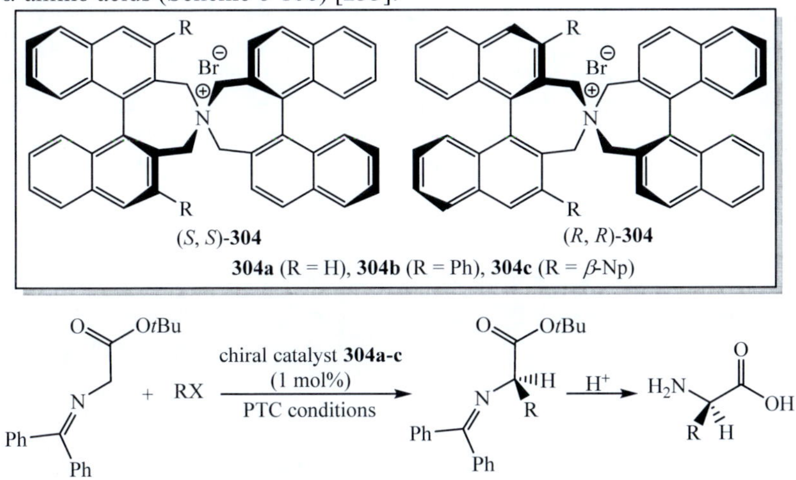

Scheme 6-106. Chemical structures of Maruoka catalysts **304** and their application in the asymmetric synthesis of α-amino acids.

The Maruoka group performed the highly enantioselective phase-transfer catalyst for alkylation and Michael reaction of β-keto esters by using a chiral quaternary ammonium salt (*S*,*S*)-**304d** as the catalyst which leads to the efficient construction of fully substituted stereogenic centers (Scheme 6-107) [234].

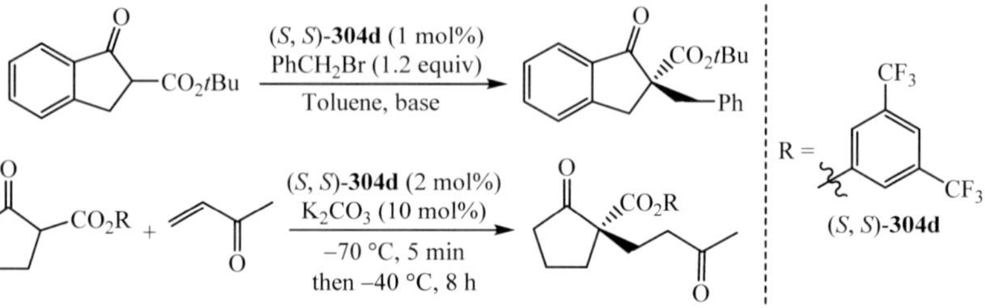

Scheme 6-107. Enantioselective phase-transfer catalytic alkylation and Michael reaction of β-keto esters.

They also illustrated that the PTC of *N*-spiro C_2-symmetric chiral quaternary ammonium bromide (*S*,*S*)-**304e** enabled the efficient, highly stereoselective conjugate addition of nitroalkanes to alkylidenemalonates under mild conditions, providing a variety of optically active γ-amino acids in excellent yields with high enantioselectivities (Scheme 6-108) [235].

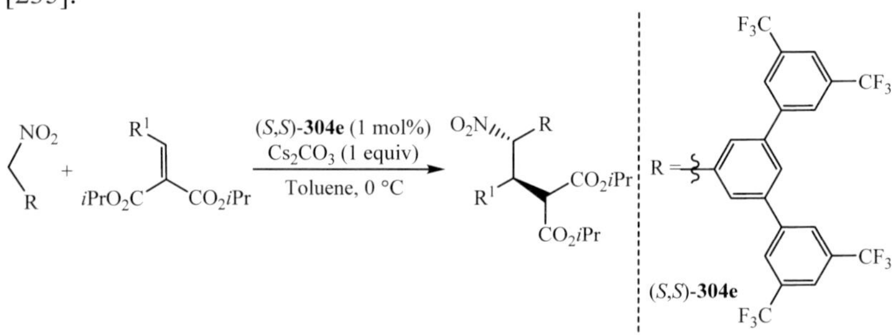

Scheme 6-108. Conjugate the addition of nitroalkanes to alkylidene-malonates.

Furthermore, Maruoka et al. prepared a new *N*-spiro chiral quaternary ammonium bromide (*R*,*R*)-**305a,b** with a 4,6-di-substituted chiral biphenyl framework in 2003 that catalyzed phase-transfer alkylation of protected glycine with high enantioselectivity and broad generality (Scheme 6-109) [236].

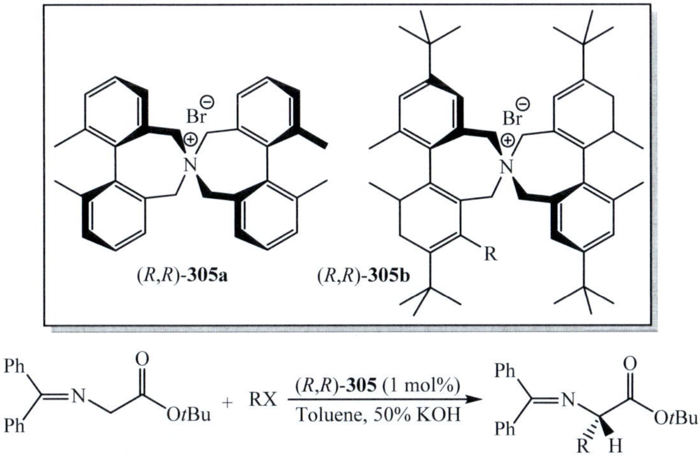

Scheme 6-109. Enantioselective phase-transfer catalytic alkylation of protected glycine derivative.

They also found that highly enantioselective Michael addition of diethyl malonate to chalcones was achieved under mild phase-transfer conditions by the successful employment of *N*-spiro C_2-symmetric chiral quaternary ammonium bromide **306** as a catalyst (Scheme 6-110). It possesses diarylhydroxymethyl functionalities as a recognition site for the prochiral electrophile [237].

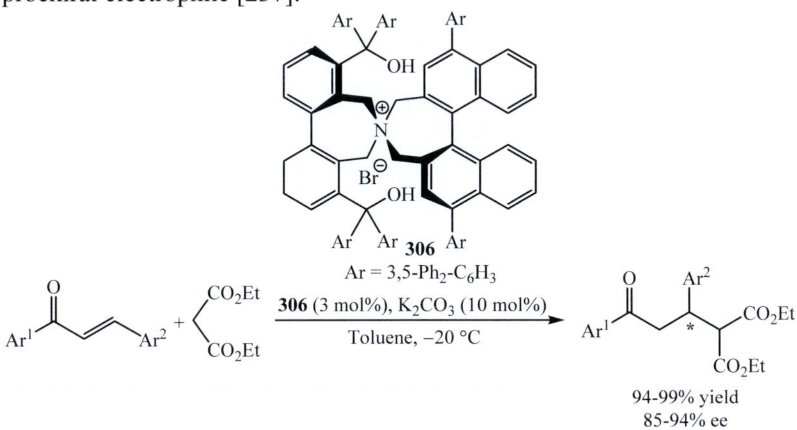

Scheme 6-110. Enantioselective phase-transfer catalytic Michael addition of diethyl malonate to chalcones.

In addition, they succeeded in designing and preparing a new, chiral PTC of type (*S*)-**307** to realize a general and valuable procedure for the asymmetric conjugate addition of various *t*-butyl α-alkyl-α-cyanoacetates to *t*-butyl propiolate (Scheme 6-111) [238].

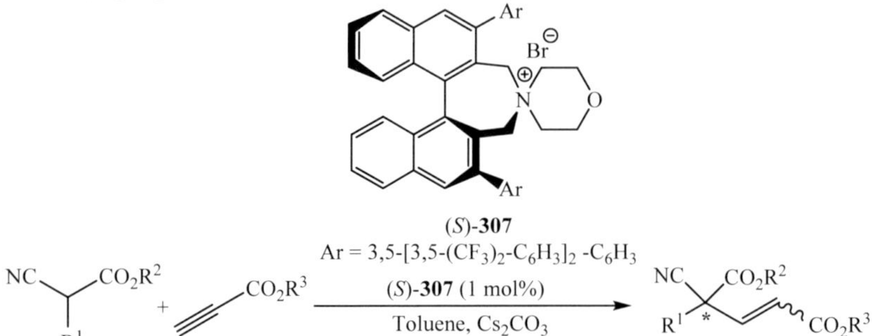

Scheme 6-111. Asymmetric conjugate addition of various *t*-butyl α-alkyl-α-cyanoacetates to *t*-butyl propiolate.

To develop new organocatalysts, Nagasawa et al. reported novel the C_2-symmetric chiral pentacyclic guanidine compounds **308a-d** as hetero-Michael reaction catalysts (Scheme 6-112). The guanidine group, a superbase, generates stabilized complex salts with anionic compounds through parallel interactions including hydrogen bonding. The structures of these catalysts **308a-d** are based on the parent skeleton of the marine natural product ptilomycalin A and related compounds. They were designed to have a C_2-symmetric chiral reaction cavity around the substrate recognition/activation site. These catalysts **308a-d** effectively catalyzed the hetero-Michael reaction depending on their cavity size. Catalysts **308a**, **308b**, and **308d** have a closed-type cavity, whereas **308c** has an open-type cavity. Thus, if the reaction proceeds at the substrate-activation site, the guanidine moiety, these catalysts lead to asymmetric induction through steric hindrance at this cavity. Optimization revealed that the closed-type cavity of **308a** is important for the high asymmetric induction of the alkylated product (*R*)-**309a-i**. Moreover, the methyl groups on the spiro ether rings of **309a** play a significant role as sterically hindering groups to control the electrophile approach, resulting in a very high level of asymmetric induction [239].

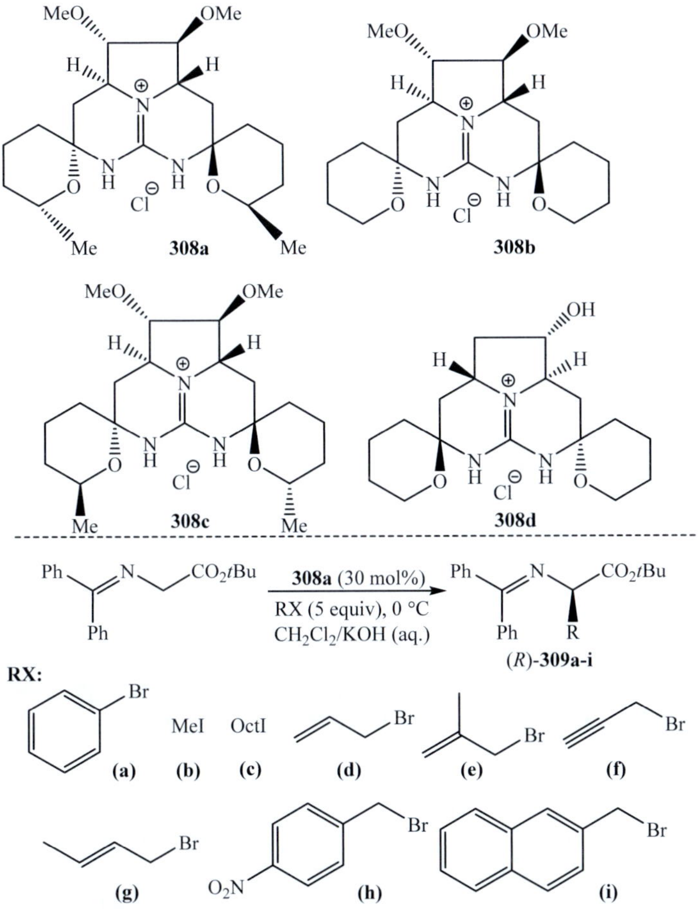

Scheme 6-112. Chemical structure of chiral pentacyclic guanidine compounds **308a-d** and their application as hetero-Michael reaction catalysts.

Likewise, Waser et al. synthesized a novel class of tartaric acid-derived *N*-spiro quaternary ammonium salts **310a-c** by starting from known TADDOLs, which consist of a dioxolane ring substituted with two mutually transoid diarylhydroxymethyl (Ar_2COH) groups (Scheme 6-113). These compounds were found to catalyze the asymmetric α-alkylation of glycine Schiff bases with good yields and high enantioselectivities [240].

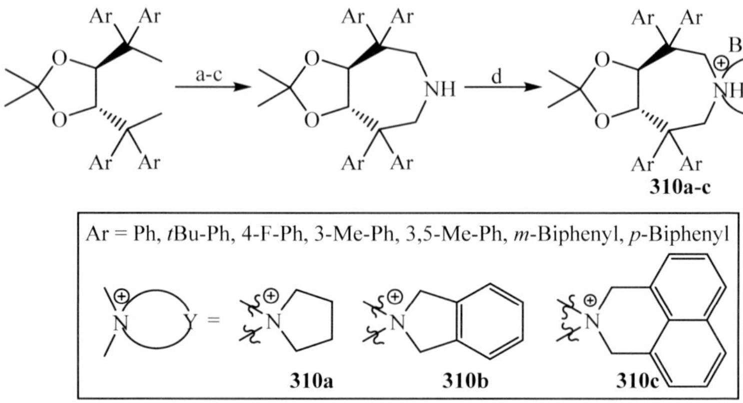

Scheme 6-113. Syntheses of tartaric acid-derived *N*-spiro quaternary ammonium salts **310a-c**. a) $SOCl_2$, Et_3N, CH_2Cl_2, 25 °C b) TMSCN, $SnCl_4$, 25 °C, 47–63% (over two steps) c) $LiAlH_4$, mesitylene, reflux, 29-43% d) Br-Y-Br, K_2CO_3, CH_3CN, reflux, 49–69%.

Shibasaki et al. reported catalytic asymmetric Michael and Mannich-type reactions of glycine Schiff bases using chiral two-center tartrate-derived diammonium salt (TaDiASs) **311a-c** [241]. Scheme 6-114 indicates the synthetic route to prepare TaDiASs **311a-c**. (*S*,*S*)-**311b** was suitable for the catalytic asymmetric Michael reaction with β-unsaturated ketones (up to 80% ee), and (*S*,*S*)-**311a** was effective in the catalytic asymmetric Michael reaction with acrylates (up to 88% ee), and the Mannich-type reaction (up to 90% ee, syn/anti up to 99:1).

In 2011, Ooi et al. demonstrated that *P*-spiro chiral tetraamino-phosphonium chloride salts **312a-c** also could function as an effective catalyst for the highly enantioselective alkylation of simple azlactones under mild solid-liquid biphasic conditions (Scheme 6-115) [242].

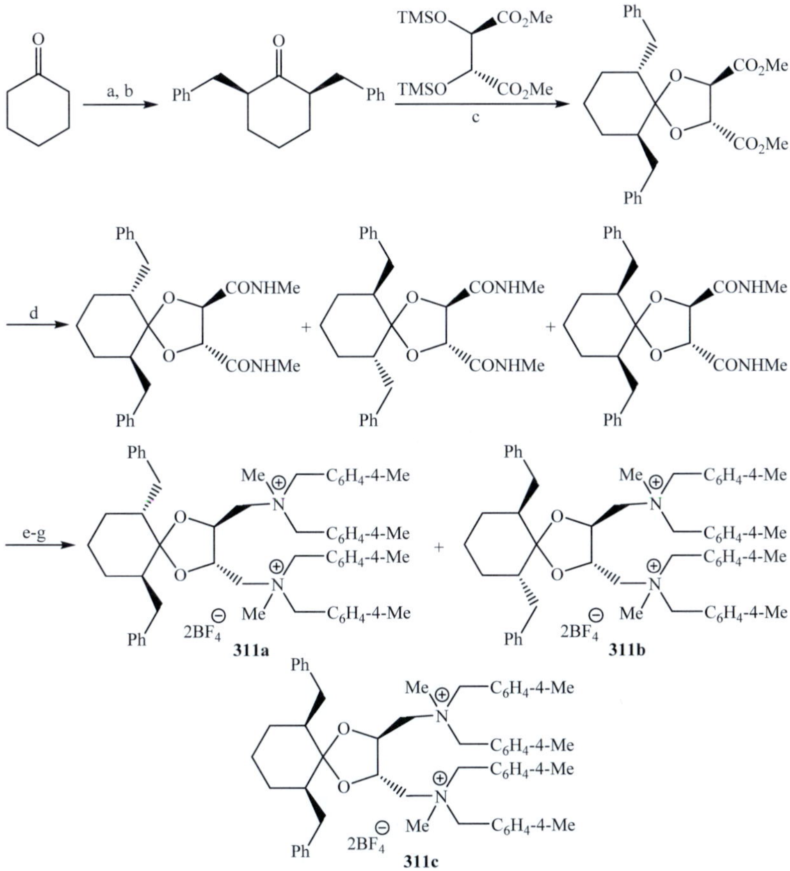

Scheme 6-114. Synthesis of (*S*,*S*)-TaDiASs **311a-c**. Reagents and conditions: a) benzaldehyde, aq. NaOH, CTACl (10 mol%), room temperature, 2 h, 90%; b) Pd/C (10 mol%), H_2, EtOAc, 4 h, 70%; c) tartrate, TMSOTf, TfOH (10 mol%), 60 °C, 48 h, quant.; d) $MeNH_2$, THF, room temperature, 5 h; e) $LiAlH_4$, THF, reflux, 2 h; f) *p*-methyl-benzyl-bromide, K_2CO_3, CH_3CN, room temperature, 2 h, 81% (two steps); g) $AgBF_4$, CH_2Cl_2, room temperature, 30 min, 92%. CTACl= hexadecyl-trimethylammonium chloride, TfOH= trifluoromethanesulfonyl, TMS= trimethylsilyl.

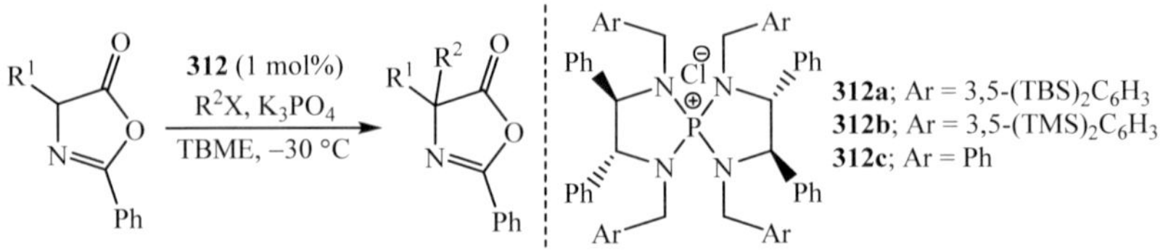

Scheme 6-115. Enantioselective alkylation of simple azlactones.

Another example is *P*-spiro chiral triaminoiminophosphorane **313**, which is used as an organic base catalyst for asymmetric Payne oxidation of *N*-sulfonyl imines with the hydrogen peroxide-trichloroacetonitrile system (Scheme 6-116) [243]. This general and scalable method indicates unique chemoselectivity, affording a stereoselective particle route to structurally diverse *N*-sulfonyl oxazidines. The mechanistic studies exhibit that the reaction proceeds via the Payne-type oxidation pathway.

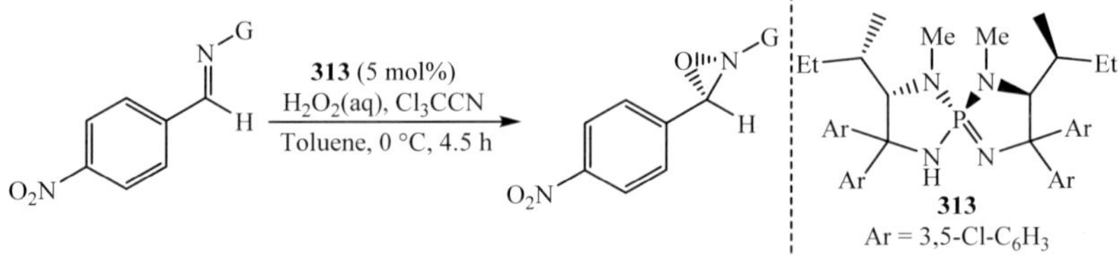

Scheme 6-116. Catalytic asymmetric Payne oxidation of *N*-sulfonyl imines.

Ooi et al. also reported a highly enantioselective synthesis of *N*-sulfonyl oxazidines using *P*-spiro chiral triaminoiminophosphorane **314**. They also presented a high regio-, diastereo-, and enantioselective 1,6-addition of azlactone to various alkenyl dienyl ketones under *P*-spiro chiral triaminoiminophosphorane **314** catalysis (Scheme 6-117). The researchers found that a chiral amino-phosphonium enolate, generated from an iminophosphorane and azlactone in a catalytic manner, recognized the direction and distance of the conjugation in alkenyl dienyl ketones, leading to a high regio-, diastereo-, and enantioselective Michael addition reaction. This reaction resulted in stereochemically functionalized enones **315** [244].

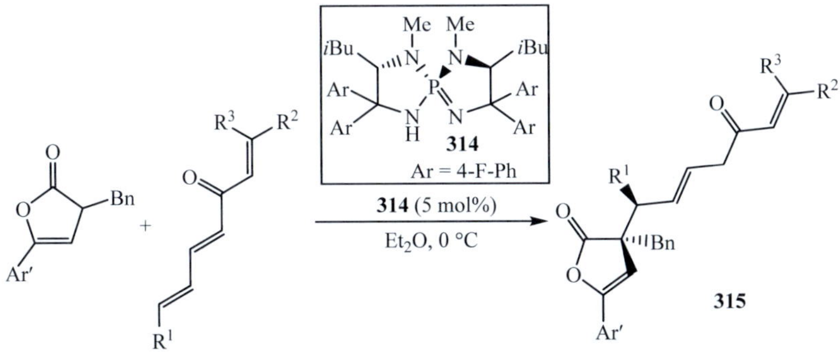

Scheme 6-117. Synthesis of stereochemically functionalized enones **315.**

The site-selectivity could be explained as follows: the catalyst can measure the distance between an activating group, such as carbonyl, and the reactive carbon center in a conjugate system of polyelectrophilic nature (Figure 6-79).

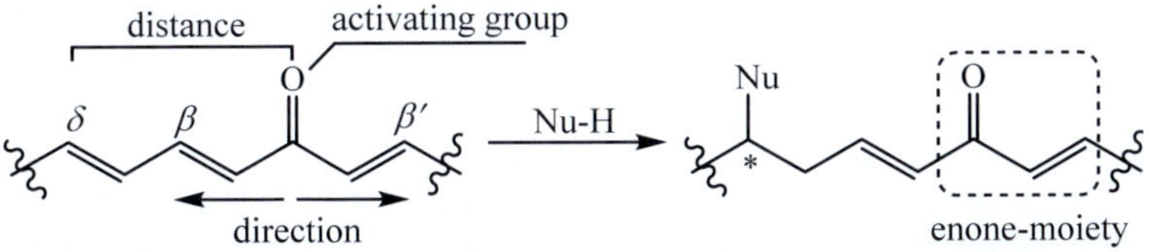

Fig. 6-79. Site-selectivity in Michael's addition to alkenyl dienyl ketones.

References

[1] Lu Yan, Wang Jin, Guo Jianyu, Tang Yihong, *et al.* 2015. "Efficient synthesis of 6-amino-2-thiaspiro[3,3]heptane hydro chloride". Heterocyclic Communications. No. 1. 1-4.

[2] Ashok Dongamanti, Mohan Gandhi Devulapally, Vikas Kumar Aamate, Srinivas Gundu, *et al.* 2016. "Microwave assisted synthesis, biological evaluation, and molecular docking of novel chroman scaffolds incorporating spirochromanone framework". Medicinal Chemistry Research. No. 12. 2882-2894.

[3] Ashok Dongamanti, Gundu Srinivas, Vikas Kumar Aamate, Mohan Gandhi Devulapally, Srinivas Reddy Malladi. 2015. "Facile ionic liquid-mediated, microwave assisted green synthesis, and anti-oxidant studies of novel indolin-2-one annulated spirochromanone conjugates". Russian Journal of General Chemistry. No. 3. 708-717.

[4] Saravana Mani Kailasam, Kaminsky Werner, Parameswaran Rajendran Subramaniam. 2018. "A facile atom economic one pot multicomponent synthesis of bioactive spiro-indenoquinoxaline pyrrolizines as potent antioxidants and anti-cancer agents". New Journal of Chemistry. No. 1. 301-310.

[5] Somsák László, Nagy Veronika, Vidal Sébastien, Czifrák Katalin, *et al.* 2008. "Novel design principle validated: Gluco pyranosylidene-spiro-oxathiazole as new nanomolar inhibitor of glycogen phosphorylase, potential antidiabetic agent". Bioorganic & Medicinal Chemistry Letters. No. 20. 5680-5683.

[6] Murugan Ramalingam, Anbazhagan Sockalingam, Narayanan S. Sriman. 2009. "Synthesis and *in vivo* antidiabetic activity of novel dispiropyrrolidines through [3+2] cycloaddition reactions with thiazolidinedione and rhodanine derivatives". European Journal of Medicinal Chemistry. No. 8. 3272-3279.

[7] Toumi Amani, Boudriga Sarra, Hamden Khaled, Sobeh Mansour, *et al.* 2021. "Synthesis, antidiabetic activity and molecular docking study of rhodanine-substitued spiro-oxindole pyrrolidines as novel α-amylase inhibitors". Bioorganic Chemistry. 104507.

[8] Gálvez Jaime, Polo Stivens, Insuasty Braulio, Gutiérrez Margarita, *et al.* 2018. "Design, facile synthesis, and evaluation of novel spiro-and pyrazolo[1,5-*c*]quinazolines as cholinesterase inhibitors: Molecular docking and MM/GBSA studies". Computational Biology and Chemistry. 218-229.

[9] Maryamabadi Ammar, Hasaninejad Alireza, Nowrouzi Najmeh, Mohebbi Gholamhossein. 2017. "Green synthesis of novel spiro-indenoquinoxaline derivatives and their cholinesterases inhibition activity". Bioorganic & Medicinal Chemistry. No. 7. 2057-2064.

[10] Hu Lei, Lan Ping, Song Qiu-Ling, Huang Zhi-Jian, *et al.* 2010. "Synthesis and antibacterial activity of C-12 pyrazolinyl spiro ketolides". European Journal of Medicinal Chemistry. No. 12. 5943-5949.

[11] S. Bhupathi Raja, Madhu Bandi, Ramana Reddy Ch. Venkata, Devi B. Rama, K. Dubey Pawan. 2017. "Ionic liquid mediated green synthesis of spiro-oxindoles from *N*-methyl quinolones and their anti bacterial activity". Journal of Heterocyclic Chemistry. No. 4. 2326-2332.

[12] S. Honmore Varsha, D. Natu Arun, M. Khedkar Vijay, A. Arkile Manisha, *et al.* 2021. "Two antibacterial spiro compounds from the roots of Artemisia pallens wall: Evidence from molecular docking". Natural Product Research. 1-8.

[13] Xiong Liang, Zhou Qin-Mei, Peng Cheng, Xie Xiao-Fang, *et al.* 2015. "Bis-spirolabdane diterpenoids from Leonurus japonicus and their anti-platelet aggregative activity". Fitoterapia. 1-6.

[14] Czopek Anna, Kubacka Monika, Bucki Adam, Siwek Agata, *et al.* 2021. "Novel serotonin 5-HT2A receptor antagonists derived from 4-phenylcyclohexane-5-spiro-and 5-methyl-5-phenyl-hydantoin, for use as potential antiplatelet agents". Pharmacological Reports. No. 5. 1361-1372.

[15] Das Prasanta, Boone Sarah, Mitra Dipanwita, Turner Lindsay, *et al.* 2020. "Synthesis and biological evaluation of fluoro-substituted spiro-isoxazolines as potential anti-viral and anti-cancer agents". RSC Advances. No. 50. 30223-30237.

[16] Das Prasanta, O. Omollo Ann, J. Sitole Lungile, McClendon Eric, *et al.* 2015. "Synthesis and investigation of novel spiro-isoxazolines as anti-cancer agents". Tetrahedron Letters. No. 14. 1794-1797.

[17] Hati Santanu, Tripathy Sayantan, Dutta Pratip Kumar, Agarwal Rahul, *et al.* 2016. "Spiro[pyrrolidine-3,3-oxindole] as potent anti-breast cancer compounds: Their design, synthesis, biological evaluation and cellular target identification". Scientific Reports. No. 1. 1-10.

[18] Safari Fatemeh, Hosseini Hajar, Bayat Mohammad, Ranjbara Ashkan. 2019. "Synthesis and evaluation of antimicrobial

activity, cytotoxic and pro-apoptotic effects of novel spiro-4*H*-pyran derivatives". RSC Advances. No. 43. 24843-24851.

[19] Lavanya Mallu, Viswambaran Asharani Indira, Thirumalai Dhakshanamurthy. 2019. "One-pot multi-component synthesis of functionalized spiropyridine and pyrido[2,3-*d*]pyrimidine scaffolds and their potent *in-vitro* anti-inflammatory and anti-oxidant investigations". Chemical Biology & Drug Design. No. 4. 464-472.

[20] Ashok Dongamanti, Madhuri Eluri Venkata Lakshmi, Sarasija Madderla, Sreekanth Sivan, *et al.* 2017. "Synthesis, biological evaluation and molecular docking of spirofurochromanone derivatives as anti-inflammatory and anti-oxidant agents". RSC Advances. No. 41. 25710-25724.

[21] Jiang Xianxing, Cao Yiming, Wang Yiqing, Liu Luping, *et al.* 2010. "A unique approach to the concise synthesis of highly optically active spirooxazolines and the discovery of a more potent oxindole-type phytoalexin analogue". Journal of the American Chemical Society. No. 43. 15328-15333.

[22] Ahmad Khan Salman, M. Asiri Abdullah. 2012. "Novel Steroidal (6*R*)-Spiro-1,3,4-thiadiazoline derivatives as anti-bacterial agents". Chinese Journal of Chemistry. No. 8. 1901-1905.

[23] Kumar Verma Anil, Bishnoi Abha, Fatma Shaheen, Parveen Huda, Singh Vineeta. 2017. "An easy synthetic access to spiro derivatives containing pyrido[1,2-*a*]pyrimidine and quinoline scaffolds and their antimicrobial activity". ChemistrySelect. No. 14. 4006-4009.

[24] Dong Yuxiang, Chollet Jacques, Matile Hugues, A. Charman Susan, *et al.* 2005. "Spiro and dispiro-1,2,4-trioxolanes as antimalarial peroxides: Charting a workable structure−activity relationship using simple prototypes". Journal of Medicinal Chemistry. No. 15. 4953-4961.

[25] Jarrahpour Aliasghar, Ebrahimi Edris, De Clercq Erik, Sinou Veronique, *et al.* 2011. "Synthesis of mono-, bis-spiro-and dispiro-β-lactams and evaluation of their antimalarial activities". Tetrahedron. No. 45. 8699-8704.

[26] Singh Divya, Fatma Shahin, Ankit Preyas, Mishra Priya, *et al.* 2013. "Microwave-induced solvent-free synthesis and antifungal screening of novel spiro heterocyclic systems containing 3,5-diphenyl-7*H*-thiazolo[3,2-*a*]pyrimidine and indolin-2-one". Synthetic Communications. No. 22. 3072-3082.

[27] R. Chavan Shrawan, S. Gavale Kishor, Khan Ayesha, Joshi Rakesh, *et al.* 2017. "Iminosugars spiro-linked with morpholine-fused 1,2,3-triazole: Synthesis, conformational analysis, glycosidase inhibitory activity, antifungal assay, and docking studies". ACS Omega. No. 10. 7203-7218.

[28] Kumar Raju Ranjith, Perumal Subbu, Senthilkumar Palaniappan, Yogeeswari Perumal, Sriram Dharmarajan. 2008. "Discovery of antimycobacterial spiro-piperidin-4-ones: An atom economic, stereoselective synthesis, and biological intervention". Journal of Medicinal Chemistry. No. 18. 5731-5735.

[29] Bharkavi Chelliah, Vivek Kumar Sundaravel, Ashraf Ali Mohamed, Osman Hasnah, *et al.* 2017. "One-pot microwave assisted stereoselective synthesis of novel dihydro-2′*H*-spiro [indene-2,1′-pyrrolo-[3,4-*c*]pyrrole]-tetraones and evaluation of their antimycobacterial activity and inhibition of AChE". Bioorganic & Medicinal Chemistry Letters. No. 14. 3071-3075.

[30] Rao Manam Rama, Teisan Sy, J. White Donald, Nicholson Benjamin, *et al.* 2005. "Lajollamycin, a nitro-tetraene spiro-β-lactone-γ-lactam antibiotic from the marine actinomycete streptomyces nodosus". Journal of Natural Products. No. 2. 240-243.

[31] Fu Chengzhang, Xie Feng, Hoffmann Judith, Wang Qiushui, *et al.* 2019. "Armeniaspirol antibiotic biosynthesis: Chlorination and oxidative dechlorination steps affording spiro[4.4]non-8-ene" . ChemBioChem. No. 6. 764-769.

[32] J. Patel Hardik, Sarra Joe, Caruso Francesco, Rossi Miriam, *et al.* 2006. "Synthesis and anticonvulsant activity of new *N*-1′,*N*-3′-disubstituted-2′*H*,3*H*,5′*H*-spiro-(2-benzofuran-1,4′-imidazolidine) -2′,3,5′-triones". Bioorganic & Medicinal Chemistry Letters. No. 17. 4644-4647.

[33] F. Zaher Ashraf, A. Khalil Nadia, M. Ahmed Eman. 2010. "Synthesis and anticonvulsant activity of new 3′-aryl-7-bromo-spiro[[1]benzothiophene-3,2′-[1,3]thiazolidine]-2,4′-diones". Oriental Journal of Chemistry. No. 4. 1241.

[34] Barmak Alireza, Niknam Khodabakhsh, Mohebbi Gholam hossein, Pournabi Hamid. 2019. "Antibacterial studies of hydroxyspiro[indoline-3,9-xanthene]trione against spiro[indoline-3,9-xanthene]trione and their use as acetyl and butyrylcholin esterase inhibitors". Microbial Pathogenesis. 95-99.

[35] Bendjedou Houaria, Barboni Luciano, Maggi Filippo, Bennaceur Malika, Benamar Houari. 2021. "Alkaloids and sesquiterpenes

from roots and leaves of Lycium europaeum L.(Solanaceae) with anti-oxidant and anti-acetylcholinesterase activities". Natural Product Research. No. 16. 2784-2788.

[36] K. Frade Silva Daiana, S. Duarte Sâmia, M. H. Lisboa Thaís, C. Ferreira Rafael, *et al.* 2019. "Antitumor effect of a novel spiro-acridine compound is associated with up-regulation of Th1-type responses and antiangiogenic action". Molecules. No. 1. 29.

[37] Wang Shengzheng, Chen Shuqiang, Guo Zhongjie, He Shipeng, *et al.* 2018. "Synthesis of spiro-tetrahydrothiopyran-oxindoles by Michael–aldol cascade reactions: Discovery of potential P53-MDM2 inhibitors with good antitumor activity". Organic & Biomolecular Chemistry. No. 4. 625-634.

[38] Gadthula Srinivas, K. Rawal Ravindra, Sharon Ashoke, Wu Dong, *et al.* 2011. "Synthesis and antiviral activity of cyclopropy l-spirocarbocyclicadenosine,(4*R*,5*S*,6*R*,7*R*)-4-(6-amino-9*H*-purin -9-yl)-7-(hydroxymethyl)spiro[2.4]heptane-5,6-diol against hepatitis C virus". Bioorganic & Medicinal Chemistry Letters. No. 13. 3982-3985.

[39] Jia Yan-Lai, Wei Mei-Yan, Chen Hai-Yan, Guan Fei-Fei, *et al.* 2015. "(+)-and (−)-Pestaloxazine A, a pair of antiviral enantiomeric alkaloid dimers with a symmetric spiro[oxazinane-piperazinedione] skeleton from Pestalotiopsis sp". Organic Letters. No. 17. 4216-4219.

[40] Vedhanarayanan Karthikeyan Subramanian, Devi Bala Balasubramanian, Paul Alex Raja Velanganni, Perumal Subbu, *et al.* 2010. "A highly atom economic, chemo-, regio-and stereo-selective synthesis and evaluation of spiro-pyrrolothiazoles as antitubercular agents". Bioorganic & Medicinal Chemistry Letters. No. 1. 350-353.

[41] Dincel Efe Doğukan, Gürsoy Elif, Yilmaz-Ozden Tugba, Ulusoy-Güzeldemirci Nuray. 2020. "Anti-oxidant activity of novel imidazo[2,1-*b*]thiazole derivatives: Design, synthesis, biological evaluation, molecular docking study and in silico ADME prediction". Bioorganic Chemistry. 104220.

[42] Czopek Anna, Byrtus Hanna, Ko1aczkowski Marcin, Paw1owski Maciej, *et al.* 2010. "Synthesis and pharmacological evaluation of new 5-(cyclo)alkyl-5-phenyl-and 5-spiroimidazolidine-2,4-dione derivatives. Novel 5-HT1A receptor agonist with potential antidepressant and anxiolytic activity". European Journal of Medicinal Chemistry. No. 4. 1295-1303.

[43] A. Hassan Alaa, F. Abdel-Latif Fathy, M. Nour El-Din Ahmed, Abdel-Aziz Mohamed, *et al.* 2011. "Synthesis of novel spiro(indolone-3,2′-[1,3,4]thiadiazol)-2-ones and evaluation of their antidepressant and anticonvulsant activities". Journal of Heterocyclic Chemistry. No. 5. 1050-1055.

[44] Rether Jan, Erkel Gerhard, Anke Timm, Sterner Olov. 2004. "Inhibition of inducible TNF-α expression by oxaspirodion, a novel spiro-compound from the ascomycete Chaetomium subspirale". Journal of Biological Chemistry. 829-834.

[45] K. Mukherjee Pulok, Maity Niladri, K. Nema Neelesh, K. Sarkar Birendra. 2011. "Bioactive compounds from natural resources against skin aging". Phytomedicine. No. 1,. 64-73.

[46] Kelidari Hamid Reza, Saeedi Majid, Hajheydari Zohreh, Akbari Jafar, *et al.* 2016. "Spironolactone loaded nanostructured lipid carrier gel for effective treatment of mild and moderate acne vulgaris: A randomized, double-blind, prospective trial". Colloids and Surfaces B: Biointerfaces. 146. 47-53.

[47] Larik Fayaz Ali, Saeed Aamer, Shahzad Danish, Faisal Muhammad, *et al.* 2017. "Synthetic approaches towards the multi target drug spironolactone and its potent analogues/derivatives". Steroids. 76-92.

[48] Palomba Martina, Rossi Leonardo, Sancineto Luca, Tramontano Enzo, *et al.* 2016. "A new vinyl selenone-based domino approach to spirocyclopropyl oxindoles endowed with anti-HIV RT activity". Organic & Biomolecular Chemistry. No. 6. 2015-2024.

[49] Bârtolo Ines, S. Santos Bruna, Fontinha Diana, Machado Marta, *et al.* 2021. "Spiro-β-lactam BSS-730A displays potent activity against HIV and Plasmodium". ACS Infectious Diseases. No. 2. 421-434.

[50] Zhang Lai-Bin, Lv Jie-Li, Liu Jun-Wei. 2016. "Phthalide derivatives with anticoagulation activities from Angelica sinensis". Journal of Natural Products. No. 7. 1857-1861.

[51] Askri Sonia, Edziri Hayet, Ben Hamouda Manel, Mchiri Chadlia, *et al.* 2022. "Synthesis, biological evaluation, density functional calculation and molecular docking analysis of novel spiro pyrrolizidines derivatives as potential anti-microbial and anti-coagulant agents". Journal of Molecular Structure. 131688.

[52] Niu Siwen, Fan Zuo-Wang, Xie Chun-Lan, Liu Qingmei, *et al.* 2017. "Spirograterpene A, a tetracyclic spiro-diterpene with a fused 5/5/5/5 ring system from the deep-sea-derived fungus

Penicillium granulatum MCCC 3A00475". Journal of Natural Products. No. 7. 2174-2177.

[53] Mustapha M. Bouhenna, Orlikova Barbora, Talhi Oualid, Schram Ben, *et al.* 2017. "Anti-proliferative, cytotoxic and NF-κB inhibitory properties of spiro(lactone-cyclohexanone) compounds in human leukemia". Anticancer Research. No. 9. 5225-5233.

[54] A. Knyazev Nickolay, V. Shmakov Stanislav, A. Pechkovskaya Sofya, S. Filatov Alexander, *et al.* 2021. "Identification of spiro-fused [3-azabicyclo[3.1.0]hexane]oxindoles as potential anti-tumor agents: Initial in vitro evaluation of anti-proliferative effect and actin cytoskeleton transformation in 3T3 and 3T3-SV40 fibroblast". International Journal of Molecular Sciences. No. 15. 8264.

[55] Ding Bo, Dai Yi, Hou Yun-Long, Yao Xin-Sheng. 2015. "Spiroalkaloids from shensong yangxin capsule". Journal of Asian natural products research. No. 5. 559-566.

[56] Gonca Ersöz, Rapposelli Simona, Darıcı Faruk, Digiacomo Maria, Yılmaz Zehra. 2016. "Antiarrhythmic activity of a new spirocyclic benzopyran activator of the cardiac mitochondrial ATP dependent potassium channels". Archives of Pharmacal Research. No. 9. 1212-1222.

[57] Czopek Anna, Kołaczkowski Marcin, Bucki Adam, Byrtus Hanna, *et al.* 2015. "Novel spirohydantoin derivative as a potent multireceptor-active antipsychotic and antidepressant agent". Bioorganic & Medicinal Chemistry. No. 13. 3436-3447.

[58] Leañez Jacques, Nuñez Jorge, García-Marchan Yael, Sojo Felipe, *et al.* 2019. "Anti-leishmanial effect of spirodihydroquinoline-oxindoles on volume regulation decrease and sterol biosynthesis of Leishmania braziliensis". Experimental Parasitology. 31-38.

[59] S. Almeida Fernanda, L.S. Sousa Gleyton , C. Rocha Juliana , F. Ribeiro Frederico, *et al.* 2021. "*In vitro* anti-Leishmania activity and molecular docking of spiro-acridine compounds as potential multitarget agents against Leishmania infantum". Bioorganic & Medicinal Chemistry Letters. 128289.

[60] Chapman Neil, Dobson Joanna, Wilson Sarah, Dahlöf Björn, *et al.* 2007. "Effect of spironolactone on blood pressure in subjects with resistant hypertension". Hypertension. No. 4. 839-845.

[61] Batterink Josh, N. Stabler Sarah, M. Tejani Aaron, T Fowkes Curt. 2010. "Spironolactone for hypertension". Cochrane. No. 8.

[62] Narender Rao Somisetti, Babu K. Sudhakar. 2011. "An improved and efficient synthesis of Irbesartan, an antihypertensive drug". Organic Communications. No. 4. 105.

[63] Satyanarayana Bollikonda, Sumalatha Yasareni, Sridhar Chaganti, Venkatraman Sundaram, Pratap Reddy Padi 2006. "New synthesis of analogues of the antihypertensive active pharmaceutical ingredient irbesartan". Synthetic Communications. No. 14. 2079-2086.

[64] Goyard David, Kónya Bálint, S. Chajistamatiou Aikaterini, D. Chrysina Evangelia, *et al.* 2016. "Glucose-derived spiro-isoxazolines are anti-hyperglycemic agents against type-2 diabetes through glycogen phosphorylase inhibition". European Journal of Medicinal Chemistry. 444-454.

[65] Goyard David, Kónya Bálint, Czifrák Katalin, Larini Paolo, *et al.* 2020. "Glucose-based spiro-oxathiazoles as *in vivo* anti-hyper glycemic agents through glycogen phosphorylase inhibition". Organic & Biomolecular Chemistry. No. 5. 931-940.

[66] Turcano Lorenzo, Battista Theo, De Haro Esther Torrente, Missineo Antonino, *et al.* 2020. "Spiro-containing derivatives show antiparasitic activity against Trypanosoma brucei through inhibition of the trypanothione reductase enzyme". PLoS Neglected Tropical Diseases. No. 5. 0008339.

[67] Uto Yoshikazu, Ueno Yuko, Kiyotsuka Yohei, Miyazawa Yuriko, *et al.* 2011. "Discovery of novel SCD1 inhibitors: 5-Alkyl-4,5-dihydro-3*H*-spir[1,5-benzoxazepine-2,4′-piperidine] analogs". European Journal of Medicinal Chemistry. No. 5. 1892-1896.

[68] Deng Yongqi, Yang Zhiwei,, W. Shipps Jr. Gerald, Lo Sie-Mun, *et al.* 2013. "Discovery of liver-targeted inhibitors of stearoyl-CoA desaturase (SCD1)". Bioorganic & Medicinal Chemistry Letters. No. 3. 791-796.

[69] Malpani Yashwardhan, Achary Raghavendra, Kim So Yeon, Jeong Hee Chun, *et al.* 2013. "Efficient synthesis of 3*H*,3′*H*-spiro[benzofuran-2,1′-isobenzofuran]-3,3′-dione as novel skeletons specifically for influenza virus type B inhibition". European Journal of Medicinal Chemistry. 534-544.

[70] Apaydın Çağla Begüm, Tansuyu Merve, Cesur Zafer, Naesens Lieve, Goktas Füsun. 2021. "Design, synthesis and anti-influenza virus activity of furan-substituted spirothiazolidinones". Bioorganic Chemistry. 104958.

[71] Li Lei, Peng Xing-Rong, Dong Jin-Run, Lu Shuang-Yang, *et al.* 2018. "Rearranged lanostane-type triterpenoids with anti-hepatic fibrosis activities from Ganoderma applanatum". RSC Advances. No. 55. 31287-31295.

[72] Luo Qi, Di Lei, Dai Wei-Feng, Lu Qing, *et al.* 2015. "Applanatumin A, a new dimeric meroterpenoid from Ganoderma applanatum that displays potent antifibrotic activity". Organic Letters. No. 5. 1110-1113.

[73] Kubota Hirokazu, Fujii Masahiro, Ikeda Ken, Takeuchi Makato, *et al.* 1998. "Spiro-substituted piperidines as neurokinin receptor antagonists. I. Design and synthesis of (±)-*N*-[2-(3,4-dichloro phenyl)-4-(spiro[isobenzofuran-1(3*H*),4′-piperidin]-1′-yl)butyl]-*N*-methylbenzamide, YM-35375, as a new lead compound for novel neurokinin receptor antagonists". Chemical and Pharmaceutical Bulletin. No. 2. 351-354.

[74] X. Qiao Jennifer, C. Wang Tammy, Ruel Réjean, Thibeault Carl, *et al.* 2013. "Conformationally constrained ortho-anilino diaryl ureas: Discovery of 1-(2-(1′-neopentylspiro[indoline-3,4′-piperidine]-1-yl)phenyl)-3-(4-(trifluoromethoxy)phenyl)urea, a potent, selective, and bioavailable P2Y1 antagonist". Journal of Medicinal Chemistry. No. 22. 9275-9295.

[75] Suzuki Mitsuko, Huu Tung Nguyen, D. Kwofie Kofi, Adegle Richard, *et al.* "New anti-trypanosomal active tetracyclic iridoid isolated from Morinda lucida Benth". Bioorganic & Medicinal Chemistry Letters. No. 15. 3030-3033.

[76] R. Cheeke Peter, Piacente Sonia, Oleszek Wieslaw. 2006. "Anti-inflammatory and anti-arthritic effects of Yucca schidigera: A review". Journal of Inflammation. No. 1. 1-7.

[77] Yang Shu-Wei, D. Ho Ginny, Tulshian Deen, Bercovici Ana, *et al.* 2014. "Bioavailable pyrrolo-benzo-1,4-diazines as Nav1.7 sodium channel blockers for the treatment of pain". Bioorganic & Medicinal Chemistry Letters. No. 21. 4958-4962.

[78] D. Ho Ginny, Tulshian Deen, Bercovici Ana, Tan Zheng, *et al.* 2014. "Discovery of pyrrolo-benzo-1,4-diazines as potent Nav1.7 sodium channel blockers". Bioorganic & Medicinal Chemistry Letters. No. 17. 4110-4113.

[79] H. Shapiro Bernared, S. Goldman Allen. 1975. "Spermatogenic arrest produced in the adult male rat by 19-norspiroxenone, a potent anti-estrogen". Proceedings of the Society for Experimental Biology and Medicine. No. 1. 226-228.

[80] Chaudhary Ashu, Sharma Neha, Sharma Poonam, Jasuja Nakuleshwar Dut, *et al.* 2008. "Contraceptive property of nitrogen/oxygen sulfur donor heterocyclic compounds". Rasayan Journal of Chemistry. 648-92.

[81] Cao Guo-An, Cheng Kin-Fai. 1996. "Antifertility agent: Stereoselective synthesis of spiro[cyclopropane-1,7′-yuehchukene]". Synthetic Communications. No. 8. 1525-1538.

[82] Luo Shi-Hong, Hua Juan, Li Chun-Huan, Liu Yan, *et al.* 2013. "Unusual antifeedant spiro-sesterterpenoid from the flowers of Leucosceptrum canum". Tetrahedron Letter. No. 3. 235-237.

[83] Hassanali Ahmed, D. Bentley Michael, M. Slawin Alexander, J. Williams David, *et al.* 1987. "Pedonin, a spiro tetranortri-terpenoid insect antifeedant from Harrisonia abyssinica". Phytochemistry. No. 2. 573-575.

[84] Zhang Zhi-xiang, Cheng Dong-Mei, Xu Han-Hong, Wu Yu-Lin, Fan Jun-Fa. 2004. "Bioactivities and mechanism of spiro enol ether analogues against Pieris rapae". Insect Science. No. 1. 19-26.

[85] Noji Satoru, Hara Yoshinori, Miura Tomoya, Yamanaka Hiroshi, *et al.* 2020. "Discovery of a Janus kinase inhibitor bearing a highly three-dimensional spiro scaffold: JTE-052 (delgocitinib) as a new dermatological agent to treat inflammatory skin disorders". Journal of Medicinal Chemistry. No. 13. 7163-7185.

[86] Bonne Claude, Raynaud Jean-Pierre. 1974. "Mode of spirono lactone anti-androgenic action: Inhibition of androstanolone binding to rat prostate androgen receptor". Molecular and Cellular Endocrinology. No. 1. 59-67.

[87] Bansal Ranju, Singh Ranjit. 2020. "Steroidal pyrazolines as a promising scaffold in drug discovery". Future Medicinal Chemistry. No. 10. 949-959.

[88] Verma Dinesh, Thompson Jacob, Swaminathan Sankar. 2016. "Spironolactone blocks Epstein–Barr virus production by inhibiting EBV SM protein function". Proceedings of the National Academy of Sciences. No. 13. 3609-3614.

[89] S. Girgis Adel, S. Panda Siva, M. Shalaby El-Sayed, F. Mabied Ahmed, *et al.* 2015. "Regioselective synthesis and theoretical studies of an anti-neoplastic fluoro-substituted dispiro-oxindole". RSC Advances. No. 19. 14780-14787.

[90] A. Mendes Joseane, Merino Pedro, Soler Tatiana, J. Salustiano Eduardo, *et al.* 2019. "Enantioselective synthesis, DFT calculations, and preliminary antineoplastic activity of dibenzo 1-

azaspiro[4.5]decanes on drug-resistant leukemias". The Journal of Organic Chemistry. No. 4. 2219-2233.
[91] Arya Kapil, Singh Rawat Diwan, Dandia Anshu, Sasai Hiroaki. 2012. "Brønsted acidic ionic liquids: Green, efficient and reusable catalyst for synthesis of fluorinated spiro[indole-thiazinones/thiazolidinones] as antihistamic agents". Journal of Fluorine Chemistry. 117-122.
[92] Lee Seong-Min, Kim Min-Sun, Hayat Faisal, Shin Dongyun. 2019. "Recent advances in the discovery of novel antiprotozoal agents". Molecules. No. 21. 3886.
[93] Zelisko Nataliya, Atamanyuk Dmytro, Vasylenko Olexandr, Grellier Philippe, Lesyk Roman. 2012. "Synthesis and anti trypanosomal activity of new 6,6,7-trisubstituted thiopyrano[2,3-*d*][1,3]thiazoles". Bioorganic & Medicinal Chemistry Letters. No. 23. 7071-7074.
[94] F. Carneiro Paula, C. R. F. Pinto Maria, K. F. Marra Roberta, de C. da Silva Fernando, *et al.* 2016. "Synthesis and antimalarial activity of quinones and structurally-related oxirane derivatives". European Journal of Medicinal Chemistry.134-140.
[95] A. Duffy Ruth, Morgan Cynthia, Naylor Robert, A. Higgins Guy, *et al.* 2012. "Rolapitant (SCH 619734): A potent, selective and orally active neurokinin NK1 receptor antagonist with centrally-mediated antiemetic effects in ferrets". Pharmacology Biochemistry and Behavior. No. 1. 95-100.
[96] Soupart Alain, Gross Peter, Legros Jean-Jacques, Alföldi Sándor, *et al.* 2006. "Successful long-term treatment of hyponatremia in syndrome of inappropriate antidiuretic hormone secretion with satavaptan (SR121463B), an orally active nonpeptide vasopressin V2-receptor antagonist". Clinical Journal of the American Society of Nephrology. No. 6. 1154-1160.
[97] Hiesinger Kerstin, Dar'in Dmitry, Proschak Ewgenij, Krasavin Mikhail. 2020. "Spirocyclic scaffolds in medicinal chemistry". Journal of Medicinal Chemistry. No. 1. 150-183.
[98] F. Abdel-Magid Ahmed. 2019. "LRRK2 kinase inhibitors as possible therapy for Parkinson's disease and other neuro-degenerative disorders". ACS Medicinal Chemistry Letters. 846-847.
[99] Zaiter Jamila, Hibot Achraf, Hafid Abderrafia, Khouili Mostafa, *et al.* 2021. "Evaluation of the cellular protection by novel spiropyrazole compounds in dopaminergic cell death". European Journal of Medicinal Chemistry. 113140.

[100] Qun-Li Zhang, Li-Jun Song, En-Si Wang. 2013. "Synthesis and antitussive effect of new hydantoin compounds". Chemical Research in Chinese Universities. No. 1. 76-81.

[101] Yang Xin-Zhou, Zhu Jian-Yu, Tang Chun-Ping, Ke Chang-Qiang, *et al.* 2009. "Alkaloids from roots of Stemona sessilifolia and their antitussive activities". Planta Medica. No. 02. 174-177.

[102] R. Shah Harshil, Ketan Savjani Jignasa. 2018. "Recent updates for designing CCR5 antagonists as anti-retroviral agents". European Journal of Medicinal Chemistry. 115-129.

[103] G. Nichols Garrett, M. Steel Helen, Bonny Tab, Adkison Kimberly, *et al.* 2008. "Hepatotoxicity observed in clinical trials of aplaviroc (GW873140)". Antimicrobial Agents and Chemotherapy. No. 3. 858-865.

[104] B. Ramachary Dhevalapally, Reddy Y. Vijayendar, Banerjee Atoshi, Banerjee Sharmistha. 2011. "Design, synthesis and biological evaluation of optically pure functionalized spiro[5,5] undecane-1,5,9-triones as HIV-1 inhibitors". Organic & Biomolecular Chemistry. No. 21. 7282-7286.

[105] Chakraborty Kajal, Antony Tima. 2021. "First report of spiro-compounds from marine macroalga Gracilaria salicornia: Prospective natural anti-inflammatory agents attenuate 5-lipoxygenase and cyclooxygenase-2". Natural Product Research. No. 5. 770-781.

[106] B. Kaye Stan. 1998. "New antimetabolites in cancer chemotherapy and their clinical impact". British Journal of Cancer. No.3. 1-7.

[107] J. Peters Godefridus, L. Van der Wilt Clasina, J. A. van Moorsel Catharina, R. Kroep Judith, *et al.* 2000. "Basis for effective combination cancer chemotherapy with antimetabolites". Pharmacology & Therapeutics. No. 2-3. 227-253.

[108] Islam Mohammad Shahidul, Al-Majid Abdullah Mohammed, F. El-Senduny Fardous, A. Badria Farid, *et al.* 2020. "Synthesis, anticancer activity, and molecular modeling of new halogenated spiro[pyrrolidine-thiazolo-oxindoles] derivatives". Applied Sciences. No. 6. 217.

[109] Zhao Juan, Chi Zhihe, Yang Zhan, Chen Xiaojie, *et al.* 2018. "Recent developments of truly stretchable thin film electronic and optoelectronic devices". Nanoscale. No. 13. 5764-5792.

[110] Xu Huihua, Yin Lan, Liu Chuan, Sheng Xing, Zhao Ni. 2018. "Recent advances in biointegrated optoelectronic devices". Advanced Materials. No. 33. 1800156.

[111] Shirota Yasuhiko, Kageyama Hiroshi. 2007. "Charge carrier transporting molecular materials and their applications in devices". Chemical Review. No. 4. 953-1010.

[112] P. I. Saragi Tobat, Spehr Till, Siebert Achim, Fuhrmann-Lieker Thomas, Salbeck Josef. 2007. "Spiro compounds for organic optoelectronics". Chemical Reviews. No. 4. 1011-1065.

[113] Bach By Udo, De Cloedt Kenny, Spreitzer Hubert, Grätzel Michael. 2000. "Characterization of hole transport in a new class of spiro-linked oligotriphenylamine compounds". Advanced Materials. No. 14. 1060-1063.

[114] Steuber By Frank, Staudigel Jörg, Stössel Matthias, Simmerer Jürgen, *et al.* 2000. "White light emission from organic LEDs utilizing spiro compounds with high-temperature stability". Advanced Materials. No. 2. 130-133.

[115] Kimura Makoto, Kuwano Seiichi, Sawaki Yasuhiko, Fujikawa Hisayoshi, *et al.* 2005. "New 9-fluorene-type trispirocyclic compounds for thermally stable hole transport materials in OLEDs". Journal of Materials Chemistry. No. 24. 2393-2398.

[116] Xie Ling-Hai, Fu Ting, Hou Xiao-Ya, Tang Chao, *et al.* 2006. "An efficient synthesis of novel spiro[[8*H*]indeno[2,1-*b*] thiophene-8,9′-fluorene] building block for blue light-emitting materials". Tetrahedron letters. No. 36. 6421-6424.

[117] Wu Yonggang, Zhang Jianyuan, Fei Zhuping, Bo Zhishan. 2008. "Spiro-bridged ladder-type poly(p-phenylene)s: Towards structurally perfect light-emitting materials". Journal of the American Chemical Society. No. 23. 7192-7193.

[118] Jiang Zuoquan, Yao Haiqing, Liu Zhongyin, Yang Chuluo, *et al.* 2009. "Bent ladder-type hexaphenylene with carbazole core and spiro linkage as stable and efficient blue emitter". Organic Letters. No. 18. 4132-4135.

[119] Cocherel Nicolas, Poriel Cyril, Vignau Laurence, Bergamini Jean-François, Rault-Berthelot Joëlle. 2010. "Dispiroxanthene-indenofluorene: A new blue emitter for nondoped organic light emitting diode applications". Organic Letters. No. 3. 452-455.

[120] Poriel Cyril, Cocherel Nicolas, Rault-Berthelot Joëlle, Vignau Laurence, Jeannin Olivier. 2011. "Incorporation of spiroxanthene units in blue-emitting oligophenylene frameworks: A new molecular design for OLED applications". Chemistry-A European Journal. No. 45. 12631-12645.

[121] Mokbel Haifaa, Poriel Cyril, Rault-Berthelot Joëlle, Eric Dumur Fred, *et al.* 2016. "A glance at violet LED sensitive photo-

initiators based on the spiroxanthene scaffold". Journal of Applied Polymer Science. No. 12.
[122] Xia Debin, Duan Chunbo, Liu Shihui, Ding Dongxue, *et al.* 2019. "Oligofluorene with multiple spiro-connections: Its and their use in blue and white OLEDs". New Journal of Chemistry. No. 9. 3788-3792.
[123] Cha Jae-Ryung, Lee Chil-Won, Gong Myoung-Seon. 2014. "New spiro[benzotetraphene-fluorene] derivatives: Synthesis and application in sky-blue fluorescent host materials". Journal of Fluorescence. No. 4. 1215-1224.
[124] Romain Maxime, Tondelier Denis, Jeannin Olivier, Geffroy Bernard, *et al.* 2015. "Properties modulation of organic semi-conductors based on a donor-spiro-acceptor (D-spiro-A) molecular design: New host materials for efficient sky-blue PhOLEDs". Journal of Materials Chemistry C. No. 37. 9701-9714.
[125] Romain Maxime, Quinton Cassandre, Tondelier Denis, Geffroy Bernard, *et al.* 2016. "Thioxanthene and dioxothioxanthene dihydroindeno[2,1-*b*]fluorenes: Synthesis, properties and applications in green and sky blue phosphorescent OLEDs". Journal of Materials Chemistry C. No. 8. 1692-1703.
[126] Ding Lei, Du Shuai, Cui Lin-Song, Zhang Fang-Hui, Liao Liang-Sheng. "Novel spiro-based host materials for application in blue and white phosphorescent organic light-emitting diodes". Organic Electronics. 108-114.
[127] Chan Chin-Yiu, Wong Yi-Chun, Chan Mei-Yee, Cheung Sin-Hang, *et al.* 2016. "Bifunctional heterocyclic spiro derivatives for organic optoelectronic devices". ACS Applied Materials & Interfaces. No. 37. 24782-24792.
[128] Sun Wei, Zhou Nonglin, Xiao Yin, Wang Shirong, Li Xianggao. 2017. "A novel spiro[acridine-9,9′-fluorene] derivatives containing phenanthroimidazole moiety for deep-blue OLED application". Chemistry–An Asian Journal. No. 23. 3069-3076.
[129] Sun Wei, Zhou Nonglin, Xiao Yin, Wang Shirong, Li Xianggao. 2018. "Novel carbazolyl-substituted spiro[acridine-9,9′-fluorene] derivatives as deep-blue emitting materials for OLED applications". Dyes and Pigments. 30-37.
[130] Qu Yang-Kun, Zheng Qi, Fan Jian, Liao Liang-Sheng, Jiang Zuo-Quan. 2021. "Spiro compounds for organic light-emitting diodes" . Accounts of Materials Research. No. 12. 1261-1271.

[131] Kim Guan-Woo, Choi Hyuntae, Kim Minjun, Lee Junwoo, *et al.* 2020. "Hole transport materials in conventional structural (n–i–p) perovskite solar cells: From past to the future". Advanced Energy Materials. No. 8. 1903403.
[132] Ning Zhijun, Fu Ying, Tian He. 2010. "Improvement of dye-sensitized solar cells: what we know and what we need to know" . Energy & Environmental Science. No. 9. 1170-1181.
[133] A. Green Martin, Ho-Baillie Anita, J. Snaith Henry. 2014. "The emergence of perovskite solar cells". Nature Photonics. No. 7. 506-514.
[134] Jung Hyun Suk, Park Nam-Gyu. 2015. "Perovskite solar cells: From materials to devices". Small. No. 1. 10-25.
[135] Pham Hong Duc, Xianqiang Li, Wenhui Li, Manzhos Sergei, *et al.* 2019. "Organic interfacial materials for perovskite-based optoelectronic devices". Energy & Environmental Science. No. 4. 1177-1209.
[136] Vivo Paola, K. Salunke Jagadish, Priimagi Arri. 2017. "Hole-transporting materials for printable perovskite solar cells". Materials. No. 9. 1087.
[137] Ganesan Paramaguru, Fu Kunwu, Gao Peng, Raabe Ines, *et al.* 2015. "A simple spiro-type hole transporting material for efficient perovskite solar cells". Energy & Environmental Science. No. 7. 1986-1991.
[138] Wang Ya-Kun, Yuan Zhong-Cheng, Shi Guo-Zheng, Li Yong-Xi, *et al.* 2016. "Dopant-free spiro-triphenylamine/fluorene as hole-transporting material for perovskite solar cells with enhanced efficiency and stability". Advanced Functional Materials. No. 9. 1375-1381.
[139] Hawash Zafer, K. Ono Luis, Qi Yabing. 2016. "Moisture and oxygen enhance conductivity of LiTFSI-doped Spiro-MeOTAD hole transport layer in perovskite solar cells". Advanced Materials Interfaces. No. 13. 1600117.
[140] Arora Neha, Orlandi Simonetta, Dar M. Ibrahim, Aghazada Sadig, *et al.* 2016. "High open-circuit voltage: Fabrication of formamidinium lead bromide perovskite solar cells using fluorene–dithiophene derivatives as hole-transporting materials". ACS Energy Letters. No. 1. 107-112.
[141] Govindan Venkatesan, Yang Kai-Cheng, Fua Yung Seng, Wu Chun-Guey. 2018. "Low-cost synthesis of heterocyclic spiro-type hole transporting materials for perovskite solar cell applications". New Journal of Chemistry. No. 9. 7332-7339.

[142] Drigo Nikita, Roldan-Carmona Cristina, Franckevicius Marius, Lin Kun-Han, *et al.* 2019. "Doped but stable: Spirobisacridine hole transporting materials for hysteresis-free and stable perovskite solar cells". Journal of the American Chemical Society. No. 4. 1792-1800.
[143] Xu Jianbin, Liang Lusheng, Mai Chi-lun, Zhang Zilong, *et al.* 2020. "Lewis-base containing spiro type hole transporting materials for high-performance perovskite solar cells with efficiency approaching 20%". Nanoscale. No. 24. 13157-13164.
[144] Akin Seckin, Bauer Michael, Uchida Ryusuke, Arora Neha, *et al.* 2020. "Cyclopentadithiophene-based hole-transporting material for highly stable perovskite solar cells with stabilized efficiencies approaching 21%". ACS Applied Energy Materials. No. 8. 7456-7463.
[145] Zhang Yan, Wu Fei, Chen Lei, Zhang Fengying, *et al.* 2020. "Molecular engineering of highly efficient dopant-free spiro-type hole transporting materials for perovskite solar cells". Solar Energy Materials and Solar Cells. 110534.
[146] M. Abdellah Islam, H. Chowdhury Towhid, Lee Jae-Joon, Islam Ashraful, *et al.* 2021. "Facile and low-cost synthesis of a novel dopant-free hole transporting material that rivals Spiro-OMeTAD for high efficiency perovskite solar cells". Sustainable Energy & Fuels. No. 1. 199-211.
[147] Abbas Mazhar, Ali Usman, Faizan Muhammad, Siddique Muhammad Bilal Ahmed. 2021. "Spirofluorene based small molecules as an alternative to traditional non-fullerene acceptors for organic solar cells". Optical and Quantum Electronics. No. 5. 1-14.
[148] Chen Yan, Wang Yanqing, Wu Di, Wu Wenfei, *et al.* 2021. "Fluorene-terminated hole transporting materials with a spiro [fluorene-9,9′-xanthene] core for perovskite solar cells". New Journal of Chemistry. No. 12. 5497-5502.
[149] Sharma Khushboo, Sharma Vinay, Sundar Sharma Shyam. 2006. "Dye-sensitized solar cells: Fundamentals and current status". Nanoscale Research Letters. No. 1. 1-46.
[150] Chiba Yasuo, Islam Ashraful, Watanabe Yuki, Komiya Ryoichi, *et al.* 2006. "Dye-sensitized solar cells with conversion efficiency of 11.1%". Japanese Journal of Applied Physics. No. 7L. L638.
[151] Nazeeruddin Mohammad Khaja, Baranoff Etienne, Grätzel Michael. 2011. "Dye-sensitized solar cells: A brief overview". Solar Energy. No. 6. 1172-1178.

[152] Singh Ajay, Radicchi Eros, Fantacci Simona, Nunzi Francesca, *et al.* 2019. "Interface electrostatics of solid-state dye-sensitized solar cells: A joint drift-diffusion and density functional theory study". The Journal of Physical Chemistry C. No. 24. 14955-14963.
[153] Wei Di. 2010. "Dye sensitized solar cells". International Journal of Molecular Sciences. No. 3. 1103-1113.
[154] Schmidt-Mende By Lukas, Bach Udo, Humphry-Baker Robin, Horiuchi Tamosu, *et al.* 2005. "Organic dye for highly efficient solid-state dye-sensitized solar cells". Advanced Materials. No. 7. 813-815.
[155] Krüger Jessica, Plass Robert, Gra1tzel Michael, J. Cameron Petra, M. Peter Laurence. 2003. "Charge transport and back reaction in solid-state dye-sensitized solar cells: A study using intensity-modulated photovoltage and photocurrent spectroscopy" . The Journal of Physical Chemistry B. No. 31. 7536-7539.
[156] H. Howie Wendy, Harris John, R. Jennings James, L. Peter Laurence. 2007. "Solid-state dye-sensitized solar cells based on spiro-MeOTAD". Solar Energy Materials and Solar Cells. No. 5. 424-426.
[157] B. Cappel Ute, A. Gibson Elizabeth, Hagfeldt Anders, Boschloo Gerrit. 2009. "Dye regeneration by spiro-MeOTAD in solid state dye-sensitized solar cells studied by photoinduced absorption spectroscopy and spectroelectrochemistry". The Journal of Physical Chemistry C. No. 15. 6275-6281.
[158] Ding By I-Kang, Tétreault Nicolas, Brillet Jérémie, E. Hardin Brian, *et al.* 2009. "Pore-filling of spiro-OMeTAD in solid-state dye sensitized solar cells: Quantification, mechanism, and consequences for device performance". Advanced Functional Materials. No. 15. 2431-2436.
[159] Melas-Kyriazi John, Ding I-Kang, Marchioro Arianna, Punzi Angela, *et al.* 2011. "The effect of hole transport material pore filling on photovoltaic performance in solid-state dye-sensitized solar cells". Advanced Energy Materials. No. 3. 407-414.
[160] Chen Dong-Yi, Tseng Wei-Hsuan, Liang Sheng-Ping, Wu Chih-I, *et al.* 2012. "Application of F4TCNQ doped spiro-MeOTAD in high performance solid state dye sensitized solar cells". Physical Chemistry Chemical Physics. No. 33. 11689-11694.
[161] B. Cappel Ute, Daeneke Torben, Bach Udo. 2012. "Oxygen-induced doping of spiro-MeOTAD in solid-state dye-sensitized

solar cells and its impact on device performance". Nano Letters. No. 9. 4925-4931.

[162] Docampo Pablo, J. Snaith Henry. 2011. "Obviating the requirement for oxygen in SnO_2-based solid-state dye-sensitized solar cells". Nanotechnology. No. 22. 225403.

[163] Dualeh Amalie, Moehl Thomas, Nazeeruddin Mohammad Khaja, Grätzel Michael. 2013. "Temperature dependence of transport properties of spiro-MeOTAD as a hole transport material in solid-state dye-sensitized solar cells". ACS Nano. No. 3. 2292-2301.

[164] H. Nguyen William, D. Bailie Colin, L. Unger Eva, D. McGehee Michael. 2014. "Enhancing the hole-conductivity of spiro-OMeTAD without oxygen or lithium salts by using spiro $(TFSI)_2$ in perovskite and dye-sensitized solar cells". Journal of the American Chemical Society. No. 31. 10996-11001.

[165] Malinauskas Tadas, Tomkute-Luksiene Daiva, Sens Rüdiger, Daskeviciene Maryte, *et al.* 2015. "Enhancing thermal stability and lifetime of solid-state dye-sensitized solar cells via molecular engineering of the hole-transporting material spiro-OMeTAD". ACS Applied Materials & Interfaces. No. 21. 11107-11116.

[166] Pozzi Gianluca, Orlandi Simonetta, Cavazzini Marco, Minudri Daniela, *et al.* 2013. "Synthesis and photovoltaic applications of a 4,4′-spirobi[cyclopenta[2,1-b;3,4-*b*′]dithiophene]bridged donor/acceptor dye". Organic Letters. No. 18. 4642-4645.

[167] Chandra Sil Manik, Fairoos Mele Kavungathodia Munavvar, Nithyanandhan Jayaraj. 2019. "Effect and position of spiro-bipropylenedioxythiophene π-spacer in donor-π-spacer-acceptor dyes for dye-sensitized solar cell". Dyes and Pigments. 313-323.

[168] Zhang Chi, Tang Wei, Zhang Limin, Han Changbao, Wang Zhong Lin. 2014. "Contact electrification field-effect transistor" . ACS Nano,. No. 8. 8702-8709.

[169] R. Tuktarov Airat, B. Salikhov Renat, A. Khuzin Artur, N. Safargalin Idris, *et al.* 2019. "Optically controlled field effect transistors based on photochromic spiropyran and fullerene C60 films". Mendeleev Communications. No. 2. 160-162.

[170] Yuvaraja Saravanan, Nawaz Ali, Liu Qian, Dubal Deepak, *et al.* 2020. "Organic field-effect transistor-based flexible sensors". Chemical Society Reviews. No. 11. 3423-3460.

[171] P.I. Saragi Tobat, Fuhrmann-Lieker Thomas, Salbeck Josef. 2005. "High ON/OFF ratio and stability of amorphous organic field-effect transistors based on spiro-linked compounds". Synthetic Metals. No. 3. 267-270.

[172] Isenberg Carolin, Reichert Thomas, P.I. Saragi Tobat. 2015. "Magnetoresistance sign change in field-effect transistors via interface doping of Spiro-TAD with F6-TNAP". Synthetic Metals. 29-34.

[173] P. I. Saragi By Tobat, Fuhrmann-Lieker Thomas, Salbeck Josef. 2006. "Comparison of charge-carrier transport in thin films of spiro-linked compounds and their corresponding parent compounds". Advanced Functional Materials. No. 7. 966-974.

[174] Ohshita Joji, Lee Kwang-Hoi, Hamamoto Daisuke, Kunugi Yoshihito, *et al.* 2004. "Synthesis of novel spiro-condensed dithienosiloles and the application to organic FET". Chemistry Letters. No. 7. 892-893.

[175] Reichert Thomas, P. I. Saragi Tobat, Salbeck Josef. 2012. "Magnetoresistive field-effect transistors based on organic donor–acceptor blends". RSC Advances. No. 19. 7388-7390.

[176] Michael Sawatzki Franz, Doan Duy Hai, Kleemann Hans, Liero Matthias, *et al.* 2018. "Balance of horizontal and vertical charge transport in organic field-effect transistors". Physical Review Applied. No. 3. 034069.

[177] Zhang Ke, B. Kotadiya Naresh, Wang Xiao-Ye, A. H. Wetzelaer Gert-Jan, *et al.* 2020. "Interlayers for improved hole injection in organic field-effect transistors". Advanced Electronic Materials. No. 6. 1901352.

[178] Lin Dongqing, Zhang Wenhua, Yin Hang, Hu Haixia, *et al.* 2022. "Cross-scale synthesis of organic high-k semiconductors based on spiro-gridized nanopolymers". Research. 9820585.

[179] Chakraborty Debarupa Dutta, Nath Lila Kanta, Chakraborty Prithviraj. 2018. "Recent progress in smart polymers: Behavior, mechanistic understanding and application". Polymer-Plastics Technology and Engineering. No. 10. 945-957.

[180] Abdollahi Amin, Roghani-Mamaqani Hossein, Razavi Bahareh. 2019. "Stimuli-chromism of photoswitches in smart polymers: Recent advances and applications as chemosensors". Progress in Polymer Science. 101149.

[181] Abdollahi Amin, Roghani-Mamaqani Hossein, Salami-Kalajahi Mehdi, Razavi Bahareh. 2020. "Encryption and authentication of security patterns by ecofriendly multi-color photoluminescent inks containing oxazolidine-functionalized nanoparticles". Journal of Colloid and Interface Science. 192-210.

[182] Viljakka Jani. 2021. "Synthesis and charactrization of indigo photoswites". Science and Engineering.

[183] Towns Andrew. 2021. “Spiropyran dyes”. Physical Sciences Reviews. No. 8. 341-368.

[184] Tajmoradi Zeinab, Roghani-Mamaqani Hossein, Salami-Kalajahi Mehdi. 2020. “Stimuli-transition of hydrophobicity/hydro-philicity in *o*-nitrobenzyl ester-containing multi-responsive copolymers: Application in patterning and droplet stabilization in heterogeneous media”. Polymer. 122859.

[185] Abdollahi Amin, Herizchi Ata, Roghani-Mamaqani Hossein, Alidaei-Sharif Hossein. 2020. “Interaction of photoswitchable nanoparticles with cellulosic materials for anticounterfeiting and authentication security documents”. Carbohydrate Polymers. 115603.

[186] Abdollahi Amin, Alidaei-Sharif Hossein, Roghani-Mamaqani Hossein, Herizchi Ata. 2020. “Photoswitchable fluorescent polymer nanoparticles as high-security anticounterfeiting materials for authentication and optical patterning”. Journal of Materials Chemistry C. No. 16. 5476-5493.

[187] Abdollahia Amin, Roghani-Mamaqania Hossein, Salami-Kalajahia Mehdi, Razavia Bahareh, Sahandi-Zangabad Keyvan. 2020. “Encryption and optical authentication of confidential cellulosic papers by ecofriendly multi-color photoluminescent inks”. Carbohydrate Polymers. 116507.

[188] Vijayamohanan Harikrishnan, F. Palermo Edmund, K. Ullal Chaitanya. 2017. “Spirothiopyran-based reversibly saturable photoresist”. Chemistry of Materials. No. 11. 4754-4760.

[189] Müller Patrick, Müller Rouven, Hammer Larissa, Barner-Kowollik Christopher, *et al.* 2019. “STED-inspired laser lithography based on photoswitchable spirothiopyran moieties”. Chemistry of Materials. No. 6. 1966-1972.

[190] M. Rowley Natalie, J. Mortimer Roger. 2002. “New electrochromic materials”. Science Progress. No. 3. 243-262.

[191] D. Reeves By Benjamin, C. Thompson Barry, A. Abboud Khalil, E. Smart Bruse, R. Reynolds John. 2002. “Dual cathodically and anodically coloring electrochromic polymer based on a spiro bipropylene dioxythiophene[(poly(spiroBiProDOT)]”. Advanced Materials. No. 10. 717-719.

[192] Villabona Marc, Benet Marina, Mena Silvia, O. Al-Kaysi Rabih, *et al.* 2018. “Multistimuli-responsive fluorescent switches based on spirocyclic meisenheimer compounds: Smart molecules for the design of optical probes and electrochromic materials”. The Journal of Organic Chemistry. No. 16. 9166-9177.

[193] Usluer Ozlem, Koyuncu Sermet, Demic Serafettin, A. J. Janssen René. 2011. "A novel high-contrast ratio electrochromic material from spiro[cyclododecane-1,9′-fluorene]bicarbazole". Journal of Polymer Science Part B: Polymer Physics. No. 5. 333-341.
[194] Xie Mo, Chen Xiao-Ru, Wu Kun, Lu Zhou, *et al.* 2021. "Pressure -induced phosphorescence enhancement and piezochromism of a carbazole-based cyclic trinuclear Cu(I) complex". Chemical Science. No. 12. 4425-4431.
[195] Luo Qing, Lv Chunyan, Sheng Haonan, Cao Feng, *et al.* 2020. "Highly bright fluorescence from dispersed dimers: Deep-red polymorphs and wide-range piezochromism". Advanced Optical Materials. No. 7. 1901836.
[196] Wilson Diane Grimard, George Drickamer Harry. 1975. "High pressure studies on spiropyrans". The Journal of Chemical Physics. No. 8. 3649-3655.
[197] Gentili Pier Luigi, Nocchetti Morena, Milianic Costanza, Favaro Gianna. 2004. "Unexpected chromogenic properties of 1,3,3-trimethylspiro(indoline-2,3′-[3*H*]naphtho[2,1-*b*][1,4]oxazine) in the solid phase: Photochromism, piezochromism and acidichromism". New Journal of Chemistry. No. 3. 379-386.
[198] Funasako Yusuke, Takaki Akihiro, Inokuchi Makoto, Mochida Tomoyuki. 2016. "Photo-, thermo-, and piezochromic Nafion film incorporating cationic spiropyran". Chemistry Letters. No. 12. 1397-1399.
[199] Meng Xiao, Qi Guangyu, Li Xue, Wang Zhangyuan, *et al.* 2016. "Spiropyran-based multi-colored switching tuned by pressure and mechanical grinding". Journal of Materials Chemistry C. No. 32. 7584-7588.
[200] He Bairon, Chang Zhengfeng, Jiang Yibin, Xu Xiaofi, *et al.* 2014. "Piezochromic luminescent and electroluminescent materials comprised of tetraphenylethene plus spirobifluorene or 9,9-diphenylfluorene". Dyes and Pigments. 87-93.
[201] Cheng Dao-Juan, Shao You-Dong. 2018. "Organocatalytic asymmetric transformations involving the cyclic imine moiety in indole and isoindole related heterocycles". Advanced Synthesis & Catalysis. No. 19. 3614-3642.
[202] Takizawa Shinobu, Kiriyama Kimiko, Ieki Kenta, Sasai Hiroaki. 2011. "A bifunctional spiro-type organocatalyst with high enantiocontrol: Application to the aza-Morita–Baylis–Hillman reactions". Chemical Communications. No. 32. 9227-9229.

[203] Takizawa Shinobu, Arteaga Fernando Arteaga, Yoshida Yasushi, Suzuki Michitaka, Sasai Hiroaki. 2014. "Enantioselective organocatalyzed formal [4+2] cycloaddition of ketimines with allenoates: Easy access to a tetrahydropyridine framework with a chiral tetrasubstituted stereogenic carbon center". Asian Journal of Organic Chemistry. No. 4. 412-415.

[204] Wang Di, Wei Yin, Shi Min. 2012. "Construction of adjacent spiro-quaternary and tertiary stereocenters through phosphine-catalyzed asymmetric [3+2] annulation of allenoates with alkylidene azlactones". Chemical Communications. No. 22. 2764-2766.

[205] Glavač Danijel, Zheng Chao, Dokli Irena, You Shu-Li, Gredičak Matija. 2017. "Chiral Brønsted acid catalyzed enantioselective aza-Friedel–Crafts reaction of cyclic α-diaryl *N*-acyl imines with indoles". The Journal of Organic Chemistry. No. 16. 8752-8760.

[206] Chung Ying Kit, C. Fu Gregory. 2009. "Phosphine-catalyzed enantioselective synthesis of oxygen heterocycles". Angewandte Chemie. No. 12. 2259-2261.

[207] Wang Qing-Gang, Zhu Shou-Fei, Ye Long-Wu, Zhou Chang-Yue, *et al.* 2010. "Catalytic asymmetric intramolecular cascade reaction for the construction of functionalized benzobicyclo [4.3.0]skeletons. Remote control of enantioselectivity". Advanced Synthesis & Catalysis. No. 11-12. 1914-1919.

[208] Liu Honglei, Liu Yang, Yuan Chunhao, Wang Guo-Peng, *et al.* 2016. "Enantioselective synthesis of spirobarbiturate-cyclohexenes through phosphine-catalyzed asymmetric [4+2] annulation of barbiturate-derived alkenes with allenoates". Organic Letters. No. 6. 1302-1305.

[209] Tamao Kohei, Nakamura Kazunori, Ishii Hiroyuki, Yamaguchi Shigehiro, Shiro Motoo. 1996. "Axially chiral Spirosilanes via catalytic asymmetric intramolecular hydrosilation". Journal of the American Chemical Society. No. 49. 12469-12470.

[210] A. Arai Midori, Kuraishi Minori, Arai Takayoshi, Sasai Hiroaki. 2001. "A new asymmetric Wacker-type cyclization and tandem cyclization promoted by Pd(II)-spiro bis(isoxazoline) catalyst". Journal of the American Chemical Society. No. 12. 2907-2908.

[211] Ebeling Gunter, S. Gruber Adriane, A. Burrow Robert, Dupont Jairton, *et al.* 2002. "A chiral spiro-oxime ligand and its organopalladium(II) coordination chemistry". Inorganic Chemistry Communications. No. 8. 552-554.

[212] Zhou Hai, Wang Wen-Hu, Fu Yu, Xie Jian-Hua, *et al.* 2003. "Highly enantioselective copper-catalyzed conjugate addition of diethylzinc to enones using chiral spiro phosphoramidites as ligands". The Journal of Organic Chemistry. No. 4. 1582-1584.

[213] Kato Takahiro, Marubayashi Kazuyoshi, Takizawa Shinobu, Sasai Hiroaki. 2004. "Development of a novel chiral spiro ligand bearing oxazoline". Tetrahedron: Asymmetry. No. 23. 3693-3697.

[214] Duan Hai-Feng, Xie Jian-Hua, Shi Wen-Jian, Zhang Qi, Zhou Qi-Lin. 2006. "Enantioselective rhodium-catalyzed addition of arylboronic acids to aldehydes using chiral spiro monophosphite ligands". Organic Letters. No. 7. 1479-1481.

[215] S. Koranne Priti, Tsujihara Tetsuya, A. Arai Midori, B. Bajracharya Gan, *et al.* 2007. "Design and synthesis of chiral hybrid spiro(isoxazole–isoxazoline) ligands". Tetrahedron: Asymmetry. No. 8. 919-923.

[216] Liu Bin, Zhu Shou-Fei, Zhang Wei, Chen Chao, Zhou Qi-Lin. 2007. "Highly enantioselective insertion of carbenoids into N−H bonds catalyzed by copper complexes of chiral spiro bisoxazolines". Journal of the American Chemical Society. No. 18. 5834-5835.

[217] Xie Jian-Hua, Zhou Qi-Lin. 2008. "Chiral diphosphine and monodentate phosphorus ligands on a spiro scaffold for transition-metal-catalyzed asymmetric reactions". Accounts of Chemical Research. No. 5. 581-593.

[218] Shu Wei, Yu Qiong, Ma Shengming. 2009. "Development of a new spiro-BOX ligand and its application in highly enantioselective palladium-catalyzed cyclization of 2-iodo anilines with allenes". Advanced Synthesis & Catalysis. No. 17. 2807-2810.

[219] Shibatomi Kazutaka, Muto Tsubasa, Sumikawa Yusuke, Narayama Akira, Iwasa Seiji. 2009. "Development of a new chiral spirooxazolinylpyridine ligand (Spymox) for asymmetric catalysis". Synlett. No. 02. 241-244.

[220] Li Shen, Zhu Shou-Fei, Xie Jian-Hua, Song Song, *et al.* 2010. "Enantioselective hydrogenation of α-aryloxy and α-alkoxy α,β-unsaturated carboxylic acids catalyzed by chiral spiroiridium/phosphino-oxazoline complexes". Journal of the American Chemical Society. No. 3. 1172-1179.

[221] Shibatomi Kazutaka, Narayama Akira, Soga Yoshinori, Muto Tsubasa, Iwasa Seiji. 2011. "Enantioselective *gem*-chloro-

fluorination of active methylene compounds". Synfacts. No. 10. 1090-1090.

[222] Narayama Akira, Shibatomi Kazutaka, Soga Yoshinori, Muto Tsubasa, Iwasa Seiji. 2013. "Copper(II)-catalyzed enantioselective fluorination of β-keto esters using chiral spiro oxazoline ligands". Synlett. No. 03. 375-378.

[223] Li Jie, Pan Wei, Wang Zheng, Zhang Xumu, Ding Kuiling. 2012. "Access to both enantiomers of α-chloro-β-keto esters with a single chiral ligand: Highly efficient enantioselective chlorination of cyclic β-keto esters catalyzed by chiral copper(II) and zinc(II) complexes of a spiro-2,2′-bischroman-based bisoxazoline ligand" . Advanced Synthesis & Catalysis. No. 10. 1980-1986.

[224] Xie Jian-Hua, Liu Xiao-Yan, Yang Xiao-Hui, Xie Jian-Bo, *et al.* 2012. "Chiral iridium catalysts bearing spiropyridine-amino phosphine ligands enable highly efficient asymmetric hydrogenation of β-aryl β-Ketoesters". Angewandte Chemie International Edition. No. 1. 201-203.

[225] Bao Deng-Hui, Wu Hui-Ling, Liu Chao-Lun, Xie Jian-Hua, Zhou Qi-Lin. 2015. "Development of chiral spiro P-N-S ligands for iridium-catalyzed asymmetric hydrogenation of β-alkyl-β-ketoesters". Angewandte Chemie. No. 30. 8915-8918.

[226] Zhang Feng-Hua, Chen, Xie Jian-Hua, Zhou Qi-Lin. 2019. "Synthesis of tridentate chiral spiro aminophosphine−oxazoline ligands and application to asymmetric hydrogenation of α-keto amides". Advanced Synthesis & Catalysis. No. 12. 2832-2835.

[227] Kraft Jochen, Golkowski Martin, Ziegler Thomas. 2016. "Spiro-fused carbohydrate oxazoline ligands: Synthesis and application as enantio-discrimination agents in asymmetric allylic alkylation" . Beilstein Journal of Organic Chemistry. No. 1. 166-171.

[228] Li Yi-Pan, Zhu Shou-Fei, Zhou Qi-Lin. 2019. "Chiral spiro phosphoramide-catalyzed sulfa-Michael addition/enantioselective protonation of exocyclic enones". Organic Letters. No. 23. 9391-9395.

[229] Luo Jun, Zhang Tao, Wang Lei, Liao Gang, *et al.* 2019. "Enantioselective synthesis of biaryl atropisomers by Pd-catalyzed C−H olefination using chiral spiro phosphoric acid ligands". Angewandte Chemie. No. 20. 6780-6784.

[230] Ooi Takashi, Maruoka Keiji. 2007. "Recent advances in asymmetric phase-transfer catalysis". Angewandte Chemie International Edition. No. 23. 4222-4266.

[231] Enders Dieter, Vinh Nguyen Thanh. 2012. "Chiral quaternary phosphonium salts: A new class of organocatalysts". Organic & Biomolecular Chemistry. No. 28. 5327-5331.

[232] J. O'donnell Martin. 2004. "The enantioselective synthesis of α-amino acids by phase-transfer catalysis with achiral Schiff base esters". Accounts of Chemical Research. No. 8. 506-517.

[233] Ooi Takashi, Kameda Minoru, Maruoka Keiji. 1999. "Molecular design of a C_2-symmetric chiral phase-transfer catalyst for practical asymmetric synthesis of α-amino acids". Journal of the American Chemical Society. No. 27. 6519-6520.

[234] Ooi Takashi, Miki Takashi, Taniguchi Mika, Shiraishi Misato, *et al.* 2003. "Highly enantioselective construction of quaternary stereocenters on β-keto esters by phase-transfer catalytic asymmetric alkylation and Michael reaction". Angewandte Chemie International Edition. No. 32. 3796-3798.

[235] Ooi Takashi, Fujioka Shingo, Maruoka Keiji. 2004. "Highly enantioselective conjugate addition of nitroalkanes to alkylidene-malonates using efficient phase-transfer catalysis of *N*-spiro chiral ammonium bromides". Journal of the American Chemical Society. No. 38. 11790-11791.

[236] Ooi Takashi, Kubota Yasushi, Maruoka Keiji. 2003. "A new *N*-spiro C_2-symmetric chiral quaternary ammonium bromide consisting of 4,6-disubstituted biphenyl subunit as an efficient chiral phase-transfer catalyst". Synlett. No. 12. 1931-1933.

[237] Ooi Takashi, Ohara Daisuke, Fukumoto Kazuhiro, Maruoka Keiji. 2005. "Importance of chiral phase-transfer catalysts with dual functions in obtaining high enantioselectivity in the Michael reaction of malonates and chalcone derivatives". Organic Letters. No. 15. 3195-3197.

[238] Wang Xisheng, Kitamura Masanori, Maruoka Keiji. 2007. "New, chiral phase transfer catalysts for effecting asymmetric conjugate additions of α-alkyl-α-cyanoacetates to acetylenic esters". Journal of the American Chemical Society. No. 5. 1038-1039.

[239] Kita Tetsuya, Georgieva Angelina, Hashimoto Yuichi, Nakata Tadashi, Nagasawa Kazuo. 2002. "C_2-Symmetric chiral pentacyclic guanidine: A phase-transfer catalyst for the asymmetric alkylation of *tert*-butyl glycinate schiff base". Angewandte Chemie International Edition. No. 15. 2832-2834.

[240] Waser Mario, Gratzer Katharina, Herchl Richard, Müller Norbert. 2012. "Design, synthesis, and application of tartaric acid derived

N-spiro quaternary ammonium salts as chiral phase-transfer catalysts". Organic & Biomolecular Chemistry. No. 2. 251-254.

[241] Shibuguchi Tomoyuki, Mihara Hisashi, Kuramochi Akiyoshi, Ohshima Takashi, Shibasaki Masakatsu. 2007. "Catalytic asymmetric phase-transfer Michael reaction and Mannich-type reaction of glycine schiff bases with tartrate-derived diammonium salts". Chemistry–An Asian Journal. No. 6. 794-801.

[242] Uraguchi Daisuke, Asai Yoshihiro, Seto Yuki, Ooi Takashi. 2009. "Asymmetric synthesis of α,α-disubstituted α-amino acids via enantioselective alkylation of azlactones under biphasic conditions using *P*-spiro chiral tetraaminophosphonium salts as a phase-transfer catalyst". Synlett. No. 04. 658-660.

[243] Uraguchi Daisuke, Tsutsumi Ryosuke, Ooi Takashi. 2014. "Catalytic asymmetric Payne oxidation under the catalysis of *P*-spiro chiral triaminoiminophosphorane: Application to the synthesis of *N*-sulfonyl oxaziridines". Tetrahedron. No. 8. 1691-1701.

[244] Yoshioka Ken, Yamada Kohei, Uraguchi Daisuke, Ooi Takashi. 2017. "Unique site-selectivity control in asymmetric Michael addition of azlactone to alkenyl dienyl ketones enabled by *P*-spiro chiral iminophosphorane catalysis". Chemical Communications. No. 40. 5495-5498.

SUBJECT INDEX